HANDBOOK OF ADVANCED SEMICONDUCTOR TECHNOLOGY AND COMPUTER SYSTEMS

HANDBOOK OF ADVANCED SEMICONDUCTOR TECHNOLOGY AND COMPUTER SYSTEMS

Edited by

GUY RABBAT

VAN NOSTRAND REINHOLD ELECTRICAL/COMPUTER SCIENCE AND ENGINEERING SERIES

VAN NOSTRAND REINHOLD COMPANY
_____ New York

Copyright © 1988 by Van Nostrand Reinhold Company Inc.
Library of Congress Catalog Card Number 87-6228
ISBN 0-442-27688-5

Printed in the United States of America

Van Nostrand Reinhold Company Inc.
115 Fifth Avenue
New York, New York 10003

Van Nostrand Reinhold Company Limited
Molly Millars Lane
Wokingham, Berkshire RG11 2PY, England

Van Nostrand Reinhold
480 La Trobe Street
Melbourne, Victoria 3000, Australia

Macmillan of Canada
Division of Canada Publishing Corporation
164 Commander Boulevard
Agincourt, Ontario MIS 3C7, Canada

16   15   14   13   12   11   10   9   8   7   6   5   4   3   2   1

Library of Congress Cataloging-in-publication Data

Handbook of advanced semiconductor technology and computer systems.

(Van Nostrand Reinhold electrical/computer science
and engineering series)
Includes index.
1. Integrated circuits—Design and construction.
2. Electronic digital computers—Circuits—Design
and construction.   I. Rabbat, Guy.   II. Series.
TK7874.A336   1987   621.395   87-6228
ISBN 0-442-27688-5

Dedicated to my wife, Elfriede

Van Nostrand Reinhold
Electrical/Computer Science and Engineering Series
Sanjit Mitra—Series Editor

Preface

Chapter 1 describes deposition as a basic microelectronics technique. Plasma enhanced chemical vapor deposition (PECVD) is a technique widely accepted in microelectronics for the deposition of amorphous dielectric films such as silicon nitride and silicon oxide. The main advantage of PECVD stems from the introduction of plasma energy to the CVD environment, which makes it possible to promote chemical reactions at relatively low temperatures. A natural extension of this is to use this plasma energy to lower the temperature required to obtain a crystalline deposit. This chapter discusses the PECVD technique and its application to the deposition of dielectric, semiconductor, and conductor films of interest to microelectronics.

Chapter 2 acquaints the reader with the technology and capabilities of plasma processing. Batch etching reactors and etching processes are approaching maturity after more than ten years of development. Requirements of anisotropic and selective etching have been met using a variety of reactor configurations and etching gases. The present emphasis is the integration of plasma etching processes into the overall fabrication sequence.

Chapter 3 reviews recent advances in high pressure oxidation technology and its applications to integrated circuits. The high pressure oxidation system, oxidation mechanisms, oxidation-induced stacking faults, impurity segregation, and oxide quality are described. Applications to bipolar and MOS devices are also presented.

Chapter 4 is a comprehensive chapter on the structures and fabrication of metal-oxide-silicon field effect transistors. The basic device structure, operation principle, and fabrication processes are described, as are fabrication technologies such as the growth of silicon dioxide films, growth kinetics, oxide charges and control of oxide charges, formation of the recessed oxide, formation of the polysilicon gate, application of ion implantation, and interconnection and contact metallurgy. Major technological advances are studied, including the application of metal silicides, sidewall technology, lightly-doped-drain (LDD) field effect transistors, and advanced device isolation techniques.

Electron beam testing is described in Chapter 5 as a tool for chip verification during the design phase and in failure analysis. A brief description is given of what the designer or failure analysis needs to know about the electron probe and signal processing. Five main modes of operation are explained in detail: voltage coding, logic-state mapping, frequency tracing, frequency mapping, and waveform measurement. The components that transform a scanning electron microscope into an electron beam tester are discussed. Electron beam test technology is treated in detail, especially integrated test systems consisting of an electron beam tester, an IC tester, and a CAD system. Three generalized examples illustrate the methodology of chip verification during the design phase. Current trends and limitations of future applications of electron beam testing are discussed.

Chapter 6 reviews thin film applications in the fabrication of contacts and interconnects for VLSI circuits. Device structures suitable for both bipolar and metal oxide semiconductor (MOS) VLSI applications tend to have shallow junction depths and contact areas (Si/metal interfaces) in the 0.2 μm to 0.5 μm and 1 μm^2 to 2μm^2 ranges, respectively; also some of the circuits require Schottky barrier doides. The use of platinum silicide and barrier layers for bipolar VLSI circuits is illustrated. Integration of a large number of components with long interconnection paths adversely impacts VLSI circuit performance, and multilevel interconnections offer an attractive solution. Application of PtSi/Ti:W/Al–Cu/SiO$_2$/Ti:W/Al film layers in the fabrication of bipolar VLSI circuits with a minimum feature size of 1.25 μm is illustrated. Three or more levels of interconnections compatible with complex packaging technologies will be required in the future. There is a growing concern that VLSI reliability may be adversely impacted by the increased current density in interconnections. Electromigration and corrosion-induced failures affecting chip reliability are examined. Areas of concern and desirable features in VLSI metallization are summarized.

In Chapter 7, the evolution of MOS integrated circuit technology is tracked from the late sixties, when the first commercially successful produce was introduced, to the eighties and the advent of VLSI, and beyond. Major advances in process technology and resulting performance and density improvements are presented. The contributions of process advances, device cleverness, and reduction in dimensions are discussed, and extrapolations to future technologies are given. Memory devices have been and continue to be the technology driver because of their predictability and effective utilization of the most advanced technology features. Memory is used throughout the chapter to illustrate major advances when applicable. Reliability and yield have been the major limiters to adoption of the most sophisticated techniques; so, where appropriate, they are highlighted.

In Chapter 8, CMOS devices are described in detail. MOS technology, through its exponential growth in the last few years, has opened up a whole new domain of applications. Cost reduction, increased integration density, and enhanced circuit performance are the three principal benefits of device scaling, and were the driving force of the 1970s, whereas CMOS has emerged as the technology of choice in the 1980s.

In Chapter 9, bipolar gate array technology is thoroughly presented. Bipolar masterslice LSIs, developed to meet the requirements of large computers, provide high performance, small system size, and high reliability. They feature low energy CML (current mode logic), a cell array masterslice, and design automation. The masterslice design has allowed the development of several hundred options with short turnaround time and at a low development cost.

The theme in Chapter 10 is the realization of high-performance computers, where system delay must be reduced to achieve a short machine cycle time. There seems to be a wide choice of hardware technologies for future high-performance computer systems. This chapter points out several factors essential for evaluating logic LSIs and packaging technologies for these computers. In comparing circuit technology, the difference in logical-function capabilities among circuits should be evaluated. Also, it is necessary to obtain the switching frequency of a circuit in actual systems for comparing CMOS circuit performance with other circuits. In order to evaluate the wire delay, the average wire length per signal net should be known. It depends on the packaging structure. Equations for estimating wire lengths in various types of two-dimensional (2-D) and three-dimensional (3 D) cell arrays are introduced.

Chapter 11 emphasizes packing for cost-performance computers. There is a significant shift in emphasis on packaging requirements for computers when absolute performance must be weighed against cost. While mainframe machines support their high performance through very dense packaging, that density is for the purpose of providing top-of-the-line capability through optimized electrical performance and the resulting high data processing throughput. With these large systems, a three-year hardware development time is tolerated, and designs are often unique or special to each manufacturer. The examples of the previous chapter show this diversity of design approach.

It is the intent of Chapter 12 on microprocessors to capture some of the flavor of current developments in the field. The first 4-bit microprocessor was introduced in 1970. Since then, the complexity and performance of these devices has increased steadily through 8-bit, 16-bit, and 32-bit architectures, with memory management and floating point support, at first on separate chips and then integrated, at clock rates of 40 MHz or more. The evolution from yesterday's unproven concept to today's standard practice has been truly astounding, as the time and resources required to implement large systems have been reduced by very large factors. At the same time, it has encouraged the growth of basic design concepts and methodologies, which in turn have provided the means to facilitate this evolution.

Programmable logic arrays (PLA), described in Chapter 13, are an orderly structure for implementing logic in minterm canonical form. This approach to organizing logic speeds the design cycle, and if proper design procedures are followed, clean economical designs for both bipolar and FET technologies are obtainable. Both combinatorial and sequential logic (finite state machines) can be designed with PLAs, using a comprehensive design procedure with Karnaugh maps. Methods to achieve low-cost solutions for designs with true-complement or two-to-four decoders are described in the chapter. The application of PLAs to a digital filter chip, and the physical layout and design automation aspects of PLA design are also covered.

Chapter 14 describes some of the techniques now in use in design for testability. Many techniques have been developed to accommodate the need for more dense and complex networks. Accompanying the presentation of some of these design techniques are some design methods of implementing specific functions—for example, counters, which sometimes appear problematic when they must be implemented in an efficient manner. In the field of design for testability, several approaches are currently being employed. An increasingly important approach is that of self-testing.

The aim of Chapter 15 is: to give a historical perspective on the development of silicon design methods (for both LSI and VLSI devices); to describe the accompanying development of CAD tools; to indicate what the requirements of a CAD system for VLSI design should be; and to attempt, in a general way, to predict the future development of this fast-moving field. Among the topics covered are: process technologies and their influence on design techniques; the rise of MOS, particularly CMOS; the use of gate arrays and standard cells; the definition of an ideal silicon compiler; the increasing use of array-based design styles; and the gradual metamorphisis of individual (largely ad hoc) CAD tools into "silicon design systems."

Chapter 16 addresses design verification at the logic design. The rapid progress of VLSI design and fabrication technologies is enabling designers to produce denser and more complex integrated circuits. As circuit integration increases, any logic design error discovered after fabricating VLSI chips becomes more time-consuming and expensive to correct. Design verification, whose goal is to ensure that the VLSI circuit functions exactly as intended before fabrication starts, is becoming vitally important. The main activities in developing an integrated circuit are functional (logic) design, physical (layout) design, and test generation. In each design activity, sophisticated verification tools are provided to confirm the correctness of the design.

Chapter 17 reviews several problems that arise in the area of design automation.

Most of these problems are shown to be NP-hard. Further, it is unlikely that any of these problems can be solved by fast approximation algorithms that guarantee solutions that are always within some fixed relative error of the optimal solution value. This presentation points out the importance of heuristics and other tools to obtain algorithms that perform well on the problem instances of interest.

Chapter 18 describes hierarchical layout design methodology, and related placement and routing algorithms that are effective for easing bottlenecks in designing several tens of thousands of custom logic VLSIs. First, a design automation configuration that is now in practical use is overviewed briefly. A layout model and a hierarchical structure that forms the basis of following discussions are presented. Then placement and routing algorithms and their evaluations are covered. Intrablock layout sets the physical size of blocks. A floor planning program can determine block locations on a chip automatically. In interblock routing, the chip may be partitioned into sets of blocks ("block assembly"), if concurrent layout design is necessary or CPU memory resources are limited. New types of block placement algorithms for automatic floor planning are presented. Objective functions and related placement algorithms for intrablock cell layout are also discussed. In addition, area comparison results of these algorithms are given. Two types of routing algorithms are applied according to the design hierarchy for multilayer wiring. An example is presented of actual VLSI chips designed by using the DA system in which these procedures are incorporated.

Chapter 19 presents layout compilation in the most thorough presentation in the industry.

Chapter 20 provides an overview of the placement function within automatic layout systems. The automatic placement problem is defined, and the data abstractions, or models, used by placement are described. The discussion divides placement algorithms into two classes: constructive and iterative. The more important algorithms in these classes, as well as applications of the algorithms within layout systems, are described. A large number of references are provided, which allow the chapter's use as a guide to placement literature.

The theme of Chapter 21 is PWBs (printed wiring boards). No matter how high the integration density of a VLSI is, and no matter how large a VLSI system grows, PWBs are still the dominant means of achieving system level interconnections. Thus development is continuing on sophisticated algorithms to raise layout performance for high density PWBs.

Chapter 22 consists of two major parts. The first part is mostly of a tutorial nature. The author gives a motivation for cell based layout methodologies, briefly discusses design methods for leaf cells, characterizes the three major cell based design methodologies (gate arrays, standard cells, and general cell approach), and finally describes methods for automated design (i.e., placement and routing) based on predefined cells. Emphasis is on the concept of "divide and conquer," used to break down the overall layout problem into a sequence of loosely coupled, more easily solvable subproblems. The second part presents new material. The author discusses the implementation of placement and routing in a specific design system for hierarchical, cell based design. Basic features of this system are fully automatic or interactively controlled top-down min-cust placement of general cells as well as automatic routing of these cells, including planar power and ground buses with variable width. Routing is done exclusively by channel routing to achieve a fast response and high completion rate. Special developments that were necessary for channel routing in a general cell environment are discussed in some detail.

Chapter 23 formalizes and discusses some issues of knowledge-based expert systems. Computer-aided design (CAD) is a collection of programs for assisting the digital system designer during each design phase. Although most CAD programs today perform tasks according to the decision-making logic of conventional programs, these programs cannot readily accommodate significant amounts of knowledge and cannot solve problems with the competence of a human expert. As knowledge-based expert system technology has advanced, providing insight into how knowledge is formalized and into its application to problems, the concept of expert systems for circuit design has begun to attract attention in the CAD community.

Chapter 24 presents the key issues and some of the solutions in the modeling and analysis of fault-tolerant multiple processor systems. Performance modeling, reliability modeling, and the combined evaluation of performance and reliability are discussed. Simple examples are used to illustrate these models. Important reference are given for further study.

Chapter 25 introduces the reader to the important area of logic programming, which has emerged as a surprisingly successful computational paradigm in the last few years, and is continuing to gain support from many fields within computer science. Typified by such successful languages as Prolog, LOGLISP, and KL0/KL1, logic programming has proved itself successful in many diverse areas, such as compiler writing, natural language processing, data bases, CAD/CAM, and expert systems.

Chapter 26 describes the organization of a highly parallel machine, with particular emphasis on the structure and function of the active memory. NON-VON is a highly parallel, non–von Neumann "supercomputer," an early prototype of which is presently operational in the Computer Science Department at Columbia University. The machine is intended to support the extremely rapid execution of a wide range of large-scale data manipulation tasks, especially those involving artificial intelligence and other symbolic applications. The general NON-VON architecture is heterogeneous in each of the three dimensions that are commonly used to classify parallel architectures: granularity, topology, and synchrony. The machine includes an "active memory" containing as many as a million simple 8-bit small processing elements, implemented using custom VLSI chips, each containing a number of processing elements. A smaller number (up to a thousand or so) more powerful large processing elements are used to broadcast independent instruction streams for execution in "multiple-SIMD" mode by the small processing elements. Tree- and mesh-structured physical connections, along with linear logical connections, are used to interconnect the SPEs, while the LPEs are interconnected through a high-bandwidth multistage interconnection network. The complete architecture also includes a secondary processing subsystem (SPS), based on a bank of intelligent disk drives and connected to the primary processing subsystem through a high-bandwidth parallel interface. Some of the most important NON-VON programming techniques are outlined, and their application to typical data processing applications are illustrated with simple examples.

Chapter 27 is concerned with advances in programmable, high-performance digital signal processor architecture. It begins with a review of several relevant signal processor architectures implemented over the past 15 years, and concludes with specific recommendations for achieving an optimal match between available technologies and high-performance signal processor design.

GUY RABBAT

Contents

1. Deposition for Microelectronics—Plasma Enhanced Chemical Vapor Deposition

Rafael Reif

Massachusetts Institute of Technology

Plasma enhanced chemical vapor deposition (PECVD) is a widely accepted technique for the deposition of dielectric films such as silicon nitride and silicon oxide. The main reason for this acceptance is the lower temperature capability of this technique compared to that of thermally driven CVD. For example, deposition temperatures of 700 to 900°C are needed to deposit silicon nitride films by thermal CVD, whereas temperatures of only 250 to 350°C are required for PECVD (Adams 1983). This lower temperature capability is due to the addition of plasma energy to the CVD environment in the form of a glow discharge, and the effective substitution of this plasma energy for thermal energy. The lowering of the deposition temperature allowed by PECVD has made it possible to deposit films of silicon nitride and/or oxide on top of wafers having a layer of aluminum or gold. In addition, PECVD is known to provide conformal coverage of steps on the substrate surface, and to yield reasonably high deposition rates.

PECVD is also being investigated as a potential technique for the deposition of crystalline films such as polysilicon (Kamins and Chiang 1982; Burger et al. 1982), epitaxial silicon (Suzuki and Itoh 1983; Shanfield and Reif 1983), gallium arsenide (Hariu et al. 1981; Pande 1983; Pande and Seabaugh 1984), and refractory metal and silicide films (Tabuchi et al. 1982). The driving force for this research is the same as that for the PECVD of dielectric films, that is, to lower the required deposition temperature while maintaining reasonable growth rates and high quality. The role of the glow discharge here, however, is somewhat more complex than in the deposition of dielectric films. In the deposition of silicon nitrides and oxides, the plasma energy replaces thermal energy mostly in the *gas*. This promotes the necessary chemical reactions at lower temperatures

1

while the deposited film is always structurally amorphous. For the deposition of crystalline films, however, the plasma energy must replace thermal energy not only in the gas but also at the wafer *surface*. This energy is needed primarily in the form of enhanced atom mobility in order to obtain a crystalline deposit.

This chapter discusses the plasma enhanced chemical vapor deposition technique, emphasizing the issues important to computer technology. It attempts to address studies not previously discussed in other reviews (Reinberg 1979a,b; Ojha 1982; Hess 1983, 1984). The chapter begins with a review of the basic physics and chemistry of nonequilibrium glow discharges, which is followed by a discussion of the effects of adding a glow discharge to a chemical vapor deposition (CVD) environment. Finally, the latter sections review the most important dielectric, semiconductor, and conductor films that have been deposited by PECVD.

NONEQUILIBRIUM GLOW DISCHARGES

A simple model of a glow discharge consists of a partially ionized gas containing equal volume densities of positive and negative charged species (mostly ions and electrons, respectively) and different volume densities of ground-state and excited species (Chapman 1980; Rand 1979). These plasmas can be generated by subjecting a gas to very high temperatures or to strong electric or magnetic fields. In thermal plasmas, the electrons, ions, and neutral gas molecules are in local thermodynamic equilibrium. In nonequilibrium or "cold" plasmas, the electrons and ions are more energetic than the background gas molecules.

Most of the glow discharges used in microelectronics are generated by the application of a radio frequency (rf) electric field to the gas, and they are nonequilibrium glow discharges (i.e., "cold" plasmas). The energy from the field initially accelerates only the few free electrons present in the gas, while any ions present are relatively unaffected because of their much heavier mass. The accelerated free electrons lose little energy in elastic collisions with atoms or molecules because of the large difference in mass. Furthermore, they initially lose little energy in inelastic collisions, such as excitation and ionization, until their energies reach the corresponding threshold energies [e.g., 11.56 eV for excitation and 15.8 eV for ionization of argon (Chapman 1980)]. Consequently, these free electrons gain energy quickly from the field.

Once the free electrons acquire sufficiently high energies, their collisions with gas species can produce excitations and ionizations, the latter generating additional electrons that in turn are accelerated by the rf field. This process quickly avalanches, creating the glow discharge. In steady state, the glow loses electrons and ions to the electrodes and to all other surfaces within the chamber, and gains a numerically equal number of electrons from ionizations and other mechanisms that produce additional electrons, such as electron emission from positive ion bombardment on the electrodes and walls.

The inelastic collisions between high-energy electrons and gas molecules give rise to reactive species, such as excited neutrals and free radicals, as well as ions and more electrons. In this manner the energy of the electrons, which is acquired from the applied electric field, is used to create highly reactive species and charged species without significantly raising the gas temperature. PECVD uses these highly reactive species to deposit thin films such as silicon nitrides, silicon oxides, and hydrogenated amorphous silicon at lower temperatures than are possible with thermally driven CVD. This is possible because the reactive species produced in the plasma have lower energy barriers to physical and chemical reactions than the parent species and, consequently, react at lower temperatures. Furthermore, the charged species in the glow discharge can be used to affect the structure (e.g., crystallinity) and other properties of the deposited films (Greene and Barnett 1982).

There are many possible inelastic collisions between electrons and reactants in a glow discharge. Examples of those believed to be significant in PECVD are listed below:

Excitation: $A + e^- \rightarrow A^* + e^-$ [1]

Ionization: $A + e^- \rightarrow A^+ + 2e^-$ [2]

Dissociation: $A_2 + e^- \rightarrow 2A + e^-$ [3]

Electron attachment: $A + e^- \rightarrow A^-$ [4]

Dissociative attachment: $A_2 + e^- \rightarrow A + A^-$ [5]

Photoemission: $A^* \rightarrow A + h\nu$ [6]

Charge transfer: $A^+ + B \rightarrow A + B^+$ [7]

where A, A_2, and B are reactants, e^- is an electron, A^* is reactant A is an excited state, and A^+, A^-, and B^+ are ions of A and B.

The rate at which these inelastic collisions create excited species, ions, free radicals, and so on, can be calculated by using a kinetic-like, reaction rate equation (Turban et al. 1979). For example, the rate at which A^* is created from reaction (1) can be given by.

$$\frac{d\,[A^*]}{dt} = k_1\,[A][e^-] \qquad [8]$$

where $d[A^*]/dt$ is the rate of formation of A^*, k_1 is the reaction rate coefficient, [A] is the concentration of species A, and $[e^-]$ is the electron concentration (Sawin 1985). A similar equation can be used to describe the reaction rates corresponding to reactions (2) through (7).

As discussed above, only high energy electrons participate successfully in inelastic collisions. This selection feature is taken into account by employing the collision cross-section concept in the definition of k_1. The cross section of a given electron-reactant inelastic collision is proportional to the probability that this inelastic collision will occur. If the energy of the electron is lower than the threshold energy required for the inelastic collision, then the collision cross section at that energy is zero. The rate coefficient k_i can be calculated by using the following equation (Turban et al. 1979):

$$k_i = \int_0^\infty \left(\frac{2E}{m_e}\right)^{\frac{1}{2}} \sigma_i(E)\, f(E)\, dE \qquad [9]$$

where E is the electron energy, m_e is the electron mass, σ_i is the collision cross section of reaction i and is a function of E, and $f(e)$ is the electron energy distribution function and gives the fraction of free electrons having a given energy. The integration is carried out over all possible electron energies. The square root term in Eq. (9) is the electron velocity.

Some collision cross-section data can be found in the literature (Brown 1966). However, most of the relevant cross sections of interest in microelectronics are not readily available. A similar situation exists with the electron energy distribution function. It is typical to assume $f(e)$ to be a Maxwell-Boltzmann distribution, that is, a distribution such that a large fraction of the electrons have energies equal to or lower than the average electron energy, with the fraction of electrons having higher energies decaying exponentially with increasing energy. However, the actual electron energy distribution function for many plasma environments is not known, and non-Maxwellian distributions have been proposed. In fact, it is expected that the reactant composition of the plasma influences $f(E)$ because the electrons lose a significant fraction of their energies through inelastic collisions with these reactants.

The glow discharges currently used in microelectronics are relatively low pressure plasmas (0.1–5 Torr). The plasma density (i.e., density of free electrons and ions) is typically about $10^{10}/cm^3$. The degree of ionization is typically $\leqslant 10^{-4}$; that is, the gas consists mostly of neutrals. A Maxwell-Boltzmann distribution is usually used to approximate the energy distribution of free electrons in the glow, with the fastest electrons having energies as high as 10 to 30 eV (Hollahan and Bell 1974).

The PECVD environment is not in thermal equilibrium. In a typical PECVD situation, the ground-state species in the gas may be slightly above 300°K, the excited species and ions may have energies corresponding to 400°K up to thousands of degrees Kelvin, and the free electrons may have energies corresponding to 10^4 to 10^5 °K (Rand 1979). Consequently, thermodynamics is of little help in predicting the outcome of a plasma deposition experiment.

POTENTIALS IN RF GLOW DISCHARGES

Several potentials are important in the glow discharges used in microelectronics: the plasma potential, the floating potential, and the sheath potentials.

The plasma potential (V_p) is the potential of the glow region of the plasma, which is normally considered nearly equipotential. It is the most positive potential in the chamber and is the reference potential for the glow discharge.

The floating potential (V_f) is the potential at which equal fluxes of negative and positive charged species arrive at an electrically floating surface in contact with the plasma. It is approximately given by:

$$V_p - V_f - \frac{kT_e}{2e} \ln \left(\frac{m_i}{2.3m_e} \right) \tag{10}$$

where T_e is the electron temperature, e is the unit electron charge, and m_i and m_e are the ion and electron mass, respectively (Chapman 1980; Vossen 1979). If the chamber walls are electrically insulated, Eq. (1) can be used to estimate the maximum energy with which positive ions bombard the walls. Most sputtering threshold energies are 20 to 40 eV (Chapman 1980). Therefore, $V_p - V_f \leq 20$ to 40 V is normally desired to avoid sputtering off the walls, which may lead to film contamination.

As mentioned earlier, the plasma potential is always positive with respect to the surface in contact with the plasma. This is so because the electron mobility in the plasma is much greater than the ion mobility, with the result that the initial electron flux to all surfaces is greater than the ion flux. Consequently, the surfaces in contact with the plasma become negatively charged, and a positive space charge layer forms near these surfaces. Because there are fewer electrons in the space charge layer, or sheath, fewer atoms are excited by electron collisions. Thus, there are fewer atoms to relax and give off radiation, and the sheath region is dark relative to the glow discharge. The sheath regions accelerate the positive ions that enter the sheaths from the glow region by random thermal motion into the electrodes and other surfaces in contact with the plasma. Similarly, the sheath regions accelerate secondary electrons emitted from surfaces (e.g., due to positive ion bombardment) into the glow region. Therefore, the sheath voltages are very important because they determine the maximum energy with which positive ions bombard the electrodes, and help estimate the maximum energy with which secondary electrons emitted from electrodes enter the glow region.

Figure 1-1(a) shows an electric circuit with an rf generator connected to the electrodes of a plasma system (Reif 1984). As indicated in Figure 1-1(b), the sheath voltages at each electrode are the same because the circuit is symmetric. However, if a "blocking" capacitor is added to the circuit (see Figure 1-2a), then negative charge from the plasma can accumulate on the capacitor, resulting in a voltage drop, V_T, across the capacitor (see Figure 1-2b). For this asymmetric circuit, V_1/V_2 and V_T are determined by the ratio of the electrode areas, that is $V_1/V_2 \approx (A_2/A_1)^n$ where $n \approx 1$ (Vossen 1979; Coburn and Kay 1972; Horwitz 1983). A theoretical model, valid only for very low pressures, suggests that $n = 4$ (Koenig and Maissel 1970), but this has not been observed experimentally.

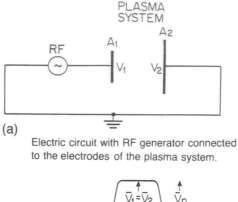

(a)

Electric circuit with RF generator connected
to the electrodes of the plasma system.

(b)

Potential distribution: $\bar{V}_1 = \bar{V}_2 = \bar{V}_p$ is the
average plasma potential.

Figure 1-1. Potential distribution in rf plasma systems without blocking capacitor (Reif 1984).
(a) Electric circuit with rf generator connected to the electrodes of the plasma system.
(b) Potential distribution: $\bar{V}_1 = \bar{V}_2 = \bar{V}_p$ is the average plasma potential.

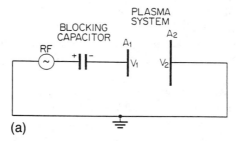

(a)

Electric circuit with RF generator connected
to the electrodes of the plasma system
through blocking capacitor.

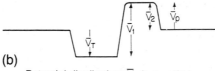

(b)

Potential distribution: $\bar{V}_T$ is a self-bias
voltage across the capacitor and $\bar{V}_1/\bar{V}_2 \approx$
$(A_2/A_1)^n$ where $n \approx 1$.

Figure 1-2. Potential distribution in rf plasma systems with blocking capacitor (Reif 1984).
(a) Electric circuit with rf generator connected to the electrodes of the plasma system through
a blocking capacitor. (b) Potential distribution: $\bar{V}_T$ is a self-bias voltage across the capacitor,
and $V_1/V_2 \approx (A_2/A_1)^n$ where $n \approx 1$.

QUALITATIVE MODEL FOR PECVD

Inelastic collisions between high-energy electrons in the plasma and gas-phase reactants create the necessary active species. The rate of these inelastic collisions is a function of the reactant partial pressure, the collision cross section (which is energy-dependent), and the electron energy distribution (Chapman 1980; Hollahan and Bell 1974). The latter is a function of plasma power and system pressure (Hollahan and Bell 1974). Besides electron-neutral collisions, other inelastic collisions are possible, such as electron-charged species, ion-neutral, ground-state-neutral, and so forth (Chapman 1980; Rand 1979). Soon a steady-state population of neutral free radicals is established. These radicals are believed to have a relatively high sticking coefficient on many substrates (Rand 1979). These absorbed radicals, because of their relatively weak internal binding, are readily broken down by ion bombardment into their constituent atoms. This enhancement of the deposition process by ion bombardment was suggested by experiments in which different film compositions were deposited on the two electrodes of an asymmetric plasma system. As indicated in the section on "PECVD of Polycrystalline Silicon Films" below, ion bombardment can also affect the crystallinity of deposited films (Greene and Barnett 1982). The temperature of the substrate also helps by providing the thermal energy required to promote surface reactions and the desorption of reaction by-products, thus lowering film contamination.

COMMERCIAL PECVD SYSTEMS

The commercial PECVD systems discussed in this section have been used successfully to deposit silicon nitride, silicon oxynitride, silicon oxide, and hydrogenated amorphous silicon films. Some of these systems are currently being used to explore the deposition of polysilicon, epitaxy, and refractory metal silicide films.

The first commercially important PECVD reactor was the Reinberg radial flow reactor introduced in 1974 (see Figure 1-3) (Reinberg 1974). The chamber design consists of two parallel, circular electrodes. The wafers are loaded onto the lower electrode, which is electrically grounded. The upper electrode is connected to an rf generator through an impedance matching network. The glow discharge is created between the two electrodes. The reactants are fed in from the gas ring and enter the deposition zone (i.e., the zone between the electrodes) at its outer edge, and flow radially in toward a pumping port at the center of the lower electrode (Sinha 1980; Kumagai 1984).

An "inverse" radial flow reactor was introduced by Applied Materials in 1976 (see Figure 1-4) (Hollahan and Rosler 1978). The gas inlet in this chamber is at the center of the lower electrode, with the gas flow directed radially outward. A magnetic drive assembly permits rotation of the lower electrode, thus ran-

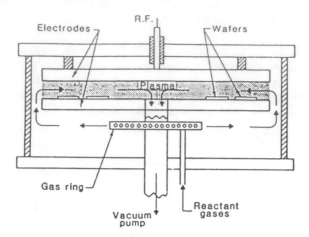

Figure 1-3. Schematic of radial flow reactor designed by Reinberg (Reinberg 1974). (Courtesy of The Electrochemical Society, Inc.)

domizing the substrate position and optimizing deposition uniformity. An improvement to this design, also introduced by Applied Materials, is shown in Figure 1-5 (Kumagai 1984). The perforated electrode in this newer design further improves deposition uniformity.

Several load-locked PECVD systems have also been developed. These systems offer better run-to-run reproducibility than earlier systems and reduced particulate formation because the reactor chamber is not cycled between atmospheric and

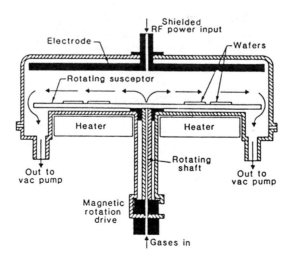

Figure 1-4. Schematic of inverse radial flow reactor (Hollahan and Rosler 1978). (Courtesy of Applied Materials, Inc.)

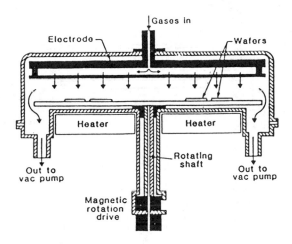

Figure 1-5.   Schematic of inverse radial flow reactor with perforated electrode (Kumagai 1904). (Courtesy of Applied Materials, Inc.)

operating pressure. Both Ulvac and LFE Corporations manufacture linear in-line systems using load-lock stations (Kumagai 1984).

A hot-wall, batch PECVD system (see Figure 1-6) was introduced simultaneously by ASM America and Pacific Western Systems in the late 1970s (Rosler and Engle 1981; Weiss 1983). The deposition chamber consists of a quartz tube placed within a resistively heated furnace. Vertically oriented graphite slabs carry the wafers in slots. Every other slab is connected to the same rf power terminal, as shown in Figure 1-6. The glow discharge is generated between adjacent electrodes. The reactants are directed along the axis of the chamber tube and between the electrodes. One of the advantages of this "tubular" design is a reduction of the collection of particulates on the wafer surface (from wafer handling and flaking off the walls), thanks to the vertical position of the wafers. In addition, good ($\pm 5\%$) film uniformity can be achieved over a large batch of wafers (e.g., ~90 125-mm wafers), and the furnace-type heating aids temperature control. Moreover, the plasma excitation technique for these systems makes it possible to pulse the rf power to the electrodes. (The rf power is applied continuously in all other systems.) Pulsing of the rf power is believed to improve deposition uniformity.

INFLUENCE OF MAJOR PECVD DEPOSITION PARAMETERS

The following general trends have been observed in typical commercial systems:

- *RF power:* An increase in rf power increases the plasma density in the glow discharge, thus increasing the degree of gas dissociation and also the substrate temperature. Consequently, increases in rf power result in increases

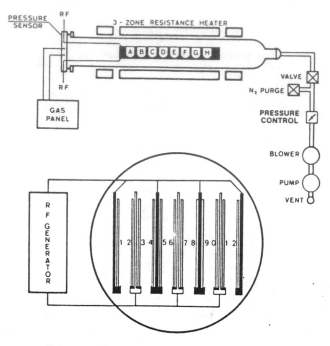

Figure 1-6. Schematic of hot-wall, batch PECVD system (Rosler and Engle 1981). (Courtesy of ASM America, Inc.)

in film deposition rate, as well as increases in reactant gas depletion. Initial transient changes in rf power have been shown to result in nonuniform stoichiometries near the film–substrate interface (Nguyen and Pan 1984), possibly due to changes in electron temperature and density before a steady state is established.

- *RF frequency:* Typical rf frequencies used are as low as 25 kHz and as high as 27 MHz. The rf frequency determines, in some cases, whether the stress in the PECVD film is compressive or tensile (Koyama et al. 1981).
- *Operating pressure:* Commercial systems normally operate in the 0.1 to 5 Torr range. Increases in system pressure result in higher deposition rates.
- *Gas-phase environment:* The partial pressures of the rate-determining reactants control the film deposition rate. Water vapor is undesirable because it inhibits the deposition of the desired film and also promotes the nucleation of particulates. The properties and stoichiometry of the films are determined by the gas-phase composition. Thickness uniformity within the same wafer and from wafer to wafer within the same run is controlled by the gas flow rate and the reactant gas depletion.
- *Substrate temperature:* Higher temperatures result in better film properties, lower film contamination, and improved film adhesion.

PECVD OF DIELECTRIC FILMS

Silicon nitride, silicon dioxide, silicon oxynitride, and silicate glasses are the most important dielectric films in microelectronics today.

Silicon nitride was the first material deposited by the PECVD technique on a large production scale. It is used extensively as a final protective coating and passivation layer for integrated circuits because it serves as an excellent diffusion barrier against moisture and alkali ions. The PECVD technique permits the low deposition temperature (250–400°C) needed to deposit this film over wafers containing an aluminum metallization layer. Furthermore, PECVD nitride films with compressive stress can be obtained, and this makes possible the deposition of relatively thick films. Silane (SiH_4) and ammonia (NH_3) are typically used as sources of silicon and nitrogen, respectively. Silane and nitrogen (N_2) can also be used, but these reactants typically yield silicon-rich films, probably due to the relatively high bond energy of N_2 (Hess 1984). These depositions are normally carried out at pressures of 0.2 to 3 Torr, which produce deposition rates of 300 to 500 Å/min. PECVD silicon nitride films contain 15 to 30 wt % of hydrogen bonded to either silicon or nitrogen (Vossen 1979; Reif 1984; Coburn and Kay 1972). This is important because film properties such as refractive index, stress, and optical absorption edge are greatly affected by the concentration and chemical distribution of hydrogen in the film.

The following general trends relating the properties of PECVD silicon nitride films and the process parameters have been observed. It is important to clarify, however, that these trends are very sensitive to reactor geometry, deposition conditions, and so on, and thus should be used only as a guide.

1. The refractive index (2.0–2.1) increases as the ratio of Si–H to N–H bond increases. The optical absorption edge shows a similar dependence (Samuelson and Mar 1982).
2. The film density (2.5 3.0 g/cm³) and stress depend on the Si/N ratio of the deposited film, while they are not affected by the Si–H/N–H bond ratio (Samuelson and Mar 1982).
3. SiH_4 and N_2 yield films with lower H content compared to SiH_4 and NH_3 (Chow et al. 1982).
4. Higher electrode temperatures yield films with lower H content (Chow et al. 1982).
5. The film etch rate increases with increasing H content (Chow et al. 1982).
6. Low excitation frequencies (e.g., 50 kHz) yield films with small compressive stress ($\sim -2 \times 10^9$ dynes/cm²) (van den Ven 1981). Higher frequencies (e.g., 13.56 MHz), however, may yield films with tensile stress ($\sim 2 \times 10^9$ dynes/cm²) (Sinha et al. 1978). Higher excitation frequencies also result in higher H content, lower film density, higher refractive index, and faster etch rate (Koyama et al. 1981).

More complete reviews of the PECVD of silicon nitride can be found in the references (Hess 1984; Sinha 1980; Zhou et al. 1985; Mar and Samuelson 1980; Classen et al. 1985).

PECVD silicon oxide films have been proposed as an interconductor dielectric material because of their low deposition temperature (250–350°C) and relatively low dielectric constant. Silane is typically used as the silicon source, while nitrous oxide (N_2O), nitric oxide (NO), carbon dioxide (CO_2), and oxygen have been used as oxygen sources. Nitrous oxide is the preferred oxidizer because of the relatively low (1.7 eV) bond dissociation energy of N–O in this molecule (Reinberg 1979). PECVD silicon oxide films are compressive ($0.07–2.4 \times 10^9$ dynes/cm^2) independent of rf frequency, and contain 2 to 9 wt % of hydrogen and less than 5 wt % of nitrogen (van den Ven 1981; Ritchie and Metz 1982). Particulate formation due to gas-phase homogeneous reactions are a typical problem in the PECVD of silicon oxide. These reactions are minimized by using relatively low power densities and high N_2O/SiH_4 ratios (van den Ven 1981). A more complete review on the PECVD of silicon oxide films can be found in the references (Hess 1984; Adams 1983; Kaganowicz et al. 1984).

PECVD silicon oxynitride films combine some of the best properties of PECVD silicon nitride and oxide films. The presence of oxygen lowers the dielectric constant of silicon nitride films, while the presence of nitrogen increases the resistance to sodium ions and moisture of silicon oxide films. Therefore, these films have found applications as interconductor dielectric layers and as surface passivation coating for integrated circuits. The deposition conditions for these films are a combination of those used for silicon nitrides and oxides (Chu et al. 1983).

Phosphosilicate glasses (PSG) and boro-phosphosilicate glasses (BPSG) can be readily deposited by PECVD in hot-wall tubular reactors (van den Ven 1981). PSG is frequently used as the interconductor layer between polysilicon and aluminum. One advantage of this material is that it provides conformal step coverage, which is accomplished by heating the wafer until the glass softens and flows. A disadvantage of this process, however, is that it requires phosphorus concentrations of 6 to 8 wt % and temperatures of 1000 to 1100°C. BPSG softens and flows at temperatures below 1000°C, and is sometimes preferred (Avigal 1983; Tong et al. 1984).

PECVD OF POLYCRYSTALLINE SILICON FILMS

Polysilicon films can be deposited by atmospheric pressure chemical vapor deposition (CVD) (Seto 1975), low pressure CVD (LPCVD) (Kamins 1980), and molecular beam deposition (MBD) (Matsui et al. 1980) at temperatures over 600°C. Polysilicon is heavily used as the gate electrode material in MOS integrated circuits, and is being studied as the semiconductor material for thin film transistors (TFTs) and three-dimensional device integration. The preferred deposition technique for integrated circuit applications is LPCVD. This technique has two basic problems: (1) the deposition rate and the film structure are very sensitive to deposition temperature, and (2) the deposition rate and thickness uniformity are affected when large quantities of dopant atoms are introduced in the film during deposition. TFTs are attractive for flat panel displays but require fabrication temperatures much lower than those used by either CVD, LPCVD, or MBD. PECVD is being studied as an alternative to LPCVD for the fabrication of gate electrodes, and as a low temperature fabrication technique for TFTs (Hirai et al. 1983). The advantages of PECVD are: (1) less sensitivity to deposition temperature, (2) potential capability of introducing large quantities of dopant atoms without affecting the deposition rate, and (3) lower deposition temperature. This section reviews the state-of-the-art of the PECVD of polycrystalline and microcrystalline silicon films. (A microcrystalline structure is defined here as an inhomogeneous phase consisting of microcrystals imbedded in an amorphous matrix.)

The temperatures required to deposit microcrystalline and polycrystalline silicon films have been found to be a strong function of deposition conditions. The influence of gas ambient, ion bombardment, and rf power has been studied in some detail.

Gas Ambient

The amorphous-to-crystalline transition temperature seems to be very sensitive to the gas ambient. Microcrystalline silicon films have been deposited at 300°C using a silane/argon plasma (Usui and Kikuchi 1979), at 80°C using a chemical transport technique in a low-pressure hydrogen plasma (Veprek et al. 1981), at 200°C using a silane/hydrogen/phosphine plasma assisted by a magnetic field (Hamasaki et al. 1980), at 400°C using a silane/helium plasma (Nagata and Kunioka 1981), and at 370°C using a silane/hydrogen/diborane direct current (dc) cathodic discharge (Carlson and Smith 1982) (see Table 1-1). Polycrystalline silicon films have been deposited at 450°C in a silane/argon plasma (Morin and Morel 1979), at 600°C in a silane/hydrogen plasma (Burger et al. 1983), and at

Table 1-1. Examples of Amorphous-to-Microcrystalline Transition Temperatures for PECVD Silicon Films (Reif 1984).

AUTHORS	DEPOSITION TEMPERATURE (°C)	GLOW DISCHARGE AMBIENT
Usui and Kikuchi (1979)	300	SiH_4/Ar
Veprek et al. (1981)	80	Chemical transport/H_2
Hamasaki et al. (1980)	200	$SiH_4/H_2/PH_3$
Carlson and Smith (1982)	370	$SiH_4/H_2/B_2H_6$

$625°C$ in a dichlorosilane/argon plasma (Kamins and Chiang 1982). Grain sizes ranging from 30 (Hamasaki et al. 1980) to 500 Å (Nagata and Kunioka 1981) and deposition rates ranging from 30 (Veprek et al. 1981) to 3000 Å/min (Usui and Kikuchi 1979) have been reported. Kamins and Chiang (1982) found their PECVD polysilicon films to be compressive. They were able to incorporate up to 5×10^{20} phosphorous atoms/cm^3 in the polysilicon film by adding phosphine (PH_3) to the gas ambient. However, they also introduced in the film up to 3×10^{20} chlorine atoms/cm^3, which were supplied by the dichlorosilane (SiH_2Cl_2) used as the silicon source.

It is difficult to assess the relative importance of the gas ambient from the literature because different researchers use different reactor geometries, operating pressures, rf powers, deposition rates, and so forth. Moreover, different researchers use different techniques to determine the amorphous-to-crystalline transition temperature (e.g., X-ray diffraction, transmission electron microscopy, electron diffraction, Raman scattering, dark-conductivity measurements, RHEED, etc.). However, several authors agree on the influence of hydrogen and dopant species in the discharge. The presence of hydrogen in the glow discharge has been found to favor the nucleation of microcrystals at lower temperatures (Hamasaki et al. 1980; Carlson and Smith 1982). This has been confirmed by optical emission spectra and IR absorption measurements, which have revealed that the concentration ratio of H/SiH radicals in the silane plasma is a key factor in initiating the nucleation of microcrystallites (Mishima et al. 1982). Similarly, the presence of either phosphine (Hamasaki et al. 1980) or diborane (Carlson and Smith 1982) in the discharge has been found to lower the amorphous-to-crystalline transition temperature.

Ion Bombardment

A flux of positive ions impinging on the film surface seems to play an important role in forming the microcrystalline film. This has been suggested by experiments in which reversing the polarity of dc discharges caused the amorphous-to-crystalline transition temperature to increase from $370°C$ to $>500°C$ (Carlson and Smith 1982).

RF Power

Silicon films deposited at a given temperature are amorphous if the power supplied to the discharge is lower than a given limit, and crystalline if higher power levels are used. For example, films deposited at $300°C$ and rf voltages ≤ 350 V were found to be amorphous, while those deposited at higher rf voltages were microcrystalline (Kamiya et al. 1981). Furthermore, the volume fraction of the crystallites in the film is increased by increasing the rf power (Mishima et al.

Table 1-2. Resistivity Change of LPCVD and PECVD Undoped Polysilicon Films Upon Annealing. The Films Were Deposited at 650°C and Annealed for 1 Hour at 1100°C in Nitrogen (Burger et al. 1983).

	LPCVD	PECVD
As-deposited resistivity (Ω-cm)	13.8×10^6	0.78×10^6
Annealed resistivity (Ω-cm)	2.1×10^6	0.96×10^6

1982; Kamiya et al. 1981). This is so probably because the ion flux to the film surface and/or the energy of ions bombarding the surface increases as the rf power increases.

The resistivities of microcrystalline silicon films are a function of the crystalline volume fraction of the material; that is, films containing a higher crystalline volume fraction exhibit lower resistivities. PECVD polysilicon films have been found to have lower intrinsic resistivities than their nonplasma counterparts (Burger et al. 1982). Moreover, the intrinsic resistivity of PECVD polysilicon films increases after a post-deposition heat treatment, while that of LPCVD films decreases (Burger et al. 1983) (see Table 1-2). This suggests a grain boundary passivation mechanism, with the passivation occurring while the film is being deposited.

PECVD OF EPITAXIAL FILMS

Silicon epitaxial layers are conventionally deposited by atmospheric or reduced (40–100 Torr) pressure CVD at temperatures of 1050 to 1200°C. The relatively high deposition temperature limits the minimum thickness and conductivity of epitaxial films deposited on heavily doped substrates because of significant dopant redistribution during growth. These limitations impose severe restrictions on the levels of device integration available for VLSI applications, and evidence the need for a technique capable of depositing silicon epitaxial layers at lower temperatures. Molecular beam epitaxy (Ota 1980), ion beam epitaxy (Zalm and Beckers 1982), ion cluster beam deposition (Takagi et al. 1976), and solid phase epitaxy (Quach and Reif 1984) have been proposed as possible low temperature alternatives. Another interesting alternative is PECVD. This section reviews the state-of-the-art of the plasma deposition of silicon epitaxial layers.

Silicon epitaxial layers have been deposited by PECVD at a temperature as low as 800°C using a horizontal water-cooled reactor (Townsend and Uddin 1973). A SiH_4/H_2 (27 MHz) discharge produced deposition rates of 200 Å/min at 800°C and operating pressures of 0.2 to 0.6 Torr. The substrate surface was in-situ cleaned by a hydrogen plasma for a few minutes at the deposition temperature immediately before the start of the deposition. The success of these experiments was attributed to this predeposition plasma cleaning step.

Silicon epitaxial layers have also been deposited at a temperature of 750°C using a vertically aligned PECVD system with rf (13.56 MHz) excitation (Suzuki and Itoh 1983; Suzuki et al. 1979). The glow discharge was confined between two mesh electrodes perpendicular to the gas flow, and the wafer was located downstream, parallel to the mesh electrodes and 150 mm below the lowest electrode. These experiments were carried out at much lower operating pressures, 1×10^{-2} and 3×10^{-3} Torr, which resulted in deposition rates of 1980 and 840 Å/min, respectively. The power level supplied to the discharge was 200 watts. The gas ambient consisted of silane, but germanium was added to the plasma at the beginning of the deposition. The addition of germanium was found to be essential, and it was proposed that it cleaned the wafer surface by removing the native silicon dioxide layer coating the substrate. The addition of germanium, however, resulted in the formation of a Si-Ge alloy in the first 1000 Å of the epitaxial film. Predeposition cleaning consisting of HCl gas (850°C, 10 min) was found unsuitable because it produced epitaxial-films with a "very smoky" looking surface (Suzuki and Itoh 1983). The films were in-situ doped by adding phosphine diluted in helium to the plasma, but they had to be annealed (1000°C, 60 min) to electrically activate most of the phosphorus in the film.

More recently, silicon epitaxial films have been deposited at temperatures as low as 650°C using a low pressure CVD system both with and without plasma enhancement using silane (Donahue et al. 1984; Donahue and Reif 1985; Reif 1984). Figure 1-7 shows a schematic diagram of this system. The chamber consists of a quartz tube sealed to stainless steel endplates by silicone gaskets. The wafer sits on a silicon-carbide-coated graphite susceptor facing a radiant heater. A 13.56 MHz generator is coupled to the gas by an external copper plate wrapped around the top third of the chamber. The rf circuit return path is through a silicon-coated stainless steel false bottom located above the bottom endplate. A dc bias can be applied between the false bottom and the graphite susceptor. By using silane during the deposition at an operating pressure of 1.5×10^{-2} Torr and at a temperature of 775°C, deposition rates of 450 and 340 Å/min were obtained with and without plasma enhancement, respectively. Twenty watts of power was supplied to the discharge during the PECVD experiments. This lower power level probably explains the lower deposition rate observed here compared to those reported in the reference (Suzuki and Itoh 1983). A predeposition in-situ cleaning of the substrate surface to remove the native silicon dioxide was found essential for achieving epitaxial growth. Surface cleaning was done by sputtering the wafer in a 50-watt argon plasma at the deposition temperature for 5 min with a dc bias of 300 V applied between the false bottom and the susceptor. A polycrystalline deposit resulted when the predeposition argon plasma or the dc bias during the predeposition cleaning was not included. Lightly doped epitaxial layers deposited on heavily doped silicon substrates by this technique

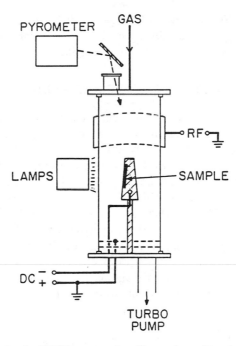

Figure 1-7. Schematic of a PECVD system for silicon epitaxy (Donahue et al. 1984; Donahue and Reif 1985).

exhibit extremely abrupt dopant concentration profiles (Donahue and Reif 1985). The deposition conditions used by different researchers to obtain silicon epitaxy by PECVD are summarized in Table 1-3 (Reif 1984).

Epitaxial layers of GaAs have also been obtained by plasma enhanced techniques (Hariu et al. 1981, Pande 1983, Pande and Seabaugh 1984). Deposition

Table 1-3. Deposition Parameters for the PECVD of Silicon Epitaxy (Reif 1984).

DEPOSITION PARAMETERS	TOWNSEND AND UDDIN (1973)	SUZUKI AND ITOH (1983)	DONAHUE, BURGER, AND REIF (1984)	
Deposition temperature (°C)	800	750	775	
Gas ambient	SiH_4/H_2	SiH_4	SiH_4	
Operating pressure (Torr)	0.2–0.6	1×10^{-2}	1.5×10^{-2}	
SiH_4 partial pressure (Torr)	—	1×10^{-2}	1.5×10^{-2}	
Discharge frequency (MHz)	27	13.56	13.56	
RF power (watts)	350	200	20	0
Deposition rate (Å/min)	200	1980	450	340

temperatures of 350°C and deposition rates of 0.2 μm/min have been reported (Hariu et al. 1981). Plasma cleaning of the substrate surface to remove native oxide layers prior to epitaxial deposition is believed to be essential.

PECVD OF REFRACTORY METALS AND THEIR SILICIDES

Refractory metals and their silicides are being investigated and evaluated as replacement materials for polysilicon. Their high temperature process compatibility along with their high conductivities [metals: $\sim$ 5 $\mu\Omega$-cm, silicides: 16–40 $\mu\Omega$-cm (Murarka 1981)] makes them attractive candidates for gate, interconnection, and contact metallization materials in VLSI circuits.

These materials can be deposited by physical vapor deposition (PVD) methods such as evaporation (Murarka 1980) and sputtering (Crowder and Zirinsky 1979), and by chemical vapor deposition (CVD) (Inoue et al. 1983). CVD offers better step coverage than PVD, and this advantage becomes particularly important as feature size becomes smaller, and sidewalls become steeper. PECVD offers the same advantage of conformal coverage as in thermally driven CVD, but, in addition, it provides more degrees of freedom, which could be used favorably to achieve the following goals:

1. To eliminate the need for post-deposition annealing. Silicides deposited by either PVD or CVD require a post-deposition annealing step (Lehrer et al. 1982) to increase and stabilize the conductivity of the films. This annealing step is needed to crystallize the films (as-deposited films are amorphous) and to achieve the desired stoichiometry [the disilicides are the most stable silicide phase (Murarka 1982)]. PECVD may make it possible to obtain a crystalline deposit with the desired stoichiometry, thus yielding a high conductivity film without the need for the post-deposition annealing treatment.
2. To control the stress of as-deposited films. By proper choice of plasma conditions, it may be possible to deposit a film with the desired stress, as has been reported for PECVD silicon nitride films (Koyama et al. 1981).
3. To lower the deposition temperature. The conductivity of chemical-vapor-deposited metal films is expected to increase with deposition temperature, partly because of increases in grain size (Chu et al. 1982). PECVD may make it possible to enhance the surface mobility of absorbed species so as to achieve a given grain size (i.e., conductivity) at lower temperatures.

PECVD has been used to deposit films of tungsten and molybdenum. Tungsten (W) films have been deposited in a radial flow reactor with rf (4.5 MHz) excitation using tungsten hexafluoride (WF_6) and H_2 (Chu et al. 1982; Tang et al. 1983). Hydrogen was needed to scavenge fluorine species, which were otherwise etching the tungsten film being deposited. Deposition rates of 40 Å/min were obtained at a substrate temperature of 350°C and an operating pressure of 0.2 Torr. Auger studies indicated the presence of silicon and oxygen in the films. The silicon contamination may have come from silicon tetrafluoride (SiF_4) detected in the WF_6 source gas. Fluorine was not detected; however, since the Auger sensitivity for fluorine is only ~ 1 wt %, fluorine contamination cannot be ruled out. The H_2/WF_6 ratio in the ambient affected the film resistivity, with higher ratios resulting in lower resistivities. This is so presumably because the scavenging of fluorine is more effective with higher hydrogen concentrations, thus lowering the fluorine contamination in the deposited film. The resistivity of as-deposited tungsten films decreased with increasing deposition temperature, probably because of larger grains and lower defect concentrations. The resistivity of the films also decreased with post-deposition heat treatments, probably because of outdiffusion of contaminants (e.g., fluorine), annealing of radiation-induced defects during deposition, or increase in the size of the grains. Films deposited at 400°C and subsequently annealed at 1100°C for 30 min had their resistivities lowered from 40 $\mu\Omega$ cm to 7 $\mu\Omega$ cm.

Molybdenum (Mo) films have been deposited by PECVD from molybdenum hexafluoride (MoF_6) and H_2 (Chu et al. 1982; Tang et al. 1983). These films were heavily contaminated with fluorine and exhibited high resistivities ($>10^{-2}$ Ω-cm for a 0.3-μm-thick film). Molybdenum films have also been deposited from molybdenum hexacarbonyl [$Mo(CO)_6$] in a dc discharge onto the cathode of a parallel plate system, but these films contained 20 to 30 wt % carbon (Okuyama 1982). A more successful attempt used molybdenum pentachloride ($MoCl_5$) diluted in H_2 to deposit Mo films by PECVD (Tabuchi et al. 1982; Inoue et al. 1983). These depositions were carried out at a pressure of 1 Torr and temperatures of 170 to 430°C. The as-deposited Mo films were amorphous, but they crystallized after high temperature annealing. Auger analysis indicated the presence of chlorine and oxygen in the as-deposited films. The chlorine concentration in the films decreased after post-deposition heat treatments. The resistivity of as-deposited Mo films decreased with increasing deposition temperature, probably because of the higher incorporation of unreacted chloride in the film at lower temperatures. The resistivity of these films decreased significantly after a post-deposition annealing at 800°C for 20 min in nitrogen, probably because of crystallization of the films and/or outdiffusion of chlorine. The lowest resistivity obtained for Mo films, after annealing, was ~ 10 $\mu\Omega$-cm.

As mentioned earlier, the high conductivities of refractory metals make them attractive candidates for gate and interconnection applications. However, these materials are chemically unstable in oxidizing ambients (Murarka 1981), and this limits their use. To overcome this problem, the silicides of these metals are being used, although their resistivities are an order of magnitude higher than those of the refractory metals. PECVD has been used to deposit silicides of molybdenum, tungsten, tantalum, and titanium.

Mo-silicide films were obtained by adding SiH_4 diluted in argon to the $MoCl_5/$

H_2 plasma ambient discussed above (Tabuchi et al. 1982). The atomic composition of the films was controlled by adjusting the composition of the gas ambient. The resistivity of the as-deposited films increased with increasing SiH_4 flow rate, possibly because of a decrease in the Mo content of the film. The resistivity of these films also increased with increasing hydrogen flow rate, presumably because of a higher concentration of unreacted chloride in the film. The as-deposited films were amorphous and crystallized into the hexagonal structure after a post-deposition annealing in nitrogen at 1000°C for 20 min. It was proposed that chlorine atoms segregated in the grain boundaries may have prevented the hexagonal structure from changing to the tetragonal structure associated with sputtered and CVD $MoSi_2$ (Inoue et al. 1983).

Tungsten silicides (W_xSi_{1-x}) have been deposited in a parallel-plate reactor with rf (13.56 MHz) excitation using WF_6 and SiH_4 diluted in helium (Akitmoto and Watanabe 1981). The substrate temperature of 230°C was relatively low, and deposition rates were about 550 to 600 Å/min. The operating pressure during deposition was kept between 0.5 and 0.7 Torr. The atomic ratio of W/Si in the film was controlled by adjusting the ratio of WF_6/SiH_4 in the gas ambient. The resistivity of the as-deposited films decreased with increasing W content. Also, the resistivity of these films decreased with post-deposition heat treatments. This is particularly interesting because the W_xSi_{1-x} films with $x \leq 0.45$ were amorphous both as deposited and after a post-deposition annealing at 1100°C for 60 min in nitrogen. Therefore, the decrease of resistivity upon annealing was attributed to outdiffusion of F and H, presumably present in the as-deposited films, and not to the crystallization of the films. The lowest resistivity obtained with these films, after annealing, was about 40 $\mu\Omega$-cm.

Tungsten-silicide films have also been deposited by thermal CVD, that is, without the assistance of a glow discharge (Brors et al. 1983). This deposition was carried out in a cold-wall, low-pressure CVD reactor by mixing silane and tungsten hexafluoride diluted with helium at 350 to 400°C and pressures between 50 and 300 mTorr. Growth rates of 500 to 1000 Å/min were obtained. These films are amorphous as deposited, but crystallize after post-deposition heat treatments, reaching resistivities as low as 35 to 40 $\mu\Omega$-cm after annealing at 1100°C for 20 min.

Tantalum-silicide films have been obtained by reacting tantalum pentachloride ($TaCl_5$) and SiH_2Cl_2 in a hydrogen ambient (Hieber et al. 1984). The films were deposited at 1.38 Torr using an inductive-coupled rf (600 kHz or 3.5 MHz) glow discharge. The films deposited at temperatures below 540°C were amorphous, while those deposited at temperatures above 580°C were crystalline and exhibited a resistivity of 70 $\mu\Omega$-cm. The film thicknesses were about 2000 to 5000 Å. After annealing at 900°C for 1 hour in an argon ambient, the resistivity of the films decreased to 55 $\mu\Omega$-cm.

Titanium-silicide films have been obtained by reacting titanium tetrachloride ($TiCl_4$) and SiH_4 in an argon plasma (Kemper et al. 1984). The depositions were carried out in a hot-wall, parallel-plate, 300 kHz plasma reactor at 1 Torr and temperatures of 300 and 350°C. The film composition was found to be very sensitive to gas-phase composition. The as-deposited films were amorphous, but crystallized after a 750°C anneal for 1 hour to yield a resistivity of 20 μΩ-cm.

Titanium-silicide films have also been deposited in a PECVD system similar to that shown in Figure 1-6 but with a modified version of the boat assembly to prevent arcing (Rosler and Engle 1984). The rf generator operates at 50 kHz. Deposition rates of 60 to 80 Å/min are obtained with an rf power of about 100 watts and deposition temperatures of 300 to 500°C. The reactants are SiH_4 and $TiCl_4$ diluted in argon. The as-deposited films are amorphous, but crystallize upon sintering above 600°C. Resistivities as low as 20 μΩ-cm were obtained after annealing at 650°C for 5 min.

More recently, Ti-silicide films have been deposited by laser-enhanced CVD (Gupta et al. 1985). In this process, an ArF excimer laser was used to photo-chemically initiate the gas phase reaction of $TiCl_4$ and SiH_4 inside a flow reactor. Argon gas is added to keep the windows clean during deposition. The operating pressure was about 1 Torr. The laser was operated at 10 kHz and produced 15-nsec pulses of energy about 30 mJ. The films were deposited at temperatures of 400 to 550°C and were amorphous, but they crystallized on annealing at 650 to 700°C. Resistivities as low as 20 μΩ-cm were obtained by this technique after annealing. Typical growth rates were 125 to 175 Å/min.

Titanium-silicide films can also be deposited in a nonplasma, cold-wall chemical vapor deposition environment (see Figure 1-8) (Tedrow et al. 1985) by reacting $TiCl_4$ and SiH_4 at temperatures of 650 to 700°C, pressures of 50 to 450 mTorr, and various $TiCl_4/SiH_4$ flow ratios. In this case, the as-deposited films are polycrystalline, and $TiSi_2$ is the predominant phase. The film thicknesses ranged from 2000 to 15,500 Å, the film resistivities were as low as 22 μΩ-cm, and Si/Ti ratios of 1.8 to 2.3 (as determined by Rutherford backscattering spec-

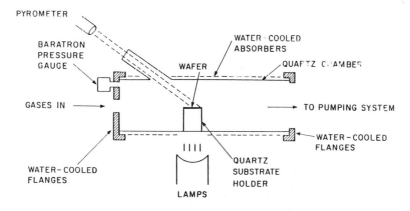

Figure 1-8. Schematic of a CVD system for titanium silicide (Tedrow et al. 1985).

Table 1-4. Refractory Metals and Silicides Discussed in the Text.

MATERIAL	GAS AMBIENT	DEPOSITION TEMPERATURE (°C)	AS-DEPOSITED STRUCTURE (A: AMORPHOUS, C: CRYSTALLINE)	MINIMUM RESISTIVITY (AFTER ANNEALING) (Ω-CM)
Tungsten (W) (Chu et al. 1982)	WF_6/H_2	350	c	7
Molybdenum (Mo) (Tabuchi et al. 1982)	$MoCl_5/H_2$	400	a	10
W-silicide (Akitmoto and Watanabe 1981)	$WF_6/SiH_4/He$	230	a	40
W-silicide (nonplasma CVD) (Brors et al. 1983)	$WF_6/SiH_4/He$	350–450	a	35
Mo-silicide (Tabuchi et al. 1982)	$MoCl_5/SiH_4/Ar/H_2$	400	a	—
Ta-silicide (Hieber et al. 1984)	$TaCl_5/SiH_2Cl_2/H_2$	<540 >580	a c	55
Ti-silicide (Kemper et al. 1984) (Rosler and Engle 1984)	$TiCl_4/SiH_4/Ar$	300–500	a	20
Ti-silicide (laser CVD) (Gupta et al. 1985)	$TiCl_4/SiH_4/Ar$	350–550	a	20
Ti-silicide (nonplasma CVD) (Tedrow et al. 1985)	$TiCl_4/SiH_4$	650–700	c	22 (as-deposited)

troscopy) were obtained. The surface smoothness of these silicide films varied from about 50 Å to about 1000 Å in the thickest film, as determined by a surface profilometer. The refractory metals and silicides discussed here are summarized in Table 1-4.

The science and technology of the PECVD of refractory metals and their silicides is very new and still in the developmental stages. Much more work is needed to optimize the deposition conditions and to study the film properties (e.g., variation in stress with plasma conditions). In particular, a deeper understanding of the relationship between deposition conditions and film properties is needed before the potential advantages of PECVD mentioned earlier can be realized.

OTHER ENERGY-ENHANCED CVD TECHNOLOGIES

There are other energy-enhanced CVD techniques that do not use a plasma and are very promising. For example, a very attractive alternative to PECVD for silicon oxide and silicon nitride films is photo-enhanced chemical vapor deposition (Peters et al. 1980). In this process, photons from a light source are used to excite reactant gases, which then deposit the desired film at temperatures as low as 100°C. One of the most significant advantages of this process is the absence of charged particles (electrons and ions) in the gas ambient, which eliminates the problems caused by radiation damage of microelectronic devices. In addition, this technique makes the selective deposition of films possible when a laser beam is used as the source of photons. Therefore, laser-enhanced CVD (Solanki et al. 1982; Bryant 1978) is potentially capable of eliminating the need for using lithographic masks to pattern films.

SUMMARY

This chapter discussed the plasma enhanced chemical vapor deposition technique and its relevance to microelectronic fabrication. In particular, fundamental aspects of nonequilibrium glow discharges were reviewed, emphasizing the issues important to thin film deposition. The voltage distribution in rf glow discharge systems was discussed, and a qualitative model for PECVD was presented. Commercial PECVD systems were briefly reviewed. The PECVD of films of present commercial importance, such as silicon nitride and silicon oxide, was discussed. The state-of-the-art in the development of the PECVD of films of potential importance in the future, such as microcrystalline silicon, epitaxial silicon, and refractory metals and their silicides, was also discussed. Finally, a brief introduction to the photo-enhanced CVD technique was included. PECVD will continue to be the preferred commercial technique for the low temperature deposition of dielectric films, and is showing great promise for the deposition of semiconductor and conductor films.

ACKNOWLEDGMENTS

The author acknowledges the MIT Center for Materials Science and Engineering (NSF-MRL Grant No. DMR81-19295-A04) for support of the PECVD of polycrystalline silicon films, the Semiconductor Research Corporation (Contract No. 83-01-033) for support of the PECVD of silicon epitaxial layers and the CVD of titanium silicide, and Analog Devices Inc. (Res. Agmt. 8-3-82) for support of the CVD of titanium silicide films.

REFERENCES

Adams, A. C., in *VLSI Technology,* S. Sze, Editor, Chap. 3, McGraw-Hill, New York, 1983a.

Adams, A. C., *Solid State Technol., 26* (4), 135 (1983b).

Akitmoto, K., and Watanabe, K., *Appl. Phys. Lett., 39,* 445 (1981).

Avigal, I., *Solid State Technol., 26* (10), 217 (1983).

Brors, D. L., Fair, J. A., Monnig, K. A., and Saraswat, K. C., *Solid State Technol., 26* (4), 183 (1983).

Brown, S. C., *Basic Data of Plasma Physics,* MIT Press, Cambridge, MA, 1966.

Bryant, W. A., *J. Electrochem. Soc., 125,* 1534 (1978).

Burger, W. R., Donahue, T. J., and Reif, R., in *VLSI Science and Technology/1982,* C. J. Dell'Oca and W. M. Bullis, Editors, pp. 87–93, The Electrochemical Society, Proc. Vol. *82-7,* Pennington, NJ, 1982.

Burger, W. R., Donahue, T. J., and Reif, R., *Proceedings* of the Fourth European Conference on Chemical Vapor Deposition, J. Bloem et al., Editors, pp. 265–272, Eindhoven Druk B.V. (1983).

Carlson, D. E., and Smith, R. W., *J. Electronic Materials, 11,* 749 (1982).

Chapman, B., *Glow Discharge Processes,* John Wiley, New York, 1980.

Chow, R., Lanford, W. A., Ke-Ming, W., and Rosler, R. S., *J. Appl. Phys., 53,* 5630 (1982).

Chu, J. K., Tang, C. C., and Hess, D. W., *Appl. Phys. Lett., 41,* 75 (1982).

Chu, J. K., Sachdev, S., and Gargini, P. A., *The Electrochem. Soc. Extended Abstracts,* Vol. *83-2,* Abs. 321, pp. 510–511 (1983).

Classen, W., et al., *J. Electrochem. Soc., 132,* 893 (1985).

Coburn, J. W., and Kay, E., *J. Appl. Phys., 43,* 4965 (1972).

Crowder, B. L., and Zirinsky, S., *IEEE Trans. Elec. Dev., ED-26,* 369 (1979).

Donahue, T. J., and Reif, R., *J. Appl. Phys., 57,* 2757 (1985).

Donahue, T. J., Burger, W. R., and Reif, R., *Appl. Phys. Lett., 44,* 346 (1984).

Greene, J. E., and Barnett, S. A., *J. Vac. Sci. Technol., 21,* 285 (1982).

Gupta, A., West, G. A., and Beeson, K. W., *The Electrochem. Soc. Extended Abstracts,* Vol. *85-1,* Abs. 268, pp. 390–391 (1985).

Hamasaki, T., Kurata, H., Hirose, M., and Osaka, Y., *Appl. Phys. Lett., 37,* 1084 (1980).

Hariu, T., Takenaka, K., Shibuya, S., Komatsu, Y., and Shibata, Y., *Thin Solid Films, 80,* 235 (1981).

Hess, D. W., in *Silicon Processing,* D. C. Gupta, Editor, p. 218, American Society for Testing and Materials, 1983.

Hess, D. W., *J. Vac. Sci. Technol. A2,* 244 (1984).

Hieber, K., Stolz, M., and Wieczorek, C., *Proceedings* of the Ninth International Conference on Chemical Vapor Deposition, pp. 205–212, The Electrochemical Society, Proc. Vol. *84-6,* Pennington, NJ (1984).

Hirai, Y., Osada, Y., Komatsu, T., Omata, S., Aihara, K., and Nakagiri, T., *Appl. Phy. Lett., 42,* 701 (1983).

Hollahan, J. R., and Bell, A. T., Editors, *Techniques and Applications of Plasma Chemistry,* John Wiley, New York, 1974.

Hollahan, J. R., and Rosler, R. S., in *Thin Film Processes*, J. L. Vossen and W. Kern, Editors, Academic Press, New York, 1978.

Horwitz, C. M., *J. Vac. Sci. Technol. A, 1,* 60 (1983).

Inoue, S., Toyokura, N., Nakamura, T., Maeda, M., and Takagi, M., *J. Electrochem. Soc., 130,* 1603 (1983).

Kaganowicz, G., Ban, V. S., and Robinson, J. W., *J. Vac. Sci. Technol. A2,* 1233 (1984).

Kamins, T. I., *J. Electrochem. Soc., 127,* 686 (1980).

Kamins, T. I., and Chiang, K. L., *J. Electrochem. Soc., 129,* 2326 (1982) and 2331 (1982).

Kamiya, T., Kishi, M., Ushirokawa, A., and Katoda, T., *Appl. Phys. Lett., 38,* 377 (1981).

Kemper, M. J. H., Koo, S. W., and Huizinga, F., *The Electrochem. Soc. Extended Abstracts,* Vol. *84-2,* Abs. 377, pp. 533–534 (1984).

Koenig, H. R., and Maissel, L. I., *IBM J. Res. Dev., 14,* 276 (1970).

Koyama, K., Takasaki, K., Maeda, M., and Takagi, M., *The Electrochem. Soc. Extended Abstracts,* Vol. *81-2,* Abs. 301, pp. 738–740 (1981). Also in *Plasma Processing,* J. Dieleman, R. G. Frieser, and G. S. Mathad, Editors, p. 478, The Electrochemical Society, Pennington, NJ, 1982.

Kumagai, H. Y., *Proceedings* of the Ninth International Conference on Chemical Vapor Deposition, McD. Robinson et al., Editors, pp. 189–204, The Electrochemical Society, Proc. Vol. *84-6,* Pennington, NJ (1984)

Lehrer, W. I., Pierce, J. M., Goo, E., and Justi, S., in *VLSI Science and Technology/1982,* C. J. Dell'Oca and W. M. Bullis, Editors, pp. 258–264, The Electrochemical Society, Proc. Vol. *82-7,* Pennington, NJ, 1982.

Mar, K. M., and Samuelson, G. M., *Solid State Technol., 23* (4), 137 (1980).

Matsui, M., Shiraki, Y., Katayama, Y., Kobayashi, K. L. I., Shintani, A., and Maruyama, E., *Appl. Phys. Lett., 37,* 936 (1980).

Mishima, Y., Miyazaki, S., Hirose, M., and Osaka, Y., *Philosophical Magazine B, 46,* 1 (1982).

Morin, F., and Morel, M., *Appl. Phys. Lett., 35,* 686 (1979).

Murarka, S. P., *J. Vac. Sci. Technol., 17,* 775 (1980).

Murarka, S. P., in *Semiconductor Silicon 1981,* H. R. Huff et al., Editors, pp. 551–561, The Electrochemical Society, Proc. Vol. *81-5,* Pennington, NJ, 1981.

Murarka, S. P., *Materials Lett., 1,* 26 (1982).

Nagata, Y., and Kunioka, A., *Appl. Phys. Lett., 38,* 142 (1981).

Nguyen, V. S., and Pan, P. H, *Appl. Phys. Lett., 45,* 134 (1984).

Ojha, S. M., in *Physics of Thin Films,* G. Hass, M. H. Francombe, and J. L. Vossen, Editors, Vol. 12, p. 237, Academic Press, New York, 1982.

Okuyama, F., *Appl. Phys. A, 28,* 125 (1982).

Ota, Y., *J. Appl. Phys., 51,* 1102 (1980).

Pande, K. P., *The Electrochem. Soc. Extended Abstracts,* Vol. *83-1,* Abs. 340, pp. 531–532 (1983).

Pande, K. P., and Seabaugh, A. C., *J. Electrochem. Soc., 131,* 1357 (1984).

Peters, J. W., Gebhart, F. L., and Hall, T. C., *Solid State Technol., 23* (9), 121 (1980).

Quach, N. T., and Reif, R., *Appl. Phys. Lett., 45,* 910 (1984).

Rand, M. J., *J. Vac. Sci. Technol., 16,* 420 (1979).

Reif, R., *J. Electrochem. Soc., 131,* 2430 (1984a).

Reif, R., *J. Vac. Sci. Technol. A, 2,* 429 (1984b).

Reinberg, A. R., *The Electrochem. Soc. Extended Abstracts,* Vol. *74-1,* Abs. 6, Spring Meeting (1974).

Reinberg, A. R., *Ann. Rev. Materials Sci., 9,* 341 (1979a).

Reinberg, A. R., *J. Electronic Materials, 8,* 345 (1979b).

Ritchie, W., and Metz, W., *The Electrochem. Soc. Extended Abstracts,* Vol *82-2,* Abs. 187, pp. 295–296 (1982).

Rosler, R. S., and Engle, G. M., *Solid State Technol., 24,* (4), 172 (1981).

Rosler, R. S., and Engle, G. M., *J. Vac. Sci. Technol., B2,* 733 (1984).

Samuelson, G. M., and Mar, K. M., *J. Electrochem. Soc.*, *129*, 1773 (1982).

Sawin, H. H., *Solid State Technol.*, *28*, 211 (1985).

Seto, J. Y. W., *J. Appl. Phys.*, *46*, 5247 (1975).

Shanfield, S. R., and Reif, R., *The Electrochem. Soc. Extended Abstracts*, Vol. *83-1*, Abs. 144, pp. 230–231 (1983).

Sinha, A. K., *Solid State Technol.*, *23*, (4), 133 (1980).

Sinha, A. K., Levinstein, H. J., Smith, T. E., Quintana, G., and Haszako, S. E., *J. Electrochem. Soc.*, *125*, 601 (1978).

Solanki, R., Boyer, P. K., and Collins, G. J., *Appl. Phys. Lett.*, *41*, 1048 (1982).

Suzuki, S., and Itoh, T., *J. Appl. Phys.*, *54*, 1466 (1983).

Suzuki, S., Okuda, H., and Itoh, T., *Jap. J. Appl. Phys.*, *19*, Supplement 19-1, 647 (1979).

Tabuchi, A., Inoue, S., Maeda, M., and Takagi, M., *Proceedings* of the 23rd Symposium on Semiconductors and IC Technology of the Electrochemical Society of Japan, p. 60 (Dec. 1–2, 1982).

Takagi, T., Yamada, I., and Sasaki, A., *Thin Solid Films*, *39*, 207 (1976).

Tang, C. C., Chu, J. K., and Hess, D. W., *Solid State Technol.*, *26* (3), 125 (1983).

Tedrow, P. K., Ilderem, V., and Reif, R., *Appl. Phys. Lett.*, *46*, 189 (1985).

Tong, J. E., Schertenleib, K., and Carpio, R. A., *Solid State Technol.*, *27* (1), 161 (1984).

Townsend, W. G., and Uddin, M. E., *Solid State Electronics*, *16*, 39 (1973).

Turban, G., Catherine, Y., and Grolleau, B., *Thin Solid Films*, *60*, 147 (1979).

Usui, S., and Kikuchi, M., *J. Non-crystalline Solids*, *34*, 1 (1979).

van den Ven, E. P. G. T., *Solid State Technol.*, *24*, (4), 167 (1981).

Veprek, S., Iqbal, Z., Oswald, H. R., and Webb, A. P., *J. Phys. C: Solid State*, *14*, 295 (1981).

Vossen, J. L., *J. Electrochem. Soc.*, *126*, 319 (1979).

Weiss, A. D., *Semiconductor International*, *6*, 88 (1983).

Zalm, P. C., and Beckers, L. J., *Appl. Phys. Lett.*, *41*, 167 (1982).

Zhou, N. S., Fujita, S., and Sasaki, A., *J. Electronic Materials*, *14*, 55 (1985).

2. Etching—Applications and Trends of Dry Etching

L. M. Ephrath and G. S. Mathad
IBM General Technology Division

The purpose of this chapter is to acquaint the reader with the technology and capabilities of plasma processing. Batch etching reactors and etching processes are approaching maturity after more than ten years of development. The VLSI requirements of anisotropic and selective etching have been met using a variety of reactor configurations and etching gases. The present emphasis is the integration of plasma etching processes into the overall fabrication sequence. As a result, etch-induced effects, including contamination by metals and insulating films and damage to silicon and silicon dioxide, are the subject of much empirical and some systematic study. The topics covered are oriented toward application—specifically to the fabrication of VLSI (micron and submicron dimension) chips. Most examples originate from the use of plasma processing to fabricate polysilicon and polycide gate FETs, but extension to the fabrication of bipolar devices is usually straightforward. A second topic of current interest is productivity. Wafer diameter has increased from 58 to 125 mm during the development of the present generation of batch etching systems. Single wafer etching is expected to provide higher productivity in the future as the wafer diameter continues to increase and batch size decreases. Even higher throughput may be possible by direct laser etching of patterns in silicon. This resistless technology is under development in several laboratories in the United States and Japan.

HISTORY OF PLASMA PROCESSING

Plasma Etching

Isotropic plasma etching of polysilicon and silicon nitride and plasma stripping of resist were introduced in the early 1970s (Shelton 1975). Etching is carried out in a tube or barrel plasma reactor in which wafers are loaded along the axis of the tube. An rf voltage is coupled into the plasma using capacitor plates or inductive coils around the outside of the reactor. The etching is isotropic because the wafer and the plasma are at approximately the same potential, with the result that the wafer is bombarded by relatively low energy ions. Commonly used etching gases are carbon tetrafluoride (CF_4) plus oxygen for polysilicon, and silicon nitride and oxygen for resist. Plasma etching reactors of this type are still used routinely for stripping silicon nitride and resist, for descumming resist patterns before etching, and for noncritical etching of low resolution patterns (greater than 5 μm). These systems are mostly manually loaded and unloaded, although recently there has been an effort in Japan to automate them. Barrel-type plasma reactors are the least expensive of the plasma reactors. However, they suffer from poor process control and poor etch uniformity.

Parallel plate or radial flow reactors evolved from a reactor designed by Reinberg for plasma deposition (Reinberg 1976). Wafers are placed on a grounded platen, and rf voltage is applied to an electrode located approximately 1.5 cm above the wafers. The etching gas is introduced at the periphery of the substrate platen and pumped through a port located in the center of the platen. The principle behind this method of introducing gases is to optimize etch uniformity by balancing two sources of nonuniformity—a nonuniform supply of reactant and a nonuniform electric field. The concentration of reactant is highest at the edge of the platen and lowest in the center because of depletion by the etching reaction. Electron density is highest in the center of the platen and lowest at the edge as a result of loss of electrons to the wall of the reactor. A schematic of a parallel plate reactor and a diagram of voltage versus position in the reactor appear in Figure 2-1 (a and b). Because the areas of the grounded and powered surfaces are approximately equal in a parallel plate reactor, the dc self bias voltage at the two surfaces is approximately equal. It can be seen from Figure 2-1(b) that a substantial voltage develops between the plasma and the grounded platen because the potential of the plasma with respect to ground is high. The plasma potential varies with rf power, pressure, and etching gas, but, in a symmetrical reactor, is typically several hundred volts. A consequence of a high plasma potential is contamination of wafers by metal sputtered from the upper electrode. The operating pressure is between 200 and 500 mTorr. The pressure cannot be reduced below this range without extinguishing the plasma. The width of the dark space increases with decreasing pressure, and a plasma cannot be sustained once the width of the dark space becomes comparable to a third of the spacing between the plates. The dependence of bias voltage on reactor geometry is discussed in detail in the references (Koenig and Maissel 1970; Vossen 1979). Polysilicon and silicon nitride are etched isotropically in $CF_4 + O_2$ plasmas; however, anisotropic etching of polysilicon has been reported using CCl_4 (Reinberg 1978) and $Cl_2 + C_2F_6$ (Mogab and Levinstein 1980). The disadvantage of this approach to fine line patterning, aside from the toxicity of the etching gases,

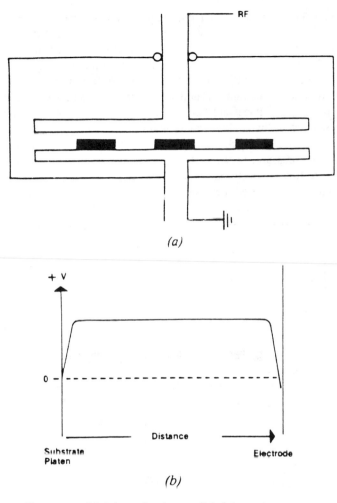

Figure 2-1.   (a) Schematic of a parallel plate reactor;
               (b) voltage vs. position in a parallel plate reactor.

is that patterns are undercut during the overetching needed to clear polysilicon over topography (Mayer and McConville 1979).

Reactive Ion Etching

Reactive ion etching (RIE), or reactive sputter etching (RSE), evolved from sputter etching reactors. RIE reactors are distinguished by an asymmetric configuration with the grounded area substantially larger than the area of the rf electrode. A consequence of this asymmetry is that almost all of the dc voltage is developed between the plasma and the powered electrode. The plasma is only slightly positive, as low as 30 V, with respect to the wall and other grounded surfaces. At these voltages, metal contamination from grounded surfaces is minimal. A schematic of an RIE reactor and a diagram of voltage versus position in the reactor are shown in Figure 2-2 (a and b). Wafers are placed on the smaller electrode (defined as the cathode) and thus subjected to bombardment by energetic ions. Typical bias voltages are between 200 and 500 V. The operating pressure is not restricted in this etching configuration because the spacing between the electrode and the top of the reactor is large compared with the width of the dark space. A low pressure, between 20 and 100 mTorr, is usually chosen because low pressure favors directional etching in a wide variety of etching gases. RIE has become a widely used technique for fine line patterning and, in addition to

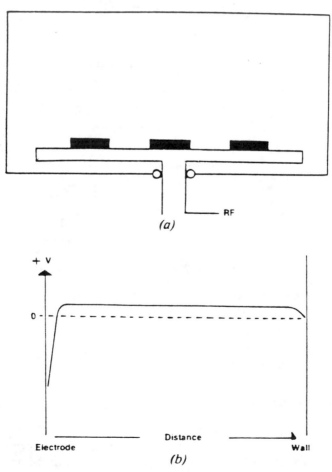

Figure 2-2.   (a) Schematic of an RIE reactor;
(b) voltage vs. position in an RIE reactor.

the original planar electrode configuration, is available commercially in the form of a vertical hexode. In the hexode reactors (Choe et al. 1985) pioneered by Bell Laboratories (U.S. Patent 4,298,443), the wafers are loaded on vertical or slightly inclined surfaces. Most of the batch reactors available today are equipped with a load lock and cassette-to-cassette automatic loading and unloading.

VLSI ETCHING REQUIREMENTS

Volatile Etch Products

A requirement for any plasma etching process is that species produced in the plasma react with the substrate to form volatile products. Appropriate combinations of substrate and etching gas exist for most of the films used to fabricate semiconductor devices. For example, silicon reacts with fluorine radicals, producing volatile silicon tetrafluoride (SiF_4). Convenient sources of fluorine radicals are Freon gases such as CF_4 and similar Freons containing one or more chlorine atoms. Silicon nitride and silicon oxide are also etched in these gases, producing volatile carbon dioxide, CN_x, COF_2, and so on. Organic films including resist are readily etched in oxygen plasmas, producing carbon dioxide and water vapor. Technologically important films that do not etch at practical rates in reactive gases include permalloy (NiFe) and aluminum oxide (Al_2O_3). This is not surprising because readily volatile compounds of nickel and iron do not exist, and, while volatile compounds of aluminum and oxygen do exist, the high bond strength of Al_2O_3 prevents etching.

Anisotropy

Anisotropic etching is required for device dimensions below approximately 2.0 μm. This requirement is introduced for small dimensions because the thicknesses of films do not scale down as rapidly as lateral dimensions. As a result, patterning by isotropic wet or plasma mode etching leads to a substantial reduction in linewidth or even to loss of the etched pattern. Anisotropic etching is observed when a voltage appears between the substrates and ions produced in the plasma. The precise role of energetic ions is not understood for all etching reactions, but their participation is evident; directly or indirectly they enhance etching by neutral species.

Selectivity

Early dry etching processes sacrificed the selectivity of wet etching to meet the original requirement for dimensional control. Applications often combined dry and wet etching steps in order to minimize undercut and, at the same time, provide etch selectivity to thin underlying layers. Much of the literature in dry etching from the mid-1970s has dealt with the discovery, application, and understanding of selective etching. Selective etching of insulators, silicon dioxide, and silicon nitride with respect to silicon and resist is needed to define isolation between devices and to open contact holes to polysilicon gate electrodes, the silicon substrate, and thin diffused regions in the silicon. The inverse selectivity is needed to pattern polysilicon gate electrodes without removing the thin underlying gate oxide. The etch rate ratio that is needed increases in both cases as devices become smaller. Higher selectivity for silicon dioxide with respect to silicon is needed because junction depth decreases faster than the thickness of the field oxide. Higher selectivity for silicon with respect to silicon dioxide is needed because the thickness of the gate oxide decreases faster than the thickness of the gate electrode. The required selectivity depends on the thicknesses of the etched and underlying films as well as on the topography produced by prior process steps. Typical requirements are between 15 and 20 to 1. An example of a calculation to determine minimum etch selectivity for patterning a polycide film over a gate oxide can be found in Wang et al. (1982). There exists a confusing variety of selective etching processes, which will be summarized in the following section.

Clean Etching

The task at hand is to integrate relatively mature etching processes into the fabrication sequence. Increased attention is being paid to the cleanliness of etching as device dimensions shrink, because full dry etching is needed at more critical levels in the fabrication process, and because cleaning steps after etching are restricted in temperature and in the amount of silicon that can be removed. Etch-induced damage and contamination can be divided into the following categories: metals sputtered from reactor components, insulating residue, lattice damage in silicon, hydrogen permeation into silicon, and radiation damage in silicon dioxide. These effects are discussed below in detail. Also discussed are device and process implications. Because RIE has become the batch etching technique of choice, the following sections on etching processes and etch-induced effects deal exclusively with etching carried out at low pressure with energetic ion bombardment.

SUMMARY OF SELECTIVE ETCHING PROCESSES

Silicon and Silicides

Selective gases for silicon and refractory metal silicides are those that produce fluorine and/or chlorine atoms in the discharge. Gases that produce fluorine include sulfur hexafluoride (SF_6) (Gdula 1979; Wagner and Brandt 1981) and nitrogen trifluoride (NF_3) (Eisele 1981; Woytek et al. 1984). The advantage of these gases is that highly selective etching of silicon with respect to oxide can be obtained. The disadvantage is that they have a tendency to undercut the etch mask and exhibit strong loading effects. It should be noted that NF_3 is toxic. The Freons—CF_4, $CClF_3$, CCl_2F_2, CCl_3F, and CCl_4—have also been used to etch silicon. These gases are neither toxic nor corrosive, with the exception of CCl_4, which is carcinogenic. Selective etching of silicon with respect to oxide increases in this set of gases as fluorine is replaced by chlorine. Selectivity in CF_4 is poor, approximately 1 to 1, but selectivity increases to approximately 5 to 1 in CCl_2F_2 (Hosokawa et al. 1974). Oxygen is often added to these Freons to further increase the selective etching of silicon with respect to oxide. RIE in CCl_3F plus oxygen was used to fabricate 1 μm dimension polysilicon gate MOSFETS (Endo and Kurogi 1980), but the reported selectivity of 5 to 1 is considered marginal. Selectivity in CCl_2F_2 has been increased to 15 to 1 in a flexible diode reactor (Pogge et al. 1983). A perforated counter electrode is added to an RIE reactor, and rf power is split between the two electrodes (Figure 2-3). In this way, quasi-independent control is gained over the flux of etching species and the energy of incident ions.

Figure 2-4 shows the etch rate of n^+ polysilicon and silicon dioxide as a function of voltage on the substrate electrode. The voltage between the electrode and ground potential is varied from zero to 400 V by varying the fraction of total rf power delivered to the two electrodes. The etch rate ratio at zero voltage is high, greater than 30 to 1, but etching is isotropic. Etching is vertical at 400 V, but the etch rate ratio is reduced to approximately 4 to 1. An intermediate voltage of 180 V produces etching that is vertical with an etch selectivity of 15 to 1. Undercutting of the etch mask occurs at voltages lower than approximately 150 V. Half micron lines and spaces etched into polysilicon are seen in Figure 2-5. A high selectivity of 20 to 1 has been reported using gas mixtures containing Cl_2 (Pogge et al. 1983). This gas is highly toxic and corrosive and is usually used only when selectivity requirements cannot be met by Freons. Some safety and environmental issues associated with the use of toxic etching gases are discussed in Herb et al. (1983).

Composite films of polysilicon and a metal silicide (polycide) are increasingly used to reduce the sheet resistance of polysilicon electrodes and interconnects (Crowder and Zirinsky 1979). The etching of tungsten silicide on polysilicon to fabricate NMOS circuits with 1 μm minimum features is described in Tsai et al. (1981). Difficulties that have been reported in etching polycide composite films include controlling etch slope (avoiding overhanging profiles), variable etch rates of the silicide film resulting from subtle variations in deposition and thermal history, and the presence of low etch rate interfacial films between polysilicon and silicide. A review of polycide etching processes appears in Chow et al. (1985).

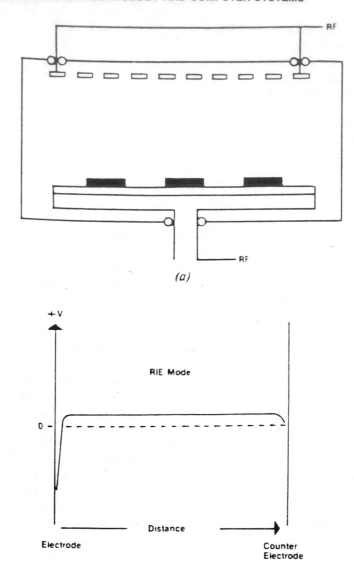

Figure 2-3.   (a) Schematic of a flexible diode reactor; (b) voltage vs. position in a flexible diode reactor.

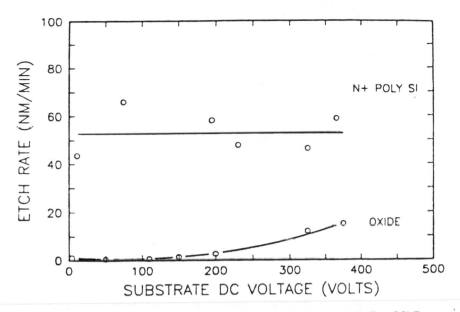

Figure 2-4.   Polysilicon and oxide etch rate vs. bias voltage in 25 mTorr CCl₂F₂.

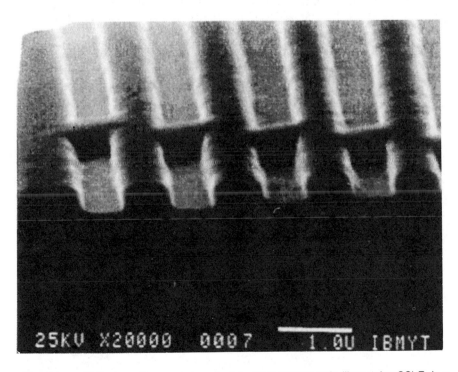

Figure 2-5.   Half micron lines and spaces etched into 250 nm polysilicon using CCl₂F₂ in a flexible diode reactor.

Silicon Dioxide

Selective etching of silicon dioxide with respect to silicon is obtained in etching gases that are deficient in fluorine. Selective gases include those that have a higher carbon to fluorine ratio than CF_4—for example, C_2F_6 (Matsuo 1980). The disadvantage of these gases is that they are polymerizing, and the tendency to form polymer on the wafers is difficult to control. The discharge can be made deficient in fluorine by adding H_2 to CF_4 (Ephrath and Petrillo 1982) or $CClF_3$ (Bennett 1982). High selectivities exceeding 40 to 1 for oxide to silicon and 30 to 1 for organic resists to silicon have been obtained in $CF_4 + H_2$ (Figure 2-6). Half micron lines and spaces etched in silicon dioxide using a PMMA etch mask are shown in Figure 2-7. Higher selectivities and larger process windows are obtained using $CClF_3 + H_2$. The selectivity obtained in the H_2 mixtures depends on the residence time, t, of gas in the discharge. The value of t (sec) can be calculated from pressure, P (Torr), the volume of the reactor, V (liters), and the flow rate, Q (Torr-liters/sec), by $t = PV/Q$. The data plotted in Figure 2-6 were generated at a residence time of 1 sec. The etch rate ratio decreases rapidly as residence time is increased. The use of H_2 requires careful attention to reactor interlocks and H_2 sensors. A popular selective etching gas for silicon dioxide is CHF_3 (Lehmann and Widmer 1978). The advantage is that no safety

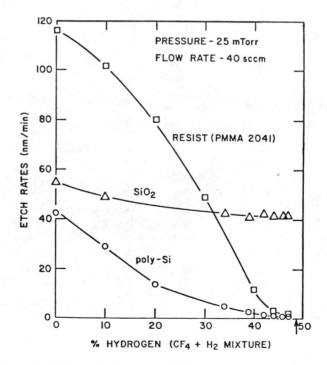

Figure 2-6. Oxide, polysilicon, and PMMA etch rates vs. percent H_2 in CF_4.

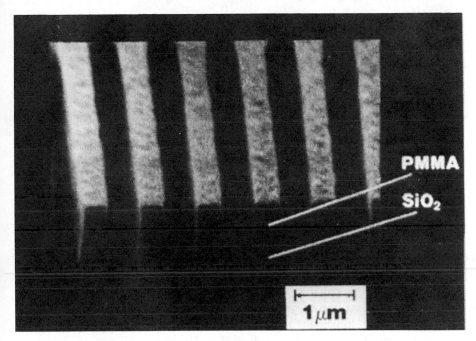

Figure 2-7. Half micron lines and space etched into 300 nm oxide using 1 μm PMMA as an RIE mask.

precautions are required, and the added complexity of controlling gas mixtures is avoided. Etch rate ratios are lower, about 10 to 1, but still adequate for many applications. Oxygen is often added to CHF_3 to reduce its tendency to polymerize (Schanfield and Hendricks 1984a and b). The addition of oxygen is, however, at the expense of selectivity. An excellent discussion of selective etching mechanisms appears in Coburn (1982).

PLASMA ETCH DAMAGE AND CONTAMINATION

Known effects of plasma processing include sputtering and redeposition of metals from reactor components onto the substrate, roughening of the substrate if the redeposited metal acts as a nonuniform etch mask, deposition of an insulating film from plasmas that tend to polymerize, hydrogen permeation into silicon, and damage to silicon and silicon dioxide.

Metal Contamination

The greatest concern is contamination by the transition metals nickel, iron, and chromium, which are constituents of stainless steel, and copper, a common alloying element in aluminum. These metals diffuse readily into silicon where they introduce energy levels near the middle of the silicon bandgap. Minority carrier lifetime is reduced as a result because these levels are efficient recombination centers. A consequence of contamination by fast-diffusing heavy metals during etching of contact holes to p-n junctions is an increase in leakage across the junction under reverse bias.

One source of these metals in an etching reactor is the rf cathode. Typical dc bias voltages are between 200 and 500 V, well above the sputtering threshold for metals of approximately 30 V (Wehner and Anderson 1970). Additional sources are reactor walls and other grounded surfaces. Metal at ground potential can be sputtered if the plasma potential exceeds the sputtering threshold. Contamination by heavy metals can be avoided by judicious choice of cathode material and reactor geometry. RF cathodes can be covered with semiconductor grade plates, for example, silicon (Bondur 1979), silica (Schwartz et al.), or polymer (Iida et al. 1982), or coated with high purity films, for example, aluminum (Ephrath and Bennett 1982). Sputtering of grounded surfaces can be avoided by increasing the area of the grounded surface with respect to the rf electrode.

A relatively simple, quantitative measure of heavy metal contamination can be obtained by measuring the retention time of MOS capacitors fabricated on etched silicon surfaces. Retention time is measured using a pulse C-V technique in which capacitors are pulsed into deep depletion (Young and Osburn 1973). The retention time, or time for full recovery to inversion, is then extrapolated from the change in capacitance over 10 sec (Figure 2-8). An acceptable retention time is considered to be 500 sec, which corresponds to a minority carrier lifetime of 25 μsec.

An example of the use of MOS retention time to monitor metal contamination is taken from Ephrath and Bennett (1982). In this study, MOS retention time was measured for substrates etched in CF_4 on aluminum substrate holders of

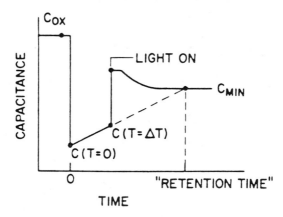

Figure 2-8.   Pulse C-V technique for measuring MOS retention time.

Table 2-1. Effect of Substrate Holder
Purity on Average
Retention Time.

ETCHING CONDITIONS	AVERAGE RETENTION TIME (SEC)
Aluminum alloy substrate holder	
RIE: CF$_4$	<10
99.999% pure aluminum substrate holder	
RIE: CF$_4$	480
Pure aluminum-coated substrate holder	
RIE: CF$_4$	>1000
Unetched control wafer	>1000

differing purity. It was found that retention time did reflect the change in purity, with values ranging from near-zero for a substrate holder fabricated from 1100 aluminum alloy to greater than 1000 sec for a substrate holder coated with 99.9999% pure aluminum (Table 2-1). An intermediate retention time of 480 sec was obtained using a substrate holder fabricated from 99.999% pure Al. The significance of these results is that device level purity silicon can be retained after exposure to energetic ion bombardment if other bombarded surfaces are sufficiently pure and are compatible with semiconductor devices. Grounded surfaces in the reactor used for these experiments were either stainless steel or aluminum alloy, but the presence of these surfaces did not result in reduced MOS retention time because the potential of the plasma with respect to ground potential was low, approximately 30 V. MOS retention time was also measured for silicon etched in different gases. It was found, using a 99.999% pure aluminum substrate holder, that the etching gas also affects retention time. Addition of H$_2$ or O$_2$ to CF$_4$ increases retention time, probably by reducing the amount of sputtered aluminum that is redeposited (Table 2-2).

The effect of various pre- and post-etch treatments on MOS retention time was also determined as part of this study. It was found that back-side gettering is effective in removing heavy metal contamination from the front of the wafer. A damaged region is created at the back of the wafer by ion implantation of argon (350 keV, 5 × 10exp15 ions/cm$_2$). Silicon wafers with and without the damaged region were etched in CF$_4$. An aluminum ally substrate holder served as the source of heavy metal contamination. MOS retention time was increased from near-zero to over 700 sec by gettering. To be effective, back-side gettering does require a high temperature step after etching to drive the metals to the damaged region. In this experiment, high temperatures were reached during the oxidation step. A post-etch cleaning step that did not involve high temperature processing was found to improve retention time for capacitors fabricated using

Table 2-2. Effect of Etching Gas on Average Retention Time.

ETCHING CONDITIONS	AVERAGE RETENTION TIME (SEC)
99.999% pure aluminum holder	
RIE: CF_4	480
RIE: $CF_4 + O_2$	690
RIE: $CF_4 + H_2$	735
Unetched control wafer	1000

the 99.999% pure substrate holder. Silicon wafers etched in $CF_4 + H_2$ were exposed after etching to an O_2 plasma followed by a BHF dip. As a result, the retention time increased to a value comparable to that of the unetched control (Table 2-3). This cleaning step grows and removes an approximately 10 nm thick plasma oxide film. Post-etch cleaning steps were ineffective for wafers etched on the aluminum alloy substrate holder. After heavy metal contamination was shown by MOS retention time to be minimal, contact holes were etched to 0.25 μm deep p-n junctions (Ting and Crowder, 1982), and junction leakage was very low (10^{-9} amps/cm^2).

The difficulty in removing heavy metal contamination has been documented by intentionally contaminating silicon with iron, nickel, and chromium and monitoring contamination levels after post-etch cleaning steps (Hosoya et al. 1985). It was found using SIMS analysis and observation of stacking fault density after oxidation that while chromium and iron are removed by standard preoxidation cleaning steps plus removal of 20 nm silicon by wet etching, nickel

Table 2-3. Effect of Post-RIE Process Steps on Average Retention Time.

ETCHING CONDITIONS	AVERAGE RETENTION TIME (SEC)
99.999% pure aluminum substrate holder	
RIE: $CF_4 + H_2$	595
RIE: $CF_4 + H_2$	
O_2 plasma + BHF	>1000
Unetched control wafer	>1000

contamination remains even after 150 nm silicon has been removed. It is speculated that nickel is redeposited during wet etching.

An example of the effect of reactor geometry on metal contamination is taken from Pang (1984). Ion bombardment of grounded stainless steel walls was reduced by decreasing the area of the rf electrode, and the effect was monitored using MOS capacitors fabricated on silicon etched in CF_4. It was found that the density of interface states decreased when the diameter of the electrode was reduced from 20 to 10 cm. In a related experiment, the stainless steel walls were coated with a film of SiO_x deposited from a $SiCl_4 + O_2$ discharge. Again, the density of interface states decreased for electrode bias voltages of 600 V and 1200 V. Pang points out that there are drawbacks to coatings, specifically pin holes and flaking.

A recently documented consequence of heavy metal contamination is the nucleation and growth of anomalously large polysilicon grains on silicon surfaces contaminated with nickel (Bendernagle 1985). Silicon surfaces were contaminated during etching in CF_4 by metal sputtered from a stainless steel fixture in close proximity to the wafer. After etching, the silicon was cleaned in acidic and basic hydrogen peroxide solutions to remove metal contamination from the surface. LPCVD polysilicon was then deposited from SiH_4. The polysilicon film contained a large number ($10exp7/cm^2$) of hemispherical bumps and whiskers. SIMS profiling showed the presence of nickel at the silicon–polysilicon interface. In supporting experiments, polysilicon was deposited on silicon wafers intentionally contaminated by evaporated nickel. Polysilicon films with large grains resulted. To rule out lattice damage as the cause of the large polysilicon grains, silicon wafers were damaged by ion implantation of argon. Smooth polysilicon was obtained on the damaged wafers. This consequence of metal contamination is technologically important because polysilicon is being used to make contact to the emitter and base regions of high-speed bipolar devices. Contact is made by depositing polysilicon onto the silicon substrate through openings etched in the insulator.

It should be noted that aluminum has not been found to affect the electrical properties of etched silicon. This is not unexpected because aluminum does not diffuse readily into silicon, and it is effectively removed by standard wet preoxidation cleaning solutions.

Surface Roughness

Redeposited metal can act as a nonuniform etch mask and cause the etching surface to roughen. An etched silicon surface that is rough is a concern when gate (or field) oxide is grown on etched silicon to form planar (or recessed) isolation, as locally high electric fields can develop at asperities and reduce the breakdown strength of the oxide. Locally high electric fields have been shown to cause the lower breakdown strength of oxide grown on polycrystalline silicon (DiMaria and Kerr 1975). Primary sources of sufficient quantities of redeposited metal to cause roughening are metal electrodes and etch mask material, which are chemically inert.

The quality of etched surfaces can be conveniently monitored by measuring the breakdown voltage of MOS capacitors (Osburn and Ormond 1972). A measure of quality, the defect density, can be calculated from a histogram of breakdown events versus applied field (Figure 2-9). Defect density is defined by the equation:

$$\text{defect density} = \ln(\text{yield})/\text{area}$$

where yield is the fraction of devices with breakdown field between a theoretical maximum, E_{max}, and 80% of E_{max}. Because of the stringent definition of yield, the calculated defect density can be high without the occurrence of low field breakdown events. An acceptable value for defect density is considered to be 100 or fewer defects/cm^2.

The defect density in oxide grown on etched silicon has been used to characterize silicon surfaces etched in CF_4 and mixtures of CF_4 with O_2 and with H_2 (Ephrath and Bennett 1982). The etch rate of silicon in CF_4 and $CF_4 + 20\%$ O_2 is 400 and 500 Å/min, respectively. The etch rate is low, approximately 30 Å/min in $CF_4 + H_2$. The substrate holder in these experiments is aluminum. The histograms of breakdown events versus applied field appear in Figure 2-10. The calculated defect density is given in Table 2-4. The defect density of oxides etched in CF_4 alone is higher than the unetched control by a factor of almost 100. The addition of O_2 or H_2 significantly decreases the defect density. The

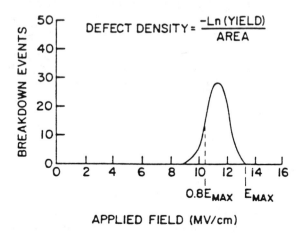

Figure 2-9 Calculation of defect density from histogram of breakdown events vs. applied field.

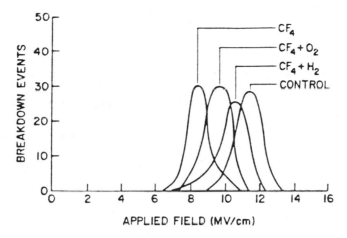

Figure 2-10.   Histograms of breakdown events vs. applied field for oxides grown on surfaces etched by RIE in CF_4, $CF_4 + O_2$, and $CF_4 + H_2$.

addition of O_2 decreased the defect density by a factor of 1.5 to 2, and the addition of H_2 decreased the defect density to a value below 100 defects/cm². The aluminum substrate holder is the source of sputtered metal and the cause of degradation in oxide quality. Aluminum was detected on all etched wafers by SIMS analysis. It is probable that the addition of O_2 to CF_4 reduces the defect density by oxidizing the aluminum substrate holder and reducing its sputter yield. The addition of H_2 to CF_4 reduces defect density to a value that would be acceptable for a control wafer by reducing the etch rate of silicon so that the silicon surface remains smooth even though aluminum is redeposited. Measurement of MOS breakdown voltage was also used in this study to evaluate cleaning steps following RIE in $CF_4 + H_2$. It was found that the growth and removal of

Table 2-4. Effect of Etching Gas on Defect Density.

ETCHING CONDITIONS	DEFECT DENSITY (DEFECTS/CM³)
99.999% pure aluminum holder	
RIE: CF_4	445
RIE: $CF_4 + O_2$	283
RIE: $CF_4 + H_2$	60
Unetched control wafer	5

a thin plasma oxide film reduced the defect density further, to a value that compared favorably with the control.

Formation of Insulating Film

Gases for selective etching of oxide form an insulating, Teflon-like polymer film on silicon. Selective etching of oxide on silicon to open contact holes results in high resistance contacts if the insulating film is not removed prior to metallization.

The presence of an insulating deposit can be conveniently detected by forming Schottky barrier diodes on etched surfaces. For example, Au contacts to silicon exposed to RIE in $CClF_3 + H_2$ show a large series resistance in the forward bias direction (Figure 2-11, from Fonash 1985), indicating the presence of an insulating layer. The presence of Teflon-like films can also be deduced from

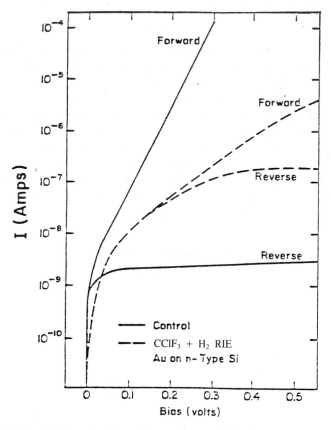

Figure 2-11. *I-V* characteristics of Au contacts on silicon etched in $CClF_3 + H_2$.

Table 2-5. Relative Auger Intensity (%).

SAMPLE	C	N	O	F	Al	Si
Control	8	—	29	—	—	63
CF_4	23	—	39	7	—	31
$CF_4 + H_2$	54	—	0	17	4	15

analysis of the etched surface by Auger electron spectroscopy (AES) and secondary ion mass spectroscopy (SIMS). Additional information on the thickness of the film and the bonding of carbon in the film is obtained by Auger depth profiling and electron spectroscopy for chemical analysis (ESCA), respectively. Silicon surfaces etched in $CF_4 + H_2$ were analyzed by these four techniques (Ephrath and Bennett 1982). For comparison, silicon surfaces etched in CF_4 and unetched silicon surfaces were also analyzed. The results are summarized in Tables 2-5 through 2-8. The surface of a sample etched in $CF_4 + H_2$ differs from that of the control in two respects. Fluorine is detected by AES on the etched surface but not on the control, and the amount of carbon is significantly higher. A sample etched in CF_4 shows intermediate levels of both elements. SIMS analysis shows the amount of hydrogen to be a factor of 5 higher after exposure to the selective etching gas. The thickness of the polymer film is approximately 20 Å. Increasing etch time from 5 to 10 min did not result in an increase in the thickness of the polymer film. The chemical bonding of carbon is consistent with that of a fluorocarbon.

AES was also used in this study to monitor the effectiveness of a number of post-etch cleaning steps. The effect of plasma and wet cleaning steps is shown in Table 2-9. The growth and removal of a thin plasma oxide produced an etched surface that was indistinguishable from that of a control. Exposure to a short etch in $CF_4 + O_2$ (removing 80 Å silicon) or to a forming gas plasma in a commercial barrel reactor removed the carbon but increased the percent oxygen on the silicon surface. A standard wet preoxidation clean is not effective in removing the polymer film.

Table 2-6. O_2 and Cs SIMS Total Integrated Processed Counts.

	O_3 SIMS				Cs SIMS			
SAMPLE	Li	B	Na	Al	H	C	N	O	F
---	---	---	---	---	---	---	---	---	---
Control	68	5.7e2	3.8e4	5.6e4	2.4e4	1.8e5	1.5e4	1.3e6	1.7e5
CF_4	125	6.4e3	1.6e4	6.6e5	2.7e4	7.4e5	3.1e4	1.6e6	3.1e6
$CF_4 + H_2$	146	6.2e3	8.8e4	3.1e7	1.5e5	1.9e7	3.0e5	2.6e6	5.5e6

Table 2-7. Auger Depth Profile.

SAMPLE	APPROXIMATE FILM THICKNESS (Å)
Control	<5
CF_4	<5
$CF_4 + H_2$	21

A forming gas plasma cleaning step was used to remove a polymer film before contacting 0.25 μm deep p-n junctions with Ti/Al metallurgy (Ting and Crowder 1982). Contact resistivity was in the low $10exp-6$ ohm-cm^2 range, and contact resistance of approximately 15 ohms/contact was measured for a chain of 1 μm dimension contact holes. It is thought that low resistance contacts are obtained in the presence of residual oxygen left by the cleaning step because of the ability of Ti to getter oxygen. High resistance contacts resulted when the cleaning step was omitted.

Silicon Damage

Ion bombardment during RIE creates a damaged layer within several hundred angstroms of the silicon surface. A damage layer created by RIE in CCl_4 (Fonash et al. 1985) and NF_3 (Ashok et al. 1983) has been shown to shift the barrier height of gold Schottky barrier diodes. The barrier height is reduced for metal to n-silicon contacts and increased for metal to p-silicon contacts. The hypothesis is that the shift is caused by positive charge associated with lattice damage. In a similar study, damage produced in silicon by RIE in $CClF_3 + H_2$ was physically characterized using reflected high energy electron diffraction (RHEED) and Rutherford backscattering (RBS) in the channeling mode, and electrically characterized using gold and platinum silicon contacts (Mu et al. in press). The

Table 2-8. Surface Elemental Compositions Estimated from ESCA Data and Expressed as Atomic Percent for Detected Elements.

SAMPLE	C—H,C—C	—CF— —C—O—	—CF$_2$—	—CF$_2$—
Control	8.3	2.7	—	—
CF_4	12.5	6.6	2.3	0.5
$CF_4 + H_2$	19.6	16.6	4.8	0.8

Table 2-9. Relative Auger Intensity (%).

SAMPLE	C	N	O	F	Al	Si
Control	8	—	29	—	—	63
$CF_4 + H_2$	54	—	10	17	4	15
$CF_4 + H_2$, forming gas plasma	10	3	47	5	8	27
$CF_4 + H_2$, peroxide clean	29	—	37	5	—	29
$CF_4 + H_2$, $CF_4 + O_2$ clean	14	—	53	—	—	33
$CF_4 + H_2$, O_2 plasma	7	—	30	—	—	63

etching process is described in Bennett (1982). The RHEED pattern obtained after etching indicates the presence of both polycrystalline and amorphous silicon. This result appears to be a general one because it is consistent with the earlier study using CCl_4. The etch rate of silicon is very different in the two gases: tens of angstroms/min in $CClF_3 + H_2$ and hundreds of angstroms/min in CCl_4. Analysis by RBS is consistent with the RHEED results. The RBS channeling spectra for silicon etched in $CClF_3 + H_2$ and a control wafer (Figure 2-12 from Mu et al. in press) shows a higher surface peak and background, indicating displaced silicon atoms for the etched silicon. Gold Schottky barrier diodes exhibited a

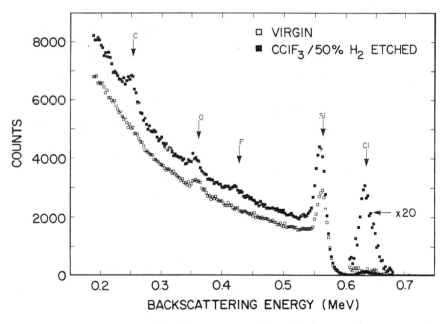

Figure 2-12.   Rutherford backscattering spectra, in the channeling mode, for silicon etched in $CClF_2 + H_2$.

high leakage current under reverse bias, presumably due to the damage-induced positive charge. Diode I-V characteristics were greatly improved by various post-RIE treatments that removed damaged silicon, but characteristics of the control were not fully restored.

RIE-induced damage to silicon has not been observed to degrade p-n junctions as shallow as 1000 Å (Ting and Crowder 1982). This is not surprising because damage is confined to within several hundred angstroms of the surface of the heavily doped silicon, well removed from the p-n junction.

Hydrogen Permeation

It has been shown by the $1\,H\,+\,15\,N \rightarrow 4\,He\,+\,12\,C\,+\,-$ nuclear resonance reaction technique that hydrogen atoms created in etching gas mixtures of H_2 permeate into silicon to a depth of approximately 500 Å (Oehrlein et al. 1985; Mu et al. 1987). TEM observations further suggest a unique defect structure in which silicon atoms displaced by hydrogen migrate several tens of angstroms before reorganizing into dislocation loops (Frieser et al. 1983). The mobility of hydrogen-induced displacement damage is attributed to the production of point defects that can diffuse during exposure to the plasma. In contrast, defects produced by heavier atoms—for example, during sputter etching or implantation of dopant atoms—are produced in stable clusters.

The presence of hydrogen or hydrogen-induced damage has not restricted the use of hydrogen-containing gases for selective etching of silicon dioxide to silicon. In a typical etching process, silicon is exposed to hydrogen only during overetching of silicon dioxide. According to Frieser et al. (1983), acceptable MOS retention time is obtained for relatively short exposure times of 5 min. Hydrogen and associated damage produced during longer exposures can be removed by etching in a nonselective gas such as CF_4.

Low energy hydrogen ions have been intentionally implanted into reactive ion etched silicon to restore I-V behavior of metal/silicon contacts (Wang et al. 1983). A correlation of restoration of I-V behavior with surface analysis data is given in Mu et al. (1985).

Radiation Damage in Silicon Dioxide

RIE and other processing environments that produce ionizing radiation create positive charge at the silicon/silicon dioxide interface and bulk neutral traps in blanket-exposed silicon dioxide. The positive charge is removed by a standard post-metallization anneal at 400°C and is not discussed further. The neutral traps are not removed by this low temperature anneal. Neutral traps are of concern because, if they are present in the gate oxide of an FET, they may become populated by hot electrons injected from silicon and cause a shift in the threshold voltage during the operation of the device. Bulk neutral traps carpeted in thermal silicon dioxide by blanket exposure to RIE have been characterized (Ephrath and DiMaria 1981). Partially etched films were incorporated into MOS capacitors, and charge was injected into the oxide by avalanche injection from silicon and internal photoemission. The build-up of trapped charge was monitored using $C–V$ and photo $I–V$ techniques. Also as part of these studies, the annealing behavior of the traps and the shielding effectiveness of process layers were determined. It was found that bulk neutral traps were produced over a wide range of rf peak-to-peak voltage. Figure 2-13 shows the trapped charge density as a function of injection time. It should be noted that the self-developed dc bias voltage in the RIE reactor used in this study was approximately one-half the rf peak-to-peak voltage. Enhanced electron trapping was observed over the range of voltage. Oxide etched at lower voltages could not be analyzed because an rf discharge could not be sustained. The trapping characteristics of silicon dioxide etched by RIE were compared to those of silicon dioxide etched by plasma etching in a high pressure barrel-type reactor. It can be seen from Figure 2-14 that enhanced trapping is absent for silicon dioxide etched in a barrel reactor

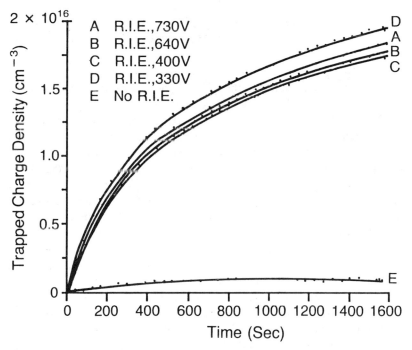

Figure 2-13.   Volume trapped charge density as a function of avalanche injection time for oxides exposed to RIE with rf peak-to-peak voltages between 730 and 330 V.

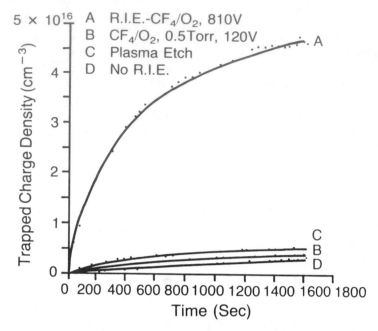

Figure 2-14. Volume trapped charge density as a function of avalanche injection time for oxides exposed to RIE in $CF_4 + O_2$, and plasma etching in RIE and barrel-type plasma reactors.

and for silicon dioxide etched in the RIE reactor under conditions that simulate plasma etching (500 mTorr, 120 V rf peak to peak).

The source of ionizing radiation in the RIE reactor is believed to be bremsstrahlung radiation created when secondary electrons from the electrode collide with grounded surfaces. This hypothesis is supported by experiments in which silicon dioxide films were positioned in the reactor so that they were protected from bombardment by other potential sources—ions and energetic secondary electrons. These films exhibited enhanced trapping. Convincing evidence comes from sputter etching experiments in which a flat counter-electrode was replaced by one that was covered with hollow tubes with open ends facing the wafers (Ryden et al. 1976). Flat band voltage shift in silicon gate MOS devices was reduced by the presence of the tubes. This result is consistent with damage caused by bremsstrahlung radiation because secondary electrons from the electrode would be ejected into a narrow angular cone about the cathode normal and so would penetrate deep within the array of tubes. X rays generated when the electrons strike a solid surface are distributed over a large angular range; so most of the X-ray flux would be absorbed by the walls of the tubes.

Experiments to evaluate the ability of process layers to shield underlying gate oxide indicate that the radiation is absorbed in several-thousand-angstrom-thick polysilicon films (Ephrath and DiMaria 1981). The trapping characteristics of oxide covered by a 350-nm-thick polysilicon film delineated by RIE and wet etching appear in Figure 2-15. Both oxides show the same low trapped charge density versus time. It has been shown more recently that damage is created by photons scattering into the gate oxide from the edges of small-dimension polysilicon gate electrodes during polysilicon etching (Sekine et al. 1984). Radiation damage from edges is not expected to occur at later etching steps after additional films have been deposited. Annealing experiments show that the bulk neutral traps are removed by a 600°C anneal in forming gas (Ephrath and DiMaria 1981). Process temperatures in excess of 600°C are reached after polysilicon electrodes are delineated; so damage introduced during this step would be removed.

Much discussion has appeared in the literature concerning the device implications of RIE radiation damage. Etching configurations have been proposed to minimize radiation damage of silicon dioxide by eliminating energetic ion bombardment. While radiation damage should be considered as device structures (absorbing films overlying the gate oxide) and processes (thermal cycles) are

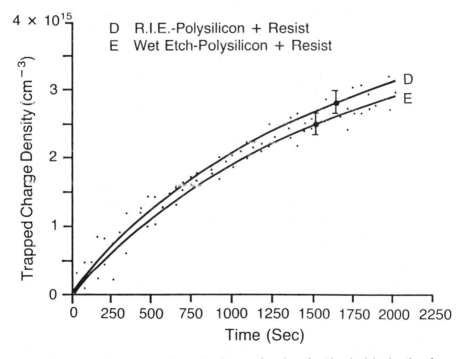

Figure 2-15.   Volume trapped charge density as a function of avalanche injection time for oxide after RIE of polysilicon electrode. Sample in which polysilicon was wet-etched serves as a control.

formulated, information to date does not indicate that the versatility or the simplicity of an etching technique should be sacrificed to eliminate damage.

RIE-induced neutral traps are not evident until they are populated by electrons injected into the gate oxide. An immediate degradation in oxide breakdown strength has been reported after RIE of polysilicon over 400 Å oxide films (Watanabe and Yoshida 1984). In this intriguing study, a high frequency of low field breakdown events was attributed to a transient voltage that occurred when the rf power supply was turned off. The source of the transient voltage was suggested to be negative charge stored in the blocking capacitor in the rf matching network and excess positive charge in the polysilicon gate electrode. One conclusion of this study was that the damage to the gate oxide could be avoided by placing an insulating material on the cathode to prevent injection of stored charge on the blocking capacitor into the wafer.

SINGLE WAFER ETCHING

Since 1982, the focus of equipment vendors has shifted to single wafer etching with vendors of single wafer reactors now outnumbering those offering batch reactors (*Semiconductor International* 1985). Commercial reactors are available with a variety of features (Broydo 1983), and many vendors claim specific process capabilities (Iscoff 1985). The attraction of this new technology lies in its *potential* to provide higher productivity, particularly as wafer size increases, and greater control over etching. Ease of automation, reduced footprint for an equivalent throughput of a batch system, relative insensitivity of throughput to wafer size, and elimination of process scale-up increase productivity. Wafer-to-wafer control over the process environment is expected to increase yield.

Currently, adequate process monitoring and diagnostics do not exist on these systems. A process mishap under these circumstances is usually detected only after the entire cassette has been processed, and the wafers have been visually inspected. This increases the risk of loss of wafers within a cassette. The situation is akin to loss of an entire batch in batch reactors. However, as the single wafer systems evolve, smart microprocessors will be able to monitor and diagnose the plasma and limit the loss to at most one wafer.

Single wafer reactors differ significantly in configuration (Mathad 1985; Wilders and Pearson 1985). The basic configuration is similar to that of planar parallel plate batch reactors, but the designs differ, depending primarily on the rage of operating pressure. The common objective of all the designs is to achieve the high etch rates needed to make single wafer etching competitive with batch etching systems.

To be competitive, the etch rate in a single wafer reactor must be approximately equal to the batch size multiple of the rate in a batch system. Most of the current single wafer systems achieve a 0.5 to 1.0 μm/min etch rate for thermal oxide compared with a 300 to 500 Å/min etch rate in batch systems (Shanfield and

Table 2-10. Techniques for Achieving High Plasma Density in Single Wafer Reactors.

	(a)	(b)	(c)
		INCREASED ELECTRON–GAS COLLISIONAL PROBABILITY BY	
Physical principle	Short mean free path	Auxiliary electron source	Increased electron path length
Pressure regime	≥ 1 Torr	0.01–0.2 Torr	≤ 0.01 Torr
Reactor type	Planar diode with low electrode gap	Triode (rf electrode)	Magnetron

Hendriks 1984a and b). The high single wafer etch rates are obtained by maximizing plasma density at reasonable power densities of approximately 5 W/cm², The configuration of the reactor depends upon the physical principles used to obtain high plasma density. Table 2-10 summarizes the three commonly used schemes. Note that each technique is optimal in a different pressure range. For this reason, the various single wafer reactors available are classified according to the pressure regime in which they operate. The objective in all these categories is to maximize ion current rather than voltage for a given power input.

High Pressure Reactors

This class of reactors is usually operated in the 1 to 10 Torr range. It is the earliest, most mature, and most widely used of the designs (Hijkata 1981; *Solid State Technology* 1982; Bersin and Gelernt 1982; Reichelderfer 1982; Mullins 1982; Lam 1982; Bithell and Peavey 1984; U.S. Patent 4,209,357). High etch rates result from the high plasma density that is achievable in this pressure range. The configuration of a practical etching reactor has required attention to several important aspects of design.

Design Considerations.   Key considerations in designing high pressure reactors have been electrode spacing, area ratio, plasma confinement, gas distribution over the wafer, excitation frequency, electrode material, and wafer cooling. Details of a reactor's design usually represent a compromise between etch rate and process capability.

Interelectrode spacing is kept quite small, usually in the range of 0.4 to 2 cm. The resulting high volume power density maximizes the degree of dissociation and ionization. At small interelectrode gaps, the area ratio between grounded and powered surfaces is close to unity. Thus, the wafers can be placed on either electrode, and the choice is based on other design considerations. Larger area ratios typical of batch reactors can be obtained by increasing the interelectrode gap (Donohoe 1985), but only at the expense of plasma density and, hence, etch rate. At small electrode gaps it is important that parallelism of the two electrodes be maintained. Schematics of two commercially available configurations are shown in Figure 2-16 (Mathad 1985).

High plasma density does not result in a high etch rate unless plasma is confined to the region over the wafer. Various gaps and pump-out openings provided in the reactor can be a significant fraction of the small reactor volume defined by the wafer diameter and the interelectrode distance. Depending upon the field strength, there is a tendency for a weak plasma to be sustained in these volumes, resulting in a reduced plasma density over the wafer. Proper confinement of the plasma increases the efficiency of the reactor (Mullins 1982) (etch rate/watt of input power) and improves etch rate uniformity, particularly at the edge of the wafer. Several schemes have been used and are shown in Figure 2-17. In Figure 2-17(a), a silica ring surrounding the lower electrode prevents the plasma from coupling with the large reactor wall area (U.S. Patent 4,324,611). A similar scheme is used in Figure 2-17(b), except that the wafer is placed within this ring (Bithell and Peavey 1984). In Figure 2-17(c), the ring performs the dual roles of providing confinement of the plasma and clamping of the wafer to the electrode (U.S. Patent 4,367,114). More complete confinement is provided in the scheme shown in Figure 2-17(d), where the confinement ring forms the dielectric wall of the reactor and provides passageways for gas exhaust (U.S. Patent 4,534,816). It should be noted that plasma confinement in these reactors results from a combination of design and operating pressure. At high pressures the mean free path is small, thus minimizing both electron loss at the reactor boundaries and lateral diffusion of the plasma.

Since collision mean free path is less than the characteristic dimension of the reactor (the radius of the reactor in this case), the flow at high pressure is viscous. A consequence of viscous flow is that the flow pattern, stagnant areas, and nonparallelism of the electrodes adversely affect etch rate uniformity. Both gas introduction and pump-out are symmetrical with the center of the wafer. Four

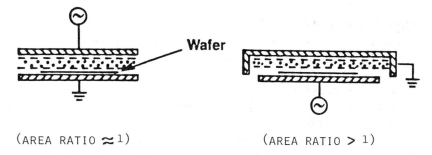

(AREA RATIO ≈ 1)   (AREA RATIO > 1)

Figure 2-16.   Schematic configuration of high pressure (≥1 Torr) SWRs (Mathad 1985).

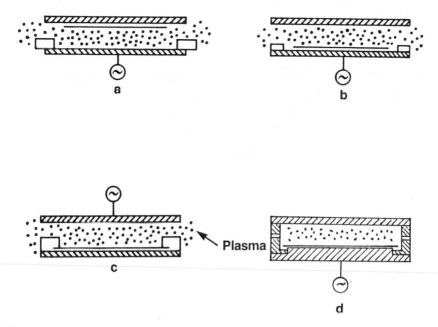

Figure 2-17.   Plasma confinement techniques used in high pressure SWRs (Mathad 1985) (U.S. Pat. 4,534,816).

distribution techniques used are shown in Figure 2-18. Distribution techniques fall into three categories: shower-head gas input and radial pump-out (Figure 2-18b, c, e), gas introduction through a porous-material electrode (U.S. Patent 4,367,114) with radial pump-out (Figure 2-18a), and shower-head input and pump-out (Figure 2-18d). The first scheme provides a uniform shower of gas through an array of small holes in the electrode, but suffers from the tendency of plasma to be sustained within some of these holes. These "plasmoids" are regions of very high plasma density, which can cause local nonuniform etch rates. The formation of plasmoids, though not well understood, is a function of gas, process variables, and hole size.

When porous material is used, as in Figure 2-18(a), gas flow is uniform with little possibility of forming plasmoids (U.S. Patent 4,367,114). The use of porous electrodes has limitations, however. When polymerizing gases (e.g., C_2F_6) are used, the electrode needs to be maintained at a high enough temperature (Bersin 1984) to minimize excessive polymer condensation and clogging of pores.

When gases are used that do not form polymers (e.g., O_2, N_2, Ar), it may be desirable to cool the upper electrode to eliminate resist hardening or to use higher power density to achieve higher etch rates. In such a case, however, it is not possible to incorporate integral cooling channels in a porous electrode.

In all of the designs discussed, the pump-out is radial, with the velocity

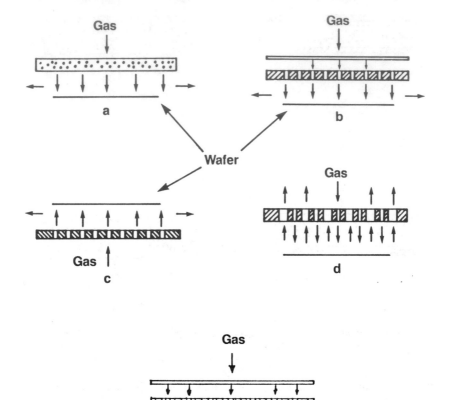

Figure 2-18. Gas distribution schemes used in high pressure SWR configurations (Mathad 1985) (U.S. Pat. 4,534,816).

increasing and reactant concentration decreasing as the gas travels toward the periphery of the wafer. This can result in center-to-edge etch rate variation, depending upon the material to be etched and process conditions. This effect will be discussed later in greater detail. The gas distribution scheme used in Figure 2-18(d) eliminates this radial effect by means of pump-out holes interspersed with gas inlet holes in the upper electrode (Donohoe 1985). However, a major drawback is the high probability that plasma will leak away through these outlets, which are larger in diameter than inlet holes. All present designs are compromises and do not offer an ideal gas distribution scheme.

Although most of these reactors are powered by 13.56 MHz generators, some use lower frequency excitation in the kilohertz range. Low frequency excitation usually is used to achieve higher ion energies and facilitate etch anisotropy (Zarowin 1983). The use of kilohertz excitation imposes certain design constraints, however. The overall capacitance is made up of contributions from the sheath, the wafer, and the gap between the wafer and the supporting electrode. The contribution of the capacitance of this gap increases with decreasing frequency and is significant in the kilohertz frequencies. A consequence is that etch rate variation can be large unless the gap at the back of the wafer is well controlled. The gap is usually minimized by clamping the wafer, thus assuring a certain degree of uniformity and repeatability of etch rate in spite of variation in wafer flatness. This results in an increase of wafer temperature, a consequence of which is that either the resist will reticulate, or the resist needs to be stabilized prior to etching.

Both the small interelectrode gap and the high degree of plasma confinement result in a high plasma potential (typically several hundred volts) (Vossen 1979). A high plasma potential causes sputtering of reactor surfaces (Peccoud et al. 1985). The nonwafer electrode is the most significant source of sputtered material because the reactor wall area is small. Any nonvolatile material sputtered onto the wafer acts as an uneven mask and causes the etching surface to roughen (Eiden et al. 1985). Many methods have been used to eliminate or minimize the problem. The use of a porous or solid graphite electrode eliminates sputtering because volatile products (CO_2, CO, CH_4, etc.) are formed. The lifetime of such a consumable electrode depends upon the plasma used. Once degraded (in terms of uniformity or particulates), the electrode must be replaced. Materials that form volatile products have been sprayed on the electrode (Mullins 1982). Some examples are plasma sprayed silicon and Teflon. With this technique, the coatings are etched, and the electrode must be replaced or recoated. A more common technique is to use an etching gas mixture that contains an easily polymerizable component (e.g., CHF_3 or C_2F_6). Such a mixture, while it etches the wafer, also passivates the upper electrode with a thin polymer film. The polymer film prevents sputtering of the metal electrode (Muraka and Mogab 1979; Corn 1984). In some configurations, wafers are held upside down on the upper electrode to prevent particulates from falling on them (Mullins 1982). Whether this approach is effective for submicron particles is uncertain. In summary, it is clear that the choice of electrode material is a compromise; the ideal material, one that forms volatile products but lasts indefinitely, is elusive.

With confined plasmas and power densities of approximately 5 W/cm^2, the demands on wafer cooling are extreme. Resist patterns reticulate or flow if the temperature during etching exceeds the post-bake temperature of the resist. One aspect of high pressure single wafer etching that tends to lower wafer temperature is gas cooling at the back of the wafer. Most designs provide adequate cooling to the wafer electrode, but, because of design constraints, the counter electrode is usually cooled only from the periphery. Thus, temperature gradients can exist, and heat can be radiated to the wafer. In one design, shown in Figure 2-18(e), temperature and temperature gradients are reduced by integrating cooling channels into the gas distribution system (U.S. Patent 4,534,816).

Performance. Three important aspects of the performance of these reactors are uniformity, loading, and anisotropy. Although high etch rates (5000 Å/min for oxide) are easily obtained, etch rate uniformity is very sensitive to many variables, some difficult to control. Sources of etch rate variation include the gap at the back of the wafer (which is a function of wafer flatness), the presence of local discharges inside gas inlet holes in the shower-head design, nonuniform clogging of pores where porous electrodes are used, and nonuniform condensation of polymers on the counter electrode from polymerizing gas plasmas. Nonuniform etching can occur if particulates are present either on the top or at the back of the wafer. Particulates can cause an intense plasma to develop locally. Etch rate uniformity of dielectrics is sensitive to gas flow rate and pressure. The profile of nonuniformity can be changed from concave (center etching slowly) to convex (edge etching slowly) by changing either flow or pressure. An example

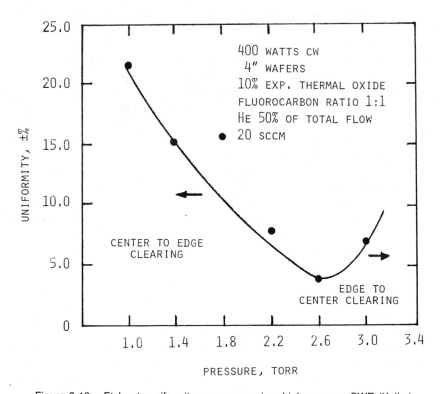

Figure 2-19. Etch rate uniformity vs. pressure in a high pressure SWR (Keller).

of this dependence is shown in Figure 2-19 (from Keller, 1984). Again, the choice of operating conditions represents a compromise. Films etched in a radial flow design usually etch more slowly at the edge because of dilution of the active species by reaction products. Very high flows minimize the variation but tend to lower the etch rate because of low residence time.

Loading effects (effect of etch rate on etching area) are observed to be more pronounced in high pressure single wafer reactors than in low etch rate batch reactors. Similarly, the difference in etch rate between blanket films and patterned films can be significant. Process performance, then, becomes dependent upon pattern factor (ratio of area etched to area masked). In the case of etching reactions that involve neutral atoms or molecules (e.g., polymer, aluminum, silicon), the etch rate increases with decreasing pattern factor. For etching reactions that depend on bombardment by ions (e.g., oxide), however, blanket etch rates exceed those of resist patterned wafers (Reinberg et al. 1982).

Anisotropic etch profiles can be obtained in high pressure plasmas, and several mechanisms have been proposed. One is that anisotropy is a function of E/P only (Zarowin 1983), where E is the field strength and P is pressure. Because the field strength in these reactors can be high, an E/P ratio similar that in low pressure batch reactors may be obtained. Probably the most significant factor in producing anisotropic etching at high pressure is sidewall inhibition or passivation (Flamm and Donnelly 1981; Robb 1985; Bruce and Malafsky 1983).

In inhibition-induced anisotropy, species adsorbed on the side walls enhance recombination of etching species and so impede lateral etching. In passivation-induced anisotropy, a thin layer of polymer is formed and maintained such that the deposition rate equals the removal rate (Bernacki and Kosick; 1984). In both cases, films do not form on horizontal surfaces because of the large ion flux. If sidewall films are to be utilized for obtaining vertical etch profiles, it is necessary to add to the etching gas a component that readily produces polymer precursors. Achieving anisotropy, then, becomes a matter of a balance between the production of etchant and passivating or recombinant species (Meith and Barker 1983). The process window for polymer-producing etching gases tends to be small. Another technique for maximizing anisotropy is use of low frequency excitation. Because ions can follow the field at these frequencies, ion energy is increased, facilitating anisotropic etching (Bruce 1981). The effect of applied frequency on ion energy is shown in Figure 2-20, taken from Donnelly et al. (1985). Disadvantages of this approach are that superimposition of a high frequency field is needed to improve uniformity, and sputtering contamination may be enhanced (Kalter et al. 1982). Anisotropic etching of organic polymers in an oxygen plasma at high pressure has not been demonstrated. Thus, this etching approach may not be suitable for image transfer applications.

In conclusion, many high pressure systems are operating satisfactorily for a wide range of materials under current lithographic ground rules. Extension of these reactors to all etching steps and to submicron patterning has yet to be

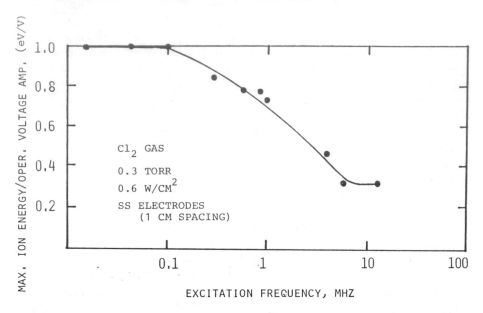

Figure 2-20.   Effect of excitation frequency on ratio of Cl_2 + ion peak energy to peak applied voltage (Donnelly et al. (1985).

demonstrated. It is these future requirements that drive the development of single wafer reactors that operate at lower pressures.

Intermediate Pressure Systems

Reactors in this category are designed to operate in the pressure range 0.01 to 0.4 Torr. The lower pressure and larger reactor volume reduce plasma density and, hence, etch rate. To increase plasma density and etch rate, these reactors provide an auxiliary source of electrons to the system. The additional source of electrons in commercially available systems is supplied through a second powered surface. The use of a hot filament has also been reported (Mantei and Wicker 1985). Disadvantages of a hot filament include a short lifetime for the filament in reactive plasmas and sputter contamination from the filament.

Design Considerations.   The object of a tri-electrode or triode reactor is to gain quasi-independent control over the ion flux and energy at the wafer surface (Chapman 1980). An additional degree of freedom is thus achieved over a diode reactor. The primary electrode is either the upper electrode, as shown in Figure

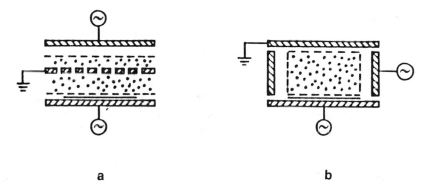

Figure 2-21.   Schematic configuration of intermediate pressure (10–400 mTorr) SWRs (Solid State Technology) (U.S. Pat. 4,464,223).

2-21(a) from *Solid State Technology* (1984), or the reactor wall, as shown in Figure 2-21(b) from U.S. Patent 4,464,223. The electrode spacing is large (4 cm), and the ratio of powered to nonpowered area is large (> 2). No effort is made to confine the plasma. Because of the relatively large reactor volume, it tends to be evenly distributed over the wafer. The gas flow is in the free molecular to viscous flow regime. Flow uniformity is important, especially at the high end of the pressure range.

Special care is taken to assure minimal sputter contamination. In the design of Figure 2-21(a), the highest sheath voltage is developed on the upper electrode to achieve high plasma density with minimal ion bombardment of the wafer. Contamination from material sputtered from the top electrode is minimized by the grid plate, which acts as a catcher plate for the sputtered material. In the design shown in Figure 2-21(b), the highest sheath voltage is developed at the wall.

More attention has been given to wafer cooling in this pressure range than was necessary in the high pressure designs. A method for cooling wafers is needed because of the absence of a naturally conducting high pressure gas film at the back of the wafer. In the design of Figure 2-21(a), clamping with helium gas cooling is employed (Bogle-Rohwer et al. 1985). In the design of Figure 2-21(b), where the wafer rests freely on the electrode, high sheath voltages cannot be used at low pressure without degrading the resist. High voltages of several hundred volts may be needed for anisotropic etching of some materials. These two designs represent trade-offs between the complexity and performance of the system.

It should be noted that the pumping package is more complex than in high pressure systems. Multiple power generators and additional matching networks further add to the complexity.

Performance.   The performance of these systems will be discussed in terms of etch rate and anisotropy and selectivity. Due to the large reactor volume (five times that of high pressure reactors), the volume power density is low. The highest etch rates are obtained when most of the voltage is applied to the wafer electrode, and when the reactor is operated at pressures greater than 0.2 Torr. Another degree of freedom that is available for increasing etch rate is the phase angle difference between the powered electrodes (Kim and Wilkinson 1984).

Anisotropy is easily achieved without invoking sidewall passivation at the low end of the pressure range when all of the power is applied to the wafer electrode. If higher pressures are used to maximize etch rate, polymerizing gases are needed to maintain anisotropy. Tailoring of etch slopes is an added capability of triode configurations. Etch profiles can be controlled by varying pressure and splitting power between the two electrodes (Bogle-Rohwer et al. 1985).

Pressures in this range allow the use of a wide range of gases to maximize selectivity. The potential of obtaining high selectivity for oxide to silicon and oxide to resist is shown in Figures 2-22(a) and 2-22(b) from (Waferetch 606). At the high CHF_3 concentrations used, gross polymerization would occur in high pressure systems, resulting in low etch rates (Castellano 1984).

In summary, the triode approach offers a method for achieving high etch rates at pressures that favor anisotropic and highly selective etching. The additional electrode also allows flexibility in controlling etch profile.

Low Pressure Systems

Low pressure systems are operated at pressures below 0.01 Torr. Plasma density and etch rate would be quite small if electrons were not trapped near the surface of the wafer to increase the number of ionizing collisions with gas molecules. Electrons are trapped in low pressure systems by applying a magnetic field, B, perpendicular to the electric field, E_0. The magnetic field causes the electrons to move in a helical path around the magnetic field lines. The resultant angular frequency, w, of the electrons is given by:

$$w = \frac{eE_o}{m_e d} + \frac{e^2 B^2}{m_e^2}$$

where e is the electron charge, m_e is electron mass, and d is the thickness of the sheath (Chapman 1980). Ions are virtually unaffected by the magnetic field because of their greater mass. In the design shown in Figure 2-23(a), additional magnets are placed on top of the anode to compress the primary magnetic field lines. The result is a flatter plasma density profile near the center of the cathode, which gives a larger uniform area (Hill and Hinson 1985). To further improve etch uniformity, the cathode is made quite large so that the wafer lies within the

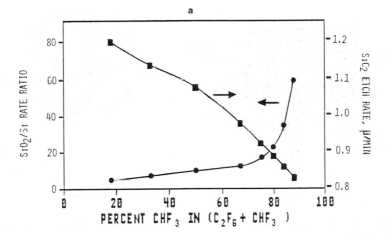

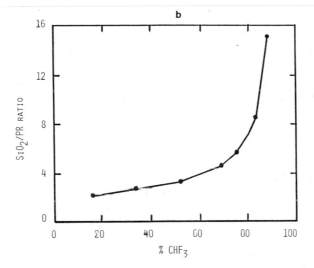

Figure 2-22.   Etch rate ratio in a triode SWR: (a) selectivity of SiO_2 to silicon; (b) selectivity of SiO_2 to photoresist (Waferetch 606—Application note).

flat portion of the field. A consequence of the large cathode is that the reactor is large, and large power inputs are necessary to achieve practical etch rates.

In the design shown in Figure 2-23(b), a series of bar magnets are arranged in a tank tread pattern (*JST News* 1984). Intense plasma bands are developed along the length of the magnet between two adjacent poles. To obtain etch uniformity, the magnet assembly is moved in a continuous loop pattern (*JST*

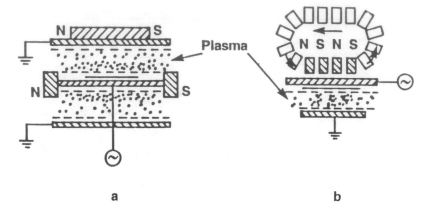

Figure 2-23.   Schematic configuration of low pressure (≧10 mTorr) SWRs (Hill and Hinson 1985; Schultheis 1985).

News 1984). Uniform etching has also been obtained by arranging fewer magnets in a horizontal plane and using a scanning motion (Okano et al. 1982). The sheath voltage at the wafer is low, approximately 40 eV (Schultheis 1985). The low voltage is a consequence of the high plasma density, which increases with increasing magnetic field strength. Note that in the design of Figure 2-23(b), the wafer is held upside down on the upper electrode to minimize particulate contamination.

High power densities of greater than 4 W/cm² are used to generate a high plasma density. To prevent excessive heating of the wafer, clamping or high pressure gas cooling at the back of the wafer is usually provided. Many clamping schemes have been used (King and Rose 1981), some of which are shown in Figure 2-24. Figure 2-24(a) shows a wafer clamped to a flat electrode by a clamp ring. Figure 2-24(b) shows similar clamping with pins. Figure 2-24(c) shows an arrangement where the wafer is clamped to a spherical electrode surface coated with a thermally conducting resilient film. The soft film provides a conformable surface, and the slight curvature of the electrode assures a uniform gap across the wafer even if the wafer is somewhat warped. In Figure 2-24(d), the electrode has a slight curvature, and gas cooling is provided (Waferetch 606). A disadvantage of mechanical clamping is that the periphery of the wafer is lost. Additional area is lost because of nonuniform etching due to the presence of the clamp. In Figure 2-24(e), clamping is accomplished by a difference in pressure on the two sides of the wafer. Finally, in Figure 2-24(f), the wafer is held by electrostatic force onto a polymer film coating the electrode (Schultheis 1985). It should be noted for the last two techniques that clamping is accomplished without loss of wafer area. As wafer diameter or process-induced warpage in-

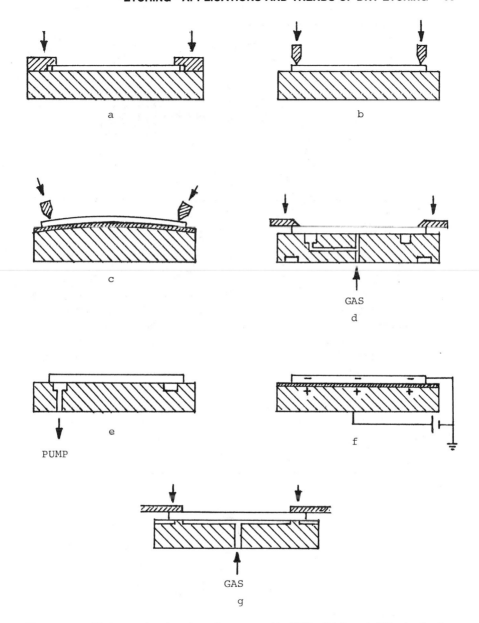

Figure 2-24.   Various wafer clamping schemes used in SWRs (Waferetch 606—Application Note).

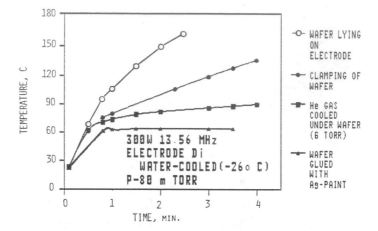

Figure 2-25. Efficiency of various wafer cooling techniques (Egerton et al. 1982).

creases, the leakage rate of high pressure cooling gas increases. Leakage can be reduced by sealing the wafer with a soft gasket or O-ring (King and Rose 1981). This technique is shown in Figure 2-24(g). A disadvantage to the use of a gasket is that it may produce a nonuniformity in the electric field at the periphery of the wafer, which would produce a nonuniform etch rate. The efficiency of some of the cooling schemes is shown in Figure 2-25, taken from Egerton et al. (1982).

Performance. The advantage of these designs is the ability to achieve high etch rates at low pressures. An example is given in Figure 2-26. The etch rates shown are from a system with the design shown in Figure 2-23(a). A greater

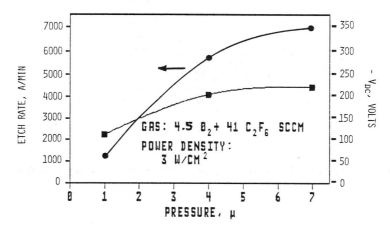

Figure 2-26. Effect of pressure on etch rate of SiO_2 in a magnetron SWR.

than 1000 Å/min etch rate is obtained for oxide at a pressure of 0.001 Torr. Anisotropic etching is easily obtained in this low pressure regime. Vertical profiles have been reported for both trilevel resist structures and for deep trenches etched into silicon. It should be noted, however, that the use of magnetic confinement of electrons to increase the etch rate tends to result in a more complex reactor, compared with those designed to operate at higher pressures.

Selective etching can be obtained by manipulating the chemical composition of the plasma as before, but an additional mechanism for obtaining selective etching has been reported. It has been observed that the magnetic field is selective in increasing the density of a particular excited species in the plasma (Lory 1984). This is shown in Figure 2-27 for xenon at a fixed rf power, with and

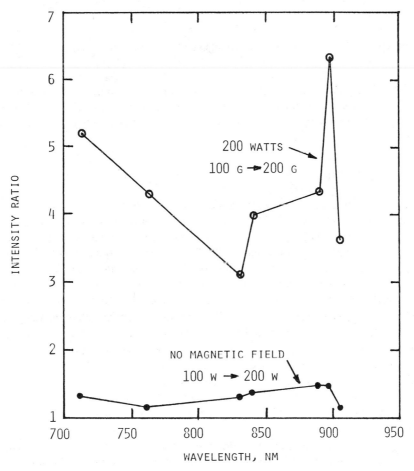

Figure 2-27.   Emission intensity enhancement of xenon atoms in a magnetic field enhanced plasma (Lory 1984).

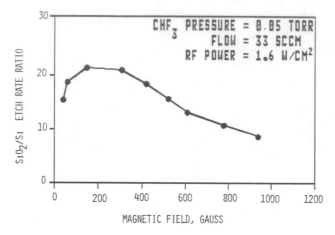

Figure 2-28.   Effect of magnetic field strength on SiO_2:Si selectivity (Okano et al. 1982).

without a magnetic field. Thus, it is possible in principle to use the combination of a particular gas and magnetic field to increase selectivity. It is shown in Figure 2-28 (Okano et al. 1982) that selectivity changes with the strength of the magnetic field.

LASER ETCHING

There has been increasing interest in the application of lasers to etching (Brewer et al. 1984a, 1985; Chuang 1983). The laser is considered an ideal source of photons because its high spectral specificity provides a mechanism for selective excitation of atoms or molecular bonds. This degree of freedom is being used in the following ways: selective absorption by gas molecules to produce etching species, selective absorption by the etching surface, and confinement of a reaction to a small region that is illuminated by the laser. The advantage of laser enhanced etching over conventional dry etching techniques is already demonstrated for some materials that have proved difficult to etch in plasma reactors, for example, copper (U.S. Patent 4,490,210). An important potential advantage lies in patterning of films without applying a mask to the wafer (Brewer et al. 1985). Laser radiation is readily focused or collimated and projected onto solid surfaces because it is monochromatic and coherent (Lacombat et al. 1980). Elimination of the conventional lithographic steps has the potential of increasing throughput and yield.

Laser etching is being pursued using three optical techniques: focused beams, nonfocused beams with contact or proximity masks, and nonfocused beams with optical projection of a mask. Focused beams have been used for direct writing of high and low resolution patterns (Ehrlich et al. 1980, 1981; Darea and Kaiser 1977) and circuit repair. While this technology has been demonstrated and is attractive for its relative simplicity, it is inherently slow. Nonfocused beams are attractive for industrial applications because of the potential for etching large area patterns. Some demonstrated applications using contact or proximity masks include etching of silicon (Sekine et al. 1983), TiC–Al$_2$O$_3$ (Chuang 1983; Brewer et al. 1984a), and polymers (Rice and Jain 1984). A disadvantage of contact or proximity masking, degradation of the mask by the corrosive etching environment, is overcome by projection laser etching. In projection laser etching, the mask is placed outside the reactor, and the pattern is projected onto the substrate (Brewer et al. 1985). Some recent applications include the patterning of glass (U.S. Patent 4,478,677; Ehrlich and Tsao 1984), GaAs (Brewer et al. 1984b), polysilicon (Horiike et al. 1984), and metals (D'Asaro et al. 1978; U.S. Patent 4,490,211).

Laser etching is under development in the United States and Japan (*The Japan Economic Journal;* Abe 1982). Its potential for etching new materials and eliminating resist masks is widely recognized, but it is also clear that formidable problems remain. Outstanding issues include etching resolution, uniformity of the laser beam, mask life, and reliability of the laser.

REFERENCES

Abe, H., *JST News, 1* (5), 11 (1982).

Ashok, S., Chow, T. P., and Baliga, B. J., *Appl. Phys. Lett., 42,* 687 (1983).

Bendernagle, R., M.S. thesis, Columbia University, 1985.

Bennett, R. S., *Electrochem. Soc. Extended Abstracts, 82-2,* 283 (1982).

Bernacki, S. E., and Kosicki, B. B., *J. Electrochem. Soc., 131,* 1926 (1984).

Bersin, R. L., *Microelectronic Manufacturing and Testing, 7* (4), 40 (1984).

Bersin, R. L., and Gelernt, B., *Optical Microlithography,* SPIE Vol. 334, p. 163, 1982.

Bithell, R., and Peavey, J., *Tegal Process Review, 1* (2), 11 (1984) (Tegal Corp., P.O. Box 10, Novato, CA).

Bogle-Rohwer, E., Gates, D., Hayler, L., Kurasaki, H., and Richardson, B., *Solid State Technol. 28* (4), 251 (1985).

Bondur, J. A., *J. Electrochem. Soc., 126,* 226 (1979).

Brewer, P. D., Halle, S., and Osgood, R. M. Jr., *Appl. Phys. Lett., 45,* 475 (1984a).

Brewer, P., McClure, D., and Osgood, R. M. Jr., *Mat. Res. Soc. Ext. Abs.,* F. Houle, T. Deutsch, and R. M. Osgood, Jr., Editors, p. 102 (1984b).

Brewer, P. D., Reksten, G. M., and Osgood, R. M. Jr., *Solid State Technol., 28* (4), 273 (1985).

Broydo, S., *Solid State Technol., 26* (4), 159 (1983).

Bruce, R. H., *Solid State Technol., 24* (10), 64 (1981).

Bruce, R. H., and Malafsky, G. P., *J. Electrochem. Soc., 130,* 1369 (1983).

Castellano, R. N., *Solid State Technol., 27* (5), 203 (1984).

Chapman, B. N., *Glow Discharge Processes,* John Wiley, New York, 1980.

Choe, D., Knapp, C., and Jacob, A., *Solid State Technol., 28* (3), 165 (1985).

Chow, T. P., Saxena, A. N., Ephrath, L. M., and Bennett, R. S., in *Dry Etching for Microelectronics,* R. A. Powell, Editor, North-Holland, New York, p. 39 (1985).

Chuang, T. J., *Mat. Res. Soc. Symp. Proc.*, *17*, 45 (1983).

Coburn, J. W., *Plasma Chemistry and Plasma Processing*, 2 (1), 1 (1982).

Corn, G., *Tegal Process Review*, *1* (2), 8 (1884) (see Wagner and Brandt 1981:201).

Crowder, B. L., and Zirinsky, S., *IEEE Trans. Electron Devices, Ed-26*, 369 (1979).

Darea, K., and Kaiser, W., *Glass Technology*, *18* (1), 19 (1977).

D'Asaro, L. A., DiLorenzo, T. V., and Fuki, H., *IEEE Trans. Electron Devices, Ed-25*, 1218 (1978).

DiMaria, D. J., and Kerr, D. R., *Appl. Phys. Lett.*, *27*, 505 (1975).

Donnelly, V. M., Flamm, D. L., and Bruce, R. H., *J. Appl. Phys.*, *58* (6), 2135 (1985).

Donohoe, K. G., in *Plasma Processing*, J. Dieleman, R. G. Frieser, and G. S. Mathad, Editors, The Electrochemical Soc., PV *82-6*, p. 306 (98).

Egerton, E. J., Nef, A., Millikin, W., Cook, W., and Baril, D., *Solid State Technol.*, *25* (8), 84 (1982).

Ehrlich, D. J., and Tsao, J. Y., *Mat. Res. Soc. Symp. Proc.*, A. Johnson, D. J. Ehrlich, M. Schlossberg, Editors, *29*, 195 (1984).

Ehrlich, D. J., Osgood, R. M., and Deutsch, T. F., *Appl. Phys. Lett.*, *36*, 698 (1980).

Ehrlich, D. J., Osgood, R. M. Jr., and Deutsch, T. F., *Appl. Phys. Lett.*, *38* (12), 1018 (1981).

Eiden, G. C., Boyer, P. K., and Hughes, J. A., in *Plasma Processing*, G. S. Mathad, G. C. Schwartz, and G. Smolinsky, Editors, The Electrochemical Society, PV *85-1*, p. 502 (1985).

Eisele, K. M., *J. Electrochem. Soc.*, *128*, 123 (1981).

Endo, N., and Kurogi, Y., *IEEE Trans. Electron Devices, Ed-27*, 1346 (1980).

Ephrath, L. M., and Bennett, R. S., *J. Electrochem. Soc.*, *129*, 1822 (1982).

Ephrath, L. M., and DiMaria, D. J., *Solid State Technol.*, *24*, (4), 182 (1981).

Ephrath, L. M., and Petrillo, E. J., *J. Electrochem. Soc.*, *129*, 2282 (1982).

Flamm, D. L., and Donnelly, V. M., *Plasma Chemistry and Plasma Processing*, *1* (4), 317 (1981).

Fonash, S. J., *Solid State Technol.*, *28* (4), 201 (Apr. 1985).

Fonash, S. J., Singh, R., Rohatgi, A., Rai-Choudhury, P., Caplan, P. J., and Poindexter, H., *J. Appl. Phys.*, *58* (2), 862 (1985).

Frieser, R. G., Montillo, F. J., Zingerman, B. N., Chu, W. K., and Mader, S. R., *J. Electrochem. Soc.*, *130*, 2237 (1983).

Gdula, R. A., *J. Electrochem. Soc.*, *126* (1979).

Herb, G. K., Caffrey, R. E., Eckroth, E. T., Janett, Q. T., Fraust, C. L., and Fulton, J. A., *Solid State Technol.*, *26* (3), 185 (1983).

Hijkata, I., *JEE*, p. 74, (Mar. 1981).

Hill, M. L., and Hinson, D. C., *Solid State Technol.* 28 (4), 243 (1985).

Horiike, Y., Sekine, M., Horioka, K., Arikado, T., Nakase, M., and Okano, H., *Mat. Res. Soc. Ext. Abs.*, F. Houle, T. Deutsch, and R. M. Osgood, Jr., Editors, p. 99 (1984).

Hosokawa, N., Matsuzaki, R., and Asamaki, T., *Japan J. Appl. Phys.*, Supplement 2, Part 1, 435 (1974).

Hosoya, T., Ozaki, Y., and Hirata, K., *J. Electrochem. Soc.*, *132*, 2436 (1985).

Iida, S., et al., Semiconductor and IC Symp., Tokyo (1982).

Iscoff, R., *Semiconductor International*, *8* (10), 48 (1985).

JST News, *3* (2), 52 (1984).

Kalter, H., Meyer, A., and Wolters, R. A. M., in *Plasma Processing*, J. Dieleman, R. G. Frieser, and G. S. Mathad, Editors, The Electrochemical Soc., PV *82-6*, p. 154 (1982).

Keller, J., *Tegal Process Review* (Tegal Corp., Novato, Calif.) (1984).

Kim, K., and Wilkinson, O., Abs. No. 394, Fifth Symp. on Plasma Processing, The Ext. Electrochemical Society (1984).

King, M., and Rose, P. H., *Nucelar Instruments and Methods, 189*, 169, North Holland Publishing Co. (1981).

Koenig, H. R., and Maissel, L. I., *IBM J. Res. Develop.*, *14*, 168 (1970).

Lacombat, M., Massin, J., Dubroeucg, G. M., and Brevignon, M., *Solid State Technol.*, *23* (8), 115 (1980).

Lam, D. K., *Solid State Technol.*, *25* (4), 215 (1982).

Lehmann, H. W., and Widmer, R., *Appl. Phys. Lett.*, *32*, 163 (1978).

Lory, E. R., *Solid State Technol.*, *27* (11), 117 (1984).

Mantei, T. D., and Wicker, T. E., *Solid State Technol.*, *28* (4), 263 (1985).

Mathad, G. S., *Solid State Technol.*, *28* (4), 221 (1985).

Matsuo, S., *J. Vac. Sci. Technol.*, *17*, 587 (1980).

Mayer, T. M., and McConville, J. H., *IEDM Techn. Digest*, 44 (Dec. 1979).

Meith, M., and Barker, A., *J. Vac. Sci. Technol.*, *Al* (2), 629 (1983).

Mogab, C. J., and Levinstein, H. J., *J. Vac. Sci. Technol.*, *17*, 721 (1980).

Mu, X. C., Fonash, S. J., Yang, B. Y., Vedam, K., Rohatgi, A., and Rieger, J., *J. Appl. Phys.* *58* (11), 4282 (1985).

Mu, X. C., Fonash, S.J., Oehrlein, G. S., Chakravarti, S. N., Parks, C., and Keller, J. H., *J. Appl. Phys.* (in press).

Mullins, C., *Solid State Technol.*, *25* (8), 88 (1982).

Muraka, S. P., and Mogab, C. J., *J. Electronic Materials*, *8*, 763 (1979).

Oehrlein, G. S., Tromp, R. M., Tsang, J. C., Lee, Y. H., and Petrillo, E. J., *J. Electrochem. Soc.*, *132*, 1441 (1985).

Okano, H., Yamazaki, T., and Horiike, Y., *Solid State Technol.*, *25* (4), 166 (1982).

Osburn, C. M., and Ormond, D. W., *J. Electrochem. Soc.*, *119*, 591 (1972).

Pang, S. W., *Solid State Technol.*, 249 (Apr. 1984).

Peccoud, L., Arroyo, J., Lassagne, P., and Peuch, M., in *Plasma Processing*, G. S. Mathad, G. C. Schwartz, and G. Smolinsky, Editors, The Electrochemical Soc., PV *85-1*, p. 443 (1985).

Pogge, H. B., Bondur, J. A., and Burkhardt, P. J., *J. Electrochem. Soc.*, *130*, 1592 (1983).

Reichelderfer, R. F., *Solid State Technol.*, *25* (4), 160 (1982).

Reinberg, A. R., *Etching for Pattern Definition*, H. G. Hughes and M. J. Rand, Editors, The Electrochemical Society, Princeton, NJ, 1976, p. 91.

Reinberg, A. R., *IEDM Techn. Digest*, 441 (Dec. 1978).

Reinberg, A. R., Dalle-Ave, J., Steinberg, G., and Bruce, R., in *Plasma Processing*, J. Dieleman, R. G. Frieser, and G. S. Mathad, Editors, The Electrochemical Society, PV *82-6*, p. 198 (1982).

Rice, S., and Jain, K., *Appl. Phys.*, *A33*, 195 (1984).

Robb, F. Y., in *Plasma Processing*, G. S. Mathad, G. Schwartz, and G. Smolinsky, Editors, The Electrochemical Society, PV *85-1*, p. 1 (1985).

Ryden, W. D., Labuda, E. F., and Clemens, J. T., in *Etching for Pattern Definition*, H. G. Hughes and M. J. Rand, Editors, The Electrochemical Society, p. 144 (1976).

Schanfield, S., and Hendricks, M., *ECS Extended Abstracts 84-2*, 561 (1984a).

Schanfield, S., and Hendricks, M., Ext. Abs. 393, Fifth Symp. on Plasma Processing, The Electrochemical Society (1984b).

Schultheis, S., *Solid State Technol.* *28* (4), 233 (1985).

Schwartz, G. C., Zielinski, L. B., and Schopen, T., in *Etching for Pattern Definition*, H. G. Hughes and M. J. Rand, Editors, The Electrochemical Society, Princeton, NJ, p. 122.

Sekine, M., Okano, H., and Horiike, Y., *Proc.* 5th Symp. Dry Processes, Inst. Electr. Eng., Tokyo, 97 (1983).

Sekine, M., Okano, H., Yamabe, K., Hayasaka, N., and Horiike, Y., *Proc.* 6th Symp. Dry Processing, Inst. Electr. Eng. of Japan, Tokyo (Oct. 1984).

Semiconductor International, *7* (13), 52 (1985).

Shelton, B. F., Kodak Microelectronic Seminar Proceedings, *Interface 75*, San Francisco, CA (Oct. 1975).

Solid State Technol., *25* (4), 79 (1982).

Solid State Technol., *27* (3), 57 (1984).

The Japan Economic Journal (Dec. 20, 1983).

Ting, C. Y., and Crowder, B. L., *J. Electrochem. Soc., 129,* 2590 (1982).

Tsai, M. Y., Chao, H. H., Ephrath, L. M., Crowder, B. L., Cramer, A., Bennett, R. S., Lucchese, C. J., and Wordeman, M. R., in *Semiconductor Silicon 1981,* H. R. Huff, R. J. Driegler, and Y. Takeishi, Editors, The Electrochemical Society, Pennington, NJ, 1981, p. 573.

Tsui, R. T. C., *Phys. Rev., 168* (1), 107 (1968).

U.S. Patent No. 4,209,357 (Assignee: Tegal Corp.).

U.S. Patent No. 4,298,443 (Assignee: AT&T).

U.S. Patent No. 4,324,611 (Assignee: Branson/IPC).

U.S. Patent No. 4,367,114 (Assignee: Perkin Elmer Corp.).

U.S. Patent No. 4,464,223 (Assignee: Tegal Corp.).

U.S. Patent No. 4,478,677 (Assignee: IBM Corp.).

U.S. Patent No. 4,490,210 (Assignee: IBM Corp.).

U.S. Patent No. 4,490,211 (Assignee: IBM Corp.).

U.S. Patent No. 4,534,816 (Assignee: IBM Corp.).

Vossen, J. L., *J. Electrochem. Soc., 126,* 319 (1979).

Waferetch 606—Application Note, GCA Corp., Bedford, MA.

Wagner, J. J., and Brandt, W. W., *Plasma Chemistry and Plasma Processing, 1,* 201 (1981).

Wang, K. L., Holloway, T. C., Pinizzoto, R. F., Sobczak, Z. P., Hunter, W. R., and Tasch, A. F. Jr., *IEEE Trans. Electron Devices, ED-29,* 547 (1982).

Wang, J. S. Fonash, S. J., and Ashok, S., *IEEE Electron Device Lett., EDL-4(12),* 432, (1983).

Watanabe, T., and Yoshida, Y., *Solid State Technol., 27* (4), 263 (1984).

Wehner, G. K., and Anderson, G. S., in *Handbook of Thin Film Technology,* L. I. Maissel and R. Glang, Editors, McGraw-Hill, New York, 1970, pp. 3–19.

Wilders, M. A., Pearson, S. M., *Tech. Proc., SEMICON/EAST,* 88 (1985).

Woytek, A. J., Lileck, J. T., and Barkanic, J. A., *Solid State Technol. 27* (3), 172 (1984).

Young, D. R., and Osburn, C. M., *J. Electrochem. Soc., 120,* 1578 (1973).

Zarowin, C. B., *J. Electrochem. Soc., 130,* 1144 (1983).

3. Oxidation Technology—The Application of High Pressure Oxidation to VLSI

Natsuro Tsubouchi

LSI Research and Development Laboratory, Mitsubishi Electric Corporation

Oxidation of silicon is one of the most important technologies in the fabrication process of semiconductor devices. Silicon oxides are widely used as field oxides, gate oxides in MOS LSIs, and isolation oxides in bipolar LSIs. Silicon oxide has other uses in the fabrication of VLSIs—for example, as a mask against implant or diffusion of dopant into silicon or as a passivation film on the surface. There are various techniques for forming silicon oxide, such as thermal oxidation, anodic oxidation, plasma oxidation, and chemical vapor deposition. From the viewpoint of oxide quality, thermal oxidation is the preferred technique because it gives rise to a low charge density between the oxide and silicon.

At the present time, advanced MOS circuits utilize a gate dielectric film of 15 to 20 nm. When the channel length of MOSFETs is scaled down to 1 μm, gate oxides of about 10 to 15 nm are required for constant scaling. The projected 1 M-bit MOS dynamic RAM will require a gate oxide of about 10 nm or less if the conventional MOS structure is used. Control and reproducibility of both the threshold voltage and the breakdown strength of the thin gate dielectrics are basic requirements for achieving high quality and reliable VLSI devices. The technique of thin gate oxide formation is one of the most important aspects of VLSI fabrication.

Thermal oxidation is performed at a high temperature for a long period of time in oxidant gases such as oxygen and steam. However, the lengthy elevated-temperature process inevitably causes crystal defects and thermal stress in the silicon substrate, resulting in degradation of device performance. So, low-temperature rapid oxidation has received increasing attention recently in an attempt to eliminate these deleterious effects.

It is well known that oxidation will occur more rapidly in pressurized ambients than at conventional atmospheric pressure. The temperature and time advantages of high pressure oxidation have the greatest potential impact in integrated circuit fabrication technology. Many problems had to be solved in order to develop such high pressure oxidation (HPO) technology for industrial use, but a great effort has been made to achieve it in the last few years. This chapter describes recent progress in high pressure oxidation technology.

HPO offers a number of advantages over conventional atmospheric oxidation processes, as follows:

1. Lower-temperature oxidation, which results in decreased impurity redistribution, shallow junction, reduction of buried collector up-diffusion, and decreased warpage or breakage.
2. Reduced process time, which means high throughput and fast turnaround time for device fabrication.
3. Suppression of oxidation-induced stacking faults, which are known to cause deleterious effects in device performance.
4. Differential oxidation rates of crystalline and polycrystalline silicon at reduced temperature.
5. Decreased temperature of phosphosilicate glass (PSG) reflow under high pressure.

Items 4 and 5 provide a unique fabrication process flow for MOS LSIs, which will be described later.

HIGH PRESSURE OXIDATION SYSTEM

The first HPO system with a safety control for production use was introduced in 1977 (Tsubouchi et al. 1977). The system was able to perform hydrogen and oxygen pyrogenic steam oxidation at pressures up to 10 atm and temperatures from 800 to 1200°C. Since then, a few manufacturers have developed HPO systems with production capability. Figure 3-1 shows an HPO system developed in Japan. It can be operated at pressures up to 10 atm at temperatures between 600 and 950°C and has a load capacity of 120 five-inch wafers with automatic load and unload mechanics. Pyrogenic steam and dry oxygen are the oxidants used for typical oxidation. Sequence and safety controls are performed with a microcomputer. Safety controls include prevention of excess hydrogen production, excess gas pressures, and overheating.

Another HPO system uses a pumped water technique, as shown in Figure 3-2 (Katz et al. 1981). The system is designed for operation at pressure up to 25 atm in a temperature range of 600 to 1000°C. Two hundred 5-inch wafers can be loaded into a 30-inch flat temperature zone.

The development of HPO systems for production use has made the technology popular in VLSI fabrication.

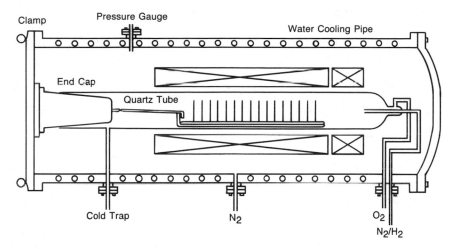

Figure 3-1. HPO system using pyrogenic steam as an oxidant.

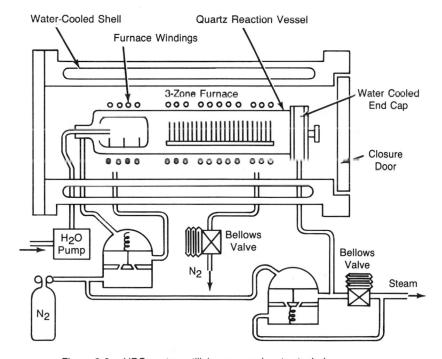

Figure 3-2. HPO system utilizing pumped water technique.

THE MECHANISM OF HIGH PRESSURE OXIDATION

Deal and Grove's model describes the kinetics of silicon oxidation (Grove 1967). The relationship is expressed as:

$$X_0^2 + AX_0 = Bt + X_i^2 + AX_i \tag{1}$$

which can be rewritten as:

$$X_0^2 + AX_0 = B(t + \tau) \tag{2}$$

where X_0 = oxide thickness, B = parabolic rate constant (μm/hr), B/A = linear rate constant (μm/hr), x_i = initial oxide thickness, and $\tau = (x_i^2 + AX_i)/B$.

The rate constants B and B/A are related to the oxidation process as follows:

$$B = 2 D_{\text{eff}} C/N \tag{3}$$

$$A = 2 D_{\text{eff}} (1/k + 1/h) \tag{4}$$

$$B/A = C/N (1/k + 1/h) \tag{5}$$

where D_{eff} is the effective diffusion coefficient of the oxidizing species, C is the equilibrium concentration of the oxidant in the oxide, N is the number of oxidant molecules incorporated into a unit volume of the oxide layer, h is the gas-phase transport coefficient, and k is a constant related to the oxide-silicon interface reaction.

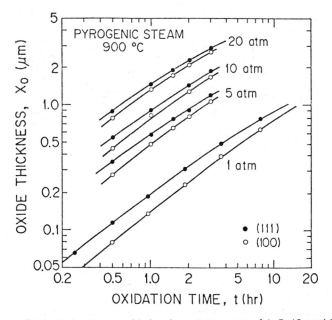

Figure 3-3.   Oxide thickness vs. oxidation time at pressures of 1, 5, 10, and 20 atm.

The general relationship, as summarized by Eqs. (1) through (5), indicates a pressure dependence of the parabolic rate constant (B) and the linear rate constant (B/A), through the dependence of the equilibrium concentration of the oxidant (C) in the oxide.

$$C - KP \qquad [6]$$

where P is the oxidant pressure and K is constant. This indicates a linear pressure dependence for both the linear and parabolic rate constants.

Figure 3-3 shows oxide thickness versus time data for steam oxidation at various pressures and 900°C (Razouk et al. 1981). Figures 3-4 and 3-5 show the results for the linear rate constant and parabolic rate constant.

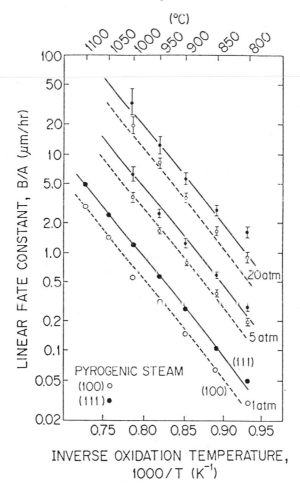

Figure 3-4. Linear rate constant B/A vs. 1000/T for (100) and (111) silicon wafers oxidized in pyrogenic steam as obtained from 1, 5, and 10 atm data.

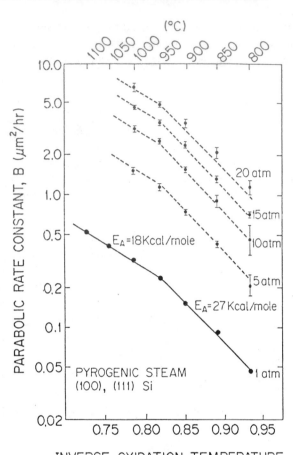

Figure 3-5.   Parabolic rate constant B vs. $1000/T$ for (100) and (111) silicon wafers oxidized in pyrogenic steam as obtained from data at 1, 5, 10, 15, and 20 atm.

Under high pressure, the oxidation time can be reduced in inverse proportion to oxidation pressure.

SUPPRESSION OF OXIDATION-INDUCED STACKING FAULT

Prolonged oxidation of silicon at high temperatures is known to result in oxidation-induced stacking faults (OSFs) in the silicon surface layer. The presence of OSFs are known to have detrimental effects on device characteristics, such as excessive junction leakage, excess noise, video defects in CCD, and so on. There has been increasing interest in the development of an oxidation method that does not generate OSFs. HCl and TCE oxidations were found to be effective in eliminating or reducing the generation of OSFs.

OSFs are also known to depend on oxidation time, temperature, and oxide thickness. Thick silicon oxides are required in a variety of integrated circuits, such as field oxides in MOS LSIs, isolation oxides in bipolar LSIs and so on. The growth of thick oxide usually generates large OSFs. However, high pressure oxidation can suppress the generation of OSFs because of the shorter times and lower temperatures involved (Tsubouchi et al. 1978).

Figure 3-6 shows length of OSF as a function of oxide thickness for high pressure oxidation and conventional wet O_2 oxidation. Solid lines indicate high pressure oxidation data, and broken lines show wet O_2 oxidation at standard

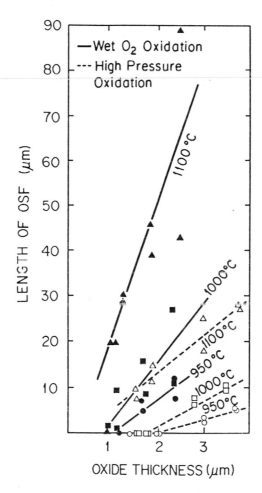

Figure 3-6. Length of oxidation-induced stacking fault as a function of oxide thickness for high pressure steam oxidation at 6.6 kg/cm² and atmospheric wet O_2 oxidation.

pressure. The length of the OSF formed during high pressure oxidation can be reduced significantly for the same temperature and oxide thickness when compared with atmospheric oxidation.

The length of the OSF can be described by the following equation in atmospheric oxidation (Murarka and Quintana 1977):

$$L = At^n \exp(-Q/kT) \qquad [7]$$

where t is oxidation time, T is oxidation temperature, and A, n, Q, and k are constants. This equation is reported to hold true for OSFs formed under high pressure oxidation as well. The constants A, n, and Q are almost equal for both

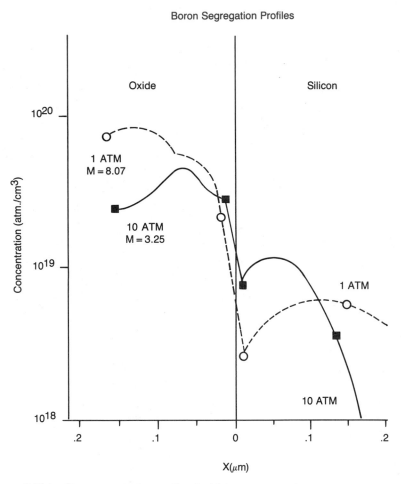

Figure 3-7(a). Boron segregation profiles for high pressure and atmospheric oxidations.

oxidations. The reduction of oxidation time under high pressure results in shorter OSFs.

IMPURITY SEGREGATION AND REDISTRIBUTION

Another important oxide interface phenomenon is the preferential movement of atomic impurities with respect to Si/SiO_2. The results of the redistribution for boron and arsenic are shown in Figure 3-7 (a and b). In both cases, the movement of impurities is greatly reduced because of low-temperature processes (Fuoss and Topich 1980).

OXIDE QUALITY

The quality of the oxide formed under high pressure is important for VLSI application. The surface state density has been reported by several researchers, who have shown the same level of surface state density in the oxide formed under high pressure as in that formed under atmospheric pressure.

The refractive index of the oxide increases with decreasing oxidation tem-

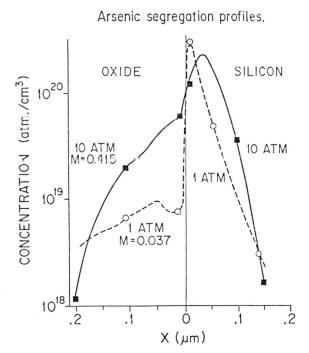

Figure 3-7(b). Arsenic segregation profiles for high pressure and atmospheric oxidations.

perature. It has been found that the refractive index depends on the temperature rather than the pressure (Hirayama et al. 1982).

The film density also depends on the oxidation temperature. A lower-temperature oxide film indicates a higher density. The relationship between refractive index and density is known to obey the Lorentz-Lorenz formula. The refractive index and density of the oxide grown by HPO are found to be consistent with this relationship. The electrical breakdown strength of oxide film depends strongly on oxide preparation and various subsequent treatments. However, the typical breakdown strength is over 8 MV/cm for thin gate oxide.

TIME-DEPENDENT DIELECTRIC BREAKDOWN OF THIN OXIDE

The dielectric breakdown of thin oxide film is strongly dependent on the time during which a bias voltage is applied to the film. The time-dependent dielectric breakdown (TDDB) of thin gate oxides is a major factor in MOS LSI reliability.

TDDB varies with the logarithm of time, and breakdown time is shortened by applying a high electric field. TDDB characteristics of thin gate oxide formed by HPO are shown in Figure 3-8 for different oxidation temperatures (Hirayama et al. 1984). Figure 3-9 shows the cumulative failure rate at 100 sec as a function

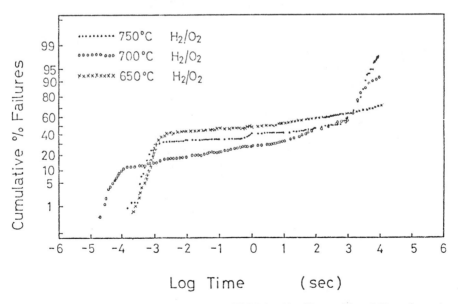

Figure 3-8. Cumulative failure percentage of TDDB for thin silicon oxides of 20 nm formed at a high pressure of 8 atm and different temperatures.

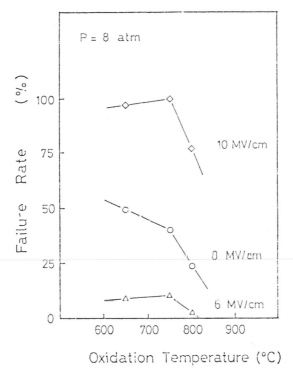

Figure 3-9. Failure rate of TDDB at 100 sec as a function of oxidation temperature under high pressure for different bias electric fields.

of oxidation temperature. The failure percentage decreases with increasing oxidation temperature.

The dependence of the cumulative failure rate on the oxidation pressure is shown in Figure 3-10. Oxidation pressure in the 2 to 8 atm range is not a dominant factor in dielectric breakdown properties.

THE APPLICATION OF HPO TO VLSI

HPO technology has gained a variety of applications in the area of VLSIs. Figure 3-11 shows the types of application for HPO systems used in Japan (Tsubouchi et al. 1983). HPO is most widely used to form the isolation oxides in bipolar LSIs. Recently, applications have been shifting toward MOS LSIs. Some HPO systems are used for general research on basic film growth mechanisms such as nitridation of silicon and oxy-nitride film. Others are used for discrete devices such as power transistors.

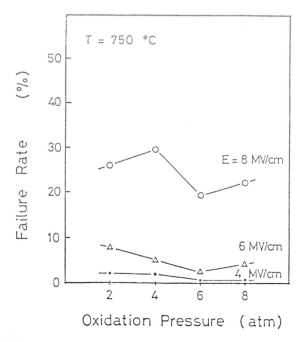

Figure 3-10.   Failure rate of TDDB at 100 sec as a function of oxidation pressure for different bias electric fields.

THE APPLICATION TO BIPOLAR LSIs

Oxide isolation with an implanted shallow junction has been widely used in high-speed bipolar LSIs. However, prolonged oxidation causes redistribution of buried collector impurities, resulting in an increase of parasitic capacitance and a decrease of breakdown voltage. HPO can reduce the impurity redistribution

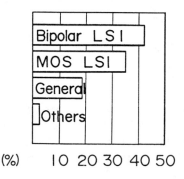

Figure 3-11.   Application fields of HPO system.

that takes place with prolonged atmospheric oxidation or at high temperature. Figure 3-12 shows the redistribution of arsenic in the buried collector during oxide isolation. The spreading resistance corresponds to the reciprocal of the arsenic concentration. The open circles are the spreading resistance profile before oxidation, and the transient profile is caused by auto-doping during epitaxial

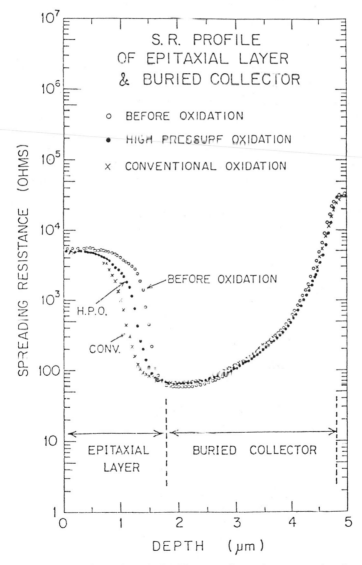

Figure 3-12. Arsenic profiles, characterized by spreading resistance as a function of depth in epitaxial layer, for oxide isolation at 5 atm and atmospheric pressure.

growth. The filled circles and the crosses are the profiles after oxidation at 1050°C under pressures of 5 kg/cm² and atmospheric pressure, respectively. The redistribution of arsenic is suppressed by HPO. The selective HPO was applied to the fabrication of emitter coupled logic gate arrays (Tsakamoto et al. 1979).

Another application to bipolar LSIs has been reported on a low-power, high-speed current mode logic using a base emitter self-aligned technology (Shimizu and Kitabayaski 1979). The oxidation of the polysilicon electrode was performed in high pressure steam to keep the base junction shallow. Figure 3-13 shows the fabrication flow.

As seen in Figure 3-13(a), the process starts with a p-type substrate and goes through n^+ buried layer formation and n epitaxial growth before conventional oxide isolation is performed. After n^+ diffusion for the collector sink, polysilicon is deposited, as shown in Figure 3-13(b). Using a photo resist mask, boron is implanted in the base area. After deposition of Si_3N_4, the polysilicon is oxidized selectively in high pressure steam. This process forms the resistor and the polysilicon electrodes of the transistor. In Figure 3-13(c), boron is implanted to form the inactive base and the resistor after the Si_3N_4 is removed. The contact windows for the emitter and collector are opened after oxide formation. Then, an n-type impurity ion, such as arsenic, is implanted and annealed. In Figure 3-13(d), all contact holes are opened. The emitter contact hole is opened by self-alignment because a thick oxide was formed around it. The polysilicon base

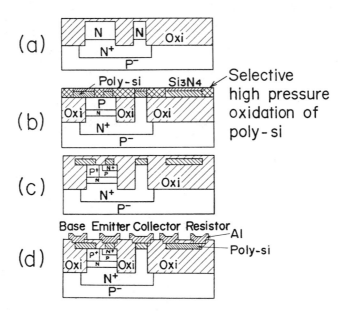

Figure 3-13.   A fabrication process flow for bipolar LSI using selective high pressure oxidation of polysilicon.

contact is offset to the distance from the emitter contact. This technology can produce a smaller device and reduce the circuit parasitic capacitance, resulting in 0.65 nsec for low power current mode logic.

MOS LSI APPLICATION

In the future, HPO application is likely to be directed toward the MOS LSI. A variety of applications have been reported so far.

Thin gate oxide characteristics are good, as mentioned above. The application to field oxides showed an improvement of refresh time in MOS dynamic RAM (Tsubouchi et al. 1979). However, recent progress in gettering technology has yielded good characteristics for refresh time without HPO.

In MOS dynamic RAM, in which a double polysilicon layer structure is widely used, a thicker silicon oxide film between two polysilicon layers is better for device performance and reliability because of the smaller parasitic capacitance and higher breakdown voltage that result.

High pressure steam oxidation at a low temperature can simultaneously grow the thin gate oxide for the second transistor and the thick intermediate oxide on the first polysilicon layer. The intermediate oxide formed by high pressure, low temperature oxidation is thicker than that formed by conventional oxidation for the same thickness of thin gate oxide (Tsubouchi et al. 1979).

With the use of differential oxidation characteristics of the silicon substrate and phosphorus doped polysilicon, another new self-aligned contact technology has been proposed (Sakamoto and Hamano 1980). The processing steps are shown in Figure 3-14. The gate oxide is defined after the tapershaping of the first polysilicon layer, shown in Figure 3-14(a). This polysilicon gate is oxidized under high pressure, as shown in Figure 3-14(b). The oxide grown on the polysilicon gate is thicker than that grown on the substrate. Subsequently, the oxide on the substrate is removed by chemical or reactive ion etching. Then, an arsenic doped polysilicon cover is deposited, and is used as a diffusion source to form the source and drain, as shown in Figure 3-14(c). As shown in Figure 3-14(d), the PSG layer is deposited, contact windows are opened, and finally metallizations are performed to complete the device. This technology has resulted in a large increase of packing density in MOS LSIs.

PSG REFLOW BY HIGH PRESSURE STEAM

Another interesting application of HPO systems to the MOS LSI fabrication process has been proposed (Mayumi et al. 1983), in which the influence of pressure, temperature, and gases on the reflow of PSG film is investigated. Figure 3-15 shows the reflow of PSG as a function of the pressure. It is found that pyrogenic steam pressure accelerates the reflow of PSG film significantly.

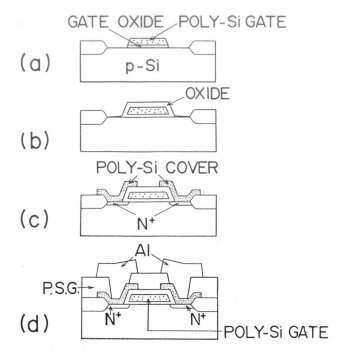

Figure 3-14.   A fabrication process flow for MOS LSI using high pressure oxidation for self-aligned contact formation.

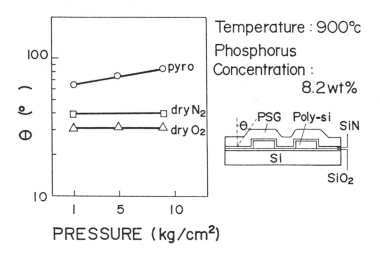

Figure 3-15.   Phosphosilicate glass reflow vs. reflow pressure for different ambient gases.

Water molecules plays an important role on the reflow mechanism. It is also discovered that the reflow temperature could be reduced by 80°C under a pressure of 8.5 kg/cm², when compared with atmospheric pressure. An N-MOS LSI fabrication process with the use of low temperature reflow of PSG has been proposed. When high pressure steam was used in the conventional PSG reflow process, some problems occurred, such as oxide grown on the contact window and the high sheet resistance of polysilicon in MOS LSIs. To eliminate these problems, silicon nitride film was used beneath the PSG.

Figure 3-16 shows the N-MOS LSI fabrication process flow using this technology. As shown in Figure 3-16(a), silicon nitride film is deposited on the surface after conventional n-channel MOS transistor fabrication. Then, PSG film, 8.0 wt %, is deposited to a thickness of 1 μm, as shown in Figure 3-16(b). Contact windows are opened by conventional photolithography after annealing of the PSG film, as shown if Figure 3-16(c). Subsequently, PSG reflow if performed at 900°C under a pressure of 9 kg/cm² for 10 min. Then, aluminum metallization is completed after etching of silicon nitride and oxide films on the contact windows. Successful results were reported for reflow at low temperature, resulting in a shallow junction depth for source and drain in MOS LSIs.

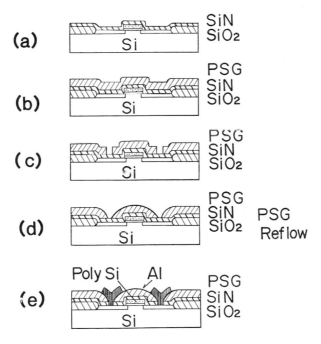

Figure 3-16.   A fabrication process flow for MOS LSI using high pressure phosphosilicate glass reflow.

Table 3-1. Applications of HPO to Bipolar LSIs.

APPLICATION	EFFECT	ADVANTAGE
Oxide isolation	• Reduction of buried collector up-diffusion • Thin epi-layer	• Low parasitic capacitance • High-speed ECL • High breakdown voltage
Polysilicon oxidation	• Shallow base junction	• Low parasitic capacitance • High-speed CML

CONCLUSION

With the introduction of HPO systems, HPO technology is finding wider applications in the semiconductor industry. In bipolar devices, oxide isolation is widely performed by HPO because of a reduction of up-diffusion of buried collector impurities with a thin epitaxial layer. HPO can improve the speed of bipolar logic because it lowers parasitic capacitance. It can also improve breakdown voltage of junctions. Applications to bipolar LSIs are summarized in Table 3-1. In MOS devices, field oxides are sometimes grown under high pressure because of reduced time and temperature. Reduction of OSGs may result in a high yield of MOS LSIs. Gate oxide formation by HPO is under investigation because a lower oxidation temperature does not always guarantee good quality. The higher oxidation rate of polysilicon than that of crystalline silicon provides a simple process to fabricate MOS LSIs with double polysilicon structure. This

Table 3-2. Applications of HPO to MOS LSIs.

APPLICATION	EFFECT	ADVANTAGE
Field oxide	Reduction of OSF	Low leakage current
Gate oxide	Less surface state density	High-quality oxide
Oxidation of polysilicon	Simultaneous growth of thin and thick oxide	Simple process of double polysilicon structure
Oxidation of polysilicon	Self-aligned contact	High packing density
PSG reflow	Reduction of impurity redistribution	Shallow source and drain

characteristic also enables a self-aligned contact technology to achieve high packing density in MOS LSIs. The accelerated PSG reflow under high pressure steam gives rise to shallow junctions in MOS LSIs. Application to MOS LSIs are summarized in Table 3-2. Some oxidation systems are being used in basic research, including silicon nitridation under high pressure. The film has a possible use as an insulator in VLSI. HPO technology could be of great importance to small device geometries in VLSI circuit fabrication in the future.

REFERENCES

Fuoss and Topich, "Heavy-Doping Effects and Impurity Segregation during High Pressure Oxidation of Silicon," *Appl. Phys. Lett.*, 275 (1980).

Grove, A. S., *Physics and Technology of Semiconductor Devices*, Wiley, New York, 1967, p. 27.

Hirayama, M., Miyoshi, H., Tsubouchi, N., and Abe, H., "High-Pressure Oxidation for Thin Gate Insulator Process," *IEEE Trans. on Electron Devices*, ED-29, 503 (1982).

Hirayama, M., Matsukawa, T., Tsubouchi, N., and Nakata, H., "Time Dependent Dielectric Breakdown Measurement of High Pressure Low Temperature Oxidized Film," *22nd Annual Proceedings, Reliability Physics Symposium*, Las Vegas, p. 146 (1984).

Katz, L. E., Howells, B. F., Adda, L. P., Thompson, T., and Carlson, D., "High Pressure Oxidation of Silicon by the Pyrogenic or Pumped Water Technique," *Solid State Technol.*, 87 (Dec. 1981).

Mayumi, S., Asahi, K., Veda, S., Fukunishi, F., Inoue, M., and Arita, S., "Application of High Pressure Oxidation Method to LSI Process," *Technical Digest on Semicon. Devices (Institute of Electronics and Communication Engineers of Japan)*, SSD-82-151, p. 19 (1983).

Murarka, S. P., and Quintana, G., "Oxidation induced Stacking faults in n- and p-type (100) silicon," *J. Appl. Phys.*, *48*, 46 (1977).

Razouk, R. R., Line, L. N., and Deal, B. E., "Kinetics of High Pressure Oxidation of Silicon in Pyrogenic Steam," *J. Electrochem. Soc.*, *128*, 2214 (1981).

Sakamoto, M., and Hamano, K., "A New Self-Aligned Contact Technology," *Technical Digest, IEDM*, 136 (1980).

Shimizu, M., and Kitabayashi, H., "BEST (Base Emitter Self-Aligned Technology): A New Fabrication Method for Bipolar LSI," *Technical Digest, IEDM*, 332 (1979).

Tsakamoto, K., Akasaka, Y., Miyoshi, Y., Tsubouchi, N., Horiba, Y., Kijima, K., and Nakata, H., "High Pressure Oxidation for Isolation of High Speed Bipolar Devices," *Technical Digest, IEDM*, 340 (1979).

Tsubouchi, N., Miyoshi, H., Nishimoto, A., Abe, H., and Satoh, R., "High Pressure Steam Apparatus for Oxidation of Si," *Jap. J. Appl. Phys.*, *16*, 1055 (1977).

Tsubouchi, N., Miyoshi, H., and Abe, H., "Suppression of Oxidation-Induced Stacking Fault Formation in Silicon by High Pressure Steam Oxidation," *Jap. J. Appl. Phys.*, *17*, Supplement 17-1, 223 (1978).

Tsubouchi, N., Miyoshi, H., Abe, H., and Enomoto, T., "The Applications of the High-Pressure Oxidation Process to the Fabrication of MOS LSI," *IEEE Trans. Electron Devices*, ED-26, 618 (1979).

Tsubouchi, N., Hirayama, M., and Nakata, H., "Present Status of Japanese Activities in Applications of High Pressure Oxidation," *The Electrochemical Society, Fall Meeting (Washington, D.C.), Technical Digest*, p. 277 (1983).

4. Structures and Fabrication of Metal-Oxide-Silicon Field-Effect Transistor

Paul J. Tsang

IBM General Technology Division

The insulated-gate field effect transistor (IGFET) has in recent years become one of the most important devices in microelectronics. The device was invented in the early 1930s (Lienfield 1930; Heil 1935), long before the invention of bipolar transistors (Bardeen and Brattain 1948; Shockley 1949), but it was not until the late 1960s that devices with commercial value were made. The long delay between conception and commercial realization can be attributed to a lack of quality substrate material and to difficulties encountered in producing a suitable material for device gate insulation and passivation. The discovery of silicon dioxide (SiO_2) provided the first breakthrough in device fabrication (Kahng and Atalla 1960; Kahng 1976). Since then, SiO_2 has been used almost exclusively as the gate insulator of the device, so the transistor is most commonly called a metal-oxide-silicon field effect transistor, or simply a MOSFET. The current conduction in a MOSFET is carried out by charge carriers of only one polarity (e.g., electrons in an n-channel FET), so the device also is often referred to as a unipolar transistor.

In the early 1960s, it was not possible to grow clean silicon dioxides. Those grown had a high density of positive oxide charge induced by contaminants and structural defects. With these "dirty" oxides, only p-channel devices were made stable enough for circuit applications (Carr and Mize 1972). Because electrical conduction in p-channel devices is carried out by holes, these devices were very slow and were not able to compete with bipolar transistors. Only in the low-end cost-performance market were p-channel MOSFET ICs able to compete with bipolar ICs, because of their simpler fabrication process and lower fabrication cost. However, by the latter part of the decade, rapid improvements in many areas of device fabrication had occurred. Especially important were a better understanding of various oxide charge mechanisms and discovery of the effect of oxide charge on device characteristics. The improvement in silicon oxidation techniques that enabled the growth of ultrapure oxide greatly accelerated the development of device fabrication, especially the fabrication of n-channel MOS-FETs. An n-channel FET was successfully made and tested by IBM in the late sixties (Critchlow et al. 1973). The device was transistor-transistor logic (TTL) compatible and had an electron channel mobility of 600 cm^2/sec-V, which is almost 40% of the electron mobility in an npn bipolar transistor.

The arrival of the fast n-channel device, together with the simpler fabrication process and the capability of achieving higher circuit integration levels, made the MOS IC as attractive as the bipolar IC, if not more so. This provoked a great surge in developmental work on MOS IC technologies in the 1970s. Progress was also made in many areas of device fabrication, including: the development of localized field oxide, the growth of thin gate oxide, polysilicon gate technology, the application of ion implantation to device fabrication, reactive ion etching techniques, and greatly improved lithographic capability.

The first n-channel MOSFET introduced by IBM in the late sixties had a channel length of about 7 μm. Today, devices with a 1.5 μm channel length are being mass-produced. In the last decade, the IC integration level of MOSFETs has increased from a few circuits to hundreds of thousands of circuits, having surpassed that of the bipolar transistor. In fact, it is fair to assert that the MOSFET is the prime driving force as well as a prime achiever in the very large scale integration (VLSI) of silicon integrated circuits. Today, MOSFETs have become the primary device for VLSI circuits such as microprocessors and semiconductor memories. The MOSFET is also an important power device. The market share of MOSFET ICs in the first part of the 1970s was insignificant. The market share today is on a par with that of the bipolar IC. Current predictions (*Electronic Market Data Book 1982*) are that in the 1990s or early 2000s, MOSFET ICs will account for more than 50% of the total production of Si ICs. The DRAM (dynamic random access memory) alone is to generate a worldwide market of more than $50 billion per year by the mid 1990s (Sunami and Asai 1985).

This chapter discusses the structure and the fabrication processes of the MOS-FET device, from its original form to today's most modern structure. The objective of this chapter is to provide newcomers to the field with an overview and a general understanding of MOSFET technology, while permitting the experienced a chance to pause, to reflect, and to design future technological endeavors.

BASIC DEVICE STRUCTURE, OPERATING PRINCIPLES, AND FABRICATION PROCESSES

Basic Device Structure and Operation Principles

A MOSFET is a switching device built on a silicon substrate, whose on–off state and current conduction are controlled by a gate voltage V_g, a drain voltage V_d, and a substrate voltage V_{bs}. Basically, as shown in Figure 4-1, the device

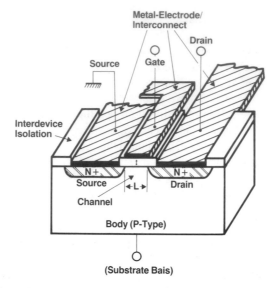

Figure 4-1. Schematic diagram of an n-channel MOSFET. (After Kahng and Atalla 1960)

has six main parts: (1) the current source and drain regions, (2) the conduction channel, (3) the device gate, (4) the device body, (5) the interdevice isolation, and (6) the interdevice connections. The device body is actually the silicon substrate on which the device is built. The device current source and drain are two diodes connected back-to-back to the device channel, whereas the device gate is placed over the device channel and is separated from it by gate insulation. Electrodes are attached to both the source and the drain and to the gate for on–off and current conduction control. A fourth electrode, the substrate bias electrode, is attached to the body of the device, that is, to the silicon substrate. The substrate bias electrode is used primarily to reduce device leakage and to modulate or to stabilize device on–off characteristics.

To construct an integrated circuit, one must place many transistors on a single substrate and connect them together to form an operational circuit. These transistors must be electrically isolated from each other for the circuit to function properly; so interdevice insulation is an integral part of device design and fabrication. In MOSFET ICs, because the device source and drain are reverse-biased during device operation, the transistors are intrinsically electrically isolated from each other in the bulk of the substrate. However, a surface leakage path may develop between adjacent devices if there is a metal interconnection line crossing over the silicon substrate between these devices. The electrical voltage carried by the interconnection line may invert the silicon substrate surface underneath it and form a conducting path, shunting the two devices. Therefore, for MOS ICs, a thick layer of oxide (so-called field oxide) is formed on the substrate surface between devices to achieve total interdevice isolation. The thickness of the field oxide is designed to provide a sufficient separation between the metal interconnection and the silicon substrate to prevent the silicon underneath the interconnection from being inverted by the voltage carried by the interconnection line during device operation.

In order to connect several devices into a functional integrated circuit, conductive (metallic) interconnection lines are formed and connected to each electrode of the device. Aluminum is used most commonly for the interdevice connections, but in recent years metal silicides and/or refractory metals have also become favored interconnection materials.

There are several types of MOSFETs. The most common means of classification is by conduction mode. When a device is fabricated on a p-type silicon substrate, the device source and drain are made of heavily doped n-type diffusion regions, and the device is called an n-channel MOSFET. On the other hand, when a device is fabricated on an n-type silicon substrate, the device source and drain are made of heavily doped p-type diffusion regions, and the device is called a p-channel MOSFET. Both devices can be further made into either an enhancement mode device or a depletion mode device by varying the relative impurity concentration in the conduction channel. An enhancement mode device is one that is not conductive at zero gate bias, and becomes conductive only when a gate-to-source voltage exceeding a certain critical turn-on voltage (also called threshold voltage) is applied. On the other hand, for a depletion mode device the reverse is true. A depletion mode device is normally conductive at zero gate bias and can be turned off only by a negative voltage equal to or larger than the critical voltage required to shut off the conduction channel.

The conduction current in an n-channel MOSFET is carried by electrons, while in a p-channel device it is carried by holes. The current conduction of an n-channel enhancement mode device is illustrated in Figures 4-2 and 4-3. The principle of current conduction of a p-channel device is the same as that of an n-channel FET except that all the voltage and charge polarities are to be reversed. Since the device source and drain are two diodes connected back-to-back, no current will flow between them unless a conduction channel is formed. The conduction channel in a MOSFET is formed by electrically inverting the channel surface with a positive gate-to-source voltage. As illustrated in Figure 4-2, the cross-section of a device channel region is essentially a structure composed of three parts: the metal (gate), the silicon dioxide (gate insulation), and the silicon (substrate or device body). The surface potential of this metal–SiO_2–Si (MOS) structure of the device channel region varies with the applied gate voltage V_g (Grove et al. 1965). When a small positive gate potential is applied, a depletion of charge carriers occurs at the channel surface. This is manifested in a bending down of the energy band of the channel MOS structure, as shown in Figure 4-2(b). There will be no conduction channel formed when the silicon surface is depleted, and the device will not conduct. An increase in the applied gate voltage

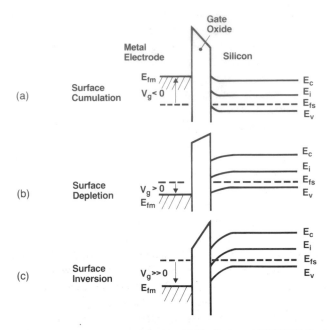

Figure 4-2.   Energy band diagram for ideal MIS diodes of MOSFET device channel when V_g = O. In the figure, (a) the surface is accumulated, (b) the surface is depleted, and (c) the surface is inverted.

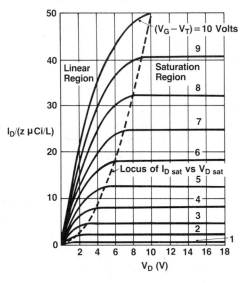

Figure 4-3.   Ideal $I-V$ characteristic of a MOSFET. The dashed line indicates the locus of the saturation drain voltage ($V_{d,sat}$). For $V_d > V_{d,sat}$, the drain remains constant.

will further increase the surface depletion but will not cause the conduction channel to form until the applied gate voltage exceeds the threshold voltage V_{th} of the device. When the gate voltage exceeds the threshold voltage, the energy band will bend so much that the intrinsic Fermi level of the MOS energy band at the surface becomes lower than the extrinsic Fermi level (see Figure 4-2c). The silicon surface is thereby inverted from a p-type to an n-type. An n-type conduction channel linking device source and drain is then formed. The device thus becomes conductive and turns on. When the device is turned on, electrons accumulate in the conductive surface channel and can be collected by the drain electrode when a positive drain voltage is applied.

Figure 4-3 shows the ideal I V characteristic of an n-channel MOSFET. Qualitatively, it is seen that when the device is turned on by a gate voltage V_g, the drain current I_d of the device first increases linearly with the drain voltage V_d (the linear region) and then gradually levels off and saturates at a value $I_{d,\text{sat}}$ when the drain voltage reaches its saturation value $V_{d,\text{sat}}$ (the saturation region). In the linear region, the drain current also increases linearly with gate voltage V_g, whereas in the saturation region, the drain current increases parabolically with the gate voltage V_g. The relationship between the drain current and gate and drain voltages of a MOSFET can be described separately for these two regions by the following first-order equations (Grove 1967):

In the linear region:

$$I_d = (W/L) \; \mu_n C_i (V_g - V_{th}) V_d, \text{ for } V_d \ll (V_g - V_{th}) \tag{1}$$

In the parabolic region:

$$I_{d,\text{sat}} = (W/L) \; m\mu_n C_i (V_g - V_{th})^2 \tag{2}$$

In these two equations, W and L are, respectively, the width and the length of the device channel, μ_n is the effective electron mobility in the device channel, C_i is the capacitance of the device gate insulation, m is a function of doping concentration and approaches 0.5 at low dopings (Brews 1981), and V_{th} is the device threshold (turn-on) voltage.

The device threshold voltage V_{th} is a function of the substrate dopant concentration N_a, the work function difference between gate and substrate ϕ_{ms}, the gate oxide charge Q_{ox}, the gate oxide thickness d_{ox}, the capacitance C_i of the device gate, and the substrate bias V_{bs} (Chero 1984):

$$V_{th} = V_{fb} + 2\psi_b + 1/C_i [2\varepsilon_s \varepsilon_o q N_a (2\psi_b - V_{bs})]^{1/2} \tag{3}$$

where $V_{fb} = \phi_{ms} - Q_{ox}/C_i$ is the flat-band voltage of the device channel MOS structure, ε_s is the dielectric constant of silicon, ε_o is the permeativity of free space, q is electron charge, and ψ_b is the bulk built-in voltage of the silicon substrate.

It is seen from these equations that the characteristics of a MOSFET are strongly affected by the physical dimensions and compositional structure (such as substrate and source/drain dopant concentration and distribution, gate material, etc.,) of the device, as well as by the device fabrication method.

Basic Fabrication Processes

For devices ranging from the rudimentary Al-gate p-channel FET to today's sophisticated submicron DRAM cell, six basic processing steps, one for each main component of the device, are needed to complete device fabrication up to the first interconnection level. These steps are the fabrications of: (1) the field isolation, (2) the gate insulation, (3) the device source and drain regions, (4) the contacts, (5) the device gate, and (6) the interconnections. The sequence of these fabrication steps depends upon the technology used in each step. In this section, the two fundamental fabrication processes, namely, the Al-gate process and the Si-gate process, are discussed.

Al-Gate Process.   For the first-generation device, in which aluminum is used as the gate and interconnection materials, the device gate and interconnection lines are made in a single step. Therefore, only five major processing steps are needed. Figure 4-4 shows the fabrication procedure for an Al-gate device. It has four photolithographic steps. The device fabrication starts with the growth of a blank layer of thick SiO_2. After this thick field oxide is grown, openings (or windows) for implementing device source and drain diffusion regions are then etched into the thick SiO_2 layer by the first lithographic etching step. Then the device source and drain regions are formed by dopant thermal diffusion. Next the device channel region is etched by the second lithographic step. After the channel region is cleaned, the gate oxide is grown. Then the contact holes to the device source and drain are opened. Finally, a layer of aluminum is evaporated onto the device substrate (or device wafer, as it is commonly called) and is subsequently etched to form the device gate and the interconnection lines that "wire up" all the devices on a chip into functional integrated circuits. At each photolithographic step, a photomask is used to transfer the device pattern image.

In the sixties and early seventies, this simple four-mask process was the workhorse used to produce millions of Al-gate MOS ICs. However, the use of aluminum as a gate material has several severe shortcomings. For example, with an Al gate, the device source and drain have to be formed before the formation of the gate. This necessitates some gate-to-source/drain overlap and thus results in a large gate overlap capacitance. The simple Al-gate structure/process was replaced by an all ion-implanted, silicon gate MOSFET structure/process in the late 1970s (Sarace et al. 1968).

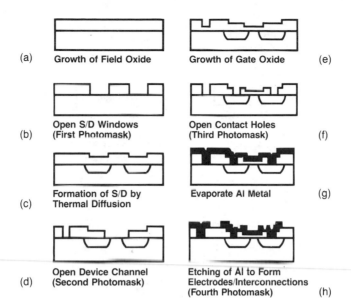

(a) Growth of Field Oxide

(b) Open S/D Windows
(First Photomask)

(c) Formation of S/D by
Thermal Diffusion

(d) Open Device Channel
(Second Photomask)

Growth of Gate Oxide (e)

Open Contact Holes
(Third Photomask) (f)

Evaporate Al Metal (g)

Etching of Al to Form
Electrodes/Interconnections
(Fourth Photomask) (h)

Figure 4-4. A1 gate process: (a) growth of field oxide; (b) etching of device source/drain windows, first photomask; (c) formation of source/drain region by thermal diffusion; (d) open channel region, second photomask; (e) growth of gate oxide; (f) open source/drain contact holes, third photomask; (g) evaporation of aluminum; (h) etching of aluminum, fourth and final photomask.

Structure and Fabrication of Polysilicon Gate FET.

In the seventies, the invention of localized oxidation (Appels et al. 1970), the development of dopant ion-implantation techniques, Osburn and Ormond 1972; Osburn and Bassous 1975), the use of polysilicon as a gate material, and the development of greatly improved photolithographic techniques brought the MOSFET fabrication process into its modern form. Localized oxidation greatly reduces the surface topography; ion implantation enables precise control of dopant distribution and concentration; and the use of polycrystalline silicon (polysilicon) gates not only greatly improves the gate insulator reliability (Gibbons 1968, 1972; Lee and Mayer 1974) but also enables the use of self-aligned device fabrication techniques. Figure 4-5 shows the structure and conventional fabrication process of a polysilicon gate MOSFET.

As shown in Figure 4-5, there are five basic, distinctive features of a polysilicon-gate FET. The first is that the gate, being made of polycrystalline silicon, is self-aligned with the device source and drain regions. This greatly reduces the gate-to-drain and gate-to-source overlap capacitances. The second such feature is a gate contact hole, which is provided for connecting the gate to the interconnection lines. The third feature that differentiates the device from an Al-gate FET is the (semi- or fully) recessed field oxide isolation (ROI). The fourth feature is that the dopant concentration in the device channel is now raised above

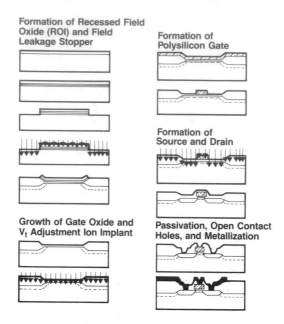

Formation of Recessed Field Oxide (ROI) and Field Leakage Stopper

Formation of Polysilicon Gate

Formation of Source and Drain

Growth of Gate Oxide and V_t Adjustment Ion Implant

Passivation, Open Contact Holes, and Metallization

Figure 4-5.   Structure of polysilicon gate MOSFET.

the substrate background level. The final distinctive feature of the device is the p+ surface leakage stopper underneath the field isolation.

As in the Al-gate process, four lithographic mask levels are needed to fabricate a basic polysilicon gate FET structure up to the first metal interconnection level. Because of the use of polysilicon high-temperature gate material and the use of ion implantation to form various diffusion regions, this fabrication process has much greater design freedom and flexibility than the Al-gate process. The process flow for the fabrication of a polysilicon gate FET is also slightly different from that of an Al-gate device. Figure 4-6 shows the processing sequence of the polysilicon gate FET. There are six major processing sequences, which are, sequentially: (1) formation of the field oxide isolation (ROI) and the surface leakage channel stopper; (2) channel doping level adjustment and the formation of gate oxide; (3) formation of the polysilicon gate; (4) formation of source and drain regions; (5) device passivation and open contact holes; and (6) formation of metal interconnections.

As shown in Figure 4-6, device fabrication starts with the formation of both the surface leakage channel stopper and the field isolation oxide. This is accomplished by field boron ion implantation and a localized oxidation process called LOCOS (localized oxidation for silicon) (Appels et al. 1970). In the process, a nitride/oxide field oxidation mask is first formed to cover the regions in which devices are to be formed; then the field ion implantation is performed; and finally

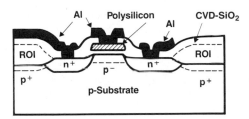

Figure 4-6. Fabrication sequence of polysilicon gate MOSFET.

the field oxide is grown. The formation of the nitride/oxide oxidation mask involves the growth of a thin "pad oxide" and the chemical vapor deposition (CVD) of silicon nitride and oxide films. After the layered oxide/nitride/oxide films are formed, they are delineated into the field oxidation pattern, which is used as the field oxidation mask. As can be seen in Figure 4-6, a photolithographic mask is used for the pattern delineation.

In the second sequence of processing steps to form the gate oxide, first the silicon wafer is thoroughly cleaned to ensure a contamination-free surface. Then the gate oxide is thermally grown in a pure or "dry" oxygen ambient. After the gate oxide is grown, a channel ion implantation is carried out to "adjust" the channel doping level so that desired device threshold voltage and threshold voltage substrate sensitivity are achieved. In addition, by choosing an appropriate type of dopant and the implant dose level, devices of either enhancement mode or depletion mode, or both, can be made at this step. In many instances, the channel dopant ion implantation is done before the growth of gate oxide. The advantage of implanting channel dopant before gate oxidation is that ion-implant–induced defects are automatically annealed out during gate oxide growth. On the other hand, implanting channel dopant before gate oxidation results in deeper channel dopant distributions, which may be undesirable.

Following the channel ion implantation, a layer of in-situ–doped (usually with phosphorus for n-channel FET) polysilicon of proper thickness is deposited by the CVD method and is subsequently delineated into the gate structure by a photolithographic step, in which an etching mask is used. These steps constitute the third processing sequence of the device fabrication. In the fourth processing sequence, the arsenic ion implant for the source and drain is performed first; this is followed by the implanted dopant activation annealing, which, in general, is carried out at temperatures ranging from 900 to 1100°C in an oxidizing ambient. After the device source and drain diffusion regions are formed, a layer of CVD silicon dioxide is deposited for surface passivation and for improving electrical insulation between the Si substrate and the metal interconnections.

In the fifth processing sequence, contact holes of the device gate and of the source and drain are etched by using the third lithographic mask. The final processing sequence uses the fourth photolithographic mask to form the metal interconnection. The most commonly used interconnection metal is aluminum with $\approx$ 2 wt% of silicon and $\approx$ 4 wt% copper added. The aluminum alloy film is deposited by a physical thin film deposition method (*Deposition Technologies* 1982). The metal interconnection pattern can be formed either by a subtractive etching method or by an additive lift-off technique (Hatzakis 1969; Hatzakis et al. 1980).

FABRICATION TECHNOLOGIES

This section discusses in some detail the four most important fabrication technologies used for MOSFETs and the effect they have on device performance. The technologies are: (1) the growth of SiO_2 films, (2) the formation of the polysilicon gate and the self-aligned source and drain, (3) the application of dopant ion implantation, and (4) the formation of interconnections and the contact metallurgy.

Growth of Silicon Dioxide Film

Silicon dioxide is one of the most important and indispensible materials for the construction of a MOSFET. It is used not only as an integral part of the device, such as the field isolation, the gate insulator, and the inter metal level insulation, but also as an operational medium in device fabrication. As we saw in the preceding section, the silicon dioxide used as field isolation and the gate insulator are formed through thermal oxidation. Precise control of film thickness and the charges in the grown oxide films are of the utmost important in device fabrication. In this section, the kinetics of oxide film growth is discussed first, followed by a discussion of the origin of oxide charge, the effect of oxide charge on device characteristics, and the means to reduce oxide charge. In the last part of this section, the procedure for forming the recessed oxide field isolation is described.

Growth Kinetics. In general, the growth of thermal silicon dioxide (SiO_2) film is carried out in a horizontal oxidation furnace in a temperature range of 700 to 1100°C in either pure (dry) oxygen or steam (wet) ambient. The oxide films thus formed are amorphous and consist of randomly arranged $Si-O_4$ tetrahedrons (see Figure 4-7). The formation of silicon dioxide is a simple chemical reaction of silicon oxidation:

$$Si(solid) + O_2 + O_2 \rightarrow SiO_2 \qquad [4]$$

and/or:

$$Si(solid) + 2H_2O \rightarrow SiO_2 + 2H_2 \qquad [5]$$

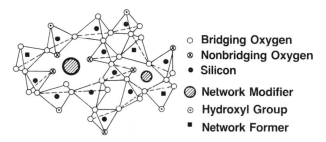

○ **Bridging Oxygen**
⊗ **Nonbridging Oxygen**
● **Silicon**

◐ **Network Modifier**
◉ **Hydroxyl Group**
■ **Network Former**

Figure 4-7. Structure of silicon dioxide.

However, the growth of silicon dioxide film from a silicon substrate involves the transport of reactant species and is more complicated than this. Existing experimental evidence has led us to believe that during film growth, silicon atoms do not leave the silicon substrate. It is the oxygen or hydroxide radical that diffuses through the already grown SiO_2 film to react with silicon at the SiO_2–Si substrate interface. The reaction rate of Si with either O_2 or OH to form SiO_2 is very great compared with the diffusion rate of oxygen or hydroxide in SiO_2. Therefore, except in the case of thin oxide films, the SiO_2 film growth rate is, in general, governed by the diffusion of oxygen or hydroxide radicals in SiO_2.

The growth kinetics of an SiO_2 film can be described by the well-known linear-parabolic law (Deal 1963; Deal and Grove 1965), in which the oxide film thickness d_{ox} is related to oxidation time and various reaction constants:

$$d_{ox}^2 + Ad_{ox} = B(t + t_i) \qquad [6]$$

Solving the quadratic equation, Eq. (6), for d_{ox} as a function of time, we obtain:

$$d_{ox} = (A/2)\{[1 + (t + t_i)/(A^2/4B)]^{1/2} - 1\} \qquad [7]$$

In the above equations, B is the parabolic rate constant, B/A is the linear rate constant, t is the oxidation time, and t_i, called the initial oxidation time, is the time that is needed to grow the initial SiO_2 film on the substrate (i.e., the native oxide film or the oxide film that was already on the substrate before oxidation). The quantity t_i is related to the initial oxide thickness d_{oi} by the following equation:

$$t_i = (d_{oi}^2 + Ad_{oi})/B \qquad [8]$$

Based on a simple phenomenological model developed by Deal and Grove (Hatzakis et al. 1980), the above rate constants can be related to the diffusivity

D of oxygen (or OH) in SiO_2, the surface reaction constant k_s of SiO_2, and the mass transport coefficient h of oxygen (and/or OH) in the oxidation ambient:

$$B = 2DC^*N^{-1} \text{ (cm}^2/\text{sec)} \qquad [9]$$

and:

$$B/A = C^*k_s h/N(k_s + h) \qquad [10]$$

where C^* is the equilibrium concentration of the oxidant in the oxide, and N is the number of oxygen atoms incorporated into a unit volume of the film.

From Eq. (7) as well as from experimental results, it is seen that there are two limiting cases of oxide film growth. For thick oxide films when $t \gg A^2/4B$, and $t \gg t_i$, oxide growth follows a parabolic law. That is, Eq. (7) becomes:

$$d_{ox}^2 = Bt \qquad [11]$$

The rate of SiO_2 growth in this case is governed by the diffusion of oxidant (i.e., O_2 or OH) in SiO_2. On the other hand, for the thin oxide films where $(t + t_i) \ll A^2/4B$, oxide growth follows a linear law:

$$d_{ox} = (B/A(t + t_i)) \qquad [12]$$

The rate of oxide growth in this case is limited by the silicon oxidation rate at the silicon–silicon dioxide interface.

Experimental results show that the temperature dependency of both the parabolic and linear rate constants can be expressed by the Arrhenius relationship:

$$B = K_p \exp(-E_{ap}/kT) \qquad [13]$$

and:

$$B/A = K_l \exp(-E_{al}/kT) \qquad [14]$$

As can be seen in Table 4-1 (Meindl et al. 1980), the activation energies E_{ap} and E_{al} of SiO_2 film growth at higher temperatures are different from those at lower temperatures. The demarcation temperature that divides the two different oxide growth mechanisms is 950°C for wet oxidation and 900°C for dry oxidation.

In general, the growth of a silicon oxide film follows quite faithfully the linear-parabolic law, except in the case of thin dry oxide films from bare silicon wafers, where significant deviation from the law exists. According to Eq. (6) and Eq. (8), extrapolating the d_{ox} versus t curve to $t = 0$ should result in a t_i value equivalent to that needed for the growth of initial oxide d_{oi}. This is observed in

Table 4-1. Rate Constants and Activation Energies of Thermal Oxidation of (100) Single Crystal Silicon.

TEMPERATURE ($°c$)	K_p ($\mu m^2/hr$)	L_1 ($\mu m/hr$)	E_{ap} (eV)	E_{al} (eV)
>950	4.2E2	—	0.78	—
<950	1.7E4	—	1.17	—
>900	—	1.77E8	—	2.05
<900	—	2.07E6	—	1.60

the wet oxides but not in the dry oxides. In dry oxides, the value of t_i obtained from data extrapolation is always larger than the calculated one. This is attributable to the fast initial SiO_2 growth in dry O_2 ambient.

Oxide Charges and Control of Oxide Charges. There are two main types of oxide charges that can be introduced into silicon dioxide during either oxide film growth or device processing: (1) the oxide charges Q_{ox}, which include the oxide fixed charge Q_f and the surface state charge Q_{it}, and (2) the sodium mobile ion charge Q_m (Deal 1980). The existence of these charges in oxide greatly deteriorates device performance and stability. In fact, this is the main problem that hampered the development of the MOSFET during the early days. Gate oxide with large positive fixed oxide charge densities causes a low threshold voltage for the n-channel device and a high threshold voltage for the p-channel device, whereas surface state charges and mobile ion (e.g., sodium ions) charges cause junction leakage and device threshold voltage instability. High oxide charges existing in the gate insulation also deteriorate device gain, or device transconductance, and cause channel leakage. The existence of oxide charge in the field oxide causes the silicon surface in the field region to invert, and leads to the formation of interdevice leakage. Therefore, it is of utmost importance to MOSFET fabrication that low oxide charges be maintained during film growth as well as throughout the entire device fabrication. Oxide films with low oxide charges can be obtained by (1) growing films in a clean system, (2) proper post-oxidation annealing, and (3) the gettering of contaminants.

Growth of Clean Oxide. Oxide charges in silicon dioxide are caused by contaminants; so the first thing one has to do to assure a stable oxide with low oxide charges is to grow oxide in a clean environment and in a clean oxidation system. The use of ultraclean gases and furnace tube, in addition to thorough cleaning of the silicon wafers, is essential to the growth of clean oxide. However, it is also found that clean oxide can be grown by introducing HCl into the oxidizing ambient during silicon oxidation (Lee and Mayer 1974; Gardiner and Shepard 1979; Gardiner et al. 1979). HCl is believed to perform two functions in the furnace during silicon oxidation. One is to clean the furnace tube; another is to clean the oxide film. The mechanism for cleaning furnace tubes by HCl is thought to be the oxidation of HCl in the HCl–O_2 mixture to form water vapor and chlorine gas. The chlorine gas thus formed will react with sodium in the furnace tube to form NaCl. Sodium chloride is volatile at high temperatures, and is swept from the furnace tube by the gas stream. By the same mechanism, HCl is considered to be effective in removing other metallic contaminants that will form volatile halides (Fe, Cr, Ni, etc.) from the furnace tube. The mechanism by which HCl cleans the oxide and reduces the fixed charges in the oxide is the formation of Na–Cl bonding, which neutralizes the sodium charges and "ties up" the mobile sodium in the oxide, thus rendering it immobile. In general, depending on the oxidation furnace, 2 to 5 mol% HCl is added to the oxidant gas flow. Oxides with fixed charges and mobile charges at a level of 10×10^{10}/cm^2 have consistently grown in the presence of HCl.

Post-Oxidation Annealing. The oxide fixed charge and surface state charge in an as-grown SiO_2 film are known to depend on oxidation temperature (Razouk and Deal 1979; Deal et al. 1967), oxidation ambient (wet or dry) (Deal et al. 1967), oxygen partial pressure (Murarka 1979), and substrate orientation (Arnold et al. 1968). Oxides grown on (100) substrates have the lowest oxide charge density, compared with oxides grown on substrates of other orientations. This is why (100) substrates are chosen for device fabrication. Deal et al. (Razouk and Deal 1979; Deal et al. 1967) have found that densities of both the oxide fixed charge and the surface state charge decrease as the oxide growth temperature increases. The oxide fixed charge density of oxide grown at 1200°C is almost an order of magnitude lower than that of oxides grown at 500°C. This growth-temperature dependence of the oxide fixed charge causes the oxide charge density to be dependent on the oxide cooling rate, if the oxide is cooled to room temperature in an oxidizing ambient. Therefore, fast cooling of the oxidized wafer is preferred to attain a low oxide fixed charge. However, for large-diameter wafers, the wafer cooling speed is limited by the occurrence of thermally induced wafer warping.

The oxidation ambient was found to affect charge density in the following way. The presence of H_2O increases charge density. Oxide grown in wet ambient usually has a higher oxide fixed charge than that grown in dry ambient. This is particularly true at high growth temperatures. On the other hand, the presence of an inert gas such as N_2 reduces oxide charges. Murarker (1979) observed a fivefold reduction of fixed charge density in dry oxide when it is grown in a nitrogen-diluted oxygen ambient.

Oxide charges can also be reduced by post-oxidation annealing in an inert gas ambient. Deal et al. (1967) showed that annealing oxide in nitrogen gas reduces oxide fixed charge density. Figure 4-8 shows the famous "Deal Q–T triangle," which summarizes the relation of Q_f to oxidation temperature and to cooling or annealing in inert ambient (N_2 or Ar). The hypotenuse of the triangle represents a Q_f versus oxidation temperature curve. The vertical bars in the diagram rep-

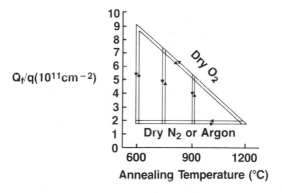

Figure 4-8. Deal et al. charge–temperature triangle showing oxide fixed charge density as function of annealing temperature for annealing times of one hour or less. (After Deal et al. 1967).

resent high-temperature anneals in inert gas, with the temperature of each of the anneals shown in the abscissa of the figure. The base of the Q–T triangle represents the cooling down of oxide in an inert gas ambient.

In view of the effect of oxidation temperature and ambient on the oxide fixed charge, and the effectiveness of inert gas annealing in reducing oxide fixed charge, the general practice of oxide growth in today's device fabrication process is to conclude with a dry oxidation followed by a short post-oxidation N_2 anneal at the oxidation temperature, and finally a cool-down to room temperature in an inert gas ambient.

To reduce the surface state charge, a low-temperature post-gate (metal) annealing in hydrogen or forming gas ambient is carried out (Deal et al. 1969; Totta and Sopher 1969). The most commonly used post-gate (metal) annealing process is done at 400 to 450°C in forming gas ambient for 20 to 30 min.

Contaminant gettering. Sodium contamination can be introduced at various stages of device fabrication. A low-cost method of eliminating sodium contamination during device fabrication and during device lifetime is to coat the oxide with a layer of phosphosilicate glass (PSG). Pliskin et al. (1967) have shown that the phosphosilicate glass acts as a getter by chemically binding sodium in an electrically neutral form. A PSG layer can be formed by the CVD method or by exposing the oxide film to P_2O_5 vapor at an elevated temperature. In most MOSFET IC fabrications, PSG is CVD-deposited over the entire circuit after gate and source/drain are implemented. It is used to getter sodium and other process-induced contaminants in the oxide, and to prevent further contamination from occurring during the device lifetime. In addition, it is also used, in some applications, as a topography modulator. Phosphosilicate glass has a sufficiently low melting point that it can be made to flow at high temperatures (>1100°C). The sharp corners of topographic steps of a MOSFET IC coated with PSG can be eliminated by a high-temperature annealing. Phosphosilicate glass used in contaminant gettering usually contains 3 to 5 mol% P_2O_5 in the film, but a higher P_2O_5 concentration is needed when it is used to modulate surface topography.

Formation of Recessed Oxide. The recessed oxide isolation (ROI) is formed by a so-called locally oxidized silicon isolation (LOCOS) method. The method was first introduced by the Philips Laboratory in the early seventies (Appels et al. 1970; Kooi and Appels 1973) for the formation of field isolation for a MOSFET IC. Since then, the recessed oxide has become the standard interdevice isolation structure for both MOSFET and bipolar VLSI ICs. Compared with planar oxide isolation, it has the advantages of reducing surface topography (vital to high density VLSI IC fabrication) and junction capacitance. In addition, in some applications such as in CMOS, it is also much superior in reducing interdevice electrical leakage.

The ROI fabrication procedure was briefly described in an earlier section (on, "Structure and Fabrication of Polysilicon Gate FET"). In this section, we will discuss the fabrication procedure in greater detail. In addition, some of the key issues encountered in its application will be discussed.

The standard procedure for ROI formation is shown in Figure 4-9(a). It involves three major steps. First, a $SiO_2/Si_3N_4/SiO_2$ trilayer film is constructed. Second, this trilayer film is delineated into the ROI pattern to cover the active device regions by either a chemical etching method or a plasma etching method.

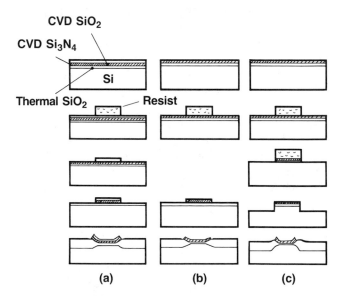

Figure 4-9. Fabrication of ROI structures: (a) "wet" process for semirecessed ROE, (b) "dry" process for semirecessed ROI, and (c) "dry" process for fully recessed ROI.

Finally, the recessed oxide isolation is grown. To form the $SiO_2/Si_3N_4/SiO_2$ trilayer oxidation mask, a thin layer of SiO_2, called pad oxide, is first grown on the silicon substrate. This is followed by deposition of the silicon nitride layer, which is done most commonly by the low pressure CVD (LPCVD) method. Finally, the overlap LPCVD SiO_2 is deposited. The silicon nitride film in this trilayer mask is the film that protects the device region from being oxidized during the growth of field oxide. Thus, it should be dense and pin-hole–free. The pad oxide provides stress relief between the nitride film and silicon substrate, while the SiO_2 overlay serves as an etching mask for the nitride film.

To delineate the trilayer film into the ROI pattern, a photoresist ROI pattern is first applied. Then, using the resist pattern as an etching mask, the overlay SiO_2 film is first etched by buffered hydrofluoric acid (BHF) solution. Next, the resist pattern is stripped off, and the wafers are cleaned. Then, using the SiO_2 overlay as an etching stencil, the ROI pattern is transferred to the silicon nitride film by hot phosphoric acid etching, after which it is ready for the growth of ROI. In the more advanced device fabrication processes, a plasma etching method is used to etch the silicon nitride layer. In this case, the resist pattern can be used as the etching mask for nitride, and the overlay SiO_2 etch mask is not needed. Thus it is eliminated from the process, as shown in Figure 4-9(b).

As the field oxide is being grown, it consumes a layer of silicon substrate, so that it is partially (about 46% of the oxide thickness) recessed ito the silicon substrate. This is the reason why the oxide isolation thus formed is called the recessed oxide isolation (ROI). In some applications, a fully recessed oxide isolation is preferred, in order to obtain, for the same ROI thickness, deeper interdevice electrical isolation and smoother surface planarity. To form the fully recessed oxide isolation, a trench with a depth equivalent to 54% of the ROI thickness is etched into the silicon substrate before the ROI growth. The fabrication sequence of a fully recessed ROI by a "dry" process is shown in Figure 4-9(c).

There are three major concerns in ROI formation. The first is the formation of a so-called brown ring on the surface of the device region underneath the trilayer dielectric mask. This brown ring is believed to be a thin oxynitride layer formed during ROI growth. Kooi et al. (1976) suggested that during ROI growth NH_3 is generated around the periphery of the oxidation mask due to the reaction between Si_3N_4 and H_2O. This newly generated NH_3 will penetrate through the pad oxide and react with the silicon substrate to form a band of oxynitride around the device region. As this oxynitride ring is generally brownish in color, it is called a brown ring. The existence of the brown ring hinders the growth of gate oxide, so that gate oxide grown underneath the brown ring has a reduced thickness, causing a premature gate oxide breakdown problem. The solution to the brown ring problem is to use a thicker pad oxide and lower field oxide growth temperature.

The second concern in ROI formation is the interface stress between the Si_3N_4

film and the silicon substrate. This stress causes oxidation-induced defect generation in the device region, leading to gate oxide deterioration (Nakajima et al. 1979) and junction leakage. The remedy for this problem is, again, to increase the pad oxide thickness.

The third concern in ROI formation is the formation of a so-called bird's head around the ROI periphery. The bird's head, as shown in Figure 4-10, includes a beak (BB) and a head body (BH) in the semirecessed ROI, while in the fully recessed ROI it also includes a crown (BC). The formation of the bird's head not only causes undesirable encroachments into device and isolation areas; it also destroys the surface planarity that a fully recessed ROI set out to achieve. Therefore, it is highly undesirable in device fabrication.

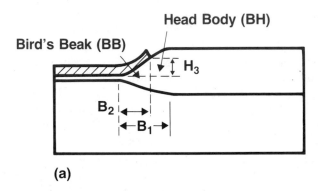

(a)

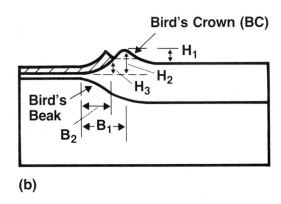

(b)

Figure 4-10. Topology of ROI's bird's head: (a) for a semirecessed ROI; (b) for a fully recessed ROI. The bird's head is characterized by a beak (BB) and a body (BH). The length of the beak is indicated by B_2 in the figure, while the head body is indicated by B_1 in the figure. In the fully recessed ROI, the bird's head is also characterized by a crown (BC). In both cases, B_2 represents encroachment into the device region, while (B_1-B_2) represents encroachment into the field region. The bird's crown in the fully recessed ROI destroys the planarity a fully recessed ROI sets out to achieve.

A study by Bassous et al. (1976) showed that the size (lateral length) of the ROI BB is strongly dependent on both the thickness of the pad oxide and that of the nitride film. They observed that the size of the ROI BB increases with the pad oxide thickness, but decreases with the nitride film thickness. From the results of Bassous et al. Figure 4-11 shows that the ROI BB can be minimized if the nitride thickness of the nitride–oxide ROI mask is thicker than 250 nm while the pad oxide thickness is kept to less than 10 nm. However, experimental

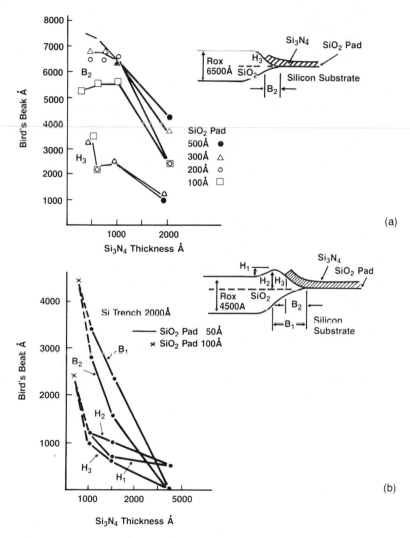

Figure 4-11.   Selected feature of bird's heads: (a) for a full ROI; (b) for a semi ROI. (After Bassous et al. 1976.)

results have also shown that, when the ROI is grown with such a thick nitride and thin pad oxide mask, the quality of the thin oxides that are subsequently grown in the active device regions (for example, the gate oxide of a MOSFET) is greatly reduced. These oxides will have a low dielectric breakdown strength and a high oxide surface state charge. The deterioration of the thin oxide grown in the active device area can be attributed to the interfacial stress of the nitride mask induced at the ROI boundary during oxide growth and the formation of the so-called brown ring around the ROI. Both can be minimized or eliminated by increasing the pad oxide thickness (Ormond and Gardiner 1980).

A desirable nitride/oxide oxidation mask for MOSFET application is one that produces ROI with the smallest ROI BB while maintaining the high quality of the thin oxide grown in the active device region. In general, this is achieved by increasing the pad oxide thickness to 30 to 60 nm and reducing the nitride thickness to below 200 nm (Nakajima et al. 1979).

The oxide charge density of the thin oxide grown after ROI in the active device region is also affected by the oxidation temperature of the ROI. The highest charge density has been observed in thin gate oxide grown on wafers whose ROI is grown at 1000°C. Both increasing and decreasing the ROI growth temperature from 1000°C reduces the oxide charge density in the gate oxide.

Formation of Polysilicon Gate

Perhaps the most crucial technological advance in MOSFET fabrication has been the development of the polysilicon gate. It brought about a radical change in device fabrication and hastened the advent of the modern VLSI era. Polysilicon possesses several advantages over aluminum: (1) It has the same high melting point and thermal expansion coefficient as the silicon substrate, so that it is possible to have high-temperature processing steps after construction of the gate structure. (2) Stable oxide can be grown on polysilicon. Therefore, a polysilicon gate has a self-passivating capability. This makes self-aligned source–drain fabrication possible. It also facilitates the implementation of two-level interconnections, thereby greatly increasing circuit wiring capability and circuit packing density. (3) The work function difference between a polysilicon gate and a device channel can be diminished by using p^{++} polysilicon gate for the n-channel device and using n^{++} polysilicon gate for the p-channel device. (4) A silicon gate provides much improved gate insulation reliability (Osburn and Ormond 1972; Osburn and Bassous 1975; Ormond 1979).

The silicon technology was first introduced to MOSFET fabrication in 1968 by Sarace et al. (Sarace et al. 1968) and Faggin et al. (Faggin et al. 1968; Faggin and Klein 1970; Vadasz et al. 1969). Many new features have since been added. The formation of a polysilicon gate has now become a standard process step in

MOSFET fabrication. It involves three major steps, which are, sequentially: (1) deposition of polycrystalline silicon, (2) formation of the gate pattern etching mask, and (3) etching of the gate pattern.

The polysilicon film is deposited by the CVD method, in which the polysilicon film is deposited on substrates through the pyrolysis of a silicon-bearing gas such as silane (SiH_4) or dichlorosilane. Usually, the film deposition is carried out in a temperature range of 600 to 800°C, either at atmospheric pressure or in a reduced pressure system. The film can either be doped in situ during film deposition or be doped extrinsically. The extrinsic doping of polysilicon is commonly done by two methods. The first one is the doped-oxide diffusion method, in which a layer of doped oxide is first deposited onto the polysilicon film; then the dopant atoms are "driven" from the doped oxide into the polysilicon by a high-temperature annealing step. Usually, the deposition of the doped oxide film and dopant drive-in are done in the same system. The second polysilicon doping method is that of ion implantation. In this process, dopant atoms of the desired dose are ion-implanted into the bulk of the polysilicon film, then they are activated and redistributed by high-temperature annealing. Diffusion of most doping elements into polysilicon is extremely fast (Kamins et al. 1972; Swaminathan et al. 1982; Coe 1977). Therefore, homogeneously doped films can be obtained quite easily. Polysilicon films deposited by the CVD method at 620 to 650°C through the pyrolysis of silane have a very fine grain structure with an average grain size of 30 to 50 nm (Anderson 1973). This may in part explain why dopant atoms diffuse so rapidly in the polysilicon film. However, a heavily doped polysilicon film has a resistivity slightly higher than that of single crystal silicon of comparable doping level, due to the grain boundary effect (Baccarani et al. 1978; Fripp 1975; Mandurah et al. 1980). In general, a sheet resistance of 30 to 50 Ω/sq can be contained in a heavily phosphorus-doped film that is 400 nm thick; this is adequate for most applications.

Several chemical solutions will etch polysilicon films. The solution that gives the best result and so is most widely used is a hot pyrocatecol solution (Fripp 1975; Mandurah et al. 1980). However, because the pyrocatecol solution attacks photoresist, a photoresist pattern cannot be applied directly on the polysilicon substrate; a SiO_2 etching mask (for etch stencil) has to be used. Therefore, the pattern delineation process of polysilicon usually consists of four steps. First, a layer of SiO_2, usually LPCVD SiO_2, is deposited. Then, the image photoresist pattern is applied by a photolithographic step. This photoresist pattern is used to etch the SiO_2 mask. After the SiO_2 mask is formed, the resist pattern is removed chemically or by oxygen plasma. Then the polysilicon film is etched into the desired gate pattern.

In a more recent development, plasma or reactive-ion-etch (RIE) processes have been used to etch polysilicon films (Irving et al. 1971; Schwartz and Schaible 1979). The most commonly used RIE etching processes for polysilicon are the

CF_4, Freon 12, Freon 13, and CCl_4 processes. In these processes, the intermediate SiO_2 stencil can be eliminated if the polysilicon film to be etched is not too thick.

A major advantage of using a polysilicon gate is the possibility of forming the device source and drain after the device gate. This is done by ion implantation. Dopants of the proper dose level and energy are implanted without an additional lithographic step. The dopants can be implanted into the device source and drain with or without a layer of "screen oxide." Implanting through a layer of screen oxide reduces the implant "channeling" effect. After dopant implantation, high-temperature annealing is carried out to activate the implanted doping atoms and to remove the ion-implant-induced defects in silicon. The device source and drain thus formed are self-aligned with the device gate, and produce a gate-to-source/drain overlap by an amount equal to the lateral diffusions of the source/drain region. The result is that the gate-to-source/drain overlap capacitance is much smaller than that found in an Al gate FET. Thus polysilicon gate FETs are much better devices, especially for use as dynamic random memories.

Application of Ion Implantation

There are three main uses for ion implantation: to form the field leakage channel stopper, to form device source and drain regions, and to modulate doping concentration in the device channel region. The advantages of ion implantation are its design flexibility and the precision that can be attained in controlling dopant concentration, doping profile depth, and minute dopant distribution changes. With ion implantation, channel doping level can be precisely set, and ultrashallow source/drain pn junctions that otherwise would not be achievable by thermal diffusion can be readily formed.

As mentioned in a previous section, arsenic rather than phosphorus is generally used to form device $n^{++}p$ source/drains, for two reasons: (1) Arsenic diffuses in silicon much more slowly than does phosphorus; therefore much shallower junctions can be formed with arsenic. (2) An arsenic-implanted junction has an overall lower defect density and a complete absence of dislocations near the metallurgical junction (Geipel and Shasteen 1980). On the other hand, a phosphorus-doped junction has a high density of dislocation at the metallurgical junction (Ikeda et al. 1974). Dislocations existing at the metallurgical junction cause a large junction leakage and thus are not desirable in device fabrication.

Source/drain ion implantation is often done through a layer of screen oxide in the hope that channeling effects can be reduced so as to facilitate the formation of shallow junctions. One side effect of ion implantation through a screen oxide is the formation of residual defects, caused presumably by the recoil implantation of oxygen (Cass and Reddi 1973; Moline and Cullis 1975; Natsusaki et al. 1977). In the case of arsenic, the residual defect region is confined within a depth of approximately half the projected ion range. Post-implant annealing in an inert gas ambient for up to 60 min at 900 to 1100°C failed to eliminate this defect region completely (Ikeda et al. 1974). An additional annealing in oxidizing ambient will cause the defects to coalesce and propagate both vertically and horizontally. However, the condition of the oxidation can be so chosen that the silicon consumption rate is approximately the dislocation propagation rate; so the remaining residual defect region still will be confined just to the surface of the n + p junction and will not penetrate into the metallurgical junction. A phosphorus-implanted junction shows a similar defect formation phenomenon, but the annealing-induced defect propagation is much faster. This is another reason for choosing arsenic over phosphorus. The minimum source/drain junction depth of an n-channel FET formed by arsenic ion implantation is about 0.15 to 0.2 μm.

The standard way to form a $p^{++}n$ source/drain for a p-channel FET is by boron ion implantation. As boron is a light atom, a heavy dose of boron implanted in silicon generates only structural defects (e.g., vacancies, interstitials), and does not cause amorphization of the silicon substrate (MacIver and Greenstein 1977). Therefore, a defect-free, low-leakage junction can be formed only by a post-implant anneal at temperatures of 900°C or higher in an inert ambient. Because of the fast diffusing characteristic of boron in silicon, a $p^{++}n$ junction formed by boron implantation and thermal activation is usually quite deep. $p^{++}n$ junctions with a depth of less than 0.3 μm can be formed by BF_2 ion implantation and/or by boron implantation into preamorphized silicon and low-temperature solid-state activation (Tsai and Streetman 1979; Crowder 1971).

Interconnection and Contact Metallurgy

The basic requirements for interconnection metallurgy in silicon integrated circuit applications are: (1) low resistivity, (2) good adhesion to silicon and/or to the silicon passivation dielectrics such as silicon dioxide and phosphorous silicate glass, (3) low contact resistance to silicon, (4) good chemical and physical stability, and (5) ease of fabrication. Aluminum and its alloys fulfill all of these requirements; so they are the most commonly used materials for interconnection metallurgy.

Aluminum has a room temperature resistivity of approximately 2.7 μohm-cm (Fraser 1983). The resistivity of its alloys such as Al–4%Cu–2%Si is about 30% higher. The contact resistivity (or the specific contact resistance) between aluminum and degenerate silicon is in the low 10^{-6} ohm-cm^2 range (Sze 1981), which is the best among many highly conductive metals. Aluminum adheres to both SiO_2 and phosphorous silicate films very well, as the heat formation of Al_2O_3 is greater than that of SiO_2. Aluminum and its alloys can be readily deposited by thermal evaporation, and can be delineated by either a wet chemical method (*Handbook of Thin Film Technology* 1970) or a chlorinated gas plasma etch (Hess 1981). In addition, Al interconnections can also be formed by a lift-off process (Grobman et al. 1979).

In MOSFET IC fabrication, because of the ultrasensitivity of the device to radiation and charging effects, physical evaporations (*Deposition Technologies* 1982) by resistance heating or by rf inductive heating are preferred, even though in some processes electron beam evaporation (Ghandhi 1983) is used because it produces cleaner films. For the same reason, etching of metal interconnections is generally done by a chemical etch or by a lift-off process.

Several problems have been encountered in using pure Al interconnections. The first problem is the formation of hillocks, that is, the formation of small surface extrusions. The existence of hillocks on the surface of Al film changes the specular nature of the film reflectivity and introduces difficulties in lithography and subsequent film coverage. Therefore, it is a major yield detractor. To circumvent this problem, a small amount of silicon is added to the Al film to cause a reduction of grain size of the film that is formed and to hinder grain growth as well. Both effects reduce the tendency to hillock formation.

The addition of silicon to an Al film also alleviates the so-called silicon-spike (or pit) problem that greatly hampered MOSFET technology in the early days. In order to promote better adhesion and to obtain low-resistance ohmic contact to Si, a post-Al metallization annealing of device wafers at about 400°C in a forming gas (10–15% H_2 in N_2) ambient is usually carried out. This low-temperature post-metal annealing is also needed to reduce the oxide surface state charge (Totta and Sopher 1969). During such low-temperature annealing, dissolution of silicon into an Al land takes place at the contact interface because of the high Si solubility in Al at those temperatures (of about 1.5 wt%). Unfortunately, silicon dissolution does not take place uniformly at the interface, but occurs along the edges and at the corners of a contact hole. It also occurs preferentially at defect sites. The result is the formation of so-called Al spikes, which penetrate deep into the silicon substrate. For a 400°C, 30-min anneal, the formation of Al spikes more than 1.0 μm in depth is commonly observed. If the device source and drain junctions are less than 1 μm in depth, post-metal annealing at 400°C for 30 min, which is most commonly done, will definitely cause a junction short. To alleviate the Al-spike formation problem, Si in an amount that slightly exceeds its solubility limit in Al at the annealing temperature is added to the Al film during film deposition (by, e.g., a co-evaporation method). This greatly suppresses the selective dissolution of silicon from the device contact area. Experimental results (Huntington and Grone 1961) have shown that with such a Si-supersaturated Al interconnection, MOSFETs with a 0.5 μm depth source/drain can be annealed at 450°C for 60 min without any detectable junction leakage deterioration and Al-spike formation.

The third problem in using pure Al interconnections is the relatively low electromigration resistance of Al. Electromigration is the material transport caused by a high-density electron flux in a conductive material, and is due to the transfer of kinetic energy from electrons, moving under the influence of the electric field applied along the conductor, to the positive metal ions (d'Heurle 1971). Any local disruption or disturbance of electron flux along the conductor causes an unbalanced metal ion transport, and results in either material accummulation or depletion at these locations. An Al interconnection that has failed in an electromigration test typically forms voids or breaks toward the cathode end, while hillocks are formed toward the anode end of the interconnection. Grain boundaries of the Al film and line width variation of the interconnection are believed to be the two most important causes of induced electron flux fluctuation and disturbance, and thus enhance the electromigration effect.

Studies have shown that the mean-time-to-failure (MTF) of a conductor can be related to the current density J in the conductor and the activation energy of low-temperature Al (self-) diffusion along grain boundaries (Black 1969), as shown in Eq. 15:

$$MTF \approx J^{-2}exp(Q/kT), \text{ for } 10^5 \leqslant J \leqslant 2 \times 10^6 \text{ A/cm}^2 \qquad [15]$$

Studies also show an increased electromigration effect in films with reduced grain size (Vaidya et al. 1980). To increase the electromigration resistance of an Al conductor, a small amount (< 4 wt%) of Cu is added to form an Al–Cu solid solution. An Al conductor containing 4 wt% Cu and 2 wt% Si shows an order-of-magnitude increase in electromigration resistance (Ames et al. 1970). The increased electromigration resistance of the Si and Cu–alloyed Al interconnection is attributed to increased grain size of the alloy and increased resistance to Al migration along grain boundaries.

STRUCTURE AND PROCESS ENHANCEMENT FOR HIGH-DENSITY, HIGH-SPEED VLSI IC

In this section, we discuss the major technological advances of the last decade or so. It is these technological advances that enable fabrication of the advanced micron and submicron devices, and make very-large-scale integration (VLSI) and ultra-large-scale integration (ULSI) possible.

Application of Metal Silicide

In the fabrication of the high-density, high-speed VLSI IC, a high-temperature metallization technology capable of matching the ever increasing intrinsic device switching speed is required. Doped polysilicon (as discussed above under "Structure and Fabrication of Polysilicon Gate FET") was the first answer to this need. However, with increased packing density and with the device dimension reduced to the submicron range, the high resistivity of doped polysilicon became intolerable. Metal silicides, mainly the noble and transition metal silicides (Murarka 1980; Chow and Steckl 1983a), with their reduced bulk and contact resistivities, could satisfy the need of, at least, the next generation of VLSI ICs, which had an integration level in the order of a hundred thousand circuits per chip (for logic) or a million bits per chip (for memory). The circuit performance improvement that can be brought about by replacing polysilicon with a metal silicide is illustrated in part by Figure 4-12 (Chow and Steckl 1983a). In the figure, the signal propagation delays of a polysilicon conduction line are compared with those of a $MoSi_2$ conduction line of the same width and length. As can be seen, more than one order-of-magnitude improvement in signal propagation speed is achieved by using the $MoSi_2$ conduction line. In addition to the lower resistivities, refractory metal silicides have the basic virtues of a high-temperature metallization system such as polysilicon. They can endure high-temperature device processing, can be delineated easily by dry etching, and can form a self-protective SiO_2 surface film. On the other hand, low-temperature-forming metal silicides offer some unique technological advantages that are not obtainable from polysilicon or Al. In this section, the application of metal silicide in the following three areas will be discussed: (1) as a contact buffer or contact barrier, (2) as a device gate, and (3) as an interconnection.

Silicide Contact Barrier/Buffer. There are two main objectives of using metal silicide in the contact area of an advanced FET: to alleviate the Al-spike penetration problem and to reduce contact resistance.

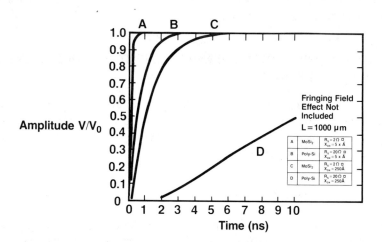

Figure 4-12. Transmission line modeling of propagation delay for 1000-μm-long $MoSi_2$ and doped polysilicon lines running over gate (250 Å) and field (5000 Å) oxide regions. (After Engeler et al., in Chow and Steckl 1983a).

Use as Contact Barrier. As a device is miniaturized, the depth of the source/drain junctions is reduced accordingly. When the junction depth is reduced to below 500 nm, doping Al with Si is no longer sufficient to ensure electrical integrity of the shallow junctions during the 400°C post-metal annealing. A metal silicide contact barrier, coupled with the use of silicon-doped Al metallurgy, was found to provide effective junction leakage protection for the 400°C post-metal annealing for junctions with depth down to 150 nm (Tsang 1976). Experimental results have shown that silicide eliminates localized silicon diffusion and acts as a diffusion barrier that hinders the interdiffusion of Si and Al. The use of a silicide contact barrier greatly reduces junction shorting caused by Al-spike formation (Hosack 1973; Kircher 1976). Silicides that have shown various degrees of efficiency in junction leakage protection are Pd_2Si, NiSi, RhSi, $TiSi_2$, $CoSi_2$, $TaSi_2$, and WSi_2.

The efficiency of leakage protection of these silicides depends upon the nature and the rate of Si and Al diffusion through them. Experimental results obtained from 0.25 μm and/or shallower n^+-p junctions have shown that, in general, the transition metal silicides provide better protection than the noble metal silicides, with RhSi (Tsang 1976) being an exception with a leakage protection capability comparable to that of $TiSi_2$ and $TaSi_2$. Table 4-2 lists the highest anneal temperature/longest annealing time a 0.25 μm silicided junction can be subjected to without developing leakage currents.

The contact barrier of noble metal silicides, and of the transition metal silicides of Ti and Co, is in general formed by a selective silicide sintering method. In the method, a blanket layer of metal of desired thickness is first deposited over the wafer that has been readied for metallization. The wafer is then subjected to an appropriate sintering anneal. During the sintering anneal, the metal film that is in the contact area will react with silicon to form a silicide, while that which is outside the contact area will remain in elemental metal form. Then the unreacted metal outside the contact areas is selectively etched away by a suitable etching solution. For platinum and rhodium silicides, etching selectivity is achieved by utilizing the possibility of forming a silicon dioxide film on the silicide surface

Table 4-2. Highest Allowable Post-Metal Anneal Temperature/Time of Silicide Contact Barriers on 0.25 μm Junction.

SILICIDE CONTACT	TEMPERATURE (°C)/TIME (MIN)
PtSi	400/30
$TiSi_2$	550/30
Pd_2Si	400/30
NiSi	550/30

but not on the surface of untreated metal during silicide sintering. After the silicide is formed, the unreacted metal outside the contact holes is selectively etched away by an etching solution that will not attack SiO_2. For silicides of Ti, Co, and Pd, there are etching solutions that will preferentially etch the unreacted metal. Table 4-3 lists the annealing procedure and the selective metal etching method for these metal silicides. Because these silicide contact barriers are formed by in-situ selective sintering, they are self-aligned with contact holes, and an additional lithographic step is unnecessary for their formation.

Owing to their higher formation temperature, contact barriers of Ta and W silicides are formed by first depositing a co-sputtered or co-evaporated silicide film, and then subetching into the contact barrier (or, alternatively, by the pattern lift-off technique). Therefore, they are not self-aligned, and an additional lithographic step (and mask) is needed in their formation. This non-self-alignment nature and the need for an extra lithographic step to form them makes Ta and W silicide barriers less attractive than the noble metal silicides and the silicides of Ti and Co, despite their superior buffering capabilities.

Use as Contact Buffer. Metal silicide is used as a contact buffer to reduce contact resistance between polysilicon interconnections themselves, between polysilicon and single crystal silicon substrates, and between polysilicon and other levels of metal interconnections. A metal silicide contact buffer also provides a viable solution to the high-contact-resistance problem often encountered in devices using Si-doped Al metallurgy, especially in those devices that have small contact openings. The high contact resistance between the Si-doped Al interconnection and the silicon substrate is caused by silicon precipitates that have segregated out from the doped Al land during post-metal anneal and/or subsequent thermal excursions. These precipitates have poor p-type conduction characteristics (Al is a deep acceptor in Si). When they have formed, whether they are

Table 4-3. Silicide Contact Formation and Selective Etch of Unreacted Metal of Four Commonly Used Silicides.

SILICIDE	SILICIDE SINTERING PROCEDURE	ETCH OF UNTREATED METAL
PtSi	550°C in Ar plus 550°C in Air	Aqua regia
Pd_2Si	280°C in Ar plus 280°C in air	KI + I aqua sol.
$TiSi_2$	600°C in N_2, strip unreacted metal, then 800°C in N_2	H_2O_2 + NH_4OH + H_2O (5:1:1 by vol.)
NiSi	400°C in forming gas	Diluted HCl

on the $p^{++}n$ junction or on the $n^{++}p$ junctions, high contact resistance between the Al conductor and the Si junction results. When these precipitates are formed on the $p^{++}n$ source/drain diffusion junction contacts of a p-channel device, they act as high resistive inclusions at these contact areas and thus effectively reduce the contact conductivity. On the other hand, when these precipitates are formed in the $n^{++}p$ source/drain diffusion junction contacts of an n-channel device, a pn junction is formed under each precipitate. For the small contacts of a high density VLSI IC, the contact area can easily be covered entirely by these precipitates to form a continuous intermediate pn junction between the conduction metal and the device source and drain. This results in an extremely high contact resistance. Most noble metal and transition metal silicides have a low contact barrier with p-type silicon. Therefore, if a silicide contact buffer is used, the silicon-precipitate-caused high-contact-resistance problem can be eliminated. Both the noble metal silicides and the transition metal silicides, described in the previous section, that have been designed for Al-spike prevention are also suitable for contact buffer applications to reduce contact resistance. Among the many silicides, the $TiSi_2$ and the $CoSi$ seem to provide the best results. Compared with others, these two silicides have the lowest bulk resistance and a contact resistance comparable to, or even lower than, that of Al–Si contacts (d'Heurle unpublished).

Silicide as Device Gate and Interconnections—Polycide Structure.

Both noble metal silicides and transition metal silicides have been used as device gates and interconnections. The transition metal silicides such as $MoSi_2$, $TaSi_2$, and WSi_2 have been favored because of their high-temperature processing capability and their capability of forming self-protective SiO_2 surface films. However, recent developments in sub-μm channel device fabrication have indicated an opposite trend, due to the successful use of the self-aligned sidewall passivation technique and the need for the lowest possible resistivity of the conduction lines.

For device gates and interconnections, silicides can be either used directly as a replacement for polysilicon or combined with polysilicon to form the *polycide* (*poly*silicon sili*cide*) layered structure, as shown in Figure 4-13 (Zirinsky and Crowder 1977; Crowder et al. 1977). The direct use of silicide has the following disadvantages, even though it is relatively simple to fabricate: (a) the not yet fully understood silicide–SiO_2 interface may introduce processing complexity; (b) a device with a pure silicide gate cannot be subject to a high-temperature oxidation process without structural degradation; and (c) post-silicide high-temperature processing steps may also result in degradation of the gate oxide. On the other hand, the polycide structure preserves the well-understood, stable $Si–SiO_2$ structure, retains the high reliability of gate oxide that is enjoyed by the polysilicon gate device, and introduces no other device processing and/or operation

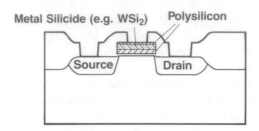

Figure 4-13. A MOSFET with polycide gate structure.

complexities. Another big advantage of the polycide structure is its capability of being oxidized to form a self-passivating SiO_2 surface film. Therefore, in general, polycide conductors are preferred, especially in gate applications.

Thin film deposition of these refractory metal silicides is generally done by either co-sputtering from dual targets or from a single target of compound or compositive material, or by dual e-beam co-evaporation. The as-deposited films are amorphous, with high resistivity, but they have little or no film stress. They are sintered at temperatures equal to or higher than 1000°C in an inert ambient to homogenize their structure and reduce their film resistivities. High-temperature sintering transfers the film into a polycrystalline structure and induces high stress, usually tensile in nature, in the film. Stress on the order of the mid to high 10^{10} dynes/cm² is commonly seen in these silicide films formed on silicon or silicon dioxide. The resistivity of these films after sintering is typically in the range of 1 to 3 Ω/sq, which is more than an order of magnitude lower than that of polysilicon. (See Table 4-4.)

To form the noble metal silicide conductors or the conductors of Ti and Co silicides, a selective sintering method similar to that used in silicide contact formation is used. In this method, polysilicon gate and/or interconnection patterns are formed first. Next, a layer of blank metal is deposited. Then, the composite structure is subjected to a silicide formation sintering heat treatment that converts the polysilicon pattern into silicide. The unreacted metal outside the polysilicon pattern is then removed by a suitable selective chemical etch. By varying the metal film thickness, either a simple silicide conductor or a polycide conductor can be formed.

Pattern delineation of refractory silicides and polycides has been carried out most often by reactive ion etch (RIE) in either radial flow on parallel plate reactors (Chow and Steckl 1983b). With some modifications, most of the RIE methods developed for polysilicon etching are also suitable for refractory silicides. The basic mechanism involved in silicide etching is similar to that for silicon. It involves the formation of volatile silicon and metal halides by the etching gases. Both fluorinated plasmas and chlorinated plasmas, as well as

Table 4-4. Resistivity of Some Commonly Used Silicides.

SILICIDE	FORMED BY/ STARTING FORM	SINTERING TEMP. (°C)	RESISTIVITY ($\mu\Omega$cm)
$TiSi_2$	metal/polySi	900	13–16
	co-sputtered	900	25
$TaSi_2$	metal/polySi	1000	35–45
	co-sputtered	1000	50–55
$MoSi_2$	co-sputtered	1000	100
WSi_2	co-sputtered	1000	70
$CoSi_2$	metal/polySi	900	17–20
	co-sputtered	900	25
$NiSi_2$	metal/polySi	900	50
	co-sputtered	900	50–60
$PtSi$	metal/polySi	600–800	28–35
Pd_2Si	metal/polySi	400	30–35

Freon plasmas, which contain both fluorine and chlorine in the molecules, have been used successfully in etching metal silicides. The most commonly used RIE method for WSi_2, $TaSi_2$, and $MoSi_2$ uses the CF_4/O_2 mixture plasma (Crowder and Zirinsky 1979; Bennett et al. 1981; Sinha et al. 1980). For WSi_2, plasmas of Freon 12 and SF_6/O_2 can also be used (Roger and Coe 1982). In fact, Freon 12 gives much better etching anisotropy and less etch undercut than CF_4/O_2 RIE. NF_3, SF_6/O_2, and CCl_4/O_2 were found to etch $MoSi_2$ (Chow and Steckl 1982; Beinvogl and Hasler 1981; Gorowitz and Saia 1982), while mixtures of SF_6–Cl_4 and CF_4–Cl_2 were found to etch $TaSi_2$ (Mattausch et al. 1983). In etching a gate structure, good etch anisotropy and high etch selectivity between silicide and SiO_2 are required. The former requirement is to assure a tight control of line width, while the latter assures timely etching termination before the thin oxide over the device source and drain regions is attacked.

In addition to those basic requirements that are common to both polysilicon and silicide gate structures, polycide gate structure etching must maintain a smooth sidewall for the etched polycide pattern. As the rates of Si and silicide etching in most reactive ion etches are different, polycide gate etching is accomplished either by a multistep etching process or by a totally anisotropic RIE with chlorinated or Freon plasma (Bennett and Chow 1985).

The Salicide Structure. In the development of advanced VLSI integrated circuits, the performance of devices and circuits has been improved, as was described in previous sections, by the use of highly conductive metal silicide for gate and interconnection applications. Further improvement in their performance can be achieved if the resistance and capacitance of device source/drain diffusion regions and the diffusion interconnections of the circuit, such as the diffusion bit lines of a dynamic memory circuit, are also reduced. This can be achieved by formation of the so-called salicide structure (Osburn et al. 1982).

Figure 4-14 shows the *salicide* (*self-ali*gned sili*cide*) structure and its fabrication procedure. As can be seen, a salicide structure is a device structure that has a polycide gate and a "silicidized" device source and drain. In a RAM IC with diffused bit lines, the bit lines are also silicidized, along with the device source/drain. In a salicide device, the device gate and source/drain are self-aligned and are silicidized in a single step. In forming the salicide structure, as shown in Figure 4-14, the device polysilicon gate and source/drain dopant implantation are first implemented. Then, using the sidewall spacer technique (Tsang et al. 1981, 1982b), the two sides of the polysilicon gate are passivated and insulated by SiO_2 sidewall spacers. Finally, using a selective silicide formation technique similar to that used to form silicide contact barriers, device gate and source/drain regions are simultaneously silicidized.

Theoretically, all the low-temperature silicides (i.e., silicides of the noble metals and the transition metals Ti and Co) are good candidates to form salicide structures. The fabrication is simple and straightforward. However, care must be taken to ensure that silicon diffusion out of the contact areas and further

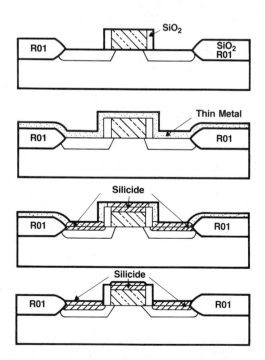

Figure 4-14.   Silicide structure and its fabrication.

propagation laterally toward the device gate during silicide sintering heat treatment are curtailed. Excessive silicon lateral diffusion from the contact area results in the formation of "silicide bridges" over the gate sidewall spacers and causes a device gate to source/drain short. Ideally, a salicide structure should be formed by the diffusion of metal only. Unfortunately, none of the silicides that are suitable for salicide application is formed through the simple mechanism of metal diffusion. These silicides are formed either by interdiffusion of both silicon and metal atoms or by silicon diffusion at the desired sintering temperatures (Poate et al. 1978). Therefore, a special silicide sintering method had to be developed to form the bridgeless salicide structure.

PtSi$_2$ and TiSi$_2$ salicides have been studied extensively and show great potential for IC application. Since both Pt and Si are diffusants in the silicide formation mechanism, the formation of a Pt silicide proceeds by initiating the silicide in an inert (e.g., forming gas) ambient until PtSi$_2$ formation begins at the Pt–Si interface. Then, sintering continues in weak oxidizing ambient to complete the salicide formation, which will suppress the occurrence of Pt silicide bridgings We lack a full understanding of how the silicide bridging is suppressed in the oxidizing ambient during Pt–silicide sintering. A possible explanation is that the fast surface diffusion of Si in Pt is totally suppressed by the formation of the thin SiO$_2$ surface film when PtSi$_2$ is formed in an oxidizing ambient (Rand and Roberts 1974). The formation of a bridge-free TiSi$_2$ salicide structure is carried out in an ultrapure nitrogen ambient (Ting et al. 1982). The use of a nitrogen ambient results in the formation of a Ti nitride surface film, both on the unreacted Ti film on the oxide sidewalls and on the Ti silicide in the contact area. The Ti nitride film probably hinders Si diffusion (or surface migration) and thus prevents bridging.

Lightly-Doped-Drain Field Effect Transistor (LDDFET) and Its Fabrication

In a scaled-down device, high field aggravations arise due to the nonscaling behavior of the resistive drops and the device subthreshold effect. To alleviate these high-field constraints, Saito et al. (1978) proposed the lightly-doped-drain FET (LDDFET), in which a narrow n$^-$ region is added between the device channel region and the device n$^+$ drain and/or source (Figure 4-15). Studies have shown (Ogura et al. 1980) that the LDD structure reduces the peak electric field in the channel and shifts the peak electric field into the n$^-$ drain region. This allows the device to operate at a higher voltage and makes the sub-μm devices operable at voltages commensurate with the loading requirements of a high density IC. In addition to the higher operating voltage, the LDD structure also reduces hot electron injection into the gate insulation and the gate-to-drain/source overlap capacitance. These all lead to a significant improvement in circuit performance.

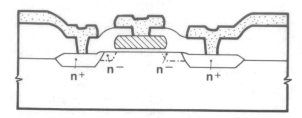

Figure 4-15. Cross section of a lightly doped-drain/source field effect transistor (LDDFET). Note the n− regions that separate the channel from the highly doped drain and source diffusions.

It is generally agreed that the LDD should be one of the standard device structures in an advanced CMOS RAM (*Electronic Market Data Book 1982*).

The fabrication procedure for an LDDFET (Tsang et al. 1982a) is shown in Fig. 4-16. In general, it is similar to that of a conventional polysilicon-gate MOSFET process, except that steps are added to construct the lightly doped drain/source extensions. As shown in Figure 4-16(a), the device is fabricated by a conventional process up to the delineation of gate polysilicon with a CVD SiO_2 etch mask. However, in LDD fabrication, special care has to be taken to obtain vertical sides for the SiO_2/polysilicon gate stack. Following the formation of the SiO_2/polysilicon gate stack, an n− phosphorus ion implantation is performed, as shown in Figure 4-16(b). Then, a layer of CVD SiO_2 of the desired thickness is conformally deposited, and, using an anisotropic and selective RIE, the planar portion of the CVD oxide is removed. As a result of this etching, an oxide sidewall spacer is left abutting the polysilicon gate stack where the deposited film is thicker in the vertical direction, as shown in Figure 4-16(c). The desired width of the oxide sidewall spacer, which determines the length of the n− region, can be precisely controlled by a set of geometric factors and the thickness of the CVD oxide (Tsang et al. 1981, 1982b). Following the formation of the sidewall spacer, arsenic-ion implantation is used to form the n+ regions. During the n+ ion implantation, the sidewall spacers effectively block the arsenic

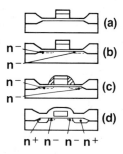

Figure 4-16. Fabrication process for LDDFET.

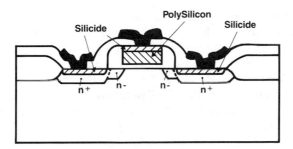

Figure 4-17. Salicide LDDFET structure.

ions from entering the two narrow n⁻ regions in the device, as shown in Figure 4-16(d). Thereafter, the device fabrication is completed by a conventional processing procedure.

Salicide LDDFET. By slightly altering the processing steps, the LDDFET can be "silicidized" to form a salicide LDDFET structure to improve the conductivity of the gate and of the device n^+ source/drain regions (Horiuchi 1985). This is shown in Figure 4-17. In fabrication of the structure (Horiuchi 1985), the polysilicon gate is etched with a multilayer resist (MLR) etch mask instead of a CVD SiO_2 mask. The MLR mask is made to have vertical sides and a thickness commensurate with the requirements of the etch rate ratio between resist and polysilicon. The SiO_2 cap of the MLR mask is designed in such a way that it can be consumed during the polysilicon gate etch but is thick enough to yield a good MLR mask. After the polysilicon gate is etched, the remainder of the MLR mask is removed by O_2 plasma, which does not attack either the oxide or polysilicon. Then, the LDD fabrication method previously described is used to form the LDD structure. Thereafter, using the method described in the salicide section, the gate polysilicon and the device drain/source are silicidized simultaneously to form the salicide-LDDFET.

Advanced Device Isolations

For high-density, high-performance VLSI applications, an interdevice isolation should have certain properties. Namely, it should be (1) capable of effectively reducing interdevice leakage, (2) planar for ease of implementing multilayer metallurgy, and (3) capable of being made small in (horizontal) size (to consume as little silicon real estate as possible). The conventional LOCOS or ROI provides good electrical isolation for shallow device regions and has a surface topography planar enough for application in ICs of low and mid-range integration. But because of the formation of the so-called ROI bird's beak and bird's crown, this simple shallow isolation becomes inadequate for high density VLSI ICs whose fabrication requires a minimum feature size on the order of 1.0 μm or smaller. In these applications, the bird's crown of the ROI makes implementation of backend-of-the-line (BEOL) multilevel metallurgy difficult, while the bird's beak encroachment causes the smallest attainable feature size to be larger than that of the lithographic capability. In addition, encroachment on the device region by the ROI bird's beak also reduces the useful silicon real estate. Several methods have been devised to reduce the ROI bird's beak. In this section, we will describe three methods that attempt to improve ROI. Later in this section, we will describe two new isolation methods that take an approach totally different from recessed oxide isolation.

Sealed-Interface Local Oxidation (SILO) Technology.   The formation of a bird's beak/crown in the ROI results primarily from lateral diffusion of oxygen along the pad oxide. Therefore, it can be completely eliminated if the pad oxide underneath the nitride mask is reduced to zero. This technique is utilized in a method called SILO (Hui et al. 1982) to form bird's-beak-free recessed oxide isolations. In the method, a nitride surface film about 10 nm thick is formed without an oxide undergrowth by using either nitrogen ion implantation or thermal (or plasma-assisted thermal) nitridization. The nitride film is then patterned into an ROI oxidation mask. The ROI formed with such a nitride mask has zero bird's beak and a bird's head length of about 40% of the ROI thickness. The bird's head can be reduced if the nitride mask is further "strengthened" by a cap consisting of a 100-nm CVD Si_3N_4 overlay and a 30-nm CVD SiO_2 pad. An ROI about 1.0 μm thick has been grown with such a mask, showing zero bird's beak and a bird's head length of only 20% of the ROI thickness. Application of SILO in the fabrication of megabit CMOS DRAMs with 1.0 μm minimum feature size has been reported (Siu 1985).

Sidewall Masked Isolation (SWAMI) Process.   This method uses a sidewall nitride mask to eliminate the bird's beak, as well as to reduce the length of the bird's head, of a fully recessed ROI (Chiu et al. 1982). Figure 4-18 illustrates schematically the fabrication sequence of the process. In the process, in addition to the normal surface nitride mask, nitride side masks are formed on the two sides of the device pedestals before ROI to provide complete oxidation protection for the device pedestal (step c in Figure 4-19). The nitride side masks are formed by the sidewall spacer technique (Tsang et al. 1981). After the ROI is grown, the nitride oxidation mask is removed by hot phosphoric acid. This leaves "sidewall ditches" along the two sides of the device pedestal. To fill these sidewall ditches, a layer of LPCVD SiO_2 is first deposited and then etched back with fluorocarbon plasma RIE. The final structure of the fully recessed ROI is planar and bird's-beak-free, and has a very short bird's head. Theoretically, a SWAMI ROI of any depth can be formed. Therefore, this is a much better interdevice isolation process for high density VLSI applications than the con-

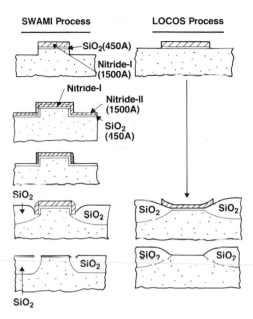

SWAMI Process   LOCOS Process

Figure 4-18. Comparison of SWAMI process and the conventional LOCOS process. (After Chiu et al. 1982)

ventional processes. However, because of the large film stress that exists in the nitride sidewalls, and the sharp bottom corner of the device pedestal step, it is highly likely that defects will be generated in the device pedestal during ROI oxidation. In fact, devices with SWAMI isolation have shown high levels of device leakage (Teng 1985). To remedy this problem, Chiu et al. (1983) have modified the process. In the modified process, as shown in Figure 4-19, the device pedestal is made with sloped sidewalls, and the nitride film covering the sidewalls is made very thin. According to Chiu et al. (1983), defect- and bird's beak/head-free SWAMI isolation has been achieved by this improved process.

Selective Polysilicon Oxidation (SEPOX) Process. Semirecessed and fully recessed ROIs provide good surface planarity at the pre-gate level. However, the good planarity is lost after the implementation of a gate structure. In addition, the ROI encounters other process difficulties such as oxidation-induced enhanced diffusion (Lin et al. 1979) and stacking-fault formation (Hu 1974). To achieve a planar post-gate structure with good substrate crystalline perfection while maintaining gate to source/drain self-alignment, a selective polysilicon oxidation (SEPOX) process was devised (Matsukawa et al. 1982). In the process, as shown in Figure 4-20, a thin pad oxide is grown first. This thin pad oxide can also be used as gate insulation. Next, a polysilicon layer of appropriate

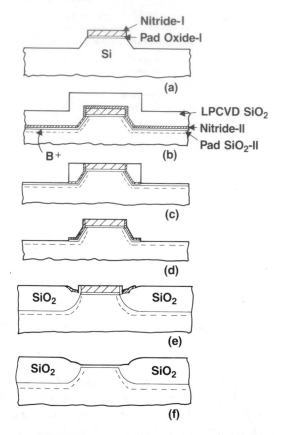

Figure 4-19. Sloped-wall SWAMI process: (a) after device patterning; (b) after nitride II and LPCVD oxide deposition; (c) after nitride/oxide RIE; (d) device pedestal structure before oxidation; (e) after field oxidation; (f) final ROI structure. (After Chiu et al. 1982.)

thickness is deposited, followed by deposition of a layer of Si_3N_4. The nitride overlay is then etched into the desired device (gate) pattern. Finally, the exposed polysilicon is completely oxidized and converted into SiO_2 isolation.

The advantages of the SEPOX isolation structure are, as stated, better planarity at the post-gate level and less disturbance of the substrate. However, the structure lacks in depth isolation capability and the flexibility of varying the device compositional structure. Therefore, its application in VLSI is limited to some special cases.

Trench Isolation. From the viewpoint of the basic requirements for an interdevice isolation (i.e., being efficient in electrical isolation and space utilization while maintaining good surface planarity), a dielectric-filled trench isolation (trench isolation) is an ideal one and surpasses all other isolation forms. The

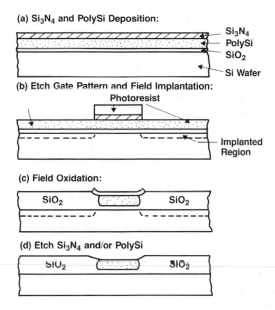

(a) Si₃N₄ and PolySi Deposition:

(b) Etch Gate Pattern and Field Implantation:

(c) Field Oxidation:

(d) Etch Si₃N₄ and/or PolySi

Figure 4-20. Process sequence of SEPOX technology. (After Matsukawa et al. 1982)

concept of using dielectric-filled trenches for interdevice isolation was conceived in the seventies, for example, in the invention of Burkhart et al. (1979). However, because of a lack of suitable fabrication technologies and tools, it was not used until the early eighties when better silicon and dielectric etching tools and etching processes became available. The need to use trench isolation in the construction of VLSI ICs to improve circuit packing density has also precipitated an industry-wide intensive effort to develop the technology.

The simplest trench isolation structure is shown in Figure 4-21 (Chiang et al. 1982), in which a trench or moat is etched around the device to be isolated and filled with CVD SiO₂, which is then etched back and planarized by, for example, an isotropic oxide RIE. There are two major problems associated with this simple oxide-filled trench isolation structure. The first one is an engineering problem; that is, it is very difficult to fill trenches without leaving a void or an insufficiently

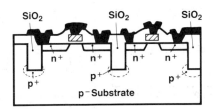

Figure 4-21. Structure of oxide-filled trench isolation for NMOSFET.

filled hole in the (upper) center of the trench ("phoenix's eye" structure), a major threat to device reliability. The second problem is the formation of defects in the device pedestal during subsequent high temperature processing steps, due to the stresses induced both by the thermal expansion difference between Si and SiO_2 and by the volume contraction of the low temperature CVD SiO_2 in the trench. Because of these problems, this simple dielectric-filled trench isolation structure sees only limited applications.

A more popular, and better, trench isolation structure is one with polysilicon refill. Much better conformal coating can be achieved by LPCVD polysilicon deposition, which eliminates the formation of phoenix's eyes. One fabrication procedure for the polysilicon-filled isolation trench is shown in Figure 4-22 (Goto et al. 1985). In the process, a trench of the desired width and depth is etched into the silicon substrate by anisotropic RIE. The trench is made with slightly outward-inclined sidewalls to assure freedom from the formation of a phoenix's eye during polysilicon refill. After the trench is formed, the wafers (silicon substrates) are cleaned, and leakage-channel-stop dopant ions are implanted into the bottom of the trench. Then a thin surface oxide is grown over the trench, followed by the deposition of a thin layer of Si_3N_4 over the entire wafer. This silicon nitride overlay presents mobile ion contamination of the device region and the field region underneath the isolation. The thin oxide underneath the nitride relieves the nitride film stress.

After the trenches are completely passivated by the $Si_3N_4SiO_2$ films, they are refilled with LPCVD polysilicon. Then, using photoresist reflow and a reactive

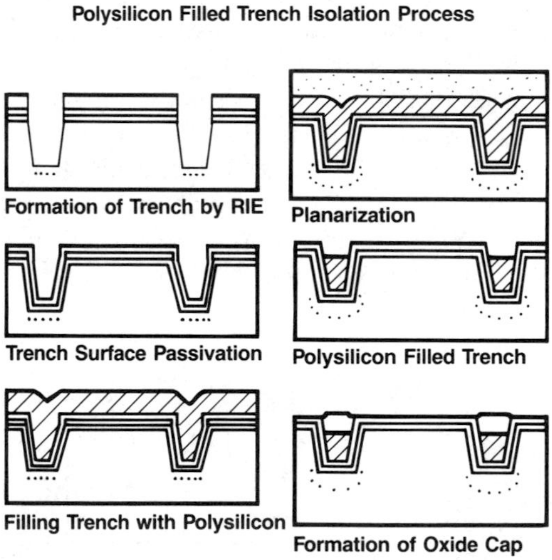

Polysilicon Filled Trench Isolation Process

Formation of Trench by RIE

Planarization

Trench Surface Passivation

Polysilicon Filled Trench

Filling Trench with Polysilicon

Formation of Oxide Cap

Figure 4-22. Fabrication procedure for polysilicon-filled trench isolation.

ion etch-back technique, the overfilled polysilicon is etched away, and the trench is co-planarized with respect to the surface of the device regions. Finally, the surface of the polysilicon in the trench is oxidized to form a layer of SiO_2 of desired thickness to "seal off" the polysilicon in the trench from the outside environment. A current assessment (*Electronic Market Data Book 1982*) has shown that polysilicon-refilled trench isolation is one of the technologies that will be essential to the success of future CMOS logic and DRAM ICs.

Isolation Achieved by Selective Silicon Epitaxial Growth.

HCl is known to etch silicon at elevated temperatures. It was found that selective silicon epitaxial growth on silicon surfaces, but not on SiO_2 or Si_3N_4 surfaces, can be achieved if a small amount of HCl is added to the Si-source gas (Voss and Kurten 1983). This selective silicon epitaxial growth is used to form oxide-filled deep trench isolation for both bipolar and MOSFET applications (Endo et al. 1982; Kurten et al. 1982). The advantage of this isolation method is that low-stress, closely spaced, oxide filled, deep narrow trenches of high aspect ratio can be readily formed. The method also provides some processing flexibility that is lacking in the normal device fabrication method.

The selective silicon epitaxial growth (SEG) isolation process, used for both bipolar transistors and MOSFETs, is illustrated schematically in Figure 4-23. The first step of the process is to form an oxide layer of a thickness equal to the isolation depth. The second step is to etch windows out of the thick oxide layer.

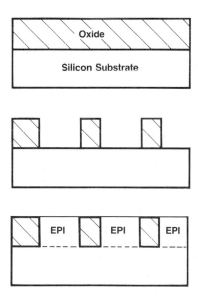

Figure 4-23. The selective silicon epitaxial growth (SEG) isolation process.

Then, at the third step, silicon is selectively grown in these windows for device fabrication.

NMOSFET fabricated on SEG silicon lands showed excellent channel mobility (long channel mobility of 630 $cm^2/V \cdot s$) (Kurten et al. 1982). SEG isolation is particularily useful to CMOS fabrication. It enables greater flexibility in the fabrication of the twin wells structure. Higher substrate resistivities and deep narrow interwell isolations are made possible for latch-up curtailment (Endo et al. 1983).

MODERN IC STRUCTURES AND THEIR FABRICATION

Based on polysilicon gate FET technology and augmented by the technology enhancements described in the previous sections, many advanced FET integrated circuits have been made. In general, the most important FET IC technologies can be grouped into three main categories: the HMOS (high performance n-MOS) technology, the DRAM (dynamic random access memory) technology, and the CMOS (complementary MOS) FET technology.

The HMOS (High-Performance MOS) Technology

The HMOS technology is essentially an advanced silicon-gate NMOS technology that produces devices of reduced dimensions to achieve higher operation speed, greater circuit packing density, and lower power consumption. Dennard et al.'s (1974) constant field scaling principle is used to miniaturize devices. In the process, both the horizontal and the vertical dimensions of a device are scaled down by a factor of $1/k$, while the device channel dopant concentration is scaled up by a factor of k. The scaled devices will be operated at voltages reduced by a factor of $1/k$ to keep the field across a device channel the same as that of the device before scaling. Because of this constant-field design, the speed of a scaled-down device is increased by a factor of k, whereas power dissipation is reduced by a factor of $1/k^2$. This results in a power–delay product reduction by a factor of $1/k^3$.

The first HMOS fabricated by Intel in 1977 (Pashley et al. 1977) has a channel length scaled down from 6.0 μm to 3.5 μm and a gate insulation scaled down from 120 nm to 70 nm. The device operates at a V_{cc} of 3 to 7 V and gives a speed–power product reduction from 4 pJ/gate to 1.0 pJ/gate.

The fabrication procedure of an HMOS is similar to that of a Si-gate FET as described earlier (under "Structure and Fabrication of Polysilicon Gate FET"), but with several device structural and process enhancements. For example, in addition to a thinner gate oxide, shallower source/drain junction, and shorter channel length, a silicon substrate of higher resistivity is used to lower the junction capacitance and to reduce the threshold voltage substrate sensitivity.

The higher dopant concentration required in the channel region, due to scaled-down device dimensions, is effected by selective ion implantation.

Different companies have developed different high-performance MOS processes to comply with their own circuit requirements. Using electron-beam lithography, IBM developed a much more advanced HMOS process to fabricate 1.0-μm single-level polysilicon MOSFET VLSI ICs (Hunter et al. 1979). The process utilizes ion implantation to form the device source/drain and to "adjust" the dopant concentration in various regions to effect the desired device characteristics. Along with the electron-beam image imprinting, a double-layer resist system (Hatzakis 1982) and reactive ion etching are used to produce 1.0-μm features and to reduce etch bias. The double-layer resist system also enables the implementation of a lift-off technique for contact and interconnection metallurgies. In order to raise the parasitic threshold of the IC's field isolation, a field etch-back procedure following the source/drain implantation is implemented. In addition, backside ion implantation is also employed for impurity gettering. Figure 4-24 shows schematically the cross section of an n-channel HMOS fabricated by this process on a 5 Ω-cm Si substrate. The device has a POCL$_3$-doped, single-level polysilicon gate, 25-nm oxide gate insulation, 35-nm deep source/drain formed by As implantation, and nominal field oxide thickness of about 450 nm.

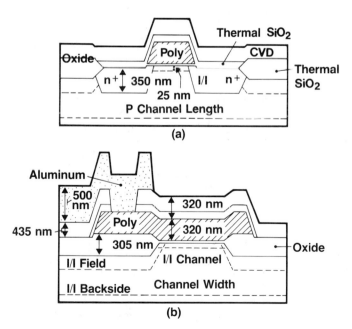

Figure 4-24.   Schematic cross sections of a half-micron channel, single polysilicon gate MOSFET. (After Hunter et al. 1979.)

Dynamic Random-Access Memories (DRAM)

The one-device memory cell and its memory (read/write) operation were invented by Dennard (1968). A combination of high packing density, moderate speed–power product, and low cost per bit made its use in forming MOSFET DRAM ICs unmatched by any other integrated-circuit memories. In the years since this memory cell's invention, its technology has progressed enormously. The density of one-device DRAM ICs has steadily increased from 1.0 kilobit in the early 1970s to today's megabits. By the end of this century, a 100 megabit/chip is predicted (Sunami and Asai 1985).

A one-device DRAM cell, as shown in Figure 4-25(a), consists of a transistor (a MOSFET switch) and a charge storage capacitor (Rideout 1979). The charge storage capacitor is connected directly in series to the source of the FET and can be charged or discharged by the MOSFET to form "1" and "0" binary states. There are, as shown in Figures 4-25(b) and 4-25(c), two ways that the lower electrode of the storage capacitor can be formed to connect to the switching device source. One is to extend the source diffusion region of the switching FET, as shown in Figure 4-25(b). Another is to form an inversion layer of

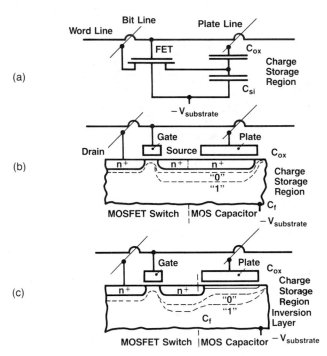

Figure 4-25. One-device memory cell illustrating (a) circuit, (b) basic structure with diffusion storage, and (c) basic structure with inversion storage. (After Rideout 1979.)

electrons by applying plate voltage, as shown in Figure 4-25(c). In either case, the storage capacitance consists of two components connected in parallel: an MOS capacitance and a junction capacitance.

In a memory array, the memory cells are attached to three array interconnection lines termed the bit lines, the word lines, and the plate lines. In general, the drain of the memory cell FET is connected to the bit line, the FET gate to the word line, and the capacitor upper electrode to the plate line. The source of the device is the lower electrode of the storage capacitor, which, as shown in Figures 4-25(b) and 4-25(c), can be either an n^+ diffusion region or a sheet of electrons of the inverted silicon surface. The word lines and the bit lines are used simultaneously to read or write the memory cells. Reading a memory cell, that is, sensing the charge state of the storage capacitor of the memory cell, is accomplished by a sense amplifier attached on the bit lines.

The performance of a DRAM is measured by the speed at which the charge state of a cell can be changed and sensed, and by how much power is needed to operate the cell. Several factors affect the performance of a DRAM, one of the most important being cell design. A figure of merit that is generally used for the design of a DRAM cell is the charge transfer ratio T (Stein and Friedrich 1973). The charge transfer ratio is the ratio of the storage capacitance of the cell to the combined capacitance of the storage capacitor and the bit line:

$$T = C_s/(C_s + nC_{bl}) < 1.0 \qquad [16]$$

where C_s is the storage capacitance, C_{bl} is the bit line element capacitance in the cell, and n is the number of cells on the bit line. A general rule of thumb for structure and process design for a DRAM cell is to make the charge transfer ratio T as large as possible.

With the gate, the bit line, and the plate of the DRAM cell made of different materials, several structures have been constructed. The simplest DRAM structure is the original one disclosed by Dennard (1968). As shown in Figure 4-26, the cell has a metal (aluminum) gate for the switching transistor. Aluminum is also used as the upper electrode (capacitor plate) of the storage capacitor, while the lower electrode of the capacitor is the extension of the diffusion region of the device's n^+ source underneath the metal plate. A simple four-mask metal-gate process is used to fabricate the device.

Several structural and process improvements have been developed to improve performance and packing density. For example, polysilicon and/or metal silicide is used for the gate of the switching transistor and for the upper plate of the storage capacitor. Using two layers of polysilicon, one for the device gate and one for the capacitor plate, precise alignment between gate and plate and between gate and device source/drain can be achieved through a self-align process that improves charge transfer efficiency. The first double-polysilicon cell with diffused bit line was successfully fabricated by Ahlquist et al. (1976). As shown

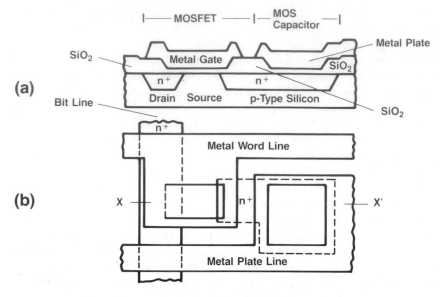

Figure 4-26.   A simple metal-gate cell. (After Rideout 1979.)

in Figure 4-27, it uses two separate layers of polysilicon: one for the device gate and one for the storage capacitor plate. The first layer of polysilicon is used to form the plate of the storage capacitor, while the second layer of polysilicon is used to form the gate of the switching FET. A staggered cell layout is employed to maximize the utilization of silicon real estate and to achieve the highest possible packing density. The use of two separate polysilicon layers for the gate and the

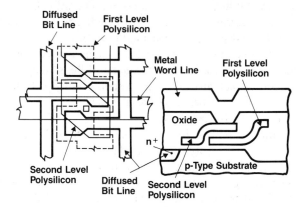

Figure 4-27.   Double polysilicon cell. Showing layout and cross section. (After Ahlquist et al. 1976.)

plate enables self-alignment of gate and plate. It increases the packing density and reduces device stray capacitances. To reduce the resistance and stray capacitances of the bit line, the diffused bit line shown in Figure 4-27 is often replaced by polysilicon and/or metal silicide.

Complementary MOS (CMOS) Technology

A CMOS circuit is constructed with both NMOS and PMOS FETs (Coppen et al. 1972). It has the advantages of lower power consumption and higher speed over NMOS and RMOS ICs (Ronen and Micheletti 1975; Ohwada et al. 1979). The construction of a DMOS IC requires the fabrication of both type FETs on the same substrate, which is achieved by forming "wells" or "tubs" of opposite doping polarity on the substrate to provide the base for the fabrication of the second type device. Depending upon the type of the wells that are formed, there are three basic types of CMOS technologies: p-well CMOS, n-well CMOS, and twin-well CMOS (also called twin tub CMOS).

Figure 4-28(a) shows the structure of a conventional p-well CMOS (Funk 1984). The p-well CMOS is a natural outgrowth of PMOS. It uses an n-type silicon substrate. To form the NMOS FET, p-type regions, or "wells," are first formed in the n-type substrate by localized diffusion. Then, following the normal device fabrication procedure, NMOS and PMOS FETs are formed in the p-wells and on the substrate, respectively.

In contrast to p-well CMOS, n-well CMOS uses p-type substrates, and n-wells are formed for the fabrication of PMOS devices. An n-well CMOS structure is shown in Figure 4-28(b) (Ohzone 1984). In the figure, n-well CMOS circuits are formed as the peripherals of the double polysilicon NMOS static RAM.

A twin-well CMOS structure is shown in Figure 4-28(c) (Schwabe et al. 1983). In building the structure, a silicon substrate of high resistivity is used. Both p-wells and n-wells are formed on the substrate by selective ion implantation and dopant drive-in. The advantage of the double-well process is that fabrication of the PMOS and NMOS FETs can be optimized separately, thus achieving the best possible CMOS structural optimization. The disadvantage of the twin well process is its process complexity.

FUTURE DEVELOPMENTS

The ultimate goals of semiconductor technology development are to achieve faster devices and higher-density circuitry. The objective of future technological development in both device structure and device fabrication will be further reduction of device size and power consumption. Current development trends, listed below, will be continued:

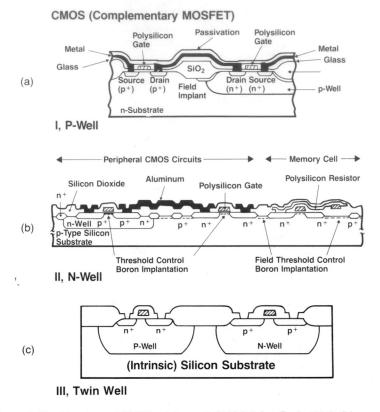

Figure 4-28. Structure of CMOSs: (a) p-well CMOS (after Funk 1984), (b) n-well CMOS used in a 64K double polysilicon NMOS RAM (after Ohzone 1984), and (c) twin well CMOS structure.

1. Down-scaling of device dimensions, both horizontally and vertically. The channel length of a device will be reduced from the current 1.0 μm to about 0.3 μm by the early to mid-1990s, with accompanying reductions in gate insulator thickness (from the current 20 nm to 5–10 nm) and source/drain junction depth (to about 100 nm) (Sunami and Asai 1985).

2. Optimization of vertical dopant profile. As device dimensions are reduced to submicron size, the vertical dopant profile will be optimized to compensate for short channel effects.

3. Greatly improved interdevice isolation. Trench isolation will be further developed and will be widely used in both DRAM and CMOS ICs.

4. Continued use of high-conductive refractory metal interconnections.

5. Structural expansion (or extension) in the z-direction. In order to save silicon real estate, extension of devices and/or circuit structure in the z-direction will be encouraged. One of the obvious and most important current

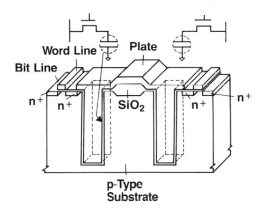

Figure 4-29. Structure of one-device dynamic memory cells with corrugated capacitor. (After Sunami et al. 1984.)

developments in utilizing the depth in silicon real estate is the use of the trench capacitor for DRAMS (Figure 4-29) (Sunami et al. 1983, 1984; Yamaguchi et al. 1985). Another development is the 3-D CMOS structure (Robinson et al. 1985), which aims to shorten the interdevice and/or the intercircuit interconnections so that signal transmission delay is greatly reduced.

6. The merging of two technologies. For example, we will see the merging of HMOS and CMOS (Yu et al. 1981), or CMOS and bipolar (Miyamoto et al. 1983), to fabricate a merged circuit with higher speed and lower power consumption, and/or special functionality.

It has been generally recognized (Sze 1983) that because of their low power consumption and less complicated device fabrication process, MOS ICs will surpass bipolar ICs in the late 1990s and become the prevailing silicon device in microelectronic application. CMOS and DRAM, the two MSO technologies that have already been playing significant roles in VLSI development, will continue to be the dominant technologies in the future. The magnitude of this prevalence is reflected in a recent assessment (Sunami and Asai 1985) that worldwide per capita DRAM usage in the year 2000 will surpass 6400 megabits, as compared with the current per capita usage of 1.0 kilobit.

REFERENCES

Ahlquist, C. N., Breivogel, J. R., Koo, J. T., McCollum, J. L., Oldham, W. G., and Renniger, A. L., "A 16,384-Bit Dynamic RAM," *IEE J. Solid-St. Circuits, SC-11*, 570–574 (1976).

Ames, I., d'Heurle, F. M., and Horstmann, R., "Reduction of Electromigration in Aluminum Films by Copper Doping," *IBM J. Res. Develop., 14*, 461 (1970).

Anderson, R., "Microstructural Analysis of Evaporated and Pyrolytical Silicon Thin Films," *J. Electrochem. Soc., 120,* 1540–1546 (1973).

Appels, J. A., Kooi, E., Paffen, M. M., Schlorje, J. J. H., and Verkuylen, W. H. C. G., "Local Oxidation of Silicon and Its Application in Semiconductor Technology," *Philips Res. Rep., 25,* 118 (1970).

Arnold, E., Ladell, J., and Abowitz, G., *Appl. Phys. Lett., 13,* 413 (1968).

Baccarani, G., Ricco, B., and Spadini, G., *J. Appl. Phys., 49,* 5565 (1978).

Bardeen, J., and Brattain, W. H., "The Transistor, a Semiconductor Triode," *Phys. Rev., 74,* 230 (1948).

Bassous, E., Yu, H. N., and Maniscalo, V., "Topology of Silicon Structure with Recessed SiO_2," *J. Electrochem. Soc., 123,* 1729 (1976).

Beinvogl, W., and Hasler, B., "Anisotropic Etching of Polysilicon and Metal Silicides in Fluorine Containing Plasmas," in *Semiconductor Silicon 1981,* H. R. Huff, F. J. Kriegler, and Y. Takeishi, The Electrochemical Society, 1981, p. 648.

Bennett, R. S., and Chow, T. P., "Polycide Etching for VLSI Applications," *Solid State Technol., 28,* 193–196 (1985).

Bennett, R. S., Ephrath, L. M., Tsai, M. Y., and Lucchese, C. J., "Polycide Etching in a Flexible Diode Reactor," The Electrochemical Society Spring Meeting, *Extended Abstracts, 81-1,* 750 (1981).

Black, J. R., "Electromigration Failure Mode in Aluminum Metallization for Semiconductor Devices," *Proc. IEEE, 57,* 1587 (1969).

Brews, J. R., "Physics of the MOS Transistor," in D. Kahng, Editor, *Applied Solid State Science,* Supplement 2A, Academic Press, New York, 1981.

Burkhart, P. J., Frieser, R. G., and Hoeg, A. J., "Process for Providing Deep Trench Silicon Isolation," *IBM Tech. Disclosure Bulletin, 22* (5), 1862 (1979).

Carr, W. N., and Mize, J. P., *MOS/LSI Design and Application,* McGraw-Hill, New York, 1972, Chap. 2.

Cass, T. R., and Reddi, V. G. K., "Anomalous Residual Damage in Si after Annealing of Through-Oxide Arsenic Implantations," *Appl. Phys. Lett. 23,* 268 (1973).

Chero, G., iec. Meeting, May 7–13, 1965. Also see Dewitt, W., Ong, *Modern MOS Technology,* McGraw-Hill, New York, 1984, 91–94.

Chiang, S. Y., Chan, K. M., Wenocur, D. W., Hui, A., and Rung, R. D., "Trench Isolation Technology for MOS Applications," in *Proc.* of the First International Symposium on VLSI Science and Technology, 1982, (Vol. *82-7*), K. E. Bean and G. A. Rozgonyi, Editors, The Electrochemical Society, 339–346 (1982).

Chiu, K. Y., Moll, J. I., and Manoliu, J., "A Bird's Beak Free Local Oxidation Technology for VLSI Circuits Fabrication," *IEEE Trans. Electron Devices, ED-29,* 536–540 (1982).

Chiu, K. Y., Moll, J. L., Chen, K. M., Lin, J., Lage, C., Angelos, S., and Tillman, R. L., "The Sloped Wall SWAMI—A Defect-Free Zero Bird's Beak Local Oxidation Process for Scaled VLSI Technology," *IEEE Trans. Electron Devices, ED-30,* 1506–1511 (1983).

Chow, T. P., and Steckl, A. J., "Planar Plasma Etching of Sputtered Mo and $MoSi_2$ Thin Films Using NF_3 Gas Mixtures," *J. Appl. Phys., 53,* 5531 (1982).

Chow, T. P., and Steckl, A. J., "Refractory Metal Silicides: Thin Film Properties and Fabrication Technology," *IEEE Trans. Electron Devices, ED-30,* 1480–1497 (1983a).

Chow, T. P., and Steckl, A. J., "A Review of Plasma Etching of Refractory Metal Silicides," in *Proc.* 4th Symp. on Plasma Processing, G. S. Mathad and G. C. Schwartz, Editors, The Electrochemical Society, 362–380 (1983b).

Coe, D. J., "The Lateral Diffusion of Boron in Polycrystalline Silicon and Its Influence on the Fabrication of Submicron MOST," *Solid-St. Electron., 20,* 985–992 (1977).

Coppen, P. J., Aubuchon, K. G., Baur, L. O., and Moyer, N. E., "A Complementary MOS 1.2 Volt Watch Circuit Using Ion Implantation," *Solid-St. Electron., 15,* 165–175 (1972).

Critchlow, D. L., Dennard, R. H., and Schuster, S. E., "Design and Characteristics of n-Channel Insulated-Gate Field Effect Transistors," *IBM J. Res. Develop., 17*, 430–442 (1973).

Crowder, B. L., "Influence of Amorphous Phase on Ion Distribution and Annealing Behavior of Group III and Group V Ions Implanted into Silicon," *J. Electrochem. Soc., 118*, 943 (1971).

Crowder, B. L., and Zirinsky, S., "1 μm MOSFET Technology. Part VII: Metal Silicide Interconnection Technology—A Future Perspective," *IEEE Trans. Electron Devices, ED-26*, 396 (1979).

Crowder, B. L., Zirinsky, S., and Ephrath, L., "The Use of Tungsten Silicide (WSi$_2$)/Polycrystalline Silicon as a Gate Material for MOS Devices," 464 RNP, *J. Electrochem. Soc., 124*, 388c (Nov. 1977).

Deal, B. E., "The Oxidation of Silicon in Dry Oxygen, Wet Oxygen, and Steam," *J. Electrochem. Soc., 110*, 527 (1963).

Deal, B. E., "Standardized Terminology for Oxide Charges Associated with Thermally Oxidized Silicon," *IEEE Trans. Electron Devices, ED-27*, 606 (1980).

Deal, B. E., and Grove, A. S., "General Relationship for the Thermal Oxidation of Silicon," *J. Appl. Phys., 36*, 3770–3778 (1965).

Deal, B. E., Sklar, M., Grove, A. S., and Snow, E. H., *J. Electrochem. Soc., 114*, 266 (1967).

Deal, B. E., MacKenna, E. L., and Castro, P. L., *J. Electrochem. Soc., 116*, 997 (1969).

Dennard, R. H., "Field-Effect Transistor Memory," U.S. Patent No. 3,387,286, June 4, 1968.

Dennard, R. H., Gaensslen, F. H., Yu, H. N., Rideout, V. L., Bassous, E., and LeBlanc, A. R., "Design of Ion-Implanted MOSFET's with Very Small Physical Dimensions," *IEEE J. Solid-State Circuits, SC-9*, 256–268 (1974).

Deposition Technologies for Films and Coatings, Rointan F. Bunshah et al., Editors, Noyes Publications, Park Ridge, NJ, 1982.

d'Heurle, F., "Electromigration and Failure in Electronics: An Introduction," *Proc. IEEE, 59*, 1409 (1971).

d'Heurle, F. M., unpublished results, IBM T. J. Watson Research Center. *Electronic Market Data Book 1982,* Electronic Industries Association, Washington, DC, 1982.

Endo, N., Tanno, K., Ishitani, A., Kurogi, Y., and Tsuya, H., "Novel Device Isolation Technology with Selective Epitaxial Growth," *IEEE IEDM 82*, 241–244 (1982).

Endo, N., Kasai, N., Ishitani, A., and Kurogi, Y., "CMOS Technology Using SEG Isolation Technique," *IEEE IEDM 83 Techn. Dig.*, 31–34 (1983).

Faggin, F., and Klein, T., "Silicon Gate Technology," *Solid-St. Electron., 13*, 1125–1144 (1970).

Faggin, F., Klein, T., and Vadasz, L., *IEEE IEDM Techn. Dig.* (1968).

Finne, R. M., and Klein, D. L., "A Water–Amine–Complexing Agent System for Etching Silicon," *J. Electrochem. Soc., 114*, 965 (1967).

Fraser, D. B., "Metallization," Chap. 9 in *VLSI Technology,* S. M. Sze, Editor, McGraw-Hill, New York, 1983.

Fripp, A., *J. Appl. Phys., 46*, 1240 (1975).

Funk, R. E., "Fast CMOS Logic Bids for TTL Sockets in MOST System," *Electronics*, 134–137 (Apr. 5, 1984).

Gardiner, J. R., and Shepard, J. F., "HCl Gettering of Chemical Vapor Deposited Layers," *IBM Tech. Discl. Bulletin, 21* 3153 (1979).

Gardiner, J. R., Riseman, J., and Shepard, J. F., "Procedure to Minimize Contamination and Defects," *IBM Tech. Discl. Bulletin, 21*, 3201 (1979).

Geipel, H. J., and Shasteen, R. B., "Implanted Source/Drain Junctions for Polysilicon Gate Technologies," *IBM J. Res. Develop. 24*, 362–369 (1980).

Ghandhi, S. K., *VLSI Fabrication Principles,* Wiley, New York, 1983, Chap. 8, 455–457.

Gibbons, J. F., "Ion Implantation in Semiconductors," *Proc. IEEE*, Part I, *56*, 295–319 (1968); Part II, *60*, 1062–1096 (1972).

Gorowitz, B., and Saia, R., "The Reactive Ion Etching of Mo and Mo Compounds for 1 μm VLSI Fabrications," General Electric TIS Rep. 82CRD249 (1982).

Goto, H., Takada, T., Nawata, and Kanai, Y., "A New Isolation Technology for Bipolar VLSI Logic (IOP-L)," *Tech. Dig. 1985 Symposium on VLSI Technology* (IEEE Cat. No. 85, CH 2125-3), Jpn. Soc. Appl. Phys., 42–43 (1985).

Grobman, W. D., Luhn, H. E., Donohue, T. B., Speth, A. J., Wilson, A., Hatzakis, M., and Chang, T. P., "1 μm MOSFET VLSI Technology: Part VI—Electron-Beam Lithography," *IEEE Trans. Electron Devices, ED-26,* 360 (1979).

Grove, A. S., *Physics and Technology of Semiconductor Devices,* Wiley, New York, 1967, Chap. 11, 317–333.

Grove, A. S., Deal, B. E., Snow, E. H., and Sah, C. T., "Investigation of Thermally Oxidized Silicon Surface Using Metal-Oxide-Semiconductor Structure," *Solid-St. Electron, 8,* 145 (1965).

Handbook of Thin Film Technology, L. I. Maissel and R. Glang, Editors McGraw-Hill, New York, 1970, 736.

Hatzakis, M., "Electron Resist for Microcircuit and Mask Productions," *J. Electrochem. Soc., 116,* 1033 (1969).

Hatzakis, M., "Multilayer Resist Systems for Lithographics," *Solid-St. Technol., 24,* 74–80 (1981).

Hatzakis, M., Canavello, B. J., and Shaw, J. M., "Single-Step Optical Lift-off Process," *IBM J. Res. Develop., 24,* 452–460 (1980).

Heil, O., British Patent No. 439,457 (1935).

Hess, D. W., "Plasma Etching of A1," *Solid-St. Technol., 24,* 189 (1981).

Horiuchi, M., "SOLID-II; High-Voltage, High Gain kA-Channel-Length CMOSFETS Using Silicide with Self-Aligned Ultra-Shallow (3S) Junction," *1985 Symp. on VLSI Technol.,* Jpn. Soc. Appl. Phys., 56–57 (1985).

Hosack, H. H., *J. Appl. Phys., 44,* 3476 (1973).

Hu, S. M., "Formation of Stacking Faults and Enhanced Diffusion in the Oxidation of Silicon," *J. Appl. Phys., 45,* 1567 (1974).

Hui, J. C., Chiu, T., Wong, S. S., and Oldham, W. G., "Sealed-Interface Local Oxidation Technology," *IEEE Trans. Electron Devices, ED-29,* 554–561 (1982).

Hunter, W. R., Ephrath, L., Grobman, W. D., Osburn, C. M., Crowder, B. L., Cramer, A., and Luhn, H. E., "1.0 μm MOSFET VLSI Technology: Part IV—A Single-Level Polysilicon Technology Using Electron-Beam Lithography," *IEEE Trans. Electron Devices, ED-26,* 353–359 (1979).

Huntington, H. B., and Grone, A. R., "Current Induced Marker Motion in Gold Wires," *J. Phys. Chem. Solids, 20,* 76 (1961).

Ikeda, T., Tamura, M., Yoshihiro, N., and Tokuyama, T., "Defects in Heavily Phosphorus-Implanted Silicon Observed after Drive-in Process," *Proc.* 6th Conf. Solid-State Devices, Tokyo, Japan, 311–316 (1974).

Irving, S. M., Lemons, K. C., and Bobos, G. E., "Gas Plasma Vapor Etching Process," U.S. Patent No. 3,615,956 (Oct. 26, 1971).

Kahng, D., "A Historical Perspective on the Development of MOS Transistors and Related Devices," *IEEE Trans. Electron Devices, ED-23,* 655 (1976).

Kahng, D., and Atalla, M. M., "Silicon–Silicon Dioxide Field Induced Surface Devices," IRE Solid-State Device Research Conf., Carnegie Inst. Tech., Pittsburgh, 1960.

Kamins, T. I., Manolin, J., and Tucker, R. N., *J. Appl. Phys. 43,* 83 (1972).

Kircher, C. J., "Contact Metallurgy for Shallow Junction Silicon Devices," *J. Appl. Phys. 47,* 5394–5399 (1976).

Kooi, E., and Appels, J. A., "Selective Oxidation of Silicon and Its Device Application," in *Semiconductor Silicon* 1973, H. R. Huff and R. R. Burgess, Editors, The Electrochemical Society, Princeton, NJ, 1973, p. 860.

Kooi, E., van Lierop, J. G., and Appels, J. A., "Formation of Silicon Nitride at a Si–SiO$_2$ Interface during Local Oxidation of Silicon and during Heat Treatment of Oxidized Silicon in NH$_3$ Gas," *J. Electrochem. Soc., 123,* 1117 (1976).

Kriegler, R. J., Cheng, Y. G., and Colton, D. R., *J. Electrochem. Soc.*, *119*, 388 (1972).

Kurten, H., Voss, H. J., Kim, W., and Engl, W. L., "Selective Low-Pressure Silicon Epitaxy for MOS and Bipolar Transistor Application," *IEEE Trans. Electron Devices, ED-30*, 1511–1515 (1982).

Lee, D. H., and Mayer, J. W., "Ion Implanted Semiconductor Devices," *Proc. IEEE*, *64*, 1241–1255 (Sept. 1974).

Lienfield, J. E., U.S. Patent No. 1,745,175 (1930).

Lin, A. M., Antoniadis, D. A., and Dutton, R. W., "The Lateral Effect of On-Boron Diffusion in 100 Silicon," *App. Phys. Lett. 35*, 799 (1979).

MacIver, B. A., and Greenstein, E., "Damage Effects in Boron and BF_2 Ion-Implanted p + n Junctions in Silicon," *J. Electrochem. Soc., 124*, 273 (1977).

Mandurah, M., Sarawat, K., Helms, C., and Kamins, T., *J. Appl. Phys., 51*, 5755 (1980).

Matsukawa, N., Nozawa, H., Matsunaga, J., and Kohyama, S., "Selective Polysilicon Oxidation Technology for VLSI Isolation," *IEEE Trans. Electron Devices, ED-29*, 561–567 (1982).

Mattausch, H. J., Hasler, B., and Beinvogl, W., "Reactive Ion Etching of Ta-Silicide/Polysilicon Double Layers for the Fabrication of Integrated Circuits," *J. Vac. Sci. Technol., B1*, 15 (1983).

Meindl, J. D., et al., "Computer-Aided Engineering of Semiconductor Integrated Circuits," Stanford Electronic Labs, TR DXG501 (July 1980).

Miyamoto, I., Saitoh, S., Shibata, H., Kanzaki, K., and Kohyama, S., "A 1.0 μ n-Well CMOS/ Bipolar Technology for VLSI Circuits," *IEEE IEDM 83*, 63–66 (1983).

Moline, R. A., and Cullis, A. G., "Residual Effects in Si Produced by Recoil Implantation of Oxygen," *Appl. Phys. Lett. 26*, 551 (1975).

Murarka, S. P., *Appl. Phys. Lett., 34*, 587 (1979).

Murarka, S. P., "Refractory Silicides for Integrated Circuits," *J. Vac. Sci. Technol., 17*, 775–792 (1980).

Nakajima, O., Shiono, N., Muramoto, S., and Hashimoto, C., "Defects in a Gate Oxide Grown after the LOCOS Process," *Jpn. J. Appl. Phys., 18*, 943 (1979).

Natsusaki, N., Tamura, M., Miyao, M., and Tokuyama, T., "Anomalous Residual Defects in Silicon after Annealing of Through-Oxide Phosphorus Implanted Samples," *Jpn. J. Appl. Phys.*, Supplement *16*, 47 (1977).

Ogura, S., Tsang, P. J., Walker, W. W., Critchlow, D. L., and Shepard, J. F., "Design and Characteristics of the Lightly Doped Drain-Source (LDD) Insulated Gate Field-effect Transistor," *IEEE Trans. Electron Devices, ED-27*, 1359–1367 (1980).

Ohwada, N., et al., *IEEE Trans. Electron Devices, ED-26*, (1979).

Ohzone, T., "64-K Static RAM Sounds NMOS Cells with CMOS Circuits," *Electronics*, 145–148 (Nov. 6, 1984).

Ormond, D. W., "Dielectric Breakdown of Silicon Dioxide Thin Film Capacitors Using Polycrystalline Silicon and Aluminum Electrodes," *J. Electrochem. Soc., 126*, 162 (1979).

Ormond, D. W., and Gardiner, J. R., "Reliability of SiO_2 Gate Dielectric with Semi-recessed Oxide Isolation," *IBM J. Res. Develop., 24*, 353–361 (1980).

Osburn, C. M., and Bassous, E., "Improved Dielectric Reliability of SiO_2 Films with Polycrystalline Silicon Electrodes," *J. Electrochem. Soc., 122*, 89– 92 (1975).

Osburn, C. M., and Ormond, D. W., "Dielectric Breakdown in Silicon Dioxide Film on Silicon," *J. Electrochem. Soc., 119*, 597–603 (1972).

Osburn, C. M., Tsai, M. Y., Roberts, S., Lucchese, C. J., and Ting, C. Y., "High Conductivity Diffusions and Gate Regions Using Self-Aligned Silicide Technology," *Proc.* First International Symposium on VLSI Science and Technology, The Electrochemical Society, *82-7*, 213 (1982).

Pashley, R., Kokonnen, K., Boleky, E., Jecmen, R., Liu, S., and Owen, W., "HMOS Scales Traditional Devices to Higher Performance Level," *Electronics*, 94–99 (Aug. 18, 1977).

Pliskin, W. A., Kerr, D. R., and Perri, J. A., "Thin Glass Films," in *Physics of Thin Films*, Vol. 4, 257–324, Academic Press, New York, 1967.

Poate, J. M., Tu, K. N., and Mayer, J. W., *Thin Films Interdiffusion and Reactions,* Wiley, New York, 1978.

Rand, M. J., and Roberts, J. F., "Observation on the Formation and Etching of Platinum Silicide," *Appl. Phys., Lett., 24,* 49–51 (1974).

Razouk, R. R., and Deal, B. E., Spring Meeting, The Electrochemical Society, *Extended Abstracts,* 363–365 (May 1979); *J. Electrochem. Soc., 126,* 1573 (1979).

Reismann, A., Berkblit, M., Chan, S. A., Kaufman, F. B., and Green, D. C., "The Controlled Etching of Silicon in Catalyzed Ethylenediamine–Pyrocatechol–Water Solutions," *J. Electrochem. Soc., 126,* 1406 (1979).

Rideout, V. Leo, "One-Device Cells for Dynamic Random-Access Memories: a Tutorial," *IEEE Trans. Electron Devices, ED-26,* 839–852 (1979).

Robinson, A. L., Antoniadis, D. A., and Maby, E. W., "Fabrication of Fully Self-Aligned Joint-Gate CMOS Structures," *IEEE Trans. Electron Devices, ED-32,* 1140–1142 (1985).

Roger, S. H., and Coe, M. E., "Low Frequency Plasma Etching of Polycide Structures in an SF_6 Glow Discharge," *Solid-St. Technol., 25,* 79 (1982).

Ronen, R. S., and Micheletti, F. B., *Solid-St. Technol., 18,* 39–46 (1975).

Saito, K., Morose, T., Sato, S., and Harada, U., "A New Short Channel MOSFET with Lightly Doped Drain," *Denshi Tsushin Rengo Taikai* (in Japanese), 220 (Apr. 1978).

Sarace, J. C., Kerwin, R. E., Klein, D. L., and Edwards, J., *Solid-St. Electron., 11,* 653 (1968).

Schwabe, U., Hersbst, H., Jacobs, E. P., and Takacs, D., "n- and p-Well Optimization for High-Speed n-Epitaxy CMOS Circuits," *IEEE Trans. Electron Devices, ED-30,* 1339–1344 (1983).

Schwartz, G. C., and Schaible, P. M., "Reactive Ion Etching of Silicon," *J. Vac. Sci. Technol., 16,* 410 (1979).

Shockley, W., "The Theory of pn Junctions in Semiconductors and pn Junction Transistors," *Bell Syst. Tech. J., 28,* 435 (1949).

Sinha, A. K., Lindenberger, W. S., Fraser, D. B., Murarka, S., and Fuls, E. N., "MOS Compatibility of High Conductivity $TaSi_2/n^+$ Poly-Si Gates," *IEEE Trans. Electron Devices, ED-27,* 1425 (1980).

Siu, W., "A 1 μm CMOS Technology Optimized for 1 Megabit DRAMs," *Digest* tech. papers, 1985, Symposium on VLSI Technology, Jpn. Soc. Appl. Phys. (IEEE Cat. No. 85 CH 2125-3), 8–9 (1985).

Stein, K., and Friedrich, H., "A 1-mil^2 Single-Transistor Memory Cell in n Silicon-Gate Technology," *IEEE J. Solid-St. Circuits, SC-8,* 319–323 (1973).

Sunami, H., Kure, T., Hashimoto, N., Itok, K., Toyabe, T, and Asai, S., "A Corrugated Capacitor Cell (CCC) for Megabit Dynamic MOS Memories," *IEEE Electron Device Lett., EDL-4,* 90–91 (1983).

Sunami, H., et al., "A Corrugated Capacitor Cell (CCC)," *IEEE Trans. Electron Devices, ED-31,* 746–753 (1984).

Sunami, Hideo, and Asai, Shojiro, "Trend in Megabit DRAM's," *Proc.* tech. papers, 1985 (2nd) International Symposium on VLSI Techn. Syst. and Appl., May 8–10, Taipei, Taiwan, 4–8 (1985).

Swaminathan, B., Sarawat, K. C., and Dutton, R. W., "Diffusion of Arsenic in Polycrystalline Silicon," *Appl. Phys. Lett. 40* (9), 795–798 (1982).

Sze, S. M., *Physics of Semiconductor Devices,* 2nd ed., Wiley, New York, 1981, 304.

Sze, S. M., "VLSI Technology Overviews and Trends," *Jpn. J. Appl. Phys. 22,* Suppl. *22-1,* 3–10 (1983).

Teng, W., Pollack, G., and Hunter, W. R., "Optimization of Sidewall Masked Isolation Process," *IEEE Trans. Electron Devices, ED-32,* 124–131 (1985).

Ting, C. Y., Iyer, S. S., Osburn, C. M., Hu, G. J., and Schweighart, A. M., "The Use of $TiSi_2$ in a Self-Aligned Silicide Technology," *Proc.,* First International Symposium on VLSI Science and Technology, The Electrochemical Society, *82-7,* 224–231 (1982).

Totta, P. A., and Sopher, R. P., *IBM J. Res. Develop.*, *18*, 226 (1969).

Tsai, M. Y., and Streetman, B. G., "Recrystallization of Implanted Amorphous Silicon Layer, I. Electrical Properties of Silicon Implanted with BF^{+2} or $Si^+ + B^+$," *J. Appl. Phys. 50*, 183 (1979).

Tsang, P. J., "RhSi Contact Barrier for Device with Shallow Junctions," paper presented at 1976 Fall Meeting, The Electrochemical Society, Oct. 1976, Las Vegas, Nevada; *Extended Abstracts*, 789–791 (1976).

Tsang, P. J., Shepard, J. F., Lechaton, J., and Ogura, S., "Characterization of Sidewall-Spacers Formed by Anisotropic RIE," *J. Electrochem. Soc. 128*, 238C (1981).

Tsang, P. J., Ogura, S., Walker, W. W., Shepard, J. F., and Critchlow, D. L., "Fabrication of High-Performance LDDFET's With Oxide Sidewall-Spacer Technology," *IEEE Trans. Electron Devices, ED-29*, 590–596 (1982a).

Tsang, P. J., Shepard, J. F., Ogura, S., and Riseman, J., "Sidewall Spacer Technology," paper presented at the 162nd national meeting, The Electrochemical Society, Detroit, Mich., Oct. 1982; *Extended Abstracts, 82-2*, 373–374 (1982b).

Vadasz, L. L., Grove, A. S., Rowe, T. A., and Moore, G. E., "Silicon-Gate Technology," *IEEE Spectrum, 6* (10), 28–35 (1969).

Vaidya, S., Fraser, D. B., and Shinha, A. K., "Electromigration Resistance of Fine Line Al," *Proc. 18th Reliability Phys. Symp.*, IEEE, New York p. 165 (1980).

Voss, H. J., and Kurten, H., "Device Isolation Technology by Selective Low-Pressure Silicon Epitaxy," *IEEE IEDM 83 Tech. Dig.*, 35–38 (1983).

Yamaguchi, K., Nishimura, R., Hagiwara, T., and Sunami, H., "Two-Dimensional Numerical Model of Memory Devices with a Corrugated Capacitor Cell Structure," *IEEE Trans. Electron Devices, ED-32*, 282–289 (1985).

Yu, K., Chwang, R. J. C., Bohr, M. T., Walkentin, P. A., Stern, S., and Berglund, C. N., "HMOS-CMOS—A Low-Power High-Performance Technology," *IEEE J. Solid State Circuits, SC-16*, 454–459 (1981).

Zirinsky, S., and Crowder, B. L., "Retractory Silicide: For High Temperature Compatible IC Conductor Lines," 463 RNP, *J. Electrochem. Soc., 124*, 388c (Nov. 1977).

5. Electron Beam Testing

E. Wolfgang
Siemens Research Laboratories

Electron beam testing refers to the measurement and mapping of electrical parameters such as voltages, currents, frequencies, and resistances with the aid of an electron probe. This discipline, originally developed from scanning electron microscopy, must now be regarded as an independent technique. It took its start from observations of contrast differences between the p and n regions, a method later designated as voltage contrast. Table 5-1 shows the principal milestones in these developments over the last three decades.

Electron beam testing is used in chip verification and failure analysis. The initial impetus for its development was in chip verification, which involves inspection of first silicon as initial samples of a new design are checked for agreement with computer simulations. The nondestructive, nonloading, and simple-to-position electron probe has proved itself eminently suited to this purpose. The methods developed for chip verification have been exploited in failure analysis, in which failures are localized after stress testing or in the field, and their causes determined.

The driving forces in the development of electron beam testing can be seen from Figure 5-1, which shows a photomontage of the tip of a mechanical probe and a number of aluminum interconnections. The 2.5 μm tip radius can hardly be further reduced; so it is clear that interconnections with widths $\leqq 1$ μm cannot be contacted for geometric reasons. Also, the probe capacitance C_p can have a very disturbing effect on the function of the circuit, since reduction of structures leads to reduced node capacitances as well. However, new active probes have been developed recently whose capacitances, C_p, have a value of only 200 fF (earlier values were in the range 2–4 pF).

Table 5-1. Milestones in the History of Voltage Contrast and Electron Beam Testing Applied to ICs.

YEAR	DISCOVERY/INTRODUCTION OF	REFERENCE
1957	Voltage contrast	Oatley and Everhart (1957)
1968	Stroboscopic scanning electron microscopy	Plows and Nixon (1968)
1968	Voltage measurement	Wells and Bremer (1968)
1975	Voltage coding	Lukianoff and Touw (1975)
1975	Internal waveforms of a DRAM	Gonzales and Powell (1975)
1976	Quantitative voltage contrast at high frequencies	Balk et al. (1976)
1977	Estimate of minimal measurable voltage in SEM	Gopinath (1977)
1979	New retarding field spectrometer	Feuerbaum (1979)
1980	Electron beam testing of microprocessors (logic-state mapping)	Crichton et al. (1980)
1980	Low energy electron probe	Kotorman (1980), Mizuno et al. (1981)
1980	Function testing of bipolar LSIs	Fujioka et al. (1980)
1982	Automated contactless SEM testing for VLSI	Thangamuthu et al. (1982)
1983	Expert systems and electron beam testing	Baille et al. (1983a)
1984	Capacitive coupling voltage contrast	Gorlich et al. (1984)
1984	Frequency mapping and tracing	Brust et al. (1984)

In recent years a great deal has been published in the field of electron beam testing. This work is scattered over many journals and the proceedings of conferences of all kinds. Here we are focusing on papers relating to the application of electron beam testing that have appeared since 1980. From a large number of review papers, five that appeared in the journal *Scanning* in 1983 deserve special mention: two by Menzel and Kubalek (pp. 103–122 and 151–171) and those of Fujioka and Ura, Feuerbaum, and Wolfgang (1983a).

This chapter concentrates on the application of electron beam testing to chip verification, and especially to voltage and waveform measurements. The sections on the electron probe and on signal processing deal with those properties that the user (e.g., a circuit designer) should be familiar with. In the section on modes of operation, five techniques are described in detail. Important modifications of scanning electron microscopes are described in the section on electron beam testers. Under electron beam test methodology, chip verification and failure analysis are explained, and concepts for integrated test systems and automated failure localization are discussed.

In the section on applications, use of the modes of operation described earlier is demonstrated by three generalized examples: signal mapping and tracing, spike detection, and comparison with computer simulation. Finally, current trends as well as the limitations of electron beam testing are discussed.

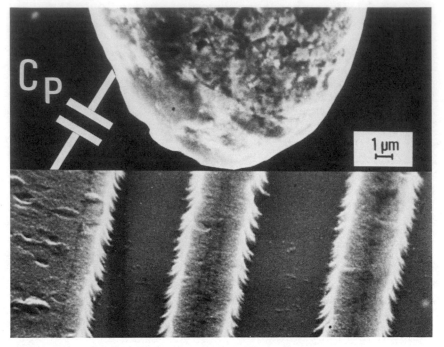

Figure 5-1. Photomontage of two scanning electron micrographs showing the tip of a mechanical tungsten probe as well as aluminum interconnections. C_p is the probe capacitance between the mechanical probe and the ground of the IC.

ELECTRON PROBE

The demands made on the electron probe and the reasons for them are listed in Table 5-2. There are several reasons for the requirement for low acceleration voltage:

1. The principal one is that irradiation damage can be avoided by using a low voltage (Wolfgang 1983a; Keery et al. 1976; Miyoshi et al. 1982).
2. Low voltages also allow passivated devices to be inspected by means of capacitive voltage contrast (Kotorman 1980, 1983; Gorlich et al. 1984; Ura and Fujioka 1983; Uchikara and Ikada 1983; Todokoro et al. 1983).
3. The current load into the circuit node can thereby be minimized (Feuerbaum 1971).

The requirement for a small spot diameter cannot be met simply, although one is accustomed to high spatial resolution from the scanning electron microscope. In electron beam testing, it should be noted that:

Table 5-2. Demands Made on the Electron Probe.

REQUIREMENT	REASON
Low acceleration voltage of primary electron beam (~ 1 kV)	To avoid irradiation damage of circuit elements and current loading of internal nodes.
Spot size 1/5 of interconnection width	To ensure that all secondary electrons that contribute to the measurement are generated at the interconnection to be measured.
Pulsed electron probe (0.1 − 1 ns)	To achieve the necessary time resolution.
Scanning electron probe	To display more than one signal.

1. A lower acceleration voltage increases the spot size. Although LaB_6 and field emission guns are superior to standard tungsten ones (Broers 1975; Swanson et al. 1983), they do, among other things, require a higher vacuum.

2. A specific number of secondary electrons is required for signal processing. Thus 6×10^6 electrons allow a voltage resolution of 100 mV to be attained. But if the pulse width is greatly reduced ($\leq$ 1 ns) because of the high time resolution aimed for, then only a few or even single electrons are generated per pulse.

3. The beam blanking systems for generating short electron pulses can lead to degradation of the spot size (Fujioka and Ura 1983; Lischke et al. 1983; Menzel and Kubalek 1979b).

In Figure 5-2, an attempt is made to present graphically the complex relationships just described. Shown on the left side are the interrelationships of the electron current i and the electron probe diameter d or the interconnection width w, as the case may be. To make sure that during measurement the electron probe remains on the interconnection, it was assumed that $w = 5d$. The electron beam diameter is about four times as large as the diameters known from the literature because of the use of beam blanking (Fazekas et al. 1978).

In the right-hand part of Figure 5-2, the interconnection width w is plotted versus the electron probe pulsed width τ_p. The sloped left-hand boundary line means that the electron pulse contains on the average one electron ($1\ e^-/\tau_p$), as can easily be calculated using the electron current i, indicated by the straight line in the left-hand part of Figure 5-2. The vertical right-hand boundary line is due to the signal processing system used, the sampling frequency of which amounts to a maximum possible pulse width, so that τ_p is 250 ns. Plotted on

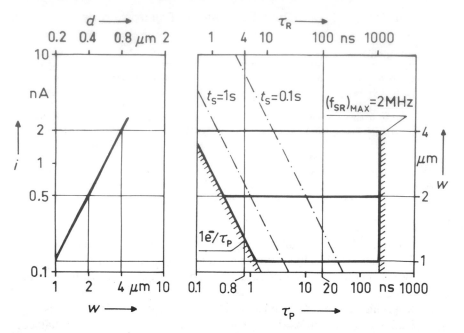

Figure 5-2.   Limitations of electron beam testing: interconnection width w, electron probe diameter $d = w/5$, probe current i, electron pulse width τ_p, rise time of pulse edge $\tau_R = 5\tau_p$, time needed for a single measurement t_s, and maximum sampling repetition frequency $(f_{SR})_{max}$. The curve in the left part represents experimental values achieved with 2.5 kV acceleration voltage, 12 mm working distance, 5 ns pulse width.

the abscissa at the upper edge of the picture is the condition for an accurate pulse edge measurement. The figure should thus be interpreted as follows: an accurate pulse edge measurement with $\tau_R = 4$ ns requires a pulse width $\tau_p = 0.8$ ns. For a measurement on an interconnection with $w = 4$ μm, there are 10 electrons/pulse; for $w = 2$ μm, there are 2.5 electrons/pulse; and for $w = 1$ μm, there is only 0.6 electron/pulse.

According to Gopinath's formula, which estimates the minimal measurable voltage by an electron probe, 6 million secondary electrons are necessary to achieve a voltage resolution of 100 mV (Gopinath 1977). Utilizing the above-mentioned relationship of the various physical parameters, the time needed for a single measuring time t_s can be estimated. Two examples, for 1 s and 0.1 s measuring time, are shown in Figure 5-2 as dashed lines.

The requirement for the scanning electron probe is the easiest to fulfill because the scanning electron microscope can routinely affect all scan modes. By using digital scan generators, all scan modes required for electron beam testing, such as area scan, line scan, or vector scan, can be realized.

SIGNAL PROCESSING

Voltage contrast results from local fields at the surface of the IC when it is electrically operated. The simplest explanation for this phenomenon is that when the interconnection is in a "high" logic state (positive voltage), it attracts back a large proportion of the resulting secondary electrons. The secondary electron collector then "sees" fewer electrons, so that these points appear dark on the CRT (Seiler 1983; Lin and Everhart 1979). On the other hand, all secondary electrons can reach the collector when the interconnection is in a "low" logic state. These points then appear bright. In addition to voltage contrast, there is a contribution of material contrast. Various methods have been suggested to isolate the voltage contrast from the material contrast (Menzel and Kubalek 1981; Fujioka et al. 1982; Sardo and Vanzi 1984).

The voltage contrast cannot be conveniently utilized for voltage measurements because it is nonlinear, and the local fields are continuously being changed by the different circuit states in the vicinity. Therefore, a series of electron spectrometers has been suggested. This approach is explained in detail in a number of review papers (Menzel and Kubalek 1983; Menzel and Brunner 1983). With increasing reduction of structural widths, the influence of local field effects should be taken into account because they can affect the accuracy of the voltage measurement (Nakamura and Sato 1983; Ura et al. 1984).

It is important for waveform measurements that the pulse-shaped secondary electron signal be processed in an optimal way. One way to ensure this is to use a boxcar generator, which has proved exceptionally effective for electron beam testing (Feuerbaum and Otto 1982).

MODES OF OPERATION

Five important modes of operation, listed in Table 5-3, are discussed in this section. They differ in their electron beam continuities, frequency relationships, scan modes, and types of signal processing. The 12 columns in the table clearly indicate that there are more than five modes of operation possible, but the others play an insignificant role in practice and will be mentioned only briefly at the end of this section. The five main modes of operation are described and explained below on the basis of schematic diagrams that also show the mode of presentation of the measured results.

Voltage Coding

Voltage coding is based on the synchronization between the line frequency of the electron beam (~ 15 kHz in TV mode) and the signal frequency of the device under test (DUT), where the IC is scanned line by line from B to E in Figure

Table 5-3. Five Main Modes of Electron Beam Testing.

ELECTRON BEAM TESTING MODES	ELECTRON PROBE			FREQUENCY RELATIONSHIP	FREQUENCY/ MINIMUM/ MAXIMUM (MHz)	SCAN MODE			SIGNAL PROCESSING		SECONDARY ELECTRON COLLECTION	
	CONTINUOUS	PULSED				AREA	LINE	VECTOR	VOLTAGE CONTRAST	VC PLUS THRESHOLD	VOLTAGE MEASUREMENT	FREQUENCY MEASUREMENT
		PHASE SHIFTED	CONSTANT PHASE									
Voltage coding	X			$f_S = n f_L$	$f_S/0.015/1.5$	X			X			
Logic-state mapping		X		$f_B = f_S/n$	$f_B/0.001/1$ $f_S/0.001/200$	X	X		X			
Frequency tracing			X	$f_B = f_S + f_{IF}$	$f_B/0.1/500$ $f_S/0.05/500$	X				X		
Frequency mapping			X	$f_B = $ varied	$f_B/0.1/500$ $f_S/0.05/500$		X	X		X		
Waveform measurement		X		$f_B = f_S/n$	$f_B/0.001/1$ $f_S/0.001/200$			X			X	X

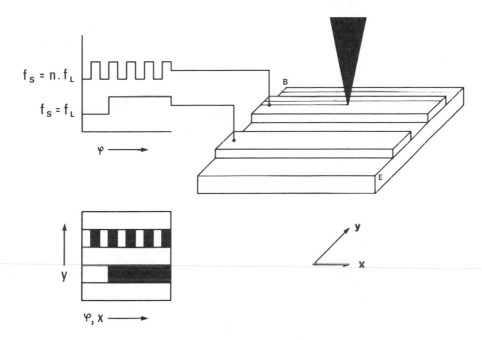

Figure 5-3. Principle of voltage coding. The unpulsed electron probe continuously scans the IC surface line by line, starting at *B* and finishing at *E*. The line and frame frequency corresponds to the TV norm. A stationary bar pattern of the logic states due to voltage contrast appears in the *x*-direction when $f_s = n\,f_L$ (line frequency $f_L = 15$ kHz).

5-3. This mode was first mentioned by Crostwait (Crostwait and Ivy 1974) and later explained in detail by Lukianoff and Touw (1975), who designated it voltage coding. Because of this synchronization, logic states along an interconnection line always change at the same point, so that the voltage contrast gives rise to a static bar pattern on the interconnections in the *x*-direction if the condition $f_s = nf_L$ is fulfilled ($n = 1, 2, 3 \ldots$). If the signal frequency corresponds to 15 kHz, then only one signal change can be seen. The maximum signal frequency of 1.5 MHz ($n = 100$) is limited by the spatial resolution of the TV monitor (625 points/column).

Logic-State Mapping

Since the majority of ICs have operating frequencies greater than 1.5 MHz, Crichton et al. (Fazekas et al. 1981) proposed logic-state mapping, in which the pulsed electron probe ($f_B = f_s/n$) repeatedly scans along the line *BE* (Figure 5-4) (Wolfgang et al. 1983; Ostrow et al. 1982). In this case f_s/n is the smallest signal frequency to be investigated, it being assumed that all other frequencies

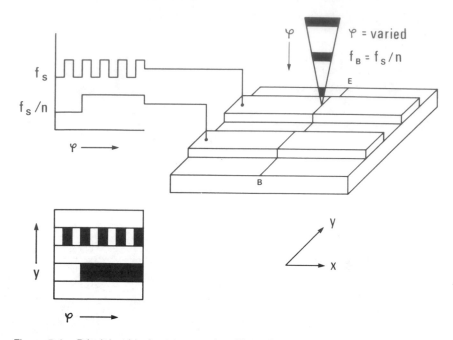

Figure 5-4. Principle of logic-state mapping. The pulsed electron probe scans along the line from *B* to *E*. After every line scan the phase is changed. The result on the CRT is a representation of distance against phase.

occurring in the IC are integral multiples n of F_B. After completion of each line scan BE, the phase ϕ is changed by a specific, freely selectable amount. The voltage contrast again results in a bar pattern along the pseudo-interconnections because the electron beam on the CRT effects an area scan. The maximum beam blanking frequency f_B is also determined by the bandwidth used for signal processing (Feuerbaum and Otto 1982). It has a typical value $\leqslant 1$ MHz. The maximum signal frequency f_s is given by the pulse width. Thus for a pulse width τ_p of 1 ns, signals with $f_s = 200$ MHz can just be represented.

An area scan may be made in place of the line scan (see Table 5-3), however. This leads to an image comparable to that obtained with voltage coding, where only the time scales differ from each other.

Frequency Tracing

Brust et al. (1984) recently presented both frequency tracing and frequency mapping (described below). They make use of the heterodyne principle in which F_B and f_s are detuned with respect to each other by a specific amount to produce an intermediate frequency f_{IF}. The electron probe pulsed with f_B executes an

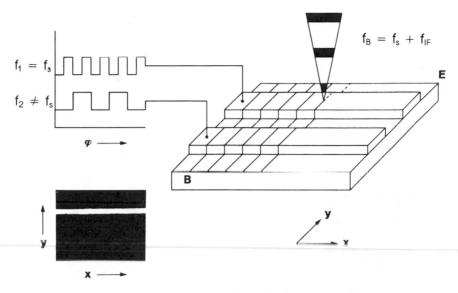

Figure 5-5. Principle of frequency tracing. The pulsed electron probe executes an area scan from B to E. The beam blanking frequency f_B differs from the signal frequency f_s. If the condition $f_{IF} = |f_B - f_s|$ is met, the secondary electron signal is amplified.

area scan from B to E (Figure 5-5). If it strikes an interconnection that carries a signal with $f_s = |f_B - f_{IF}|$, then this signal passes through the filter and is amplified. If it exceeds a prespecified threshold, the CRT is unblanked. In contrast to other techniques, frequency tracing and frequency mapping do not require any synchronization between the DUT and the electron probe.

Frequency Mapping

This mode permits the measurement of signals of unknown frequency f_s. The pulsed electron probe repeatedly scans a line from B to E (Figure 5-6). After every line scan, the frequency f_B is varied by a specific freely selectable amount so that the secondary electron signal is amplified, as in the above case, if the condition $f_{IF} = |f_B - f_s|$ is met. The electron beam of the CRT executes a line scan so that the distance across the interconnections is shown in the y-direction, and the frequency f_B along the x-direction. Thus, a pair of bright bars occurs, lying at a distance f_{IF} to the left and right of the desired frequency f_s.

In both techniques, frequency tracing and frequency mapping, the lower boundary of f_B and f_s is fixed by the frequency $f_{IF} = 50$ kHz, and the upper boundary is limited by the beam blanking system, with a present typical value of 500 MHz.

If a vector scan is carried out in place of a line scan, the line pair is replaced by a point pair with a spacing of $2f_{IF}$ in the x-direction.

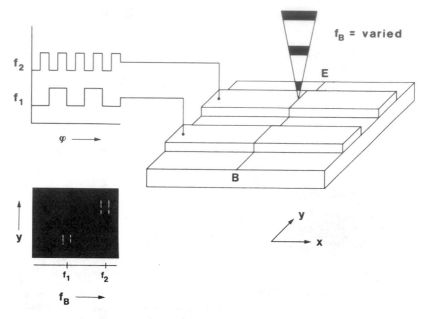

Figure 5-6. Principle of frequency mapping. The pulsed electron scans along the line from B to E. After every line scan, the beam blanking frequency f_B is changed by a specific amount. If the condition $f_{iF} = |f_B - f_s|$ is met, the CRT is unblanked. The result is a bar pair to the right and left of the sought-for frequencies f_1 and f_2.

Waveform Measurement

In contrast to the four electron beam testing modes already described, the waveform mode does not utilize voltage contrast, as it is nonlinear. Instead, the voltage is measured with the aid of a secondary electron spectrometer (Feuerbaum 1979; Menzel and Kubalek 1983).

In this mode, the electron probe pulsed with $f_B = f_s/n$ is kept pointed at an interconnection (Figure 5-7). As the measurement duration increases, the phase ϕ is changed, as in logic-state mapping, and the resulting waveform appears after a few seconds. As in logic-state mapping, f_B is the smallest signal frequency that should be measured. Here again it is assumed that the other signal frequencies are integral multiples of f_B.

After completion of a waveform measurement, the electron probe is directed at the next interconnection to be measured. The frequency relationship is the same in this case as for logic-state mapping.

Other Electron Beam Testing Modes

The first mode used was static voltage contact, which is today important only in the analysis of hard errors. Plows and Nixon (1968) introduced the stroboscopic

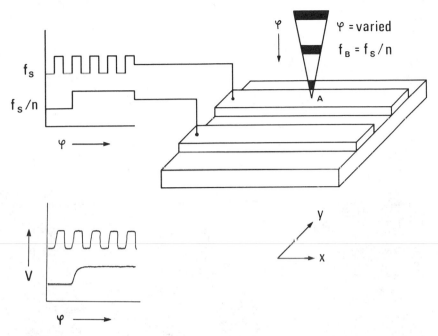

Figure 5-7.   Principle of the waveform mode. The pulsed electron probe remains directed at a point (*A*) during the waveform measurement. The beam blanking frequency f_B corresponds to the lowest signal frequency to be measured. During the measurement, the phase ϕ is shifted by a specific amount. The voltage value results from a feedback loop that may contain a retarding field spectrometer, a secondary electron collector, a photomultiplier, and amplifiers.

voltage contrast method, which has now been replaced by logic-state mapping, as the latter includes the information content of many stroboscopic voltage contrast micrographs.

The timing diagram method was proposed by Fazekas et al. (1981), and it was extended by Ostrow et al. (1982) to a real-time logic state analysis. However, this procedure requires such a high probe current that it is not suitable for interconnections that are becoming ever narrower.

ELECTRON BEAM TESTERS

The instruments known today as electron beam testers (EBT) are modified scanning electron microscopes. This section will not deal with the four or five EBTs available on the market, but will describe a number of fundamental properties that every EBT should or can have.

A schematic configuration of a standard EBT is shown in Figure 5-8(a). It consists of an SEM, an IC adapter, the signal feed, and an IC drive unit. Not marked in the diagram are the beam blanking unit and the special signal pro-

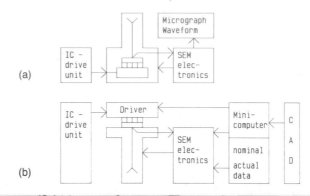

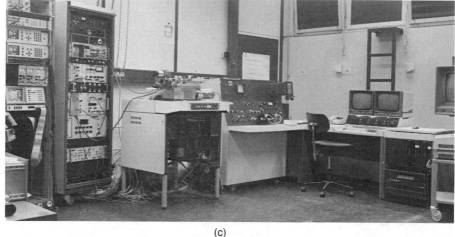

(c)

Figure 5-8. Electron beam testers. (a) Standard configuration. (b) Automated electron beam tester (Kollensperger et al. 1984). (c) Photograph of the automated electron beam tester.

cessing unit, which may also contain a secondary electron spectrometer. Figures 5-8(b) and 5-8(c) show a more advanced EBT whose special feature is that its electron optical column has been rotated by 180° (Kollensperger et al. 1984). This arrangement permits an optimal signal feed to be attained, as shown in Figure 5-9. The ICs, which are mounted in ceramic packages, are vacuum-sealed onto the adapter plate by means of a flat rubber seal (Figure 5-9a). This ensures that only the cavity containing the chip is under vacuum, while the pins are open to the air. The specimen chamber can thus be kept small (minichamber in Figure 5-9), and the signal feed can be kept short.

For investigations of the chips on the wafer, wafer probes were set up in a vacuum (Fazekas et al. 1978; Menzel and Kubalek 1979a). Figure 5-10(a) is a schematic diagram of a large-area specimen chamber. The wafer is placed on two x-y stages, one of which can be rotated around an angle α. The adapter card

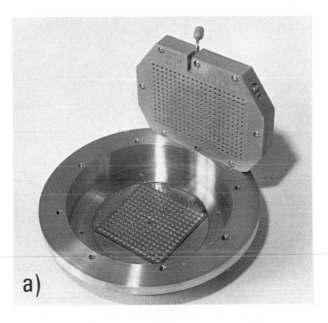

Figure 5-9.   Signal feed for packaged ICs in the case of Figures 5-8(b) and 5-8(c). (a) Adapter plate for a 241-pin grid array and Textool socket. (b) *x-y* stage in which the adapter plate is mounted.

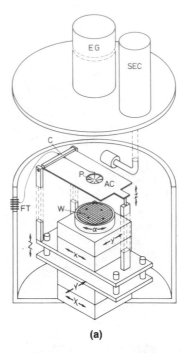

(a)

Figure 5-10.   Large-area specimen chamber with wafer probes. (a) Exploded view of chamber with two *x-y* stages, wafer *W*, adapter card *AC*, probes *P*, connector *C*, feedthroughs *FT*, electron column *EG*, and secondary electron collector *SEC*. (b) Photograph of two *x-y* stages and wafer. (c) Photograph of chamber and adapter card for a packaged device under test. Signal drives are located in front of and behind the DUT.

AC with the probes *p* is lowered onto the pads with the aid of the SEM image. The lower *x-y* stage allows the contacted chip to be moved with respect to the electron optical-axis (Figure 5-10b), so that magnified images of all regions of the chip can be obtained.

Many articles on beam blanking systems are covered in the previously mentioned review papers (Fujioka and Ura 1983; Menzel and Kubalek 1979b). An electron beam tester with a particularly high time resolution was recently presented by Goto et al. (1984).

Chip verification of first silicon often requires that circuit parts be separated and subjected to mechanical probes, a supply voltage, and/or signals (Wolfgang 1983a). The separation can be effected by means of laser surgery (Dallmann et al. 1985). To make an effective comparison with the simulation, the DUT must be analyzed at the same temperatures at which the simulation was carried out. For this purpose, the DUT must be either cooled or heated, by applying Peltier elements (Wolfgang 1983a) or by means of an air stream (Kollensperger et al. 1984).

(b)

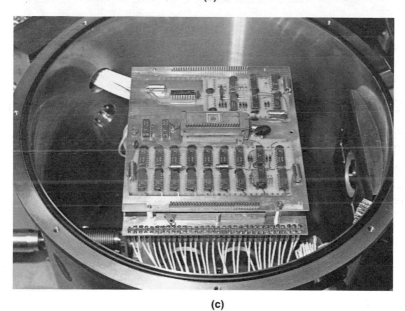

(c)

Figure 5-10.   *(continued)*

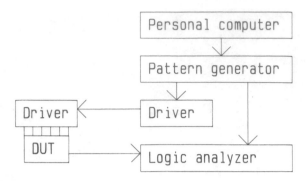

Figure 5-11.   Block diagram of an IC drive unit that is used in connection with the automated electron beam tester (Figure 5-8).

An important element of the EBT is the IC drive unit. A suggested solution is shown schematically in Figure 5-11. The pulse sequence for a program loop of minimum duration and frequency f_2/n is generated with the aid of a personal computer and transmitted to a pattern generator. This requires drivers to be fitted at both ends of the feed cable and arranged as close as possible to the pins on the DUT side. A logic analyzer is used to check whether the prespecified pulse sequences are really present at the DUT, and whether this actually functions during the measurement. The analyzer can be applied in such a way as to allow an actual/nominal comparison to be made (Kollensperger et al. 1984).

ELECTRON BEAM TEST METHODOLOGY

Electron beam testing has two areas of application: chip verification and failure analysis (Wolfgang 1983b; Fazekas et al. 1983). Table 5-4 shows the tasks for which electron beam testing is used, what types of defaults are detected, and finally what nominal data are available to provide a comparison with the actual situation.

Chip Verification

The procedure is outlined in six stages in Table 5-5. The first stage is very important: the provision of internal test pads on the upper metallization level. Since they can be made as small as the via holes, they take up no extra space.

The second state involves the usual determination of operating margins and measurement of output waveforms using automated test equipment (ATE) (Krampl et al. 1983).

The third is the preparation stage. During the design phase an open ceramic package can be selected in most cases. Passivation layers must then be removed

Table 5-4. Applications of Electron Beam Testing.

	CHIP VERIFICATION	FAILURE ANALYSIS
Tasks	To determine quality of: • models for simulation • design rules • CAD system • chip To shorten design period	To determine modes and causes of failure
Application	During design phase	During production and testing, in the case of: • losses of yield • stress testing • failures in the field
Type of fault	Design-related	Material- and process-related
Nominal values	Computer simulation	"Golden device"

by methods such as plasma etching. The IC operating conditions must be adjusted to simulate the electrical conditions of the automatic test equipment and to reproduce the faulty situation (Wolfgang 1983a).

At the fourth stage, the fault is localized with the aid of logic-state and frequency mappings. The recommended procedure is, if possible, to operate the IC at constant frequency and temperature and to modify the supply voltage so

Table 5-5. Chip Verification During the Design Phase.

PROCEDURE	TEST EQUIPMENT USED	ACTIONS
Design for testability		Provide internal test pads at the metallization level.
Failure detection and characterization	ATE	Determine margins of operation. Measure output waveforms.
Preparation		Set up operating conditions that disclose the failure or weakness (voltage, frequency short instruction sequence loop). Open package and remove passivation layer.
Localization	EBT	Record logic-state mappings within and outside the margins of operation.
Quantification	EBT	Measure waveforms at internal nodes (actual data).
Verification	EBT	Display actual data and nominal data (computer simulation).

ATE, automatic test equipment; EBT, electron beam testing.

that correct functioning is initially obtained and is then followed by the malfunction.

After location of the positions within the IC that exhibit disagreement between the logic state mappings in functioning and faulty modes, waveform measurements must be made at individual internal nodes.

The actual data are then compared in the final stage with the calculated nominal data to find the reason for the design weakness (Feuerbaum and Hernaut 1978).

Failure Analysis

The four stages of the procedure are listed in Table 5-6. Electron beam testing is only one of many analytical methods used in the clarification of technology-related failures. Some of the nondestructive processes available are described in the standard failure analysis row. Destructive material analyses, in the fourth row, are applied after the defect has been successfully localized (Doyle and Morris 1982; Marcus and Sheng 1983; Oppolzer and Huber 1983).

To localize the defect, it is sufficient in most cases to use qualitative electron

Table 5-6. Failure Analysis Procedure When Electron Beam Testing is Used for Defect Localization and Characterization.

PROCEDURE	TEST EQUIPMENT USED	ACTIONS
Failure detection and characterization	ATE	Locate failing IC within a printed circuit board. Determine margins of operation. Carry out electrical measurements at inputs and outputs.
Standard failure analysis		Inspect for gross defects (X-ray, hermeticity, etc.). Open package and inspect visually for defects (light microscope, scanning electron microscope).
Localization and electrical characterization	EBT	Record and compare logic-state mappings of failing and "golden" device. Locate defective area. Characterize local electrical malfunction.
Defect identification and material analysis		Local optical and SEM inspection. Further preparation to access lower parts of the chip. Preparation of IC for X-ray microprobe analysis, Auger microprobe analysis, transmission electron microscopy, etc.

beam testing methods only (Beall et al. 1980; Yuasa et al. 1980; Bertsche and Charles 1982; Wolcott 1982; Ura et al. 1982; Davidson 1983b; Baille et al. 1983b). In this stage it is usually not advisable to remove the passivation layers to avoid any alteration of the defect. If passivation layers are present, electron beam testing at low primary electron energies ($\leq$ 1 keV) allows voltage contrasts to be observed through the passivation layers (Kotorman 1980, 1983; Gorlich et al. 1984; Ura and Fujioka 1983; Uchikara and Ikada 1983; Todokoro et al. 1983).

Since often—especially in the case of external suppliers—circuit diagrams, layouts, and nominal logic states at internal nodes are not available, it is useful to compare the failing device with a corresponding good device ("golden device").

Integrated Test System

In examining the procedure used in chip verification and failure analysis, which involves initial failure detection and characterization by means of an automated test equipment, at first glance the integration of EBT and ATE seems an obvious step. For chip verification, however, the chip layout (point of measurement) data as well as logic and transient simulations are also required. An integrated system thus should resemble the configuration shown in Figure 5-12 (Kollensperger et al. 1984). Other integrated systems are described in Thangamuthu et al. (1982),

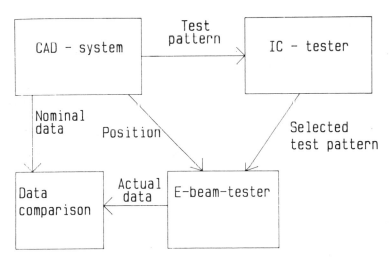

Figure 5-12.   Integrated test system for chip verification where the automated electron beam tester (Figure 5-8, Kollensperger et al. 1984) is prepared for connection to a CAD system and a pin of an IC-tester.

Walter et al. (1982), May et al. (1984), Kuji et al. (1984), and Hosoi et al. (1984).

It should be noted that the IC tester (ATE) is faster than the EBT by many orders of magnitude. A linkup to the CAD system must therefore be regarded as more important than one to the pattern generator (Kollensperger et al. 1984; Yamaguchi et al. 1984).

Automated Failure Localization

The testability of ICs is evaluated by the ease of control and observation from the external pins, which are accessible to the IC tester. The use of electron beam testing will greatly improve observability because the internal node signals are, in most cases, present on the upper-level metallization. If this is not secured, then optimal test points may be calculated (Breuer 1983). The faulty regions can also be delimited by sensitizing/desensitizing test vectors (Berger-Sabbatel and Courtois 1983), and a signal processing system, ROMUALD, used to evaluate the static voltage contrast. A comparable technique uses the FINDER system, which compares voltage contrast micrographs with artificial voltage contrast patterns generated with the aid of the CAD system (Kuji et al. 1984). A very effective system was developed in the Intel company (May et al. 1984): a signal processing system processes stroboscopic voltage patterns, which are generated in every clock phase of the program loop by a faulty operational state (e.g., with two different operating speeds). The good micrographs are then subtracted from the faulty ones—if no fault has yet occurred in this clock phase, then nothing can be seen on the processed micrograph. If the clock is increased stepwise, then the point of origin of the fault can be seen. As the clock continues counting, the fault propagation can be observed, and a fault tree can be set up.

The integrated test system described above (Kollensperger et al. 1984) was developed for the guided probe approach. If the test points are known, they can be located and checked automatically by transferring the coordinates from the CAD system.

A suggestion that points far into the future is that of automatic problem solving. Its key is a knowledge base containing everything that is known about the IC. Clauses help defaults to be automatically localized at the circuit block level (Berger-Sabbatel and Courtois 1983).

APPLICATIONS

In this section, a review is initially made of the literature on EBT applications. Two principal trends can be distinguished: small structures and low analog signals in RAMs and more complex circuits containing greater numbers of transistors for microprocessors. For this reason, papers published prior to 1980 are not listed here. For memories, which may be simply operated with short loops, three types of measurement are of principal importance: determination of the analog sense signal, measurement of the excess voltage for bootstrap nodes, and exact timing of RAS (row address strobe) and CAS (column address strobe) chains (Lukianoff et al. 1982; Burghard et al. 1983; Sato and Saito 1983; Yamaguchi et al. 1984; Sawada et al. 1984).

Analyses of microprocessors are often very difficult to describe because of their complexity (Crichton et al. 1980; Fazekas et al. 1981; Wolfgang et al. 1983; Garth and Nixon 1982; Davidson 1983a). This may explain why so many fewer analyses have been published on this subject than on RAMs. The potential applications of EBT must, however, be regarded as being more important for logic ICs, and especially for full custom ICs, than for RAMs.

For this reason, three examples of full custom logic are presented here. In the first example, three out of the five modes of operation, namely, logic-state mapping, frequency tracing, and frequency mapping, are used at the same point of an 8086 microprocessor. The second example shows the detection of spikes, also in the 8086 microprocessor, with the aid of logic-state mapping and waveform measurement. The third example shows a comparison between measurement and simulation of a telecommunications IC.

Signal Mapping and Tracing

Figure 5-13(a) shows a small section of an 8086 microprocessor. The stroboscopic voltage contrast method is used; so some interconnections appear dark and others bright. Figure 5-13(b) shows a logic-state mapping in area scan mode, where the topography of the interconnections is maintained. The bar pattern of the logic states on the different interconnections is completely visible, however, only when the interconnections run parallel. On short or vertical interconnections, only parts of the program loop are mapped.

Figure 5-13(c) shows a logic-state mapping that was recorded by line scan along the line b–b (indicated in Figure 5-13a). The complete program loop is now mapped on all lines. It comprises 48 clocks (lowest interconnection) and lasts 8 μs. The clock frequency f_{CL} in the case shown is 6 MHz.

Figure 5-13(d) shows the frequency mapping along the line scan a–a above the region 100 kHz to 3.5 MHz. The bar pair of the clock CL can be clearly seen below, to the right around 3.33 MHz and at a spacing of $\pm f_{IF}$ (50 kHz). Then in the same horizontal line the second and third harmonics of the preliminary electron probe frequency f_B detect f_s, with spacings of the bright bars of $2f_{IF}/2$ and $2f_{IF}/3$, respectively. The four interconnections, whose bit pattern is repeated after 48 clocks, produce the pattern at the upper left edge of the figure. It

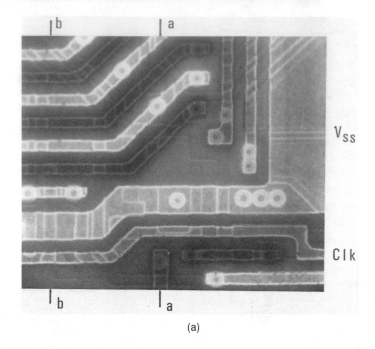

(a)

Figure 5-13. Signal mapping and tracing, demonstrated on a section of an 8086 microprocessor. (a) Stroboscopic voltage contrast. The letters a–a and b–b show the position of the line scans for figure parts (d) and (c), respectively. (b) Logic-state mapping with a duration of the program loop of 48 clocks or 8 μs, using an area scan. (c) Logic-state mapping under the same conditions as in part (b), using a line scan b–b. (d) Frequency mapping along the line a–a over a frequency range from 100 KHz to 3.5 MHz. (e) Frequency tracing at 3.38 MHz. Only the interconnections carrying the clock signal are visible. (f) Frequency tracing at 119 kHz. Note that interconnection lines that are covered with an oxide layer are visible in this mode (arrow).

contains one of the bar pairs of the fundamental frequency (69.3 + 50 = 119.3 kHz), one of the bar pairs of the second harmonics (138.6 + 50 = 188.6 kHz), and the bar pairs of the third and fourth harmonics.

Finally, Figure 5-13(e) and 5-13(f) show the frequency tracings for the frequencies 3.38 MHz (3.33 + 0.05 MHz) and 119 kHz (3.33 MHz/48 + 50kHz). It is worth noting in Figure 5-13(f) that in the middle of the micrograph (arrow), too, an interconnection is visible at the lower level that is covered by a SiO_2 layer about 0.8 μm thick. The brightness differences between the interconnections in the frequency tracing originate from the different proportions of high and low states in the program loop (see Figure 5-13b).

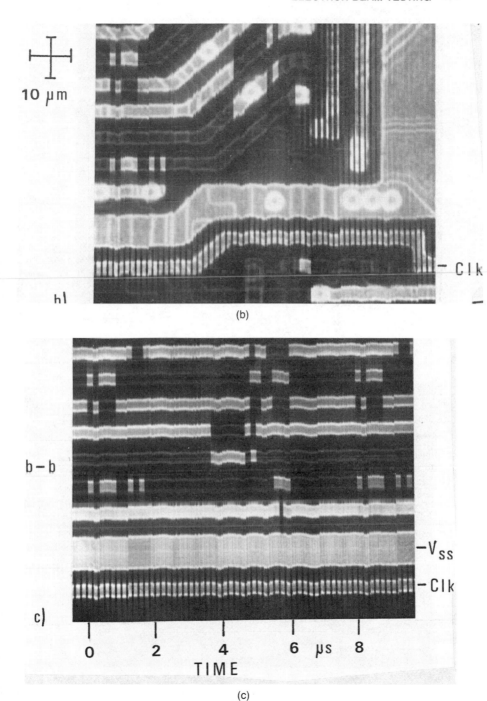

(b)

(c)

Figure 5-13.   *(continued)*

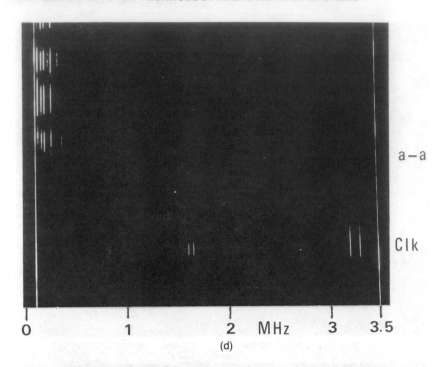

a−a

Clk

0   1   2   MHz   3   3.5

(d)

FREQUENCY

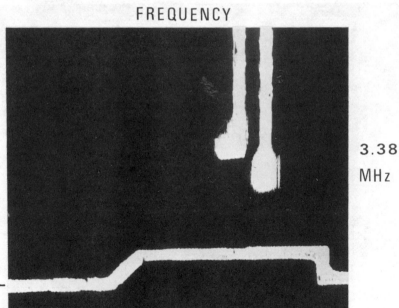

3.38
MHz

(e)

Figure 5-13.   *(continued)*

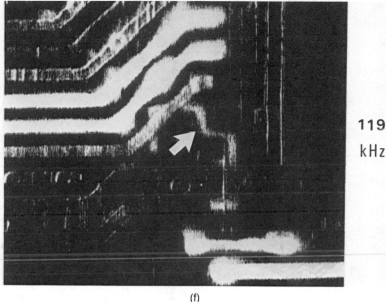

119
kHz

(f)

Figure 5-13. *(continued)*

Spike Detection

Figure 5-14 illustrates how voltage spikes can be detected and quantified. Figure parts (a) through (d) represent a series of logic-state mappings recorded in line scan mode. The time scales in the mappings were varied: Figure 5-14(a) shows the entire region of 48 clocks or 8 μs. Three spikes are already detectable at both upper narrow interconnections as fine dark bars, already clearly visible in Figure 5-14(b). The time scale covered in Figure 5-14(b) is from clock 11 to clock 21. At the fifth interconnection from above, a fine bright bar appears at the right edge of the micrograph (arrows), which is shown magnified in Figure 5-14(c) (from clocks 21–22) and magnified again in Figure 5-14(d). The width of the spikes is now 5 ns, and can be easily read. Clearly recognizable in Figure 5-14(b) are the rising and falling edges of the other signals, which are recorded in Figure 5-14(e) as waveforms. The spike has a half-width of 5.5 ns and leading and trailing pulse edge slopes of 2 V/ns and 1 V/ns, respectively.

Comparison with Computer Simulation

As described in the preceding chapter, propagation delay measurements require a comparison of measured actual data with simulated nominal data. Figure 5-15 shows this for three signals of a delay stage. Waveform *A* represents the

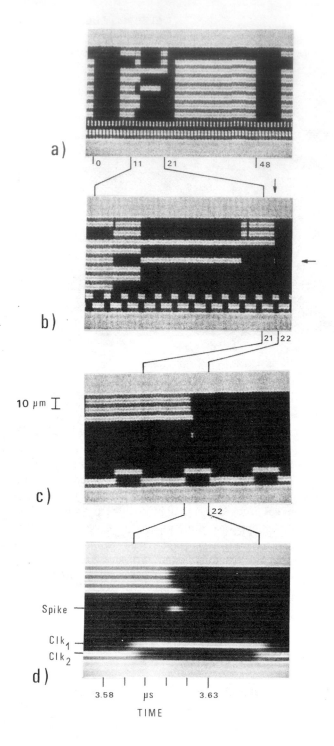

a)

b)

c)

10 μm

d)

Spike

Clk$_1$

Clk$_2$

3.58   μs   3.63

TIME

Figure 5-14.

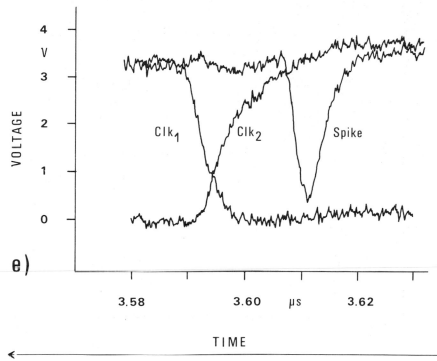

e)

Figure 5-14.   Spike detection within an 8086 microprocessor. (a–d) Series of logic state mappings (line scan) with different time scales. The numbers indicate the clock sequence within the program loop consisting of 48 clocks. (e) Spike detection within an 8086.

1 MHz clock. Waveform B activates, with its falling edge, an RC element. At a 2.5 V transition switching voltage of signal C, signal B goes to the high stage. A comparison between the measured and simulated waveforms shows that the rising edge of B is delayed by 55 ns. This behavior is due to parasitic capacitances, which make the capacitance C in the RC element larger than had been assumed.

CONCLUSIONS

Three decades after the discovery of voltage contrast (Oatley and Everhart 1957), electron beam testing has become a recognized and highly developed tool for chip verification and failure analysis. This can be seen, for example, from the fact that 20 contributions on this subject were presented at the 1984 electron beam testing symposium in Osaka, and 8 contributions at the 1985 SEM conference in Las Vegas. But it is also apparent because of the five commercial electron beam testers that have recently appeared on the market. Advances in this field due to the rapid development of ICs clearly foretell an extremely good future for such testing.

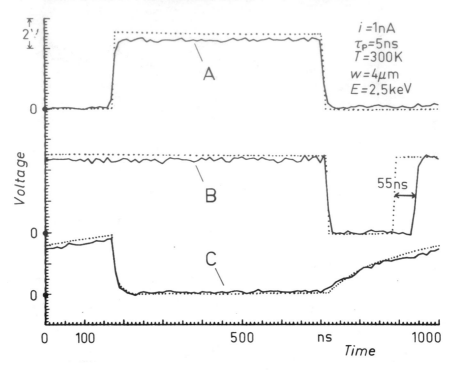

Figure 5-15.   Propagation delay measurement at a delay stage of an advanced CMOS IC. The dotted curves represent SPICE simulations.

These developments also show us the limitations of electron beam testing, however. Of particular significance is the fact that interconnection lines are becoming even narrower, with those on the 1 Mbit DRAM already down to between 1 and 1.2 μm in width. Closely linked to these developments is a reduction in the rise and fall times of the pulse edges and the need to carry out measurements at low acceleration voltages in order to avoid electron beam irradiation damage. Because all three effects lead to fewer primary electrons being present in the pulsed electron probe, either accuracy suffers or very long measurement times must be accepted as a tradeoff. A way out of this situation was shown in Plies and Otto (1985): by applying a mathematical deconvolution to the measured waveform, electron pulses having a pulse width 20 times longer may be used, which naturally also means 20 times as many secondary electrons available for signal processing.

The packaging of ICs can also be very inconvenient for internal measurements, especially when the chips are mounted face down in the package. The signal feed problems that occur in such cases make test preparation very inconvenient and time-consuming. As yet no solution is in sight for those ICs whose transistors are buried in silicon (three-dimensional transistors). But analysis can also be made much more difficult where information about chip layout and simulation data are lacking, since it is impossible to decide for complex ICs without reference data whether the circuit functions correctly at internal nodes or not.

ACKNOWLEDGMENTS

I would like to thank my colleagues W. Argyo, H. D. Brust, F. Fox, F. Gaal, J. Kolzer, J. Otto, and E. Plies for their helpful advice; L. Reidt for her photographic work; M. Hof for typing the manuscript; and H. J. Pfleiderer for his general support.

REFERENCES

SEM = Scanning Electron Microscopy, Proceedings of the annual conference, (1967–1977) ITTRI Chicago, (1978–1985) SEM Inc., AMF O'Hare, IL, U.S.A.

ISTFA = Proceedings of International Symposium for Testing and Failure Analysis.

Baille, G., Courtois, B., Laurent, J., and Rubat du Merac, K. C., "E-beam Testing: Prospective of Methodologies for Microprocessors and Memories Failure Analysis," *Journee D'Electronique*, Presses polytechniques romandes, Lausanne, Suisse, 299–311 (1983a).

Baille, G., et al., "Testing for Failure Analysis: New Tools and New Test Methods," *Proc.* 13th Fault-Tolerant Computing Symposium, Milano, Italy (1983b).

Balk, L. J., Feuerbaum, H. P., Kubalek, E., and Menzel, E., "Quantitative Voltage Contrast at High Frequencies in the SEM," *SEM* IV, pp. 615 624 (1976).

Beall, J. R., Wilson, D. D., and Echols, W. E., "SEM Techniques for the Isolation of Failures in Memory Circuits," *ISTFA*, pp. 1–8 (1980).

Berger-Sabbatel, G., and Courtois, B., "E-Beam Testing Strategies for VLSI," *Proc.* Sixth European Conference on Circuit Theory and Design, Sept. 6–8, Stuttgart, pp. 82–88 (1983).

Bertsche, K. J., and Charles, H. K., "The Practical Implementation of Voltage Contrast as a Diagnostic Tool," 20th Annual *Proc.* Reliability Physics, pp. 167 177 (1982).

Breuer, M. A., "Automatic Design for Testability Based on a Cost Measure," *IEEE 83 Autoteston*, pp. 138–142 (1983).

Broers, A. N., "Electron Sources for Scanning Electron Microscopy," *SEM* III, pp. 661–670 (1975).

Brust, H. D., Fox, F., and Wolfgang, E., "Frequency Tracing and Mapping: Novel Electron Beam Testing Methods," *Microcircuit Engineering 84*, Berlin, *Proc.* Int. Conf. Microlithography (1984).

Burghard, R. A., Brabenac, C. J., and Ward, C. M., "Applying 1 Nanosecond Stroboscopic Voltage Contrast with an Energy Analyzer to Dynamic Random Access Memory Measurements," *SEM* IV, pp. 1611–1618 (1983).

Crichton, G., Fazekas, P., Wolfgang, E., "Electron Beam Testing of Microprocessors," *Digest* of papers 1980 IEEE Test Conference, pp. 444–449 (1980).

Crostwait, D. L., and Ivy, F. W., "Voltage Contrast Methods for Semiconductor Device Failure Analysis," *SEM*, pp. 935–940 (1974).

Dallmann, A., Menzel, G., Weyl, R., and Fox, F., "Failure Analysis of ECL Memories by Means of Voltage Contrast Measurements and Advanced Preparation Techniques," *Proc.* International Reliability Symposium, Orlando, FL, 1985.

Davidson, S. M., "SEM Voltage Contrast Techniques for CMOS Device Assessment," *Inst. Phys. Conf. Ser.* No. 67: Section 9, Paper presented at Microsc. Semicond. Mater. Conf., Oxford, Mar. 21–23, pp. 415–420 (1983a).

Davidson, S. M., "Latch-up and Timing Failure Analysis of CMOS VLSI Using Electron Beam Techniques," *Proc.* IEEE International Reliability Physics Symposium, pp. 130–137 (1983b).

Doyle, E., Morris, B., *Failure Analysis Techniques—A Procedural Guide*, General Electric Company, Electronics Laboratory, Syracuse, NY, 1982.

Fazekas, P. et al., "On Wafer Defect Classification of LSI Circuits Using a Modified SEM," *SEM* I, pp. 801–806 (1978).

Fazekas, P., Feuerbaum, H. P., and Wolfgang, E., "Scanning Electron Beam Proves VLSI Chips," *Electronics, 14*, 105–112 (1981).

Fazekas, P., Wolfgang, E., Gerling, W., and Goser, K., "Electron Beam Testing Methodology," *Curriculum for Test Technology*, IEEE Catalog Number 83 CH 1978-6, pp. 26–31 (1983).

Feuerbaum, H. P., "VLSI Testing Using the Electron Probe," *SEM* I, pp. 285–296 (1979).

Feuerbaum, H. P., "Electron Beam Testing: Methods and Applications," *Scanning, 5*, 14–24 (1983).

Feuerbaum, H. P., and Hernaut, K., "Application of Electron Beam Measurements for Verification of Computer Simulations for VLSI Circuits," *SEM* I, pp. 795–799 (1978).

Feuerbaum, H., and Otto, J., "Improved SE Signal Processing for Waveform Measurements," *SEM* IV, pp. 1501–1505 (1982).

Fujioka, H., and Ura, K., "Electron Beam Blanking Systems," *Scanning, 5*, 3–13 (1983).

Fujioka, H., Nakamae, K., and Ura, K., "Function Testing of Bipolar ICs and LSI's with the Stroboscopic Scanning Electron Microscope," *IEEE J. Solid-State Circ., SC-15*, 177–183 (1980).

Fujioka, H., Tsujitake, M., and Ura, K., "Voltage Contrast Isolation by Frame-by-Frame Subtraction in the Scanning Electron Microscope," *SEM* III, pp. 1053–1060 (1982).

Garth, S. C. J., and Nixon, W. C., "Inspection of a TMS 9995 16 Bit Microprocessor VLSI Using Videotape Recording from a Scanning Electron Microscope," *Microcircuit Engineering 82*, Int. Conf. Microlithography, Grenoble, France, Oct. 5–8, 183–186 (1982).

Gonzales, A. J., and Powell, M. W., "Internal Waveform Measurements of the MOS Three Transistor, Dynamic RAM Using SEM Stroboscopic Techniques," *Technical Digest* of IEDM, New York, pp. 119–122 (1975).

Gopinath, A., "Estimate of Minimum Measurable Voltage in the SEM," *J. Phys. E: Sci. Instrum., 10*, 911–913 (1977).

Gorlich, S., Herrmann, K. D., and Kubalek, E., "Basic Investigations of Capacitive Coupling Voltage Contrast," *Microcircuit Engineering 84*, Berlin, *Proc.* Int. Conf. Microlithography (1984).

Goto, Y., et al., "Electron Beam Prober for LSI Testing with 100 ps Time Resolution," *Proc.* Int. Test Conf., Philadelphia, pp. 543–549 (1984).

Hosoi, H., Nikawa, K., and Ooiwa, T., "Electron Beam Tester Directly Connected with Multipurpose LSI Tester," *Proc.* Symposium on Electron Beam Testing, Nov. 9–10, Osaka, pp. 27–30 (1984).

Keery, W. J., Leedy, K. O., and Galloway, K. J., "Electron Beam Effects on Microelectronic Devices," *SEM* I, p. 507–514 (1976).

Kollensperger, P. et al., "Automated Electron Beam Testing of VLSI Circuits," *Proc.* Int. Test Conf., pp. 550–556 (1984).

Kotorman, L., "Noncharging Electron Beam Pulse Prober on FET Wafers," *SEM* IV, pp. 77–84 (1980).

Kotorman, L., "Low Energy Electron Microscopy Utilized in Dynamic Circuit Analysis and Failure Detection on LSI–VLSI Internal Circuits," *IEEE Trans. on Components, Hybrids, and Manuf. Techn.* CHMT-6, pp. 527–536 (1983).

Krampl, G., Gaal, F., van der Molen, O., and Fazekas, P., "Modern Techniques for Failure Localization and Identification in VLSI Circuits," *Journee D'Electronique*, Presses polytechniques romandes, 1015 Lausanne, Suisse, 271–281 (1983).

Kuji, N., Tamana, T., and Yano, T., "On-Line Stroboscopic SEM," *Proc.* Symposium on Electron Beam Testing, Nov. 9–10, Osaka, pp. 19–22 (1984).

Lin, Y. C., and Everhart, T. E., "Study on Voltage Contrast in SEM," *J. Vac. Sci. Technol., 16*, 1856–1860 (1979).

Lischke, B., Plies, E., and Schmitt, T., "Resolution Limits in Stroboscopic Electron Beam Instruments," *SEM* III, pp. 1177–1185 (1983).

Lukianoff, G. V., and Touw, T. R., "Voltage Coding: Temporal versus Spatial Frequencies," *SEM*, pp. 465–471 (1975).

Lukianoff, G. V., Wolcott, J. S., and Morissey, J. M., "Electron-Beam Testing of VLSI Dynamic RAMs" *Automatic Testing 82 & Test Instrumentation*, Paris, France, Sept. 21 23, pp. 117–130 (1982).

Mangir, T. M., "Interconnection Technology Issues for Testing and Reconfiguration of WSI," *Proc.* IEEE International Conference on Computer Design: VLSI in Computers, Oct. 8–11, Port Chester, NY, pp. 127 131 (1985).

Marcus, R. B., and Sheng, T. T., *Transmission Electron Microscopy of Silicon VLSI Circuits and Structures*, Wiley, New York (1983).

May, T. C., Scott, G. L., Meieran, E. S., Winer, P., and Rao, V. R., "Dynamic Fault Imaging of VLSI Random Logic Devices," *Proc.* IEEE Int. Reliability Physics Symposium, pp. 95–108 (1984).

Menzel, E., and Brunner, H., "Secondary Electron Analyzers for Voltage Measurements," *SEM* I, pp. 65–75 (1983).

Menzel, E., and Kubalek, E., "Electron Beam Test System for VLSI Circuit Inspection," *SEM* I, pp. 297–304 (1979a)

Menzel, E., and Kubalek, E., "Electron Beam Chopping Systems in the SEM," *SEM* I, pp. 305–318 (1979).

Menzel, E., and Kubalek, E., "Electron Beam Test Techniques for Integrated Circuits," *SEM* I, pp. 305–322 (1981).

Menzel, E., and Kubalek, E., "Fundamentals of Electron Beam Testing of Integrated Circuits," *Scanning*, 5, 103–122 (1983).

Menzel, E., and Kubalek, E., "Secondary Electron Detection Systems for Quantitative Voltage Measurements," *Scanning*, 5, 151–171 (1983).

Miyoshi, M., Ishikawa, M., and Okumura, K., "Effect of Electron Beam Testing on the Short Channel Metal Oxide Semiconductor Characteristics," *SEM* IV, pp. 1507–1514 (1982).

Mizuno, F., et al., "Failure Analysis of LSI Using a Scanning Low-Energy Electron Probe," *J. Vac. Sci. Technol.*, 19, 1019–1023 (1981).

Nakamura, H., and Sato, Y., "An Analysis of the Local Field Effect on Electron Probe Voltage Measurements," *SEM* III, p. 1187–1195 (1983).

Oatley, C. W., and Everhart, T. E., "The Examination of p n Junctions with the Scanning Electron Microscope," *J. Electron*, 2, 568–570 (1957).

Oppolzer, H., and Huber, V., "Details in Microstructure and Geometrical Configuration of Integrated Circuits Studied by Transmission Electron Microscopy," *Curriculum for Test Technology*, IEEE Catalog Number 83 CH 1978-6, pp. 461–466 (1983).

Ostrow, M., et al., "IC-Internal Electron Beam Logic State Analysis," *SEM* II, pp. 525–530 (1982).

Plies, E., and Otto, J., "Voltage Measurements inside ICs Using Mechanical and Electron Probes," *SEM* (1985).

Plows, G. S., and Nixon, W. C., "Stroboscopic Scanning Electron Microscopy," *J. Phys. E: Scient. Instr.*, 1, 595–600 (1968).

Sardo, A., and Vanzi, M., "Digital Beam Control for Fast Differential Voltage Contrast," *Scanning*, 6, 122–127 (1984).

Sato, M., and Saito, S., "Electron Beam Testing Techniques for Dynamic Memory," *Extended Abstracts* 15th Conf. Solid State Devices and Materials, Tokyo pp. 273–276 (1983).

Swada, K., Sakurai, T., Oktani, T., Isobe, M., and Iizuka, T., "EB Tester Evaluation of 256 kb CMOS SRAM," *Proc.* Symposium on Electron Beam Testing, Nov. 9–10, Osaka, pp. 35–38 (1984).

Seiler, H., "Secondary Electron Emission in the Scanning Electron Microscope," *J. Appl. Phys. 54*, R1–R18 (1983).

Shaver, D. C., "Techniques for Electron Beam Testing and Restructuring Integrated Circuits," *J. Vac. Sci. Technol. 19*, 1010–1013 (1981).

Swanson, L. W., Tuggle, D., and Li, J. Z., "The Role of Field Emission in Submicron Electron Beam Testing," *Thin Solid Films, 106*, 241–255 (1983).

Thangamuthu, K., Macari, M., and Cohen, S., "Automated Contactless Digital Test System for VLSI," *Digest* of papers, 1982 International Test Conference, Philadelphia, Nov. 15–18, pp. 634–638 (1982).

Todokoro, H., Fukuhara, S., and Komoda, T., "Stroboscopic SEM with 1 keV Electrons," *SEM* II, pp. 561–568 (1983).

Uchikara, Y., and Ikada, S., "Inspection of Microdevices over the Passivation Top Layer Surface Using the Low Voltage SEM," *Jpn. J. Appl. Phys., 22*, L645–L647 (1983).

Ura, K., and Fujioka, H., "Function-Testing of Passivated LSI's with Stroboscopic Scanning Electron Microscope," *Proc.* 14th Conf. Solid State Devices, Tokyo (1982); *Jpn. J. Appl. Phys., 22-1*, 541–546 (1983).

Ura, K., Fujioka, H., Nakamae, K., and Ishisaka, M., "Stroboscopic Observation of Passivated Microprocessor Chips by Scanning Electron Microscopy," *SEM* III, pp. 1061–1068 (1982).

Ura, K., Fujioka, H., and Nakamae, K., "Reduction of Local Field Effect on Voltage Contrast," *SEM* III, pp. 1075–1080 (1984).

Walter, M. J., Eldering, C. A., Krevis, K. M., and Haberer, J. R., "Internal Node Testing by Tester Aided Voltage Contrast," *Proc.* ISTFA 1982 Int. Symposium for Testing and Failure Analysis, San Jose, CA, Oct. 25–27, p. 156–161 (1982).

Wells, O. C., and Bremer, C. G., "Voltage Measurement in the SEM," *J. Phys. E., 1*, 902 (1968).

Wolcott, J. S., "Electron Beam Testing for Verification of Voltage Distribution on VLSI Circuits," *ISTFA* pp. 149–155 (1982).

Wolfgang, E., "Electron Beam Testing: Problems in Practice," *Scanning, 5*, 71–83 (1983a).

Wolfgang, E., "Tracing of Design Weakness in VLSI Circuits Using the Electron Probe," *Inst. Phys. Conf. Ser.* No. 67: Section 9, Paper presented at Microsc. Semicond. Mater. Conf., Oxford, Mar. 21–23, pp. 407–414 (1983b).

Wolfgang, E., Fazekas, P., Otto, J., and Crichton, G., "Electron Beam Testing of Microprocessors," *Hardware and Software Concepts in VLSI*, Guy Rabbat, Editor, Van Nostrand Reinhold, New York, 1983, pp. 296–329.

Yamaguchi, K., Kanetami, K., Todokoro, H., and Akimoto, K., "Waveform-Observation on Bipolar RAM with Electron Beam LSI Tester," *Proc.* Symposium on Electron Beam Testing, Nov. 9–10, Osaka, pp. 23–26 (1984).

Yuasa, Y., Fujita, M., and Manabe, N., "SEM Stroboscopic Techniques to Failure Analysis of LSI's," *ISTFA*, pp. 9–14 (1980).

6. Metallization for Very Large-Scale Integrated Circuits

P. B. Ghate

Texas Instruments Incorporated

Progress in patterning technologies and computer-aided circuit designs has brought us to the threshold of very large scale integrated (VLSI) circuits with a million or more devices to be integrated on a silicon chip (Block 1981; Heilmeier 1984; Asai 1984). The ever increasing demand for high performance (speed) and high density (large number of components) circuits is continuously met by new circuit design, new device structures fabricated with innovative variations and combinations of the basic bipolar and field effect transistor (FET) technologies, and shrinking the geometries. Some of the device technologies, with their acronyms, considered suitable for VLSI applications are listed: I^2L (integrated injection logic), STL (Schottky transistor logic), ISL (integrated injection Schottky logic), NMOS (n-channel metal oxide semiconductor) CMOS (complementary MOS), MESFET (metal silicon FET), SOS (silicon on sapphire), SOI (silicon on insulator), and so on. New device structures and process flows continue to emerge. Once a device technology is selected for VLSI applications, the metallization must provide silicon-to-metal contacts, gate material if needed, interconnections (i.e., conducting paths between devices on the chip for charge transfer), and compatible bond pads for packaging and assembly operations. Because integration of a large number of components on a chip consumes a relatively large chip area for interconnections, multilevel (conductor/insulator/conductor/ . . .) interconnections are required to minimize signal delays due to long interconnection paths, to improve packaging density of components by reducing chip size, and to achieve circuit performance goals (Ghate et al. 1977; Ghate 1982a,b,c).

The impact of interconnection resistance on circuit performance is illustrated in Figure 6-1. Here, the access time versus feature size for a 4K static memory with 20 $\Omega/\square$ polysilicon gate and 1 $\Omega/\square$ refractory gate material shows that

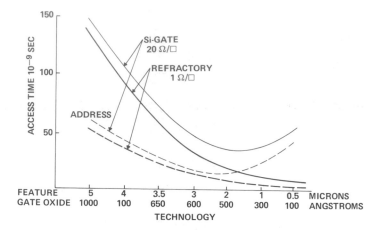

Figure 6-1. Impact of interconnection resistance on the performance of 4K-bit static memory.

polysilicon gate technology begins to degrade the access time for an approximately 2 μm device feature size solely because of high interconnection resistance. A larger chip area with a high packing density of smaller geometry appears to be the normal trend for both bipolar and MOS technologies (Figure 6-2).

The metallization requirements vary with device technologies, and no unique metallization solution compatible with all VLSI circuits can be defined. However, all VLSI metallizations share some common concerns, such as forming Si/metal contacts to shallow junction devices, Schottky contacts, barrier layers, deposition and definition of gate interconnections, step coverage, metal and insulator layers for multilevel interconnections, packaging considerations, and reliability. All these concerns are highly interdependent, and process optimization is sought within the framework of VLSI circuit performance goals.

In this chapter, thin film applications in the fabrication of contacts and interconnections for silicon VLSI circuits and some of the failure mechanisms limiting the reliability of interconnections will be reviewed. The order of topics presented in the following pages follows the normal sequence of process steps, and the discussion will be limited to highlight the problems and solutions. Later, future trends in VLSI metallization are reviewed.

METAL/SEMICONDUCTOR CONTACTS

In VLSI processing, three types of contacts are encountered: (1) metal/semiconductor contacts, (2) metal/insulator contacts (e.g., gate electrodes, capacitors), and (3) metal/metal contacts (e.g., vias and bond pads). Formation of reproducible metal/semiconductor contacts is a first major step in VLSI metallization. In this

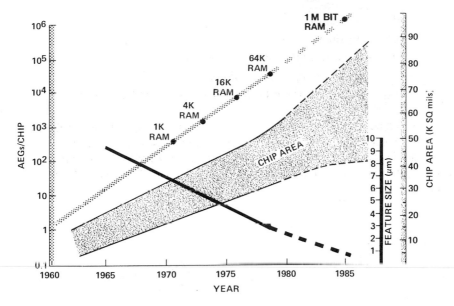

Figure 6-2.   Trend in chip size and minimum feature size for silicon integrated circuits.

section, barrier heights of metal/semiconductor contacts, current–voltage characteristics of metal/semiconductor contacts, and ohmic contacts are briefly reviewed.

The Schottky Barrier

When a metal makes an intimate contact with a semiconductor, the Fermi levels in the two materials must be coincident at thermal equilibrium, and as a result a potential barrier is formed between the two. Current transport in metal/semiconductor barriers is mainly due to majority carriers, in contrast to p-n junctions, where the minority carriers are responsible. An expression for the current–voltage characteristic developed in the framework of the thermionic emission (TE) model is as follows:

$$J = A^*T^2 \exp\left(-q\phi_B/kT\right) \cdot \exp\left(q(V + \phi_B)/kT\right) \text{ for } V > 3kT/q$$
$$= J_{ST} \exp\left(qV/nkT\right)$$

where:

A^* = effective Richardson constant = 120 amp/cm^2/$^\circ$K^2

ϕB = barrier height

ϕB = lowering of the barrier height due to image forces

V = bias voltage

k = Boltzmann constant

n = parameter in the range of 1 to 1.3

T = absolute temperature in degrees Kelvin

J_{ST} = saturation current for a reverse bias = $A*T^2 \exp(q\phi_B/kT)$

The TE model satisfactorily explains the current–voltage characteristics for a lightly doped semiconductor (10^{18}cm^{-3}) in contact with a metal. The specific contact resistance R_c (expressed in ohms-cm^2) for a Schottky barrier diode (SBD) is found to be temperature-dependent, as can be seen from the following relationship:

$$R_c = \frac{\partial V}{\partial j} \text{ as } j \to \text{o}$$

$$= \frac{k}{qA^{**}T} \exp(q\phi_B/kT)$$

For a disc contact of radius r, the spreading resistance R_s is given by:

$$R_s = p/4r$$

where p is the resistivity of the semiconductor. Hence for small contacts, the voltage drop across an SBD is governed not only by R_c but also by R_s.

In many semiconductor device applications, metal contacts are made to semiconductors whose doping density lies in the range of 10^{18} to 10^{20} cm^{-3}. As the doping density starts increasing, the depletion width W starts decreasing as (doping density)$^{-0.5}$. For highly doped semiconductors (10^{20} cm^{-3}) the depletion width W drops to a range of 1 to 3 nm. For metal–semiconductor contacts with such small depletion widths, the current transport across the interface is dominated by quantum mechanical tunneling through the barrier rather than by thermionic emission. For intermediate doping density, the current voltage characteristics are developed in the framework of a thermionic field emission (TFE). It is sufficient to note here that once the current transport is dominated by tunneling, the specific contact resistance is primarily governed by the shape (width and height) of the barrier, and is almost independent of temperature.

Experimentally measured barrier heights for different metals on n-type silicon are presented in Table 6-1.

The SBD is a majority carrier device with inherent fast response in contrast to a p-n junction, where the current is transported by minority carriers and the response time is longer. Because of the fast response of the SBD, it is used to shunt excess-base current through the diode as a majority-carrier current in a

Table 6-1. Typical Schottky Barrier Heights for Metal–n-Type Silicon.

METAL	ϕN (eV)
Al	0.65
Au	0.79–0.80
Mo	0.57–0.59
Ni	0.66
Pd	0.75
Pt	0.80–0.87
W	0.65–0.67

Note: $\phi_{BN} + \phi_{Bp} = E_g = 1.1$ eV for silicon.

clamped transistor. The measured saturation time can be reduced to about 10% of that of the original transistor, and it is on the order of a few nanoseconds. SBDs are widely used in bipolar ICs to improve circuit performance.

Ohmic Contacts.   For most applications, a low impedance ohmic contact is desirable. Specific contact resistance R_c is defined as the reciprocal of the derivative of current density with respect to voltage and is expressed in units of Ω-cm^2. It is an important figure of merit for transport characteristics across a metal/semiconductor interface. For an ohmic contact, the current–voltage characteristic is linear. In practice, a contact is considered ohmic if the voltage drop across the device is small and does not affect the device characteristics significantly.

In general, ohmic contacts to silicon are achieved by establishing metal contacts to heavily doped n (or p) regions. In such a metal/semiconductor contact, the depletion width in the heavily doped semiconductor region is small, and the current transport takes place by a tunneling mechanism. For a given metal–silicon system, specific contact resistance depends on surface concentration [$\log R_c$ varies as $(N_D)^{-0.5}$, where N_D is the surface concentration]. Under simplifying assumptions (neglecting current crowding, perimeter effects, and so on), the contact resistance r_c for a contact of area A is given by $r_c = R_c A$. Typical test structures used (Blair and Ghate 1977) for contact resistance measurements and the schematic of the series combination of contact resistance and diffused resistors are shown in Figure 6-3. This methodology is similar to the one proposed by Berger (1972). Experimental data (Ghate 1982a,b,c) for Si/platinum silicide and Si/aluminum contacts are presented in Figure 6-4.

HILLOCK FORMATION — DUE TO COMPRESSIVE STRESS

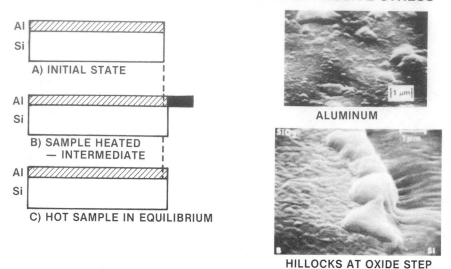

Figure 6-3. Test structures and schematic of the series combination of contact resistance and diffused resistors. (A) Initial state. (B) Sample heated—intermediate. (C) Hot sample in equilibrium.

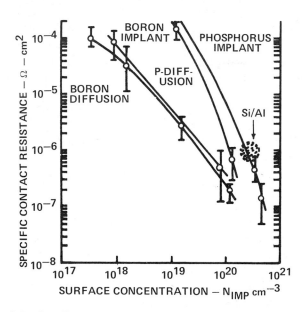

Figure 6-4. Specific contact resistance data for Si/PtSi and Si/Al contacts.

FABRICATION OF METAL/SEMICONDUCTOR CONTACTS

Preparation of clean surfaces is a prerequisite in the formation of contacts. Several techniques, such as ultrasonic cleaning, chemical etching, and others prior to loading the substrates in film deposition equipment, as well as plasma glow discharges, sputter etching, ion milling, and so on inside the film deposition equipment prior to film deposition, have been employed. In general, metal films are vacuum-deposited by several methods, such as electron beam (EB) evaporation, RF sputtering, and dc-magnetron sputtering onto clean semiconductor surfaces. The metallized substrates, after appropriate patterning steps, are heating to a temperature slightly below the metal–silicon eutectic temperature to form the contacts. During this annealing step, normally referred to as contact sintering, a thin layer of metal silicide or silicon-rich metal compound is formed due to interdiffusion. This diffused layer has a metal-like behavior, and the metal–silicon interface is no longer sharp. However, the depletion width in the semiconductor at the contact remains thin enough for the electrons to tunnel through, and the primary objective of forming ohmic contacts is achieved.

Si/Al Contacts

Aluminum has been the most widely used metal for contacts and interconnections in both bipolar and MOS ICs. The ohmic contacts to highly doped p+ and n+ type silicon contacts are easily made by removing or consuming the native silicon dioxide, permitting the deposited metal film to come into intimate contact with the silicon surface. This is most commonly effected by annealing the Al metallized substrate at some temperature between 450°C and 525°C for a period of 15 to 30 min in N_2 or H_2 ambience. Typical specific contact resistance is on the order of 1 to 10 $\times$ 10^{-6} Ω-cm^2.

Erosion of Si from the contact windows has been a common observation for ICs fabricated with Al metallization, as Si from the contact window diffuses into Al to satisfy the solid solubility at the sintering temperature; however, this has not been a major yield-limiting factor in the fabrication of ICs of emitter junction depths > 1 μm. Si atoms are found to migrate as far as 24 to 40 μm, and the depth of erosion in contacts (referred to as contact pitting or spiking) varies from 100 to 500 nm. A typical example of Si/Al contacts is illustrated in Figure 6-5. It is generally observed that Si erosion does not proceed uniformly downward. Possibly because of a thin oxide layer or some barrier layer inhibiting interfacial interaction, the erosion pattern spreads horizontally in the shape of a triangle for (111) substrates, and as rectangular pits on (100) substrates; and on cooling, the dissolved Si precipitates in the contact windows and on the SiO_2 surface.

Use of Al–Si films has been proposed as a solution to minimize contact pitting (Figure 6-6). The solid solubility of Si in Al at the contact sintering temperature

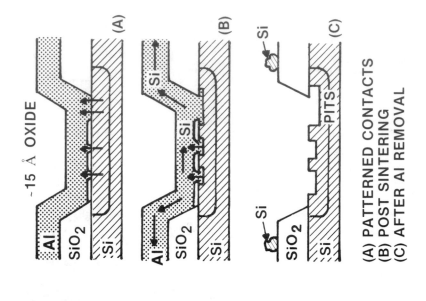

(A) PATTERNED CONTACTS
(B) POST SINTERING
(C) AFTER Al REMOVAL

(111) SILICON

(100) SILICON

Figure 6-5.   Erosion of silicon from (111) and (100) substrates used for Si/Al contacts.

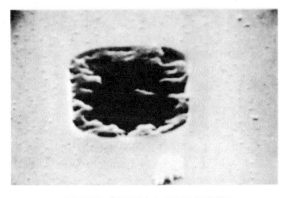

AFTER METAL STRIPPED   (a)

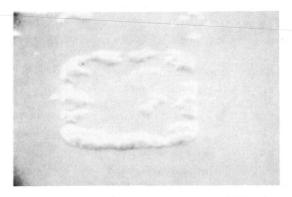

METAL AND OXIDE STRIPPED   (b)

Figure 6-6. Si/Al contacts. (a) Metal removed from contact area: (b) metal and oxide removed from contact area. Several temperature excursions will result in a regrown Si-layer. The thickness of the regrown silicon layer will affect contact resistance.

is approximately 0.6 to 1.0 wt%. Several methods, such as dual e-gun, RF induction heated source (IN-source), or dc-magnetron sputtering, have been employed to deposit Al–Si films. Also, Al–Cu–Si ternary alloy films have been used to minimize contact pitting on shallow ($<$ 1 μm) junction devices and to utilize the beneficial effects of Cu for improving electromigration resistance of interconnections. Excess Si in Al–Si or Al–Cu–Si films is known to precipitate on Si and SiO_2 surfaces. The Si precipitates in contact windows affect the characteristics of rectifying (Schottky Si/M) contacts (Card 1975). Also, these precipitates cause variation in contact resistance (Ghate 1981a). The Si/Al-alloy interface has a specific contact resistance of 0.36 to 1.45 $\times$ $10^{-6}\Omega$ $-cm^2$. Use of Al–Si films does not seem to pose major problems for ICs with 3 $\times$ 3 μm^2

contacts with junction depths > 350 nm; however, for contacts $1 \times 1 \ \mu m^2$ or smaller in size with junction depths on the order of 100 to 300 nm, application of Al–Si films is less desirable.

Si/PtSi Contacts

Advances in the design of high performance bipolar (e.g., Schottky clamped transistor-transistor logic) circuits have demanded not only ohmic contacts to n^+ and p+ type Si areas on the chip, but also Schottky contacts to high resistivity (1 to 10 Ω-cm) n-type epitaxial layers. A number of near noble (Pt, Pd, Ni) and refractory (Cr, Ti, Mo, W) metals have been explored for contact formation. Near noble silicides are readily formed below 600°C, and the refractory silicides are formed at temperatures in excess of 600°C. Furthermore, consumption of Si from the contact windows during metal silicide layer formation is controllable by a preselected metal layer thickness. The silicide layer formation process offers an alternate approach for eliminating contact pitting encountered with Si/Al contacts. In general, a suitable barrier layer between the metal silicide contacts and Al film conductor metallization is required for reliable operation.

   Among the several metal silicides explored for bipolar circuit applications, platinum silicide layers have found widespread use in forming Schottky and low-resistance ohmic contacts (Lepselter and Andrews 1969). Platinum silicide layers are formed by vacuum deposition of 20 to 100 nm thick Pt films onto Si substrate with clean surfaces in contact windows, and by annealing the metallized substrates in N_2 and/or air ambience at temperature in the 400 to 600°C range for a period of 10 to 30 min. The native oxide layer on a chemically cleaned surface is on the order of 1 to 3 nm, and it has not been a limiting factor in platinum silicide formation; however, in-situ sputter cleaning of Si contacts is preferred. During contact sintering, Pt reacts with Si to form Pt_2Si, and once all the Pt layer is converted to Pt_2Si, growth of PtSi starts at the Si interface, and the PtSi layer formation is completed. When hot substrates are pulled into air (or $N_2 + O_2$ ambience), Si atoms on the PtSi surface react with O_2 to form SiO_2. It is the oxide layer that protects PtSi from dissolution during subsequent stripping of unreacted platinum on the oxide surface.

   Formation of a stable and reproducible Si/PtSi interface is one of the critical steps in IC fabrication. Electrical characteristics of the Si/PtSi interface (Schottky or ohmic contacts) depend on the dopant concentration and surface cleanliness (presence or absence of oxide layers and impurities) prior to silicide formation. For example, in some bipolar applications, the forward voltage drop V_F of a PtSi Schottky diode at some prescribed current level is one of the critical circuit parameters. An appropriate diode area is designed on the chip. When contacts are clean, Pt films are deposited under good vacuum conditions, and silicide layers are formed under controlled ambience, sharp Si/PtSi interfaces of the desired electrical characteristics are realized. On the other hand, an oxide layer (or some barrier layer) on the order of 5 nm or greater in the contact windows interferes with the formation of a uniform platinum silicide layer. A 5-nm oxide layer tends to be discontinuous, and Pt reacts with Si through the discontinuities. The final Si/silicide interface is rather rough (less sharp), and the effective area of the contact deviates considerably from the designed value. Impurities at the Si/Pt interface interfere in the transport of Pt and Si atoms to form Pt_2Si first and then PtSi. Note that Pt_2Si and PtSi have orthorhombic and tetragonal crystal structures, respectively. It is a general observation that impurities interfere in solid-state transformations, and it is to be anticipated that impurities at the Si surface will inhibit the transformation of Pt_2Si into PtSi.

In a Si/PtSi contact Study (Ghate 1982a,b,c), the distribution of V_F (the forward voltage drop) at 10 μA of more than one hundred 25 × 25 μm^2 area Schottky diodes uniformly spaced on 7.5-cm-diameter (100 n-type substrates was monitored. Data analysis showed three types of distribution: (1) V_F distributed around 450 mV, (2) V_F distributed around 425 mV, and (3) a random distribution of V_F ranging from 350 to 460 mV. The first two distributions were reconciled with the barrier heights of PtSi (0.85 V) and Pt_2Si (0.82 V), respectively; and the third distribution was considered to result from a mixture of PtSi and Pt_2Si with a rough (nonsmooth) Si/silicide interface. The composition of these three types of silicide layers as determined by Auger electron spectroscopy in conjunction with depth profiling is presented in Figure 6-7. The profiles (a) and (b) correspond to PtSi and Pt_2Si films, respectively. The profile (c) clearly shows a rather rough (not sharp) Si/silicide interface, a result that is consistent with the transmission electron microscopy observations of exceedingly rough Si/PtSi interfaces in the cross-sectional specimens (Foll et al. 1981).

Formation of low-resistance ohmic contacts to n+ and p+ contacts on the substrate is accomplished by following a process sequence similar to the formation of Si/PtSi Schottky diodes. In fact, if the IC design requires PtSi Schottky diodes, these Schottky diodes and ohmic contacts are effected in a single step. The contact resistance of the Si/PtSi interface is a critical parameter and is monitored with test structures described earlier. Scaled versions of test structure have been used in the process development of VLSI circuits with 1.25 μm feature size. In Figure 6-8, we present a Schottky test structure (Shockley 1964; Yu 1970) with 1.25 × 7.6 μm^2 (0.05 × 0.3 mil^2) contacts fabricated with electron-beam patterning techniques. The PtSi contacts are made to an As implanted Si surface (20 Ω/□ with an approximately 0.35 μm junction depth). The measured contact resistance is on the order of 0.5 and corresponds to a specific contact resistance of $0.5 × 10^{-7}$ Ω-cm^2.

As the junction depths of devices suitable for VLSI applications approach the 100 to 300 nm range, the surface preparation of Si in contact windows and the silicide layer thickness require stringent process controls to minimize Si consumption from the contact areas. Recently, deposition of alternate thin layers of Pt and Si has been proposed to fabricate Si/PtSi, ohmic contacts, and Schottky diodes with minimum Si consumption in the contact windows (Tsaur et al. 1981).

There is considerable development activity to tailor the barrier heights of

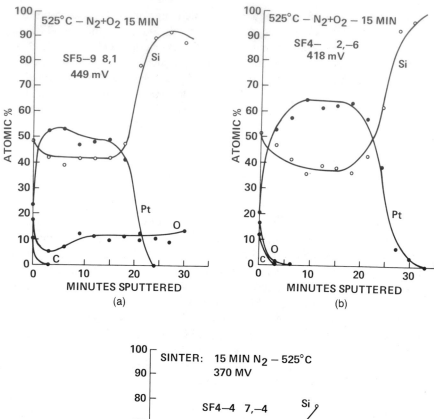

Figure 6-7. Composition depth profiles of platinum silicide layers in three different Schottky contacts: (a) V_F = 449 mV, (b) V_F = 418 mV, and (c) V_F = 370 mV.

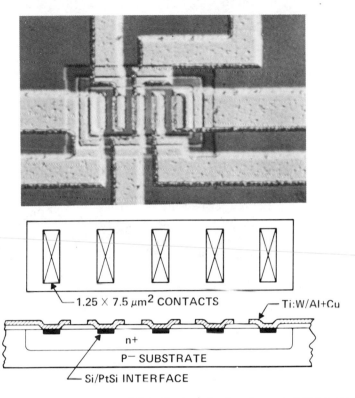

Figure 6-8. Electron beam-patterned Schottky test structure for contact resistance measurements (contact area 1.25 × 7.62 μm²).

Schottky diodes. Low energy ion implantation (antimony) has been employed to adjust charge density in surface layers adjacent to the metal–silicon interface and thereby increase the effective barrier height of Ni–Si diodes on p type Si substrates (Shannon 1974). Also, adjusting the barrier heights by adjusting the Pt and Ni composition in Pt_xNi_y films on n-type Si substrates has been attempted (Terry and Saltich 1976). Variation in barrier heights as a function of the variable composition of Pt_xSi_y films and the stability of such films obtained by code position of Pt and Si atoms have been studied (Eizenberg et al. 1981).

It may be noted here that STL VLSI circuits require two types of Schottky diodes on the same chip (Sloan 1979). The circuit configuration is such that the signal voltage swing ΔV related to the change in logic state from 1 to 0 corresponds to the difference in forward voltage drops of these two Schottky diodes (Figure 6-9). Under simplifying assumptions, if the emission coefficients and areas of the diodes are set equal to each other, then ΔV becomes simply the difference between the barrier heights. For example, if a PtSi Schottky diode with a barrier height of 850 mV is selected as one of the diodes, and if 200 mV

* SCHOTTKY DIODES FOR STL

Figure 6-9.   Schematic of voltage swing in STL circuits using two types of Schottky diodes.

is the required voltage swing, then a second type of diode with 650-mV barrier height needs to be incorporated in the circuit. As the circuit is normally expected to operate over a prescribed temperature range, the *I-V* characteristics of these two types of diodes need to track with each other. The STL technology mentioned here is one of the many examples where VLSI circuit designs will continue to challenge process development in tailoring the properties of Si/metal contacts.

BARRIER LAYER

In microelectronics, multilayer film applications continue to be explored for meeting device fabrication and performance goals, and the concept of a diffusion barrier in VLSI contacts and interconnection technology is not new (Lepselter 1966; Cunningham 1965; Cunningham et al. 1970). Simply stated, a layer of material *B,* interposed between layers *A* and *C,* is a barrier layer if it inhibits the mixing of *A* and *C* and thereby permits the exploitation of the desirable properties of *A* and *C* (Figure 6-10). During device fabrication and later in the device use condition, some intermixing (interdiffusion) at the *A/B* and *B/C* interfaces is to be expected. The choice of barrier layer is dictated by the desired physical properties. The "effectiveness of a barrier" can be measured only under prescribed conditions.

For ICs fabricated with PtSi contacts, the subsequent film interconnection technology must maintain the integrity of the Si/PtSi interface and be compatible with chip packaging technology. Bilayer and trilayer metal films with gold as the primary conductor have been used for interconnections; for example, Mo/Au, W/Au, Cr/Au, Ti:W/Au, Ti/Pt/Au, and so on. (Notation: *X/Y* denotes a two-layered structure composed of film *X* deposited first closest to the silicon substrate, followed by a second layer film *Y* deposited on top of *X*.) Since Au adhesion to SiO_2 is poor, film layers of Mo, W, Cr, and Ti:W (on the order of 150 to 250 nm) provide the necessary adhesion and also serve as a diffusion barrier between the contacts and Au. In the case of Ti/Pt, Ti (50 to 100 nm) provides the adhesion, and the composite Ti/Pt layer (with Pt = 150 to 350 nm) serves as a diffusion barrier—that is, it acts as a diffusion barrier for transport

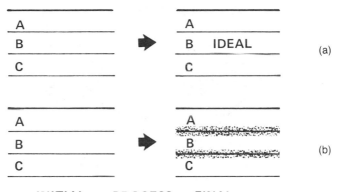

Figure 6-10.   Schematic of a barrier layer *B* interposed between layers *A* and *C*. (a) Ideal barrier with no intermixing. (b) A realistic barrier layer with slight intermixing at the interfaces.

of Au not only to the Si/PtSi interface but also to the SiO_2/Ti interface; a Au-free SiO_2/Ti interface is a key factor in maintaining the integrity of the packaged chip where Au wires are attached to the Au bond pads. Also note that this composite Ti/Pt diffusion barrier is compatible with beam lead technology where 12 to 25 μm thick Au is electroplated on Au bond pads to form the beam leads.

Multilayer films such as Cr/Ag/Au and Cr/Cu/Au have been used on top of the Al bond pads in flip-chip technology where Pb–Sn solder or Au bumps are to be used in subsequent chip attachment operations (Chen et al. 1981).

There are several classes of diffusion barriers wherein layers *C* and *B* interdiffuse to form compounds (Nicolet 1978; Nicolet and Bartur 1981). The growth kinetics of these compounds determine the useful lifetime of the diffusion barriers. In order to eliminate contact pitting with Al metallization, a layer of Ti under Al is used to provide the contacts to Si and also to serve as a diffusion barrier between Si and Al. Here Ti and Al will interdiffuse to form compounds; and when the Ti layer is consumed, Si/Al contacts are established, and the effectiveness of the barrier is lost.

In the early seventies, a pseudoalloy of titanium and tungsten, Ti:W (10:90 wt%), was introduced as a barrier layer in the PtSi/Ti:W/Au metallization (Cunningham et al. 1970). The Ti:W layer was process-compatible, displayed superior corrosion resistance, and provided the necessary adhesion and barrier properties. The use of the Ti:W layer offered process simplification with the replacement of sequential sputter depositions of two layers such as Ti/W with a single deposition from one target.

A cost-effective alternative to PtSi/Ti:W/Au metallization was to replace Au and Al. The new metallization scheme offered one of the solutions for eliminating contact pitting observed with Al metallization (Figure 6-11). The effectiveness of the Ti:W barrier layer (100 to 200 nm) and application of PtSi/Ti:W/Al

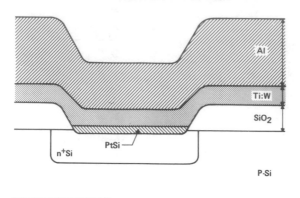

BARRIER LAYER Ti:W IS EXPECTED TO MAINTAIN THE INTEGRITY OF
PtSi CONTACTS (Si/PtSi INTERFACE) THROUGHOUT THE USEFUL LIFE OF
THE DEVICE

Figure 6-11.   Schematic cross section of PtSi/Ti:W/Al metallization to ICs.

metallization in IC fabrication have been reported (Ting and Wittmar 1982).
Also, the Ti:W layer has been successfully applied as a barrier layer between
PtSi contacts and Al–Cu interconnection metallization on a VLSI circuit with
junction depths on the order of 300 to 350 nm.

Since the pseudoalloy Ti:W film is sputter-deposited, the film composition
depends not only on the target composition but also on film deposition conditions.
The vacuum ambience has profound effects on the barrier properties of Ti:W
films, and deposition parameters have to be optimized.

Recently, reactively sputtered TiN films have received considerable attention
as diffusion barriers in VLSI metallization. DC magnetron sputtered films with
resistivity as low as 75 $\mu\Omega$-cm and their process compatibility with a lift-off
technique have been reported. However, TiN films have not found widespread
use as diffusion barriers in multilayered films because of the process complexity
of etching individual layers. The TiN films appear to be effective diffusion
barriers between silicided contacts and conductors up to 600°C; their application
as effective diffusion barriers in VLSI circuits has yet to be demonstrated (Nicolet
1978; Ho 1982a).

Diffusion barriers will continue to be used in VLSI metallization schemes.
Development of a suitable barrier layer will depend on the availability of data
on interdiffusion and growth kinetics of compounds in multilayered films. This
is a very fruitful area of research, and more work is needed (Nowicki and Nicolet
1982). As there is no unique VLSI metallization, a unique barrier layer does
not exist, but a cost-effective solution for the barrier will continue to be developed
and implemented.

INTERCONNECTIONS

Single-Level Interconnections

A large number of metal films in single, bilayer, and trilayer configurations have been explored for IC interconnections:

Single layer: Al, Al–Si, Al–Cu, Al–Si–Cu, Cr, W, etc.
Bilayer: Mo/Au, Cr/Au, Ti:W/Au, Ti:W/Au
Trilayer: Ti/Pd/Au, Ti/Pt/Au, Cr/Ag/Au

(Note: $X–Y$ denotes an alloy film of elements X and Y of a composition to be specified later.)

Any metallization system to be considered for IC interconnections has to satisfy certain minimum requirements, some of which are listed below:

1. Good ohmic contacts to both n and p type silicon.
2. Adaptability to practical methods of film deposition to provide low resistive interconnection paths ($3–10$ $\mu\Omega$-cm).
3. Adherence to both silicon and silicon dioxide.
4. Patternability.
5. Compatibility with bonding and packaging techniques.
6. Reliability under normal operating conditions.
7. Cost-effectiveness.

Aluminum films have found widespread use in the IC industry because of their process compatibility and cost-effectiveness. Aluminum has a high conductivity, and films can be vacuum-deposited within 5 to 10% of bulk resistivity (2.76 $\mu\Omega$-cm). A film of 1 μm nominal thickness has a sheet resistance of 0.03 $\Omega/\square$, and interconnection resistances (as compared to contact resistances) are generally low.

Bilayer and trilayer films with Au as the principal conductor are also used for interconnections where process and reliability considerations outweigh cost factors. Since Au films adhere poorly to SiO_2, refractory metals such as Mo, Ti:W, and Ti/Pt provide the necessary adhesion to the SiO_2 and act as a barrier layer between the Si contact areas on gold films. Metal systems other than Al and Au as primary conductors have been used in special applications; however, these two metals satisfy most of the IC interconnection needs.

Aluminum has been the workhorse—up to 90% or more of the IC interconnections are accomplished with vacuum-deposited Al films. As the demand for high-performance circuits continued to grow and device processing technologies advanced, junction depths began to approach the 1-μm range, and Al–Si films were introduced to solve contact pitting problems. In the mid-1960s, electro-migration in Al film conductors was identified as one of the primary failure mechanisms limiting IC reliability. (Interconnection reliability will be discussed in a later section.) A search for reliability improvement led to the introduction of Al–Cu films (d'Heurle and Ho 1978). With advances in film deposition technologies, Al–Cu–Si films were introduced to utilize the beneficial effects of Cu for electromigration resistance and Si for minimizing the erosion of contacts. Even though other Al-alloy films have been examined for IC interconnection applications, Al–Si, Al–Cu, and Al–Cu–Si films are favored in terms of IC process compatibility. The Al-alloy composition, resistivity, and microstructure of these films strongly depend on film deposition techniques and affect the reliability of IC contacts and interconnections (Ghate 1981a,b).

It was recognized very early in the development of small-scale and medium-scale (SSI and MSI) bipolar circuits that PtSi ohmic and Schottky contacts have to be used to improve circuit performance. Also, it was noted that the PtSi layer (in the PtSi/Al bilayer system) is a poor barrier for preventing Al from reaching the Si/PtSi interface. As a result, PtSi contacts have been used in conjunction with a barrier layer/conductor metallization scheme. The PtSi/Ti:W/Al metalli-zation has been widely used in bipolar ICs. Aluminum film conductors have been replaced by Al–Cu film conductors to minimize hillock growth and improve electromigration resistance (Ghate 1982a,b,c).

Interconnection requirements differ for bipolar and MOS ICs. Bipolar IC designs impose stringent guidelines that the interconnection sheet resistance be less that 0.1 $\Omega/\square$ because they rely on relatively large current levels and low voltage drops. Also, they rely on shallow junction devices, narrow base widths, and sometimes Schottky diodes. In early bipolar-IC development, diffused tun-nels, with relatively large sheet resistance, were used for interconnects, but in recent years these tunnels have been abandoned in favor of low resistance metal film interconnects. As a general rule, all bipolar ICs' single and multilevel interconnections are fabricated with sheet resistances less than 0.1 $\Omega/\square$ and at temperatures below 500°C to maintain device integrity.

It is conceivable that future bipolar IC designs may employ relatively high resistance (1–3 $\Omega/\square$, feasible with refractory and metal silicides) and conventional low resistance (0.1 $\Omega/\square$) films for short and long interconnects, respectively, in alternate layers of multilevel metallization schemes.

Gate/Interconnects for MOS ICs

MOS IC designs demand self-aligned gate structures, short channel lengths, thin gate oxides, relatively shallow source and drain junction depths, and a compatible contact and interconnection scheme. The evolution of gate/interconnect tech-

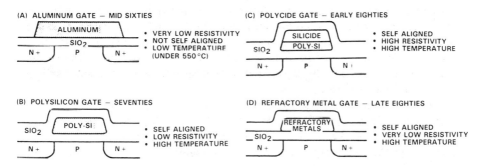

Figure 6-12. Evolution of MOS gate technologies, mid-sixties through late eighties.

nology for MOS ICs is shown in Figure 6-12 (McGreivy and Pickar 1982). Though early MOS ICs started with Al as the gate and interconnect material, it was soon realized that Al had to be replaced with some material compatible with high temperature processing. As current levels in MOS IC interconnects are relatively low (compared to bipolar ICs), and MOS IC designs can tolerate relatively high sheet resistances for interconnects, refractory metals such as Mo and W were attempted in the late sixties. Refractory gate materials were either electron-beam- or RF-sputter-deposited, and these processes made it more difficult to control the thresholds of MOS transistors. Chemical vapor deposition of tungsten films as gate/interconnection material was also attempted. These approaches did not reach production status.

About the same time, advances in ion implantation technology provided a tool for threshold adjustments of MOS transistors by introducing a controlled amount of impurities at the Si/SiO$_2$ (Si/gate oxide) interface. High temperature ($>600°C$) ion implantation damage annealing was required in the process sequence after the gate interconnection and patterning steps. The compatibility and availability of chemically vapor deposited (CVD) polysilicon films as gate and interconnect materials provided a major breakthrough for MOS LSI circuits with single and double polysilicon layers as multilevel interconnections. The polysilicon layers provided process flexibility in (1) achieving self-aligned gate structures, (2) annealing ion implantation damage, and (3) growing thermal oxides on these layers as part of an interlevel insulator.

Furthermore, reflow of CVD phosphorus doped SiO$_2$ on polysilicon interconnects smoothed the topography for metal coverage on oxide steps. As device geometries continued to shrink and circuit designs required lower interconnection resistances, reduction in polysilicon sheet resistance from 100 $\Omega/\square$ to 20 $\Omega/\square$ was accomplished by introducing appropriate dopants in polysilicon layers. The CVD polysilicon was an "almost ideal" material for MOS VLSI circuits.

For VLSI applications, as feature size reduction toward 1 μm and integration of 100,000 more devices on large chip areas continued, it was recognized that polysilicon had to be replaced with low resistance materials. Figure 6-1 illustrated the impact of interconnection resistance—for 20 $\Omega/\square$ and 1 $\Omega/\square$ polysilicon gate and refractory gate materials, respectively—on the access time of a 4K static memory. For this device, polysilicon technology began to degrade the access time for an approximately 2 µm device feature size solely because of high interconnection resistance.

The semiconductor industry is actively exploring the application of refractory metals and silicides such as Mo, W, $TiSi_2$, $TaSi_2$, $MoSi_2$, WSi_2, and so on, as MOS gate/interconnection materials to achieve sheet resistances in the 1 to 3 $\Omega/\square$ range (Murarka 1983; Saraswat and Mohammadi 1982). Table 6-2 summarizes the resistivity and deposition methods of some of these materials. Process sequences, similar to those of polysilicon, are attempted with refractory metals/silicides, and a number of process-related problems are being solved for successful implementation of these materials. Some of the problems and methods used to overcome them are outlined in the following list:

1. Resistivity of physically vapor deposited (electron beam or sputtering) films is sensitive to vacuum ambience during the film deposition. Annealing steps are needed to lower the resistivity.
2. Adhesion and internal stresses of these films depend on deposition parameters and film thicknesses. Co-evaporation of metal and silicon to achieve silicon-rich films and deposition from silicide sources are attempted to overcome the stress problem. Bilayers of silicon/metal and silicon/metal silicide are reacted at a high temperature in an inert atmosphere to form a silicon/silicon-rich silicide. This composite layer is commonly referred to

Table 6-2. Metal Silicides, 500 mm Nominal Thickness.

SILICIDE	PROCESS	RESISTIVITY, $\mu\Omega$-cm
$MoSi_2$	Cosputter E-beam	≈ 100
Si/Mo	E-beam	
$W Si_2$	Cosputter	≈ 70
W	CVD	—
$TaSi_2$	Silicon/metal cosputter	35–45 ~ 50–55
$TiSi_2$	Silicon/metal cosputter	13–16 ~ 25
PtSi	Silicon/metal	35–40
Pd_2Si	Silicon/metal	30–35

as "polycide" and appears to be a promising solution to overcome the adhesion problem.

There is a growing concern about the role of high series source/drain (and contact) resistances on device performance as the scaling trend continues to shrink device geometries and require shallow junction depths (Scott et al. 1982). Lowering of the series resistance in the source/drain regions seems to be possible with silicided junctions. The silicided junctions could be formed by the reaction of refractory metal in contact regions in the same process step of polycide formation by the reaction of polysilicon/refractory metal bilayers.

3. The oxidation process of these refractory silicides and bilayers of silicon/metal results in volumetric changes and causes changes in internal stress levels.

4. These refractory metal and silicide layers are patternable with plasma, sputter, and/or reactive ion etching techniques. Pattern definitions with anisotropic etching resulting from the dry processes are preferred, in contrast to isotropic etching from wet chemicals. $TiSi_2$ and $TaSi_2$ are etchable in buffered HF solutions.

5. The polycide/Al multilayered structures appear to be stable up to 500°C and pose no serious problems to device packaging processes that are carried out below 500°C. The Al–Si films continue to meet the contact (up to 300 nm junction depths) and top layer interconnection requirements. The issue of long interconnection time delays on VLSI chips with increasing chip areas and decreasing feature size has been examined as a function of resistivities of interconnection materials (Saraswat and Mohammadi 1982). Results of this study indicate that silicides are less desirable materials for interconnections on VLSI chips of areas greater than 2 mm^2 and feature sizes < 1 μm and smaller. It has been concluded that conductor materials with low resistivities of 3 to 5 μΩ-cm are needed for VLSI applications.

Multilevel interconnections

Evolution of MSI and LSI circuits to meet customer requirements necessitated the use of multilevel (at least two) interconnections to achieve higher packing density, reduced propagation delays, and reduced chip size (Ghate and Fuller 1981). The basic elements of a two-level interconnection scheme are crossovers (a second-level interconnect crossing the first-level interconnect separated by an insulator layer) and vias (locations for level-to-level connections). Because early IC industry growth was dominated by bipolar technology, multilevel interconnection schemes with Al and Au film conductors were developed to meet the demands of leading edge bipolar ICs. Some metal–insulator combinations employed to achieve multilevel interconnections are:$Al/SiO_2/Al$, $Al/Al_2O_3/Al$,

Al/polyimide/Al, Mo–Au–Mo–SiO$_2$/Mo–Au, and Ti:W–Au–Ti:W/SiO$_2$/Ti:W–Au. Aluminum and gold films provide the desired low resistive interconnections, and SiO$_2$, Al$_2$O$_3$, or polyimide provides the necessary insulation between layers.

Hillocks. Because Al films were most widely used for single-level interconnections, two-level interconnections with Al as the conductor and deposited SiO$_2$ as the insulation layer were the logical choice. However, hillock formation in pure Al films and metal coverage on oxide steps posed serious problems in implementing this Al/SiO$_2$/Al two-level interconnection scheme. As the aluminum metallized substrate is heated to a temperature in the 100 to 180°C range for curing photoresist used in the patterning step, or to a temperature of 450°C for contact sintering, and/or to a temperature of 300 to 400°C for insulation deposition, the Al film is subjected to a compressive stress due to the difference in coefficients of thermal expansion of Al and Si substrate.

This compressive stress in the Al film is relieved by hillock formation (Figure 6-13). The height of the hillock may equal the metal thickness. It has also been observed that the room temperature internal stress in Al films subsequent to annealing at temperature T_A (450°) corresponds to a thermal stress equivalent to a temperature difference of 125°C. This observation suggests that the thermal stress retained in Al films depends on the recrystallization temperature (T_R). The addition of impurities to Al films is known to raise T_R (e.g., from the 100–125°C range to 175–200°C) by suppressing grain boundary motion; so the use of doped films (deposited under optimum conditions) is expected to produce hillock-free surfaces. As the film in tensile stress is heated to T_A, the stress level crosses zero value at T_R and becomes increasingly compressive up to T_A; then the film relaxes to a zero stress level by grain boundary motion and hillock growth. Besides addition of impurities, optimization of substrate temperature and vacuum ambience during film depositions have been tried to suppress hillock growth. Alloy films and optimization of film deposition parameters have been successfully tried to achieve hillock-free films for Josephson junction devices. Our experience has shown that addition of approximately 2 wt% Cu to Al films adequately suppresses hillock formation for multilevel interconnections (Ting and Wittmar 1982). Also, the resistivity of Al–Cu (2 wt%) films is on the order of 3.3 $\mu\Omega$-cm and well within the 15 to 20% of bulk resistivity of pure Al.

Recently, suppression of hillock growth by ion implantation has been reported (Kamei et al. 1984).

Crossovers. The crossovers must meet the requirements of insulation between levels and continuity of the crossing interconnection. Defect-free (pinhole) interlevel insulation layers and hillock-free Al surfaces are desirable to eliminate interlevel shorts. Topography of the deposited insulation (SiO$_2$) layer covering first-level interconnects (e.g., reentrant folds developed over vertical edges of first-level interconnects) has been a major factor affecting the integrity of

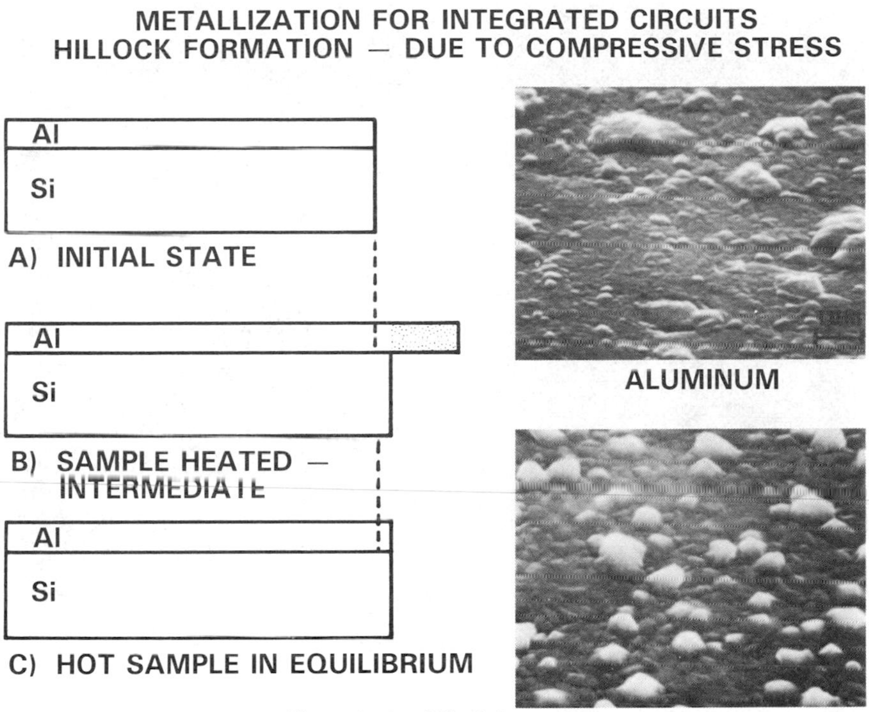

METALLIZATION FOR INTEGRATED CIRCUITS
HILLOCK FORMATION — DUE TO COMPRESSIVE STRESS

A) INITIAL STATE

ALUMINUM

B) SAMPLE HEATED —
INTERMEDIATE

C) HOT SAMPLE IN EQUILIBRIUM

Figure 6-13.   Hillock formation.   **Al + Si**

crossovers because of poor metal coverage on these oxide steps (see Figure 6-14). Over the years, many schemes have been employed to meet the crossover requirements.

An anodic two-level Al/Al$_2$O$_3$/Al system, which produces an almost flat topography for the second-level metal, was introduced to solve the crossover problem. In the anodic process, photomasking of aluminum metallized substrate is used to selectively protect portions of the metal film from electrolytic conversion to an oxide, just as it is used in conventional processing to protect against chemical etching. The resultant structure is an interconnect pattern that is embedded in and overcoated with the anodic oxide of aluminum.

This process could provide pillar structures in the vias and minimize metal coverage problem at the via steps. The anodic process was an elegant solution at the time it was introduced.

Complete conversion of Al to Al$_2$O$_3$ between interconnects over the entire substrate requires that a suitable current path be provided for completion of the anodic process. Anodic oxidation, like many other corrosion processes, is grain-boundary-enhanced; hence, the average grain diameter should be considerably less than the film thickness for good anodization results. As the complexity of ICs continued to increase and the lead widths and spacing continued to shrink,

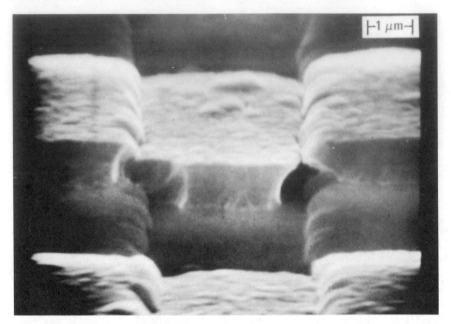

Figure 6-14. Defect in second-level metal at reentrant oxide step.

process controls to achieve high packing density interconnects became too stringent for the anodic process using wet chemicals, and the anodic two-level Al/Al$_2$O$_3$/Al interconnection scheme was abandoned.

Since the mid-1960s, sputtered quartz has been explored for multilevel insulator application. Topography of the sputtered quartz layer at crossover strongly depends on edge profiles of the underlying metal, and the resulting reentrant folds at crossovers have been problem areas for achieving continuity of second-level metal leads. Planarization of a sputtered quartz layer has been attempted by partial resputtering during quartz deposition. This process exploits enhanced sputter etch rates at sharp corners, and it has been used in multilevel interconnection processing with some degree of success. Relatively thicker oxide and metal layers have been employed.

Another dielectric film that has been widely used in multilevel interconnection processing is the CVD SiO$_2$ layers. Plasma excited depositions of SiO$_2$ layers at a relatively lower temperature (250–350°C) as compared to thermally excited CVD depositions (400–500°C) have the beneficial effect of lowering hillock density on Al films and thereby minimizing interlevel shorts. Plasma deposited SiO$_2$ films have been employed as interlevel insulators on a variety of bipolar ICs.

In recent years, polyimides have been proposed as multilevel insulators and protective coatings for ICs (Evans et al. 1980; Blech et al. 1978). In this process, polyimides in liquid form are applied like photoresists to metal patterned substrates, and then cured to remove the solvent to convert them into a solid form. These polyimides provide an excellent planarization of the surface for second-level metal interconnections. However, widespread use of polyimides as multilevel insulators has been rather slow because of a number of process-related problems such as purity, adhesion, absorption and/or retention of moisture, controlled via etching, and so on. As the quality of polyimides has improved, there has been renewed interest in the application of polyimides as multilevel insulators in VLSI processing.

Vias.   Multilevel interconnect systems must provide for interlevel contact through holes or vias in the interlevel insulation layers. In the design of a conventional two-level interconnection system, a "nested via" is employed. Expanded first-level metal pads are used for two reasons: (1) to provide a tolerance for level-to-level registration at the via definition step and (2) to provide an etch step so that the passivation layer beneath the first-level metal is not attacked while the insulator (SiO$_2$) is etched for via formation. Also, the second metal overlaps the via to protect the first level during the second-level etching. In the conventional Al/SiO$_2$/Al system, with nested vias, the Al pad is larger than the via on all four sides, and it serves as an etch stop for oxide etchants. Such use of an expanded metal pad increases the effective spacing between first-level interconnects, and lowers the packing density. Also, the placement of nested vias for a complex LSI circuit make the chip layout operation somewhat more cumbersome. The absence of oversize pads for via locations makes them more attractive for computer-aided designs of interconnection layouts, and allows the maximum packing density of first-level interconnections permissible with patterning capabilities. "Oversized" via configurations with the proper selection of an insulator (oxide) etch stop such as silicon nitride under the first-level Al eliminate the oversized Al pads and permit maximum packing density of first-level interconnects (see Figure 6-15). The use of an oversized via scheme in the fabrication of two-level interconnections for a VLSI circuit (the 16-bit I^{2}L microprocessor circuit) has been reported (Ghate 1982a). As device geometries continue to shrink, via etching and metal coverage in the vias make this oversize via design somewhat undesirable.

If level-to-level registration, via sizing accuracy, and interlevel contact resistance tolerances of the circuit permit, then fully nested vias placed on the nominal width of the first-level interconnection without expanded metal pads can be employed. An I^{2}L 4-bit microprocessor circuit fabricated with electron-beam patterning uses a nest–via scheme for the two-level interconnections (Evans et al. 1980). Here, a 1.1-$\mu\Omega$-m wide via is placed on a 2.5-$\mu\Omega$-m wide first-level interconnection.

The etching of vias is a critical step. Sloped oxide contours for the vias are desirable for achieving continuity of the second-level interconnections. The prob-

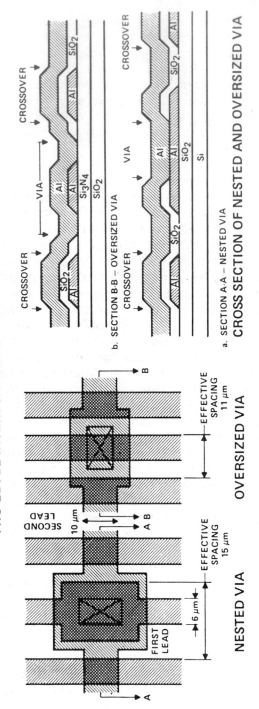

Figure 6-15.   Schematic cross section of nested and oversize vias.

lem of nonfunctional ICs due to high interfacial resistance ($\approx 10^{-2}$ Ω-cm^2) in vias is directly related to the incomplete removal of the insulator layer in the vias; also impurities introduced on metal surfaces in cleaning steps produce a barrier layer. In-situ cleaning of vias by ion milling or by sputter etching appears to be a viable solution to overcome the via resistance problems. With in-situ sputter cleaning, an interface resistance in the range of 10^{-8} to 10^{-7} Ω-cm^2 is observed for Al/Al interfaces in vias.

METAL COVERAGE ON OXIDE STEPS

Metal coverage on oxide steps is an area of major concern in IC processing. As long as metal films are vacuum-deposited either from a molten source or by sputtering, the evaporated or sputtered atoms follow line-of-sight trajectories, and microcracks develop in the films near the oxide steps due to shadowing effects (Blech et al. 1978). Innovative toolings with different source-to-substrate configurations and moving substrate holders have evolved to attack the step coverage problem. Other methods include film depositions under high background pressure, ion plating, and so on. These methods have not found widespread application in VLSI interconnection processing.

We have carried out a model calculation to determine metal thickness as a function of step angle for substrates metallized in a conventional canted domed planetary system with concentric source and a horizontal rotating substrate over a circular magnetron (Fuller and Ghate 1979). A cosine distribution for the surface sources is assumed, and proximity effects of the adjacent steps are ignored. Results of these model calculations, presented in Figure 6-16, are consistent with our experimental observations. A thinning factor of 50% or greater should be expected for steps exceeding a 60° angle. Experimental observations on the impact of via size on step coverage, presented in Figure 6-17, clearly show that proximity effects due to adjacent steps need to be comprehended in circuit layouts. As VLSI designs continue to shrink device geometries, and vertical edge profiles are demanded for contact windows and interconnections, metal coverage problems must be solved within the framework of deposition tooling and device performance goals.

APPLICATION OF TWO-LEVEL INTERCONNECTIONS

In the fabrication of two-level interconnections, the desirability of sloped first-level interconnects to generate smooth oxide contours for second metal coverage has been well recognized.

A metallization process scheme compatible with LSI and VLSI bipolar ICs can be as follows:

PtSi/Ti:W/Al–Cu/SiO$_2$/Al

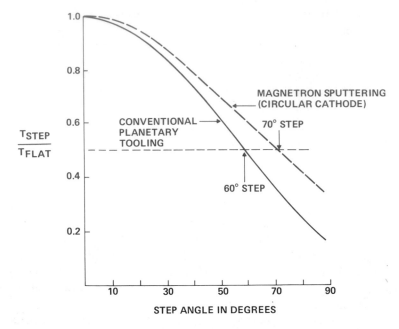

Figure 6-16. Metal coverage on oxide steps; results of a model calculation for two types of film depositions.

where PtSi forms the ohmic and Schottky contacts, Ti:W acts as a barrier layer, Al–Cu films provide first-level interconnections, SiO_2 (deposited in a plasma reactor) serves the purpose of an interlevel insulator, and Al acts as a second-level interconnection. Such a metallization process scheme has been applied in the fabrication of two-level interconnections with nested vias, and a portion of this circuit is shown in Figure 6-18(a). An "oversized" via scheme has been employed in the fabrication of two-level interconnections for a VLSI circuit (the 16-bit I^2L microprocessor circuit—SBP9900) fabricated with conventional optical patterning techniques, and a portion of this circuit is shown in Figure 6-18(b). The width and spacing of first-level interconnections on the early design of SBP9900 were 6.25 μm (0.25 mil) and 5.0 μm (0.2 mil), respectively. This circuit has more than 25,000 Si/metal contacts (3.7 × 2.5 μm^2), 15,000 crossovers and 10,000 vias. The cumulative total length of first- and second-level interconnections is approximately 6 meters. The chip area is on the order of 0.58 cm^2 (90K mil^2); approximately 6 meters. 65% of the chip area is occupied by interconnections. Oversize vias provide maximum packing density of first-level interconnections on this chip; also level-to-level registration between the first-level leads and vias is less demanding. If level-to-level registration, via sizing accuracy, and interlevel contact resistance tolerances of the circuit permit, then

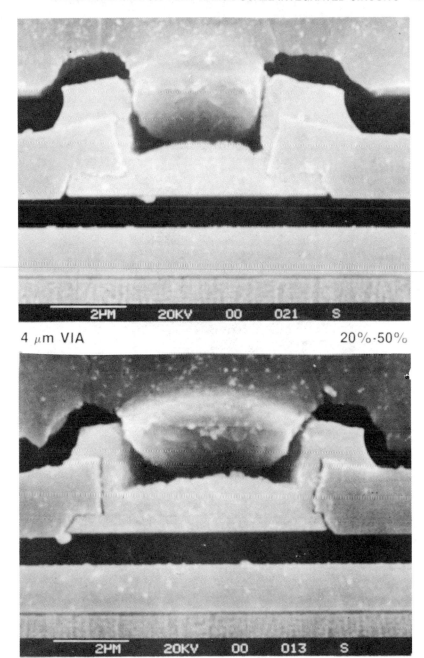

4 μm VIA   20%-50%

6 μm VIA   38%-54%

Figure 6-17.   Step coverage as a function of via size.

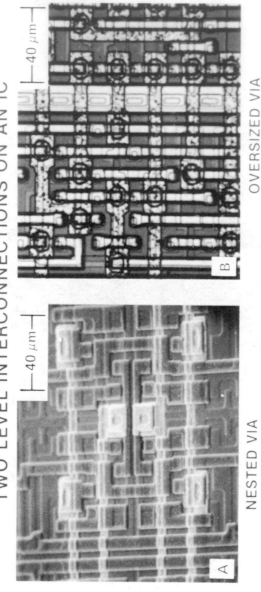

TWO LEVEL INTERCONNECTIONS ON AN IC

NESTED VIA

OVERSIZED VIA

Figure 6-18. Two-level interconnections. (a) Nested vias on a bipolar memory circuit. (b) Oversize vias on the 16-bit I²L microprocessor circuit SBP9900.

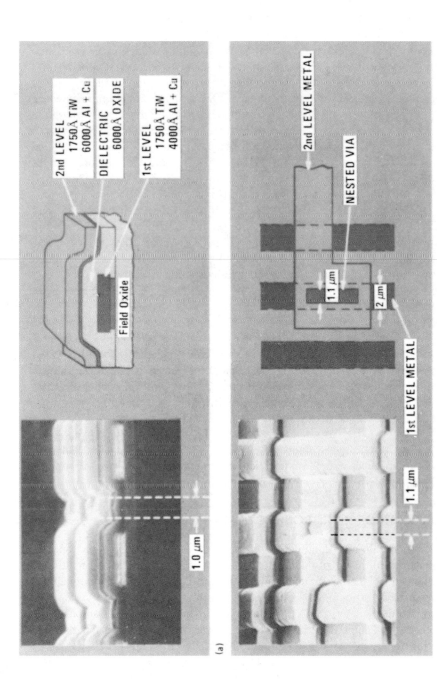

Figure 6-19. Crossovers and nested vias of two-level interconnections on electron beam-patterned VLSI circuit SBP9989E.

fully nested vias placed on the nominal width of the first-level interconnections without expanded metal pads can be employed.

Recently the PtSi/Ti:W/Al–Cu/SiO$_2$/Ti:W/Al–Cu metallization has been successfully extended to a VLSI class bipolar circuit (a revised version of a 16-bit microprocessor—SBP9989E) fabricated with the application of electron-beam lithography. Level-to-level registration is ± 0.2 μm. The minimum feature size on this circuit is 1.25 μm. Junction depths of devices on this circuit are on the order of 0.25 μm. PtSi layers are used for ohmic contacts to n$^+$ (As implanted) collector regions. The test structure described earlier has been used to determine the contact resistance of 1.25 × 7.5 μm^2 contacts, and the measured values correspond to a specific contact resistance of 0.5 × 10^{-7} Ω-cm^2. Sputtered Ti:W/Al–Cu bilayers were used for both first- and second-level interconnections. The Ti:W layer in the first level acts as a diffusion barrier between PtSi and Al–Cu films; use of the Ti:W layer in the second level provides the flexibility of reworking the second-level interconnections if the need arises (Ti:W acts as an etch stop for the Al–Cu interconnection definition step). The metal pitch (width + spacing) of first-level interconnections is 5 μm with a 2.5-μm design value for the metal width. Interconnection definitions were accomplished with wet etching, and the width of first-level interconnections was close to 2.0 μm. Metal thicknesses of 170/500 nm and 170/700 nm for the first- and second-level interconnections were chosen to maintain reasonable aspect ratios. A 600-nm-thick plasma deposited SiO$_2$ was used as an interlevel insulator. Excellent level-to-level registration of the electron-beam patterning equipment enabled the use of nested vias 1.1 μm in width placed on a 2.0-μm-wide first-level interconnection. Vias in plasma oxide were etched in a parallel plate reactor using a fluorocarbon gas. A cross section of crossovers and the placement of a via are shown in Figure 6-19. Also, in Figure 6-20, the full size SBP9989 chip fabricated with optical patterning is compared with an SBP9989E chip fabricated with electron beam direct writing on the slice; the minimum feature sizes for optical and e-beam patterned chips are 4.5 μm and 1.25 μm, respectively.

RELIABILITY OF INTERCONNECTIONS

As the VLSI chip area continues to grow, with a relatively large portion (approximately two-thirds) of the chip area being utilized by single and multilevel interconnections, there is an increasing awareness that interconnection reliability will be a major factor limiting VLSI chip reliability. Furthermore, system designers are demanding that VLSI chips should maintain, if not surpass, the reliability of LSI chips. Today's VLSI reliability goals are 100 FITs under normal operating conditions. (1 FIT = 1 failure in 10^9 device hours.)

Interconnection reliability is likely to be limited by three classes of problems: circuit designs and layouts, process controls and manufacturing defects, and basic materials properties. Here, the discussion will be limited to the failure

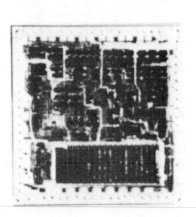

E-BEAM DIRECT WRITE SBP9989E

- AREA 22,000 MIL^2
- PERFORMANCE: 10 MHZ MAX AT 280 MW (65 MW FOR 5.5 MHZ)
- MINIMUM FEATURE 1.25 MICRONS

OPTICAL SBP9989

- AREA 57,000 MIL^2
- PERFORMANCE: 5.5 MHZ MAX AT 600 MW
- MINIMUM FEATURE 4.5 MICRONS

Figure 6-20.   Comparison of optically patterned SBP9989 chip with electron beam-patterned SBP9989E.

modes dependent on basic material properties. Literature surveys suggest that failure modes, limiting interconnection reliability, may be classified in three broad categories: electromigration, corrosion, and others.

Electromigration

Reliability studies, since the mid-sixties, have identified electromigration (EM) induced failures as one of the primary failure mechanisms limiting the reliability of IC interconnections (d'Heurle and Ho 1978; Ho 1982a; Ghate 1982a,b,c; Black 1982). Mass transport under the influence of an impressed dc electric field is referred to as electromigration. Depletion and accumulation of matter due to flux divergences along the conductor paths are responsible for the failures.

An example of depletion of matter/void formation in an Al–Si interconnect of an MOS IC is shown in Figure 6-21(a); accumulation of matter and whisker growth (seen at random) in a Ti:W/Al interconnect of a bipolar IC is shown in Figure 6-21(b).

Most of the electromigration testing has been carried out with Al and Al-alloy film conductors, as they are most widely used on integrated circuits. A general expression for the median time to failure (MTF) is as follows:

$$\text{MTF} = Aj^{-n} \exp{(Q/kT)}$$

where:

A = a parameter depending on geometry, physical characteristics of the film, protective coating, and substrate

j = the current density of A cm^{-2}

n = the exponent: $1 < n < 3$

Q = the activation energy

T = the average temperature of the conductor

k = Boltmann constant

Mass transport in films is governed by grain boundary diffusion, and Q is a suitably averaged activation energy for the film. MTF data are observed to obey log-normal distribution with a dispersion parameter σ. Consider the case of plasma SiO$_2$-passivated Ti:W/Al and Ti:W/Al–Cu (2 wt%) film interconnections. Reported electromigration life test data on test samples (9 μm wide, 1.14 mm long, and 170 nm Ti:W/800 nm Al or Al + Cu film conductors on oxidized silicon substrates followed by approximately 1-μm-thick plasma SiO$_2$ passivation) subjected to a current density of 1×10^6 A cm^{-2} in the 150 to 270°C temperature range are shown in Figure 6-22. Time-to-failure data are found to

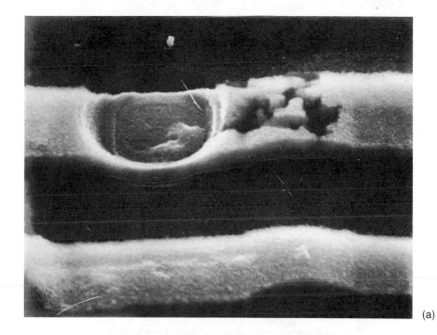

(a)

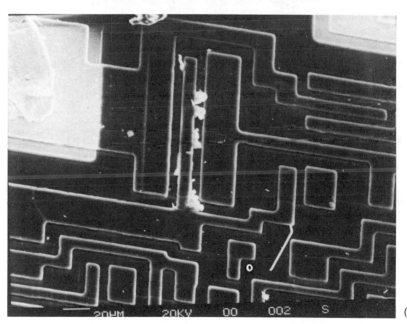

(b)

Figure 6-21. (a) Migration-induced failures in Al film conductors (MOS devices). (b) Electromigration-induced failures in Ti:W aluminum interconnects (bipolar devices).

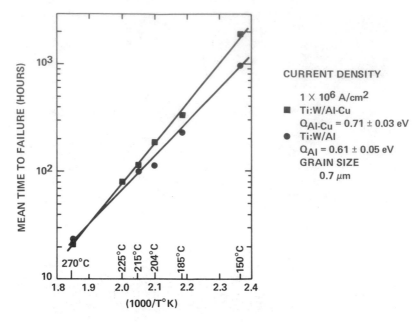

Figure 6-22.   TI MTF data for Ti:W/Al and Ti:W/Al–Cu.

obey log-normal distribution with an average dispersion parameter of σ = 0.43 nm, and are consistent with activation energies of 0.61 ± 0.05 eV and 0.071 ± 0.03 eV for Ti:W/Al and Ti:W/Al + Cu film conductors, respectively (Ghate and Blair 1978). Based on these life test data, one may proceed to calculate the failure rates of these film conductors subjected to current densities of $j = 5 \times 10^5$ A cm^{-2} and 2×10^5 A cm^{-2} at 85°C operating temperature.

The exponent $n = 1.5$ is used for MTF extrapolation to 85°C. The predicted failure rates for these conductors are displayed in Figure 6-23.

Examination of the failure rate curves suggests that glass-passivated Ti:W/Al film interconnections operating at 85°C and at a current density of 5×10^5 A cm^{-2} will experience a failure rate at 50 FITs (1 FIT = 1×10^{-9} failures/hr) after 9K hours usage at a 100% duty cycle (or after 18K hours usage at a 50% duty cycle). If the current density were 2×10^5 A cm^{-2}, the same interconnections would experience failure rate of 50 FITs after approximately 46K hours at a 100% duty cycle (or 92K hours at a 50% duty cycle). Because interconnects carry pulse currents, one can use the "on-time" approximation for failure rate predictions as a function of usage time.

Electromigration studies in thin films suggest that interconnection fabrication procedures (film properties, microstructure, composition) and chip operating conditions (temperature, current density, duty cycle) have a profound impact on failure rates. For example, replacement of pure Al by Al–Cu film interconnects

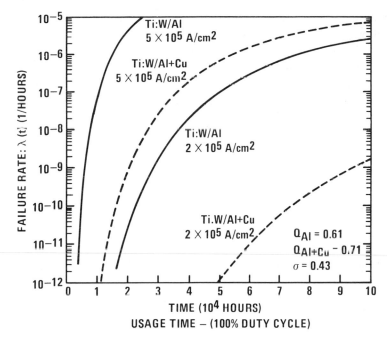

Figure 6-23.   Predicted failure rates of Ti:W/Al and Ti:W/Al + Cu film conductors as a function of usage time at 85°C.

reduces failure rates for a specified IC design layout and operating conditions; or, for the same metallization, lowering the operating temperature improves reliability.

Since electromigration induced failure rates depend on the activation energy Q, gold film interconnects with $Q = 0.90$ eV can be expected to display a far superior electromigration resistance in comparison to Al interconnects (Blair et al. 1972; Denning et al. 1984). Reported life test data on Ti:W/Au interconnects suggest a failure rate of 10 FITs for $j = 2 \times 10^5$ A cm^{-2} at 150°C operating temperature after 10 years of continuous usage.

Recently, it has been reported that MTFs of Al and Al-alloy film conductors decrease with decreasing width in the 8 to 2 μm range and suddenly turn around and start increasing below a width of 2 μm (Kinsbron 1980; Vaidya et al. 1980). EB-evaporated Al–Cu (0.5 wt%) films are reported to be superior to sputter-gun evaporated films. The superior performance of EB-evaporated Al–Cu (0.5 wt%) films has been rationalized on the basis of a "bamboo" type grain structure [(111) textured films with grain boundaries normal to the substrate].

In another study on Al–Cu–Si film conductors, superior performance was attributed to a very uniform grain size of 0.25 μm and to uniform Cu distribution throughout the films (Ghate 1981). This study supports the thesis that electromigration-induced failures resulting from flux divergences due to microstructural inhomogeneities can be avoided by employing uniform grain size distribution across the conductor width and thereby enhancing VLSI interconnect reliability.

Use of plasma (or reactive ion) etching of Cu doped Al films in Cl_2-containing gases to form narrow (1 to 3 μm wide) interconnects has increased the potential for corrosion-induced failures. Because Al–Cu films have demonstrated their superior electromigration, researchers are busy developing proprietary recipes to overcome the corrosion problem. Concurrently, alternate Al-alloy metallization systems are explored. One study suggests that Al–Si–Ti films are compatible with dry etching techniques without the corrosion problem, and that their electromigration resistance is equivalent to that of Al–Cu–Si (Black 1970).

VLSI Interconnection Statistics

In the VLSI circuits, some of the signal paths tend to be fairly long. Interconnection technology is constantly challenged to respond with smaller metal pitches (width + space) and multilevel interconnections. A histogram of the signal paths of a VLSI chip with two-level interconnections is shown in Figure 6-24 (Ghate 1981). Note that the majority of the signal paths are relatively short, whereas some of the signal paths are long (10 mm). The current density distribution is complex. The interconnection distribution suggests that time-to-failure data collected on a "relatively long" test current should provide an asymptotic MTF value and thus adequately describe the interconnection failure rates. Because the shorter interconnections, under equivalent current density, have longer MTFs, longer interconnections will have a significant impact on VLSI reliability.

Contacts

Electromigration failures are observed not only in the interconnections but also in Si/Al contacts with Al as the contact/interconnection metallization. Experimental studies on contact chains indicate that Si atoms are transported in the direction of electron flow from contact windows, eroding the contacts; and these Si atoms are later deposited onto the Si contacts where electrons leave the Al interconnections (Black 1970). Erosion of contacts leads to increased leakage currents and device failures. These failures depend on junction depth and circuit layouts (i.e., distance of contact window edge from the diffusion window). Experimental data on test structures have been collected to predict contact failure rates versus interconnection failure rates (Vaidya and Sinha 1982). However, it must be noted that Si erosion and growth of precipitates do not proceed uniformly in contact windows, and furthermore they depend on Si surface preparation of contact windows prior to metal deposition. A certain amount of caution needs to be exercised in failure rate prediction.

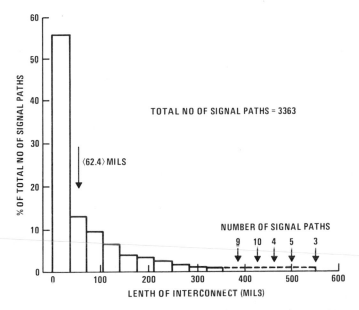

Figure 6-24.   Histogram of signal path lengths (first and second level M1 + M2).

Reliability of shallow junction devices is a major concern, solved by interposing either a silicide/barrier layer or refractory silicide/barrier layer between the contacts and interconnections.

Corrosion

The corrosion of IC metallization has been a frequent reliability problem (Kolesar 1974; Paulson and Kirk 1976; Bhide and Eldridge 1983). A large number of variables such as metallization, encapsulation, ionic contamination, temperature, humidity, and applied bias affect corrosion failures. The semiconductor industry has employed hermetic packages to protect device metallizations from corrosion by sealing the devices in an inert ambient with minimum (less than 3000 ppm) moisture inside the package. The current trend is to replace hermetic packages with plastic packages to reduce cost. Unlike hermetic packages, these packages are readily penetrated by moisture and contaminants. Furthermore, small amounts of ion contaminants, mostly chlorides, are present in the encapsulating (molding) materials. As moisture and contaminants reach the device metallization, corrosion reactions set in that eventually lead to device failures. Examples include bond pad corrosion and internal metallization corrosion due to the penetration of moisture through the protective coating.

Corrosion reactions may be classified in three broad categories.

1. Galvanic: Two dissimilar metals set up an electrochemical cell and corrosion occurs at the anode. An example is the corrosion of the Mo layer in Mo/Au metal systems; this metal system has been essentially abandoned in favor of a corrosion-resistant Ti:W/Au metallization.
2. Electrolytic or concentration: An example is the corrosion of Al leads in the presence of moisture with phosphorus or chloride ions.
3. Simple anodic: An example is corrosion of the Al/NiCr system. Here an applied bias is necessary to initiate the corrosion reactions.

The corrosion reactions are complex, and it is sufficient to note that in the presence of moisture, contaminants either introduced during device fabrication or transported to the metallization subsequently during device operation with or without applied bias are potential sources of corrosion-induced metallization failures.

It has been noted that Al is the most widely used metallization for ICs. It is a common practice in bipolar IC fabrication to use phosphosilicate glass (PSG) layers as passivation for devices to tie up the mobile Na ionic contamination. Also, in MOS LSI fabrication, a phosphorus-doped CVD SiO_2 layer (4 to 10 wt% phosphorus) is deposited and then reflowed at a high temperature to produce smooth topography for metal coverage on oxide steps. The Al and Al-alloy film conductors are formed on this PSG layer. Furthermore, it is not uncommon to find a CVD SiO_2 layer deposited on top of these interconnects as a protective coating. This SiO_2 layer is doped with phosphorus to reduce internal stresses and thereby eliminate cracks in the layer. Data in the literature lead to the logical conclusion that corrosion of Al and Al conductors decreases with a decreasing amount of phosphorus in the PSG layer in the presence of moisture, and that failure rates are sensitive to the applied bias on these conductors.

The corrosion susceptibility of plastic encapsulated ICs is assessed by accelerated testing of a small sample of devices at 85% relative humidity (RH) and at 85°C, referred to as 85/85, for periods ranging from 25 to 1000 hours (in some cases 2000 hours) with or without bias. The cumulative failures are measured at specified intervals up to 1000 hours. These data are reported to indicate the reliability of these devices against corrosion (e.g., 1% failures at 2K hours at 85/85).

Attempts have been made to develop models to use 85/85 data to predict failure rates for use conditions. Recently, accelerated 85/85 test data on RTV silicon rubber-coated aluminum metallized NMOS 4K DRAMs were analyzed by Striny and Schelling (1981). The interconnections on these devices were 6 μm wide with 8 μm spacing. These plastic encapsulated devices showed a 4% cumulative failure after 9312 hours. The data were analyzed in the framework of the Eyring model. The MTF was represented by the expression:

$$MTF = A \exp \frac{Q}{kT} + \frac{\beta}{H}$$

where:

A = constant

Q = activation energy for temperature T

B = activation energy for humidity H

T = temperature

H = relative humidity

The published aluminum corrosion data on ICs were analyzed to predict the acceleration factor (AF) from 85/85 data for typical (central office) operating condition of 55% RH at 38°C. Striny and Schelling predicted an AF = 1000; that is, devices exhibiting a MTF of 1 year at 85/85 would exhibit a MTF of 1000 years at 55% RH at 38°C. With a MTF = 20,000 hours at 85/85 and a conservative estimate of AF = 100, Striny and Schelling predicted that for devices operating under high humidity conditions (in Baton Rouge, LA) a failure rate of 100 FITs would be reached after 40 years. (Note: The failure rate is increasing with time.)

There is some concern about the corrosion susceptibility of Al–Cu films, which were introduced to improve electromigration resistance, as compared to that of pure Al interconnections. The Al–Cu and Al–Cu–Si metallization systems have been in use since the early seventies, and they have been selectively implemented on bipolar and MOS ICs for single and multilevel interconnections (Ghate et al. 1977; Chen et al. 1981; Gardiner and Halley 1981). Process parameters have been optimized to achieve a homogeneous alloy with low (0 to 4 wt%) Cu in these Al–Cu and Al–Cu–Si films. When Al and Al–Cu film deposition and fabrication processes are carried out under strict process controls, there is no measurable difference in the corrosion-induced device failure rates of those films under normal operating conditions. Recently it has been reported that there is no noticeable difference in the corrosion susceptibility of wet etched versus dry etched Al–Cu–Si and Ti–W/Al–Cu film conductors (Bukhman et al. 1985). The SC industry is constantly searching for alternate test methods to reduce the 85/85 test time. Highly accelerated stress tests (HAST) with high temperature and humidity are being explored, and efforts are under way to relate HAST data to 85/85 data.

In summary, the corrosion processes are complex, and reliability modeling is far from satisfactory. There is a general consensus about the factors affecting corrosion failures. Making material selections according to basic principles and using process controls to avoid ionic contamination have been the best ways to minimize corrosion failures and thereby achieve cumulative failures of less than

0.5% after 1000 hours at 85/85. These data correspond to an average failure rate of 5 FITs under normal operating conditions.) Since VLSI processing involves phosphorus-doped SiO_2 and dissimilar metals from contacts through packaging, the most stringent process controls to eliminate ionic contamination from the surface of the chips and moisture from the package appear to be the best remedies against corrosion-induced interconnect failures.

Others

Since the total interconnection scheme encompasses contacts, barrier metals, gate materials, conductors, barrier layers between conductors, and bumps for packaging, a number of process-related problems affect interconnection reliability. No attempt will be made to discuss the other reliability issues here except for a brief description of an open metal failure mode recently observed with Al–Si interconnections on MOS devices (O'Donnell et al. 1984).

As interconnection width continues to decrease, the size of Si precipitates (nodules) approaches the width dimension. Because the current in these interconnections bypasses the Si nodules, the current density increases by a factor of two to three above design guidelines. Early failures are noticed in electromigration testing; also these open metal or void failures are noticeable in temperature cycling (-65 to $150°C$) tests. It appears that this failure mode results from internal stresses in the film.

The Al–Si films are in tension, and the magnitude of this stress is affected not only by Al–Si processing but also by the nature of the internal stresses in the protective coating. Some of the Si nodules grow at the expense of other small Si precipitates within the diffusion length and give rise to a variable distribution for Si nodules in narrow interconnections. It is conceivable that different lengths along a long interconnection are pinned by Si contacts and these Si nodules, and that these individual segments of interconnections sustain high internal stresses (2–3×10^9 dynes/cm^2, a factor of 2 to 5 above the breaking strength of bulk Al). Like all other systems under stresses, these interconnections will try to attain a stress-free state. It appears that metal voids develop because of creep phenomena. [Plastic deformation (dislocation motion is required) under constant stress is called creep.] The early failures in electromigration experiments are manifestations of these voids that develop because of tensile stresses in the films (Curry et al. 1984).

FUTURE TRENDS

The increasing demand for high-performance VLSI circuits will continue to drive both bipolar and MOS technologies to innovate device structures with shrinking geometries. Optical patterning will be approaching its limit at a minimum feature size of 1 μm in manufacturing environments.

Electron beam and X-ray lithography techniques continue to be developed for pattern definitions below the 1 μm feature size. Chip size continues to grow with increased densities of components. The complexity of interconnections suggests that a 0.5 μm feature size appears to be the lower limit for device structures for silicon VLSI circuits.

Whether the device is bipolar or MOS, the metallization processes must be compatible with shallow (about 100–150 nm) junction devices and multilevel interconnections., In bipolar applications, the circuits may demand, besides ohmic contacts to n^+ and p^+ regions, two types of Schottky contacts with prescribed barrier heights.

The interconnection requirements of bipolar and MOS VLSI circuits will differ in several respects. For example, bipolar designs will demand that interconnection sheet resistances be 0.1 Ω/□ or lower, whereas MOS designers will see significant circuit performance improvement when polysilicon gate interconnections (20 to 30 Ω/□) are replaced by refractory metal silicides with a 1 to 5 Ω/□ sheet resistance. CVD metal systems may be needed to solve the problems of metal coverage on steep oxide steps. Some MOS devices require ion implantation of dopants after the first-level gate interconnection process step has been completed. Process flexibility is required to anneal the ion implantation damage in these devices.

Differences of this type need to be comprehended in the design of the metallization scheme. As custom VLSI circuit design costs continue to soar, computer-aided design tools find widespread use, and customization of VLSI circuits is accomplished with gate arrays. A master slice with a large number of gates is produced by one of the bipolar or MOS technologies. A computer generates the necessary multilevel interconnection layouts on the chip, and the master slice is processed through the multilevel interconnections to interconnect the gates. It is very likely that three or more levels of interconnections will be required to meet circuit performance goals. Also, the number of input–output pins may be so large (e.g., more that 100) that the normal mode of gold wire bonding to aluminum pads located near the periphery of the chip may have to be abandoned in favor of gang bonding with gold or gold-coated copper bumps. The use of Pb–Sn solder bumps arranged in the form of an array on the chip has been reported (Ting and Wittmar 1982). A VLSI metallization process scheme with silicide contacts/three levels of interconnections/barrier layer/bumps may be visualized as follows (Figure 6-25):

$$PtSi(Ti:W)/(Al–Cu)/I_1/(Ti:W/(Al–Cu)/I_2/(Ti:W)/(Al–Cu)/I_3$$

$$Cr/Ag/Au/(plated\ gold\ bumps)\ or\ Cr/Cu/Au/(Pb–Sn\ solder\ bumps)$$

This process scheme has (1) one silicide layer of contacts to silicon, (2) three barrier layers of Ti:W, the first layer to keep the Al–Cu away from the PtSi and

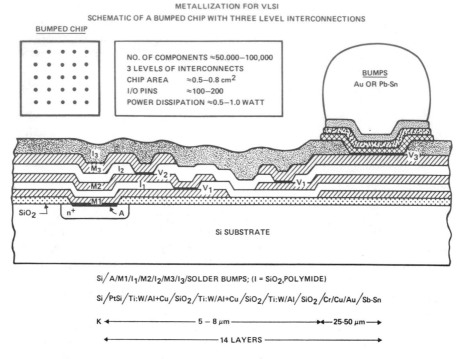

Figure 6-25. VLSI metallization process scheme.

the other two as etch stops, (3) three layers of Al–Cu films for interconnections, (4) three layers of insulators (I_1, I_2, and I_3) through which vias and bond pads have to be etched for interlevel interconnections, and (5) an adhesion layer of chromium, a composite bilayer of Cr/Ag as a barrier layer under gold bumps, or similar Cr/Cu/Au layers under Pb–Sn bumps, with a cumulative total of 14 layers. It is imperative that the individual layers in the 14-layer composite, normally dubbed the VLSI metallization, maintain the integrity of silicon devices on the chip and provide the prescribed signal paths.

Everyone engaged in VLSI efforts intends to "design in" reliability by using carefully agreed upon current density design guidelines and process technologies. Furthermore, stringent process controls and accelerated test data on test structures are required to ensure that the metallization processes continue to meet VLSI chip reliability goals.

The areas of concern/desirable features in VLSI metallization may be summarized as follows:

1. Impurities at the silicon–metal interface and their impact on ohmic and Schottky contacts.

2. Reliable barrier layers for contacts and packaging operations.
3. Metal coverage on steps.
4. Microstructure (grain size) and impurity control of film interconnections for reliability improvement.
5. Deposition and patterning of refractory metal silicides for MOS VLSI circuit applications.
6. Defect-free relatively thin interlevel insulator layers capable of producing planar surfaces for subsequent metallization.
7. Internal stresses in multilayered structures.
8. Dry processing (sputter, plasma, reactive ion etching) for metal lead definitions and via etching in insulators.
9. Stable silicon/metal and metal/metal interfaces through multilevel and packaging operations.
10. Low cost.

Metallization is one of the last critical steps performed in VLSI processing, and deposition and patterning of multilayer metal and insulator films must provide a cost-effective solution for interconnecting devices on the chip. The process scheme must be compatible with chip attachment requirements. Data on interdiffusion in thin films and surface characterization tools are essential for successful implementation of a reliable VLSI metallization scheme.

ACKNOWLEDGEMENTS

The author wishes to thank J. C. Blair, Clyde Fuller, Lloyd Crosthwait, Steve Evans, and Frank Morris for numerous helpful discussions over the years.

REFERENCES

Asai, S., "Trends in Megabit DRAMs," *IEDM Digest Tech. Papers*, San Francisco, pp. 6–9 (1984).
Berger, H., "Models for Contacts to Planar Devices," *Solid State Electronics, 15*, 145–158 (1972).
Bhide, V., and Eldridge, J. M., "Aluminum Conductor Line Corrosion," *IEEE 21st Annual Proc. Rel Phys.*, pp. 44–51 (1983).
Black, J. R., "RF Power Transistor Metallization Failure," *Ieee Trans. Electron Devices, 17*, 800–803 (1970).
Black, J. R., "Current Limitations of Thin Film Conductors," *IEEE 20th Annual Proc. Rel. Phys.*, pp. 300–306 (1982).
Blair, J. C., and Ghate, P. B., "The Effects of Vacuum Ambience on Si/Al Contacts," *J. Vac. Sci. Technol., 14*, 79–84 (1977).
Blair, J. C., Fuller, C. R., Ghate, P. B., and Haywood, C. T., "Electromigration Induced Failures in, and Microstructures and Resistivity of, Sputtered Gold Films," *J. Appl. Phys., 43* (2), 307–311 (1972).

Blech, I. A., Fraser, D. B., and Haszko, E. M., "Optimization of Al Step Coverage through Computer Simulation and Scanning Electron Microscopy," *J. Vac. Sci. Technol., 15,* 13–19 (1978).

Bloch, E., "VLSI Trends," in *Semiconductor Silicon 1981, Proceedings,* H. R. Huff, R. J. Kriegler and Y. Takeishi, Editors, *81-5,* 20–32, The Electrochemical Society, Pennington, NJ (1981).

Bukhman, J., Goodner, R., and Hulseweh, T., "Corrosion Evaluation of Metal Films Patterned by Reactive Ion Etching," Abstract No. 221 of a paper presented at the Electrochemical Society Meeting in Toronto, Ontario, Canada, *J. Electrochem. Soc., 132,* (3), 132 (1985).

Card, H. C., "Si/Al Schottky Barrier Diodes," in "Metal–Semiconductor Contacts," *Inst. of Phys. Conf. Ser.* No. 22, Manchester, England, pp. 129–137 (1974); *Solid State Commun., 16,* 87–90 (1975).

Chen, J. Z., Chin, W. N., and Jen, T. S., "A High Density Bipolar Logic Master Slice for Small Systems," *IBM J. Res. Develop. 23,* 145–152 (1981).

Cunningham, J. A., "Expanded Contacts and Interconnections to Monolithic Silicon Integrated Circuits," *Solid State Electron. 8,* 735–745 (1965).

Cunningham, J. A., Fuller, C. R., and Haywood, C. T., "Corrosion Resistance of Several Integrated Circuits Metallization Systems," *IEEE Trans. Rel., 19,* 182–187 (1970).

Curry, J., Fitzgibbon, G., Guan, Y., Muollo, R., Nelson, G., and Thomas, A., "New Failure Mechanisms in Sputtered Aluminum–Silicon Films," *IEEE 22nd Annual Proc. Phys.,* pp. 6–8 (1984).

Denning, D. C., LaCombe, D. J., and Christan, A., "Reliability of High Temperature I^2L Integrated Circuits," *IEEE 22nd Annual Proc. Rel. Phys.,* pp. 30–36 (1984).

d'Heurle, F. M., and Ho, P. S., "Electromigration in Thin Films," in *Thin Films: Interdiffusion and Reactions,* J. M. Poate K. N. Tu and J. W. Mayer, Editors, Wiley, New York, 1978.

Eizenberg, M., Foll, H., and Tu, K. N., "Formation of Shallow Contacts to Si Using Pt–Si and Pd–Si Alloy Films," *J. Appl. Phys., 52,* 861–868 (1981).

Evans, S. A., Morris, S. A., Arledge, L. A., Englade, J., O., and Fuller, C. R., "A 1-μm Bipolar VLSI Technology," *IEEE Trans. Electron Devices, ED27,* 1373–1379 (1980).

Fisher, F., and Neppl, F., "Sputtered Ti-Doped Al–Si for Enhanced Interconnect Reliability," *IEEE 22nd Annual Proc. Rel. Phys.,* pp. 190–192 (1984).

Foll, F., Ho, P. S., and Tu, K. N., "Cross Sectional Transmission Electron Microscopy of Silicon–Silicide Interfaces," *J. Appl. Phys., 52,* 250–255 (1981).

Fuller, E. R., and Ghate, P. B., "Magnetron Sputtered Aluminum Films for Integrated Circuits Connections," *Thin Solid Films, 64,* 25–37 (1979).

Gardiner, K. M., and Halley, S. R., "Manufacturing High Density Memory Chips," *Solid State Technol. 24* (10), 117–122 (1981).

Ghate, P. B., "Aluminum Alloy Metallization for Integrated Circuits," *Thin Solid Films, 83,* 195–205 (1981a).

Ghate, P. B., "Electromigration Testing of Al-Alloy Films," *IEEE 19th Annual Proc. Rel. Phys.,* pp. 243–250 (1981b).

Ghate, P. B., "Deposition Techniques and Microelectronic Applications," in *Deposition Techniques for Films and Coatings,* R. F. Bunshel et al., Editors, Noyce Publications, Park Ridge, NJ, pp. 514–547, 1982a.

Ghate, P. B., "Electromigration Induced Failures in VLSI Interconnects," *IEEE 20th Annual Proc. Rel. Phys.,* pp. 292–299 (1982b).

Ghate, P. B., "Metallization for Very Large Scale Integrated Circuits," *Thin Solid Films, 93,* 359–383 (1982c).

Ghate, P. B., and Blair, J. C., "Electromigration Testing of Ti:W/Al and Ti:W/Al–Cu Film Conductors," *Thin Solid Films, 55,* 113–123 (1978).

Ghate, P. B., and Fuller, C. R., "Multilevel Interconnections for VLSI," in *Semiconductor Silicon 1981, Proceedings,* H. R. Huff, R. J. Kriegler, and Y. Takeishi, Editors, *81-5,* 680–693, The Electrochemical Society, Pennington, NJ (1981).

Ghate, P. B., Blair, J. C., and Fuller, C. R., "Metallization for Microelectronics," *Thin Solid Films, 45,* 69–84 (1977).

Ghate, P. B., Blair, J. C., Fuller, C. R., and McGuire, G. E., "Application of Ti:W Barrier Metallization for Integrated Circuits," *Thin Solid Films, 53,* 117–128 (1978).

Heilmeier, G. H., "Microelectronics: End of the Beginning or Beginning of the End," *IEDM Digest Tech. Papers,* San Francisco, pp. 2–5 (1984).

Ho, P. S., "General Aspects of Barrier Layers for Very Large Scale Integration Applications, I: Concepts," *Thin Solid Films, 96,* 301–316 (1982a).

Ho, P. S., "Basic Problems For Electromigration in VLSI Applications," *IEEE, 20th Annual Proc. Rel. Phys.,* pp. 288–291 (1982b).

Kamei, Y., Kameda, M., and Nakayama, M., "Ion Implanted Double Level Metal Process," *IEDM Digest Tech. Papers,* pp. 183–141 (1984).

Kinsbron, E., "A Model for the Width Dependence of Electromigration Lifetimes in Aluminum Thin Film Stripes," *Appl. Phys. Letters, 36,* 968–970 (1980).

Kolesar, S. C., "Principles of Corrosion," *IEEE 12th Annual Proc. Rel. Phys.,* pp. 155–167 (1974).

Lepselter, M. P., "Beam Lead Technology," *Bell Syst. Techn. J.,* 45, 233–243 (1966).

Lepselter, M. P., and Andrews, J. M., in *Ohmic Contacts to Semiconductors,* B. Schwartz, Editor, The Electrochemical Society Softbound Symposium Series, New York, 1969, pp. 159–186.

McGreivy, D. J., and Pickar, K. A., Editors, *VLSI Technologies through the 80's and Beyond,* IEEE Computer Society, IEEE Catalog No. EHO 192-5, 1982, pp. 185–209.

Murarka, S. P., *Silicides for VLSI Applications,* Academic Press, New York, 1983.

Nicolet, M. A., "Diffusion Barriers in Thin Films," *Thin Solid Films, 52,* 415–425 (1978).

Nicolet, M. A., and Bartur, M., "Diffusion Barriers in Layered Contact Structures," *J. Vac. Sci. Technol., 19,* 786–793 (1981).

Nowicki, R. S., and Nicolet, M. A., "General Aspects of Barrier Layers for Very Large Scale Integration Application II: Practice," *Thin Solid Films, 96,* 317–326 (1982).

O'Donnell, S. J., Bartling, J. W., and Hill, G., "Silicon Inclusions in Aluminum Interconnects," *IEEE 22nd Annual Proc. Rel. Phys.,* pp. 9–16 (1984).

Paulson, W. H., and Kirk, R. W., "The Effects of Phosphorous Doped Passivation Glass on the Corrosion of Aluminum," *IEEE 12th Annual Proc. Rel. Phys.,* pp. 172–179 (1976).

Saraswat, K. C., and Mohammadi, F., "Effect of Scaling of Interconnections on the Time Delay of VLSI Circuits," *IEEE Trans. Electron Devices, ED29,* 645–650 (1982).

Scott, D. B., Hunter, W. R., and Shichijo, M., "A Transmission Line Model for Silicided Diffusions: Impact on the Performance of VLSI Circuits," *IEEE Trans. Electron Devices, ED29,* (4), 651–661 (1982).

Shannon, J. M., "Reducing the Effective Height of a Schottky Barrier Using Low-Energy Ion Implantation," *Appl. Phys. Letters, 24,* 269–371 (1974).

Shockley, W., "Research and Investigation of Inverse Epitaxial VHF Power Transistors," Final Technical Report No. AL-TDR-64-207, 113, Air Force Atomic Laboratory, Air Force Systems Command, Wright-Patterson Air Force Base, OH. (Sept. 1964).

Sloan, B. J., "STL Technology," *IEDM Digest Tech. Papers,* Washington, DC, pp. 324–327 (1979).

Striny, K. M., and Schelling, A. W., "Reliability Evaluation of Aluminum-Metallized MOS Dynamic RAMs in Plastic Packages in High Humidity and Temperature Environments," *IEEE Trans. Components, Hybrids and Manu. Technol., CHMT-4* (4), 476–481 (1981) (Proc. of the 31st Electronic Components Conf. 1981, Atlanta, GA, May 11–13, 1981.)

Sze, S. M., *Physics of Semiconductor Devices,* 2nd ed., Wiley, New York, 1981, Chap. 5, pp. 245–311.

Terry, L. E., and Saltich, J., "Schottky Barrier Heights of Nickel–Platinum Silicide Contacts on N-Type Si," *Appl. Phys. Letters, 28,* 229–231 (1976).

Ting, C. Y., and Wittmar, M., "The Use of Titanium-Based Contact Barrier Layers in Silicon Technology," *Thin Solid Films, 96,* 327–245 (1982).

Tsaur, R. Y., Silversmith, D. J., Mountain, R. W., Huang, L. S., Lau, S. S., and Sheng, T. T., "Shallow PtSi–Si Schottky Barrier Contacts Formed by a Multilayer Metallization Technique," *J. Appl. Phys., 52,* 5243–5246 (1981).

Vaidya, S., Fraser, D. B., and Sinha, A. K., "Electromigratiron Resistance of Fine Line Aluminum for VLSI Applications," *IEEE 18th Annual Proc. Rel. Phys.,* pp. 165–170 (1980).

Vaidya, S., and Sinha, A. K., "Electromigration Induced Leakage at Shallow Junction Contacts Metallized with Aluminum/Polysilicon," *IEEE 20th Annual Proc. Rel. Phys.,* pp. 50–54 (1982).

Yu, A. Y. C., "Electron Tunneling and Contact Resistance of Metal–Silicon Contact Barriers," *Solid State Electronics, 13,* 239–247 (1970).

7. MOS Technology Advances

Youssef A. El-Mansy
Intel Corporation

William M. Siu
Intel Corporation

In this chapter, the evolution of MOS integrated circuit technology is tracked from the late sixties, when the first commercially successful product was introduced, to the eighties and the advent of VLSI, and beyond. Major advances in process technology and resulting performance and density improvements are presented. The contributions of process advances, device cleverness, and reduction in dimensions are discussed, and extrapolations to future technologies are given. Memory devices have been and continue to be the technology driver because of their predictability and effective utilization of the most advanced technology features. Memory will be used throughout this chapter to illustrate major advances when applicable. Reliability and yield have been the major limiters to adopting the most sophisticated techniques; so, where appropriate, they will be highlighted

Figure 7-1 shows the growth in components per chip versus the year of introduction. As can be seen, the number of components per chip increased by a factor of more than a hundred over the 1975–85 decade. This was accomplished through reduction in feature size, growth of chip area, and device innovations (Moore 1975). The relative contribution of these three areas is illustrated qualitatively in Figure 7-2. Throughout this evolution the technology has been continuously changing, not only as represented by the above-mentioned indicators, but also in terms of the primary device. For technical reasons, the workhorse technology started as p-channel in the early days, changing to n-channel technology; and most recently CMOS is emerging as the mainstream technology. Another subtle change that has taken place over time is the change in emphasis from aggressive transistor scaling to interconnect optimization. The chips have evolved from an area dominated by transistors to an interconnect-limited density.

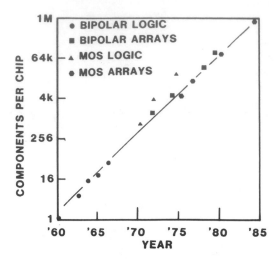

Figure 7-1. Growth in components per chip.

Also, as the chips have grown larger and become more densely populated, power dissipation has emerged as a real performance limiter and reliability concern. A reduction of power supply voltages and the move to CMOS will help somewhat, but chip heating promises to be one of the key limiters to device count growth in chips.

We start the chapter by tracking the evolution of process technology over the

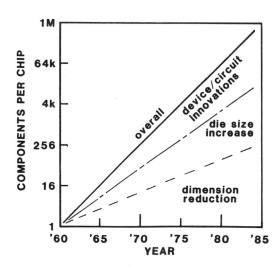

Figure 7-2. Composite effect of device/circuit innovations, die size increase, and dimension reduction.

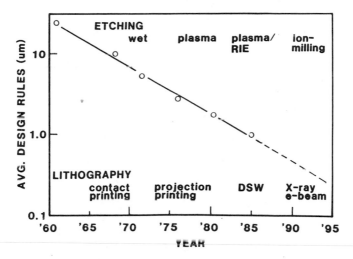

Figure 7-3. Evolution of process technology over the past two decades as measured by design rules, and projection into the 1990s.

last two decades (see Figure 7-3), followed by a description of progress in device and memory cell optimization. Some key technological advances are then presented, as well as factors that will likely affect the rate of progress in the near future.

EVOLUTION OF PROCESS TECHNOLOGY

In the early sixties, MOS devices took a major step closer to becoming reality with the advent of thermal oxidation of the silicon surfaces (Kahng and Attalla 1960). This made these devices stable and reproducible for the first time since their conception some 25 years earlier.

The making of a MOS integrated circuit requires a minimum of five major steps:

1. Definition of active gate areas.
2. Isolation around the devices.
3. Access or contacts to the various active device regions.
4. Interconnections among the active devices.
5. Passivation and scratch protection over the chip.

The photomasking sequence of the earliest MOS ICs followed the above requirements very closely. As Figure 7-4 shows, that process used five masking steps, aluminum gates, and p-channel devices. This is basically an aluminum

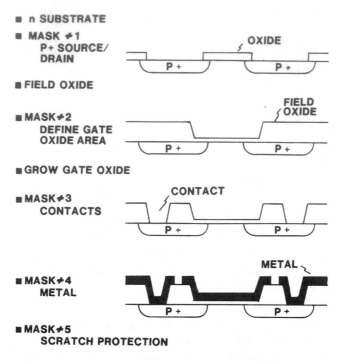

■ n SUBSTRATE

■ MASK ≠1
    P+ SOURCE/
    DRAIN

■ FIELD OXIDE

■ MASK≠2
    DEFINE GATE
    OXIDE AREA

■ GROW GATE OXIDE

■ MASK≠3
    CONTACTS

■ MASK≠4
    METAL

■ MASK≠5
    SCRATCH PROTECTION

Figure 7-4.   The basic five-mask, metal gate PMOS process.

gate technology in which p-channel transistors were used for the superior isolation capability on n-type substrates. This technology was utilized in making the earliest shift registers and logic chips (see Table 7-1). It had the following drawbacks, however:

1. Only aluminum and p^+ diffusions were available as interconnect levels.
2. The last high temperature cycle is the gate oxide growth cycle, so that high temperature gettering is not feasible.
3. It uses p-channel devices, which are inherently slower than n-channel devices.
4. Device gates need to be aligned to the source and drain, resulting in large device areas, high overlap capacitance, and poor device matching.

These were the major driving forces for further process improvements. In the late sixties, a breakthrough took place when polysilicon gates were successfully used in the making of the 256 bit MOS RAM (Faggin and Klein 1970). This was still a p-channel technology, but polysilicon replaced aluminum for MOS device gates. Also, an extra step was introduced to enable a direct contact between the polysilicon gate and the substrate, which was referred to as a buried contact.

Table 7-1. Process Evolution, a Chronology of Process and Product Advancement.

YEAR	PROCESS	PRODUCT
1962	Metal gate PMOS	Shift registers, logic chips
1969	Si gate PMOS	256 bit RAM
1971	Si gate NMOS	Static 1K RAM
1974	Si gate NMOS	4K RAM
1976	Si gate NMOS, double poly	16K RAM
1979	Si gate NMOS	64K RAM
1982	Refractory gate NMOS	256K RAM
1984	CMOS	256K RAM

The ability to make source and drain diffusions self-aligned with the gate, as well as the buried contacts, produced significant area savings. Figure 7-5 shows the major steps for that early process. In addition to density improvement, the use of polysilicon gates allowed the use of high temperature gettering up to opening contacts, and flowed glass as an insulator between polysilicon and metal to improve topography. The first commercially viable RAM was introduced using that process (see Table 7-1).

An NMOS version of the polysilicon gate process was introduced shortly thereafter (*International Electron Devices Meeting* 1975). The isolation problem inherent in NMOS processing is caused by a positive ionic charge in surface layers that depletes isolation areas between devices in the p-substrate. This problem was solved with the use of ion implantation, in which a masked boron implant was introduced in the isolation areas. Even though this step solved the device isolation problem, the resulting n-channel devices were inherently less stable because of their sensitivity to positive ion drift, primarily sodium ions, which causes an electric field change in the isolation as well as device areas. The fact that two masking steps were used for the field oxide and field implant made for cumbersome isolation rules, which are sensitive to the registration accuracy between the two layers. In spite of this, that process was used to introduce a 1K SRAM memory product (see Table 7-1).

The next major milestone was the invention of LOCOS isolation (Appels and Paffen 1971), which combined the two masking steps into one and produced self-aligned implants to the field oxide. The basic steps needed in making that process are illustrated in Figure 7-6. As can be seen, this process provides improved isolation, density, and topography. However, an extra chemically vapor deposited (CVD) nitride film was needed, which contributed to yield and reliability problems. The most significant among these problems was what is commonly known as the white ribbon effect, depicted in Figure 7-7, where a

- N SUBSTRATE
- FIELD OXIDE
- MASK #1
  DEVICE WELL
- GATE OXIDE
- MASK #2
  BURIED
  CONTACT
- POLY
  DEPOSITION
- MASK #3
  POLY GATE
- SELF-ALIGNED
  SOURCE/DRAIN
  DIFFUSION
- CVD OXIDE
- MASK #4
  CONTACT
- MASK #5
  METAL
- MASK #6
  SCRATCH PROTECTION

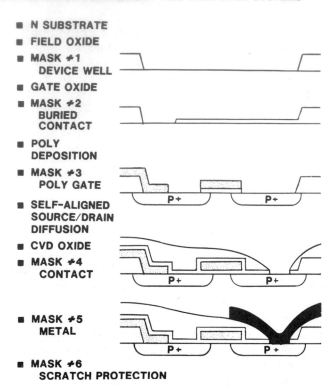

Figure 7-5.   The basic six-mask, silicon gate PMOS process.

silicon nitride ribbon is formed in the region between device and isolation during field oxide growth, and causes a thin region during subsequent gate oxide growth. This problem was eventually solved by proper removal of that ribbon prior to transistor gate oxide growth. This basic NMOS process with LOCOS isolation was the mainstay for making commercially successfull products in the mid-seventies. The key steps for such a process, which were used for the 4K dRAM, are shown in Figure 7-8. The basic structure of that process stayed essentially stable and formed the cornerstone of subsequent technologies. The major drive and motivation continued to be density and performance. The next major step came with the utilization of ion implantation for threshold tailoring to produce multiple threshold devices within the same process, using the slow-diffusing arsenic instead of phosphorous for source and drain diffusions. The result was a high performance transistor, so the process was referred to as HMOS (Pashley et al. 1977) (Figure 7-9). The process provided four different types of transistors by using two extra masking steps for threshold adjust implants. Such a high performance technology for the first time provided MOS device performance that was competitive with bipolar capability, yet at the much higher integration

LOCOS ISOLATION

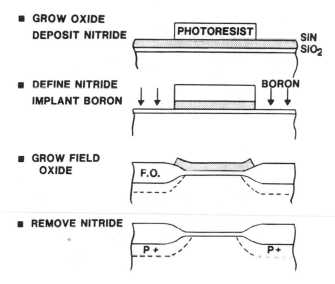

Figure 7-6. Basic process flow for LOCOS. A self-aligned field implant region is produced.

levels and lower cost afforded by MOS processing. This basic process and subsequently scaled generations of it, referred to as HMOS II (Jecmen et al. 1979) and HMOS III (Liu et al. 1982), were the principal vehicles for high performance and high density products during the late seventies and into the eighties. Even in 1985, the majority of products were still made using this technology or variations of it to suit specific product requirements.

The HMOS process structure was adapted in the early eighties to produce a CMOS process. By utilizing the vast manufacturing experience that existed for

PROBLEMS WITH LOCOS

- **REQUIRES CVD NITRIDE**
- **STRESS IN SUBSTRATE**
- **WHITE RIBBON EFFECT**

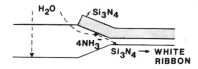

Figure 7-7. White ribbon effect in LOCOS processing.

SI GATE NMOS (4K RAM PROCESS)

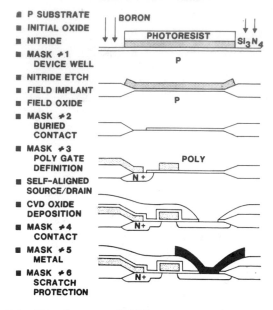

- ■ P SUBSTRATE
- ■ INITIAL OXIDE
- ■ NITRIDE
- ■ MASK #1
  DEVICE WELL
- ■ NITRIDE ETCH
- ■ FIELD IMPLANT
- ■ FIELD OXIDE
- ■ MASK #2
  BURIED
  CONTACT
- ■ MASK #3
  POLY GATE
  DEFINITION
- ■ SELF–ALIGNED
  SOURCE/DRAIN
- ■ CVD OXIDE
  DEPOSITION
- ■ MASK #4
  CONTACT
- ■ MASK #5
  METAL
- ■ MASK #6
  SCRATCH
  PROTECTION

Figure 7-8.   The silicon gate, NMOS process used for the 4K dRAM.

HMOS

- ■ HIGH RESISTIVITY P SUBSTRATE
- ■ INITIAL OXIDE
- ■ NITRIDE
- ■ MASK #1
  DEVICE WELL
- ■ NITRIDE ETCH
- ■ FIELD IMPLANT
- ■ FIELD OXIDE
- ■ NITRIDE REMOVAL
- ■ MASK #2
  DEPLETION IMPLANT
- ■ MASK #3
  ENHANCEMENT IMPLANT

- ■ MASK #4
  BURIED CONTACT
- ■ POLY DEPOSITION AND DOPING
- ■ MASK #5
  POLY DEFINITION
- ■ MASK #6
  CONTACT
- ■ MASK #7
  METAL
- ■ MASK #8
  SCRATCH PROTECTION

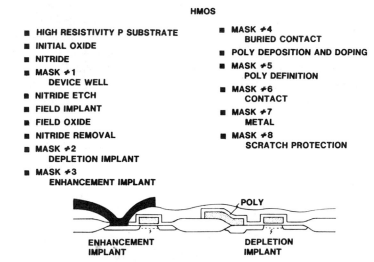

ENHANCEMENT
IMPLANT

DEPLETION
IMPLANT

Figure 7-9.   HMOS process flow and resultant process cross section. Four transistor types, including enhancement and depletion mode devices, are produced.

making HMOS, a variation for making p-channel transistors within the same technology resulted in a CMOS technology that has comparable density and performance to the HMOS technology but runs at much lower power dissipation levels (Yu et al. 1981). The basic process flow for such a technology is shown in Figure 7-10. As can be seen, with the exception of the well needed to make the p-channel transistor, the elements of the process are virtually the same as the HMOS process; so both technologies can share the same manufacturing line. Even though the viability and advantages of such a technique were demonstrated

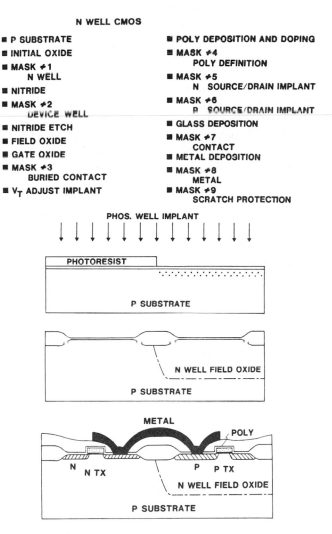

N WELL CMOS

- P SUBSTRATE
- INITIAL OXIDE
- MASK #1
  N WELL
- NITRIDE
- MASK #2
  DEVICE WELL
- NITRIDE ETCH
- FIELD OXIDE
- GATE OXIDE
- MASK #3
  BURIED CONTACT
- V_T ADJUST IMPLANT
- POLY DEPOSITION AND DOPING
- MASK #4
  POLY DEFINITION
- MASK #5
  N SOURCE/DRAIN IMPLANT
- MASK #6
  P SOURCE/DRAIN IMPLANT
- GLASS DEPOSITION
- MASK #7
  CONTACT
- METAL DEPOSITION
- MASK #8
  METAL
- MASK #9
  SCRATCH PROTECTION

PHOS. WELL IMPLANT

PHOTORESIST

P SUBSTRATE

N WELL FIELD OXIDE

P SUBSTRATE

METAL

POLY

N   N TX

P   P TX

N WELL FIELD OXIDE

P SUBSTRATE

Figure 7-10.   N-well CMOS technology process flow and cross sections.

in the early eighties, it was only in the mid-1980s that a major conversion occurred of existing process capability and new products to the new CMOS technology.

This does not imply that CMOS was introduced only recently. In fact, CMOS has existed ever since the days of metal gate p-channel MOS technology. In the early days the CMOS process was based on PMOS technology (White and Cricchi 1966). However, the relative complexity of CMOS compared to either PMOS or NMOS was so high that it was only used for special low power applications. Only recently, as NMOS technology has evolved and its complexity increased, has the relative complexity gap, as measured by the number of processing steps, narrowed from as much as a factor of two to a difference of only 20%. This is probably the single most important reason for the momentum that CMOS is gaining, which promises to make CMOS the dominant technology of choice across all products by the late eighties. Figure 7-11 depicts the evolutionary path of various MOS technologies during the seventies and eighties. As can be seen, two major approaches are emerging for VLSI application of CMOS. First, there is the evolutionary path for the standard p-well CMOS, which adopts the advanced capability of process tools and is mainly used in high density, high performance static memory because of its compatibility with the high resistance poly load cell and buried contacts. Second is the outgrowth of HMOS technology into a high performance CMOS technology by building upon the majority of process steps and adding the few extra steps needed to make the n-well. This approach is finding extensive use in dynamic memory (dRAMs), electrically programmable memory (EPROMs), and some logic applications. The mainstream technology for logic applications continues to build on the available memory technology, and no trend has evolved yet because of its relationship to the preference and existing capability of the manufacturers.

Now that the complexity gap between NMOS and CMOS has narrowed, the explosive popularity of CMOS can be traced to the following factors:

1. The inherent low power capability of CMOS circuits.
2. Their simpler circuit design.
3. The availability of special features in CMOS technology.

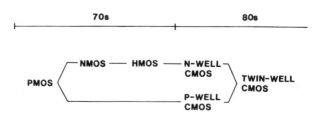

Figure 7-11. Evolutionary path of various MOS technologies from PMOS to CMOS.

Table 7-2. Design Ease With CMOS.

	NMOS	CMOS
64K dRAMS	i2164	i51C64
# Clocks	40	14
# Random Tx	1600	1100
Trac (nsec)	100	150
Special Features	—	Static column

An example of two 64K dRAMs designed in NMOS and CMOS to illustrate the above-mentioned points is shown in Table 7-2 (Chwang et al. 1983). As can be seen, the CMOS version uses approximately 30% fewer transistors than the NMOS. Most of the savings are due to the static nature of CMOS and are reflected by the dramatic reduction in the number of clocks used in the chip

DEVICE AND MEMORY CELL EVOLUTION

The process advance described in the previous section has manifested itself in a dramatic improvement in circuit performance. The key element in the advance has been the transistor, which started as a metal gate p-channel, evolved into a silicon gate p-channel and then a silicon gate n-channel, and finally led to the high performance n- and p-channel transistors of today.

The MOS Transistor

Figure 7-12 illustrates cross sections of n-channel and p-channel transistors, their schematic representation, and the transfer current–voltage characteristics. For an n-channel device the current carriers are electrons, whereas for a p-channel device the carriers are holes. Four separate regions or terminals exist in an MOS transistor: source, drain, gate, and substrate. For normal operation, the source, drain, and gate voltages measured with respect to substrate are positive for an n-channel and negative for a p-channel.

A key parameter in determining transistor operation is the threshold voltage (V_T), which marks the boundary between conduction and no conduction in the device. The expression for V_T is given by:

$$V_T = \phi_{gs} + 2\phi_f + \frac{1}{C_o\,(Q_{ss} + Q_b)}$$

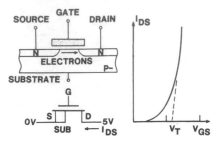

N-CHANNEL

- CURRENT CARRIERS ARE ELECTRONS
- FOR NORMAL OPERATION, SOURCE, DRAIN, AND
  GATE VOLTAGES ARE POSITIVE w.r.t. SUBSTRATE

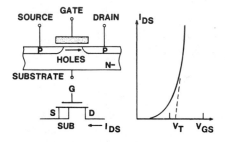

P-CHANNEL

- CURRENT CARRIERS ARE HOLES
- FOR NORMAL OPERATION, SOURCE, DRAIN, AND
  GATE VOLTAGES ARE NEGATIVE v w.r.t. SUBSTRATE

Figure 7-12.   Conduction behavior of n- and p-channel MOS transistors.

where: V_T is the threshold voltage; ϕ_{gs} is the gate–substrate work function difference; ϕ_f is the substrate Fermi level; C_o is the gate oxide capacitance per unit area; Q_{ss} is the fixed surface state charge density per unit area; Q_b is the surface depletion charge density per unit area at the onset of conduction.

The source–drain current in this case can be written as follows:

For $(V_D - V_S) < (V_G - V_T)$,

$$I_D = \mu \cdot C_o \cdot \frac{W}{L} \cdot F_s \left[(V_G - V_T)(V_D - V_S) - \frac{1}{2} V_D^2 + \frac{1}{2} V_S^2 \right]$$

For $(V_D - V_S) \geq (V_G - V_T)$,

$$I_D = \frac{\mu}{2} C_o \frac{W}{L} \cdot F_s \left[V_G - V_T - V_S \right]^2$$

where: I_D is the source–drain current; μ is the electron (hole) mobility; W and L are the width and length of the device; F_s is a small device parameter (El-Mansy 1982); V_G, V_S, and V_D are the gate, source, and drain voltages with respect to the substrate.

This equation is plotted for the same size n- and p-channel devices in Figure 7-13. As can be seen, even for this scale the n-channel transistor has about a factor of two current-drive advantage over the p-channel transistor. In the past decade, MOS transistors have been systematically scaled down in dimensions, in both the horizontal and vertical directions. Rules for this scaling were originally formulated in the early seventies (Dennard et al. 1974), and have since been

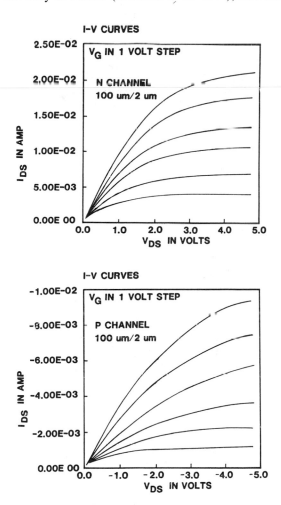

Figure 7-13.   Comparison of the drive capability of n- and p-channel transistors of the same size.

SCALING EXAMPLES

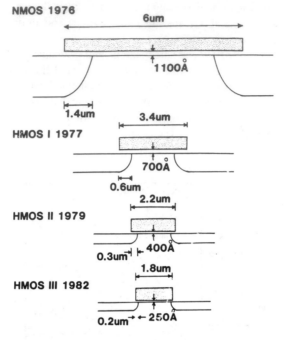

NMOS 1976
6um
1100Å

1.4um
3.4um
HMOS I 1977
700Å

0.6um
2.2um
HMOS II 1979
400Å
0.3um

1.8um
HMOS III 1982
0.2um 250Å

SCALING THEORY

Scaled Parameters	*Scaling Factor*	
Device Dimension: L, W, Xo, Xj, Xl	1/S	
Substrate Doping: Na	S	
Supply Voltage	1	1/s
Affected Parameters		
Device Current (I)	S	1/S
Miller Capacitance (C)	1/2	/S
Gate Delay (VC/1)	1/S^2	1/S
Power Dissipation (VI)	S	1/S^2
Speed × Power Produce (V^{2}C)	1/S	1/S^3

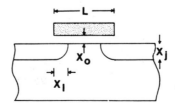

L
X_o X_j
X_l

Figure 7-14.   Examples of MOS transistor scaling.

studied and improved upon. Figure 7-14 summarizes the scaling theory as originally presented. If both lateral and vertical dimensions are reduced by a scaling
factor S (<1), the device performance as measured by the speed–power product
improves by the same factor. Further improvement can be achieved by reducing
the operating voltage by the same factor. However, for practical and standard-
ization reasons, the supply voltage was not reduced at the same pace as the other
parameters, and, in general, scaling followed the constant voltage column in the
table. Figure 7-14 shows a sequence of MOS device examples introduced during
the last decade or so. Device dimensions during this period were reduced by
about a factor of five, and the resulting speed improvement was in excess of 20-
fold (see Figure 7-15).

The pace of scaling devices down has been constrained by a number of factors
that affect yield, manufacturability, and reliability. Briefly, we may note that:

1. As the gate length shrinks, the source to drain separation becomes small,
   and punch-through takes place. The gate loses control over the drain, and
   the device fails to operate as an MOS transistor. Junction and substrate
   doping adjustments and profiling have been used successfully to reduce
   this problem.
2. At short channel lengths, high source drain electric fields generate hot (high
   energy) electrons. Some of these electrons are injected into the gate oxide,
   resulting in trapped negative charge and subsequent shifts in threshold

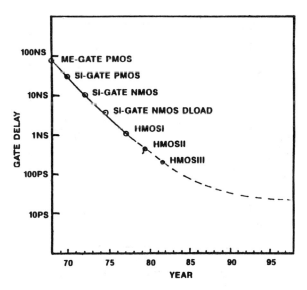

Figure 7-15.   Gain in performance over the past two decades as the MOS transistor is
scaled.

voltage. Again, tailoring of source/drain doping profiles has been used to reduce the peak electric fields, but this is considered an interim solution. A more viable solution to that problem is to improve the gate oxides in order to reduce their trapping efficiency.

3. As gate oxides get thinner, the ability to control defects is drastically reduced. This results in a severe yield impact and subsequent reliability and wear-out problems. The development of high-quality, low-defect oxides, and possibly the use of composite layers for gate dielectrics, are showing promise.

4. As junctions are scaled down, the diffusion resistance gets higher and starts to impact the available current drive of the device (Figure 7-16). Plating source–drain diffusion with a low-resistance material such as silicides or refractory metals is one of the most promising techniques (see Figure 7-17).

5. Critical dimension control is a key requirements as geometries shrink. A device with a 3μm gate length can tolerate 0.5 μm variation but at 1 μm gate length, the variation needs to be less than 0.2 μm. Advances in lithography and etching control are the pacesetters for this issue.

6. Interconnect parasitics are becoming dominant in determining performance in today's circuits. As dimensions are reduced, the fringing capacitance between adjacent lines is dominating the total node capacitance (see Figure 7-18). The value of such capacitance is increasing at a fast pace and promises to be the major performance limiter. For larger chips, the inter-

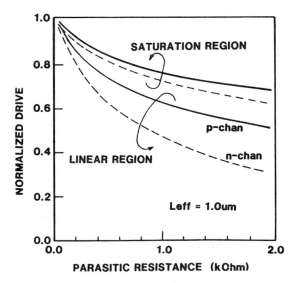

Figure 7-16.   Effect of source/drain resistance on the drive capability of n- and p-channel transistors.

CMOS PROCESS WITH SELF-ALIGNED SILICIDE GATE AND SOURCE/DRAIN

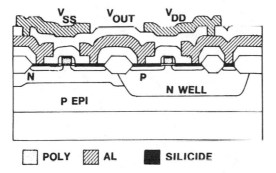

Figure 7-17.   Plannting source/drain regions with low-resistivity silicide.

connect resistance becomes critical, and the resulting time constant requires some changes in chip architecture or pad configuration. The interconnection timely delay is shown in Figure 7-19 for some of the commonly used wiring systems. Also shown in the figure is the gate delay T_G of the driver

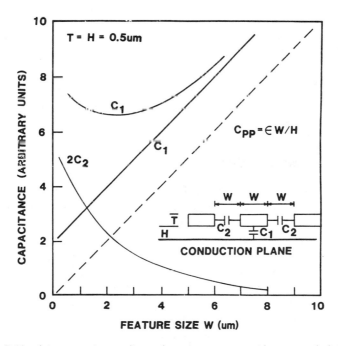

Figure 7-18.   Interconnect capacitance increases as geometries are scaled down.

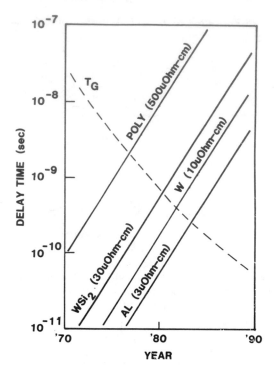

Figure 7-19.   Delay times of common interconnect systems and evolution during the seventies and eighties.

circuit. As can be seen, to utilize the high drive and intrinsic speed capability of scaled devices, migration to a lower-resistance interconnect system must take place.

Evolution of Memory Cells

As process technology evolved and devices were scaled, circuit techniques saw a parallel change that made a significant contribution to the performance and density improvement of integrated circuits. The progression of random access memory (RAM) cells is presented to illustrate that point.

The basic RAM cell is a six-element (Burns et al. 1966) cell consisting of a cross-coupled latch that stores the data in a locked-in mode and does not require refresh. As shown in Figure 7-20, four of the elements are transistors, provided by the technology, and two are load elements. The load elements have gone through a changeover time, but the basic configuration of the memory cell is still in use for static memory. That cell is very stable and provides high performance, but its biggest disadvantage is that it consumes a large chip area. The

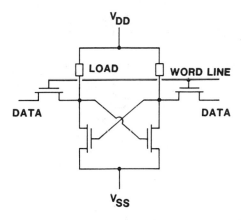

Figure 7-20.   Six-element static RAM cell schematic.

load element changed from a resistor to an n-channel transistor and back to a resistor load with varying implementation. The most advanced implementation utilizes double polysilicon technology where one of the polysilicon layers is used for the load. This provides for the smallest cell because the load can be folded over the other active elements. For the highest density, this type of memory cell is commonly used even in CMOS technologies. For micropower applications, however, a six-transistor cell is used where p-channel devices are used as load elements because control of the load resistor is not very precise. For dynamic memory cells, the evolution took a different path toward reducing the area requirements for the memory cell. The first major change was the reduction of the number of transistors required from six to three devices (Regitz and Karp 1970), which were used in making 1K dRAMs in the early seventies (see Figure 7-21). This cell, however, needed refreshing of the data periodically and every time a memory cell was accessed, which resulted in complicating the circuit design and degrading the performance although the area saving was significant.

The major breakthrough came with the invention of the one-transistor (1T) memory cell (Dennard 1968), shown in Figure 7-22. This cell, in reality, comprises two elements: an access transistor and a storage capacitor. The two select lines for the cell control the gate and source terminal of the access transistor. The 1T cell was first used in 4K dRAMs in the mid-seventies. The use of this basic cell has spread to all dynamic memory chips ever since, and it is still the only viable candidate to date, including the 1MB generation. It has gone through numerous layout and processing configurations (see Figure 7-23), but the basic principle is still the same. The 1T cell saw a dramatic reduction in size over the last decade, which was primarily driven by technology advances. Figure 7-24 illustrates cell size change as a function of year of introduction for increasing-density generations of various types of memory products. A ratio of 4 to 6 existed

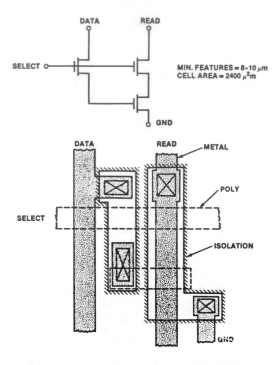

Figure 7-21.   3T cell used in the 1K dRAM.

between static and dynamic memory cell size over the years, depending on the load type used in the static memory.

BASIC TECHNOLOGY BUILDING BLOCKS FOR ADVANCED IC PROCESSES

The advances in IC technologies illustrated earlier in this chapter were fueled by innovations in every element in the process technology. Advances in thin film technology, dry etching, metallurgy, oxidation/diffusion, and lithography have all helped to pace the rapid progress witnessed in the industry. The basic physics and chemistry of most of these elements have been reviewed in the previous chapters. In this section, we will focus on how these building blocks may be integrated harmoniously to form a high performance process, but first, let us examine the goals of process integration.

Process Integration

The goal of process integration, the act of putting a process technology together from basic technology building blocks, has remained relatively constant over

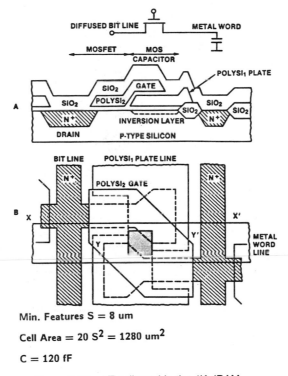

Min. Features S = 8 um

Cell Area $= 20\ S^2 = 1280\ um^2$

C = 120 fF

Figure 7-22.   1T cell used in the 4K dRAM.

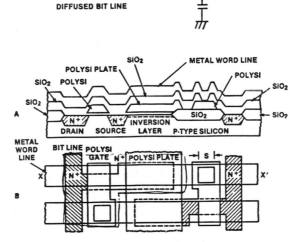

Figure 7-23.   Double poly 1T Cell used in modern dynamic RAMs.

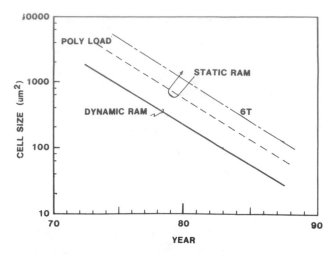

Figure 7-24.   Decrease in cell size for static and dynamic RAM cells over the last 15 years.

this period of time. It was, and still is, to build a high performance IC technology as measured by more functions per unit area and higher performance (usually measured by speed and power) per function. Many of the key advances in IC process technology have revolved around a basic set of building blocks. To understand what these building blocks are and their relative importance in a typical IC process technology, let us begin by examining the layout of a basic active device, the transistor.

Figure 7-25 shows a typical transistor layout. The drive capability of such a transistor depends on the channel width (W) to channel length (L) ratio, as well as other technology parameters such as gate dielectric thickness, gate electrode resistivity, source/drain technology, and metallization. Also noted in this figure is the significant overhead area taken up by the diffusion and the isolation. Given the stated objective of incorporating the most powerful transistor in the smallest layout area, the basic objective of process integration is to find the proper mix of basic technology building blocks to realize this goal.

A key measure of the prowess of a given technology is its design rules or layout rules. This set of parameters measures how closely the components that make up the transistor (such as contacts, isolation, etc.) can be placed together— hence, the packing density of the technology. The two key factors that determine the design rules are the minimum feature size and electrical constraints. The minimum feature size, in turn, is controlled largely by two basic technology building blocks: lithography and etching. Figure 7-26 depicts the role of these two key elements in determining the minimum feature size on a given layer.

In the remainder of this chapter we will focus on some key building blocks that have fueled the advances in IC process technology in the past 15 years;

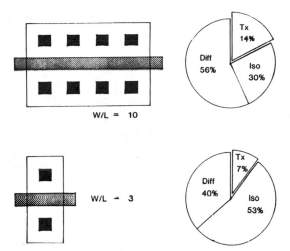

W/L = 10

W/L ≈ 3

Figure 7-25. Typical layout of a transistor showing the "overheads" in isolation and diffusion

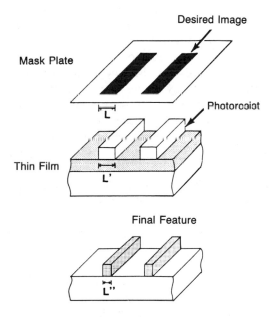

Desired Image

Mask Plate

Photoresist

Thin Film

Final Feature

Figure 7-26. The desired image, with dimension L, is printed onto the photoresist with dimension L'. After etching, the final feature has a dimension L".

namely, lithography, etching, isolation, gate electrodes, source/drain technologies, and metallization.

Lithography

The cornerstone of the advances in IC technology is lithography. The density of IC technologies depends so much on lithography that the generations of the technologies are usually referenced to lithographic capabilities (e.g., 2 μm technology, 1 μm technology, etc.).

The important role of lithography in IC technology can be readily appreciated when one considers that in a modern IC process typically 10 to 15 "masks" or lithographic steps are employed to define the many layers of images that make up the transistor and interconnect structures. With few exceptions, the minimum size geometry that can be created is determined by the resolution capability of the lithographic tool. Thus, the density of a given technology is governed largely by the available lithographic process.

Figure 7-27 depicts the evolution of lithographic tools used in the IC industry over the past decade. In all of these processes, a thin layer (usually 1–3 μm) of light-sensitive polymer (photoresist) is applied by spin coating on the surface of the wafer, followed by exposure of the photoresist layer through a mask pattern on a glass plate. Depending on the type of photoresist used, the exposed region will be dissolved away during the developing process (positive photoresist), or vice versa (negative photoresist). The resultant polymer image left on the wafer is then used to transfer the desired pattern onto the underlying substrate by means of etching, deposition, or ion implantation.

In the early days of IC processing, the lithographic tool most commonly used was contact printing. It is relatively simple with a glass mask plate placed directly over the photoresist coated wafer. Upon exposure of the pattern, the wafer is removed, and the image is formed by developing. Because there are no optics involved in the projection of the image, contact printing is capable of very high resolution, down to the submicron range. Its minimum resolution is limited only by mask dimensions, exposure wavelength, and diffraction limitations. Contact printing lost favor in the industry primarily because of its practical limitations. The contact process can damage the photoresist and cause degradation in the resist pattern as well as in the mask.

The next step in the industry was the projection aligner. Unlike contact printers, the projection aligner uses mirror optics to project an image on a master plate onto the wafer. In most designs a slit system is used to minimize aberration limitations. The wafer/master plate is scanned to transfer the entire image. The projection aligner overcame many of the practical limitations of contact printing and became the workhorse of the industry at the 2 μm level. However, with the demand for ever higher resolution approaching the 1.50 μm level, projection

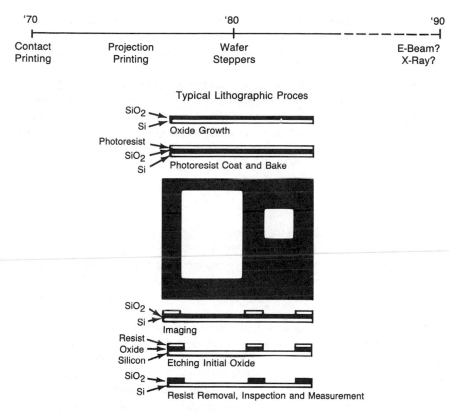

'70   '80   '90

Contact Printing   Projection Printing   Wafer Steppers   E-Beam? X-Ray?

Typical Lithographic Proces

SiO₂, Si — Oxide Growth

Photoresist, SiO₂, Si — Photoresist Coat and Bake

SiO₂, Si — Imaging

Resist, Oxide, Silicon — Etching Initial Oxide

SiO₂, Si — Resist Removal, Inspection and Measurement

Figure 7-27. Evolution of photolithographic tools used in IC processing. Also shown is a typical lithography process.

aligners, limited by aberration and diffraction, gave way to another type of projection lithographic tool. the direct step on wafer (DSW), commonly called the wafer stepper.

Similar to projection aligners, DSW systems also project an image onto the wafer from a master plate. These systems differ from projection aligners in the use of a lens system to accomplish this, as opposed to a mirror system. Also, unlike a projection aligner, which scans a whole wafer during one exposure, the DSW system "steps" across the wafer and makes a matrix of single exposures. These two differences (namely, lens vs. mirror optics and whole wafer scan vs. step-by-step exposure) have a significant impact on the ultimate design rules achievable. First, the use of a lens system allows for a significantly higher numerical aperture (NA) for the DSW—typically 0.15 to 0.20 for a projection aligner and 0.30 or greater for a DSW—hence a higher image resolution. This, in turn, translates into a smaller feature size capability. Second, the step-by-step exposure arrangement of the DSW system minimizes the across-the-wafer distortion problem inherent in full-wafer-scan projection aligners, yielding an improved layer-to-layer registration and corresponding improvement in the design rules.

The direction of future lithographic tools beyond DSW is not clear. Electron-beam direct write, optical-beam direct write, and X-ray projection have all been touted as the wave of the future. All of these techniques offer improvements in resolution and alignment outside of the R&D laboratory, but the choice of future lithographic tools will always involve a trade-off between capability and economics.

Etching

The importance of etching in the advances of IC process technology ranks with that of lithography. In the early days of IC process technology, wet chemical etching was the mainstay. The past few years have witnessed extensive migration from wet etching to dry etching. Let us now examine the underlying reasons behind this migration and the role etching technology plays in the advancement of process technology.

By its nature, wet chemical etching, except for the special case of crystallographic plan etching, is isotropic. As such, the lateral etching component is usually in the same order as the vertically etched thickness. The resultant difference between the resist image and the final thin film image (see Figure 7-28), the etch bias, will make the feature size smaller than it was intended. Although this may sound like a good way to reduce the feature size, the overall pitch does not decrease. Furthermore, practical constraints of control of the image size makes wet etching of small features infeasible.

The advances in dry etching were discussed in detail in Chapter 2. The key advantage dry etching offers is control of small image sizes. In addition, RIE dry etch processes can give true anisotropic etch profiles in multilayer thin film stacks, such as polycide films, a key requirement in the integration of advanced high density processes.

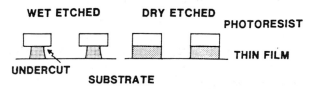

Figure 7-28.   Comparison of isotropic wet etching and anisotropic dry etching. The former results in undercutting and a smaller final feature. However, control of the dimension and pitch of the feature is compromised.

Isolation

The important role of the isolation process in design rule overheads was discussed in Chapter 4. Early MOS technologies use a thick field oxide for isolation, taking advantage of the inherently higher threshold voltage to isolate between two diffusion regions. "Moats" are then etched into the field oxide to allow the formation of diffusion regions. The LOCOS process, described earlier in the chapter, represents a major advance in isolation technology through incorporation of a self-aligned channel stopper implant and a more planar topography. Thus, the LOCOS process allows a thinner field oxide to be used and places fewer constraints on the doping of the underlying substrate. These improvements opened avenues for optimization of the transistors for improved performance as well as improved packing density.

One major drawback of the LOCOS process lies in the formation of "bird's beak" at the edge of the isolation (see Figure 7-6). Typically, this beak is in the 1.0 μm range. Although this "encroachment" does not present a serious problem on older technologies (2 μm or above), its limitations are becoming more serious in advanced technologies.

The drive to reduce the bird's beak in the isolation has been aided by dynamic RAM development. In a typical dRAM memory cell (Figure 7-29), the encroachment of the bird's beak results in a direct reduction of the storage area. In the early 1980s many investigators (Hui et al. 1982; Endo et al. 1983; Shibata et al. 1983) were engaged in the development of a viable technique to reduce

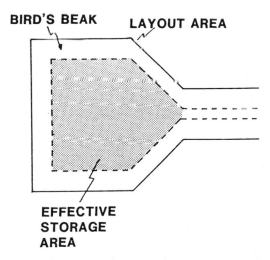

Figure 7-29.   A typical dRAM cell storage area. The effective storage area is less than the layout area because of the encroachment from the LOCOS bird's beak.

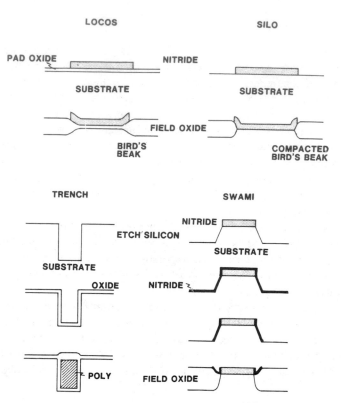

Figure 7-30. Conventional LOCOS and new reduced bird's beak isolation techniques. SILO uses direct nitridation of the silicon surface, while the TRENCH and SWAMI processes require silicon etching.

the bird's beak. Key advances were made in this area, resulting in a variety of isolation technologies. As depicted in Figure 7-30, these technologies ranged from trenches etched into the silicon to simple modification of the time-tested LOCOS process.

Gate Electrodes

Aside from acting as a control electrode to turn the transistor on and off, the gate electrode affects the propagation delay of a signal down its length. Also, the type of material used for the gate electrode can affect the source/drain technology, as well as subsequent processing.

The transition from aluminum gates to polysilicon gates was described earlier in this chapter. Polysilicon gates offer the distinct advantages of self-aligned source/drains as well as high temperature processing subsequent to gate electrode formation. Limited by grain boundary conduction, for practical thicknesses, polysilicon sheet resistivity is limited to 15 to 40 ohms/square. For stand-alone transistors this does not place a severe limitation on the gate delay. However, in memory applications where the gates of hundreds of transistors are connected together (as in the word line), propagation delay along the gate electrode becomes significant. In the past decade many workers in IC process technology have been trying to find the perfect gate electrode material. The necessary attributes are low resistivity, ability to withstand high temperature processing, and the absence of negative reliability effects on the process.

Two viable technologies, polycide gate electrodes and refractory metal gate electrodes, are illustrated in Figure 7-31.

Polycide gate electrodes utilize polysilicon as the base electrode, coated by a refractory metal silicide overcoat. The use of tungsten silicide, titanium silicide, tantalum silicide, and molybdenum silicide have all been reported in the literature. The merit of this approach lies primarily in the use of the well-known, well-characterized polysilicon layer as a buffer between the low-resistivity, high-stress silicide top layer. This configuration provides for good threshold voltage control (as in silicon gate technologies) and is easily integrated into existing silicon gate technologies. Resistivities in the range of 1 to 3 ohms/square have been reported. This represents an order of magnitude reduction in sheet resistivity compared to polysilicon gate electrodes and has made the basic silicon gate technology viable at the 1.0 μm generation.

Looking into the next decade (see Figure 7-17), the demand for even higher gate speeds (<100 psec) will again make polycide technologies obsolete. Waiting in the wings are refractory metal gate technologies and multilayer metal technologies. Multilayer metal technologies are simple extensions of polysilicon gate technologies using low resistivity interconnects (such as aluminum) to reduce the propagation delay along long lines. The merits of the multilayer metal technology lie in the known characteristics and reliability of silicon gate technology.

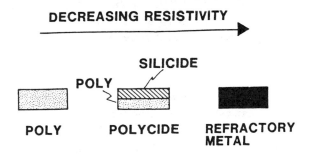

Figure 7-31.   Three common types of gate electrodes: conventional polysilicon, a polycide gate electrode consisting of a base poly layer and a silicide coating, and a refractory metal gate electrode.

Refractory metal gate technologies, on the other hand, will necessitate a complete replacement of the silicon gate process but offer a potentially simpler process sequence and enjoy the layout advantages of a low-resistance gate technology.

Source/Drain Technologies

The consequences of scaling on source/drain resistance were discussed above, in the section on "Device and Memory Cell Evolution." For the 1 μm technology and beyond, relatively high source/drain resistances will become a limiting factor in the performance of the technology. This effect was illustrated in Figure 7-16.

The most promising technique in reducing the source/drain resistivity is the use of a self-aligned silicidation process, shown in Figure 7-32. This approach can effectively reduce the source/drain resistivity by over an order of magnitude, thereby improving the performance of the technology without further device scaling and its associated reliability problems.

Metallization

The initial metallurgy used in the IC industry was pure aluminum. Aluminum was chosen for its low resistivity, ease of deposition, etchability, and ability to form ohmic contacts with silicon. As the technology advanced, the scaling of junction depths exposed one major problem with aluminum, that of aluminum interdiffusing through the shallow junctions causing spiking failures.

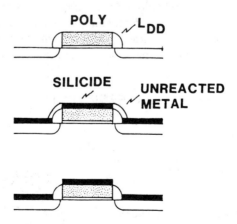

Figure 7-32.   After formation of the L_{DD} spacer, refractory metal is deposited and reacted with the silicon at an elevated temperature, resulting in the formation of a silicide layer over the poly and the source/drain regions. The unreacted metal over the spacer region is removed.

Thus, silicon (1–2%) was added to the aluminum to form a saturated alloy to prevent this phenomenon. The Al–Si metallurgy has been an industry standard for many years. However, as technology advances further, the ever increasing number of components per chip leads to a higher power density dissipation. This in turn places a significant demand on the metallurgy, and electromigration-induced failure becomes a real threat.

The migration to a new metallurgy is met with many conflicting requirements. Low resistivity, etchability, corrosion resistance, ability to make ohmic contacts, and so forth, make the choices rather limited. Although the use of other metallurgies, such as tungsten, has been reported, aluminum compounds are likely to remain the mainstay of VLSI interconnects in the foreseeable future.

ACKNOWLEDGMENT

The authors express their gratitude to Leanna Woodall for her untiring effort in preparing this manuscript.

REFERENCES

Appels, J. A., and Paffen, M. M., *Philips Research Report*, p. 157 (1971).

Burns, J. R., et al., *ISSCC Digest of Technical Papers*, pp. 118–119 (1966).

Chwang, R., et al., *ISCC Digest of Technical Papers*, pp. 56–57 (1983).

Dennard, R. H., U.S. Patent 3, 387, 286 (1968).

Dennard, R. H., et al., *IEEE Journal of Solid State Circuits*, SC-9, 256 (Oct. 1974).

El-Mansy, Y., *IEEE Trans. Electron Devices*, pp. 567–573 (1982).

Endo, N., et al., *Technical Digest IEDM*, pp. 31–34 (1983).

Faggin, F., and Klein, T., *Solid State Electronics*, 13, 1125–1144 (1970).

Hui, J., et al., *Technical Digest IEDM*, pp. 220–223 (1982).

International Electron Devices Meeting, pp. 295–301 (1975).

Jecmen, R. M., et al., *ISSCC Digest of Technical Papers*, pp. 100–101 (1979).

Kahng, D. H., and Attalla, M. M., IRE-AIEE Solid State Device Research Conference, Pittsburgh, PA (June 1960).

Liu, S. S., et al., *ISSCC Digest of Technical Papers*, pp. 234–235 (1982).

Moore, G. E., *International Electron Devices Meeting*, p. 11 (1975).

Pashley, R., et al., *ISSCC Digest*, pp. 22–23 (1977).

Regitz, W. M., and Karp, J., *ISSCC Digest of Technical Papers*, pp. 42–43 (1970).

Shibata, T., et al., *Technical Digest IEDM*, pp. 27–30 (1983).

White, M. H., and Cricchi, J. R., *Solid-State Electronics*, pp. 991–1008 (1966).

Yu, K., et al., *ISSCC Digest of Technical Papers*, pp. 208–209 (1981).

8. CMOS Devices

Pallab Chatterjee

VLSI Design Laboratory Texas Instruments Inc.

MOS technology, through its exponential growth in the last few years, has opened up a whole new domain of applications. Cost reduction, increased integration density, and improved circuit performance are the three benefits of device scaling. They combined to be the economic and technological driving force of the 1970s. The decade of the seventies was dominated by nMOS technology, and the linear scaling of device structures (Dennard et al. 1974) was adequate, as shown in Table 8-1. The switching speed of the active device determined circuit performance, and device isolation and interconnection were not a major problem.

In the eighties, the demand for reliability and system level power considerations focused attention on chip power dissipation. CMOS emerged as the undisputed technology of choice of VLSI level integration.

Specific applications demanded that particular features be incorporated in the technology. These application-oriented requirements rendered the prescription of linear scaling of active devices impractical as we reached the 1 μm geometry range. A "discontinuity" in device structure and the materials needed for IC fabrication, as well as the techniques used for device and circuit design, developed. The isolation and interconnection of devices assumed a more important role in determining circuit performance.

This chapter will briefly survey the state-of-the-art in CMOS technology and some of the device and structural changes that are being made to serve specific applications of VLSI.

TABLE 8-1. Constant Field Scaling.*

PARAMETER	SCALE FACTOR
Lateral dimensions (L, W)	$1/K$
Vertical dimensions (t_{ox}, X_j)	$1/K$
Doping concentration (N_A)	K
Voltage	$1/K$
Current	$1/K$
Capacitance	$1/K$
Delay	$1/K$
Power	$1/K^2$
Power density	1
Power–delay product	$1/K^3$

*Nonscaled quantities: resistance, subthreshold slope, built-in junction potential.

ANATOMY OF A BASIC CMOS PROCESS

The fabrication of CMOS devices requires the integration of an n-channel and a p-channel transistor isolated from each other in a common substrate with an interconnect system that allows metallic connections to both polarity devices. Implementation of the CMOS fabrication process in the seventies was based on a large number of masking operations to separately form the n-channel and p-channel devices. This made the process complex and reduced the yields compared to nMOS processes. The technology used for the fabrication of CMOS devices blossomed in the late seventies, borrowing heavily from the learning obtained from nMOS technology. The use of self-aligning techniques reduced the masking steps needed for CMOS, whereas special power-saving circuit design techniques increased the masking steps used to generate multiple threshold devices for nMOS. The fabrication of CMOS in a bulk substrate involves a parasitic SCR structure that is not present in nMOS. Thus, the focus of technology development work in translating from nMOS to CMOS has centered on avoidance of a possible latch-up of the SCR.

These developments in process and device technology, coupled with the ease of designing static CMOS, have been the main driving forces in making CMOS the key technology choice for VLSI in the eighties.

A cross section of fabrication steps in the formation of a basic CMOS structure is shown in Figure 8-1. The choice of the substrate material is governed by isolation and latch-up considerations. In order to lower the substrate impedance and thus increase the current required to sustain latch-up, a low conductance p$^+$ substrate with a thin p-epitaxial layer is preferred. The effective doping in the

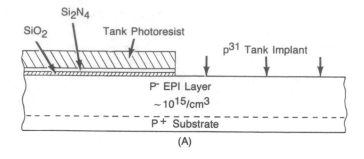

(A)

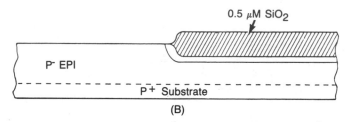

(B)

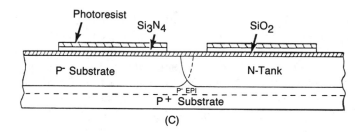

(C)

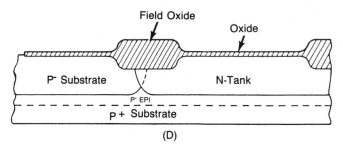

(D)

Figure 8-1. A typical set of cross-sections during a CMOS fabrication sequence (labeled A–F).

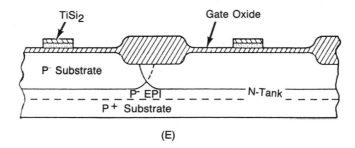

(E)

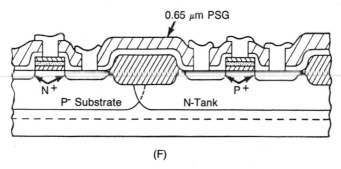

(F)

Figure 8-1. *(Continued)*

active device areas is set by ion implantation for both n-channel and p-channel devices. These implantations form the lateral isolation for the p-channel and n-channel devices. These regions are variously called "tank," "tub," or "well." The definition of both p-tank and n-tank with ion implantation is referred to as a twin tank process. The implantation is performed with a single mask step, using a self-aligning technique that grows a protective oxide on the first side so that the second side can then be implanted. The tank is typically several microns deep to avoid being punched through when biased. A long impurity diffusion step is necessary to achieve the necessary tank depth. During this step, impurities move laterally, causing a weakening of isolation near the tank boundary. This limits how close the n-channel and p-channel devices may be placed to each other.

The next step in fabricating a CMOS device is to isolate the transistors from each other, since the tank provides a boundary grouping all p-channel and all n-channel devices. This is done by growing a thick oxide region selectively, just as is done in nMOS. The active areas are protected from oxidation using a silicon nitride oxidation barrier. This is defined in a single masking step for both p-channel and n-channel devices. Normally it is necessary to implant boron into the regions that will separate the n-channel transistors from each other, so that low-leakage isolation is provided. If the n-tank concentration is low, this implant needs to be masked from the n-tank region. However, the n-tank concentration in sub 2 μm CMOS is such that it is not necessary to mask this implant. This reduces one masking step in the implementation of CMOS.

After the tank and isolation oxide are formed, it is necessary to form the n-channel and p-channel MOSFETs. For state-of-the-art CMOS, the gate of both of these MOSFETs is formed with n^+ polysilicon. Thus, the flat band voltage is one band gap different for the n-channel, compared to the p-channel. As a result, the p-channel device has a threshold voltage of nearly -2 V, if no threshold-adjusting implants are performed. The implantation of boron to create a buried channel type p-channel is needed. It is also necessary to increase the threshold voltage of the n-channel device, unless the p-tank doping is already high enough. This also requires a boron implant. It is possible to achieve both these purposes with a single implant, thereby saving another masking step.

The gate electrode is patterned and etched as in nMOS technology. The use of a sandwich structure of polysilicon clad with refractory metal silicide is common, and this reduces the sheet resistance of the gate electrode so that it can be used as an effective layer of interconnect.

The formation of the source and drain regions in CMOS would normally require two masking steps. Recently, it has been shown that the use of counterdoping is possible to fabricate both types of source and drain regions using one mask. The penalty that is paid in the counterdoping scheme is that the sheet resistance of the diffusion regions is increased. For near 1 μm feature size, a self-aligned silicide process has been used that allows the gate and source drain regions of both the n-channel and p-channel regions to be metal-clad at once. This reduces the effective sheet resistance of all the levels.

The fabrication of the metal system and the contacts is generally similar to the process sequences used in nMOS technology. The only significant difference is that the contacts are made to both polarities of devices. This requires that some precautions be taken to protect the shallow junctions from the dopants in the reflow glass layer used to smooth the surface. The glass is typically doped with phosphorus and boron.

The structure described above is basic CMOS technology that can be applied to some simple logic applications. However, specific requirements of different VLSI applications necessitate that the device design be modified, or that other device and interconnect facilities be added to this basic flow. Most of these fabrication tradeoffs have to be reexamined as the feature size is scaled down. We will discuss these scaling-related issues later in this chapter.

DESIGNING WITH CMOS TECHNOLOGY

The most attractive feature of CMOS is that for the classical static CMOS implementation, as shown in Figure 8-2. Each logic gate is a series combination of an n-channel and a p-channel device. Regardless of the state of the input,

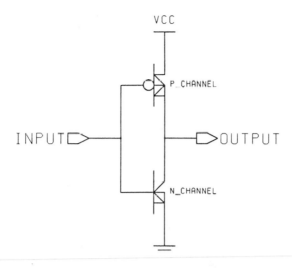

Figure 8-2. Circuit schematic of a static CMOS inverter.

one of the two devices will be off. This results in a standby power in static CMOS circuits that is very low. Current flows in the circuit only when a transition of state is occurring. This feature makes it extremely easy to manage the power dissipation budget in CMOS designs. Further, the output is always connected to one of the power supply rails because at any given state only one transistor is on, and the other is off. This guarantees that the logic swing will be determined by the power supply voltage only, and not by the ratio of the effective impedance of the devices, as is the case with static nMOS design. If there were no density or performance penalties, static CMOS would be the ideal way to implement digital circuits.

Although many circuits use static or classical CMOS, where one p-channel and one n-channel device are used for each logic switch, the density of such layouts is poor compared to nMOS. There are two reasons for this loss in density. The first is that the spacing between a p-channel and an n-channel device is not easy to scale. The second is that the static CMOS requires many connections to realize the series parallel device structures for multi-input NOR gates (Figure 8-3).

In order to work around the power dissipation and other problems of static nMOS, dynamic nMOS circuits were invented in the seventies. The design style of a modern CMOS is to some extent derived from knowledge gathered from nMOS technology. Thus, various new styles have been proposed. These styles trade layout complexity for clocking complexity. Domino CMOS (Figure 8-4) realizes a logic function with nMOS logic style and uses a clock and a buffer to perform the logic function in two stages. When the clock is low, the p-channel pull-up is on, and the output node is precharged. At this time the inputs are set

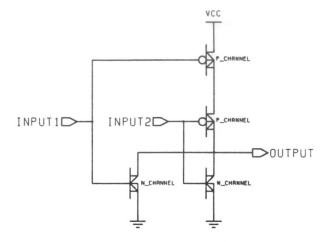

Figure 8-3.   Circuit schematic of a static CMOS two-input NOR gate.

up on the n-channel devices. When the clock goes high, the output node falls if there is a path through the logic to discharge it. This realization of logic is more efficient in layout while requiring a clock to be propagated. It also comes at some expense of noise margin compared to static CMOS.

It is certain that other forms of CMOS circuits will evolve in the future. The driving force behind them is the desire to perform a given logic function with nMOS density at static CMOS power level with no clocks and no sacrifice in speed or noise margin.

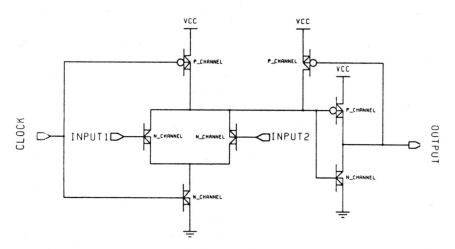

Figure 8-4.   Circuit schematic of a domino CMOS two-point NOR gate.

CUSTOMIZING THE TECHNOLOGY: APPLICATION CONSTRAINTS

CMOS technology has been applied to realize VLSI circuits that perform memory and logic functions. Each of these applications has some special requirement that could best be served by customizing the technology. On the other hand, the economy of scale and the continuous learning and cost reduction achieved in the industry are based on designing different circuit applications using an identical technology. The evolution of technology and the structural changes that are incorporated are largely the result of a tradeoff of customization against volume and cost.

Memory applications require the largest volumes of chips and are very cost-sensitive. Thus, the technology used for memory generally is heavily customized. The additions to the basic process required to perform the memory functions are all dictated by the memory cell itself.

The dynamic RAM is realized with a memory element that consists of a capacitor on which charge is stored and a transistor that isolates it from the bit line. The cell is addressed by turning on the transistor. The major problems in the scaling of a dRAM cell have involved the amount of charge stored, which depends on the size of the capacitor and the voltage swing. As the density of this kind of memory is increased, the area of the cell must decrease. The minimum amount of charge to be stored has been limited by noise sources such as alpha particles emitted from residual radioactive elements in the package materials. In order to scale the cell area without losing charge, it is necessary to maintain the voltage swing and increase the capacitance per unit area by thinning down the gate oxide. The voltage swing at the capacitor can be maintained at the power supply level only if the gate voltage can be bootstrapped a threshold voltage above the power supply. Thus, all isolations and transistors of dRAM must withstand high voltages. CMOS provides two additional hooks for dRAMs. As the noise source is an alpha particle that generates charge up to 20 μm into the silicon, it is possible to fabricate the memory cells in the tank region so that the pn junction can form a barrier that keeps some of the charge away from the cell. This makes CMOS implementation attractive. The second reason for using CMOS in dRAMs is system-driven. Most of the applications of dRAMs require that bits be retrieved from addresses near the first address in a sequence. In computer architecture jargon, this is referred to as a locality of reference. If we could provide a way to allow a large number of bits with "nearby" addresses to be accessed very rapidly after the first bit was accessed, the system speed could be increased. This is easily done if all the cells accessed on one row of the dRAM can be held temporarily in a latch from which they can be randomly accessed very quickly. This function, referred to as a "static column mode," is implemented with a reasonable power penalty only if CMOS technology is used.

The electrically programmable ROM (EPROM) application requires the availability of a thin insulator layer through which one can transport small amounts of charge in order to change the threshold voltage of a transistor that acts as the memory element. This is realized with a transistor that has two gates, one stacked on top of the other. The transport of current between these two gates requires the use of a very thin dielectric, generation of high voltages to allow this current to flow when programming, and making sure that there is little enough leakage through the thin dielectric at lower voltages so that the charge stored in the floating gate cannot escape for years.

The static RAM application requires the fabrication of a dense CMOS latch as the basic cell. This is best done by having a buried contact from the gate to the n^+ and p^+ layers. Other than this requirement (which is a density tradeoff), static RAMs can be fabricated with the basic CMOS process described above (in "Anatomy of a Basic CMOS Process"). For this reason, static RAMs have been popular in technology development.

Logic applications, typically led by microprocessors and peripherals, require the hooking up of a large number of random gates. This is best accomplished by having two levels of metal available for wiring. The density of logic circuits is generally limited by the pitch on these two metal levels. The trend in today's microprocessors is toward reduced instruction set (RISC) architecture. This allows processing to be done extremely fast, so that the main performance bottleneck is the time to fetch the data from memory for the next operation. This bottleneck can be broken if a large amount of data and an instruction cache are available on the chip. This would lead to the inclusion of large static RAMs on the chip with a microprocessor. This synergism agrees very well with the fact that the technology needs for sRAMs and microprocessors are very similar. Another emerging application area is field programmable logic. This would require that EPROM and microprocessor technology be merged together.

These applications, as well as the continued trend toward smaller feature sizes, are resulting in some interesting changes in the way CMOS technology is evolving. In the rest of the chapter we will discuss these issues.

THE POWER SUPPLY DILEMMA

The scaling laws of the seventies were based on a constant electric field premise, which required that as the channel length was scaled down, the operating voltage should be scaled as well. The constant electric field would eliminate the effects of velocity saturation and carrier heating during device operation.

Circuit and system level constraints, however, favored a constant power supply voltage over many generations of ICs. The dominance of TTL devices has currently set the power supply standard at 5 V $\pm$ 10%. However, some additional circuit and system level constraints must be understood before device design issues are discussed.

First, the statistical fluctuation of process parameters is a significant problem at submicron feature sizes because scaling manufacturing tolerances is very expensive and technologically complex. Second, the requirement for ICs to operate over a large range of temperatures results in a tolerance requirement at the circuit level that can comprehend the temperature coefficients of the threshold voltage and mobility. The third circuit level constraint is based on noise margins, both internal and external to the chip. This stems from cross-talk and coupling problems.

All these tolerances add up to voltage margin requirements from a circuit design point of view. These voltage margins dictate that power supply voltages cannot be scaled down at the same rate as device geometries. Indeed, for the largest volume IC application today (dRAMS), signal-to-noise considerations have constrained the power supply to be maintained at 5 V. JEDEC has established 3.3 V $\pm$ 10% as the next power supply standard, but its adoption is not expected until <0.7 μm geometries are used.

In order to solve the device design problems that arise from these circuit constraints, a lightly doped drain (LDD) transistor structure was invented. We will focus on performance limitations and reliability problems associated with nonscaling of the power supply and the use of LDD structures.

PERFORMANCE LIMITATIONS DUE TO VOLTAGE NONSCALING

In digital circuits the active device is assumed to be an ideal switch with a precisely determined switching threshold, below which it conducts no current, and above which it has a very low on impedance and high drive current. The MOSFET is not an ideal switch, and at submicron geometries the threshold voltage, off current, and drive current all depend on the voltage operation.

THRESHOLD VOLTAGE

Of interest in the formulation of the threshold voltage for small geometries is the length and width dependence, which determines the fluctuations that may be expected in the processing of these devices. An important aspect of the measurement of threshold voltage for small geometry is dependence on the drain voltage. This is a direct manifestation of the degree of two-dimensional influence on the surface potential and may, indeed, be construed as a figure of merit for device design. As the channel length of the MOSFET gets shorter, control of the threshold voltage becomes more difficult because of the short channel effect caused by drain-induced barrier lowering (DIBL) (Troutman 1979). The problem is further aggravated if the controllability of the gate length in the fabrication does not improve with the channel length such that $\Delta L/L$ is constant.

The reduction in threshold voltage at short channel lengths as shown in Figure 8-5 is a direct manifestation of the effects of two-dimensional charge sharing and drain-induced barrier lowering.

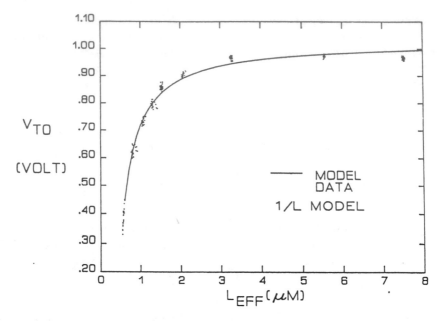

Figure 8-5. Dependence of threshold voltage on channel length, demonstrating two-dimensional charge sharing and DIBL.

There are two-dimensional fields associated with reduction of the width dimension also. The lateral field at the edge of the channel is determined by the shape of the transition region oxide. This field is generally small for LOCOS processes. However, there is a significant lateral encroachment of dopants, which causes an additional amount of depletion charge to be associated with the gate (Kers et al. 1982; Noble and Cottrell 1976).

In both commercial and military IC applications, it is important to monitor the temperature dependence of the threshold voltage. The temperature coefficient of the threshold voltage is about 2 mV/°C at 1 μm, and it increases slightly at submicron dimensions.

Since the threshold voltage is directly dependent on the value of ϕ_{MS}, the choice of the gate electrode determines the level of channel doping required to obtain a given threshold voltage. The use of n^+ polysilicon gate is common for scaled n-channel technology. However, the channel doping required for a given threshold voltage would be much less if p^+ polysilicon were practical. This is normally not possible because the gate oxide is not a good diffusion barrier for boron. Recent work on oxynitrides shows that they may be able to provide the desired diffusion barrier (Sodini et al. 1982). This would reduce the temperature sensitivity of the threshold and further reduce the body effect at the cost of increased DIBL. The choice of the gate electrode work function is even more interesting for CMOS circuits. A ϕ_{MS} electrode such as molybdenum has been proposed (Yasuda et al. 1982) to make design for p-channel and n-channel devices more symmetrical and less prone to geometry effects.

Drive Current Limitation

In digital circuits the figure of merit of a MOSFET switch, when turned on, is its ability to deliver current when both gate and drain to source voltage are equal to V_{cc}. Mobility in the linear region, together with velocity saturation, and effects in the saturation region determine this current.

The impact of nonscaling of voltage on the internal operation of the MOSFET can be understood in terms of increases in the vertical and lateral electric field.

Scaling devices requires increasing the doping in the channel region. It has been shown that this increases the lateral electric field and decreases the effective carrier mobility. Figure 8-6 shows the effects of the increased vertical electric

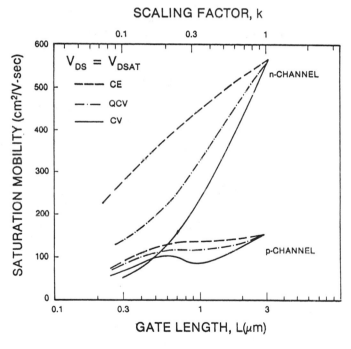

Figure 8-6. Reduction of effective mobility in scaled devices due to vertical electrical field. Three cases are shown: *CV* is for constant voltage, *QCV* allows voltage scaling as the square root of feature size scaling, and *CE* is based on a constant electric field.

field on mobility, for both p-channel and n-channel devices. As the gate length is scaled, three scenarios of voltage scaling are shown, and it is clear that constant voltage scaling results in a very rapid degradation of device performance.

The nonscaling of voltages also has a major impact on device performance if the junction depth is scaled down (Scott et al. 1982). Figure 8-7 shows an increase in sheet resistance as a function of decreased junction depth. Increased source/drain resistance decreases the effective transconductance of the transistors, as shown in Figure 8-8.

In order to solve this problem, self-aligned silicide cladding techniques have been developed. The use of $TiSi_2$ has allowed the sheet resistance of the source/drain region to be as low as 1 ohm/square for 0.2 μm deep junctions. This is a major discontinuity in the use of new material in ICs.

The lightly doped drain structure fabricated using a sidewall oxide is emerging

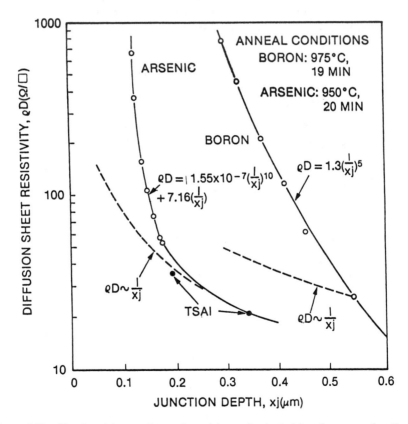

Figure 8-7. Sheet resistance of arsenic and boron implanted junctions as a function of junction depth.

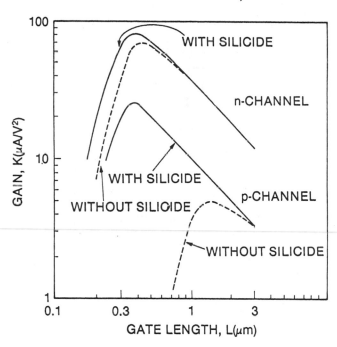

SCALING FACTOR, k

Figure 8-8. Device gain as a function of gate length for scaled devices. Note the impact of the silicided source/drain on p-channel devices.

as the source/drain structure of choice for submicron CMOS. This structure is shown in Figure 8-9. Note that the n region under the sidewall oxide contributes a high sheet resistance section, and this does reduce the available transconductance. In addition, it is important that the contact resistance between the silicide cladding and the source/drain diffusion be low in order to avoid further degradation in transconductance.

Subthreshold Current

Subthreshold or weak inversion conduction in MOSFETs is of growing concern in the operation of small-geometry MOSFET circuits. This region of MOSFET currents affects dynamic circuits, CMOS standby power, and the precision of analog circuits. Furthermore, the requirements of adequate subthreshold turn-off are contradictory to the requirements of device scaling for above-threshold operation.

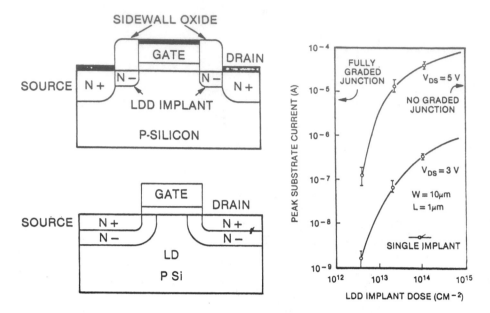

Figure 8-9.   Dependence of substrate current on LDD implant dose for an LDD transistor structure.

A convenient measure of the subthreshold turn-off characteristic is the gate bias swing S needed to reduce the subthreshold current by one decade (Brews 1979). By this definition:

$$S = \log_{10}\frac{dV_{GS}}{d(\ln I_D)}$$

[1]

Subthreshold conduction is by diffusion over the potential barrier set by the surface potential in the channel. Thus, it depends on channel doping, oxide thickness, and substrate bias (Brews 1979). In order to maintain a small gate swing S for scaled-down MOSFETs, the substrate doping level needs to be low, and the oxide thickness needs to be small. The low doping is opposite to that needed to scale down MOSFETs with low drain induced barrier lowering. Figure 8-10 shows that the application of substrate bias can reduce G for a given oxide thickness and channel doping. This points to the use of substrate bias for submicron CMOS. It is important to note that the subthreshold slope is proportional to RT so that operating at higher temperatures causes an increase in standby currents.

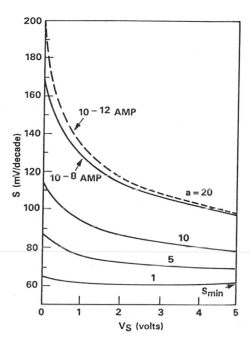

Figure 8-10. Subthreshold slope s as a function of substrate bias. The parameter $\alpha = 2\ \varepsilon_s t_{ox}/\varepsilon_{ox}\ L_D$, where L_D is the Debye length.

RELIABILITY PROBLEMS DUE TO VOLTAGE NONSCALING

The increase in electric field near the drain end of scaled-down MOSFEs causes low-level impact ionization, which results in a majority carrier current to the substrate. The carrier temperature can be quite high, and for electrons, injection into the insulator is likely. This produces long-term degradation of the device. The hot carriers can also emit photons through mechanisms such as Brehmstrallung or interband transition. These photons can cause errors in dynamic memories built in the CMOS. A circuit level solution to these problems is to use a series device in each inverter, which is wasteful of density.

The maximum electric field near the drain determines the extent to which hot carrier effects occur. The maximum field occurs between the switch-off point and the drain, and for abrupt drain junctions is represented as:

$$E_m = \frac{V_d - V_{dsat}}{l}, \quad l \propto t_{ox}^{\frac{1}{3}}\ x_j^{\frac{1}{2}} \qquad [2]$$

Thus, reduction of the electric field would require increases in gate oxide thickness and junction depth, both of which are contrary to the scaling laws.

It is possible to increase l by using a lightly doped drain junction. Figure 8-9 shows the dependence of the substrate current on the LDD implant dose, demonstrating that it is possible to reduce the peak electric field. The position of the peak electric field is also important in determining the degradation mechanism. If the maximum field occurs under the sidewall oxide, the charge injected into the oxide will be trapped over the n⁻ region. If the doping of the n⁻ region is light, this can cause a depletion of the n⁻ region, thus increasing the series resistance and decreasing the transconductance of the device. If the peak field is under the gate, the trapped charge manifests itself as a shift in the threshold voltage. Figure 8-11 shows the cumulative effect of stressing an LDD device where the initial degradation is due to trapped charge in the oxide. As the degradation continues, the maximum field moves under the gate, and a shift in V_t occurs.

Another significant reliability issue is the time-dependent dielectric breakdown

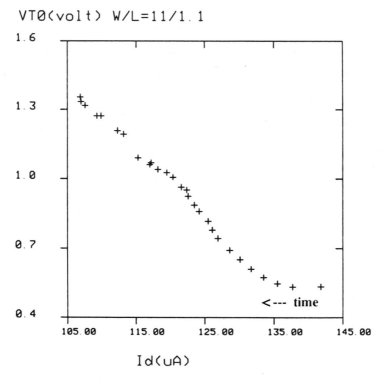

Figure 8-11.   Change in device current vs. the change in V_t during hot electron stress of LDD devices.

in thin dielectrics. If the supply voltage is not scaled, the Fowler Nordheim transport through thinner oxides causes trapped charge buildup that finally results in failure of the gate oxide. This is very severe in dRAM capacitors and EPROMs. In the case of capacitors, the solution to this problem is to reduce the oxide field by holding the storage plate at $V_{cc}/2$. The endurance specifications of nonvolatile memories are governed by this mechanism.

ISOLATION SCALING—SO NEAR, YET SO FAR

The scaling of integrated circuits assumes that the active devices can be densely packed on a common substrate with adequate isolation between the devices. The problems of device isolation can be classified in three categories.

The first problem relates to the common substrate. Work on the traditional measure of lifetime and crystal defects is progressing adequately to reach the goal of submicron VLSI. However, the electrical noise induced in the substrate, as well as the tank of CMOS devices, increases as the switching speed increases. There are excess carriers induced in the surface by various mechanisms, which act as sources of coupling between devices in MOS circuits, especially when on-chip substrate pumps are used. The use of epitaxial substrates is an acceptable solution to this problem.

The second problem in isolation is based on the loss of device density due to large transition regions in the LOCOS isolation. The elimination of this encroachment has been the major emphasis of a large variety of novel isolation structures (Wang et al. 1982; Chiu et al. 1982; Kurosawa et al. 1981).

The isolation of p-channel devices from n-channel devices in CMOS circuits is probably the most significant isolation problem. Typical p^+ to n^+ isolation design rules in state-of-the art CMOS are 2.5 times the minimum design rules. Two problems are encountered in the scaling of this isolation. The first is the design of adequate field thresholds near tank boundaries, and the second is providing latch-up immunity. The use of twin tank CMOS has reduced the field threshold problem by defining a sharp tank boundary.

The latch-up problem in submicron CMOS has been studied in detail. Figure 8-12 (a and b) shows the most accepted solution (Hall et al. 1984; Taur et al. 1984) to the latch-up problem. The scaling of the tank depth causes a significant increase in the bipolar gain of the vertical p-n-p device. If a p/p^+ epitaxial substrate is used, most of the current induced due to a forward bias in the tank goes to the substrate, leaving a small fraction that can turn on the lateral p-n-p device. This increases the holding current. The use of a silicided butted contact in the tank further increases this effect. The holding voltage in latch-up is now determined by the sheet resistance of the tank, the shallow epi layer, and the distance of the contact from the tank edge. The holding voltage can be larger than V_{cc}, as shown in Figure 8-12(b). This allows the circuit to recover from

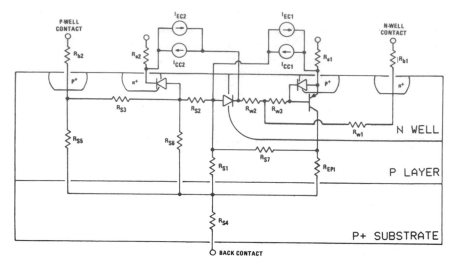

Figure 8-12(a). Cross section of CMOS tank shows elements pertinent to latch-up.

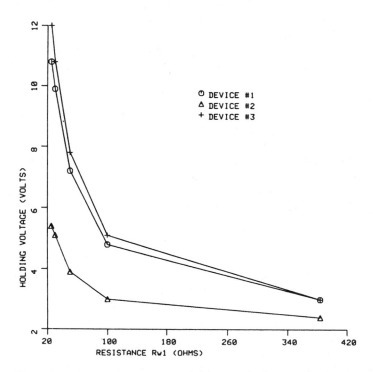

Figure 8-12(b). Holding voltage as a function of contact spacing in tank.

latch-up. The use of low impedance substrate bias can reduce latch-up significantly. However, high impedance substrate bias generated on the chip may be prone to latch-up (Hall et al. 1984). SOI technology is the ideal solution to this problem.

THE INTERCONNECT BOTTLENECK

The scaling of CMOS devices to micron geometries has resulted in an explosion of interconnect requirements to hook up these devices to perform a circuit function. As the integration of more functions on a single chip continues to expand, the number of interconnects needed is predicted to grow more than linearly. This growth in interconnect requirements has resulted in the total functionality of the chip being dominated by the electrical response of the interconnects. In addition, technologists have found that scaling the interconnects is limited by the physical properties of the metal that is used, as well as the electrical properties. These requirements are discussed below.

The most significant improvement in integrated circuit performance in the last decade derived from the reduction of wiring capacitance as dimensions were scaled down. Because wire dimensions were significantly larger than insulator thickness, the capacitance of the wires could be calculated by considering them to be parallel plate capacitors. The wiring needed to connect submicron devices cannot be viewed as ideal parallel plate capacitors with "thin plates." As the geometries scale down, the wires must carry larger current densities in order to carry power to different parts of the circuits. In addition, the signals that must be transmitted through these wires are switching much faster, as device speed is increasing with down-scaled dimensions. Because the chip sizes that are being built today are actually increasing to allow larger functions to be integrated, the average length of the wire is large, as is the capacitance of the wire. Therefore, larger current densities must be carried through these tiny wires, which produce a phenomenon called electromigration where the electron current flowing through the wire can cause metal ions to move by "pushing" the ions as they go by. If the metal is formed of uneven grains, they can cause voids to form, or metal ions to pile up in places. The result is that it is not reliable to scale down the thickness of the wires along with the down-scaling of the width of the wires. Thus, the aspect ratio of the wires tends to be large in submicron circuits. As the devices are being placed closer together to increase density, there is an increase in the unevenness of the surface over which the wires must run. These current density limitations, as well as topological constraints, produce the interconnect environment shown in Figure 8-13. This results in a large capacitive coupling area between wires and from wires to devices. As a result, circuit performance slows down. Figure 8-14 shows the total capacitance of a parallel

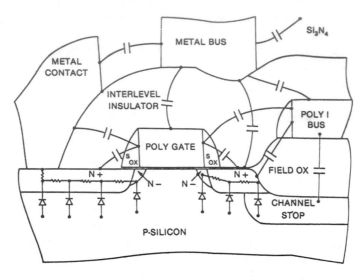

Figure 8-13.   Scale drawing of the parasitic environment in 1 μm MOS technology.

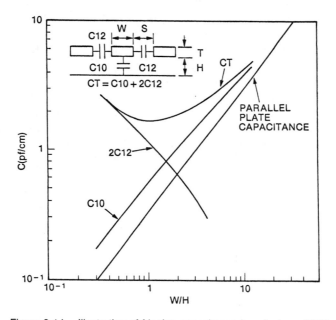

Figure 8-14.   Illustration of fringing capacitance in submicron VLSI.

bus line as the pitch is reduced. The fringing components are large, and mutual capacitance dominates at the submicron pitch so that the total capacitance per unit length of wire increases. This increase could be more than a factor of two, and thus affects speed and functionality of some circuits drastically. As the functional integration level has increased, the chip size has been growing, so that on-chip bus lines can run the length of the chip. Speed improvement thus requires that functional architectures be available to allow very short distance signal propagation.

In today's application of CMOS technology to microprocessors, the use of two levels of wiring has seemed essential. As integration density is increased, multilevel interconnection is desirable. The problems of achieving true multilevel metallization are similar to the concerns with two levels. At each level, the vias to the lower level must be reliably opened, and the surface must be planarized to ensure that the wiring layer can be patterned without breaking or shorting out due to filament formation along rough edges. The current density and resistivity limitations imposed by available metal layers are already limiting the speed of circuits at 1 μm. Speeds must be so maintained that the propagation delays through the interconnect are of the order of the switching delays through the transistors. Advances in multilevel metallization will include methods to planarize the surface of the integrated circuit before a metal layer is deposited and patterned. Currently, many techniques are under development for obtaining this planarization. The most promising techniques use the deposition of thick layers of oxide, followed by the deposition of a resist layer that fills up the hills and valleys. This composite layer is then etched back with a plasma that chemically attacks the oxide and the resist at the same rate. This is a critical etch step that must be timed correctly to etch away all the resist and much of the oxide, leaving a smoother surface behind. This kind of planarization technique has been shown in the laboratory to provide smoothing that can be repeated to get up to five layers of metal interconnection.

Power distribution noise is a severe problem in CMOS. Because each transition of a CMOS inverter requires a spike of current, inductive noise may limit the total performance in CMOS. Figure 8-15 shows typical line inductance on the chip as well as the noise generated by a single inverter. Power distribution branches may support many hundreds of inverters in addition to I/Os that switch even larger current, so this noise problem is very severe in submicron CMOS. There are many ways of minimizing the problem. Large pin count chips can afford to have many pads devoted to introduction of power and ground. This partitions the problems of power distributions. It is desirable to have power pads available, not only at the periphery of the chip, but inside the chip as well, so that the inductive effects of distribution of power on the chip can be minimized. Packaging systems that use flip chip bonding are well suited to this kind of power distribution.

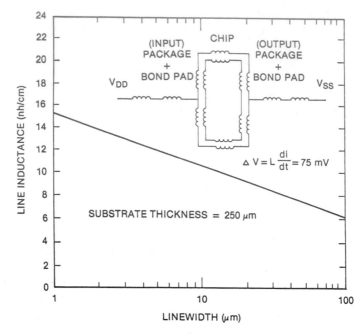

Figure 8-15. The impact of line inductance on power supply noise. Inset shows noise due to typical switching of a CMOS driver.

Another major consideration in the development of interconnections is the ability to withstand the operating environment. Most VLSI chips are subjected to both temperature cycles and residual moisture in the package. The temperature cycles can generate very large stresses, which produce metal fatigue and cracking during repeated temperature cycles. Corrosion occurs because traces of moisture interact with phosphorus doping in the interlevel oxides. The doping is provided to make the oxides soften and flow at low temperature so they can cover the topography smoothly. The reaction with moisture forms phosphoric acid, which can etch away the aluminum metallization commonly used in ICs. These problems are generally very acute for chips that are housed in plastic packages. Much effort is required to minimize these reliability problems, and most of the solutions to the reliability issues require a tradeoff of the obtainable electrical properties of the interconnection layers.

The importance of solving these problems of interconnects seems today to overshadow the problems of the active devices themselves. The rate of progress of the technology is definitely being paced by our ability to solve these problems.

THE THIRD DIMENSION

Although scaling of devices to submicron geometries is the principal thrust of technology development today, the use of the third dimension to increase packing density is being very actively pursued as well. Density considerations are most prominent in memories, and the main effort in the use of the third dimension has been aimed toward memories.

The use of the third dimension may be compared to the development of the construction industry. Subways and underground structures may allow a function to be performed, or skyscrapers may, where land is costly. In a similar way, structures have been proposed that dig into the silicon surface to make active devices. There have also been structures proposed that build multiple layers of devices on top of the silicon substrate. The two major developments in 3D technology are the use of trench capacitor structures, which utilize the substrate material, and the use of stacked structures, which are overlaid on top of the normal active device layer.

The trench capacitor is an elegant solution to the dRAM problem where the charge storage area could not be scaled. The most significant concern with trench capacitors is the possibility of punch-through (Sunami et al. 1984; Elahy et al. 1984). Figure 8-16 (a and b) shows the minimum doping needed to scale the trench-to-trench distance. It is seen that below 0.75 μm, trench-to-trench spacing cannot be scaled. Many versions of a megabit dRAM have been fabricated by integrating the trench capacitor with the basic CMOS process. Trenches that are typically used in this kind of application are about 3 μm deep. However, 8 to 9 μm deep trenches have been demonstrated to be feasible, and the 4-megabit dRAM generation will probably have these deep trenches.

The stacked CMOS structure shown in Figure 8-17 was developed on the basis of hydrogen passivated polysilicon transistors (Shichijo et al. 1984). It has been shown that these transistors can be made to have very low leakage. This feature makes them suitable for load devices in dRAMs (Shah et al. 1984) where the transistor has to act like a switchable impedance. The high impedance state is needed to reduce the power consumption per bit. However, if an ionized particle hits the circuit, the excess charge generated must be quickly removed so that it does not cause an error. This is done by dynamically switching to the low impedance state.

Attempts are under way to stack many layers of active devices on top of each other so that the circuit elements and the wiring can all be distributed over the various layers. This allows maximum flexibility in connecting the devices. It also allows a different function to be performed at each layer. For example, one may perform an imaging function at the top layer, then have an analog to digital conversion at a lower layer, and do some processing at still lower layers.

Broad application of these 3-D concepts is likely in VLSI chips of the 1990s.

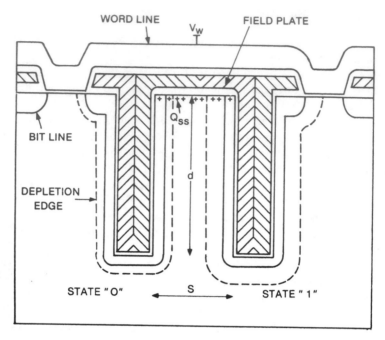

Figure 8-16(a). Trench capacitor cells using 3-D capacitors.

THE ECONOMICS OF VLSI

In past decades, development of VLSI technology has been aided by the growth of new applications. It has been feasible to provide the large investment required in the evolution of VLSI technology because of the known markets that the technology provides. Markets and development costs have increased exponentially. Continued evolution of the technology into the submicron regime will be very expensive and can be funded only if potential markets continue to attract investors.

The costs of developing the design, process, and manufacturing technology are increasing as the feature size is scaled down. Applications develop slowly, and the total cycle may be six to eight years long. A major cash-flow problem results because of the delayed return on investment. This situation will slow the progress of the total technology unless the volume of chips required by the current generation of the technology can provide the cash flow needed to sustain the development costs. These considerations may dominate the rate at which the technology evolves.

The traditional concept of the "learning curve" has been used to create volume demands in the semiconductor industry; the prices of the chips drop as the volume

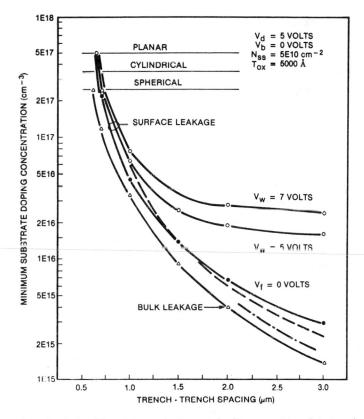

Figure 8-16(b).   Analysis of the minimum doping required to prevent trench-to-trench punch-through.

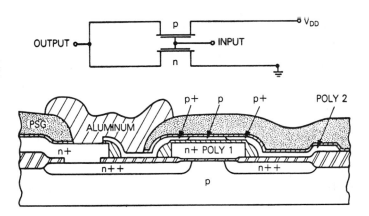

Figure 8-17.   Cross section of stacked CMOS structure.

increases so that economies of scale can be used to fuel larger volumes. In the last few years, however, the semiconductor industry has been unable to find high-volume markets in VLSI, other than memories. This situation has put a severe pressure on memory chips to absorb the development cost of the technology for all applications. It is expected that other high-volume applications of VLSI will emerge in the next few years that will continue to fuel the exponential growth of the industry.

SUMMARY

CMOS technology issues that affect the continued growth of the semiconductor industry have been discussed. Although the gain in performance is slowing as we go into the submicron regime, the economic incentives now available will carry us to the half-micron regime. The use of large chips and 3-D integration provide a continued growth path. However, current device techniques are not ideal for exploiting features below 0.3 μm, and new device and architecture ideas are necessary for continued growth.

REFERENCES

Brews, J. R., *IEEE Trans. Electron Devices, ED-26*, 1282 (1979).

Chiu, K. Y., Moll, J. L., and Maroliu, J., *IEEE Trans. Electron Devices, ED-29*, 536 (1982).

Dennard, R. H., Gaensslen, F. H., Yu, H. N., Rideout, V., Bassons, E., and LeBlanc, A. R., *IEEE J. Solid State Circuits, SC-9*, 256 (1974).

Elahy, M., Shichijo, H., Chatterjee, P. K., Shah, A. H., Benerjee, S. K., and Womack, R. H., *IEDM Tech. Digest*, p. 248 (1984).

Hall, J. E., Seitchik, J. A., Arledge, L. A., Yang, P., and Fung, P. K., *IEDM Tech Digest*, p. 292 (1984).

Kers, L. A., Beguwala, M. M. E., and Custode, F. Z., *IEEE Trans. Electron Devices, ED-28*, 1490 (1982).

Kurosawa, K., Shibata, T., and Iizuka, H., Device Research Conference, Santa Barbara, CA (1981).

Noble, W. P., and Cottrell, P E., *IEEE IEDM Tech. Digest*, p. 583 (1976).

Scott, D. B., Hunter, W. R., and Shichijo, H., *IEEE Trans. Electron Devices, ED-29*, 651 (1982).

Shah, A. H., et al., 1984 Symposium on VLSI Tech., San Diego, CA (1984).

Shichijo, H., et al., *IEDM Tech. Digest*, p. 228 (1984).

Sodini, C. G., Wong, S. S., Ekstedt, T. W., and Oldham, W. G., Device Research Conference (June 1982).

Sunami, H., Kure, T., Yagi, K., Yamaguchi, K., and Shimizu, S., *IEDM Tech. Digest*, p. 232 (1984).

Taur, Y., Chang, W. H., and Dennard, R. H., *IEDM Tech. Digest*, p. 398 (1984).

Troutman, R. R., *IEEE Trans. Electron Devices, ED-26*, 461 (1979).

Wang, K. L., Saller, S. A., Hunter, W. R., Chatterjee, P. K., and Yang, P., *IEEE Trans. Electron Devices, ED-29*, p. 541 (1982).

Yasuda, H., Hashimoto, K., Nozawa, H., Ochii, K., and Kohyama, S., 1982 Symposium on VLSI Technology, Oiso, *Digest* of Tech. Papers, p. 14 (1982).

9. Bipolar Gate Array Technology

Seiken Yano

Systems LSI Development Division NEC Corporation

Demand is increasing for new data processing systems that offer higher performance, smaller system size, lower cost, and higher reliability. High-speed LSIs are indispensable in meeting these requirements. However, many problems are revealed as their integration level is increased.

First, existing high-speed circuits, such as emitter coupled logic (ECL), can meet performance objectives, but they dissipate too much power. When many ECL circuits are integrated on a chip, increased power dissipation necessitates the use of an expensive special cooling system. High-speed circuits with low power dissipation are needed for LSIs, but they, in turn, demand advanced bipolar technology to realize the low power dissipation and high performance.

Second, as the circuits in a chip are increased by advanced LSI technology, standard LSI products cannot satisfy design requirements relating to their performance and functionality. Therefore, high-speed, system-oriented LSIs are needed for high-performance systems. These LSIs, however, require a large variety of parts and, in turn, are very costly to develop. Moreover, the development time for these LSIs eventually occupies a large portion of the system's development time. Therefore, it is desirable to develop system-oriented LSIs with a short turnaround time and a low development cost.

Many efforts have been made to achieve such high-performance system-oriented LSIs. One approach uses bipolar gate arrays or masterslices. However, existing gate arrays and masterslices present problems in the areas of performance realization and power dissipation. To overcome these disadvantages, new masterslice LSIs have been developed that feature low-energy CML (current mode logic), a cell array masterslice, and design automation, with application to large computers. In these LSIs, transistor utilization efficiency and power dissipation can be improved by the use of low-energy CML with a series-gated structure, collector dotting, and emitter dotting.

CIRCUIT REQUIREMENTS

Performance improvement for data processing systems depends mainly on advances in circuit delay time and chip integration level. Figure 9-1 shows the relation between circuit delay and chip integration level, corresponding to the energy level per gate. The energy levels shown in the figure assume that allowable power dissipation per chip is limited to one watt. The figure shows that a small computer with 10,000 gates can be integrated on a chip only when gates with 10 ns delay can be realized with 0.1 mW operational power (i.e., 1 pJ). The allowable power dissipation per chip cannot be increased much more because of cooling system limitations. Therefore, low-energy circuits are required to realize very-large-scale integration. The low-energy CML (LCML) circuit can

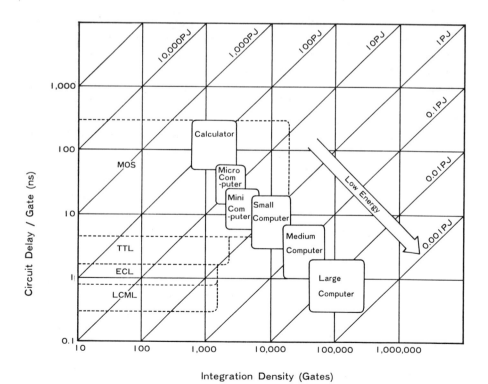

Integration Density (Gates)

Figure 9-1. Relation between circuit delay and chip integration level.

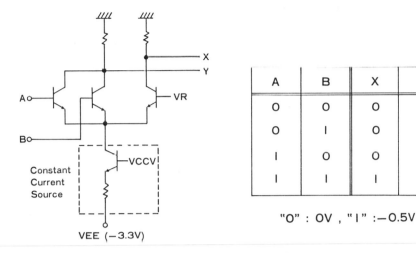

A	B	X	Y
O	O	O	I
O	I	O	I
I	O	O	I
I	I	I	O

"O" : OV , "I" :—0.5V

Figure 9-2.   Low energy CML.

satisfy this requirement (Kimura et al. 1974; Akazawa et al. 1978; Takahashi et al. 1979; Yano et al. 1980; Ouchi et al. 1983).

The basic LCML circuit, shown in Figure 9-2, provides NAND/AND function in negative notation. It features a small logic swing (500 mV) and low power supply voltage. The circuit is free of transistor saturation because of its small logic swing. The power supply voltage can be reduced to −3.3 V from the conventional −5.2 V, keeping all the circuit transistors from saturation. The basic internal circuit dissipates 1.4 mW with 0.35 ns delay, and the power–speed product (i.e., energy) is reduced to less than 0.5 pJ.

A constant current circuit, which is constructed by a transistor and a resistor, is provided for each internal and external LCML circuit to ensure a sufficient noise margin for the LCML with a small logic swing. The constant current circuit is driven by a constant voltage source, VCCV, which is also integrated on the chip. The circuit can minimize logic level variations due to power supply, temperature, and process parameter variations. This results in a small logic swing for the LCML.

The reference voltage, VR, can be either supplied from an external power supply unit or generated on the chip from the power supply VEE. In the former case, circuit noise immunity can be easily tested by changing the external reference voltage source. In the latter case, the on-chip reference voltage generator is constructed by the same constant source and two parallel-collector resistors as used in logic circuits, so that the reference voltage is exactly in the middle of high and low logic levels. When the power supply voltage and/or temperature causes a logic level change, the reference voltage is also changed in the same direction as the logic level and provides a good symmetric noise immunity.

EMITTER DOTTING

The LCML with emitter-follower has an output level that is one diode voltage drop of the basic LCML circuit output. The emitter-follower output has a low output impedance and results in high-speed operation at a heavy load condition, such as many fan-outs and/or a long line drive. The reference voltage, VR, must also be shifted to be referred to by the emitter-follower output signals from other gates.

Emitter dotting, which is a connection of emitter-follower outputs for two or more circuits, can provide the AND function in negative notation. For example, a NAND/AND function, equivalent to an AND-OR-INV function, can be realized at the emitter-follower output for NAND circuits. Therefore, emitter dotting can improve not only transistor usage efficiency but also system performance by reducing logic stage numbers, as shown in Figure 9-3.

COLLECTOR DOTTING

Collector dotting, where one collector resistor is shared by two or more logic circuits, can provide the OR function in negative notation. By collector dotting, the AND-OR function can be realized at the output of one logic stage. This also results in a reduction in transistor and logic stage numbers required to construct a function capability, as shown in Figure 9-4.

In collector dotting, the possibility exists that two or more circuit currents would flow in the collector resistor. This causes a large logic swing and transistor

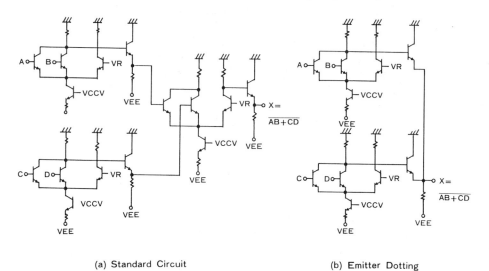

(a) Standard Circuit   (b) Emitter Dotting

Figure 9-3.   Emitter dotting. (a) Standard circuit; (b) emitter dotting.

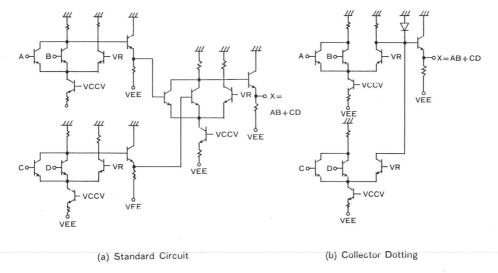

(a) Standard Circuit   (b) Collector Dotting

Figure 9-4. Collector dotting. (a) Standard circuit; (b) collector dotting.

saturation. To avoid this state, a diode clamp is applied paralleling the collector register. If the logical condition assures that two or more circuits do not switch on simultaneously, collector dotting without a diode clamp can be applied.

SERIES-GATED STRUCTURE

The LCML circuit with a two-level gated structure has cascade current switches between ground and a constant current course, as shown in Figure 9-5. The circuit can provide two output types: the collector output and the emitter follower output. The first voltage level, the collector output, has a 500 mV logic swing from 0 V (logic "0") to −500 mV (logic "1"). The second voltage level, the emitter-follower output, is one diode voltage drop of the first voltage level and is restricted to internal use only. The upper and the lower current switches have the first and the second voltage level inputs, respectively.

The two-level series-gated structure has many advantages over the single-level current switch, whose function is limited to the AND/NAND gate. By interconnecting transistors in different patterns, the circuit can be changed to furnish many usable functions. These functions include the exclusive OR gate, multiplexer, decoders, and latches. Each circuit has one current source, and therefore, has a single-gate power dissipation. The propagation delay for a series-gated circuit with typical fan-outs and line length can be expected to be less than one-and-a-half times that for the basic LCML circuit.

By using more than one series gate, more complex functions can be easily realized, including master-slave flip-flop (two current sources), full adder (two

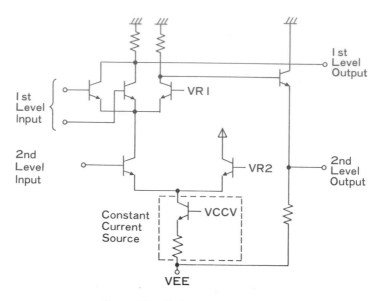

Figure 9-5. Series-gated structure.

current sources), and register files. To obtain these functions with a small number of transistors, collector dotting and emitter dotting also can be applied, combining with a two-level series-gated structure. Collector dotting without a diode clamp can be applied only when two or more collector-dotted circuits may not be switched on simultaneously. A full adder is one example of the application of these techniques (see Figure 9-6).

If signal transfer between function blocks is realized by using both the first and the second voltage level, the same performance and efficiency in resource utilization will be achieved as with custom design circuits. For example, a 3-to-1 multiplexer, constructed with ten transistors, has three data inputs (D0, D1, and D2) at the first voltage level and three control inputs (S0, S1, and S2) at the second voltage level, as shown in Figure 9-7. These second voltage level signals are ordinarily generated as the control logic output and are distributed to several similar multiplexers, depending on bit width. Therefore, compared to a conventional multiplexer with a decoder, a multiplexer without a decoder improves resource utilization efficiency and performance.

MASTERSLICE

The significance of the masterslice approach has been widely recognized since gate arrays, or masterslice approaches, began to be extensively used in mainframe computers (Blood 1978; Hively 1978; Offerdahl 1978; Braeckelmann et al. 1979;

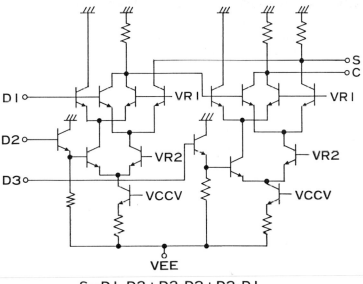

$$S = DI \cdot D2 + D2 \cdot D3 + D3 \cdot DI$$
$$C = DI \oplus D2 \oplus D3$$

Figure 9-6. Low-energy CML with series-gated structure, collector dotting, and emitter dotting.

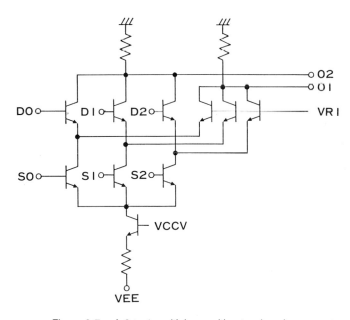

Figure 9-7. A 3-to-1 multiplexer without a decoder.

Blumberg and Brenner 1979; Nakaya et al. 1981; Lee and Bass 1982; Suzuki et al. 1983; Sato et al. 1983; Takahashi et al. 1983). This approach is valuable because individual mainframe models are manufactured only in numbers ranging from several hundred to several thousand; so the full custom design approach is not justifiable economically, in spite of the substantial speed improvement it would offer.

The masterslice approach allows high-volume production during the diffusion stages with a single set of masks. The common diffusion set can be customized by a variety of metal masks. This method results in short turnaround times and low development cost.

The layout time is also shortened because the need to design the diffusion layers for each option development is eliminated. Once a layout is complete, sample quantities can be obtained in a short time, compared to the time required for a full custom design. This is possible mainly because wafers can be stockpiled in a completely diffused status, with first layer metal deposited. The low development cost is due to the shortened development time and the small quantity of custom masks required.

One of these approaches uses ECL gate arrays, which have obvious advantages over full custom design circuits in regard to turnaround time and development cost. However, ECL gate arrays now on the market have high-power consumption, which requires a special cooling system.

The ECL masterslice approach has improved the efficiency of current source usage by utilizing the advantages of a series-gated structure. This results in low power dissipation per equivalent gate. There have been several attempts to reduce chip power dissipation by utilizing low-power circuits for internal logic. However, LSI power dissipation cannot be decreased much more, as long as a high-voltage ECL is applied.

CELL ARRAY MASTERSLICE

A new LCML masterslice, which is constructed by a cell array substrate, has been developed to overcome the disadvantages of existing ECL masterslices. Cells are the most elementary devices that are stepped and repeated to form the logic gate portion of the chip. The cell array masterslice allows a flexible function block layout and high resource utilization.

The LCML masterslice chip is 7.6 × 7.3 mm in size and consists of two parts: an internal cell array logic section and an external buffer section. The internal cell array logic section is 6.3 × 5.9 mm in size and has 832 cells arranged in a 26 by 32 matrix, as shown in Figure 9-8. Each cell contains eight small-size transistors for current switches and constant sources, four medium-size transistors for emitter-followers, and 26 resistors, as shown in Figure 9-9. This arrangement allows a flexible function block layout. Each internal cell can contain 2.5 equivalent gates, on average.

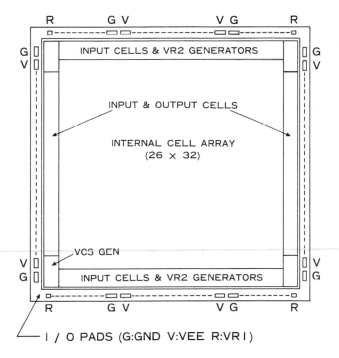

Figure 9-8. Cell array masterslice.

The external buffer area surrounds the chip. There are 56 output buffers located on opposite sides of the chip, 112 input buffers, six current source voltage (VCCV) generators, and 52 internal reference voltage (VR2) generators. The large transistors of output buffers are designed to handle the currents necessary for driving off the chip. The external buffer circuit features an 8.47 mA current

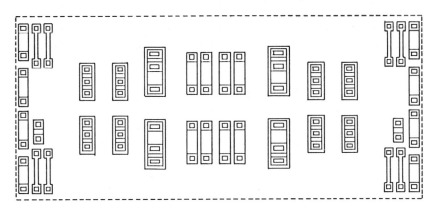

Figure 9-9. Cell with 12 transistors.

switch, a 500 mV logic swing, and 59 ohms output impedance. It has one- or two-input AND/NAND functions.

The chip has 132 external pads for the film carrier bonding system. Each pad has a double-bumped structure to avoid edge contact between lead and substrate without any lead formation. These pads include eight pads for ground (GND), eight pads for the -3.3 V power supply (VEE), and four pads for the reference voltage supply (VR1). The power distribution and pad location were carefully designed to reduce the ground line voltage drop, which has a direct influence on the high output level of external buffers.

The masterslice has two metallized layers of aluminum: the first layer of aluminum and the second layer of aluminum covered with gold. After function block allocation, the aluminum-metallized layer, used for function blocks, is processed in preassigned patterns by the DA system. The first and the second aluminum metallized layers have 320 channels with 8 μm pitch and 370 channels with 10 μm pitch, respectively.

To maintain high-speed LCML operation, a high density packaging system with small lead frame capacitance is required. This is the reason why a film carrier bonding system, which features batch bonding and high reliability, is adapted to the masterslice. Lead frames are made by photoresist etching of copper foil, which is laminated onto a 35-mm-width film carrier, and by gold plating. The chips are assembled on the film carrier by thermal compression bonding between the gold pads on the chips and lead frames. Chips with lead frames are punched off the film carrier and assembled on a ceramic substrate. Each chip occupies about 10 mm^2 of the substrate, which results in a high packing density.

DESIGN AUTOMATION

A DA (design automation) system is indispensable for designing LSIs with a short turnaround time and for developing many LSIs with limited resources. The system supports all the activities from function block design to testing the prototypes, as shown in the flow chart in Figure 9-10. These activities include logic stimulation, function block placement, routing, delay analysis, test pattern generation, and fault simulation (Kurobe et al. 1977; Yoshizawa et al. 1979, 1982; Kato 1983; Sasaki 1980).

Before each circuit in an LSI family is designed, the masterslice substrate and function block design have been accomplished. A masterslice substrate consists of an array of internal cells, routing areas, and external buffer areas, as shown in Figure 9-8. Physical information on the masterslice is complied in the structure library and the pattern library. Power supply lines are usually fixed on the substrate and processed in preassigned patterns by the DA system.

After block layout design, the block circuits are simulated by a circuit analysis

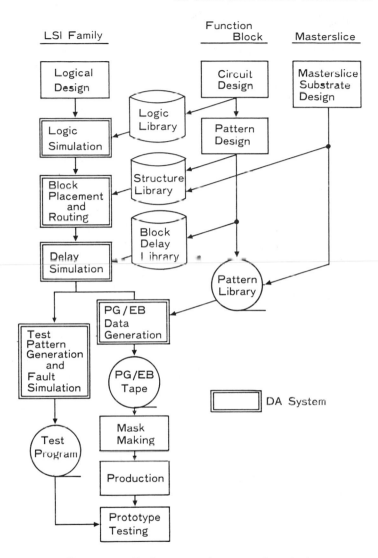

Figure 9-10.   Design automation system flow chart.

program to determine the function block delay time. Logical information and physical information regarding the function blocks are checked by the DA system to ensure their coincidence. Logical, physical, and electrical information on the function blocks is compiled in the block libraries. These libraries are referred to by the DA system at every stage in the chip design.

LOGIC SIMULATION

The first step in LSI family design is logic simulation. The designer inputs logical interconnection information between function blocks and a functional description of the chip using functional description language. The mixed level logic simulator verifies the logical operation of the two input files, called the logical interconnection file and the functional description file. In this step, two input files are simulated in parallel using automatically generated simulation patterns. Then the simulated results are compared and checked. The logical interconnection file is used for chip layout design, and the functional description file is used for system simulation in the next design step.

The block logical functions are designed as macro gates, which are described by basic gates such as AND, NAND, OR, NOR, and DFF. Circuit delay times also are assigned to each of several basic gates, which are used to construct a function block. Thus, the logic simulator can also simulate the signal propagation delay in the LSIs by using function block libraries.

LAYOUT DESIGN

The next step is layout design, which consists of automatic two-dimensional placement, routing, and delay simulation. At this stage, based on a set of signal nets between function block terminals, the cell location for each block and the pad location for each external terminal are determined, and the wiring patterns for each net are routed.

Placement

Placement is used to assign function blocks in cell arrays on the substrate so as to facilitate routing. The final layout design goal is to achieve 100 percent routing. However, it is too difficult to consider the final goal itself in the placement stage. Therefore, function blocks are allocated to minimize the total routing length of all signal nets.

Routing

The routing procedure is divided into two stages: global routing and detailed routing. After the whole chip areas are partitioned into portions vertically and horizontally, global routing determines, for each net, portions through which the net will be routed. Global routes for all nets, expressed as a set of portions, are obtained so that the wire density in each portion does not exceed the wire capacity for the portion. Then the precise positions of wiring patterns are decided in the detailed routing, while being constrained from the above global routes.

The detailed routing tries to find the wiring patterns for any given pair of terminals, or a pin-pair. The pin-pairs for each net are defined and ordered for routing. A line search method can be adopted with a few modifications when three-layer wiring patterns are required for a specific masterslice.

Specific items related to the LCML circuit are also processed by the layout system. They include checking signal voltage level coincidence for two-level signal transfer among function blocks, selection of an emitter-follower resistance, line-width modification to reduce the series resistance between emitter-follower sources, current density and line-length check, power calculation, and other circuit design rule checks.

Delay Simulation

After layout design, the detailed media delay, which includes line delay and fan-out delay for each net, is calculated by using basic delay parameters in the block delay library. The media delay and circuit delay are merged into the same database. Detailed ac characteristics can be estimated by delay simulation.

The logic simulator is used for delay simulation again. In this case, the designer can input the test patterns to activate a special delay path. Sometimes another delay time calculation tool is used, which traces all net delay without any test pattern. It calculates minimum/maximum delay and clock skew for the clock distribution net and detects race conditions between flip-flops on a chip. After all ac characteristic requirements are satisfied, the layout system generates the pattern generator tape or electron beam tape for individual mask making.

TEST-PATTERN GENERATION AND FAULT SIMULATION

Device test patterns are generated automatically by using the scan path technique. This technique is very useful for accomplishing a reduction in computing time and test steps, with small amounts of logic circuit increase. The test patterns used in logic simulation also can be used as device test patterns.

A fault simulator generates the comparison vector forecast by input test patterns and checks fault coverage. If necessary, manual test patterns will be added to attain the detection rate design goal. A standard device test program is automatically modified by the individual data for each option development. Finally, the generated test patterns are converted to the LSI tester language.

DESIGN FOR TESTABILITY

In LSI family development, design for testability is important in order to test the device completely and efficiently. The scan path technique is applied to all the LSI devices, assigning three special pads for testing: shift mode control, shift data input, and shift data output. These preassigned pads allow the devices to be tested by the same tester pin assignment, which results in a low-cost LSI tester with special serial scan function limited to these three tester pins. The technique also makes automatic test pattern generation easier (Funatsu et al. 1975, 1978; Kawai et al. 1980).

In the scan path technique, all the flip-flops in an LSI are connected serially and form a shift register. Each flip-flop has an input data selector, which is controlled by the shift mode control input. This structure allows the flip-flops to be used as ordinary flip-flops in the normal mode and as shift registers in the test mode. In the normal mode, normal data are loaded to the flip-flop. In the shift mode, output from the previous flip-flop in the shift register chain is loaded. The output signal for each flip-flop drives both the following flip-flop and other gates.

The shift register approach enables the sequential logic circuit to be transformed to a combinational one because all the states of the circuit are completely determined by the shift input data and other input terminal data. After the flip-flops are set (by advancing the clock) to the new states, they can be read out by shifting out registers through the shift data output. Therefore, shift registers in the circuit can be treated as equivalent input/output terminals.

Generally, every sequential logic can be separated into combinational logic and flip-flops (register). In the scan path technique, the input stage to each flip-flop is a selector that selects normal input data to the flip-flop or adjacent flip-flop data, according to the shift mode control. In the shift mode, the flip-flops make a shift register. The first stage input for the shift register is the shift data input, and the final stage output is the shift data output.

The LSIs are tested by the scan path technique as follows. First, primary input data ①, shown in Figure 9-11, is applied to input terminals under the normal mode. Second, test data ② is shifted into shift registers by changing the shift mode control from the normal mode to the shift mode. Then, primary output data ③ is compared with the forecast test vector. After the mode is set back to normal, output pattern ④ from combinational circuits is loaded to the register by advancing a clock. The shift mode control is then set to the shift mode, and individual data bits from the shift data output are observed after each clock advance.

When flip-flops are combined in register files, a modified scan path technique is required. In this case, the LSI must contain extra support hardware: an address counter that increments only in shift mode, a write data selector, and read and write address selectors, as shown in Figure 9-12. In shift mode, the address selectors select the counter output, and the write data selector selects the shifted read data and shift in data. With the support of this hardware, the register files can be included in the same scan path as ordinary flip-flops. As a result, the scan path technique can be successfully applied to all LSIs developed.

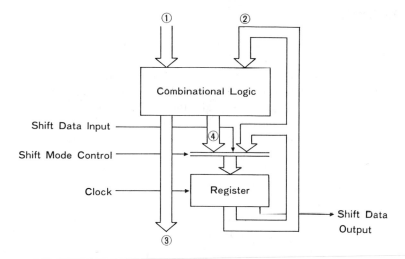

Figure 9-11.   An LSI with shift path.

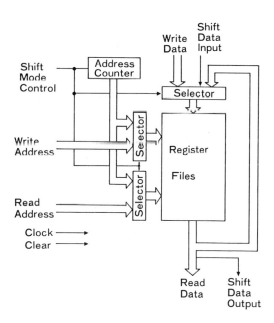

Figure 9-12.   Register files with shift path.

APPLICATION

The masterslice has been adopted to meet the demands of system-oriented LSIs. The first members of the LSI family are selected to have higher chip counts per chip type in a system and/or among systems. They include the multiplier, 144-bit contents addressable memory, 32-bit right/left shifter, and channel controllers.

Representative design data for the LSI family show that they utilize 80 percent of the internal cells and average the equivalent of 1660 gates. These chips contain an average of 710 networks with 2.5 fan-outs, which means 1755 interconnections are routed on the chip. The total line length is about 1400 mm. The typical power dissipation for the chip is 3.8 watts. It can be efficiently cooled by a forced-air cooling system.

One example of these LSIs is a 32-bit right/left shifter intended for general-purpose usage. A block diagram is shown in Figure 9-13. This chip has 64-bit data inputs, nine control inputs, 32-bit data outputs with four parity bits, and several status bits. The control inputs can select right/left shift, shift values, and modes. The delay time from data input to data output is 4.4 ns. The total gate count is 1800 gates, and power dissipation is 4.0 watts. Figure 9-14 is an LSI microphotograph.

CONCLUSION

The low-energy CML with -3.3 V power supply and 500 mV logic swing was adopted for a basic masterslice circuit to realize both high speed and low power. The constant current source, which is provided for each LCML, can minimize logic-level variations due to environmental change. Emitter dotting, collector dotting, and the series-gated structure can offer usable means for transistor reduction and logic stage number reduction.

The cell array masterslice can offer a short turnaround time in the development of LSI family members. The design automation system supports all activities from logical design to testing the prototype. The system has improved the design turnaround time and design quality.

The LSIs developed on the masterslice have an average of 1660 gates and utilize 80 percent of the internal cells. Typical power dissipation for these LSIs is 3.8 watts. They can be cooled by a forced-air cooling system. The LSI family members will be increased to achieve higher performance, smaller system size, and better reliability.

ACKNOWLEDGMENT

The author wishes to express his thanks to the many hardworking individuals contributing to LCML integrated circuit development at NEC. In particular, he wishes to thank Dr. H. Kanai, T. Takemura, K. Shimizu, and H. Murano for their valuable suggestions and management efforts.

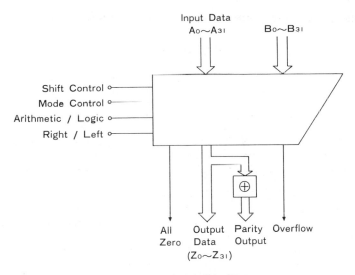

Figure 9-13. A right/left shifter.

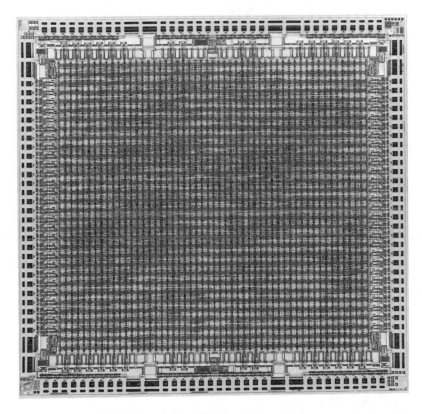

Figure 9-14. LSI microphotograph.

303

REFERENCES

Akazawa, Y., Kodama, H., Sudo, T., Takahashi, T., Nakamura, T., and Kimura, K., "A High-Speed 1600-Gate Bipolar LSI Processor," *ISSCC '78, Digest of Technical Papers*, pp. 208–209 (1978).

Blood, B., "High-Speed Gate Arrays Are the Building Blocks for Modern Computer Systems," *1978 WESCON*, 1/3, pp. 1–5 (1978).

Blumberg, R. J., and Brenner, S., "A 1500-Gate, Random Logic, Large-Scale Integrated (LSI) Masterslice," *IEEE J. Solid-State Circuits, SC-14* (5), 818–822 (Oct. 1979).

Braeckelmann, W., et al., "A Masterslice LSI for Subnanosecond Random Logic," *IEEE J. Solid-State Circuits, SC-14* (5), 829 (Oct. 1979).

Funatsu, S., Wakatsuki, W., and Arima, T., "Test Generation Systems in Japan," *Proc.* 12th Design Automation Conference, pp. 114–122 (June 1975).

Funatsu, S., Wakatsuki, N., and Yamada, A., "Designing Digital Circuit with Easily Testable Consideration," *1973 Annual Test Conf.*, pp. 98–102 (Oct. 1978).

Hively, J. W., "Subnanosecond ECL Gate Array," *1978 WESCON*, 3/3, pp. 1–10 (1978).

Kato, S., "FDL: A Structural Behavior Description Language," *Proc.* 6th CHDL, pp. 137–152 (May 1983).

Kawai, M., Funatsu, S., and Yamada, A., "Application of Shift Register Approach and Its Effective Implementation," *Proc.* 1970 Test Conference, pp. 22–25 (Nov. 1980).

Kimura, K., Shimizu, K., Shiba, H., Nakanura, T., and Yano, S., "LSI Using Low-Energy, High-Speed CML Circuits," *Colloque International sur les Circuits integres complexes*, pp. 278–287 (Dec. 1974).

Kurobe, T., Nemoto, S., Shikata, Y., and Kani, K., "LSI Logic Simulation System; LOGOS2," Monograph of Technical Group on DA of Information Processing Society of Japan, DA31-2 (1977).

Lee, S., and Bass, A. S., "A 2500-Gate Bipolar Macrocell Array with 250 ps Gate Delay," *IEEE J. Solid-State Circuits, SC-17* (5), 913–918 (Oct. 1982).

Nakaya, M., et al., "A Bipolar 2500/Gate Subnanosecond Masterslice LSI," *IEEE J. Solid-State Circuits, SC-16* (5), 558–562 (Oct. 1981).

Offerdahl, R. E., " High Utilization of Masterslice LSI Array," *1978 WESCON*, 3/5, pp. 1–11 (1978).

Ouchi, Y., Sano, M., Sato, F., Nakashiba, H., Nakamae, M., Shiraki, H., and Yoshimura, N., "Sub-Picojoule Bipolar LSI with a 350 ps Gate Delay and 2000 Gates," *Proc.* ICCD '83, pp. 97–100 (Oct. 1983).

Sato, F., Takahashi, T., Misawa, H., Misawa, H., and Kimura, K., "A Subnanosecond 2000-Gate Gate Array with ECL 100K Compatibility," *Proc.* CICC '83, pp. 23–26 (May 1983).

Sasaki, T., "MIXS: A Mixed-Level Simulator for Large Digital System Verification," *Proc.* 17th Design Automation Conf., pp. 626–633 (June 1980).

Suzuki, M., Horiguchi, S., and Sudo, T., "A 5000-Gate Bipolar Masterslice LSI with 500 ps Loaded Gate Delay" *ISSCC '83, Digest of Technical Papers*, pp. 150–151 (1983).

Takahashi, Y., Hagiwara, N., Ito, Y., Matsuhiro, K., Nakashima, T., Nakamura, T., Yano, S., and Kimura, K., "A 1200-Gate, Subnanosecond Masterslice LSI," Paper Tech. Group, IECE, Japan, SSD79-49 (Oct. 1979).

Takahaski, T., Nakashiba, H., Nakamura, T., and Kimura, K., "A High-Speed 16-Bit Expandable Multiplier Using 5000-Gate ECL Gate Array," *Proc.* ICCD '83, pp. 264–267 (Oct. 1983).

Yano, S., Ouchi, Y., Kimura, K., and Ito, Y., "A Masterslice Approach to Low-Energy High-Speed LSIs," *Proc.* ICCC '80, pp. 203–206 (1980).

Yoshizawa, H., Kawanishi, E., Goto, S., Kishimoto, A., Fujinami, Y., and Kani, K., "Automatic Layout Algorithms for Masterslice LSI," *Proc. IEEE* (1979).

Yoshizawa, H., Nomura, M., and Nakao, M., "A CAD System for Gate Array Automated Design," *Proc.* 4th CICC '82, pp. 260–262 (1982).

10. Packaging Technologies for High-Performance Computers

Akira Masaki

Central Research Laboratory, Hitachi, Ltd.

To realize high-performance computers, system delay (i.e., the sum of packaging delay and on-chip delay per circuit stage) must be reduced to achieve a short machine cycle time. Components of system delay, t_{sd}, are basic delay, loading delay, and wire delay. Also, it is necessary to reduce the power dissipation per circuit to make it possible to increase the number of circuits per LSI and per system.

There seems to be a wide choice of hardware technologies for future high-performance computer systems. This chapter will point out several factors that are essential to the evaluation of logic LSIs and packaging technologies for high-performance computers.

When comparing circuit technologies, one must ask how many circuits are required in all, and how many circuit stages the critical path consists of in a specific functional block implemented by each circuit technology. What actually should be minimized is not the power and delay of the unit circuit, but that of the functional block. In other words, the difference in logical-function capabilities among circuits should be evaluated (Masaki and Chiba 1982).

CMOS, which does not dissipate power except in the switching operation, is well recognized as an effective technology for decreasing total system power dissipation. However, the power of a CMOS circuit increases with operating frequency. Therefore, it is necessary to obtain the switching frequency of a circuit in actual systems for comparing CMOS circuit performance with other circuits.

In order to evaluate the wire delay, the average wire length per signal net should be known. It is written as:

$$l_w = n_{LD} \times l_{pp} \qquad [1]$$

where n_{LD} is the average number of loads, and l_{pp} is the average connection (pin-to-pin) length. The number of loads depends on the circuit type. Apparently, the connection (or wire) length depends on the packaging structure. Equations for estimating wire lengths in various types of two-dimensional (2-D) and three-dimensional (3-D) cell arrays will be introduced, and examples of applications of these equations will be described in the latter part of this chapter.

CONSIDERATIONS IN CIRCUIT COMPARISON

Logical-Function Capability of the Circuit

It is not easy to obtain a general solution for this problem. There exists no effective means of studying it other than carrying out a logical design for each circuit type. The results depend strongly upon the logical characteristics of the functional block, as well as upon the differences in individual abilities of the designers. Nevertheless, differences in logical-function capabilities of circuits can never be neglected.

For implementing a particular functional block, the required number of circuits, N_{CT}, and the number of circuit stages in the critical path of the block, NB_{CP}, are dependent on the circuit type. Using the numbers for a "standard" circuit, N_{CTO} and N_{CPO}, the following normalized values are defined:

$$n_{CT} = N_{CT}/N_{CTO} \qquad [2]$$
$$n_{CP} = N_{CP}/N_{CPO} \qquad [3]$$

Then, system delay and power for a normalized circuit can be written as:

$$t_{sdn} = n_{CP} \cdot t_{sd} \qquad [4]$$
$$P_n = n_{CT} \cdot P_C \qquad [5]$$

where t_{sd} and P_C are system delay and power for the evaluated circuits. Table 10-1 shows n_{CT} and n_{CP} for various circuits obtained by carefully designing a 4-bit arithmetic and logical unit (Masaki et al. 1984). In this table, the logical-function capability is represented by the inverse of the product of N_{CTO} and

Table 10-1. Logical-function capability of circuits.

	SG-ECL	CD-ECL	ECL	CD-CML	CML	EF-NTL	NTL	CMOS*
n_{CT}	0.58	1	1.30	1.28	1.37	1.37	1.72	0.67
n_{CP}	0.83	1	1.28	1.83	1.76	1.76	2.10	1.28
$1/(n_{CT} \cdot n_{CP})$	2.09	1	0.47	0.61	0.40	0.42	0.28	1.18
Equivalent circuits						BFL	DCFL	

N_{CPO}. As shown in the table, the capability of the logically most powerful circuit is more than seven times larger than that of the least powerful one. However, differences in actual power dissipation and delay are usually not so large, because the basic delay and the number of loads are large for logically more powerful circuits.

Switching Frequency of Circuits in Systems

To estimate the performance of a CMOS circuit that dissipates power only during switching operations, the switching frequency of the circuit in a system environment needs to be known. This problem can be more comprehensively analyzed by introducing the quantity:

$$k = T_s/t_{sd} \qquad [6]$$

where T_s is the average switching period for a circuit in system operation, instead of using the operating frequency of the circuit, f.

The value of k depends solely upon the logical structure of the system, as long as the performance of the circuit is fully utilized within it. From this definition, the power of a CMOS circuit can be written as:

$$P_C = E/(k \cdot t_{sd}) \qquad [7]$$

Here, E is the switching energy of a CMOS circuit, including the wire and loading capacitances. Several ideas for obtaining the value of k have been considered (Masaki and Chiba 1982): (1) investigate the logical structure of the system; (2) measure the switching frequency of logic signals in system operation; (3) measure power dissipation in systems built by using CMOS circuits; and (4) count the number of times the circuit switches by simulating the logical operation of the system.

Table 10-2 shows the result obtained by applying these ideas. It is found that the value of k is within the range 20 to 200 for most computer systems.

Table 10-2. Estimated and measured values of k.

METHOD / SYSTEM	MINI-COMPUTER A	MINI-COMPUTER B	LARGE-SCALE COMPUTER	COMPLETELY PIPELINED SYSTEM
(1) Investigation of logical structure of systems	180	120	60–20	~4
(2) Measurement of switching freq. of signals in system	200			
(3) Measurement of power of CMOS computer		170–130		
(4) Logic simulation			80–30	

EQUATIONS FOR ESTIMATING WIRE LENGTH

Examples Of 2-D and 3-D Packaging Structures

In some machines, modules of LSIs are mounted on printed circuit boards, with the boards then arranged to form a large plane. The single-plane structure can be classified as 2-D packaging.

On he other hand, many machines utilize a 3-D-like packaging structure. In these machines, LSIs are mounted on plug-in cards, which are then mounted on backboards using edge connectors.

Strictly speaking, this structure is not a complete 3-D picture because there is no plane-to-plane connection, and the wires from one card to another should go through edge connectors. Therefore, this structure is often termed stacked 2-D or 2.5-D. In this chapter, however, it will be classified as a variant of 3-D for mathematical reasons.

An example of a complete 3-D structure is the 3-D wafer computer (Grinberg et al. 1984). A stack of silicon wafers makes up the computer, and the adjacent wafers are connected face to face using the microbridge interconnection technique.

The purpose of adopting a 3-D or 3-D-like packaging structure is to obtain a high packaging density. It is also generally expected that wire length in a 3-D or 3-D-like structure is shorter than that in a 2-D structure. Relative to this, theoretical comparisons will be made in the following sections.

Equations For Estimating Wire Lengths

Various equations have been reported for estimating wire length. However, most of them are applicable only to two-dimensional, square arrays. In this chapter, equations for estimating wire lengths in various types of 2-D and 3-D system packaging structures will be introduced (Masaki 1984). In this section, a method for obtaining the equations is described, and the equations themselves are given in the chapter appendix, in a form suitable for computer programming.

2-D Arrays. To begin with, packaging structures are mathematically modeled. Figure 10-1(2) illustrates 2-D arrays. Here, the dots represent cells, and each cell corresponds to an LSI if the array is a plug-in card, or to a circuit or a circuit block if the array is an LSI chip. It is assumed that there are 2^L by 2^M cells in the array, and the average center-to-center spacing of the cells is given by x or y.

Complete 2-D Partitioning. In a cell array, the wires that connect cells run in all directions. If we partition this array hypothetically into four blocks, we will find that the number of wires that go out of a block is the value obtained by Rent's rule. The portional relationships of two neighboring blocks are shown

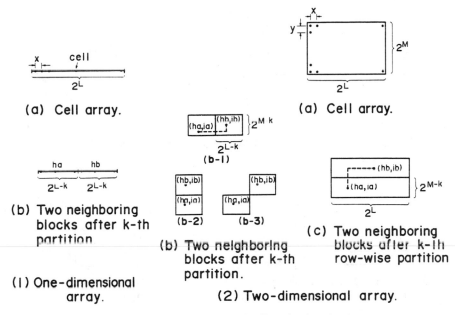

Figure 10-1.   Models for 1-D and 2-D packaging structures.

in Figure 10-1(2-b). The average wire length between the blocks can be obtained by averaging all of the interblock cell-to-cell distances, taking into account the number of wires. Therefore, the average length of all of the wires that exist in this array can be obtained by repeating this hypothetical partitioning.

Although such partitioning is hypothetical, it is related to actual partitioning and placement design. In an actual placement, logically tightly coupled cells are grouped into a block, so as to decrease the total wire length. This itself is the partitioning mentioned above.

When M is smaller than L, the blocks obtained after M-th partitioning are treated as one-dimensional arrays, as shown in Figure 10-1(1).

Row-wise Partitioning. The manner of partitioning described above applies to the case when either x is not very different form y, or L is not very different from M. If x is considerably smaller than y, it is advantageous to adopt the row-wise partitioning shown in Figure 10-1(2-c), to reduce the total wire length. This applies in practice to several gate array LSIs.

3-D and 3-D-like Arrays.

Figure 10-2 illustrates models for 3-D and 3-D-like cell arrays. It is assumed that there are 2^L by 2^M by 2^N cells in the array, as shown in Figure 10-2(1), and that the average center-to-center spacings of the cells are x, y, and z.

If completely three-dimensional wiring is possible, the partitioning will be

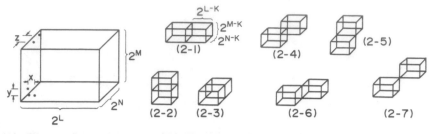

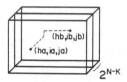

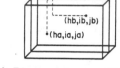

(1) Three-dimensional   (2) Two neighboring blocks
    cell array.   after k-th partition.

(3) Two neighboring blocks   (4) Plane-wise partition,
    after k-th plane-wise partition.   edge-connection.

Figure 10-2.   Models for 3-D and 3-D-like packaging structures.

done in a manner where the array is divided into eight blocks, that is, 2 by 2
by 2. In this case, the positional relations between neighboring blocks are shown
in Figure 10-2(2).

Plane-wise Partitioning with Face-to-Face Connection. With currently available 3-D technology, the interplane wiring pitch is far larger than he intraplane
wiring pitch. Therefore, in most actual cases, the plane-wise partitioning shown
in Figure 10-2(3) is advantageous.

Plane-wise Partitioning with Edge Connectors (Stacked 2-D). Figure 10-2(4)
indicates a model for the stacked 2-D structure that utilizes edge connectors.
Mathematically, the equation for this structure is similar to that for Figure 10-
2(3), except that the length of the route via the edge connector should be taken
into account. For this reason the stacked 2-D structure is classified as a 3-D
variant in this chapter.

These structures may have:

• One edge connector per plane.
• Two edge connectors on opposite sides of the plane.
• Edge connectors on three or four sides of the plane.

Equations can be derived for stacked 2-D structures when connectors are
located on only one edge of the plane, or when connectors are located on two

edges on opposite sides of the plane. This is done by adding the length of the route via the edge connector to the length obtained in the previous section.

However, if connectors are located on three or four sides of the plane, deriving equations that g. 'e the shortest routes is very complicated. A numerical method should be used to estimate wire lengths in these cases.

Comparison of Theoretical and Experimental Values

Figure 10-3 compares theoretical wire lengths with experimental values for 2-D arrays (Masaki et al. 1984). These values were obtained by analyzing wiring data for 12 parts of four types of bipolar and CMOs gate arrays. The experimental values lie between 50% and 100% of the theoretical values. It should be noted that both the x and y components also fall in this range. The reason the experimental values are smaller than the theoretical is that in actual placement design, cell locations in a block are decided with respect to cell relationships in the neighboring blocks.

This result indicates that good estimation values can be obtained by multiplying the theoretical values by 0.75. Although the result is for 2-D arrays, there is no reason why estimation cannot be done for 3-D arrays in the same manner.

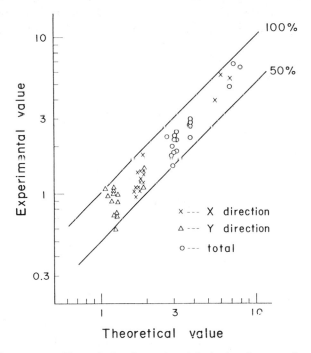

Figure 10-3.   Theoretical and experimental wire length comparison.

EVALUATION EXAMPLES

Comparison of LSI Performances

Figure 10-4 shows a comparison of on-chip performance for various circuits, obtained by considering the factors described above and by assuming the level of device technologies currently used (Masaki 1984). For this comparison, a CD-ECL (ECL with collector-dotting function) is taken as the "standard" circuit. It can be concluded that CMOS is about three to five times slower than the fastest bipolar circuits.

Prospect for CMOS VLSI Performance

Figure 10-5 shows the prospect for CMOS VLSI-ULSI performance (Masaki et al. 1984), estimated as a result of the factors described above and assuming classical scaling theory. The cases were s (scaling factor) = 1, 2, and 4, corresponding to L_g (gate length) = 2 μm, 1 μm, and 0.5 μm, respectively. The n_1 in the figure denotes the number of signal layers on the chip. Undoubtedly, a low-power, fairly fast VLSI-ULSI will be realized using CMOS in the future. However, it can also be concluded that the achievement of subnanosecond t_{sdn} will not be easy, even if L_g 1 μm technologies are adopted.

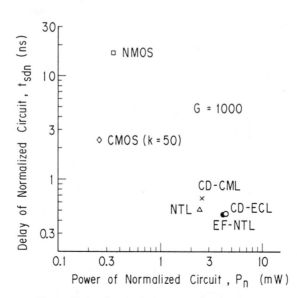

Figure 10-4. Comparison of circuit performance.

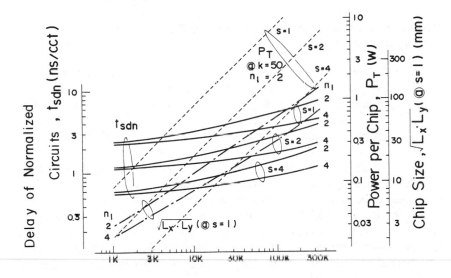

Figure 10-5. Performance prospect for CMOS VLSI, where s is the scaling factor, and n_1 is the number of the signal layers.

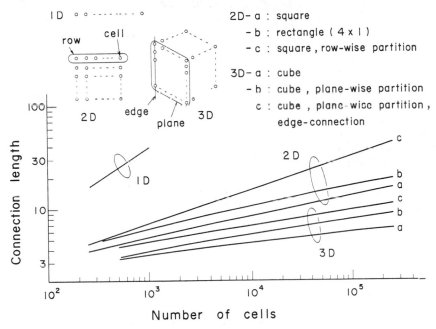

Figure 10-6. Average connection length in terms of cell pitch in various cell arrays ($L = M = N$ and $x = y = z$).

Comparison of Wire Lengths in Various 2-D and 3-D Packaging Structures

Figure 10-6 shows theoretical wire lengths for various cell arrays obtained using the equations presented above. Here it is assumed that $L = M = N$ and $x = y = z$. As can be seen from the figure, the wire lengths for 3-D arrays are shorter than for 2-D arrays, and the ratio increases with the total number of cells. However, this does not directly apply to actual cases, as in such cases L, M, and N are not equal, nor are x, y, and z.

Table 10-3 shows an estimation and comparison of wire lengths in various types of cardboard packaging structures. The total number of LSIs for all cases is assumed to be 900. For the 2-D structure, it is assumed that a large 1.2 m by 1.2 m card is used, and LSIs are mounted as a 30 by 30 array. The LSI pitch on the card is 4 cm for all of the cases. For 3-D and 3-D-like arrays, it is assumed that 50 LSI are mounted on a 40 cm by 20 cm card, with 18 cards mounted on a backboard at a pitch of 2 cm. The estimated wire lengths are listed in the power portion of the table.

The x component of the wire, l_{ppx}, the y component l_{ppy}, and the z component

Table 10-3. Wire lengths in various types of cardboard packaging structures.

				3-D AND 3-D-LIKE				
				STACKED 2-D				3-D WITH
			2-D	1 EDGE PER CARD	2 EDGES PER CARD	3 EDGES PER CARD	4 EDGES PER CARD	FACE-TO-FACE CON.
TOTAL NO OF LSIs			900	900				
NO OF LSIs PER CARD			30 × 30	10 × 5				
NO OF CARDS PER BACKBOARD			—	18				
CARD SIZE (cm × cm)			120 × 120	40 × 20	←	←	←	←
BACKBOARD SIZE (cm × cm)			—	40 × 40				
LSI PITCH	x	(cm)	4	4				
	y	(cm)	4	4				
CARD PITCH	z	(cm)	—	2				
AVERGAE LSI TO LSI WIRE LENGTH								
X - COMPONENT	lppx	(cm)	7.61	6.65	6.65	6.76	6.80	6.65
Y -	lppy	(cm)	7.61	4.75	3.85	3.64	3.52	2.84
Z -	lppz	(cm)	—	0.94	0.94	0.94	0.94	0.94
TOTAL	lpp	(cm)	15.22	12.33	11.43	11.34	11.26	10.42

Table 10-4. Wire lengths in 3-D wafer computer packaging structures.

		PLANE-WISE PARTITION	COMPLETE 3-D
TOTAL NO OF CHIPS		900	
NO OF CHIPS PER WAFER		10×10	
NO OF WAFERS PER STACK		9	
WAFER SIZE (mm)		70×70	
STACK SIZE (mm)		$70 \times 70 \times 10$	←
CHIP PITCH x (mm)		7	
y (mm)		7	
WAFER PITCH z (mm)		1	
AVERAGE CHIP TO CHIP WIRE LENGTH			
X - COMPONENT	lppx (mm)	9.30	6.58
Y -	lppy (mm)	9.30	6.58
Z -	lppz (mm)	0.21	0.81
TOTAL	lpp (mm)	18.82	13.97

l_{ppz}, are obtained separately. The total length, l_{pp}, for the 2-D is 20 to 50% longer than that of the 3-D and 3-D-like structure. For stacked 2-D structures, increasing the number of edge connectors is effective for decreasing the y-component of the wire length. Also, face-to-face connection between cards, which characterizes the 3-D structure, is effective for decreasing the y-component.

Among the various packaging systems, the 3-D wafer computer structure is regarded as one of the most sophisticated packaging technologies. The results of a case study for estimating average wire length is such packaging are outline in Table 10-4. The total number of LSI chips is 900, which is equal to the cardboard systems mentioned above. It is assumed that 100 chips are fabricated on a 7 cm by 7 cm wafer with 7 mm center-to-center spacing, and that nine wafers are stacked 1 mm apart.

If the pin pitch for the interwafer connector is not very fine, the plane-wise partitioning is applicable, and the wire length is about 2 cm. On the other hand, if the interwafer connection technology is quite sophisticated, the complete 3-D partitioning can be applied, with the length decreasing to 75% of the plane-wise partitioning.

REFERENCES

Grinberg, J., Nudd, G. R., and Etchells, R. D., "A Cellular VLSI Architecture," *IEEE Computer*, *17*(1), 69–81 (Jan. 1984).

Masaki, A., "Wire Length Equations in Various Types of 2-D and 3-D System Packaging," in *Proc.* 1st IEEE CHMT Symposium, CHMT, Tokyo, pp. 147–152, Oct. 1984.

Masaki, A., and Chiba, T., "Design Aspects of VLSI for Computer Logic," *IEEE Trans. Ed, ED-29*, (4), 751–756 (Apr. 1982).

Masaki, A., Yamada, M., Asano, M., Tanaka, H., Itoh, H., and Hashimoto, N., "Perspectives on Hardware Technologies for Very-High Performance Computers," in *Proc.* 1984 International Conference on Computer Design, Rye Town, New York, pp. 561–564, Oct. 1984.

APPENDIX

Notations:

L	Number of cells in x direction (logarithm with the base 2)
M	Number of cells in y direction (logarithm with the base 2)
N	Number of cells in z direction (logarithm with the base 2)
R	Exponent in Rent's rule equation
NPP	Number of wires in the array (for obtaining actual number of wires, NPP should be multiplied by a suitable coefficient)
NPP3	Number of wires that appear in three-dimensional partitioning (including plane-wise partitioning)
NPP2	Number of wires that appear in two-dimensional partitioning (including row-wise partitioning)
NPP1	Number of wires that appear in one-dimensional partitioning
LPP	Average wire length in the array
LPPX	x-component of LPP
LPPY	y-component of LPP
LPPZ	z-component of LPP
LPP3	Average length of wires that appear in three-dimensional partitioning
LPPX3	x-component of LPP3
LPPY3	y-component of LPP3
LPPZ3	z-component of LPP3
LPP2	Average length of wires that appear in two-dimensional partitioning
LPPX2	x-component of LPP2
LPPY2	y-component of LPP2
LPP1	Average length of wires that appear in one-dimensional partitioning
LPPX1	x-component of LPP1 ($=$ LPP1)

Equations

Complete 2-D partitioning:

$$\text{NPP2} = 2\hat{}(L+M)*2\hat{}((L-M)*(R\text{-}1))*(1-2\hat{}(2*(R\text{-}1)*M))$$
$$\text{NPP1} = 2\hat{}(L+M)*(1-2\hat{}((R\text{-}1)*(L-M)))$$
$$\text{NPP} = \text{NPP2} + \text{NPP1}$$
$$\text{LPPX2} = 1/9*(1-2\hat{}(2*(R-1)))/$$
$$(1-2\hat{}(2*(R-1)*M))*(7*2\hat{}(L-M)$$
$$*(2\hat{}((2*R-1)*M)-1)*M)-1)/(2\hat{}(2*R-1)-1)-2\hat{}(M-L) \quad (R \ne 1/2)$$
$$*(1-2\hat{}((2*R-3)*M))/(1-2\hat{}(2*R-3)))*X$$

$$LPPY2 = 1/9*(1 - 2\hat{}(2*(R - 1)))/$$
$$(1 - 2\hat{}(2*(R - 1)*M))*(7*(2\hat{}((2*R - 1)$$
$$*M) - 1)/(2\hat{}(2*R - 1) - 1) - (1 - 2\hat{}((2*R - 3)*M))/ \qquad (R \neq 1/2)$$
$$(1 - 2\hat{}(2*R - 3)))*Y$$

$$LPPX2 = 1/9*(1 - 2\hat{}(2*(R - 1)))/$$
$$(1 - 2\hat{}(2*(R - 1)*M))*(7*2\hat{}(L - M)$$
$$*M - 2\hat{}(M - L)*(1 - 2\hat{}((2*R - 3)*M))/(1 - 2\hat{}(2*R - 3)))*X \qquad (R \neq 1/2)$$

$$LPPY2 = 1/9*(1 - 2\hat{}(2*(R - 1)))/$$
$$(1 - 2\hat{}(2*(R - 1)*M))*(7*M - (1 - 2\hat{}((2*R - 3)$$
$$*M))/1 - 2\hat{}(2*R - 3)))*Y \qquad (R \neq 1/2)$$

$$LPPX1 = (1 - 2\hat{}(R - 1))/$$
$$(1 - 2\hat{}((R - 1)*(L - M)))*(2\hat{}(R*(L - M)) - 1)/$$
$$(2\hat{}R - 1)*X$$

$$LPPX = (LPPX2*NPP2 + LPPX1*NPP1)/NPP$$
$$LPPY = LPPY2*NPP2/NPP$$
$$LPP = LPPX + LPPY$$

Row-wise partitioning:

$$LPPX2 = 1/3*(2\hat{}L - 2\hat{}(-L))*X$$
$$LPPY2 = (1 - 2\hat{}(R - 1))/(1 - 2\hat{}((R-$$
$$1)*M))*(1 - 2\hat{}((R - 1)*M))*(1 - 2\hat{}(R*M))/(1 - 2\hat{}R)*Y$$
$$LPPX1 = (1 - 2\hat{}(R - 1))/(1 - 2\hat{}((R - 1)*L))*(2\hat{}(R*L) - 1)/$$
$$(2\hat{}R - 1)*X$$

NPP2, NPP1, NPP, LPPX, LPPY, and LPP are identical to those expressions in "Complete 2-D partitioning."

3-D and 3-D-like Arrays.

Complete 3-D partitioning:

$$NPP3 = 2\hat{}(L + M + N)*2\hat{}((L + M - 2*N)*(R - 1))*$$
$$(1 - 2\hat{}(3*(R - 1) *N))$$

$$NPP2 = 2\hat{}(L + M + N)*2\hat{}((L - M)*(R - 1))*(1 - 2\hat{}(2*(R - 1)*$$
$$(M - N)))$$

$$NPP1 = 2\hat{}(L + M + N)*(1 - 2\hat{}((R - 1)*(L - M)))$$
$$NPP = NPP3 + NPP2 + NPP1$$

$$LPPX3 = 1/7*(1 - 2\hat{}(3*(R - 1)))/$$
$$(1 - 2\hat{}(3*(R - 1)*N))*(5*2\hat{}(L - N)$$
$$*(2\hat{}((3*R - 2)*N) - 1)/ \qquad (R \neq 2/3)$$
$$(2\hat{}(3*R - 2) - 1) - 2\hat{}(N - L)*(1 - 2\hat{}((3*R - 4)$$
$$*N))?(1 - 2\hat{}(3*R - 4)))*X$$

$$LPPY3 = 1/7*(1 - 2\hat{}(3*R - 1)))/$$
$$(1 - 2\hat{}(3*(R - 1)*N))*(5*2\hat{}(M - N)$$
$$*(2\hat{}((3*R - 2)*N) - 1)/(2\hat{}(3*R - 2) - 1) - 2\hat{}(N - M)$$
$$*N))/(1 - 2\hat{}(3*R - 4)))*Y \qquad (R \neq 2/3)$$

$$\text{LPPZ3} = 1/7*(1 - 2\hat{}(3*(R-1)))/$$
$$(1 - 2\hat{}(3*(R-1)*N))*(5*(2\hat{}((3*R-2)$$
$$*N)-1)/(2\hat{}(3*r-2)-1)-(1-2\hat{}((3*R-4)*N))/ \qquad (R \ne 2/3)$$
$$(1 - 2\hat{}(3*R-4)))*Z$$

$$\text{LPPX3} = 1/7*(1 - 2\hat{}(3*(R-1)))/$$
$$(1 - 2\hat{}(3*(R-1)*N))*(5*2\hat{}(L-N)$$
$$*N - 2\hat{}(N-L)*(1 - 2\hat{}((3*R-4)*N))/(1 - 2\hat{}(3*R-4)))*X \qquad (R \ne 2/3)$$

$$\text{LPPY3} = 1/7*(1 - 2\hat{}(3*(R-1)))/$$
$$(1 - 2\hat{}(3*(R-1)*N))*(5*2\hat{}(M-N)$$
$$*N - 2\hat{}(N-M)*(1 - 2\hat{}((3*R-4)*N))/(1 - 2\hat{}(3*R-4)))*Y \qquad (R \ne 2/3)$$

$$\text{LPPZ3} = 1/7*(1 - 2\hat{}(3*(R-1)))/(1 - 2\hat{}(3*(R-1)*N))*$$
$$(5*N - (1 - 2\hat{}((3*R-4)*N))/(1 - 2\hat{}(3*R-4)))*Z \qquad (R \ne 2/3)$$

$$\text{LPPX2} = 1/9*(1 - 2\hat{}(2*(R-1)))/$$
$$(1 - 2\hat{}(2*(R-1)*(M-N)))*(7*2\hat{}(L-M)*$$
$$(2\hat{}((2*R-1)*(M-N))-1)/$$
$$(2\hat{}(2*R-1)-1)-2\hat{}(M-L)*(1 - 2\hat{}((2*R-3)*$$
$$(M-N)))/(1 - 2\hat{}(2*R-3)))*X \qquad (R \ne 1/2)$$

$$\text{LPPY2} = 1/9*(1 - 2\hat{}(2*(R-1)))/(1 - 2\hat{}(2*(R-1)*(M-N)))$$
$$*(7*(2\hat{}((2*R-1)*(M-N))-1)/ \qquad (R \ne 1/2)$$
$$(2\hat{}(2*R-1)-1)-(1-2\hat{}((2*R-3)$$
$$*(M-N)))/(1 - 2\hat{}((2*R-3)))*Y$$

$$\text{LPPX2} = 1/9*(1 - 2\hat{}(2*(R-1)))/$$
$$(1 - 2\hat{}(2*(R-1)*(M-N)))*(7*2\hat{}(L-M)$$
$$*(M-N) - 2\hat{}(M-L)*(1 - 2\hat{}((2*R-3)*(M-N)))/ \qquad (R \ne 1/2)$$
$$(1 - 2\hat{}(2*R-3)))*X$$

$$\text{LPPY2} = 1/9*(1 - 2\hat{}(2*(R-1)))/(1 - 2\hat{}(2*(R-1)*(M-N)))$$
$$*(7*(M-N) - (1 - 2\hat{}((2*R-3)*(M-N)))/(1 - 2\hat{}(2*R-3)))*Y \quad (R \ne 1/2)$$

$$\text{LPPX1} = (1 - 2\hat{}(R-1))/$$
$$(1 - 2\hat{}((R-1)*(L-M)))*(2\hat{}(R*(L-M))-1)/$$
$$(2\hat{}R-1)*X$$

$$\text{LPPX} = (\text{LPPX3*NPP3} + \text{LPPX2*NPP2} + \text{LPPX1*NPP1})/$$
$$\text{NPP}$$

$$\text{LPPY} = (\text{LPPY3*NPP3} + \text{LPPY2*NPP2})/\text{NPP}$$
$$\text{LPPZ} = \text{LPPZ3*NPP3/NPP}$$
$$\text{LPP} = \text{LPPX} + \text{LPPY} + \text{LPPZ}$$

Plane-wise partitioning with face-to-face connection:

$$\text{LPPX3} = 1/3*(2\hat{}L - 2\hat{}(-L))*X$$
$$\text{LPPY3} = 1/3*(2\hat{}M - 2\hat{}(-M))*Y$$
$$\text{LPPZ3} = (1 - 2\hat{}(R-1))*(2\hat{}(N*R)-1)/(1 - 2\hat{}(N*(R-1)))/$$
$$(2\hat{}R-1)*Z$$
$$\text{LPPX2} = 1/9*(1 - 2\hat{}(2*(R-1)))/$$
$$(1 - 2\hat{}(2*(R-1)*M))*(7*2\hat{}(L-M)$$

$$*(2^{\wedge}((2*R-1)*M)-1)/ \qquad\qquad (R \neq 1/2)$$
$$(2^{\wedge}(2*R-1)-1)-2^{\wedge}(M-L)*(1-2^{\wedge}((2*R-3)*M))/$$
$$(-2^{\wedge}(2*R-3)))*X$$
$$\text{LPPY2} \;=\; 1/9*(1-2^{\wedge}(2*(R-1)))/$$
$$(1-2^{\wedge}(2*(R-1)*M))*(7*(2^{\wedge}((2*R-1)*M)-1)/$$
$$(2^{\wedge}(2*R-1)-1)-(1-2^{\wedge}((2*R-3)*M))/(1-2^{\wedge}(2*R-3)))*Y \quad (R \neq 1/2)$$
$$\text{LPPX2} \;=\; 1/9*(1-2^{\wedge}(2*(R-1)))/$$
$$(1-2^{\wedge}(2*(R-1)*M))*(7*2^{\wedge}(L-M)$$
$$*M-2^{\wedge}(M-L)*(1-2^{\wedge}((2*R-3)*M))/(1-2^{\wedge}(2*R-3)))*X \quad (R \neq 1/2)$$
$$\text{LPPY2} \;=\; 1/9*(1-2^{\wedge}(2*(R-1)))/(1-2^{\wedge}(2*(R-1)*M))$$
$$*(7*M-(1-2^{\wedge}((2*R-3)*M))/(1-2^{\wedge}(2*R-3)))*Y \qquad (R \neq 1/2)$$
$$\text{LPPX1} \;\; (1-2^{\wedge}(R-1))/$$
$$(1-2^{\wedge}((R-1)*(L-M)))*(2^{\wedge}(R*(L-M))-1)/$$
$$(2^{\wedge}R-1)*X$$

NPP3, NPP2, NPP1, NPP, LPPX, LPPY, LPPZ, and LPP are identical to those expressions in "Complete 3-D partitioning."

Plane-wise partitioning with edge connectors (stacked 2-D):

With one edge connector per plane:

$$\text{LPPX3} \;=\; 1/3*(2^{\wedge}L-2^{\wedge}(-L))*X$$
$$\text{LPPY3} \;=\; 2^{\wedge}M*Y$$
$$\text{LPPZ3} \;=\; (1-2^{\wedge}(R-1))*(2^{\wedge}(N*R)-1)/(1-2^{\wedge}(N*(R-1)))/$$
$$(2^{\wedge}R-1)*Z$$
$$\text{LPPX2} \;=\; 1/9*(1-2^{\wedge}(2*(R-1)))/$$
$$(1-2^{\wedge}(2*(R-1)*M))*(7*2^{\wedge}(L-M)$$
$$*(2^{\wedge}((2*R-1)*M)-1)/$$
$$(2^{\wedge}(2*R-1)-1)-2^{\wedge}(M-L)*(1-2^{\wedge}((2*R-3)*M))/$$
$$(1-2^{\wedge}(2*R-3)))*X \qquad\qquad (R \neq 1/2)$$
$$\text{LPPY2} \;=\; 1/9*(1-2^{\wedge}(2*(R-1)))/$$
$$(1-2^{\wedge}(2*(R-1)*M))*(7*(2^{\wedge}((2*R-1)*M)-1/$$
$$(2^{\wedge}(2*R-1)-1)-(1-2^{\wedge}((2*R-3)*M))/(1-2^{\wedge}(2*R-3)))*Y \quad (R \neq 1/2)$$
$$\text{LPPX2} \;=\; 1/9*(1-2^{\wedge}(2*(R-1)))/$$
$$(1-2^{\wedge}(2*(R-1)*M))*(7*2^{\wedge}(L-M)*M-2\,(M-L)$$
$$*(1-2^{\wedge}((2*R-3)*M))/(1-2^{\wedge}(2*R-3)))*X \qquad (R \neq 1/2)$$
$$\text{LPPY2} \;=\; 1/9*(1-2^{\wedge}(2*(R-1)))/(1-2^{\wedge}(2*(R-1)*M))$$
$$*(7*M-(1-2^{\wedge}((2*R-3)*M))/(1-2^{\wedge}(2*R-3)))*Y \qquad (R \neq 1/2)$$
$$\text{LPPX1} \;=\; (1-2^{\wedge}(R-1))/$$
$$(1-2^{\wedge}((R-1)*(L-M)))*(2^{\wedge}(R*(L-M))-1/$$
$$(2^{\wedge}R-1)*X$$

NPP3, NPP2, NPP1, NPP, LPPX, LPPY, LPPZ, and LPP are identical to those expressions in "Complete 3-D partitioning."

With two edge connectors on the opposite sides of the plane:

$$LPPX3 = 1/3*(2\char94 L - 2\char94(-L))*X$$

$$LPPY3 = 1/3*(2*2\char94 L - 2\char94(-L))*X$$

$$LPPY3 = 1/3*(2*2\char94 M + 2\char94(-M))*Y$$

$$LPPZ3 = (1 - 2\char94(R-1))*(2\char94(N*R)-1)/(1 - 2\char94(N*(R-1)))/$$
$$(2\char94 R - 1)*Z$$

$$LPPX2 = 1/9*(1 - 2\char94(2*(R-1)))/$$
$$(1 - 2\char94(2*(R-1)*M))*(7*2\char94(L-M)$$
$$*(2\char94((2*R-1)*M)-1)/ \qquad (R \neq 1/2)$$
$$(2\char94(2*R-1)-1) - 2\char94(M-L)*(1 - 2\char94((2*R-3)*M))/$$
$$(1 - 2\char94(2*R-3)))*X$$

$$LPPY2 = 1/9*(1 - 2\char94(2*(R-1)))/(1 - 2\char94(2*R-1)*M))$$
$$*(7*(2\char94((2*R-1)*M)-1)/ \qquad (R \neq 1/2)$$
$$(2\char94(2*R-1)-1) - (1 - 2\char94((2*R-3)*M))/$$
$$(1 - 2\char94(2*R-3)))*Y$$

$$LPPX2 = 1/9*(1 - 2\char94(2*(R-1)))/$$
$$(1 - 2\char94(2*(R-1)*M))*(7*2\char94(L-M)$$
$$*M - 2\char94(M-L)*(1 - 2\char94((2*R-3)*M))/(1 - 2\char94(2*R-3)))*X \qquad (R \neq 1/2)$$

$$LPPY2 = 1/9*(1 - 2\char94(2*(R-1)))/(1 - 2\char94(2*(R-1)*M))$$
$$*(7*M - (1 - 2\char94((2*R-3)*M))?(1 - 2\char94(2*R-3)))*Y \qquad (R \neq 1/2)$$

$$LPPX1 = (1 - 2\char94(R-1))/$$
$$(1 - 2\char94((R-1)*(L-M)))*(2\char94(R*(L-M))-1)/$$
$$(2\char94 R - 1)*X$$

NPP3, NPP2, NPP1, NPP, LPPX, LPPY, LPPZ, and LPP are identical to those expressions in "Complete 3-D partitioning."

11. Packaging Cost-Performance Computers

John W. Balde

Interconnection Decision Consulting, Flemington, NJ

There is a significant shift in emphasis on packaging requirements for computers when absolute performance must be weighed against cost. While mainframe machines support their high performance through very dense packaging, that density is for the purpose of providing top-of-the-line capability through optimized electrical performance and the resulting high data processing throughput. With these large systems, a three-year hardware development time is tolerated, and designs are often unique or special to each manufacturer. The discussion of 3-D packaging of the previous chapter shows this concern for maximum speed and density.

It is this emphasis on high performance that makes the design emphasis for high end computers concentrate on such issues as decoupling requirements, termination of interconnection lines in characteristic impedance, low capacitance requirements for MOS circuitry, and the future demands of gallium arsenide semiconductors. Designers of cost-performance machines must also understand system design methods and the requirements for good electrical performance; however, the emphasis is not on the maximum performance possible, but rather on reasonable performance at a low cost. Market demands require a much shorter design cycle, mandating the use of as many standard parts as possible. Packaging designs in the cost-performance arena are often very similar, with only a few deviations from the current "standard" packaging approach. This chapter will concentrate on possible differences in approach, in the full understanding that the most successful designs are most apt to be widely imitated. First, however, we should explore the reasons why unique packaging designs are not very likely in this market.

Until recently, the computer market was principally the mainframe market.

In the early seventies the personal computer (PC) market was virtually non-existent, as shown by a comparison of mainframe versus PCs. Even in 1980 the PC market demands on semiconductor vendors were negligible compared with mainframes. The market for office terminals, identical in design and manufacturing requirements to desktop microcomputers, was also small compared to that for mainframes. Today, however, the market for small-scale, cost-driven computer hardware is already equal in dollar volume to the mainframe market, and it is expected to grow to $20 billion by 1990.

Similarly, the minicomputer market was a small, though not insignificant, fraction of the mainframe market. Minis accounted for approximately $1 billion in 1981, with their market expected to grow to $5 billion in 1987, and perhaps $15 billion in 1990. Sales of both of these cost-performance categories, minis and PCs, will far outstrip the mainframe volume.

Manufacturers of mainframe systems claim otherwise, predicting an upturn in mainframe systems on the theory that increased sales of PCs will spawn a demand for mainframe capability as PC users look to bigger machines for data processing of centralized information. This at best a big "maybe." There is a strong trend toward individual organizations owning their own PCs and mini-computers within big corporations, so that those organizations are independent of the traditional centralized computer. Thus a major new class of PC and minicomputer users is created, which has no relationship to larger machines and provides convincing evidence that the predicted sudden upturn of the mainframe market may indeed be wishful thinking.

This emergence of the cost-performance machine will make a profound difference in the semiconductor and packaging markets. There are many players—many companies that produce or will produce cost-performance machines. Because the players are relatively small, new designs are often turned around in only a few months, and sales shift rapidly to newer machines that offer more performance at similar or lower prices. This market sees an interval from design start to volume sales of less than a year, and a rate of design change that makes it unlikely that a new design two years in the making can satisfactorily compete with one conceived a year later. To summarize, the cost-performance market has many players, short design-to-sale intervals, and stringent cost constraints. The principal solution is the use of more and more standard parts and standardized approaches to physical design.

To understand the design and marketing realities of the cost-performance computer, consider the following:

- Four or five VLSI chips can provide up to 90% of the central (CPU) performance
- Design of a special chip, custom or standard cell, can cost $250,000 and take three or four months. That is tolerable for a few devices but would be excessive if all the functions of a computer were to be designed in special

high-density chips. Amdahl could design a computer in the seventies, bread-board it in standard DIP technology, and then translate that design to a few dozen LSI chips and achieve a dense system design with good performance and competitive cost. The Amdahl approach, with its use of surface attach chip carriers on conventional printed circuit boards, is a good model for small computer design—except that there is seldom if ever enough time or manpower available to design a new VLSI-based small system in just a few months. Instead, a small-system designer is usually limited to only a few special chips plus some standard gate arrays, depending on off-the-shelf ICs for the balance of the circuitry.

- The amount of circuit concentration in the few permissible VLSI devices is so great that over 90% of the computer can be on just a few chips, but the remainder of the circuitry that ties the computer together can constitute over 90% of the board area. Recent minicomputer CPU circuit boards form Digital Equipment Corporation illustrate this dramatically. The circuit core of the central processor is on four chips in the center of the board, yet the board is so full of DIP-packaged devices for the rest of the circuitry that is the DIPs that dictate the nature of the board packaging.

The 9300 minicomputer made by National Cash Register (see Figure 11-1) is another telling example. The attaché case–sized minicomputer with over 1 MIP capability is a four-board machine, yet the VLSI of the central processor is just four chips. With four 15″ × 20″ boards, or 1200 square inches of board real estate, the VLSI content of the central processor covers only 4 square inches. Even allowing for the increased surface area of sockets for the VLSI devices, the VLSI-dominated board area would still be just 8 square inches, or less than 1% of the total board area.

The IEEE VLSI Workshop held in Clearwater, Florida in January 1984 featured a spirited discussion of this phenomenon. That conference was largely attended by those whose engineering lives are devoted to trying to increase the density, performance, and capability of the individual VLSI chip. That is important for the mainframe market, yet for cost-performance computers, the dominant marketplace of the late eighties, there is little concern for such an increase in raw chip capability. Consider the NCR minicomputer: if VLSI capability were to increase tenfold, the board area devoted to the VLSI portion of the circuit would decrease below the present 1%, but the effect on the overall size of the system would be entirely undetectable! It is the size of the off-the-shelf, standard "glue" or interface circuits that determines the size of the computer.

Why, then, is there no shift to reduce the size of the "glue" circuits? First the functional role of the devices needs to be understood. These ICs chips provide, among other functions, the peripheral drivers, the hex-to-decimal convertors, and the line drivers to give the power and/or characteristic impedance matching needed to send signals over great distances. They are interspersed throughout

Figure 11-1. The 9300 minicomputer is a typical cost-performance computer in that most of its four PC boards carry non-VLSI circuitry. (Photo courtesy of NCR)

the board in locations determined by the circuit architecture and functions required for a particular system. These functions are highly standardized, and semi-conductor manufacturers can find a sufficient volume of sales to make them available at an attractive, very low price. Combining these functions into super chips would not only takes design time and effort, but would tend to produce chips that are specific to a particular system manufacturer.

A shift to such chips specificity is counterproductive in several ways: First, it reduces the volume of sales and increases the cost of the design applied to each sale unit, often making it impossible to sell to other potential buyers. Second, no other manufacturer would want to buy large chip circuits identical to such a product. (Where, then, would a system manufacturer's uniqueness and innovation be rewarded? Where could one find a competitive edge?) Third, although designers could shift to the greater use of standard cell, semicustom, or ASIC chips, that would require that designers delegate their design responsibility to others—something no one wants to do. Designers would much rather chose standard chips for the glue circuits, so that they stay in control.

The result of all this is that PC and minicomputer designs can be expected to have only a few unique VLSI chips, surrounded by a rather large number of standard off-the-shelf chips providing rather standard functions.

Given that, where is the opportunity for innovation and packaging design with some competitive edge? Let us consider the options:

1. The simplest option is to try to stay with DIP IC packaging and the comfortable world of through-hole mounting of both ICs and passive components— use wave soldering, use component insertion equipment, and continue to design with double-sided circuit boards. Do not underestimate the viability of this approach. Most computer products are still made this way, and a continued life for this type of design is possible in some markets. The use of the larger VLSI devices is precluded because semiconductor manufacturers will not package such large chips in a DIP. The number of system manufacturers willing to deal with a 68- or 84-pin DIP package 6 inches long is too small to make that solution economically feasible, and even the small computer manufacturer cannot afford to throw away all the board space and circuit performance advantages of high density VLSI circuits.

2. The next option is more popular: stay with DIPS for the "glue" circuits, but use more advanced packaging for the VLSI portion of the circuit. Here more choices become available. One can order the VLSI chip in a pin grid array package like the one in Figure 11-2. If the rest of the board uses DIPs, then a board with two signal planes with added ground and power planes will probably be adequate. If a pin grid array package with a fully populated array of pins is used, the size of the package might be too small to have an adequate chip cavity.

Figure 11-2.   The pin grid array package is often advocated as a means of integrating VLSI devices into through-hole PC boards.

More important, the cavity would have to be on the top side of the package, requiring that the heat of the chip be carried away through the leads to the printed circuit board. Since printed circuits board are poor conductors of heat, that solution is usually unacceptable. It is better to construct the chip package with the cavity down, providing for upward heat removal through a surface-bonded heat sink. This also has the advantage that the population of pins on the bottom of the package is reduced per unit area, often to nearly the level of wiring density that can be supported by a double-sided board or, at most, four signal layers.

THE PIN GRID ARRAY DILEMMA

Now a big question arises: If mounting the pin grid array package is easy, inserting its pin in through-holes on the board on already-standard 100-mil centers, and if soldering it is easy, using wave soldering, what is the possibility of package removal? Unfortunately, it is not good. These packages can be desoldered from a PC board, but either special spring tension extractors must be provided to prevent excessive removal forces, or much care must be used. If there is one instance of less-than-adequate solder melting—rather likely considering the number of pins involved—a via barrel can be pulled from the board, destroying the entire assembly.

One alternative is to put the pin grid array in a socket, which provides full printed circuit board compatibility. But if a socket can provide board compatibility, why use the pin grid array as the chip package? Why not take advantage of the smaller package outline, the shorter, more uniform circuit path lengths, and the lower path resistances of chip carriers, and their much lower cost?

Socketing greatly expands the range of packaging choices, including:

- Ceramic leadless packages in bottom contact sockets with hold-down retainers. If the heat sink gets in the way, it is even possible to mount the heat sink on the top spring retainer.
- Ceramic leadless packages with edge clips that effectively add pin leads. These packages can also be socketed in the same type of socket used for the pin grid array, and have the advantage that there is no need for a top hold-down retainer and no restriction on the heat sink. There is a slight problem with the add-on clips—they are somewhat prone to handling and insertion damage, and their insertion into the socket is slightly more difficult than that of the pin grid array. Along with their adaptability to the pin grid array socket, the principal advantage of edge clips is that they permit the use of ceramic packages with their generally good crosstalk and characteristic impedance control.
- Postmolded plastic leaded packages, inserted into sockets that press the sides of the package and leave the top available for heat sinks. Multiple sources exist for the sockets, and the package is both small and rugged. The unique

advantage of this package and socket combination is that both can be surface-attached for all-surface-attach boards, permitting the shift to surface attach without change of part.

ALL SURFACE ATTACH OPTIONS

The next alternative is to try to go "all surface attach." This has been the approach of the military, who have used chip carriers quite successfully. Here again, options and trade-offs abound.

One choice is to use leadless chip carriers, either ceramic or epoxy-glass like the EPIC package used by British Telecom. This epoxy-glass package is designed to be mounted directly on the circuit board, matching thermal expansion coefficients. Here the principal problem is one of how to cope with the differential expansions of the package and the board due to differential heating. The all-ceramic construction is not usually considered suitable. Even all-ceramic constructions have a solder joint cracking problem with the temperature changes encountered from storage to turn-on. The expansion-controlled equivalents to ceramic—polyimide, kevlar, quartz, Ultem, epoxy on copper-Invar-copper boards—are usually too expensive for this cost-sensitive market. Although vendors perceive that these products would be of interest, users are nearly unanimous in not wanting this type of solution in a cost-performance product.

The situation is worse than that. Preventing the horizontal expansion of materials that must increase in size forces an expansion in the Z direction. That puts severe stresses on the vias—stresses that cause almost 100% via cracking during temperature cycling if the use of copper-Invar-copper constrains X-Y expansion to provide a good match for the leadless ceramic packages. The original report by Gary Beane of TI is restricted, but a discussion of the problem is available from the IPC in a videotape.

There is another sound reason for lack of interest in controlled-expansion materials: they can provide only an approximate match for packages of uniform wattages, and computers do not have uniform wattages on their chips. There may be 3 to 5 watts on the VLSI chips and 100 milliwatts on a hex-to-decimal convertor. The concept of tailoring the expansion of a board to be different at different portions of the board is suspect, requiring much faith. Imagine a checkerboard with one square without the Alloy 42 or Cu-In-Cu backing so it can expand while the world around it stays tightly temperature-controlled. The excess material must go to Z-axis buckling!

An alternative solution is to cool well, keeping the temperature constant. Unfortunately, although mainframes and telephone central processors exist in a commercial world of air conditioned rooms and a consistent thermal environment, the cost-performance products may not. The only way to cope with the differential expansion problem in these designs is to keep the wattages low, which means small packages, small chips, and giving up the advantages of VLSI.

SURFACE ATTACH
and the use of
Chip Carriers and Surface Mount Components

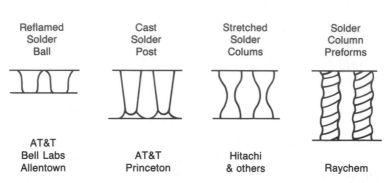

Figure 11-3. Specially shaped solder bumps can help designers cope with some of the thermal expansion problems of leadless packages.

A more practical solution is to use leaded packages, or employ very tall solder bumps to provide the compliance that can permit leadless packages on a printed circuit board. The solder ball technology of Bell Laboratories Allentown is not good enough for printed circuit substrate use. The Hitachi hourglass stretched bumps are better, increasing the solder height to 18 to 20 mils. Providing cast bumps can increase the height to 30 or even 40 mils, and the Raychem tall solder attachment columns can provide up to 100 mils. Various solder bump geometries are shown in Figure 11-3. This kind of approach provides enough compliance of accommodate differential X-Y expansion.

The rub here is that boards also flex in the Z dimension. The lack of uniformity of copper (or the deliberate allocation of horizontal stripes to one layer and vertical to the other) causes the board to warp as its temperature changes. There may be an initial warp as the copper is removed during circuit etch, and there will be more warping as the board goes through the soldering operation and returns to room temperature. Additional changes in board shape occur as the board is inserted and removed from a card cage or slot, and as the computer is turned on and off with the accompanying temperature changes. The Z-axis flexure of ceramic is virtually zero; typical printed circuit boards flex 50 mils in a 10-inch board, or 5 mils to the inch. For a 1-inch package, that is $2 \frac{1}{2}$ mils Z-axis pull, assuming a 20-mil solder leg. This may be tolerable if the solder height is actually 20 mils, but is troublesome if the solder is just a few mils thick.

Leadless chip carriers on submodule boards, mounted in turn on larger printed circuit boards, have become an established historic solution for the telecommunications and industrial equipment areas—so why not use them for the cost-performance market? The design in question may have a mixture of high density surface mount packaging with other portions of the circuit that use older, low density techniques—so why provide the additional layers and higher technology to surface mount all of the circuit if a significant area of the board surface does not need higher interconnection density? What may seem like a good idea here will not hold up to closer examination. If a cost-performance system is to compete in tomorrow's market, it must go to surface attach, with all the automation that implies. Handling two substrates and remaining with interfaces that are through-hole when the rest of the technology is becoming surface attach is an unattractive option, to say the least.

When all the various surface attach options are considered, the best solution seems to be to use leaded chip carriers as the dominant package for future cost-performance boards. There are some problems involved in making sure that the leads are adequately compliant, and there are difficulties in optimizing the soldering process, but these problems will be resolved. Once that issue is decided, the next major question is the packaging efficiency—the gain in board real estate and system volume.

BOARD AREA—THE DEFINITIVE CRITERION FOR PACKAGING EFFICIENCY

The criterion for deciding whether a design is efficient is the ultimate measure of packaging density, the percentage of the board surface area occupied by active silicon. The definitive paper on this subject, by Knausenberger and Schaper (1984) of Bell Laboratories, shows that the move from DIP packaging to 28 leaded chip carrier surface mounted packages increases the silicon density from 0.6% to just under 1%—not an overwhelming leap. Using 68 lead packages achieves a 5% jump in density—a major improvement, but one denied to the cost-performance market because the vendors cannot take the time for all-new chip designs.

Should we accept this situation? Yet, but only partially. The design criterion of Bell Laboratories for telecommunication apparatus is rooted in mainframe design philosophy: design to cope with changes because they will inevitably occur—a good philosophy, to be ignored at one's peril. But higher packaging densities are possible if one mounts the chip carrier packages as close together as possible, with aisle spaces of 100 mils or less. Texas Instruments, in the paper by Garth and Gray (1983), shows boards with silicon densities of over 2% delivering the 300% packaging density improvement widely touted for chip carriers over DIPs. But there is a price: reduced ability to inspect, reduced ability to socket, and no ability to add change wiring unless one is willing to tolerate the "stick the wire end into the solder fillet" changes often used by the military—a workshop "kluge," a solution suitable for prototype manufacture, but one that should provoke considerable and justifiable anxiety in a production setting.

THE FOUR-CHIP MULTICHIP PACKAGING OPTION

In devising an alternative, one might well begin with Amdahl's approach to cache memory packaging in its 580 computer, as shown in Figure 11-4. There the manufacturer uses four ceramic chip carriers on a small platform. The individual packages could never be mounted on the big board with that close a spacing; they could not be adequately inspected, and certainly never satisfactorily removed. Instead, the four-chip assembly is tested and stockpiled, then mounted and removed as if it were another VLSI package. This is encouraging, but there may be an even better way.

Both Bauer (1984) of RCA and Stafford (1980) of Bell Laboratories have advocated the use of small hybrids—four chips maximum; and Stuhlbarg (1983) of Hughes also supports this position. There is a considerable advantage to the use of such small multichip packages. Wire bonding yields are still quite good for this size assembly, so the failure rate during assembly is tolerable, and the package size can often be identical to the package selected for the VLSI circuit chips. This uniformity of package size is a powerful advantage in printed circuit board design. The computer aided design technology for chips has shortened the design time for ICs to 4 to 6 weeks, but printed circuit board design time is often over 9 weeks. Use of the same design philosophy currently applied to chips can be facilitated by the use of a uniform "cell size" on the printed circuit board, just as it has facilitated IC chip design on silicon.

Figure 11-4. The Amdahl 580 employs platforms holding four chip carriers, with each platform treated as a single VLSI device. (Photo courtesy of Amdahl)

Figure 11-5. Memory circuitry, with its extreme regularity, is unlikely to require changes and thus is a good candidate for close package spacing. (Photo courtesy of Texas Instruments)

Sperry has similarly used larger hybrids, possible because its design for the computer in question used a quite large package footprint for the VLSI packages. Balde (1984a) discussed the relative advantage of this compared to the use of smaller hybrids.

What happens if this four-chip hybrid philosophy is analyzed using the Schaper–Knausenberger criterion? Balde (1984b) examines this design using that criterion. Apparently the density can be improved to about 4% and still retain the wider aisles needed to facilitate design changes and improved testing. The closer packing density of the Texas Instruments approach shown in Figure 11-5 may be satisfactory for memory chip mounting, where the regularity of the wiring makes changes extremely unlikely.

Why not replace these four-chip packages with single LSI designs? It is not practical to do so because the topological configurations within a system use many different permutations and combinations of standard chips, and it is initially unlikely that the same chip set would be used in exactly the same topological configurations. Wafer scale integration efforts had to cope with that same problem; each cluster of functions has many identical cells, only a small fraction of which were selected to provide the interconnection of functions required for a desired circuit. The advantage of the four-chip package approach is that it can deliver only those chips needed at a given spot on the board.

This much said, one might be tempted to interconnect the chips internally in the package to provide the required function. This is a very bad idea because if the chips were connected internally, the package would immediately become nonstandard, and within a short period of time there would be thousands of alternative package wiring patterns. If these packages are to serve the cost-performance market, they must be readily available off the shelf, packed with equally off-the-shelf standard chips. Furthermore Bauer (1984) Has found that the use of internally interconnected multichip packages decreases reliability because they can be functionally tested only as a group. The assembly's overall reliability is reduced if the chips cannot be checked individually. Bauer finds that penalty to be 2:1; use of the internally connected chips without external access to the individual chip functions doubles the failure rate. One can sometimes invent patterns, such as those shown in Figure 11-6, that permit some external testing and still provide internal interconnections, but such a design would be counterproductive for the cost-performance market.

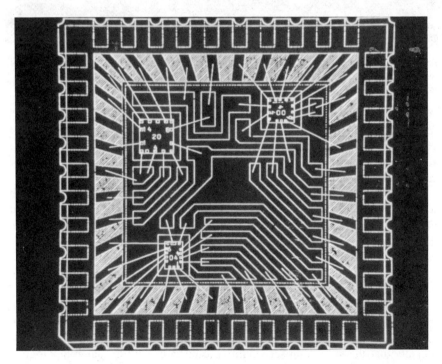

Figure 11-6. Multichip packages that combine internal interconnections with external test points are generally not feasible in cost-performance designs.

Figure 11-7. A memory packaging approach by Hitachi uses an open package and gel chip protection to enhance reliability economically. (Photo courtesy of Hitachi)

Should the multichip packages be postmolded? Only if the yield penalty is not excessive. Using an open or premolded package permits testing before the package is sealed. Recent advances in understanding of the performance of silicone gel encapsulants have led to the use of an open package with gel chip protection, which permits superior reliability performance at a low cost. The recent Hitachi memory package shown in Figure 11-7 is a good example of that solution.

Assuming that the multichip package concept becomes accepted, it is unlikely that resulting devices will remain in multichip form. Once one has a four-chip package with a particular set of functions, there will be the inevitable tendency to reuse the same set of functions elsewhere on the board, perhaps rotating the package to meet the needs of the new location. If their usage then increases, some of these multichip packages will have sales in the millions, and the LSI designer will become interested. Soon a single chip package will be produced, somewhat lowering the cost. If the package is the same size as the earlier package and follows the established pinout, it can be sued without change in the board or circuit design, with no disruption to the spare parts pipeline.

This is standard evolution. Given sufficient demand, LSI designs always replace hybrids, but then the hybrid manufacturer simply makes another hybrid with the new LSI chips. This evolution allows the use of the multichip approach to offer potential densities that were previously thought possible only with wafer scale technology.

SIPS

Any discussion that includes memory packages and multichip hybrids would not be complete without recognizing the single in line package. The SIP (see Figure

Figure 11-8.   The single inline (SIP) platform for chip carriers is particularly useful for memory packaging on throughhole PC boards.

11-8) is still an important alternative for memory packaging. At a penalty of some vertical height, one can use this multichip package to provide quite high memory density. In fact, the memory density produced by a combination of SIP and chip carrier is three to four times what is possible with the same packages mounted flat on a board, at a penalty of three or four in board-to-board spacing. The advantage is less backplane or board interconnection wiring, an interesting trade-off that is more of a manufacturing implementation decision than one of system design. The printed circuit board can remain in two-sided plated-through hole wave solder technology, with the reflow soldering confined to small assemblies. This is an interesting transitional design approach of some usefulness.

RESISTORS AND CAPACITORS

All of the discussion so far has concerned active circuit elements. Computer technology typically uses fewer passive components than are found in a consumer apparatus, but there are still some to deal with: characteristic impedance terminators, decoupling capacitors, and some discrete resistors and capacitors used for other purposes. These components can be mounted in a number of ways:

1. They can be mounted on the top or IC surface of the board, in the space between ICs and/or multichip hybrid components. This is a feasible approach, but one that decreases the circuit density to the extent that the aisle spacing is increased.
2. They can be mounted under the chip carrier packages, using the clearance space provided under the plastic body. This is a particularly appropriate solution for decoupling capacitors.
3. They can be mounted in small outline J-leaded packages (SOJ) or chip carriers and treated as if they were just another IC package. This promises to be a major and useful package to mix with leaded chip carriers. The SOJ package is nearly as small as chip carriers of low lead count, and their circuit boards are easier to route than those of the leaded chip carriers.
4. They can be mounted on the other, noncomponent side of the circuits boards, and wave-soldered—perhaps the most attractive and cost-effective solution.

HIGH-SPEED/HIGH-PERFORMANCE

Like other areas of computer technology, cost-performance computers see increases in the speed of operation as VLSI capability increases and as circuit board size decreases with increased packaging density. For the most part, however, cost-performance systems tend not to introduce gallium arsenide chips or use the highest possible performance silicon circuits. The techniques of high performance are pioneered elsewhere. That disclaimer notwithstanding, many of the physical design considerations will be found to be identical for cost-performance and high performance computers.

FUTURE INCREASES IN CIRCUIT DENSITY

Increased performance of the VLSI chips will be accompanied by an increase in the I/O lead count for each package. Lead counts will rise to 168, 198, and 260, demanding chips carriers with 25-mil pad/lead spacing, with 20-mil spacing not far behind. The increased chip size of the VLSI chips, and the increase in the size and I/O count of their packages will be reflected in similar increases in packing density of the "glue" circuits. This means a general increase in circuit density.

PROBE POINTS AND CHANGE PADS

When the IC packages have I/Os on 50-mil centers, the question of aisle spacing depends principally on the ability to test, and to provide the possibility of change. Fanning of aisle spacing is principally dependent on the ability to test, and to

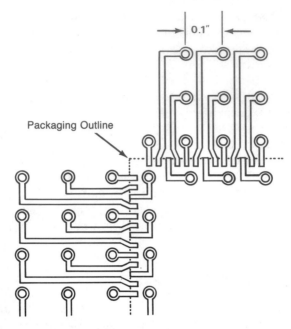

Figure 11-9. Fanout to test pads on 100-mil centers and cuttable links for ease of modification can be provided with little penalty in circuit density.

provide the possibility of change. Fanning out to vias on 100-mil spacing is generally straightforward, and provision of cuttable links is possible without too much of a penalty in density. Figure 11-9 shows one possible arrangement of the I/O pads and the vias and change pads. With increased I/O and reduction in pad spacing, fanout to 100-mil test pads and vias is still possible; but it begins to consume much more space than the board surface required to mount the chips, so something has to give.

One of the best alternatives is to reduce the number of I/O leads. The series resistance of the traces in a chip carrier can be tolerated for the signal leads, but can be a source of excessive common impedance coupling for power and ground leads. The typical solution has been to make multiple parallel connections for power and ground, with the result that the power and ground paths often consume as much as 25% of the total I/O count. If ceramic chip packages with tungsten or thick film metallization are used, this is a necessary solution. With leaded packages, however, a look at the relative resistivity of the alternate circuit materials is in order. A shift to copper lead material can so reduce the common impedance coupling that the 25% share of the device's I/O count is rendered unnecessary.

This reduction of lead count with copper package circuitry can be much more easily implemented in the chip carrier format than in a pin grid array. I/O reduction can make the package smaller, reducing the need for the excessive lead counts typical of the pin packages. For cost-performance systems, it is better to go to a smaller package and use the reduction in board footprint to provide aisle space for circuit change and test.

This shift is beginning. AMP is readying a ceramic package with copper metallization with 10-mil contact spacing around the periphery. Such a tight spacing cannot be soldered to the board, and it cannot be implemented in leads, but sockets can and have been designed to contact the surface pads of the ceramic chip carrier with that spacing. The result is a very important contribution to the packaging of high I/O devices, and the good news for the cost-performance designer is that the packages can be surface-mounted among the other surface amount packages.

Fanning chip carrier I/Os to 100-mil test pads can increase the aisle spacing to 500 mils between packages. Reduction to 50-mil spacing of the vias and test pads is too costly in circuit routing density; even 13-mil vias in 25-mil pads provide only 25 mils between via pads, requiring 5-mil line and space to get two circuit stripes. That is not only costly in terms of printed circuit yield, but it also produces problems with excessive crosstalk. The alternative, putting one trace between pads, doubles the number of layers of the PC board, and increases cost by reducing yield because of the number of layers. The oft-cited yield–risk graph of Figure 11-10 presents a realistic picture of the problem.

Another problem is that, until recently, 50-mil vias and pads have been difficult to probe-test. Although the hollow needles of the probe guides can be less than 50 mils O.D., it is difficult to make actuators that spring-load the needles to achieve the contact force and still have the necessary resilience. A cluster of test probes on 50-mil centers can be spread out to 100-mil centers to get more reliable actuators, but a 10-inch-square board cannot provide the long, sinuous needle length required to fan out the pin points of the board contact to an array of actuator ends in a 20-in-square pattern. Those leads would have such high inductance and crosstalk that no working circuit could be exercised, and the long bent needles would make even simple continuity testing unreliable.

Fifty-mil spacing may not be the answer. Instead, the best possibility is 72-mil spacing—the use of the interleaved row of 100-mil spaced vias at the center of the squares on a 100-mil grid. This doubles, rather than quadruples, the via density, but the result can be probed with the newer probe points, and the circuit board can use reasonable stripes with high layer circuit density. With 13- or 16-mil holes in 30-mil pads, the between-pad spacing of over 40 mils can provide 8-mil line and space, or better yet 7-mil stripes with 9-mil spaces, 6-mil traces, and 10-mil spaces. This is manufacturable circuitry with a reasonable yield because only a few layers are needed. Four signal layers can meet most circuit density requirements, with greater density possible either by an increase to six

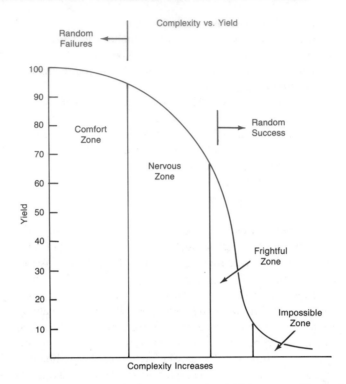

Figure 11-10.   As PC board complexity increases, the yield decreases. The implication for cost-performance systems is clear.

signal layers, or a reduction of the pad diameter to 25 mils and the use of 4-mil lines and 7-mil spaces, a choice largely dependent on the available production capability.

ALTERNATE SUBSTRATE MATERIALS

Most substrates for cost-performance systems have been and will continue to be rigid epoxy glass laminate. However, two other possibilities are available for this market.

Phenolic boards have been used extensively in Japan for consumer products. Their surface insulation resistance is adequate for the low voltages encountered in computers, and the ability to make satisfactory copper traces on them has been established. Where phenolic fails is in Z-axis expansion and in via performance. Vias in phenolic can and do crack and snap if the thickness of the phenolic is greater than 30 mils or so. Use of multilayer boards in phenolic has been generally ruled out because of his barrel cracking, and also because the body resistivity of phenolic boards can be very erratic. Internal leakage paths can be formed as the material becomes loaded with plating chemicals, creating such a severe problem that vias are to be avoided with this material, at least with conventional plating methods.

Phenolic is not entirely ruled out, however, because the use of increasing quantities of surface attach passive components can so thoroughly eliminate the use of vias for crossovers that single-sided boards can be designed for some less complicated interconnection needs. The use of zero-ohm resistors can provide crossovers without vias, as shown in Figure 11-11, and increase the possibilities for such interconnection boards, although generally not enough to permit that solution for cost-performance computers. Part of the problem is that digital circuits just do not use enough resistors to provide the necessary crossover paths as a fallout of circuit manufacture, and the high density IC circuits do not provide room on the top surface of the board for the required wiring.

Far more important are the new dielectric materials for multilayer boards. Use of the Gore pre-preg materials with Gore-Tex instead of fiberglass reduces the dielectric constant to 2.7, providing a significant improvement in propagation delay and reduced circuit capacitance. Even if the characteristic impedance is to be kept constant, the ability to reduce the spacing to the ground plane can

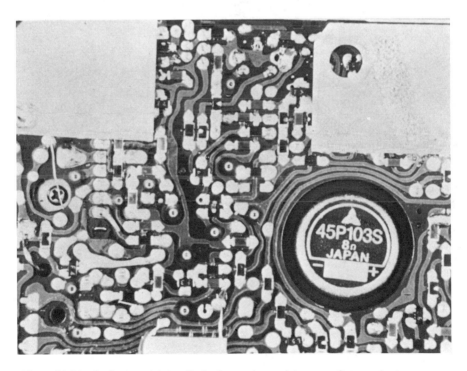

Figure 11-11.   In Japan, surface-attached zero-ohm resistors are often used as crossovers in order to stay with single-sided PC boards.

increase the density of the circuit. Rogers has an equivalent material, fiberglass-reinforced PTFE, also with a 2.7 dielectric constant. Even newer constructions based on liquid crystal polymer are being actively explored. The cost-performance world can use these materials to improve performance without resorting to the exotic co-fired multilayer hybrid package constructions.

FLEXIBLE SUBSTRATES

The use of flexible circuitry bonded to metal plates, shown in Figure 11-12, was originally advocated by Vernon Brown of Bell Laboratories Denver. This arrangement has also been used by Burroughs, where it is called SLIC (single layer interconnection circuits), and it is advocated and marketed by Welwyn in the United Kingdom. In fact, this approach works very well as long as the flexible circuit can be implemented as a two-sided design.

That design was suitable for the interconnection densities of the early eighties, producing a capability of up to 125 inches of circuitry per inch of circuit board. An increase to packages of 68 leads or greater as the norm requires four signal layers, and raises severe questions for the approach. By the time one has a four-

Figure 11-12.   Use of flexible circuitry bonded to metallic backing can be a useful approach is two circuit layers are sufficient to the task at hand.

layer interconnection substrate, and includes the top pad layers and possible interposed ground planes, the circuit is far from flexible. Laminating it to a metal plate is no different from laminating a rigid board. To make the technique even more problematic, the metal support and control plate must provide a near match to the TCE of the substrate in an attempt to control expansion.

CONNECTORS FOR COST-PERFORMANCE SYSTEMS

Connections from the circuit boards of these systems can be conventional two-part connectors, or even edge card types. In larger systems, the increase of circuit density in each board of a system generates a high I/O density at the card edge for the leads that must communicate with the other boards of a system. For cost-performance systems, increases in circuit density may provide so large a per centage of the total system on one board that the number of leads exiting to other boards becomes quite modest, with remaining board I/O confined to communications links to other autonomous system functions such as peripheral drivers and graphics generators. This situation will often permit the use of card edge connectors. Card edge is often deemed inferior to the use of two-piece connector systems, and with the requirement for gold-plated tabs with card edge, the costs of the systems can be nearly comparable. A more likely reason for choosing the two-piece approach is that the I/O density from the board requires the equivalent of 50-mil circuit lead spacing at the communications edges of the board, suggesting at least the use of a two-row, 100-mil spacing of 25-mil-square posts interfacing with a box connector to flat cable. Given that solution, certain options must be considered:

1. The connector may have conventional solder tails, requiring holes in the board. If the major portion of the board has shifted to surface attach, the soldering of the connector may be the only operation that is not accomplished using vapor phase soldering. This suggests additional cost.

Beyond this cost disadvantage, the use of a solder tail connector can introduce signal routing problems. It can be very difficult to get the traces to the outer row of contacts through the "picket fence" of the inner row of pads around the holes.

2. Another alternative is the use of surface attached solder tails. These can permit connector mounting and connection by vapor phase, but only if the plastic of the connector can withstand the temperature of the soldering vapor for the requisite time. Most connector materials cannot, with Ryton being one of the exceptions. It is possible to mold the connector in epoxy or in a thermoset plastic, but that would so increase the time in the molding press that there would have to be a tenfold increase in the press overhead charged to the connector—a costly answer.

Surface-attached solder tail connectors can be used without resorting to high-temperature plastics, however, if the tails are arranged in a straight line row to permit the use of hot mandrel soldering, focused infrared, or laser reflow. Each of these methods works, but all add considerably to the soldering assembly cost.

3. A third alternative is to rely on wave soldering of the board assemblies, even though surface vapor phase soldering might be the first method contemplated. Wave soldering is usually thought to be out of the question for high I/O lead density on the chip packages, and indeed it is for inverted "packages down" soldering. But there is another possibility: wave solder on the bottom of the board, connecting the resistors and capacitors, and let the heat from the solder wave conduct up through the vias to melt the screened solder paste on the top side. Surface attach by reflow soldering of screened paste need not be vapor phase or even infrared; it can be done with conducted heat from a solder wave on the bottom.

At this point, soldering alternatives are well worth reviewing. The choice of soldering method influences the design of the board, not only in the choice of connectors and the placement of parts, but in the use and placement of vias. Hot plate and hot belt soldering is usually a military equipment choice, possible because of the use of ceramic substrates. Hot oven, hot air, and infrared are reasonable alternatives, with the choice dependent on the amount of special design and process characterization needed to ensure satisfactory performance. Infrared may be used for board preheat before vapor phase or hot air soldering, and indeed it may have a particularly necessary place in ensuring satisfactory surface attachment of components that are to be wave-soldered through conducted heat.

MULTIPLE PIN PACKAGES REVISITED

Even if the maximum number of chip I/Os can be held within 168, 198, or 260 for the near term, permitting chip carriers, is it true that future packages will have to be pin grid array? I think not, like many others working in the field. All of the pin grid array packages that have higher lead counts today are multichip packages, with considerable design time and high cost. This solution does not appear practical for cost-performance systems.

But what about the assumption that VLSI chips will require a greater lead count, even for an individual chip? Another large "maybe" is the answer; probably the increased integration of function on a chip will reduce the I/O so far below the Rent's rule projection that there will be little or no need for I/O counts greater than 260. It is too early to be certain, but the board penalty that comes with the use of high I/O pin grid arrays suggests that cost-performance systems cannot afford this luxury. There will have to be a considerable base of high-performance use of parts that become standard and available before designers of cost-driven systems can choose this option, if they ever can.

WHAT OF TAB?

Tape automated bonding can be used to attach the chips internally to multichip hybrids, and that option can be independently pursued if many multichip hybrids are chosen. Direct TAB mounting on an epoxy-glass board poses problems of automation, or accuracy of lead registration, and the use of mixed sizes of ICs also suggests problems. The VLSI chips with their high I/O counts pose great problems in registration and dimensional control, but even the use of TAB to mount the "glue" circuits would require tab mounting of chips on boards 10 or 12 inches square. Pick and place machines to locate and push the chip carriers into the paste on boards of that size are possible, but accurate placement and attachment of the more fragile leads of TAB is another matter. Even in Japan, Sharp—a leading TAB practitioner—is contemplating the use of the butt-joined leaded chip carrier proposed by John Walker of Northern Telecom.

BUTT-JOINT PACKAGES

The butt-joint leaded packages (see Figure 11-13), called CIRQUAD by Northern Telecom and L-lead by Sharp, are a practical alternative to the J-leaded chip carriers and SOJ leaded packages. They are placed in solder paste, just like the J-leaded units, and reflow-soldered by vapor phase, infrared, or hot air methods. These joints have been found quite acceptable if the solder does not wick ex-

Figure 11-13.   The butt-joint leaded package is a practical alternative to surface-attached J-leaded chip carriers. (Photo courtesy of Northern Telecom)

cessively up the lead, and Z-axis flexing of the substrate is constrained. Limiting Z-axis bending is a good idea in any case; so this package may be a real alternative to the J-leaded chip carriers.

However, the only sockets available for the butt-leaded packages are of the "dead bug" variety. These sockets require that the package be entered into the socket with the leads up, which itself is not a problem but necessitates a complete redesign of the board if the socket is to be eliminated, because the leads and connections would then occur at different points on the package. This solution, then, is not generally satisfactory, but it is possible for small assemblies for which socketing is not contemplated except for burn-in.

In the previous discussion, the use of copper-Invar-copper substrates was deemed inappropriate to the commercial cost-performance market, because of the high cost and problems with TCE match and delamination. That does not rule out the possibility of some kind of metal reinforcement of printed circuit boards to prevent bending and temperature flexing. If the printed circuit board is approximately 16 ppm/C, the metal reinforcing plate can be copper or other metal alloy at 16 ppm/C also. That reinforcement can vastly improve the reliability of the solder joints, for both butt-joint and J-leaded packages. Assuming adequate insulation, the adhesive bonding of a board to a metal substrate also suggests the possibility of using flex material.

CONCLUSIONS

Cost-performance packaging requires the selection of solutions that can be implemented quickly with stock parts and produced in high volume using a high level of automation. Choosing among all the options does not rule out the use of pin grid arrays, ceramic multilayer substrates, matched temperature coefficient boards, and hermetic ceramic packages, but there seem to be more appropriate possibilities.

In particular, the use of leaded plastic packages with silicone gel encapsulation, with similar packages for the single chips and for multichip "glue" circuit assemblies, seems to offer a very high density solution that is quite inexpensive. Conventional epoxy-glass boards with modest line and space requirements seem likely, with bonding to a metal stiffener advisable in some cases.

The packages can be gull-wing flat-pack-like types, J-lead or butt-joint. Socketing requirements are the principal consideration for that decision.

The spacing on the board can be made very close, but only at the risk of reduced assembly yield and nearly impossible broad repair and component replacement. Even greater is the risk of poor testability because of difficulty in probling and troubleshooting.

The alternative—slightly reduced spacing—permits testing pads, cutable links for circuit change and repair, and use of sockets. If the density is greatly increased through the use of multichip packages, the price of this "insurance" may not be excessive. It is a practical, cost-effective compromise.

No matter what the system requirements, there are so many trade-offs that there will be room and rationale for many solutions that meet the individual needs of the designing companies and their applications.

REFERENCES

Balde, J. W., "As Packaging Density Increases, Focus Shifts to SSI and MSI Chips," *Electronics,* *57* (11), 103 (1984a).

Balde, J. W., "New Packaging Strategy to Reduce System Costs," *IEEE Transactions on Components, Hybrids and Manufacturing Technology, CHMT-7* (3), 257 (1984b).

Bauer, J., Presentation during Surface Mount Seminar, ISHM, Dallas, Texas (Sept. 1984).

Garth, E., and Gray, F., "The Impact of Plastic Leaded Chip Carriers (PLCC's) on Multilayer Design and Fabrication," *Proc.* Technical Conference, Third Annual International Electronics Packaging Conference, p. 535, Oct. 24–26, 1983.

Knausenberger, W. H., and Schaper, L. W., "Interconnection Costs of Various Substrates—The Myth of Cheap Wire," *IEEE Transactions on Components, Hybrids and Manufacturing Technology, CHMT-7* (3), 261 (Sept. 1984).

Stafford, J. W., "Chip Carriers: Applications and Future Directions," *Electronic Packaging & Production,* p. 135 (July 19, 1980).

Stuhlbarg, S. M., "Hermetic Chip Carrier Packaging, from JEDEC to VLSI," *Proc.* Technical Conference, Third Annual Electronics Packaging Conference, p. 104, Oct. 24–26, 1983.

12. Microprocessors

Kenneth C. Smith
and
Mark A. Scott
University of Toronto

TERMINOLOGY

In considering the origins and history of the development of microcomputers, it is useful to distinguish three devices: the microprocessor, the microcomputer, and the microcontroller. In the usual perspective, these devices differ in the nature, quantity, and location of memory and peripheral interfaces. Although some of the distinctions made are not dramatic, they may serve to make the description that follows a more orderly one.

The term "microprocessor" will serve to describe the basic computational/logical element of a larger system. Here, originally, the prefix "micro" referred to both physical size and capability. However, as technology advances, reduced capability is no longer an issue; witness the Intel iAPX 432 (Lattin et al. 1981) and HP 32-bit (Beyers et al. 1981) products utilizing hundreds of thousands of transistors on a chip and having a very high level of computing ability (Gupta and Toong 1983).

The term "microcomputer" will generally serve to describe a complete, probably general-purpose, system. Such a system will include a relatively large memory in which RAM is far more predominant than ROM. By virtue of its size this memory normally is not packaged with the processor. Also, a microcomputer will include means to access peripherals usual to a general-purpose system.

The term "microcontroller" generally serves a more specific role. Whereas the included processor may be quite generally applicable, the completed controller usually is not. As a result, its memory and I/O requirements are relatively restricted. Also, ROM dominates RAM. As a result of these limitations, single-chip implementations incorporating a small RAM, a larger ROM, and a flexible I/O system are common. Such is the MCS 48 series (Barry 1985; Wakerley 1979), one of the world's best-selling controller families.

GENERAL ISSUES

An Overview

Since the initial development of microprocessors about 1970, the driving force behind microprocessor design has changed dramatically. The first microprocessor (the Intel 4004) was designed to replace a large amount of fixed random control logic (Noyce and Hoff 1981). The explosive expansion of microprocessors, beginning with the development of many 8-bit microprocessors during the 1970s, was fueled initially by the same requirements. Later, in the latter half of the decade, their use as microcomputers expanded extensively. Currently, 16-bit and 32-bit microprocessors cover a huge range of applications, extending to the computational power of recent mainframe computers. The primary use of current larger microprocessors is in computer systems rather than as replacements for fixed control logic.

However, the microprocessor is only one component in an integrated system. In the same time frame, equally dramatic reductions in memory cost and access times have had a profound effect on microprocessor development. If the other system components, such as memory, had not undergone major changes in performance and cost, they would have severely restricted possible microprocessor architectures.

Although it is never certain which system component provides the driving force for the refinement of other system components, making improved integrated systems possible, memory is quite special in this regard. In fact, as noted below in the section on "Memory and Microprocessor Development," it is instructive to examine the effect that advances in memory technology have had on microprocessor architecture. It seems that the somewhat predictable rate of advance in memory technology can be prognosticative of developments in microprocessor architecture.

The development of the first microprocessors was driven primarily by hardware constraints. Current microprocessor development, while still correspondingly limited, is driven as much, if not more, by software considerations (De Prycker 1983).

The creation of a successful microprocessor is becoming increasingly dependent on the availability of software and its development costs. Thus it is necessary to evolve microprocessor families that can extend the life of the microprocessor software well beyond the few years of the hardware development cycle, by facilitating improved hardware designs that remain compatible with existing software. Also, a well-conceived extended microprocessor family can spread software development costs over a wider user base by covering a number of market segments, from complete small microcontroller systems with memory and microprocessor integrated on the same chip to large multichip microcomputer systems.

Product-Line Continuity

As noted, the success of a microprocessor design is increasingly influenced by software factors and constraints. To ensure a large pool of available software, many manufacturers have developed software-compatible families of microprocessors. Whereas each member of the family serves a different market niche (from small single-user, single-task, single-processor systems such as device controller to large multiuser, multitask, multiprocessor systems such as a generalpurpose computer), each member of the family executes the same instruction set, in at least an upward compatible manner (Toong and Gupta 1981).

The size of the pool of available software can also include in the system design, without initial implementation (at least as viewed from external pins), predicted architectural enhancements such as those induced by predictably decreasing memory prices.

. The Intel 8086 and 8088 processors provide two examples of the points made above. First, the 8086 processor has a 20-bit address bus and a 16-bit data bus, while the 8088 processor has a 20-bit address bus and an 8-bit data bus (Intel 1979). Apart from the bus control logic, the internal architectures of the two processors are essentially identical. They provide the same register organization, they execute the same instruction set, and they support the same system organization. With two such microprocessors chips, the manufacturer is able to provide processors that serve different market segments without enduring the development costs of separate software. The user benefits as well, being able to migrate from one processor to another, as processing requirements necessitate larger bus widths and throughput, without having to make software changes.

Second, within both the 8086 and 8088 chip families there is also a selection of output bus signals, ranging from minimum to maximum mode. A minimummode system generally consists of a single microprocessor, whereas a maximummode system is generally considered to have one or more microprocessors or coprocessors associated with the primary microprocessor, as well as bus arbiters and controllers. Although the maximum-mode signals are necessary in a multiple processor environment, the overhead of their generation and handling is eliminated in a simple single-processor design. The implementation of a variety of external-pin signals on a chip is a minor factor in terms of the overall complexity of the microprocessor chip; yet it also allows the manufacturer to cover a larger market area, while allowing the user to transfer between architectures. An important example of this idea is the addition of an 8087 floating-point coprocessor (Intel 1979), as requirements warrant, without a rewriting of the software.

In fact, the Intel 8086 or 8088 supports a number of coprocessors, including the 8087 floating-point coprocessor and the 8089 I/O coprocessor. These coprocessors perform functions that could be performed by the main processor, but they do so much more efficiently than the main processor. They allow higher performance in a system that includes them, while they need not be included in simple low-cost systems. The main processor has certain operation codes that are reserved for the coprocessors. When such an operation code is to be executed, the main processor performs no operation apart from calculating the effective address. The coprocessor utilizes the data bus to load or store operands in response to such coprocessor operation codes. The coprocessor then completes the operation while the main processor also continues its operations. While 8087 floating-point coprocessor operations can be emulated in software, they will be several times slower and, as well, will require memory for the emulation code. However, in a system where floating-point operations are rare, software emulation is a common choice. Furthermore, as system requirements increase, it is not difficult to change from emulation to coprocessor operation. The 8089 I/O coprocessor supports a number of I/O operations, which, while not very complex in themselves, have a high cost due to the overhead of interrupt servicing on the main processor. In general, a coprocessor architecture also allows the easy development of processor enhancements after the development of the initial processor, particularly in situations where market forces prevent inclusion of the enhancements in a redesign of the processor chip (Intel 1979).

The Zilog 8001 and 8002 are examples of upward compatability in a microprocessor family. The Zilog 8002 utilizes a 16-bit address space and 16-bit data bus. The Zilog 8001 is similar in basic architecture, but it also includes segmentation facilities (7 bits) to greatly increase the address space. To allow for software compatibility between the two members of the product line, the 8001 includes an 8002-emulation mode where the segmentation is disabled (Titus et al. 1981).

Finally, the section on the Motorola 68000 provides an interesting example of foresight in the area of potential product-line extension. Although the initial processor offering supported only a 16-bit data bus and a 24-bit address bus, the internal registers were 32 bits wide. At the time the 68000 was developed, memory costs were such that a 24-bit address space was more than adequate, and 16-bit data buses were considered large for microprocessors. However, the wisdom in the choice of the full 32-bit internal implementation has been shown by the recent introduction of the 68020. This processor, having a full 32-bit internal and external implementation, was destined to appear, as other factors such as memory costs made such a development feasible (Motorola 1980; Stritter and Gunter 1979).

Flexible Design

Historically, a motive for the creation of a new microprocessor design, in the face of the large associated costs, has been the search for a truly special general-purpose device, one offering true flexibility and adaptability to a reasonably large subset of perceived customer needs. The result of this process is apparent: it is the creation, by a large number of manufacturers at considerable cost and reduced profit margins, of a very large number of microprocessor and microcontroller products, whose prices may be argued to be higher than necessary because of the generality they support. The approach is normally justified by the desire to spread the high cost of product creation over a wide user base.

This problem is not a new one. In the search for its solution, there has been a vast increase in emphasis on flexible design methods, design tools, and techniques such as standard-cell and modular-cell (Takeda et al. 1985) layout. Included in the approaches taken is the identification of a variety of modularizing ideas—attempts to create adaptable precepts from which larger systems, such as an entire microprocessor IC, could be assembled. Early basic examples are seen in the use of ROMs, PALs, and bus structures in usual designs. More recently, one sees products that generalize the approach somewhat, providing standard modules such as register files and code-compressed ROM-based features that undergo economical product tuning at a late preproduction stage (Maejima et al. 1985).

Note that this trend apparently can be viewed also as a logical extension of the semicustom-array idea to a macro-module environment, within the setting of a manufacturer's internal product–production process.

SOFTWARE ISSUES

The Optimal Instruction Set

A recurring theme in the history of computer architecture has been the search for an optimal instruction set (Fairclough 1982). This search, in a very real sense, has been the driving force behind, first the computer revolution, and then the microcomputer insurrection.

Implicit in the search has been the desire to find an instruction set that is more complete, more flexible, truly more "general-purpose" (Patterson and Seguin 1982). At every stage, the extent of a new development was balanced by the cost of the technology available to support it. Successful products provided a marked improvement in performance at a marginal increase in price. Yet each such success guided and motivated the improvements in technology needed for the cycle to begin anew, to provide again the capacity to add generality at low cost.

But are the products of this cyclic incremental process really what they seem to be—truly general-purpose processors? It is the view, for example, of the proponents of reduced instruction set architectures that such is not the case.

There is considerable evidence that they may be correct. One need only look at the object code compiled by the compiler products of reputable vendors of microprocessor software, to find inelegant and wasteful consequences of trying to cope with the potential generality of the user's need with a manageable subset of the processor's capability. Although one might conclude that the software provider is less than perfect, could it be that the environment in which the vendor works is unduly muddled?

Among various analyses of the statistics of use of machine instructions, one reported in 1984 (Supnik 1985), on instruction counts in a "typical" job mix, is particularly relevant. It is reviewed below in the section on the MicroVax 32, in the context of an overview of the development of the MicroVax 32. Of the 304 instructions available in the DEC VAX 11 architecture, slightly fewer than 20%, accounting for 60% of the required microcode, represent only 0.2% of the instructions accessed, and account for only about 1% of the job mix execution time. While, in a particular implementation, replacement of these instructions by software (macrocode) degraded the overall performance of a single chip implementation by only 4%, it provided other, remarkable benefits. The control store was reduced by a factor of five, and the active chip area was reduced by nearly one-half. Such a reduction in complexity improves the speed performance for all instructions and simplifies design and test activities; and it probably reduces development time, ensuring earlier market entry.

Recently, several NMOS single-chip VLSI processors have been designed with the search for a more fundamental simple architecture in mind. The RISC I, RISC II (Sherburne et al. 1984a,b), and MIPS (Rowen et al. 1984) architectures all employ low-level instructions, each executing in one machine cycle. Regular execution timing is shown to simplify the implementation of pipelining and to retain conceptual simplicity. In contrast to commercial single-chip processors, where half the die area (Sherburne et al. 1984b) is devoted to control of the data path, a reduced instruction set computer (RISC) requires far less instruction decoding and control logic. In RISCs, control can utilize small fast programmed-logic arrays, which reduce control delays so that they no longer constitute the critical performance limiting path. The area freed up can be used for other purposes, such as expanded general register files that reduce the rate of main memory access. More will be said about this in the section on "Register File Machines."

Microprogramming vs. Macroprogramming

The design of a sophisticated microprocessor instruction set is usually implemented by microcode contained on the processor chip. The need to create a family of microprocessors of various capabilities, which all execute the same instruction set, can be dealt with by making various choices concerning the separation between microcode and normal user code (macrocode). In any instruction set, a few instructions are executed very infrequently. In addition, it is not unusual for the least-used members of the instruction set to be the most demanding in terms of microcode resources (e.g., double-precision floating-point divide).

Thus a lower-level member of a microprocessor design family could be implemented with only frequently used instructions microcoded, while infrequently used instructions would produce a processor trap. When desired, the unimplemented instructions could be emulated by the trap-handling macrocode software.

On the other hand, a higher-level member of a microprocessor design family would have all of the instruction set in microcode. The higher-level member could be implemented in a larger chip, quite different from the lower-level member, with additional microcode to handle the previously emulated instructions. Alternatively, the additional microcode could be implemented through the addition of a coprocessor to the lower-level member. Such a coprocessor would support only the previously emulated instructions.

Orthogonal vs. Amorphous Instruction Sets

Instruction set encoding is an area in which the diverse needs of manufacturers, basic machine users, and compiler writers are in apparent conflict. In their respective searches for efficiency, particularly in short-word-length machines, two extremes may be identified: the amorphous and the orthogonal approaches to instruction set design.

An orthogonal instruction to some extent sacrifices the number of useful instructions that may be implemented with a given number of opcode bits, but in return it provides an instruction set that is generally easier to learn, and in which it is easier to write high-level language compilers. With an orthogonal instruction set, the opcode bits may be considered to be divided up into a number of fields. Each field controls one aspect of the instruction execution in a manner that is constant over the whole, or at least a large subset, of the instruction set. On the other hand, with an amorphous instruction set, no particular interpretation of the opcode bits is intended, or can be made.

For example, an 8-bit opcode provides 256 possible operations, which, if not organized in any particular way, would probably be difficult to learn and use. However, an 8-bit opcode, divided into two 4-bit fields, also provides 256 possible operations, but by virtue of the need for choice from 16 possible values for the first field, and 16 possible values for the second field, implies there are only 32 items to learn to use, rather than 256. It may be true that some of the 16-by-16 combinations will not be useful instructions, but if the proportion of useful instructions is high enough, then the technique is considered justifiable.

No microprocessor manufacturer has an instruction set that is either completely amorphous or completely orthogonal, but there is considerable variation between manufacturers in the number of orthogonal subsets of which their instruction set is composed.

High-Level Language Constructs

Recent microrprocessor designs include a number of instructions intended to allow efficient implementation of high-level language constructs. Examples include array bounds checking instructions and decimal arithmetic instructions. However, even more steps can be taken to improve the implementation of high-level languages. The strong type constraints of Pascal and Ada provide examples of such needs. Here, explicit checking by macrocode for uninitialized or invalid values will in many cases be more expensive in terms of computing resources than the actual computations that result. Yet instructions incorporating implied checking in association with actual computation could be implemented in microcode with considerable reduction in the amount of processing required.

Multiuser/Multitasking

Until recently, most microprocessors were designed only for a single-user, single-task mode of operation. However, as the computing power of microprocessors grows, the demand for their utilization in multiuser, multitask environments increases. A successful microprocessor design must address its intended role in ways that satisfy the needs of a multitasking environment without unduly impacting its use stand alone.

User-Supervisor Modes, Privileged Instructions.

In a multiuser environment, a number of processor operations should be performed only by the operating system. These operations include input and output operations, halts, and control of interrupts, as well as changing of virtual memory maps. A microprocessor that is to support such an environment must have a means of disabling such instructions when a user is running. Typically this facility is implemented by providing both user and supervisor operating states. In the supervisor state, all instructions can be used, while in the user state, certain instructions such as those detailed above, usually termed privileged instructions, are disabled, or alternatively cause traps (that is, software interrupts). A processor can only enter the supervisor state through an interrupt or trap.

Virtual Memory and Segmentation.

In the multiuser environment, it must be possible to protect users from each other. Specifically, a user should not be able to read or corrupt the memory occupied by another user. This requires an approach, termed virtual memory, where the parts of the physical memory not occupied by the running process can be excluded from that part assigned to a user by means of his virtual memory map. In addition, the operating system and input–output areas (each memory-mapped) would normally be excluded from a user's virtual memory map.

The Test-and-Set (Noninterruptible) Instruction.

In a multitasking environment, there normally will be interaction between the various tasks. Such concurrent programming structures as semaphores, resource locks, monitors, and so on, will be used to structure this interaction. At the lowest level, all of these concurrent programming structures require a noninterruptible test-and-set instruction for implementation.

In this instruction, the memory address or register that controls access to the concurrent programming structure is tested, and the condition flags are set, based on the result of the test. In the same instruction, the controlling memory address or register is also set to a particular value. This will typically alternate between two values, representing availability or unavailability of the concurrent programming structure. The set part of the instruction will set the memory address or register to the value indicating unavailability; the test part of the instruction will set the condition flags to indicate the prior status of the referenced memory address or register. After the test-and-set instruction is executed, the condition flags are then tested. If they indicate available, then the task continues. If they indicate unavailable, then the task must wait until it is awakened by the task that currently controls the concurrent programming structure (when that task exits the structure).

The test-and-set must be a single noninterruptible instruction because if it were implemented as a test instruction followed by a set (or load) instruction, a logical difficulty would arise. This occurs under the condition that one task, having executed the test instruction, and having found the structure available, is about to execute the set instruction, when it is interrupted by a timer interrupt that causes another task to be given to the processor. In this event, the second task will find the concurrent programming structure available when the task tests the structure, and thus will set the memory address or register to indicate unavailability. If a subsequent timer interrupt should result in the first task being given back to the processor while the second task is still in the concurrent programming structure, then a major concurrency error will occur. This happens because the first task, having already executed its test instruction and concluded that the structure is available, will merely set the memory address or register to indicate unavailability, and proceed into the concurrent programming structure. Because the major purpose of concurrent programming structures is to ensure exclusive access to critical sections of code, providing access to two tasks at the same time could have disastrous consequences.

The test-and-set instruction is sometimes implemented by first disabling interrupts, then doing a test instruction, and then doing a set instruction. However, this approach has a high overhead because the interrupt flag of the first task must be saved and afterward restored. Thus five separate instructions (save interrupt flag, clear interrupt flag, test, set, restore interrupt flag) would be required to implement test-and-set on a processor that did not support it as a unified processor instruction.

Fast Context Switching. In a multiuser or multitasking environment, concurrency is simulated by allowing each user or task access to the processor for a short period of time, and then switching to another user or task. Each time the processor is switched from one user to another, all the registers and flags of the first must be saved, and those of the second must be restored. Such task switching must be completely transparent to the user, who has no idea when switching will occur. In a typical multiuser environment, task switching will occur every 10 to 100 msec. Thus the overhead associated with context switching would be quite large. For example on a processor with a typical register set of 16 registers, it would be quite substantial, requiring the push of 15 registers and the flags onto the stack, saving the stack pointer register in a fixed location, loading the stack pointer register from another fixed location, and the pop of 15 registers and the flags from the new stack.

Some processors have special instructions to reduce the overhead of context switching. An instruction that will transfer multiple registers to the stack is an example of this approach. Another, very fast solution to the context-switching overhead problem is to have multiple register sets in the processor and switch between the sets, obviating the need to offload the register contents from chip to memory.

In addition to the overhead of switching the processor context, a virtual memory system will have to deal with the overhead of switching the virtual memory map.

ARCHITECTURAL ISSUES

Increasing Processor Performance (Sherburne et al. 1984b)

A fundamental limit to computer performance is reached when memory traffic reaches the I/O bandwidth supported by the processor chip. This bandwidth limit is set in turn by area and power constraints: parallelism is limited by space for I/O pads and drivers, while the speed of I/O drivers with external capacitive loads is bounded by the delay–power product. On the other hand, multiplexing pads for several transactions per cycle requires faster settling and even more power.

Memory traffic consists of two components, instructions and data; and several options are available to reduce either component separately. Traditional architectures create powerful instruction constructs equivalent to many simple instructions. At the other extreme, reduced instruction set computer (RISC) architectures adopt a reduced instruction set in order to minimize instruction encoding and control, and to increase speed through simplicity. Either scheme can use an instruction cache, at least to speed up the main-memory side of the interaction. Also, either could employ a data cache for the same reason. In fact, the greatest reduction in chip I/O bandwidth can be made by transforming the external cache to an on-chip stack for either program or data or both. However, particularly for data, a simple stack is perhaps not an optimal choice. A more general-purpose large register file can be of greater value in a subroutine-dominated environment. That, in any case, is the view of RISC advocates, as will be seen in the sections on "Register File Machines" and RISC II.

Efficient Stack Access

Virtually all microprocessor architectures support a stack to some extent. The stack is used most frequently for parameter and return-address passing to subroutines. It is also used for variable storage with reentrant code. It is very important that particular attention be paid to the optimization of stack access during the design of a processor. There is much room for improvement in this area of processor architecture.

It is common for a high-level-language optimizing compiler to produce nonoptimal code during subroutine calls. For each call, local high-frequency operands contained in chip registers must be stored back in their normal memory locations, then pushed onto the stack if they are subroutine parameters. Later, after the call, local high-frequency operands used within the subroutine must be loaded from memory (or stack) back into a chip working-register; then when the subroutine returns, they must be stored back in their normal memory locations. When the calling program resumes, local high-frequency operands once again must be loaded from memory back into a chip register.

The overhead described above puts a high price on code written with numerous subroutine calls. Although the overhead can be justified in support of the essential ideas of separate compilation of calling and called programs, and of standard interfaces, it is so high for small subroutines that it is common for high-level-language compilers to have a separate register-based parameter-passing mechanism for system subroutines about which the compiler is aware.

There are, however, direct means by which processor architecture can reduce the overhead described above. Register-based parameter passing has been used, for example, in RISC architectures. Another technique—used, for example, in the FORTH machine architecture (see section on "Special-Purpose Architectures")—is to have the top part of the stack contained in chip registers (Gold et al. 1985).

Memory vs. Accumulator Organizations

Early microprocessor architectures tended to have few memory reference instructions, often simply a load and a store. Other operations, such as arithmetic,
comparison, and logical instructions, were performed utilizing a limited set of
processor registers dominated by an accumulator. Many instructions used implied
addressing (of the accumulator register) for at least one of the operands. While
these techniques made processor design and architecture simpler, they created
a number of bottlenecks that affected the implementation of efficient software
on microprocessors.

The most significant and enduring bottleneck in microprocessor architecture
has been the chip boundary. Operations performed completely within the chip
are much more efficient than those that require access to an external bus. The
time required to drive the much higher loads of an external bus, the pin-count
limitation that often leads to multiplexing different signals on the same external
pin, and the equivalent factors in the devices connected to the bus, all combine
to make bus accesses much longer than internal access times.

Another bottleneck in microprocessor design has been the limited number of
on-chip registers provided. Efficient software, particularly optimizing high-level-
language compilers, will always attempt to bring local high-frequency operands
into a register during the period of need, and to store the final register value to
its memory location only at the end of the period. However, the number of
usable internal registers can put a severe limitation on this kind of optimization.
With too few registers, critical operands must be moved back and forth from
memory more often than desired, simply because there are not enough registers
to hold all those identified.

Current microprocessor architectures support a larger set of internal registers
than earlier designs, generally at least 16. However, the number of internal
registers actually usable for local high-frequency operands is far fewer because,
at any time, a number of registers must be dedicated to operand addressing.

Current microprocessor architectures also support a much larger range of
memory-accessing instructions than simply load and store. However, it is still
useful to distinguish organizations as being of accumulator or memory type,
based upon whether both operands of an instruction may be memory addresses,
or at least one must be an internal or chip register.

Memory-based processor architectures are usually supersets of accumulator-
based processor architectures. Each processor organization allows instructions
where both operands are chip registers. However, memory architectures, and
current implementations of accumulator architectures, both allow instructions
where one operand is a chip register, and the other operand is an external memory
location. But only the memory architecture allows instructions where both operands are external memory locations. Current implementations of accumulator
architectures, however, allow the external memory operand to be either source
or destination.

The implementation of software (particularly high-level) is simplified in a memory-based organization. Although instructions where both operands are memory locations are slower than those where at least one operand is a chip register, they are faster than the equivalent in an accumulator architecture, that is, loading the source operand from memory into a chip register and performing the operation with the chip register and the destination operand. In a memory-based organization, instructions having both operands in memory will generally be used for operands of local low-frequency usage, while chip registers will be used for operands of local high-frequency usage. An additional advantage over accumulator-based architectures is that no chip register containing the latter must be freed up to load the former, less frequently used operand.

Reduction of Memory Access

As implied already, the reduced instruction set computer is an object of current research with considerable impact on microprocessor development. In contrast, current microprocessor architectures, which by comparison may be characterized as complex instruction set computers, use a microprogram-based architecture with a number of complex instructions, such as block move, to attempt to limit the number of memory references, and thus increase the speed of common high-level-language operations. Such an approach is motivated by the fact that a limiting factor in the speed of a microprocessor architecture is very often the number of memory bus operations required both to fetch the instruction and its data and then to store the result.

Using the example of a block move, a nonmicroprogrammed register-based microprocessor architecture would require a machine-language subroutine to load the accumulator register based upon the source index register, increment the source index register, store the accumulator based upon the destination index register, increment the destination index register, decrement the block count, and repeat until the count reaches zero. While the memory references required to load and store the data may be considered essential, they form only a fraction of the total memory accesses, primarily for instruction fetches, that are necessary to execute such a machine-language subroutine.

Various steps are possible to reduce the number of memory references to only those essential to the desired process. With a microprogrammed microprocessor architecture, it is convenient to build complex instructions that perform a number of functions. Thus the increment of the index register may be performed by the same instruction that loaded or stored the accumulator register, based upon the address data from the index register. Indeed, in current microprocessor architectures, the entire block move may be implemented as a single instruction, with only a single instruction fetch as overhead beyond the memory references essential to performing the block move itself.

While this approach has been taken by several current commercial microprocessor products, it results in about half the microprocessor chip being taken up by microprogram storage. The result, in current architectures, is the availability of space for only a limited number of general-purpose registers. This consequence is particularly undesirable because, although register-to-register instructions are much faster than memory-to-memory instructions, the latter, with memory-to-register transfers, are often the limiting factor in the overall speed of the processor.

The RISC approach (Sherburne et al. 1984b) has been to give up the large microprogram storage area on a chip in return for a large number of general-purpose registers. With more general-purpose registers a larger number of variables can be kept in registers, and the overhead in moving the variables between registers and memory is greatly reduced. Furthermore, in a register-rich environment, techniques such as passing subroutine parameters in registers, rather than via a memory stack, become feasible.

Register File Machines (Sherburne et al. 1984b)

As noted, register-based architectures can store frequently used operands in a fast register file. However, register allocation, as organized by the usual compiler, is often done independently for each procedure call. The result is that register contents must be swapped off-chip to make room for the next procedure. For many machines this makes subroutine call and return a very time-consuming operation, leading to a major performance degradation when typical high-level-language-originated programs are executed. One may conclude that performance could be greatly improved by better compiler design. However, more relevant to this discussion, one may also conclude that this task could be simplified by larger on-chip register files and enhancement of the mechanisms by which they are accessed. One such attempt is described below.

A Larger Register File Organization (Sherburne et al. 1984a,b). In order to reduce the overhead in memory access associated with procedure calls, the RISC II microprocessor provides a large on-chip local memory, organized in multiple overlapping register banks called windows. Each bank supports a different level of the dynamic procedure-calling hierarchy. The active procedure has access to a total of 32 registers: 10 global registers accessible to any procedure, 10 local to the present procedure, and 8 high and 6 low, shared by adjacent procedure levels. The shared registers are used for parameter passing without explicit data movement. The overlapping banks are organized within a contiguous block of 138 registers. Of these, 128 implement a circular buffer of 8 banks, of which 22 are singled out at any time by a window pointer.

This pointer moves in steps of 16 for each call or return. When procedure nesting exceeds the buffer, an overflow occurs, and one or two banks are swapped to main memory. They are restored when underflow occurs.

The choice of bank number and window size was made on the basis of architectural simulation and performance studies indicating over/underflow for only a small percentage of procedure calls with eight windows, each of width (6 + 10 + 6). This choice can be a critical one, for as register file size increases, performance degrades as access time increases with the requirement for longer buses.

I/O—Memory Mapped vs. Separate Bus or Control

In the early days, of computers in general and then of microcomputers, processor architectures kept memory and I/O separate, utilizing special instructions and different address spaces. However, an alternative, called memory-mapped I/O, is now quite common. With this technique, a portion of the memory address space is allocated to I/O device registers. To the software, these registers appear as normal memory locations, in that all instructions that work with a memory location may be used with a device register. However, a device register will differ in several ways from a normal memory location: it may be read-only, or write-only, and it may have side effects, such as outputting a character when written to, or automatically changing value when a device becomes ready.

The Motorola 68000 uses memory-mapped I/O exclusively, while the Intel 8086 supports both memory-mapped I/O and separate I/O bus and control designs. This implementation of separate I/O has a number of disadvantages. It has two register bottlenecks: one register being used for data transfer and one register being used for port addressing (only the first 256 ports can be addressed with an address contained in the operation code). Only a single load or store is possible with each I/O instruction. Also, a number of memory instructions, such as block move, logical, and arithmetic instructions, cannot be applied directly to device registers in a system with separate I/O architecture.

The intention of separate I/O bus and control designs is to keep external device interfaces simple. This is done by allowing only certain forms of access rather than the more general access possible in memory-mapped I/O. However, with the increasing integration of device interfaces, the increased complexity generated by separate I/O and memory buses (often multiplexed over the same physical bus), and the need for the processor to support I/O as well as memory instructions, the general trend now is toward memory-mapped I/O.

Special-Purpose Architectures

As remarked elsewhere, many of the past and potential developments in microprocessor architecture relate to improved performance in the implementation of high-level-language software. In this regard, it is instructive to consider in particular the consequences of a design targeted at FORTH, arguably the most high-level language (Gold et al. 1985).

The FORTH language is a high-level language that is often used in real-time control applications where performance and size are considerable constraints. The language consists of a number of words (commands). Programming in FORTH is done by defining additional words in terms of words previously defined, until the desired program is defined as a single word. Only the most basic words are generally coded in machine language. The language is highly stack-oriented, and the code generated is highly threaded. It functions primarily by pushing data on the stack and then calling a previously defined subroutine (that is, a word). This process repeats itself down to words at the lowest level that perform most of the actual computation.

NC4000 FORTH Microprocessor (Gold et al. 1985). The execution of FORTH code has a number of unusual characteristics that make it a candidate for a specialized rather than general-purpose microprocessor architecture. FORTH requires frequent stack access for both data and return addresses, the stacks for data and for return addresses are logically separate, and the overhead associated with subroutine calls must be kept to a minimum.

One design, the NC4000 architecture (Gold et al. 1985), provides three memory buses, all with 16 bits of data. Apart from the conventional memory bus to the main memory, where program and variables are stored, there are two buses with 8-bit address spaces for the return-address stack and the data stack. The top two elements of the data stack are internal registers within the processor chip.

A subroutine call instruction implemented as a single-word opcode (having 1 bit of opcode and 15 bits of subroutine address) is fetched in a single main-memory bus cycle. The storing of the return address on the return-address stack takes place in parallel with the fetch of the subsequent instruction. A subroutine return is executed without an additional bus cycle, as each of the processor instructions (except subroutine call) includes one opcode bit that specifies a subroutine return at the end of the instruction. The retrieval of the return address from the return address stack takes place in parallel with operations on the data stack and main memory required by the instruction.

Most FORTH operations deal with the top one or two elements of the data stack, using them as inputs to the computation while returning the result to the data stack. Because the top two stack elements are processor registers, push and pop between the memory-based and processor-based portions of the stack can be performed in parallel with the computation on its top two elements. For example, the operation "add" removes the top two elements of the stack, adds them, and returns the result to the stack. Thus only the first processor stack register will remain filled, while the second processor stack register must be refilled from the memory-based portion of the stack. However, because that value is not required for the computation, both the add and the transfer can be performed in parallel.

Microprogrammed Bit-Slice Processors (Alexandridis 1978)

For most applications of a fixed-instruction-set, single-chip, general-purpose VLSI microprocessor will suffice. However, for special needs requiring very high performance or unusual instruction processing, a user-microprogrammed bit-slice processor will often provide a better solution.

A single-chip fixed microprocessor combines both data processing and control functions on a single chip. With the increasing complexity of general-purpose microprocessor instruction sets, control functions will usually be implemented with microcode. However, because this aspect is fixed and hidden, it does not affect microprocessor functionality as perceived by the user. The various architectural characteristics of single-chip fixed microprocessors, such as word length, instruction set, and bus architecture, are unchangeable. On the other hand, the architectural characteristics of a microprogrammed bit-slice microprocessor can (and must) be defined by the user.

A bit-slice microprocessor design has two main components, the register arithmetic logic unit (RALU) and the microprogram control unit (MCU). The RALU unit generally represents a slice (usually 2 to 8 bits wide) of a complete RALU. A number of RALU units are connected together to provide an architecture with the required word size. The most widely used bit-slice family, the AMD2900 series, has a 4-bit RALU complete with slices of 16 registers on the chip. Thus a 16-bit processor would be constructed with four such RALU units. The RALU has a number of inputs that control its various stages—the selection of register or bus inputs to the ALU, ALU operation, shifter operation, selection of register or bus outputs from the shifter, and so on. These control inputs would normally be generated by a microprogram provided by the MCU and its associated memory. The MCU also will normally contain processor flags and will control next-address generation in the micrprogram.

A number of interconnection architectures can be used to link the RALU units with the MCU and the microprogram memory. The simplest arrangement is one in which all three components operate synchronously. The MCU provides an address to the microprogram memory, which returns the control inputs to the RALU and MCU, and then the MCU generates another address for the microprogram memory. With additional components, this process can be speeded up by pipelining the microprogram memory fetches with the RALU and MCU next-address generation cycles. More complex architectures, where there may be several independent RALU units and MCU units operating in parallel, can produce even greater processing speed. However, such architectures are complex and costly to develop, and can present a formidable microprogramming task.

The increasing complexity of general-purpose microprocessor instruction sets and the availability of special-purpose coprocessors for such applications as floating-point calculations will reduce the need for microprogrammed bit-slice processors in the future. Although they are flexible, their flexibility comes at a much higher development cost than an architecture based on a general-purpose microprocessor.

IMPLEMENTATION ISSUES

The Expansion of the Product Design Cycle

As the trend to integrate increasingly powerful and complex processors on a single chip continues, concern mounts over the rapidly rising complexity facing the designer. Tens of man-years, extending over years of elapsed time, are required to transform a 32-bit architectural specification into a salable product. Perhaps most distressingly, if the project extends beyond a few years, the rapid pace of technological change ensures that many of the assumptions originally made are no longer valid.

In recognition of the nonlinear costs of delay, chips are partitioned into modules and assigned to several design teams to minimize elapsed time. The design of each module proceeds within constraints identified by the global team managers. Regrettably this divide-and-conquer approach, the tradition in the industry, is prone to errors of assumption and omission. Moreover, by virtually requiring that no team be globally aware, the process actually ensures that intermodule or global optimization is not likely, nor even possible.

This problem is not new. It has been, and continues to be, endemic in the process of creation of large complex systems of any sort.

Other solutions have been sought. The creation of increasingly sophisticated CAD tool suites, that is, design and simulation packages sharing a common data base and management control system (Johnson 1984), is one approach. But it, itself, and its design team, perhaps exemplify the very problem it seeks to resolve. Its basis lies in the assumption of the need to coordinate and interrelate the processes we already know. Unfortunately, yet understandably, this is done in ignorance of the ones we do not know!

Perhaps it is time to step back and recognize, especially in the area of microprocessor design, that it may not be best simply to proceed in isolation with more complete and more complex general-purpose processors on a single chip. In a more general sense, recognition that the global optimum is not simply the optimal combination of local optima may be an essential step in the process of evolution of integrated microprocessors. This may be an underlying motive of the proponents of reduced instruction set computer (RISC) architectures (see above, the section on "Register File Machines")—an attempt is to identify a system that, while powerful, is small enough to be designed well, and in a reasonable time.

Technology Trends (Bell 1984; Burger et al. 1984)

Concomitant with conceptual developments, reduction in feature size has been, and continues to be, extremely important in microprocessor-related developments. Thus MOS semiconductor memory, a harbinger of microprocessor development, has exhibited a density increase of about 60% per year. Most importantly, as increased density reduces system costs, it provides increased performance, both through reduction of delay within the chip and through reduction in the number of interconnected chips in a system. Thus MOS-based microprocessor performance is likely to continue to evolve at the current rate of about 50% per year (Bell 1984). This, in contrast with the 20% performance growth rates of TTL- and ECL-based larger systems, helps explain the obvious trend to microprocessor-based design. By way of example, one can note that since 1975, the cycle time of the AMD 2901 bipolar slice family has decreased by roughly a factor of 4 (about 15% per year) to 50 nsec, while in the same period, that of the Motorola 6800/68000 family has decreased by a factor of 50, to about 60 nsec.

An important effect of this same technological trend can be seen in the related increase in the performance/price ratio of microprocessor-based commercial products. Thus for Fortran benchmarks, an IBM 3081C, a DEC VAX-11/780, and an IBM PC (utilizing an Intel 8087 floating-point coprocessor), costing 4, 0.2, and 0.006 $M, respectively, and running at about 2, 0.3, and 0.04 M flops (million floating point operations per second), respectively, deliver 0.6, 1.7, and 6.7 floating point operations per second per dollar, respectively (Bell 1984).

But what is the significance of all of this to the current discussion? Certainly it indicates a very strong force directed toward greater use of relatively smaller machines. However, at the same time, one sees regularly an increasing need to solve problems of ever greater size and complexity. One consequence of these effects will surely be an increasing emphasis on hardware and software to support "cooperative multiprocessing." Here one may envision a group of identical processors, normally engaged in providing separate simultaneous services, such as conventional multi-tasking, generalized coprocessor implementation, multi-user access, dynamic graphics, or enriched I/O, being reconfigured upon special need, to act as one large coordinated multiprocessor on very large problems (which arise characteristically, albeit less frequently than proponents of nonreconfigurable arrays would hope).

CMOS, an Ascending Technology

Of the many attributes of CMOS, its potential for low-power operation has always been one of the most attractive. Until quite recently, however, this property seemed to be overshadowed by the increase in processing complexity and chip area that CMOS implied. As a result, CMOS has been relegated generally to special applications where zero standby power is critical, such as in portable equipment or where power failure immunity is needed.

Now the situation is changing, for several reasons. One of these appears to be the increasing desire for increased chip functionality. The apparent trend to 32-bit architectures is a particular example. Obviously, an inherently low-power technology, such as CMOS, is of interest in this context.

There are currently several 32-bit CMOS processors at various stages of development (Zorpette 1985). Included are the AT&T WE32100, the Motorola MC68020 (see the section on the Motorola 68000), the National NS32C032, and the Intel 80386, recently delivered or nearly deliverable. It is interesting to note that, in each case, the CMOS processor is a version of an earlier NMOS product that offers improved performance at equal or greater functionality, with execution rates well in excess of 1 million instructions per second (MIPS). National has also been pursuing the NS32C532, a half-million-transistor CMOS chip with on-board memory management, floating point arithmetic, and other features (Zorpette 1985).

In addition, there has been a general increase in the number of CMOS equivalents of smaller processors, one of them the MC68HC11A8 by Motorola (Zorpette 1985), incorporating a large amount of RAM, ROM, and EEPROM and a number of I/O interfaces including an 8-bit analog-to-digital converter (see the section on the Motorola 68000). A recent development has been the introduction of radiation-hardened functionally compatible CMOS products by Harris—the 80C85RH, equivalent to the Intel 8085, with an Intel 8086 equivalent to be available shortly (Zorpette 1985).

Memory and Microprocessor Development

As indicated elsewhere, semiconductor memory can be viewed as the bellwether of microprocessor development (Bell 1984). An interesting perodicity in the announcement of memory products, as well as a marked similarity in the pricing evolution of each such product, has been noted (Lepselter and Sze 1985). Since 1971, and the introduction of a 1K chip, quadrupling of chip capacity has occurred at intervals of about 3 to 4 years. After product introduction, the price per chip at first falls rapidly, slowing to a rate of $1 per year at a cost per chip of roughly $3. Extrapolation of these data indicates that the $3 level may be reached for 256K chips in 1988 and IM chips in 1991.

As implied above, one can observe a synchronization of microprocessor development with the evolution of memory-chip size (Bell 1984). Thus, a correspondence can be seen of the 8080 with the 4K chip; the Z80 and 6502 with the 16K chip; the 8086 and 68000 with the 64K chip; and the 80286, 68020, and 32032 (MacGregor 1984) with the 256K chip. On the basis of these observations, one may only surmise the nature of microprocessor developments to come.

There are other bases as well from which conjectures can be made on the memory–processor symbiosis. Reflect only for a moment on the history of electronic computing, as flagged by the following tuples intended to relate selected attributes of memory hierarchy and processor technology: (1) drum, vacuum tube; (b) drum, core, vacuum tube; (c) core, transistor; (d) core, cache, transistor; (e) core, IC cache, IC; (f) IC memory, IC cache, IC; (g)———on-chip cache and registers; (j)?

What is the relevancy of this? Of much that can be said, note only a few highlights; (1) the recurrence of a matching element between logic functions operating at disparate speeds [the core in (b), the cache in (d), (e), and (f), and the registers in (g) and (h)]; (2) the tendency to unify to a single technology (currently silicon); and (3) the closer coupling (and even merging) of the memory and processing parts. Certainly item (1) represents a trend that apparently continues. Recent chip designs include more and more speed-matching attributes.

Note, however, that observation (2) is probably too narrow a view. While the example presented does not properly illustrate it, there is other evidence (for example, the early use of CRT tube memories in vacuum tube machines) that suggests the idea that (2) is part of a longer cycle of unification and fracture. Expect to see some new directions in high-speed processing technology in the near term.

Finally, one sees several explicit examples of the processor–memory merger. Commercial single-chip processors are obvious direct examples. More subtle examples are the RISC machines. Memory-management features of modern designs are a related development.

The Increasing Use of (E)EPROM (Barry 1985)

Included in a general trend to incorporate on-board memory in new chip designs, for applications ranging from small controllers to large computers, is the appearance of electrically programmable read-only memory (EPROM) and electrically erasable and programmable read-only-memory (EEPROM) (Bagula and Wong 1983). Both Intel and Motorola are including EEPROM capability in their microcontroller product lines. Motorola recently sampled its MC6805K2 and MC6805K3 "applications-specific" microprocessors incorporating peripherals customization, which include up to 256 bytes of EEPROM. In addition, Motorola expects to replace the ROM on existing chips with EPROM to produce the MC1468705F2 and MC68701F4 with 1K and 4K bytes, respectively. Motorola has recently introduced the most highly integrated circuit yet produced by the company, the MC68HC11, with 512 bytes of EEPROM, 8K bytes of ROM, 256 bytes of RAM, an 8-bit analog-to-digital converter, and other peripheral interfaces.

Packaging (Farrell 1984; Barry 1985)

In the nearly two decades of the microprocessor age, no truly clear-cut packaging standard has been accepted. While the dual-in-line package (DIP) has come close, being used for 90% of all integrated circuits shipped, it suffers from two major problems. On the other hand, for some applications, particularly in the consumer area, both its cost and size become critical, each overwhelming the corresponding aspect of the IC itself. On the other hand, for high performance systems, its pin limitation, in effect to 64 pins, is very restrictive.

The reasons for the popularity of the DIP are several. It is a stiff-leaded package, inherently strong and available in different materials, that has few electrical quirks and is suited both to manual and automatic insertion. Lead spacing at 0.1-inch centers provides a good compromise between circuit and wiring density and accessibility. While 90% of the DIPs shipped are plastic at relatively low cost, the availability of the more robust, rugged and reliable, ceramic, compatible package has been important.

Flatpacks with 0.05-inch-centered leads have less than 1% of the market, being popular with the military and a small number of special users. There is some indication that this market share will drop with the advent of the surface-mounted, small-outline (SO) package, which, however, is seriously limited to 24 or fewer pins. Chip-on-board (COB) packaging in which the IC chip is bonded directly to the PCB pads and epoxy-coated, used for the last few years for consumer applications such as watches and more recently in the games market, has less than a 3% market share.

More recent alternatives are the chip carrier and the quad surface mount (QSM) plastic packages, both having from 16 to 156 pins at 0.05-inch centers. These offer a reduction in PCB area occupied, while maintaining a reasonably low cost for both the package and its installation. The QSM has appeared in response to the need for a small cost-effective package to match the decreasing cost and potential ubiquity of 8-bit microprocessors. For these reasons, it may present a highly reliable alternative to COB packaging, particularly in the microcontroller area. A third recent alternative is the pin grid array (PGA), suited for several hundred pins at 0.1-inch centers but at an increased cost of package, installation, and board layout complexity.

EXEMPLARY SYSTEMS

The Evolution of an 8-Bit Family—The MC6800

In 1975 the MC6800 became the first single-supply, 5-volt microprocessor available on the market. The MC6800, utilizing NMOS transistors operating at a 1 MHz clock, includes CPU control circuitry, a basic ALU, two 8-bit accumulators, a 16-bit stack pointer, a 16-bit program counter, and facilities for several addressing modes (Farrell 1984). Subsequent versions with processing improvements operate up to 2 MHz.

Later developments diverged along two distinct paths. One, beginning with the MC6802/08, was to add peripherals to the same basic architecture. The other, beginning with the MC6809, was to enhance the basic processing capability.

The MC6809, available in 1978, was an early indicator of an important trend. While an 8-bit machine as seen from its external bus, internally the 6809 handled 16-bit words. Six 16-bit registers were provided, as well as an enhanced instruction set including multiply, and addressing modes suited to efficient high-level-language design. For example, indirect addressing made position-independent coding easy, thereby encouraging the use of ROM-based software.

The MC6802, available in 1977 and incorporating 11,000 transistors, represented the first step by Motorola in the transition from microprocessor to microcomputer, retaining the MC6800 internal software architecture but aiming to reduce the need for additional components in system applications. The MC6802 contains its own oscillator and 128 bytes of RAM, 32 of which can be retained at low power in standby mode, for example, in the case of power failure.

Subsequently, the MC6801/03 evolved, becoming available in 1979. While maintaining the MC6800 instruction set, it includes some additional instructions such as multiply. Following the trend toward completeness as a microcomputer, the MC6801—in addition to an internal clock generator, 128 bytes of RAM, and a ROM—has the ROM capacity increased (to 2K bytes) with facilities to utilize external ROM, and has a divide-by-four clock output, a serial communications interface, and a 16-bit multifunction timer.

The success of the MC6801 pointed to the need for a family of relatively complete microcomputers, focused on a range of microcontroller tasks, including a variety of types, each produced at relatively low cost. The result was the evolution of a standard building-block approach leading to the MC6805 family, which includes more than two dozen members. For true application specificity, both dynamic NMOS and static CMOS versions are available. All of the processors have the same core, a CPU with control and ALU, having for the most part a common instruction set. The exception is that the CMOS products (designated 146805), having a static implementation, allow two additional instructions, namely, stop and wait. The instruction set is designed for byte-efficient program storage, striving for the greatest amount of program function in a given on-chip ROM. This includes the ability to nest subroutines and to do true bit test and manipulation, as well as utilize multifunction instructions and versatile address modes.

Associated with the CPU core, all MC6805 types contain an 8-bit accumulator or working register, an 8-bit index register, a 5- to 6-bit stack pointer, a 5-bit condition code register, and an 11- to 13-bit program counter (Farrell 1984).

Each types also contains a self-check ROM that permits self-test of vital functions without expensive test equipment.

Although all 6805 family members share the same CPU core and register, the amount of RAM and the amount and type of ROM provided to each vary. The major differences among members lie in the area of peripherals. All have I/O ports, but the numbers and types of ports vary. Although many have timers, the type of timer varies. One member has a four-input multiplexed 8-bit analog-to-digital converter with external reference; another provides direct LCD drive; another incorporates a phase-lock loop; some have EPROM for low volume production and protyping.

One sees in the MC6805 family an interesting lesson: There is an alternative to a ready-made, ready-to-go, completely general-purpose, costly, multipin product, a majority of whose potential is never used. The alternative advanced by the MC6805 concept is specificity at the very final stages of design through a building-block process.

The concept suggests yet another alternative, which also operates on the principles of minimal package size, reduced pin count, and reduced cost with maximized flexibility, but also reduces the costs of stocking, replacement, and so on. This approach is to utilize a richer subset of peripheral components included within a somewhat more standardized chip whose pin connections are finally customized through EPROM or even EEPROM.

The MC68HC11 (Sibigtroth 1984). This new microcomputer represents another kind of evolution. It is a super-eight-bit microcomputer utilizing a 52-lead QSM (quad surface amount) package in which the route to flexibility and versatility is high pin count, while economy is maintained by new low-cost packaging.

The MC68HC11 uses HC-type CMOS for high speed at low power with static operation. Reduction in power improves both the reliability of the chip itself and that of its power supply, particularly in small systems where redundant supplies may be feasible. Nonvolatile memory is up to 8K bytes of ROM (or HCMOS EEPROM, depending on the product variation) plus 512 bytes of HCMOS EEPROM. Included also are 256 bytes of static RAM, which uses very little power in standby. The architecture, based on the MC6801, contains the entire 6801 instruction set plus enhancements such as a second index register, which, by facilitating block moves, reduces code complexity. An enhanced 16-bit timer includes three capture registers, five output-compare registers, a prescaler, and improved interrupt capability. In addition, a watchdog timer, real-time interrupt, and pulse-accumulator feature implement a great number of systems-timing needs with a minimum of software and no additional hardware. Besides four parallel 8-bit I/O ports, the unit contains two independent serial I/O subsystems, and an eight-channel multiplexed 8-bit successive-approximation analog-to-digital converter.

The augmented MC6801 instruction set, totaling 91 instructions in all, includes divide and compare, as well as bit manipulation and test. Multiple addressing modes extend over the entire 64K (65,536) byte memory map, from which the 4K-byte ROM can be removed by EEPROM programming of a separate register.

The MC68HC11 microcomputer can be seen to operate in one of four modes, depending on the state of dedicated pins: (a) a single chip mode with address and data lines unavailable outside, (b) an expanded multiplexed mode in which it will operate as a microprocessor and access off-chip memory, (c) a bootstrap mode for EEPROM programming and self-test of both processor and system, and (d) an expanded test mode for factory testing. A bootstrap ROM, invisible to the system in operating mode, is available in the other modes under control of the mode-select pin.

The pins of the MC68HC11 are interestingly multipurpose. Port A comprises three inputs, five outputs, and a timer I/O. Port B comprises eight outputs with strobe capability and also multiplexes the eight high-order address bits in the expanded mode. Port C comprises eight I/O lines and/or latch inputs that are capable of strobed and handshake operation, but also multiplex the eight low-order address bits and the data lines. Port D has six I/O lines that are general-purpose or for serial peripheral or communications interfaces, and two dedicated to serial I/O. Port E has eight lines for digital inputs or for analog inputs to the internal multiplexer.

A 16-bit multifunction timer is an enhanced version of that available in the MC68001. A four-stage prescaler allows division by 1, 4, 8, or 16. Three input-capture registers operate on either the rising or falling input edges. Five output-compare registers are provided.

The serial communication subsystem is a full-duplex asynchronous non-return-to-zero system of eight or nine bits operating at eight basic baud rates with prescaler extension. It can be used easily with rudimentary peripherals such as shift registers, or even with other similar processors on a master-slave basis.

The MC68000 Family

While the development of the MC6809 proceeded, Motorola was creating a 16/32 bit family, the MC68000 (Farrell 1984). Introduced in 1979 as the first general-purpose 16-bit microprocessor, it unfortunately suffered from problems that were continued in subsequent versions in 1980 and 1981.

The MC68000 was intended as an efficient vehicle for high-level languages. The architecture supports multiusers and multitasks and is conveniently operated in multiprocessor configurations. It was designed as a clear, flexible, general-purpose architecture to support expansions of instruction set and other features for future products. It operates in either user or supervisor modes and employs an asynchronous processor bus.

The MC68000 contains seventeen 32-bit registers, of which eight are general-purpose for data and nine somewhat specialized for address. Of the latter, one is the user-stack pointer, and another points to the supervisory stack.

Several product variations have appeared. For example, in 1982 the MC68008 was introduced in a smaller package with compatibility to an 8-bit I/O system. A virtual memory version, the MC68010, plug-in compatible with the MC68000, was also introduced in 1982. Also available is the MC68012, which can address up to 2 Gbyte (2^{31} bytes) and accommodates read/modify/write operation.

Recently introduced (MacGregor et al. 1984) is a full 32-bit internal and external implementation, the MC68020, utilizing 200,000 transistors operating at clock rates up to 16.7 MHz, which has all the features of both the MC68000 and the MC68010. It also can accept a coprocessor interface, and has an expanded address space, an instruction cache for increased performance, and an enhanced instruction set. For testing, it utilizes a judicious mixture of structured and functional testing (Myers 1984), designed to reduce test time and dedicated testing hardware, while maximizing test coverage. Test logic area overhead is less than 3% in the MC68020, partly as a result of the internal bus structure and the use of a 16-bit signature register and a few additional multiplexers for partitioning.

MicroVAX 32—A Vax 11 Single Chip (Plus) Implementation (Beck et al. 1984; Supnik 1984)

Announced early in 1984 (Beck et al. 1984) was a 32-bit microprocessor chip that emulates the instruction set and demand-paged virtual memory management of the DEC VAX 11 superminicomputer. Implemented in 3-μm dual-metal NMOS, containing 101,000 installed transistors in 125,000 transistor sites, the chip operates at 20 MHz with full 32-bit internal and external data paths. The 8.7 × 8.8 mm chip is mounted in a 68-pin surface-mounted leaded chip carrier, and dissipates less than 3 watts from a single 5 V supply with on-chip back-bias generator. It operates from an internal clock with a 200-nsec microcycle, supporting a 400-nsec I/O cycle, and thereby permitting DRAM access without wait states.

The microprocessor chip provides instruction-set, data-type, and memory-management compatibility with the VAX 11 family. Correspondingly, the user-visible machine consists of 16 general-purpose registers, a processor-status word (PSW), and 18 privileged registers. The instruction-set architecture, incorporating overlapped instruction prefetch and parallel instruction decode, implements 304 instructions and 14 data types: 175 instructions and 6 data types in the microprocessor chip itself, 70 floating instructions and 3 data types in a companion floating point chip, 27 instructions in microcode-assisted macrocode, and 32 (decimal and string) instructions entirely in macrocode. Nine data types are supported: byte, word, longword, and quadword integers; variable-length bit fields; variable-length character strings; single, double, and extended-double precision floating-point numbers.

The on-chip, tightly integrated, memory-management architecture supports demand-paged virtual-memory management. Virtual memory is 4 Gbyte, divided into four 1-Gbyte regions of 2^{23} 512-byte pages each. Two regions are mapped through double-level page tables, with the third mapped through a single-level page table and the fourth reserved. The physical address space is 1 Gbyte. The chip utilizes a pipelined microprogrammed parallel 32-bit implementation having eight principal sections (Beck et al. 1984). The chip is organized into three major strips. The left strip contains the data paths for instruction prefetch, instruction execution, and memory management. The middle strip contains the corresponding control logic. The right strip contains sequencing and clocking logic for both internal and external operations. The data path (E Box) contains the 16 user-visible general-purpose registers, 20 microcode scratch registers, a 32-bit arithmetic logic unit (ALU), and a 32-bit barrel shifter (utilizing a pass-transistor network). The ALU employs 4-bit carry lookahead with nibble ripple utilizing dual rail logic for speed. Special hardware is provided to assist integer multiplication. Two registers can be read, an ALU or shift operation performed, and a result written to a register, within one 200-nsec microcycle.

The memory management unit (MMU or MBox) contains three address registers, two for data and one for instructions, and a fully associative translation buffer utilizing true LRU replacement and access checking logic. The MMU overlaps address translation with data-path operations. Virtual addresses are looked up in the translation buffer. If found, the corresponding physical address is sent off-chip. In this case, the overhead in address translation is less than 25 nsec. If the lookup is unsuccessful, special microcode is brought to bear to update the translation buffer.

The instruction-decode logic (IBOX) contains an 8-byte prefetch queue and three instruction-decode PLAs. It fetches instruction data during unused bus cycles and decodes operation codes and operand specifiers in parallel with other operations. The IBox can parse an opcode, a specifier, and an address displacement in one 200-nsec microcycle.

The data and logic interface (DBox) contains memory-data read-and-write rotators, a longword-write buffer, external-operations sequencer, and pin-control logic. It controls all external operations, performing unaligned memory references across long word boundaries, independently of regular microcode, overlapping memory writes with other operations.

The control store contains sixteen hundred 39-bit words of microcode implemented as dynamic virtual-ground ROM having an access time of less than 100 nsec.

In addition, there is an instruction-decode PLA that provides 19 bits of information relating to each instruction, a microsequencer to access the control store, interrupt logic to mediate four vectored and three nonvectored external interrupts, and clock logic to generate eight internal two-phase clocks from an external double-frequency source (at 40 MHz).

Essentially, all chip sections operate concurrently yet independently. Thus, while the EBox performs an arithmetic operation, the control store accesses the next microinstruction, the microsequencer prepares the address of the following microinstruction, the MBox is translating an address, the IBox is decoding an instruction or operand specifier and prefetching further instruction data, and the DBox is initiating or completing an external access.

As a result of the parallelism described, macroinstructions proceed to completion as follows: register to register in 400 nsec, memory to register in 800 nsec, register to memory in 600 to 1200 nsec, conditional branch (not taken) in 200 ns, and conditional branch (taken) in 800 nsec.

The partitioning of the instruction set in the MicroVax 32 represents an interesting study of the hardware–firmware–software exchange. Of the native VAX instructions, the MicroVax 32 implements 57.6% by count but 98.1% by frequency of use, using only 20% of the microcode bits required for a complete implementation. In most workloads, those instructions implemented by macrocode emulation, 19.4% by count, account for only 0.2% by frequency of use and about 1% by time for execution.

Even with performance assists by the chip, macrocode emulation takes about four times longer than the direct microcode implementation, so overall performance degradation through omitted instructions is only about 4%. However, by virtue of the decision to balance implementation with emulation, the control store has been reduced by a factor of five, and the active chip area by nearly half.

In the MicroVax 32, design for test was limited by available silicon. The principle test features are serial shift register with feedback for reading out the control store, the IPLA, and the microsequencer outputs; a test mode in which normal sequencing is replaced by external microaddresses; and dedicated microcode to optimize static readout in the special test mode.

A VLSI Superminicomputer Chip Set—The VLSI VAX 11-780

Announced early in 1984 (Johnson 1984) was a 32-bit eight-component chip set utilizing four custom chips to implement the full functionality and comparable performance of the VAX 11-780 superminicomputer, but with a 400:1 reduction in integrated circuit count and a 225:1 reduction in printed-circuit-board area. In all, 1,220,550 transistors implement the full 304-element instruction set, employing all 17 data types in the 4-Gbyte virtual memory architecture.

The basis of the CPU is a 54,760-transistor instruction fetch and execution (IE) chip containing three complex parallel-operating micromachines: the instruction prefetch and decode unit (IBox); the execution and microsequencer unit (EBox); the virtual memory management and bus controller unit (MBox).

The MBox contains the first level of the chip set's two-level address translation buffer (TB), namely, the totally associative mini-translation buffer (MTB) containing one instruction-stream and four data-stream entries. If the MTB access "hits," the physical address is presented to the external macroaddress/data bus, and cache data is returned, all in one 200 nsec microcycle. If the MTB access "misses," the virtual address is presented off-chip to the backup translation buffer (BTB) located in the memory/peripheral subsystem (M) chip, which returns a new valid entry for the MTB, ensuring a hit for an MTB retry in the succeeding cycle.

The memory/peripheral subsystem (M) chip, comprising 54,670 transistors, contains the controller and address tag array for the 512 entry BTB, as well as the controller and tag array for the CPU's 2K × 32-bit data cache, enabling the CPU to perform both address translation and data fetch in the same 200-nsec microcycle. As well, the M chip incorporates twenty-four 32-bit general-purpose registers and the MOS clock generator for the entire chip set, as well as several peripheral support functions usually requiring separate support chips, such as time-of-day clock, interval time, interrupt controller, and four independently baud-rate-programmable UARTs. Note that large commodity static RAMs are used for actual data storage in the large address-translation and data caches.

An optional 42,520-transistor floating point accelerator (F) chip uses parallel 67-bit fraction and 13-bit exponent data paths and 100-nsec cycle to implement 61 floating point instructions on one single and two double-precision data types, as well as integer, multiply, and divide. Internal microcode allows this chip to operate as a true coprocessor. For systems without the F chip, the corresponding operations are supported by microcode using the IE chip data paths.

The 16K × 40-bit microprogram store comprises five patchable control store (PS) chips mounted on a single ceramic hybrid. Each 207,950-transistor PS chip contains a mask-programmable 16K × 8-bit ROM array for microprogram storage, a 1K × 8-bit RAM array to hold ROM patches, and a 32 × 14-bit CAM for trapping patch addresses, which are stacked to provide the system with 150 patch addresses in all. The system thereby allows a software-supported alternative to field-retrofit in the event that microcode bugs are detected.

A 24,330-transistor 32-bit bus interface (BI) chip couples the IE chip CPU to the 13.3 Mbyte/sec external system bus. The chip handles all bus protocols including interrupt, DMA, and multiword transfers in both uniprocessor and multiprocessor environments.

An interface to various system buses is provided by a multiport bus controller made of custom CMOS gate arrays. All chips in the VAX chip set communicate via two half-cycle multiplexed buses—the 32-bit data/address lines (DAL) and a 40-bit microinstruction/microaddress bus (MIB). There are also dedicated signal lines between chips for status and control information.

As in the case of the MicroVax 32, the success of the design rested on the development of a comprehensive CAD tool suite (Johnson 1984); Johnson et al. 1984; Supnik 1984; Curtis 1985; Kessler and Ganesan 1985). This was assembled in parallel with the development of process technology. The interaction was important; for example, actual NMOS circuit test chips showed the need for a double-metal interconnect topology with a minimum circuit feature of 3 μm (drawn).

The RISC II Processor (Patterson and Seguin 1982)

The RISC II CPU was designed with a strong emphasis on optimizing its architecture for the intended application, that is, to run compiled-C programs in a UNIX environment, within the special constraints of VLSI implementation. Layout generally used a rectilinear subset of the Mead and Conway design rules; however, critical iterated modules in the register file and ALU were carefully optimized for density and low power. Implementation was targeted at the "standard" 4-μm Si-gate NMOS process available from several "foundries." Subsequently a 3-μm version was created by direct 4:3 scaling, which operated with a 50% increase in both speed and power.

Performance, equivalent to many popular processors such as the Intel 80286, is seen to be greatly enhanced by the simplification inherent in the reduced instruction set and by the use of the space freed by this for a large internal register file. This 138 × 32b memory, organized in eight overlapping register banks, has been demonstrated to alleviate the chip I/O bottleneck by retaining frequently used operands on-chip.

Successful operation of both the original and scaled chips in first silicon, following 2.8 man-years of design time, is attributed to extensive simulation at several levels made possible by the tool-intensive environment encouraged at Berkeley.

The RISC II architecture uses a small, but carefully selected, set of simple instructions that support the most frequently occurring operations. It relies on an orthogonally partitioned 32-bit instruction format and regular instruction timing. Control complexity is reduced correspondingly to occupy less than 10% of the chip area, from the 40 to 70% typical of current full-instruction-set microprocessors. Regularity of instructions permits the realization of a simple, yet effective, pipeline. In both RISC II implementations, the machine cycle time (either 500 or 330 nsec) is limited by the data path rather than by the control path within a large microstore.

Chip area is dominated by the highly regular data path. Nearly half the data path, and more than one-third of the chip area, is occupied by a large register file of 138 words of 32 bits. Control circuitry consists of a few small PLAs utilizing at most a small percentage of the chip. For example, an opcode decoder, consisting of an AND plane and single OR row, and occupying only 0.5% of the chip area, suffices for the 30-element instruction set.

One implementation of RISC II, in 5-mask NMOS with $\lambda = 1.5$ μm (and corresponding gate length of 3 μm) on a 4.3 nm $\times$ 7.7 mm die and operating at a 12 MHz clock rate, provides an instruction cycle of 330 nsec and a peak throughput of 3 million instructions per second (MIPS) at a power level of less than 2 watts from a 5 V supply (with no substrate bias). On the basis of analysis, using integer C programs compiled with a standard UNIX compiler and raw performance data, it seems that RISC II can outperform popular processors such as the M68000, HP9000, Intel 80286, and VAX-11/780.

ACKNOWLEDGMENTS

The support of the Natural Sciences and Engineering Research Council of Canada through Operating Grant A1753 is gratefully acknowledged.

REFERENCES

Alexandridis, N. A., "Bit-sliced Microprocessor Architecture," *Computer, 11* (6), 69–92 (June 1978).

Bal, S., Kaminker, A., Lavi, Y., Menachem, A., and Soha, Z., "The NS16000 Family—Advances in Architecture and Hardware," *IEEE Computer, 15* (6), 58–68 (June 1982).

Bagula, M., and Wong, R., "A 5V Self-Adaptive Microcomputer with 16Kb of E^2 Program Storage and Security," *IEEE ISSCC Digest,* pp. 34–35, New York, Feb. 23–25, 1983.

Barry, R., "8-Bit Microprocessor Chips," Special Edition Report, *Electronic Design, 33* (6), 84–100 (Mar. 14, 1985).

Beyers, J. W., Dohse, L. J., Fuxetola, J. P., Kochis, R. L., Lob, C. G., Taylor, G. L., and Zeller, E. R., "A 32b VLSI CPU Chip," *IEEE ISSCC Digest,* pp. 104–105, New York, Feb. 18–20, 1981.

Beck, J., Dobberpuhl, D., Doherty, M. J., Dorenkamp, E., Grondalski, B., Grondalski, D., Henry, K., Miller, M., Supnik, B., Thierauf, S., and Witek, R., "A 32b Microprocessor with On-Chip Virtual Memory Management," *IEEE ISSCC Digest,* pp. 178–179, San Francisco, Feb. 22–24, 1984.

Bell, C. G., "The Mini and Micro Industries," *IEEE Computer, 17* (10), 14–30 (Oct. 1984).

Burger, R. M., Cavin, R. K., Holton, W. C., and Sumney, L. W., "The Impact of ICs on Computer Technology," *IEEE Computer, 17* (10), 88–95 (Oct. 1984).

Cain, J. T., Editor, *Selected Reprints on Microprocessors and Microcomputers,* Selected reprints from *Computer* and *IEEE Micro,* 3rd Ed., IEEE No. EHO 214–7, 1984, 366 pp.

Curtis, W., "Designing the Micro-Vax Using Silicon Compilation," COMPCON '85, San Francisco, Feb. 25–28, 1985.

DePrycker, M., "A Performance Comparison of Three Contemporary 16-Bit Microprocessors," *IEEE Micro, 3* (2), 26–37 (Apr. 1983).

Fairclough, D. A., "A Unique Microprocessor Instruction Set," *IEEE Micro, 2* (2), 8–18 (May 1982).

Farrell, J. J., "The Advancing Technology of Motorola's Microprocessors and Microcomputers," *IEEE Micro, 4* (5), 55–63 (Oct. 1984).

Gold, J., Moore, C., and Brodie, L., "Fast Processor-Chip Takes Its Instruction Directly From FORTH," *Electronic Design,* pp. 127–138 (Mar. 21, 1985).

Gupta, A., and Toong, M. D. H., "An Architectural Comparison of 32-Bit Microprocessors," *IEEE Micro, 3* (1), 9 22 (Feb. 1983).

Intel, "8086 Family Users Manual," Intel Corporation, 1979.

Johnsen, D., Budde, D., Carson, D., and Peterson, C., "Intel iAPX 432—VLSI Building Blocks for a Fault-Tolerant Computer," AFIPS Conference *Proc. 52,* 531–538, NCC (1983).

Johnson, W. N., "A VLSI Superminicomputer CPU," *IEEE ISSCC Digest,* pp. 174–175, San Francisco, Feb. 22–24, 1984.

Johnson, W. N., Herrick, W. V., and Grundmann, W. J., "A VLSI VAX Chip Set," *IEEE J. Solid-State Circuits, SC-19* (5), 663–674 (Oct. 1984).

Kessler, A. J., and Ganesan, A., "Standard Cell VLSI Design: A Tutorial," *IEEE Circuits & Devices Magazine, 1,* (1), 17–33 (Jan. 1985).

Kozernchak, E. B., and Cheney, G., "The AT&T WE 32100 Microsystem," COMPCON '85, San Francisco, Feb. 25–28, 1985.

Lattin, W. W., Bayliss, J. A., Budde, D. L., Colley, S. R., Cox, G. W., Goodman, A. L., Rattner, J. R., Richardson, W. S., and Swanson, R. C., "A 32b VLSI Micromainframe Computer System," *IEEE ISSCC Digest,* pp. 110–111, New York, Feb. 18–20, 1981.

Lepselter, M. P., and Sze, S. M., "DRAM Pricing Trends—The π Rule," *IEEE Circuits & Devices Magazine, 1* (1), 53 54 (Jan. 1985).

MacGregor, D., "The MC68020—Corrections and Comparisons" *IEEE Micro, 4,* 3–4 (Oct. 1984).

MacGregor, D., Mothersole, D., and Moyer, B., "The Motorola MC68020," *IEEE Micro 4* (4), 101–118 (Aug. 1984).

Maejima, H., Kida, H., Kihara, T., Baba, S., and Akao, Y., "A CMOS Microprocessor with Instruction-Controlled Register File and ROM," *IEEE ISSCC Digest,* pp. 12–13, New York, Feb. 13–15, 1985.

Motorola, "Motorola MC68000, 16 bit Microprocessor Users Manual," Motorola Semiconductor Products Inc., 1980.

Myers, W., "Testing of 200,000 Transistor-Chips Kept Manageable," *Micronews, IEEE Micro, 4* (5), 68 (Oct. 1984).

Noyce, R. N., and Hoff, Jr., M. E., "A History of Microprocessor Development at Intel," *IEEE Micro, 1* (1), 8–22 (Feb. 1981).

Patterson, D. A., and Seguin, C. H., "A VLSI RISC," *IEEE Computer, 15* (9), 8–22 (Sept. 1982).

Rowen, C., Przbylski, S., Jouppi, N. P., Gross, T. R., Shott, J. D., and Hennessy, J. L., "A Pipelined 32b NMOS Microprocessor," *IEEE ISSCC Digest,* pp. 180–181, San Francisco, Feb. 22–24, 1984.

Sherburne, Jr., R. W., Katevenis, M. G. H., Patterson, D. A., and Seguin, C. H., "A 32b NMOS Microprocessor with a Large Register File," *IEEE ISSCC Digest,* pp. 168–169, San Francisco, Feb. 22–24, 1984a.

Sherburne, Jr., R. W., Katevenis, M. G. H., Patterson, D. A., and Seguin, C. H., "A 32-Bit NMOS Microprocessor with a Large Register File," *IEEE J. Solid-State Circuits, SC-19* (5), 682–689 (1984b).

Sibigtroth, J. M., "Motorola's MC68HC11: Definition and Design of a VLSI Microprocessor," *IEEE Micro, 4* (1), 54–65 (Feb. 1984).

Stritter, E., and Gunter, T., "A Microprocessor Architecture for a Changing World: The Motorola 68000," *IEEE Computer, 12* (2), 43–52 (Feb. 1979).

Supnik, R. M., "MicroVAX 32, A 32 Bit Microprocessor," *IEEE J. Solid-State Circuits, SC-19* (5), 675–681 (Oct. 1984).

Supnik, R., "Microprocessors and Floating Point Processors," Session overview, *IEEE ISSCC Digest,* p. 11, New York, Feb. 13–15, 1985.

Takeda, K., Ishino, F., Ito, Y., Kasai, R., and Nakashima, T., "A Single Chip 80b Floating Point Processor," *IEEE ISSCC Digest,* pp. 16–17, New York, Feb. 13–15, 1985.

Titus, C., Titus, J., Baldwin, A., Hubin, W., Scanlon, L., *16 Bit Microprocessors,* Blacksburg Continuing Education Series, Howard W. Sams, 1981, 350 pp.

Toong, M. D. H., and Gupta, A., "An Architectural Comparison of Contemporary 16-Bit Microprocessors," *IEEE Micro, 1* (2), 26–38 (May 1981).

Wakerley, J. F., "The Intel MCS-48 Microcomputer Family: A Critique," *IEEE Computer, 12* (2), 107–116 (Feb. 1979).

Zorpette, G., "Microprocessors," *IEEE Spectrum, 22* (1), 53–55 (Jan. 1985).

13. Programmable Logic Arrays (PLAs)— Design and Application

John Wyn Jones

IBM United Kingdom Laboratories Limited

Paul Lowy

IBM Corporation

Programmable logic arrays (PLAs) were originated in the early days of LSI in an effort to reduce the long design cycle and turnaround time required to obtain chips. The first PLAs were single chips mounted in 16- or 24-pin modules, and consisted mainly of two arrays of devices plus peripheral circuits. The devices within each array were selectively connected to one another to implement logical functions, a procedure that remains the same today. Wafers for these chips were processed through all manufacturing steps needed to build the transistors but not their interconnections. That is, they were processed through manufacturing using all masks up until the metal layers; these layers were to be "programmed" or given a personality when the chip function was designed. The wafers were stockpiled so that, after design, completed modules were available in just a few weeks. Today PLAs are used both as prototype hardware—fusible link PLAs can be personalized in minutes—and in final designs. In the latter case they are likely to appear as a subportion of a larger chip, and in this case they are referred to as PLA macros. Logic design with PLAs is relatively straightforward and requires less documentation than conventional design, yet the design procedure may be made quite rigorous. Designs are race-free, and there is little physical design. PLAs can be interconnected to perform any combinatorial logic function within the bounds of the number of devices, inputs, and outputs of the PLA, as will now be shown.

BASIC PRINCIPLES

Any set of logic functions may be expressed in canonical minterm form, that is, $F1 = AB\overline{C} + \overline{A}BC + \ldots$, $F2 = \overline{A}\overline{B}C + AB\overline{C} + \ldots$, etc., where $AB\overline{C}$

379

is a minterm or product term. Therefore, this form represents each function as the sum of product terms (the OR of ANDs), where each product term is the AND of either true or inverted inputs. The PLA is the realization of this logic form. In a PLA, the AND outputs are known as the *product terms* or AND *word lines*. The outputs of the OR circuits that form the sum of the product terms are sometimes called sum terms, but more frequently just OR *words* or OR *outputs*. Both AND and OR circuits are physically arranged as arrays, and their connections are determined by the logic designer based on the function to be implemented (See Figure 13-1.)

The true and inverted signals are formed by decoding the inputs. A 1-bit decoder is an inverter with both its input and its output brought out and perhaps buffered. Each AND circuit forms a row of the AND array, and each OR circuit forms a column in the OR array. The complete PLA is shown schematically in Figure 13-2.

The darkened circles in the AND array represent connections to AND inputs, and the horizontal lines are product terms. Similarly, in the OR array the darkened circles indicate the connections to OR inputs, and vertical lines are sum terms. Any input in an AND array row can be connected to either the true or inverted output of the corresponding decoder, or else not connected. Similarly, any input in an OR array column can be connected or not to the product term driving it. Figure 13-2 is often implemented physically just as is shown in the schematic, that is, with decoder outputs orthogonal to AND inputs and with AND outputs orthogonal to OR inputs. The figure illustrates the connections—the personality— of the PLA for the functions $f1 = A\overline{B}\overline{C} + \overline{A}B$, and $f2 = \overline{A}B + \overline{B}C$.

For a PLA having 1-bit decoders, there are two inputs to the AND array for each PLA input (the true and inverted of each input) and the maximum number of product terms for n PLA inputs is 2^{2n}. However, since the product of the true and complement of the same variable is always 0, this number is never implemented. The maximum number of inputs to the OR array is the number of

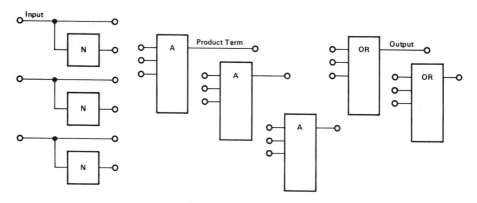

Figure 13-1. PLA definitions.

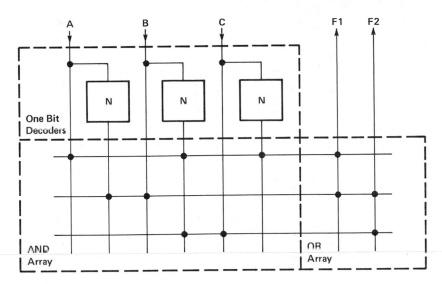

Figure 13-2. PLA schematic.

product terms; the number of outputs is the number of OR circuits and also the maximum number of sum terms that can be implemented. In practice, the physical implementation, that is, the available silicon area and the number of available PLA I/Os, limits the size of the PLA. Example: A PLA chip with 16 inputs, 8 outputs, and 48 product terms will have 1536 AND array devices. A maximum of 8 functions with 48 minterms per function is achievable. Even though there are 32 inputs to the AND array, each minterm can only have a maximum of 16 variables (the product of the true and complement of the same variable is always 0), so the maximum number of used devices in the AND array is half the total crosspoints, or 768 devices.

PHYSICAL IMPLEMENTATION

PLAs have been manufactured in both bipolar and FET technologies. The circuit schematic of a bipolar PLA that utilizes diodes as the device element for the arrays is illustrated in Figure 13-3.

In the AND array, all of the decoder lines that are connected to a row must be positive (greater than + V minus the voltage drop across the diode) for that word line to go positive or, to use the common term, be selected. In the OR array, since the direction of the diodes is reversed with respect to the load device, any product term that is connected to an OR input will cause the corresponding output to be positive when that input is positive (+ V minus the voltage drop across the AND word's load device). The determination of whether to connect the diodes or not is usually done in the contact or cut processing step, for example,

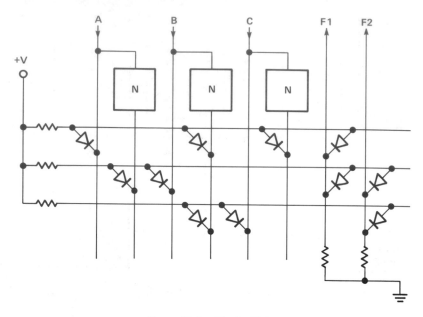

Figure 13-3.   Bipolar PLA.

with the mask aperture that connects the metal product wire to the p junction of the diode.

A particularly useful PLA for prototyping is a "field programmable" PLA such as the Signetics 82S100 (Signetics Corporation). This bipolar PLA has a NiCr fusible link in series with each device at the crosspoint. Crosspoints of this array with personalized and unpersonalized devices are shown in Figure 13-4.

Decoder circuitry (not shown) is included on the chip for personalization. To personalize, unwanted crosspoints are selected by the decoder and removed by passing high current through the NiCr link, opening the diode. The other diodes are left intact. Because the chip is fabricated with every crosspoint connected, it can be tested thoroughly before it is personalized.

Table 13-1 shows the specifications for several field programmable bipolar PLAs.

PLAs are implemented efficiently in FET technologies such as NMOS (n-channel metal oxide semiconductor), which is a simpler process than that used in bipolar designs. Both AND and OR arrays are made from NOR circuits. Consider an NMOS PLA used to obtain the function $f = A\overline{B}\overline{C} + \overline{A}B$. Let $P1 = A\overline{B}\overline{C}$ and $P2 = \overline{A}B$. By De Morgan's rule, $P1 = (\overline{A} + B + C)$ and $P2 = (A + \overline{B})$, which are NOR expressions and, therefore, map into the FET AND array. Now let $Q = P1 + P2$. The NOR of $P1 + P2$ yields $\overline{Q}$, so to obtain Q the output of the OR array must be inverted. This may be done at the output of the PLA or elsewhere in the design. Figure 13-5 shows a schematic of an unclocked NMOS PLA.

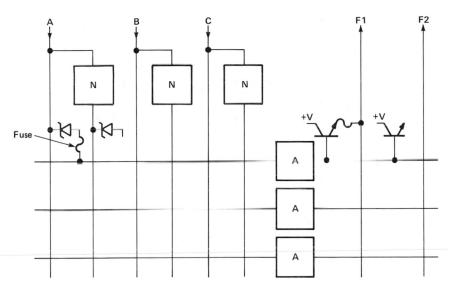

Figure 13-4.   Fusible link PLA crosspoints.

PLAs can also be implemented in the CMOS technology. An NMOS NOR circuit used for a product or a sum term is shown in Figure 13-6(a), and the corresponding CMOS circuit in Figure 13-6(b).

It is seen that an n-way NOR in CMOS requires n n-channel devices and n p-channel devices, which renders a PLA design of this type too costly in terms of area. However, if a clocked or "precharge" approach (Wood 1979) for the AND array (Figure 13-7) is used, the overhead is considerably less.

The purpose of device $Q1$ is to precharge the word line node to a "1" state so that pullup devices are not required. Transistor $Q2$ is turned off to prevent inputs that switch during precharge time from discharging the word line node. Since $Q1$ is a p device and $Q2$ an n device, a single clock pulse for the entire AND array suffices; that is, $Q1$ is on when $Q2$ is off, and vice versa. The product

Table 13-1. Field programmable bipolar PLAs.

	INPUTS	OUTPUTS	PRODUCT TERMS	FEEDBACKS	CYCLE TIME (NS)
Signetics 82s100 (Signetics)	16	8	48	0	50
NEC uPB450D (NEC)	24	16	72	16	200
Monolithic Memories PAL64R32 (Monolithic Memories)	64	32*	256	0	55

*OR array not fully programmable.

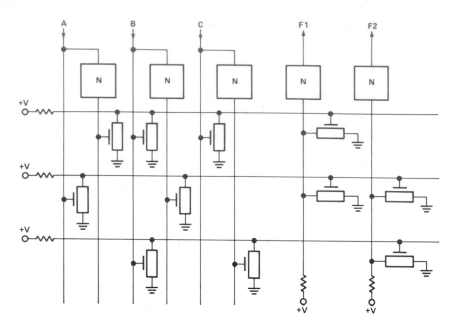

Figure 13-5.   NMOS PLA.

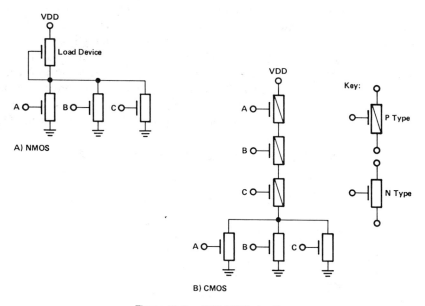

Figure 13-6.   FET NOR circuits.

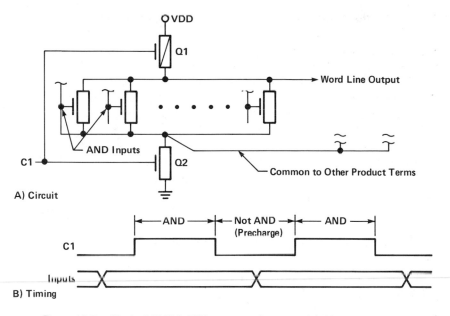

A) Circuit

B) Timing

Figure 13-7.   Clocked CMOS AND array product term. (a) Circuit. (b) Timing.

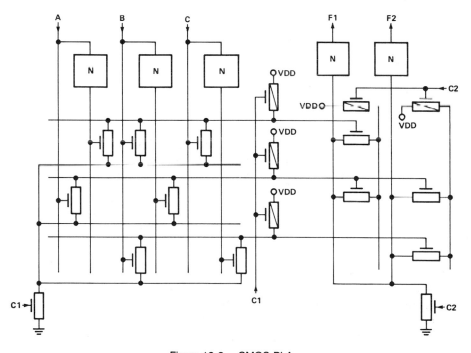

Figure 13-8.   CMOS PLA.

terms are charged to VDD during "not AND time" through $Q1$, while $Q2$ is gated off to prevent charge from flowing to ground. There is a timing requirement that the inputs to the AND array be finished switching before $C1$ turns device $Q2$ on, or else there will not be enough time to gate the OR array. If proper timing is employed, then if any input becomes positive, the word line goes to logical "0"; otherwise if all inputs are "0," the word line remains at a "1" (selected). This is precisely the same logic function produced by the NMOS NOR configuration, the "static" PLA. The OR array may be designed identically to the AND array, but the outputs, like the NMOS version, must be inverted to form the true of the desired function. Because proper operation depends on nodes holding their charge for part of the cycle, coupling between arrays must be avoided. This may be accomplished by isolating the arrays with buffer circuits. A schematic (buffers omitted) is shown as Figure 13-8.

CODED TABULAR LOGIC

The logic designer will find it convenient to treat the PLA as a logic table into which desired functions are coded. One need not be concerned with circuit or implementation details other than knowing the input and output polarities, performance, and power. We suggest a matrix format for the table where the columns are the AND array inputs and OR array outputs, and the rows are the product terms. The symbols for coding the logic and their definitions are:

1. In the AND array:
   a. I—match on a logical one.
   b. O—match on a logical zero.
   c. .—match on either a logical one or zero (don't care state).
2. In the OR array:
   a. I—set output to logical one if any associated product term is selected; or to logical zero if no associated product term is selected.
   b. .—take no action (the OR output is independent of this product term).

When the functions of the PLA are increased—by adding additional decoder capability, for example—additional symbols representing the new functions are included in the table.

By using these conventions, any inversion of signals within the PLA may be ignored; that is, the table represents the logical function of the PLA regardless of the technology used, so that it is easier to understand the design and to minimize it. Figure 13-9 is our example in tabular form, with applied inputs and resultant outputs shown on top.

In the figure, product term 3 is selected, which results in output f_2 (column 5) going to 1.

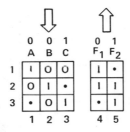

Figure 13-9.   PLA in tabular form.

EXTENDING PLA FUNCTION

Partitioning

The function of the PLA can be enhanced by increasing the function of the decoders. For a PLA with two inputs each driving a 1-bit decoder, how many useful product terms can be formed in the AND array? Since there are four input rails to the AND array, there can be 2^4 or 16 possible product terms formed. These are shown in Figure 13-10 for a diode AND array; note that only 9 of the

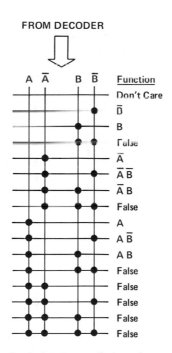

Figure 13-10.   Logic functions—diode and array, 1-bit decoding.

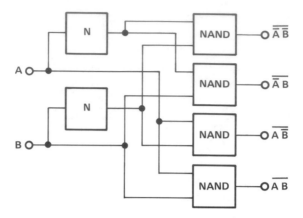

Figure 13-11. A 2-bit decoder.

16 terms are useful. However, if the same two inputs are partitioned together—
that is, they are decoded two at a time by a 2-to-4 decoder (Figure 13-11)—the
useful combinations in the AND array are increased from 9 to 15 (Figure 13-
12).

Now the product terms can include factors of two variables, not just one. This

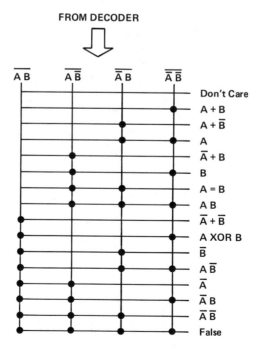

Figure 13-12.   Logic functions—diode AND array, 2-bit decoding.

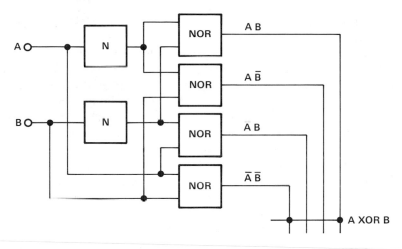

Figure 13-13.   A 2-bit decoder—NMOS implementation.

means that a PLA term can include expressions such as $(A \text{ XOR } B)$ and $(C + \overline{D})$. A typical 2-bit decoder (2-*bit partitioner*) for an NMOS PLA is shown in Figure 13-13. The personality for the exclusive OR function and the decoded outputs in the figure differ from those of the diode AND array of Figure 13-12 because the NMOS PLA and the decoder are implemented with NOR circuits.

Earlier we stated that a tabular format frees the designer from implementation details, and this is true as well for 2-bit decoding. Figure 13-14 shows the symbols and their meaning.

Logical Operation	A B ↓ ↓ 1 Bit Decoder	A B ↓ ↓ 2 Bit Decoder
Don't Care	. .	. .
B	. I	. I
A	I .	I .
$\overline{B}$	. O	. O
$\overline{A}$	O .	O .
$\overline{A}$ B	O I	O I
A $\overline{B}$	I O	I O
A B	I I	I I
A = B	—	EE (E̲qual)
A XOR B	—	UU (U̲nequal)
A OR $\overline{B}$	—	PN (A P̲ositive or B N̲egative)
$\overline{A}$ OR B	—	NP (A N̲egative or B P̲ositive)
$\overline{A}$ OR $\overline{B}$	—	NN (A N̲egative or B N̲egative)
A OR B	—	PP (A P̲ositive or B P̲ositive)

Figure 13-14.   Symbols for the AND array.

The cost difference between a PLA with 2-bit decoders and one with 1-bit decoders is determined by comparing the overhead of the 2-bit decoders with the benefits gained by the extra six functions: Exclusive-Or-Not (EE), Exclusive-Or (UU), and the OR operators (PN, NP, NN, PP). That is, functions such as the Exclusive-Or require two product terms with single-bit decoding, but only one product term with 2-bit partitioning. Another noteworthy point is that there is no penalty in the AND array for forming pairs of inputs; that is, two 1-bit decoders require four physical columns. One may inquire if the benefits of 2-to-4 decoders accrue for higher-order decoders, that is, 4-to-16. The answer is no because the 4-to-16 decoder requires 16 columns for 4 inputs versus 8 for the 2-bit partitioner. Also, the number of available functions only increases from 225 (15^2) to 255 ($16^2 - 1$). Therefore, if inputs are to be partitioned, 2-bit partitioning is judged optimum.

Storage Elements

PLA function can also be extended by adding storage elements, that is, latches of various types on the outputs. If the PLA is a macro, the latches may be part of the macro or remote from it in another macro. If the PLA is a chip, then the latches and driver circuits are located next to the OR outputs. When the latch outputs at the OR output of the PLA are connected to the input decoders, a finite state machine for accomplishing sequential logic is formed. A schematic diagram is shown in Figure 13-15.

Types of latches used with PLAs include set-reset, JK, toggle, polarity hold with gated clock, and polarity hold preceded by Exclusive-Or circuit. An OR array that drives set-reset or JK latches requires two output columns or rails for each latch, one for the set and one for the reset. Later we will illustrate the use of latches in a state machine with a design example.

It is more important to provide increased function for a PLA chip than for a PLA macro because the former is limited by a fixed number of I/Os and product

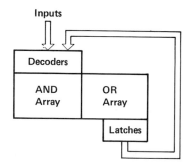

Figure 13-15. PLA finite state machine.

terms. An example is the NEC uPB450d PLA chip (NEC), which includes additional features such as 2-bit partitioning, JK latches feeding back to inputs, and polarity hold latches on outputs.

DESIGN REQUIREMENTS

Logic design with PLAs is accomplished by specifying the solution as the sum of products (AND-OR). The design should have the lowest-cost solution, that is, the minimum number of product terms, sum terms, and devices. This will free up more function on a PLA chip, and will minimize area and delay for a PLA macro. Generally, as a designer becomes more familiar with the design, he or she will build a library of PLA functions. Such a library will include functions such as incrementers, decrementers, and logical and arithmetic operators. From this base, it is relatively easy to modify one of these functions to satisfy specific requirements (e.g., change a counter from modulo 8 to modulo 10). A physical PLA may contain many such functions, and the problem of minimization to achieve the lowest cost must be solved. Minimization should be considered during the generation of each function, but global minimization of the total PLA should be left until the design is complete and is functionally correct because it is difficult to perform design changes on minimized PLAs that have been combined with other functions. Only after the PLA has been minimized and thoroughly simulated should physical design begin.

DESIGN METHODOLOGY

A conventional method of expressing the output as a sum of products of the input variables is to write out all the minterms for the function, and then select only the prime implicants necessary to cover these minterms. A formal procedure for this is the Quine-McCluskey method, which, although it does lend itself nicely to a computer program solution, is a rather tedious manual process (hence error-prone) and thus not the best method for initial PLA design. Another approach is to write the truth table (see Figure 13-16), and then for each output write the corresponding Karnaugh map (Veitch diagram) (see Figure 13-17). We have found this technique easy to manage for PLA designs. At this point, it may be helpful for the reader to review these basic approaches and their nomenclature (Chu 1962; Hellerman 1967).

   These approaches, used properly, guarantee the minimum solution for single-valued functions. However, PLAs normally have multiple outputs representing multiple functions, so the minimum solution cannot always be achieved by using Karnaugh maps. For example, the methods above would not indicate to a designer that in a PLA with two outputs, each driven by identical product terms, an output and a product term could be removed. Another restriction of Karnaugh maps is that relations between functions of more than four variables are difficult to

Inputs	Outputs
W X Y Z	w x y z
0 0 0 0	0 0 0 1
0 0 0 1	0 0 1 0
0 0 1 0	0 0 1 1
0 0 1 1	0 1 0 0
0 1 0 0	0 1 0 1
0 1 0 1	0 1 1 0
0 1 1 0	0 1 1 1
0 1 1 1	1 0 0 0
1 0 0 0	1 0 0 1
1 0 0 1	1 0 1 0
1 0 1 0	1 0 1 1
1 0 1 1	1 1 0 0
1 1 0 0	1 1 0 1
1 1 0 1	1 1 1 0
1 1 1 0	1 1 1 1
1 1 1 1	0 0 0 0

Figure 13-16. Truth table for modulo 16 incrementer.

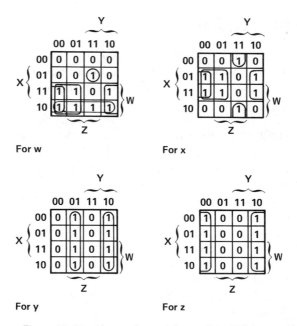

Figure 13-17. Karnaugh maps for modulo 16 incrementer.

visualize. Thus a function of many variables has to be partitioned by factoring some of the variables; for example, replacing $f\,(ABCDE)$ with $f(ABCZ)$ where Z is $f(DE)$. Admittedly this process may become difficult to manage, and again one may not be able to obtain the minimum solution. Therefore the designer will have to do as much as practical and depend on computer minimization tools to make up the difference.

A four-step design method based on the conventional truth table and Karnaugh map approach for 1-bit (true-complement) decoders is:

1. Write the truth table.
2. Write the Karnaugh map for each output.
3. Obtain the elementary areas from the maps and write the corresponding equations.
4. Map the equations into the PLA table.

This procedure is illustrated by the following example, the design of a 4-bit incrementing function (a modulo 16 incrementer) using 1-bit decoding.

The first step is to write the truth table for the incrementer, and the second step is to create a Karnaugh map for each output in turn.

The third step minimizes the solution. Determine the elementary areas (shown by the curved lines) and write the corresponding expressions:

$$w = W\bar{Y} + W\bar{Z} + \bar{W}XYZ + W\bar{X}$$
$$x = X\bar{Y} + \bar{X}YZ + X\bar{Z}$$
$$y = Y\bar{Z} + \bar{Y}Z$$
$$z = \bar{Z}$$

The fourth and last step of the procedure is to map these expressions directly into a PLA using a tabular format (Figure 13-18).

The design of the modulo 16 incrementer is now complete. The function may be part of a larger PLA, or be a macro as shown in Figure 13-19.

W X Y Z	w x y z	
I . O .	I . . .	$w = W\bar{Y}$
I . . O	I . . .	$+ W\bar{Z}$
O I I I	I . . .	$+ \bar{W}XYZ$
I O . .	I . . .	$+ W\bar{X}$
. I O .	. I . .	$x = X\bar{Y}$
. O I I	. I . .	$+ \bar{X}YZ$
. I . O	. I . .	$+ X\bar{Z}$
. . I O	. . I .	$y = Y\bar{Z}$
. . O I	. . I .	$+ \bar{Y}Z$
. . . O	. . . I	$z = \bar{Z}$

Figure 13-18.   PLA personalization for modulo 16 incrementer.

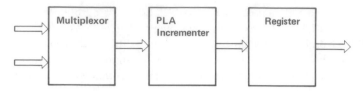

Figure 13-19. PLA in a macro design.

Designing with 2-Bit Partitioners

One approach to determine if 2-bit decoders improve the design is to write the equations for them directly from the Karnaugh map and evaluate the results. The technique is to first seek configurations of three 1's in a row or column (OR operators), or two 1's spaced apart by a single row or column (XOR or XORN operators). However, there is an inconsistency between this technique and that for a 1-bit decoder design. If a minterm is combined with another to form a function of two variables, then the largest elementary area that covers that minterm is not necessarily required for a minimum solution. In other words, because 2-bit partitioning yields expressions other than the product term canonical form, don't care terms that would have been removed during the conventional design process with Karnaugh maps must now be put back to achieve a minimum solution! This is the case for w in the example below, where the Karnaugh maps of Figure 13-17 for the 4-bit incrementer are analyzed for a 2-bit decoder design.

For w: Instead of including $W\overline{X}YZ$ in the area for the fourth row, combine $W\overline{X}YZ$ with $\overline{W}XYZ$ to give $(W \text{ XOR } X) YZ$. The third and fourth rows (omitting $W\overline{X}YZ$, of course) are adjacent OR terms, so the six minterms in these two rows are captured by $W (\overline{Y} + \overline{Z})$.

For x: Combine $\overline{W}XYZ$ with $W\overline{X}YZ$ to give $\overline{X}YZ$. The other six minterms are captured by $X (\overline{Y} + \overline{Z})$.

For y: The eight minterms are given by $Y \text{ XOR } Z$.

For z: The eight minterms are given by $\overline{Z}$.

These expressions can now be mapped directly into a PLA as shown in Figure 13-20.

```
W X Y Z      w x y z
I . N N      I . . .      w = W (Ȳ + Z̄)
U U I I      I . . .        + (W XOR X) YZ
. I N N      . I . .      x = X (Ȳ + Z̄)
. O I I      . I . .        + (X̄) YZ
. . U U      . . I .      y = Y XOR Z
. . . O      . . . I      z = Z̄
```

Figure 13-20. PLA personalization for incrementer.

The choice of *WX* and *YZ* for the Karnaugh map axes of Figure 13-17 was a good one for the 2-bit decoder design. In practice, some amount of trial and error is normally required to get the best pairing. We realize too that it may be difficult at first to determine the best covers for a 2-bit decoder design. Therefore, a second, more methodical approach is followed to exploit the use of 2-bit decoders for a PLA originally designed with single-bit decoders. The procedure is as follows:

1. Reorder the product terms that drive the same OR word so that they are adjacent, and follow the next steps for terms common to the same OR word.
2. Adjust the position of the words so that terms that differ by two variables only are adjacent, and call these a group. Note that at this stage there should be no terms that differ by one variable if the 1-bit decoder solution was minimum.
3. The pair of variables that are different from term to term can now be combined as inputs to the 2-bit decoder. However, it may be advantageous to expand the original product terms before combining them.

One should first attempt to find solutions containing OR operators (PP, PN, NP, NN) rather than Exclusive-Or (XOR, UU) and Exclusive-Or-Not (XORN, EE) operators. The reason for this is that three terms can be combined into one with the OR terms and only two with XOR and XORN. The advantage of choosing OR operators first is demonstrated in the following design example for a hypothetical function. The 1-bit decoder solution exists, and the terms have been ordered:

A	B	C	OUTPUT
.	O	I	I
.	I	O	I
O	I	I	I
I	O	O	I

Partitioning *B* and *C* and using XOR for the first two terms reduces the four terms to three:

A	B	C	OUTPUT
.	U	U	I
O	I	I	I
I	O	O	I

Now let us go back and expand the terms to see if the OR expressions can be used. The expanded terms (don't cares included) are:

A	B	C	OUTPUT
O	O	I	I
I	O	I	I
O	I	O	I
I	I	O	I
O	I	I	I
I	O	O	I

Rearranging the terms so that the set for $A = 0$ are in adjacent rows, we obtain:

A	B	C	OUTPUT
O	O	I	I
O	I	O	I
O	I	I	I
I	O	I	I
I	I	O	I
I	O	O	I

If B and C are paired, a solution of only two terms is obtained:

A	B	C	OUTPUT
O	P	P	I
I	N	N	I

Therefore, one must be prepared to expand the terms to the full set before the best solution can be realized, and to utilize OR operators first.

Another example to demonstrate the application of this procedure is the 3-bit comparator design specified by the truth table of Figure 13-21.

Input A			Input B			Output
W	X	Y	W'	X'	Y'	F
0	0	0	0	0	0	1
0	0	1	0	0	1	1
0	1	0	0	1	0	1
0	1	1	0	1	1	1
1	0	0	1	0	0	1
1	0	1	1	0	1	1
1	1	0	1	1	0	1
1	1	1	1	1	1	1

Figure 13-21. Truth table for the compare function.

Input A			Input B			Output
W	X	Y	W'	X'	Y'	F
0	0	E	0	0	E	1
0	1	E	0	1	F	1
1	0	E	1	0	E	1
1	1	E	1	1	E	1

Figure 13-22. Truth table for compare function after partitioning Y and Y'.

All the terms have been expanded (there are no don't cares) and placed in optimum order for defining groups. That is, terms that differ by two variables only are adjacent. By definition, the first two terms are the first group. Pairing Y and Y' allows the XORN function to reduce them to one term. After this trial it can be seen that when the remaining groups are combined, the truth table of Figure 13-22 results.

It can now be seen from Figure 13-22 that Y and Y' no longer differ, and by inspection X and X' may be paired and the four remaining terms reduced to two. Finally, W and W' are paired, yielding only one term in the PLA table as shown below:

WW'	XX'	YY'	F
EE	EE	EE	I

Clearly, the best pairing has been achieved, albeit by trial and error.

The question of whether there is a better way to obtain optimum partitioning for 2-bit decoders remains. That is, given inputs A, B, C, and D, is the best pairing obtained from AB and CD, AC and BD, or AD and BC? Or, is it better to simply not pair them and instead use 1-bit decoders? We know that for a PLA designed with 1 bit decoders, the Karnaugh map indicates exactly those minterms that differ by only one variable, and one looks for a particular topology, that is, the rectangles representing the elementary areas. Similarly, terms differing by two variables also have a particular topology, and we take advantage of this fact to determine the best partitioning. A two-dimensional truth table, akin to a Karnaugh map, indicates quickly if a useful pattern exists for a particular set of

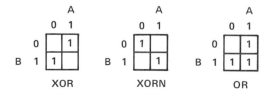

Figure 13-23. PLA personalization for logical operators.

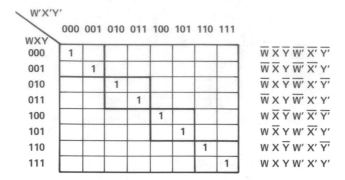

Figure 13-24. Two-dimensional truth table for the compare function.

input pairs. If no useful pattern exists, one may proceed iteratively and redraw the truth table. Of particular interest are the patterns for these functions of two variables: XOR, XORN, and the OR operators ($A + B$, $A + \overline{B}$, $\overline{A} + B$, and $\overline{A} + \overline{B}$). These functions reduce the number of product terms, for the case of 1-bit decoding, from two or three to only one. The two-dimensional truth table patterns for the logic operators XOR, XORN, and OR are shown in Figure 13-23.

The procedure for determining the best pairing is to assign the input order, write the two-dimensional truth table, scan for useful patterns, and repeat if necessary. With experience it is possible to write the PLA table for certain functions such as comparators directly from the two-dimensional truth table. To demonstrate, the comparator will be designed using this procedure. First, the two-dimensional truth table (Figure 13-24) is created from Figure 13-21.

One observes from the table that the XORN pattern reoccurs for Y and Y', indicating that our choice of input order was fortuitous, and that Y and Y' should be partitioned together. At this point, Y XORN Y' could be factored out of the truth table and called EE, as we did previously with the one-dimensional truth table (Figure 13-22). Then the two-dimensional truth table could be rewritten, and from the pattern for X and X' we would factor out X XORN X', and so

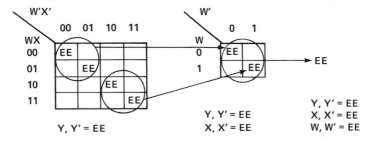

Figure 13-25. Deriving the comparator PLA table graphically.

forth. However, there is a much simpler graphical technique available. Notice that the code points along the edges of the truth table of Figure 13-24 are in increasing binary order, and that if the patterns for XORN are replaced by the PLA table symbol EE, a new pattern set is formed, shown in the left-most pattern of Figure 13-25. The trick now is to treat the diagonal pattern of two EE's for X and X' just as we treated the diagonal pattern of 1's for Y and Y'. By so doing, we can derive the next pattern in Figure 13-25, and by applying the same reasoning once again, we arrive at the symbol EE for W, W'.

We have now derived the PLA table for the compare function:

WW'	XX'	YY'	F
EE	EE	EE	I

This is a neat implementation indeed for a PLA that has 2-bit decoding, because it is the single product of XORN terms. In fact, more inputs can be added, and the result is still a single product term!

A5A6A7A8 \ A1A2A3A4	0000	0001	0010	0011	0100	0101	0110
0000		1	1		1		
0001	1			1		1	
0010	1			1			
0011		1	1				
0100	1						

a) Corner for All 8 Bits

A5A6A7 \ A1A2A3	000	001	010	011	100	101	110	111
000	UU	EE	EE	UU	EE	UU	UU	EE
001	EE	UU	UU	EE	UU	EE	EE	UU
010	EE	UU	UU	EE	UU	EE	EE	UU
011	UU	EE	EE	UU	EE	UU	UU	EE
100	EE	UU	UU	EE	UU	EE	EE	UU
101	UU	EE	EE	UU	EE	UU	UU	EE
110	UU	EE	EE	UU	EE	UU	UU	EE
111	EE	UU	UU	EE	UU	EE	EE	UU

b) A4 Partitioned with A8

Figure 13-26. Two-dimensional truth table for parity generator. (a) Corner for all 8 bits. (b) *A*4 partitioned with *A*8.

Let us now see if this graphic technique will also give us a neat solution for functions of many logic levels such as adders and parity generators. We choose the latter for our design example. The parity generation over 8 bits of data can be expressed as follows: $P = A1$ XOR $A2$ XOR $A3$ XOR $A4$ XOR $A5$ XOR $A6$ XOR $A7$ XOR $A8$. Another way to express P is that $P = 1$ if the number

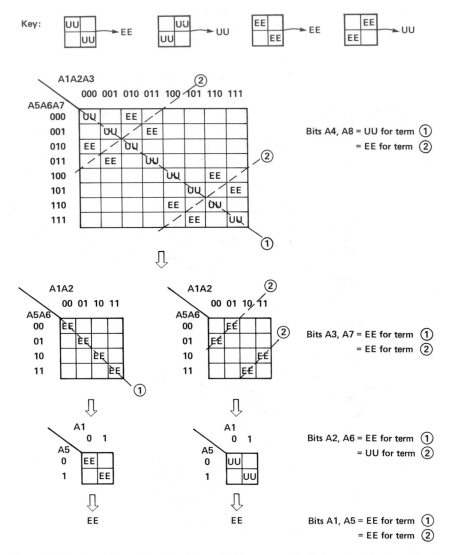

Figure 13-27.   Deriving parity generator product terms graphically from the two dimensional truth table.

A4	A8	A3	A7	A2	A6	A1	A5		P
U	U	E	E	E	E	E	E		I
E	E	U	U	E	E	E	E		I
E	E	E	E	U	U	E	E		I
E	E	E	E	E	E	U	U		I
E	E	U	U	U	U	U	U		I
U	U	E	E	U	U	U	U		I
U	U	U	U	E	E	U	U		I
U	U	U	U	U	U	E	E		I

Figure 13-28.   PLA personalization for parity generator.

of 1's in the 8 bits of data is odd; otherwise, $P = 0$. Therefore, the truth table will yield a 1 for half of the possible 256 combinations of the 8 bits.

Even though only one corner of the two-dimensional truth table is shown in Figure 13-26(a), the patterns for XOR and XORN may be observed repeating in a symmetric way, and we use this fact to draw the next truth table. Figure 13-26(b) is the truth table for the parity generator with the patterns for A4 and A8 replaced with EE or UU. One must now, in a fashion similar to the comparator example, determine for sets of either EE or UU, the corresponding function for bits A3 and A7, then bits A2 and A6, and finally bits A1 and A5. Instead of a wordy explanation, we present Figure 13-27 to demonstrate this technique for two of the eight product terms.

The entire PLA is shown in Figure 13-28. Term 1 from our example of Figure 13-27 is in row one, and term 2 is in row three.

We observe that if the parity generator were implemented as a 1-bit decoder PLA, there would be 128 product terms because every product term differs from every other in at least two bit positions and therefore cannot be minimized. We can quickly check the dimensions of our table by noting that each occurrence of UU or EE represents two terms, and that each row of four pairs of inputs represents 2^4 or 16 product terms, so that the PLA table does indeed represent 128 minterms.

It is emphasized once again that this graphical technique applies to the OR operators $(A + B, A + \overline{B}, \overline{A} + B,$ and $\overline{A} + \overline{B})$ and the XOR and XORN functions, and that the OR operators provide a three-to-one reduction in product terms.

MORE DESIGN EXAMPLES

In the following examples, a truth table and Karnaugh maps (showing the elementary areas for the 1-bit decoder design) for each output variable have been created, and the PLA for both 1- and 2-bit decoders is shown in tabular format.

W X Y Z	w x y z
0 0 0 0	0 0 0 1
0 0 0 1	0 0 1 0
0 0 1 0	0 0 1 1
0 0 1 1	0 1 0 0
0 1 0 0	0 1 0 1
0 1 0 1	0 1 1 0
0 1 1 0	0 1 1 1
0 1 1 1	1 0 0 0
1 0 0 0	1 0 0 1

Figure 13-29.　Truth table for modulo 10 incrementer.

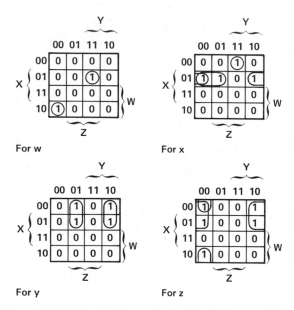

Figure 13-30.　Karnaugh maps for modulo 10 incrementer.

W X Y Z	w x y z
I 0 0 0	I . . I
0 I I I	I . . .
0 I . 0	. I . .
0 I 0 .	. I . .
0 0 I I	. I . .
0 . I 0	. . I .
0 . 0 I	. . I .
0 . . 0	. . . I

$$w, z = W\overline{X}\overline{Y}\overline{Z}$$
$$w = \overline{W}XYZ$$
$$x = \overline{W}X\overline{Z}$$
$$+ \overline{W}X\overline{Y}$$
$$+ \overline{W}\overline{X}YZ$$
$$y = \overline{W}Y\overline{Z}$$
$$+ \overline{W}\overline{Y}Z$$
$$z = \overline{W}\overline{Z}$$

Figure 13-31.　PLA personalization for modulo 10 incrementer.

W X Y Z   w x y z

$$
\begin{array}{|cccc|}
\hline
I & 0 & 0 & 0 \\
0 & I & I & I \\
0 & 0 & I & I \\
0 & I & N & N \\
0 & . & U & U \\
0 & . & . & 0 \\
\hline
\end{array}
\qquad
\begin{array}{|cccc|}
\hline
I & . & . & I \\
I & . & . & . \\
. & I & . & . \\
. & I & . & . \\
. & . & I & . \\
. & . & . & I \\
\hline
\end{array}
$$

$$w, z = W\overline{X}\,\overline{Y}\,\overline{Z}$$
$$w = \overline{W}XYZ$$
$$x = \overline{W}\,\overline{X}YZ$$
$$\quad + \overline{W}X\,(\overline{Y} + \overline{Z})$$
$$y = \overline{W}\,(Y \text{ XOR } Z)$$
$$z = \overline{W}\,\overline{Z}$$

Figure 13-32.   PLA personalization with 2-bit decoder.

Example: Modulo 10 Incrementer

The truth table and the Karnaugh maps are shown in Figures 13-29 and 13-30. The expressions obtained from the elementary areas are:

$$w = W\overline{X}\,\overline{Y}\overline{Z} + \overline{W}XYZ$$
$$x = \overline{W}X\overline{Z} + \overline{W}X\overline{Y} + \overline{W}\overline{X}YZ$$
$$y = \overline{W}\overline{Y}Z + \overline{W}Y\overline{Z}$$
$$z = \overline{W}\overline{Z} + \overline{X}\overline{Y}\overline{Z}$$

When the last minterm of z, that is, $\overline{X}\overline{Y}\overline{Z}$, is expanded to $\overline{W}\overline{X}\overline{Y}\overline{Z} + W\overline{X}\overline{Y}\overline{Z}$, it is seen that $\overline{W}\overline{X}\overline{Y}\overline{Z}$ is covered by the term $\overline{W}\overline{Z}$, and $W\overline{X}\overline{Y}\overline{Z}$, is included in the same minterm required for w. These expressions can now be mapped directly into a PLA, as shown in Figure 13-31.

Now the modulo 10 incrementer is designed using a 2-bit decoder, and the solution, Fig 13-32, is most easily obtained by scanning the maps of Figure 13-30. There is an OR operator in the second row of the map for x, and an XOR operator in the map for y.

Example: 2-Bit Add Function

The Karnaugh map is taken for each output in turn. (See Figures 13-33 through 13-35.) The axes have been changed for the best partitioning.

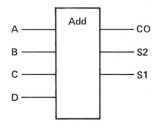

Figure 13-33.   Block diagram of a 2-bit add function.

A	B	C	D	CO	S2	S1
0	0	0	0	0	0	0
0	0	0	1	0	0	1
0	0	1	0	0	1	0
0	0	1	1	0	1	1
0	1	0	0	0	0	1
0	1	0	1	0	1	0
0	1	1	0	0	1	1
0	1	1	1	1	0	0
1	0	0	0	0	1	0
1	0	0	1	0	1	1
1	0	1	0	1	0	0
1	0	1	1	1	0	1
1	1	0	0	0	1	1
1	1	0	1	1	0	0
1	1	1	0	1	0	1
1	1	1	1	1	1	0

Figure 13-34.   Truth table for a 2-bit add function.

Expressions for each variable from the elementary areas on the maps are:

$$S1 = \overline{B}D + B\overline{D}$$
$$S2 = \overline{A}\overline{B}C + AB\overline{C} + \overline{A}C\overline{D} + A\overline{C}\overline{D} + \overline{A}B\overline{C}D + ABCD$$
$$C\theta = AC + BCD + ABD$$

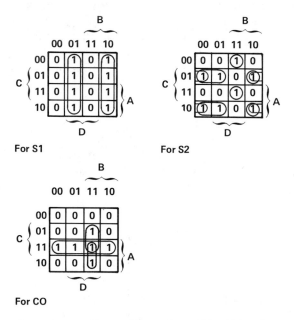

For S1   For S2

For CO

Figure 13-35.   Karnaugh maps for a 2-bit add function.

A	B	C	D		CO	S2	S1		
.	O	.	I		.	.	I		S1 = $\overline{B}D$
.	I	.	O		.	.	I		$+\ B\overline{D}$
O	O	I	.		.	I	.		S2 = $\overline{A}\overline{B}C$
I	O	O	.		.	I	.		$+\ A\overline{B}\overline{C}$
O	.	I	O		.	I	.		$+\ \overline{A}C\overline{D}$
I	.	O	O		.	I	.		$+\ A\overline{C}\overline{D}$
O	I	O	I		.	I	.		$+\ \overline{A}B\overline{C}D$
I	I	I	I		.	I	.		$+\ ABCD$
I	.	I	.		I	.	.		CO = AC
.	I	I	I		I	.	.		$+\ BCD$
I	I	.	I		I	.	.		$+\ ABD$

Figure 13-36.   PLA personalization for 2-bit add function.

These expressions can now be mapped directly into the PLA table (Figure 13-36).

The 2-bit decoder version is written from the Karnaugh maps (Figure 13-37). The map for $S2$ is interesting because it contains two OR operators that can be combined. That is, the minterms of rows two and four can be expressed as $\overline{A}C(\overline{B}+D) + A\overline{C}(\overline{B}+D) = (A\ \text{XOR}\ C)(\overline{B}+D)$. Graphically, an expression of this type is represented by two rows (or two columns), each with the same pattern, separated by a vacant row (or column).

Example: Two's Complement Function

The expression derived from the Karnaugh maps (see Figures 13-38 and 13-39) are:

$$A' = \overline{A}C + \overline{A}D + \overline{A}B + A\overline{B}\overline{C}\overline{D}$$
$$B' = \overline{B}D + \overline{B}C + B\overline{C}\overline{D}$$
$$C' = C\overline{D} + \overline{C}D$$
$$D' = D$$

Figure 13-40 is the 1-bit decoder version, and Figure 13-41 the 2-bit decoder version of the PLA.

A	B	C	D		CO	S2	S1		
.	U	.	U		.	.	I		S1 = B XOR D
U	N	U	N		.	I	.		S2 = (A XOR C) $(\overline{B}+\overline{D})$
E	I	E	I		.	I	.		$+$ (A XORN C) BD
I	.	I	.		I	.	.		CO = AC
U	I	U	I		I	.	.		$+$ (A XOR C) BD

Figure 13-37.   PLA personalization for 2-bit add function with 2-bit decoder.

A	B	C	D	A′	B′	C′	D′
0	0	0	0	0	0	0	0
0	0	0	1	1	1	1	1
0	0	1	0	1	1	1	0
0	0	1	1	1	1	0	1
0	1	0	0	1	1	0	0
0	1	0	1	1	0	1	1
0	1	1	0	1	0	1	0
0	1	1	1	1	0	0	1
1	0	0	0	1	0	0	0
1	0	0	1	0	1	1	1
1	0	1	0	0	1	1	0
1	0	1	1	0	1	0	1
1	1	0	0	0	1	0	0
1	1	0	1	0	0	1	1
1	1	1	0	0	0	1	0
1	1	1	1	0	0	0	1

Figure 13-38.   Truth table for two's complement function.

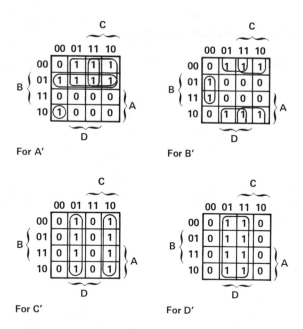

Figure 13-39.   Karnaugh map for two's complement.

A	B	C	D		A'	B'	C'	D'	
.	.	.	I		.	.	.	I	$D' = D$
.	.	I	O		.	.	.	I	$C' = C\bar{D}$
.	.	O	I		.	.	I	.	$+\ \bar{C}D$
.	O	.	I		.	I	.	.	$B' = \bar{B}D$
.	O	I	.		.	I	.	.	$+\ \bar{B}C$
.	I	O	O		.	I	.	.	$+\ B\bar{C}\bar{D}$
O	.	I	.		I	.	.	.	$A' = \bar{A}C$
O	.	.	I		I	.	.	.	$+\ \bar{A}D$
O	I	.	.		I	.	.	.	$+\ AB$
I	O	O	O		I	.	.	.	$+\ A\bar{B}\bar{C}\bar{D}$

Figure 13-40. PLA personalization for two's complement.

SEQUENTIAL LOGIC DESIGN

Our discussion of storage elements briefly mentioned how a PLA could be configured as a finite state machine (Figure 13-15), and we will now discuss this in more detail. The current state of the machine is stored in the latches connected between OR and AND arrays (the feedback register), while the combinatorial logic for determining the next state—the transition algorithm—is the personality of the PLA itself. The product terms are functions of both external inputs (new state information) and feedback register outputs (current state information). The sum terms are personalized to change the state of the feedback register and the PLA output.

We recommend that the feedback register be an LSSD, or level sensitive scan design, configuration (Eichelberger and Williams 1977). LSSD requires two latches per register for test pattern generation purposes, and to provide a race-free design. The first latch of the pair, the L1 latch, captures the data from the PLA's output, and the second, the L2 latch, receives the data from the first so that it is available at the input of the PLA. Typical clock timing is: CLOCKA loads L1, and CLOCKB transfers its contents to L2. These two clocks are nonoverlapping (Figure 13-42), and the period between CLOCKB and CLOCKA is sufficient for the signal to propagate through L2 and the PLA. In this way there can never be a race between new state and old state information.

A	B	C	D		A'	B'	C'	D'	
.	.	.	I		.	.	.	I	$D' = D$
.	.	U	U		.	.	I	.	$C' = C\ \text{XOR}\ D$
.	O	P	P		.	I	.	.	$B' = \bar{B}\,(C + D)$
.	I	O	O		.	I	.	.	$+\ B\bar{C}\bar{D}$
O	.	P	P		I	.	.	.	$A' = \bar{A}\,(C + D)$
U	U	O	O		I	.	.	.	$+\ (A\ \text{XOR}\ B)\ \bar{C}\bar{D}$

Figure 13-41. PLA personalization for two's complement with 2-bit decoder.

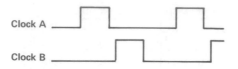

Figure 13-42. Clock timing for feedback register.

The design procedure for a state machine PLA is the same as that for a combinatorial logic PLA with the addition of these prerequisite steps:

1. Generate a state diagram.
2. Convert the state diagram to a state table.
3. Convert the state table to a transition table, which is really just the familiar truth table.

Example: Design of a 3-Bit Up-Down Counter (Modulo 8 Counter) Using Polarity Hold Latches

Counter operation is described by the following rule: Step the count value up if the input I is 1, and down if I is 0. The design is LSSD, and the input I does not switch during CLOCKA time. (See Figures 13-43 through 13-48.)

The PLA configuration and the design diagrams are shown. Please note that the Karnaugh map of Figure 13-47 shows the areas for the 2-bit decoder design, and not the largest elementary area.

Other latches besides polarity hold latches are useful in sequential logic applications, and the PLA personality is programmed to suit the type of latch used in the feedback register (i.e., polarity hold, JK, toggle, etc.). For a set-reset latch the transition conditions are: no change, set to 1, and reset to 0. The JK latch has a fourth condition, toggle, and this latch's operation is described below.

JIN	KIN	OUTPUT	PLA OR SYMBOL
0	0	Previous state	· (no action
1	0	1	S (Set)
0	1	0	R (Reset)
1	1	1 if previous state = 0, 0 if previous state = 1	T (toggle)

The toggle latch, a special version of the JK latch, only performs the last transition condition above. Since the J and K inputs are a pair of adjacent OR rails, the toggle latch is implemented as a JK latch with its J and K inputs connected so that a single OR rail may be used.

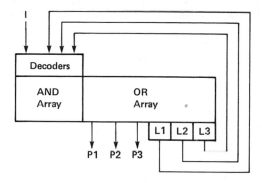

Figure 13-43. PLA configured as an up-down counter.

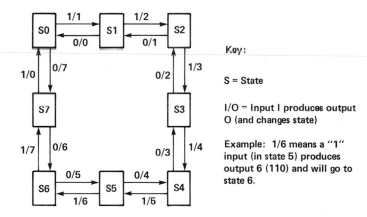

Key:

S = State

I/O – Input I produces output
O (and changes state)

Example: 1/6 means a "1"
input (in state 5) produces
output 6 (110) and will go to
state 6.

Figure 13-44. State diagram of the up-down counter.

| State | Input(I) | |
L1L2L3	0	1
S0	S7,7	S1,1
S1	S0,0	S2,2
S2	S1,1	S3,3
S3	S2,2	S4,4
S4	S3,3	S5,5
S5	S4,4	S6,6
S6	S5,5	S7,7
S7	S6,6	S0,0

Output
Next State

Figure 13-45. State table for the up-down counter.

409

I	L1	L2	L3	I1	I2	I3	P1	P2	P3	
0	0	0	0	1	1	1	1	1	1	0/7,7
1	0	0	0	0	0	1	0	0	1	1/1,1
0	0	0	1	0	0	0	0	0	0	0/0,0
1	0	0	1	0	1	0	0	1	0	1/2,2
0	0	1	0	0	0	1	0	0	1	0/1,1
1	0	1	0	0	1	1	0	1	1	1/3,3
0	0	1	1	0	1	0	0	1	0	0/2,2
1	0	1	1	1	0	0	1	0	0	1/4,4
0	1	0	0	0	1	1	0	1	1	0/3,3
1	1	0	0	1	0	1	1	0	1	1/5,5
0	1	0	1	1	0	0	1	0	0	0/4,4
1	1	0	1	1	1	0	1	1	0	1/6,6
0	1	1	0	1	0	1	1	0	1	0/5,5
1	1	1	0	1	1	1	1	1	1	1/7,7
0	1	1	1	1	1	0	1	1	0	0/6,6
1	1	1	1	0	0	0	0	0	0	1/0,0

Next State — Output — Input

Figure 13-46.   Transition table for the up-down counter.

Example: Design of a 4-Bit Incrementer (Modulo 16 Counter) Using an LSSD JK Latch

There is no input to the PLA; that is, the count is always stepped every clock cycle. The design procedure is the same as that of the up-down counter. (See Figures 13-49 through 13-51.)

The Karnaugh maps produce the following relationships for the toggle condition:

$$\text{Toggle } (JKz) = \text{always true}$$
$$\text{Toggle } (JKy) = Z$$
$$\text{Toggle } (JKx) = YZ$$
$$\text{Toggle } (JKw) = XYZ$$

The first product term may look peculiar until one realizes that the low-order bit toggles each clock cycle regardless of the state of the counter. There are no bits in input W's column because W is the highest-order bit of the counter.

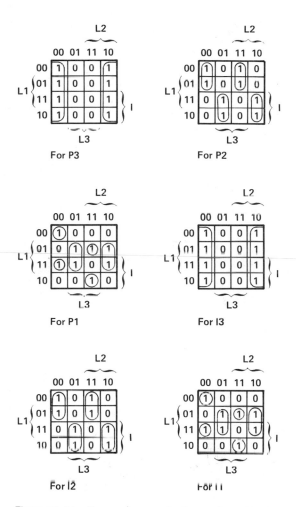

Figure 13-47. Karnaugh maps for the up-down counter.

I	L1	L2	L3		I1	I2	I3	P1	P2	P3
.	.	.	O		.	.	I	.	.	I
I	.	U	U		.	I	.	.	I	.
O	.	E	E		.	I	.	.	I	.
E	E	O	O		I	.	.	I	.	.
.	I	U	U		I	.	.	I	.	.
U	U	I	I		I	.	.	I	.	.

Figure 13-48. PLA personalization for the up-down counter.

411

W X Y Z	JKw JKx JKy JKz
0 0 0 0	0 0 0 T
0 0 0 1	0 .0 T T
0 0 1 0	0 0 0 T
0 0 1 1	0 T T T
0 1 0 0	0 0 0 T
0 1 0 1	0 0 T T
0 1 1 0	0 0 0 T
0 1 1 1	T T T T
1 0 0 0	0 0 0 T
1 0 0 1	0 0 T T
1 0 1 0	0 0 0 T
1 0 1 1	0 T T T
1 1 0 0	0 0 0 T
1 1 0 1	0 0 T T
1 1 1 0	0 0 0 T
1 1 1 1	T T T T

Figure 13-49. Truth table for JK increment function.

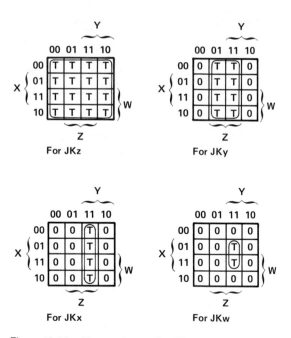

Figure 13-50. Karnaugh map for JK increment function.

412

```
W X Y Z        JKw JKx JKy JKz
. . . .        .    .    .    T
. . . |        .    .    T    .
. . | |        .    T    .    .
. | | |        T    .    .    .
```

Figure 13-51.   PLA personalization for JK increment function.

REMOVING REDUNDANCIES

We said earlier in our design requirements section that the design should have the minimum number of product terms, sum terms, and devices. A necessary although not sufficient requirement for a minimum design is that there be no extra or redundant devices. Regardless of the care taken, redundancies can creep into a design. This occurs when the designer fails to see an elementary area on a Karnaugh map, or perhaps fails to choose the optimum personality for a 2-bit decoder function. Redundancies also arise when a function has too many variables to manage accurately with a Karnaugh map. Therefore, it is desirable to run the PLA tables through a program that will detect redundancies. A device that is redundant is one that can be removed without altering logical function, and if the device is left in the design, a test pattern will not be able to tell whether the actual device in the manufactured part is bad or not. (If it is bad, the function will still work properly.) It follows that a program for generating test patterns for a PLA such as that described by Eichelberger and Lindbloom (1979) can also flag redundancies. However, for reasons of clarity and "to know what is going on," removal of redundancies should be left to the designer, and not be part of the program. A program such as the one mentioned above is very useful because it is difficult to detect redundancies visually. For example, it may not be apparent that the function $Q = \overline{A}CDEF + ABCDEF$ is redundant. However, once the program has indicated that Q has a redundant bit, the redundancy can be verified by factoring the equation as follows: Let $Z = CDEF$. Then $Q = \overline{A}Z + ABZ = Z(\overline{A} + AB) = Z(\overline{A} + B)$. Therefore, bit A should be removed from the $ABCDEF$ product term.

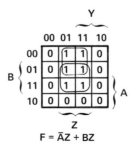

F = ĀZ + BZ

Figure 13-52.   Karnaugh map for $F = \overline{A}Z + ABZ$.

Alternatively, the Karnaugh map (Figure 13-52) can be used to verify bit removal.

If redundancies are indicated for functions of 2-bit decoders, the method of expanding product terms that we suggested for designing with 2-bit decoders applies. A common oversight is to code a function of this form:

A	B	C	OUTPUT
I	U	U	I
I	O	O	I

To correct, the UU term is expanded, and both words are replaced by:

A	B	C	OUTPUT
I	N	N	I

CHIP DESIGN WITH PLA MACROS

Our final example is chosen to demonstrate the design and application of PLAs and to tie together some of the concepts we have discussed.

A digital algorithmic unit, implemented on a large-scale integrated circuit chip as six programmable logic arrays and other circuits, provides concurrent digital filtering for four sampled servo control loops. The transfer functions of the digital filters determine the necessary phase and amplitude compensation for the servos and are defined by stored coefficients. Filter output is calculated by first multiplying a sample input value (representing the current state of the servo) by a set of stored coefficients, then multiplying stored values (representing the previous state of the servo) by another set of stored coefficients, and finally adding these matrix products.

The major data paths in the unit, controlled hierarchically, are shown schematically in Figure 13-53.

The set of stored values (or partial results) for each of the four servos, A, B, C, and D, is written to and read from the RAM under control of the LSAD (local storage address) PLA. The COEF (coefficient) PLA is a storage facility that serves the same purpose as a ROS by storing each set of coefficients. The coefficients are labeled a_i and b_i; the a coefficients are the zeros, and the b coefficients are the poles of the filter's transfer function. The CNTL (control) PLA is designed as a finite state machine (its latches are not shown in the figure) and it executes several self-contained sequences or control subroutines (e.g., vector sum and vector product subroutines). It decodes the address bus to identify the servo associated with the input sample on the data bus, and then broadcasts this address, a subroutine identifier, and a sequence count to the LSAD, DATA,

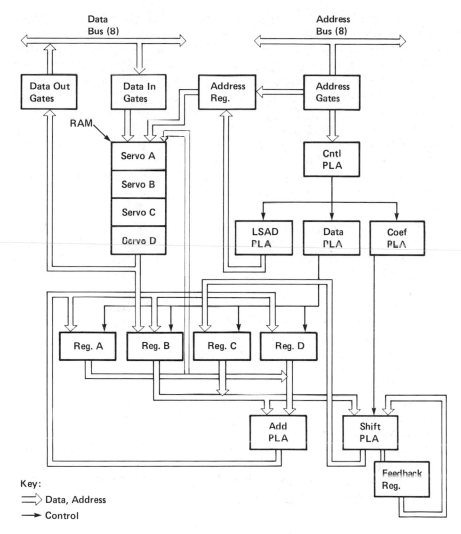

Figure 13-53.   Digital algorithmic unit data flow.

and COEF PLAs. The address information indicates which of the four sets of partial results must be read from the RAM by the LSAD PLA. The LSAD PLA generates the read and write addresses for the RAM by maintaining the contents of the address registers; the DATA PLA generates control lines to gate data to the various registers; and the COEF PLA outputs the proper coefficients to the SHIFT PLA for multiply operations. The ADD PLA is a combinatorial logic functional element that adds the 8-bit values in the storage registers to the corresponding 8-bit values in the RAM. The corresponding bits of each bus are paired, and are the inputs to 2-bit decoders. The ADD PLA is of the type invented by Weinberger (1979), and in this design, the OR outputs of the PLA are paired and drive XOR circuits. These are designed physically so that their inputs align with the OR outputs. Because physical layout normally necessitates that each latch be at least as wide as the pitch of two OR array columns, the XOR function fits nicely between the OR array output and a parity hold latch, and therefore conserves area. The SHIFT PLA is an example of both combinatorial and sequential logic. It shifts data and detects overflow and underflow conditions as required for the multiply operation. Four bits of a long data string (typically 16) are received and may be shifted any number of bits left or right. Then the control lines from the DATA PLA direct the shifted 4-bit result to the proper register. The design of this PLA is presented in detail as an example of a typical PLA macro.

The input control bits of the SHIFT PLA are divided into two fields. The first field of three bits selects the function, while the second field of two bits specifies the length of the shift, that is, 0, 1, 2, or 3 bit positions. The bits of the first field are coded so that any one of the following functions may be selected:

- Shift data a specified number of places into the feedback register.
- Gate out the contents of the feedback register while inhibiting the input data from being gated out.
- Analyze the input data for overflow or underflow conditions.
- Gate out the results of the analysis above.

The design procedure was to draw a map of the functions required, and then arrange the functions so that the control code combined as many functional tables as possible.

To explain the choice of a control code, we will digress for a moment. Instead of our problem, suppose the functions AND, XOR, and OR are to be coded. It is beneficial to code AND and XOR with don't cares so that the OR function need not be coded. A suitable code is:

$$AND \quad O.I$$
$$XOR \quad OI.$$

Now if the OR operation is assigned 011, the OR function does not require coding because the inclusive OR is the logical sum of the AND and XOR functions. This technique would save one product term for every bit on the bus to be OR'd.

The control assignment for the SHIFT PLA was derived with the previous example in mind and is shown in Figure 13-54.

Figure 13-55 is the SHIFT PLA table showing the feedback register. The control inputs are columns 1 through 5. Columns 1 through 3 select the function,

1＼23	00	01	11	10
0		(a) Shift Input Data to Outputs and to Feedback Registers	(b) Gate Feedback Register to Outputs (If Shift Count=0 and High Order Data Is Unsigned (i.e. Col. 6 = 0)	
1	(c) Analyze the Data for Overflow or Underflow with Shift Count=0	(a) (c) Shift Input Data to Outputs and Feedback Register Analyze the Data for Overflow or Underflow with Shift Count=0	(d) Gate Out Result of Overflow and Underflow Analysis	(d) Gate Out Result of Overflow and Underflow Analysis

Figure 13-54.   Control assignment for SHIFT PLA.

and 4 and 5 the shift amount. Columns 6 through 9 are the data inputs, and 10 through 13 are the inputs from the feedback register. The remaining columns comprise the OR array. Columns 14 through 17 are connected to the feedback register, and columns 18 through 21 are the four bits of output. The functions of the SHIFT PLA can be ascertained by reading the code points for the control function in columns 1, 2, and 3, and then finding their definition in the map (Figure 13-54). For example, the code for product terms 1 through 16 is .OI, corresponding to the shift operation. Terms 1 through 4 pass data through un- shifted, while 5 through 8 shift data one bit, and so on.

Another interesting aspect of this PLA is the feature of extending the sign for arithmetic right shift operation; that is, the sign bit is inserted into the positions to its right according to the count. The control code for this is OII and operates on words 17 to 20 and 39 to 42. Words 17 to 20 gate the most significant bits of the shifted data from the feedback register to the output, and words 39 to 42 extend the sign bit (column 6) the amount specified by the count field. This extension of the sign is not obvious because redundant bits have been removed; therefore, the personality must be analyzed in order to understand this function.

PHYSICAL LAYOUT

The goal of the layout designer is to achieve the necessary performance in the minimum area without violating any of the mask shape constraints needed to guarantee that the product is manufacturable. At the same time, he or she should bear in mind that the regular structure must be maintained, or else physical design automation will be too difficult.

The spacing of input to input, product term to product term, and output to

```
                    ┌──────┐
                    ▼      │
   INPUT      FEEDBACK OUTPUT
              1111 1111 1122
   123456789  0123 4567 8901

 1 .OIOOI... ....  .... I...    1 SHIFT 0 BIT 1  ┐
 2 .OIOO.I.. ....  .... .I..    2 SHIFT 0 BIT 2  │
 3 .OIOO..I. ....  .... ..I.    3 SHIFT 0 BIT 3  │
 4 .OIOO...I ....  .... ...I    4 SHIFT 0 BIT 4  │
 5 .OIOII... ....  ...I         5 SHIFT 1 BIT 1  │
 6 .OIOI.I.. ....  .... I...    6 SHIFT 1 BIT 2  │
 7 .OIOI..I. ....  .... .I..    7 SHIFT 1 BIT 3  │
 8 .OIOI...I ....  .... ..I.    8 SHIFT 1 BIT 4  │
 9 .OIIOI... ....  ..I. ....    9 SHIFT 2 BIT 1  ├ a
10 .OIIO.I.. ....  ...I ....   10 SHIFT 2 BIT 2  │
11 .OIIO..I. ....  .... I...   11 SHIFT 2 BIT 3  │
12 .OIIO...I ....  .... .I..   12 SHIFT 2 BIT 4  │
13 .OIIII... ....  .I.. ....   13  SHIFT 3 BIT 1 │
14 .OIII.I.. ....  ..I. ....   14 SHIFT 3 BIT 2  │
15 .OIII..I. ....  ...I ....   15 SHIFT 3 BIT 3  │
16 .OIII...I ....  .... I...   16 SHIFT 3 BIT 4  ┘
17 O.I...... I...  .... I...   17 FEED BACK BIT 1 ┐
18 O.I...... .I..  .... .I..   18 FEED BACK BIT 2 ├ a, b
19 O.I...... ..I.  .... ..I.   19 FEED BACK BIT 3 │
20 O.I...... ...I  .... ...I   20 FEED BACK BIT 4 ┘
21 IO...OPP. O...  ..I. ....   21 OVER            ┐
22 IO...O..I O...  ..I. ....   22 OVER            │
23 IO...INN. O...  .I.. ....   23 UNDERFLOW       │
24 IO...I..O O...  .I.. ....   24 UNDERFLOW       │
25 IO...I... O...  ...I ....   25 SIGN            │
26 I........ O...  I... ....   26 SEQ             ├ c
27 IO...O... I..I  .I.. ....   27 UNDERFLOW       │
28 IO...I... I..O  ..I. ....   28 OVERFLOW        │
29 IO....... I.I.  ..I. ....   29 MAINT           │
30 IO....... II..  .I.. ....   30 MAINT           ┘
31 II....... ..I.  .... .III   31 OVERFLOW        ┐
32 II....... II..  .... I...   32 UNDERFLOW       │
33 II....... O.I.  ..I. I...   33 MAINTAIN/OVERFLOW
34 II....... OI..  .I.. ....   34 MAINTAIN        │
35 .I...I... .OO.  .... I...   35 PASS            ├ d
36 II....I.. .O..  .... .I..   36 PASS            │
37 II.....I. .O..  .... ..I.   37 PASS            │
38 II......I .O..  .... ...I   38 PASS            ┘
39 OI.OOI... ....  .... ...I   39 PROP − SIGN
40 OI.O.I... ....  .... .II.   40 PROP − SIGN
41 OI..OI... ....  .... .I..   41 PROP − SIGN
42 OI...I... ....  .... I...   42 PROP − SIGN
43 OIO..I... ....  .... .III   43 PROP − SIGN
```

Figure 13-55.   SHIFT PLA.

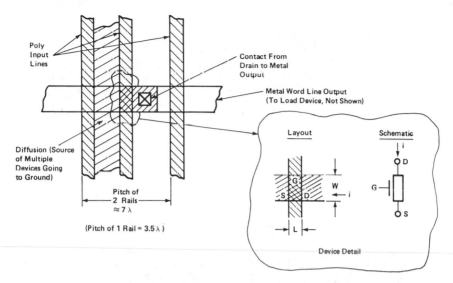

Figure 13-56.   Layout of NMOS PLA crosspoint.

output should be at the minimum technology spacing if possible. For an NMOS technology utilizing polysilicon as the gate material, for example, one of the factors that determines the minimum pitch of the input lines is the length of the device. Figure 13-56 is a typical layout for a static array crosspoint.

The minimum technology dimension, called λ (lambda), is also the device length. If a typical minimum device length were 2 μm, then the input pitch, which is about 3.5 λ, would equal 7 μm. Therefore, as soon as the number of inputs, outputs, and product terms is known, the PLA's area can accurately be predicted. For an NMOS design, the performance of the PLA is governed by the current (i) that the devices sink, and the capacitive load they drive. The current is directly proportional to the device's width, and inversely proportional to its length; capacitance is directly proportional to PLA size. The larger the PLA is (or the more prevalent its use), the more important it is to have an optimum layout to conserve chip area and reduce delay.

An advantage of PLAs—and this advantage is independent of the technology used—is that the performance of each PLA is calculable in a straightforward manner. When the analysis of one PLA is known, the delay of any other of the same type can be parameterized as a function of words, personalized devices, and so on.

FOLDING

Folding is a physical design technique that minimizes area by rearranging words and outputs. Its effectiveness is dependent on the particular bit patterns of the

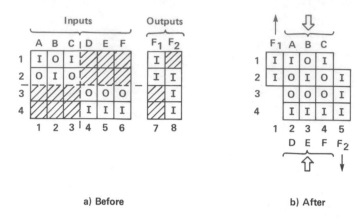

a) Before

b) After

Figure 13-57. Folding a PLA. (a) Before. (b) After.

design, and it is most effective when there are open areas in the arrays. A simple example of a candidate PLA suitable for folding is shown in Figure 13-57(a).

One observes that there are vacant areas in both AND and OR arrays, which have been shaded. Another observation is that because term 2 is the input to both sum terms, if this PLA were divided into two smaller arrays and separated, term 2 would have to be repeated. To avoid adding the extra term, the array is folded along the dashed lines. The subarrays are shifted as shown in Figure 13-57(b). By folding, rather than separating, the OR bits at 2, 1 and 2, 5 in Figure 13-57(b) are driven by the same AND term. Folding has been used for mapping designs into a PLA chip (Wood 1979), and for reducing the area of PLA macros (Rinaldi 1983).

DESIGN AUTOMATION

The design process for PLAs is generally well automated. It is straightforward to write design programs because the PLA is both ordered and regular. Programs have been written for documentation, design verification, test pattern generation, and mask generation. Work station manufacturers and software houses such as Seattle Silicon Technology (Concorde Software) offer programs of this type. We will discuss briefly the concepts of these tools based on the programs we have used.

Documentation

Once the designer has devised the logic table, the symbols are conveniently entered with a full screen editor. A program creates a skeleton file with row and column numbers on the periphery of a matrix of points. These are displayed, and the designer replaces points with the proper symbols. Another program syntax checks the data and creates a hard copy. It also expands and reformats the data for use by the simulator, test pattern generator, and physical design generator programs.

Minimization

It is desirable to analyze the PLA with a computer program after the final design step is completed. Since the number of combinations of variables can be large, the time taken to produce the best solution can be very long, so this process should be attempted only after the PLA has been verified to be functionally correct, and no more changes are expected. A suitable algorithm for PLAs with 1-bit decoders is one that searches the minterms common to each output of the OR array, and indicates when bits differ in one position only. The product terms in a 2-bit decoder design can be expanded (e.g., UU is expanded to 10 and 01) and then treated like any other. The difficulty is that for terms of many factors such as a comparator term, the number of expanded terms grows as 2^n where n is the number of input pairs. The detection of redundancies by running test pattern generation programs in the manner discussed earlier is also recommended.

Simulation

Simulator programs require a subroutine for each type of entity that is simulated (e.g., NAND, NOR, REGISTER, ROS, etc.). Therefore, a routine for the PLA is also provided. The logical data base includes a single block representation and a personality table for each PLA used. During simulation each PLA is modified by its associated personality. That is, a design with five PLAs would use the same subroutine, but there would be five copies of the PLA in the compiled input to the simulator.

Test Pattern Generation

At least two approaches for modeling PLAs have been used. These approaches assume that PLA designs do not have any failure mechanisms different from random logic. This has proved true for our PLA experience consisting of the bipolar, NMOS, and CMOS designs we have described. One method of handling PLAs is to expand the bit patterns into equivalent circuits. Figure 13-58 shows how an NMOS PLA is expanded into equivalent NOR circuits so that standard test pattern techniques may then be applied.

A second method, used on PLA chips, is to write a program that analyzes the bit pattern directly and generates patterns to test for each fault (Eichelberger and

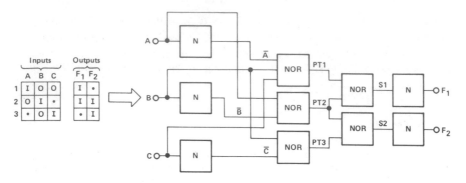

Figure 13-58.   NMOS PLA exposion.

Lindbloom 1979). This is an efficient way of generating the patterns, but it is difficult to implement when there is a mix of PLAs and random circuits.

Physical Design Generation

If engineers were to generate all the mask shapes for a PLA manually, they would probably use an interactive graphic computer design system such as CALMA (Calma Company) or IGS (Carmody et al. 1978). The rectangles, polygons, and

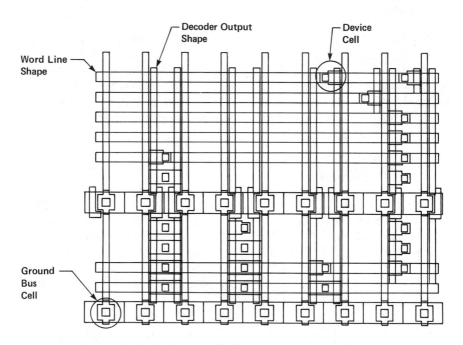

Figure 13-59.   Plot of part of a PLA generated automatically.

lines that are entered are stored on the system in a graphics language for later conversion to optical data. The PLA AND inputs, product terms, and OR output lines are constructed one at a time with required dimensions. It is normal practice for an engineer using a graphics system to design circuits or elements that are used repeatedly first, and store them as named graphic entities called "structures" or "cells." For a PLA, cells are used for the device at the crosspoint, the partitioning circuits, power bus rails, and so on. The designer would determine where devices should be placed and add device cells accordingly. To automate this procedure, the shapes for the AND inputs, and so forth, must be generated algorithmically using program-calculated dimensions and locations, and these numbers in turn are determined by the number and pitch of array rows and columns. In a similar fashion to manual operation, the program calculates the position to which a needed cell (i.e., the device at the crosspoint) should be placed, and codes the reference (or "call") to that cell at that position. The map of where the devices are to go—the expanded version of the PLA table—is the input that tells the program where to place the devices within the array. Figure 13-59 is a plot of a portion of an NMOS PLA mask set that was generated automatically.

FUTURE USE

We envision continued use for PLAs as macros on large chips in both NMOS and CMOS designs, and also as chips for prototyping VLSI designs. However, we believe stand-alone PLA chips will have difficulty competing with gate array chips because the latter generally provide more design flexibility than PLA chips at a moderate physical design effort. With the advent of synthesis or mapping techniques, prototypes can be designed, built, and debugged with PLAs efficiently, and then converted to gate arrays or PLA macros on custom chips. Another approach to prototyping is the PAL, which utilizes an AND array plus random logic and is being manufactured in fusible link technology, so that with both PALs and PLAs the user has reasonable flexibility. This is a good short-term solution for prototyping systems of fewer than 20,000 gates. Eventually, fusible link PLAs will be replaced by electrically alterable PLAs (Wood et al. 1981) so that the prototype can be changed without destroying the chip. Another possibility is to use newer and denser technologies to build PLAs with storage elements at the crosspoint, and produce a software-alterable chip.

SUMMARY

This chapter has described PLAs and a design methodology that we believe will lead to improved designer productivity and high-quality designs. Fusible link PLA chips are very useful for prototyping VLSI designs on the bench or in a system environment, and require no physical design. PLAs provide a rigorous environment for design, yet logic design is easy, and documentation of logic in tabular form is easy to understand. Minimization, simulation, and test pattern generation are automated, and algorithmic PLA generators exist for the physical design of PLA macros for VLSI chips. Alternatively, PLA designs can be converted by synthesis, if desired, to a more efficient form for a particular technology. Therefore, the design discipline described in this chapter is highly relevant to VLSI designs of 100,000 circuits or more.

ACKNOWLEDGMENTS

The authors are indebted to J. C. Logue, under whose leadership much of the PLA work at IBM began, and who originally suggested that a technical report on this subject be written. Initially H. Fleisher, L. I. Maissel, N. F. Brickman, F. Howley, and W. W. Wu made significant contributions to PLA design. We thank T. J. Charest, K. A. Chen, G. C. Luckett, D. R. Rozales, and R. D. Taylor for their contributions to this chapter. We are grateful to E. Holton for her valuable suggestions, and to R. L. Ryf for his assistance in preparing the manuscript for publication. W. D. Benedict, C. O. R. Dodman, and R. A. Proctor provided encouragement and support, and we especially acknowledge D. E. Cuzner for his continuing support of the development and application of PLAs.

REFERENCES

Calma Company, 501 Sycamore Drive, Milpitas, CA 95035-7489.

Carmody, P., Barone, A., Morrell, J. and Lovejoy, C., "An Interactive Graphics System for Large Scale Integration Design," *Proc.* International Conference on Integration Techniques in Computer Aided Design, Bologna, Italy, pp. 281–294, Sept. 1978

Chu, Y., *Digital Computer Fundamentals,* McGraw-Hill, New York, 1962, pp. 136–147, 153–155.

Concorde Software, Seattle Silicon Technology Inc., 12356 Northrup Way, Bellevue, WA 98005.

Eichelberger, E. B., and Lindbloom, E., "A Heuristic Test-Pattern Generator for Programmable Logic Arrays," *IBM J. Res. Devel., 24,* 15–22 (1979).

Eichelberger, E. B., and Williams, T. W., "A Logic Design Structure for LSI Testability," *Proc. 14th Design Automation Conference,* New Orleans, LA, June 1977.

Hellerman, H., *Digital Computer System Principles,* McGraw-Hill, New York, 1967, pp. 181–183.

Monolithic Memories, 2175 Mission College Blvd., Mail Stop 9-14, Santa Clara, CA 95054. PAL® is a registered trademark of Monolithic Memories.

NEC, Nippon Electric Company, Ltd., Tokyo, Japan.

Rinaldi, M., "8 Bit PLA Adder," *IBM Technical Disclosure Bulletin, 26,* (6), 2792–2793 (Nov. 1983).

Signetics Corporation, 811 E. Arques Avenue, P.O. Box 3409, Sunnyvale, CA 94088-3409.

Weinberger, A., "High-Speed Programmable Logic Array Adders," *IBM J. Res. Dev., 23,* 163–168 (1979).

Wood, R. A., "A High Density Programmable Logic Array Chip," *IEEE Transactions on Computers, C-28* (9) (Sept. 1979).

Wood, R., Hsieh, Y., Price, C. and Wang, P., "An Electrically Alterable PLA for Fast Turnaround Time VLSI Development Hardware," *IEEE Journal of Solid State Circuits, SC-16* (5) (Oct. 1981).

14. Design for Testability

T. Williams
IBM Corporation

Many design techniques have been developed to accommodate the need for more dense and complex networks. This chapter will describe some of the techniques that are still in use.

Accompanying the presentation of some of these design techniques will be some design methods of implementing specific functions—for example, counters, which sometimes appear problematic when they must be implemented in an efficient manner.

In the field of design for testability, three approaches are currently being employed. The ad hoc approach is used when the designer is not capable of changing the silicon on chips but can do something at the board level. With the structured approach, or structured design techniques, the designer can influence the design of the chip and therefore control certain attributes, not only at the chip level but also at the board and system level. These techniques are usually associated with design automation tools, as mentioned in prior chapters, which will facilitate test generation, fault simulation, and possibly design verification.

The third approach to design for testability, which is becoming increasingly important, is the self-testing approach. Three different areas in self-testing will be explored in this chapter. First is the microprocessor self-stimulated test, in which the microprocessor executes some code and stimulates other devices on the board. The second area is in-situ self-testing, in which system latches are used to generate patterns, as well as to subsume patterns in a way that is deemed very efficient in terms of both "time" and "test generation/fault stimulation." The third area of self-testing includes the ex-situ techniques, where latches are used external to the system function. Again, these latches are used to generate test patterns, as well as to subsume test patterns.

AD HOC TECHNIQUES

In this section three techniques will be explored: in-circuit testing, bus structured design at the board level, and signature analysis. All three are predicated on the same assumption, that the designer is not able to change the internals of a module that is brought off the shelf in standard available sets. Furthermore, in general, the logic models for these networks are not available, especially logic models for microprocessor designs. Even if some of these microprocessor logic models were available, they would be, in most cases, sequentially so complex that to do test generation, even if it were manual test generation or fault simulation, would be far too expensive in terms of computer resources. Therefore, some techniques must be put in place to solve this collection of problems.

In-Circuit Testing

The in-circuit testing technique has become very popular and is still actively used. Basically, the technique makes the assumption that the components that are being put onto a board have just been tested by incoming inspection or some in-house testing. As these components have been tested and thus are assumed good at the board level, all that needs to be done is to make sure that the manufacturing process was carried out completely. The basic objective is to get rid of solder splashes, which could be manifested as shorts, or opens, where the soldering did not take effect correctly, or a component is put into the wrong place.

The mechanism for in-circuit testing is as follows: Basically, the board is powered up so that all components are receiving their power. Next, a node is probed between two modules on the board, as shown in Figure 14-1. If the tester can drive enough current into this node that is being contacted between the primary input of one module and the primary output of another module, the node can be driven up to a plus level, that is, not drawing any current from $+V_{cc}$, so that the primary input of the module to the right will perceive that net as being the logic one. This is called overdrive.

The purpose of the overdrive function is to give external control over this net so that the tester now has control of the primary input of the module on the right. Similarly, if the tester has dotted into the output of the module, it can observe the results of this module to input stimulus (assuming that no other device is dotted into it). Thus, a module can be tested while it is on the board as if it were an independent module whose inputs and outputs the tester controls. The drawback is that when one uses overdrive, there is more than the normal amount of current flowing through devices, which causes junction heating. As a result, reliability problems can be incurred. Thus, the number of patterns— that is, the length of time that overdrive can be used for a module—is usually restricted an approximate millisecond in duration. This has implications on the

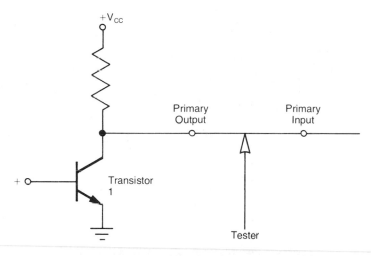

Figure 14-1.   In-circuit probing of net between two modules.

test length that can be applied to a given module, especially if the module is a microprocessor. Clearly, all test patterns necessary to test the internal devices of this model cannot be applied, but a sufficient test-set can be applied to make sure that all the primary inputs and primary outputs are not stuck, and to give reasonable confidence that the correct module is in the correct place—in the knowledge, of course, that these modules have been tested just prior to insertion.

Bus Structured Approach

The next ad hoc method is the bus structured approach at the board level. Microprocessors are designed with a bus-structured architecture (i.e., a data bus and an address bus play an essential role in the operation of the microprocessor), and this structure is usually preserved at the board level, as shown in Figure 14-2. The access this provides to the primary inputs and primary outputs associated with the bus and with other units on the board facilitates testing of the board. However, much attention is paid to testing the microprocessor on the micro-computer board, and little attention seems to be directed to the support chips, when, in fact, the complexity of some of the support chips used in microcomputer boards is equal to or greater than that of microprocessors themselves. Of course, the testing function at the board level is predicated again on functional rather than structural patterns because the logic models for the microprocessor and, most likely, the support chips *are not known*.

Without the bus structure at the board level, testing of the microcomputer board would be extremely difficult. Here use of the ad hoc approach would not solve the testing problem.

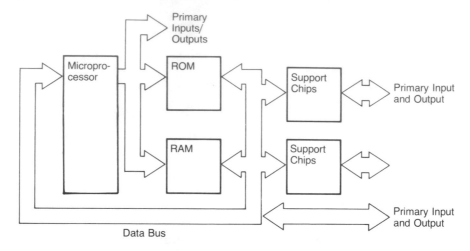

Figure 14-2. Bus structured microcomputer board.

The bus-structured architecture is a form of partitioning—a way of isolating one module from another—so that testing can be done. For example, the RAM and the ROS can be tested—which, had they been embedded in the logic, would have been very difficult to test. That is, it would be necessary to go through the address bus, as well as to be able to read and write any kind of data values into the RAM. This technique takes a number of control lines, and faults on the bus are difficult to isolate.

Signature Analysis

The signature analysis approach is an unusual ad hoc technique, in that the technique really requires planning at the board level. Care must be taken to follow a set of design rules at the board level so that the board can be tested is best suited to bus-structured architectures, as mentioned before, and, in particular, those associated with microprocessors. This will become more apparent shortly.

The core of the signature analysis approach is to use linear feedback shift registers, which are incorporated into the test equipment and are not part of the design. Figure 14-3 shows a 4-bit linear feedback shift register.

The shift register is composed of polarity hole latches, labeled L1's and L2's. The L1 latches are controlled by the Shift A clock, and the L2 latches are controlled by the Shift B clock. The output of the third and fourth shift register, S_3 and S_4, respectively, will be exclusive OR'ed and fed around to the input of the first shift register on the left. The term "linear" is used because exclusive OR's are used which preserve linear operations, and "feedback" refers to the

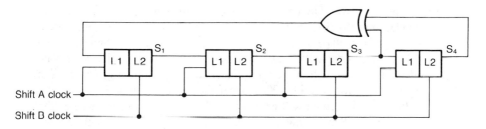

Figure 14-3. Linear feedback shift register.

fact that outputs of shift registers are fed back to the input of the shift register; hence, the name, linear feedback shift register.

Thus, to shift this linear feedback shift register, an A clock is turned on and off, followed by a B clock being turned on and off, and each bit in the shift register will move to the right one position, and the exclusive OR of S_3 and S_4 will now be stored in L1 and L2 latches of the first shift register.

A shift register with a single clock could be used, but for simplicity a double-clock shift register will be described here. Figure 14-4 shows the working mechanism of the linear feedback shift register shown in Figure 4-3. If the initial value for the four shift register positions happens to be all zeros, Figure 14-4(a), then when the A and B clocks are applied, the three bits in S_1, S_2, and S_3 will not appear in S_2, S_3, and S_4, respectively; that is, they will still be all zeros.

	S_1	S_2	S_3	S_4
If initial value =	0	0	0	0

(a) Absorbing state

	S_1	S_2	S_3	S_4
If initial value =	0	0	0	1
	1	0	0	0
	0	1	0	0
	0	0	1	0
	1	0	0	1
	1	1	0	0
	0	1	1	0
	1	0	1	1
	0	1	0	1
	1	0	1	0
	1	1	0	1
	1	1	1	0
	1	1	1	1
	0	1	1	1
	0	0	1	1

$$2^n - 1 = 2^4 - 1 = 15$$

(b) Count of fifteen which is maximal for a four bit linear feedback shift register.

Figure 14-4. Counting capabilities of a linear feedback shift register.

However, the exclusive OR of zero with zero will yield another zero, which will be loaded into the first shift register position, S_1, and this one will take on the new value of zero. The conclusion is that when this particular value is loaded into the shift register, it will always stay there; that is, it is an absorbing value or an absorbing state. If, however, the linear feedback shift register is started at any other value—for example, 0001—then the count shown in Figure 14-4(b) will be followed. This count will be a count of 15. *Note:* After the values 0011 are obtained for S_1, S_2, S_3, and S_4, respectively, the next value will be 0001, which is exactly the value that was used in the beginning. This means that this linear feedback shift register will count in a somewhat strange way, by a count of 15. This counter is a maximal-length linear feedback shift register. It cannot count to any value higher than 15 with four shift register positions and using linear elements. The count is shown to be $2n - 1$, with n being the number of shift registers. In this case, there are four, so that number comes out to be $16 - 1$, which is equal to 15. If a maximal-length linear feedback shift register is desired for longer than four bits, a table can be consulted that will tell the configuration of the exclusive OR bits. It may take more than one exclusive OR gate to contain a maximal-length linear feedback shift register, and usually many choices can be made to implement a linear feedback shift register of maximal length.

The signature analysis register differs from the linear feedback shift register in that an extra exclusive OR is used, as shown in Figure 14-5. Thus, if a stream of bits, Y, can be sampled synchronously with the shifting of the linear feedback shift register, some very interesting results can be obtained.

For example, let's assume that the initial value in the linear feedback shift register is 0001, and that the stream of bits going from right to left for the vector Y is basically 1011110. If these seven bits are sampled synchronously with the shifting of the shift register, the results shown in Figure 14-5 will occur. After the first shift, the pattern will be 1000; after the second shift, it will be 1100

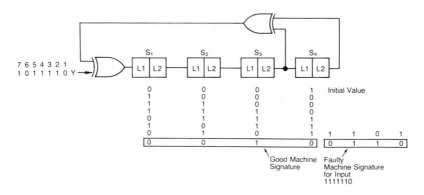

Figure 14-5. Signature analysis register.

(remember that a shift basically occurs with the result of the A and B clocks, as shown in Figure 14-3). Note that at the end of the seventh shift, the values stored in the four bits of the shift register are 0010 for S_1 through S_4, respectively. This value is called the signature. It is the signature associated with that given initial value in the shift register and that particular stream of bits 1 through 7, in exactly that order. If the sixth bit in the stream of bits coming into the Y input was not a zero and was a one, then the output sequence after the seventh bit has been shifted in would equal 0110, as shown in Figure 14-5. This would be a signature that one would obtain if this were a faulty machine response. This would be known as the faulty machine signature, and the signature 0010 would be the good machine signature. Thus, one can see how this technique can be used to distinguish the good machine response from the faulty machine response. The question is: how accurate is it? The probability of a stream of vectors coming in and having the exact same good machine signature is approximately $\frac{1}{2}n$ where n is equal to the number of shift register positions. By using a 16-bit linear feedback shift register, the probabilities can be made extremely small. (There is considerable controversy about how to calculate the exact probability, and to date this has not been done because the error statistics are not well known.)

The environment in which the signature analysis tool is used is shown in Figure 14-6. The signature analysis tool is external to the board under test; it is part of the test equipment.

First, the board is initialized so that all values of latches on the board are fixed to given values. If there were any X values—that is, unknown values—then the signatures could not be collected properly because X exclusive OR'ed with anything is X. The result would soon be all X's.

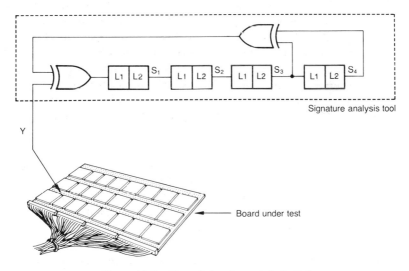

Figure 14-6.   Use of signature analysis tool.

Once the board is initialized, the signature analysis tool is initialized as well through a human value, just as in Figure 14-5 the shift register was initialized to 0001. Once the board and signature analysis tool have been initialized, clocking can begin.

Typically, there is a microprocessor on the board. The microprocessor will start to execute a program that is stored in ROM on the board or, possibly, even on the test equipment. The microprocessor will then go through this set of patterns, which will stimulate the entire board. With each clock pulse that is applied to the microprocessor, and A–B shift pair of clocks will be applied to the signature analysis tool to cause a sampling of the inputs from the probe, which will constitute the stream of Y bits as in Figure 14-5.

A fixed number of clock pulses will be applied to the board and to the linear feedback shift register. Let's assume that the number is 40. At the end of 40 clock pulses, assuming that we begin with a golden board (i.e., a perfectly good board), the results that remain in the linear feedback shift register latches will be the good machine signature for that given node. This process is repeated over and over again so that all the signatures for every node on the board for precisely 40 patterns are known. With the next board to be tested, the board will be initialized, the signature analysis tool will be initialized, and the sample will be taken on a given node. If that node's output after 40 clock pulses leaves a signature different from the good machine's signature in the signature analysis tool, then that module's output is bad, and all its inputs will be checked for proper signatures. If proper signatures are obtained, then the defect is likely contained within that module or, possibly, some soldering or other defects associated with that module. If some input on that module is defective, it is further backtraced to its inputs, and the next module is checked to determine if it has good inputs with bad outputs, to see if it is defective. This process is continued until a module is obtained with a bad output and all good inputs, or one returns to the same point from which he/she initially started. If one came back to the same node—that is, followed a loop of, say, three modules—there could be no diagnosis of where the actual fault occurred. The problem is that these three modules are in a closed loop. Thus we arrive at more of the design rules that are necessary for signature analysis at the board level to be a good diagnostic tool. These rules are as follows:

1. The board must have some way of being stimulated, either by the tester or self-stimulated by the use of a microprocessor.
2. The board must be initializable—that is, you cannot have a situation where there are many unknown values in latches, as this would give rise to X's. Once an X got into the linear feedback shift register, the results would be unknown.
3. The board needs to be strongly clocked—that is, it cannot be an asynchronous network. If it were asynchronous, then synchronization of that network with a linear feedback shift register would be virtually impossible.

4. All global feedbacks on the board must be broken to achieve good diagnosis. If this latter rule is not followed, then the diagnostic resolution will be degraded.

All the design for testability techniques mentioned have a given overhead, including this one. Certainly, there is some overhead for the ROM space to execute the self-stimulated tests. Furthermore, there may be some overhead to break feedback loops, and there may be some overhead to make the design strongly clocked rather than asynchronous. Once this has been paid, then the testing procedure is fairly straightforward.

It is also important to note that the test patterns for this type of testing technique are essentially functional test patterns; structural test patterns based on stuck faults are not available by virtue of the fact that the logic models are not available for these networks. Signature analysis may be the best that can be done for this particular board with the given inputs that the designer has.

STRUCTURED DESIGN TECHNIQUES

As integration levels increased in the late 1960s, test generation—that is, automatic test generation and fault simulation for sequential networks—became more and more costly. As a result, other techniques were looked into, such as faster test generation algorithms and faster fault simulation algorithms. No matter what was looked at, the end result showed that something had to be done to the designs in order to make them testable. This gave birth to the idea of design for testability: changing the logic design at the chip level to have a positive influence on the testing of not only the chip but subsequent packaging levels as well.

The concept that seemed to be employed by most people involved is a well-known one, predicated from the structure that was put forth by Huffman in 1954. He basically stated that any sequential network can be represented by a combinational logic network with primary inputs and primary outputs, and memory elements fed by the combinational logic whose outputs feed back into the combinational logic itself. However, these elements could be memory elements, or they could be delays or a number of different descriptions. This form is shown in Figure 14-7.

Given this structure of a sequential machine, if one wanted to reduce the test generation problem to one of testing combinational logic rather than sequential logic, as combinational logic is far more straightforward than sequential logic, all one must do is be able to control and observe these memory elements, whatever they may be. If one can do this, then the network shown in Figure 14-8 is all that needs to be used where test generation and fault simulation are concerned. The latch elements (or memory elements) can be treated as pseudoprimary inputs and pseudoprimary outputs. The next question is: what is the best way to control and observe these latch elements?

One possibility would be to cut all the feedback loops feeding from the memory

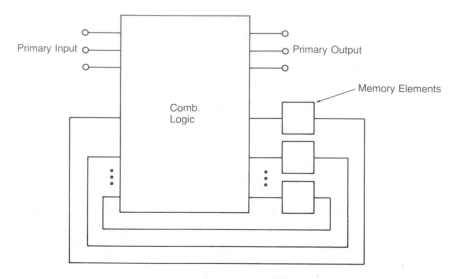

Figure 14-7. Basic structure of the Huffman model.

elements back to the combinational logic, thus having direct access to the board and subsequent packaging levels. Clearly, this is not a viable technique because it would take many I/Os and, in particular, two pins for each memory element in the design. The most valuable resource in VLSI is pins, so this is too costly, and thus unsatisfactory.

Another approach is to treat the memory elements as a RAM, addressing the RAM, reading and writing it. This would give controllability and observability, but the system path could not be that cumbersome. (An approach that does use this technique is called Random Access Scan; it will be discussed shortly.)

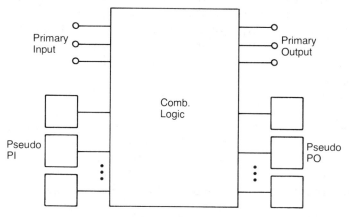

Figure 14-8. Transformation of the Huffman model where memory elements are controllable and observable.

Another approach to this problem would be to control the memory elements by having a shift register function—some way of shifting values into the memory elements and shifting them out once you have captured some results in them. This technique is fundamentally taken from the scan concepts, of which there are several. NEC, or Nippon Electric Company, Ltd., uses a technique called Scan Path; Sperry Corporation uses one called Scan Set; and IBM uses one called LSSD, just to mention a few. Since the announcement of these particular design techniques was made in the mid- to late 1970s, many different variations of these techniques have been proposed. The LSSD technique will be reviewed here, along with Random Access Scan. This will give a good overview of the technique, particularly for LSSD, where some design examples will also be given to show some idea of how these techniques can be used in logic design.

Along with these structured design approaches, there are: a design automation system that will analyze the designs to ensure that they follow the design rules; programs to automatically generate the test patterns; programs to fault simulate and verify that the test patterns do, in fact, cover the assumed fault set; and a program to create the data file, which is consumable by a tester. In essence, then, the structured design approach is used for design for testability; it is really a design methodology, and there is a design automation system to support it so that the overall process will be very fast and cost-effective.

Level Sensitive Scan Design, LSSD

This section will describe the LSSD approach to designing sequential logic. We will begin with definitions of level sensitive design, and then the scan approach, followed by a set of design rules, as well as some examples.

There are two basic concepts in LSSD, and the design rules reflect this. The first concept is that the network operates correctly and is not dependent on rise time, fall time, or minimum delay time, on the individual logic gates. The only dependency is that the total delay through a number of levels should be less than some known value. This method is called level sensitive design.

The second concept is that all internal storage elements other than memories, ROMs, and RAMs are implemented as part of shift registers. Given that they are part of shift registers, their controllability and observability are straightforward via the shifting mechanism.

The definition for level sensitive is as follows:

A network is level sensitive if and only if the steady-state response to any of the allowed input changes is independent of the gate and wire delays in that subnetwork.

In essence, this definition means that certain pins are allowed to change at certain times. The ones that are most critical in level sensitive design are the clocks, to be discussed shortly. Other primary input signals are not constrained to the same degree as clocks.

Given that the objective is to have a level sensitive design, the latches that are used should also be level sensitive. Consequently, the basic storage element should not contain hazards or races. The hazard-free polarity latch shown in Figure 14-9 is an example of a latch whose latching characteristics are independent of the rise and fall times of the System Clock C. This attribute is obtained by Gate 3. This gate essentially implements a redundant term; that is, testing the output of Gate 3, Stuck At 1 is untestable in the DC sense. However, given that this fault does not exist, this latch will latch up irrespective of fall time of the System Clock C—that is, when C is going from a 1 to a 0.

Just for clarity, D is the system data port; C is the system clock input; and L is the latch output. As a matter of fact, except for polarity inversion problems, Gate 1 can either feed Gate 4 or Gate 2 with the appropriate change in polarity on the system clock. What this says is that this latch will work irrespective of the relative delays when the clock obtain arrives at Gates 2 and 4, or when they arrive at Gate 5. Also included in Figure 14-9 is an exitation table for the latch and the equation for the latch (which also apply for the hazard-free latch). Note that the term $1D$ is the term for the hazard-free portion of the latch and is redundant. Figure 14-10 shows a block or symbolic representation of a polarity hold shift register latch, SRL, along with a hazard-free implementation using NAND gates.

As was mentioned earlier, the clocks play a central role in the operation of the sequential elements in an LSSD environment. In particular, there are three clocks that enter this SRL: C, the system clock; A, a shift clock; and B, another shift clock. Only one clock will be activated at any one time.

The "off" values for the three clocks are zero. Let's assume that these three

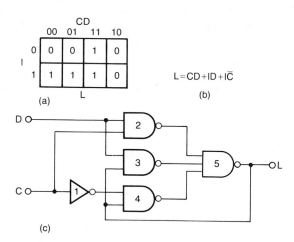

(a)

	CD			
I	00	01	11	10
0	0	0	1	0
1	1	1	1	0

L

(b)

$$L = CD + ID + I\overline{C}$$

(c)

Figure 14-9. Hazard-free polarity-hold latch. (a) Excitation table. (b) Exicitation equation. (c) Logic implementation.

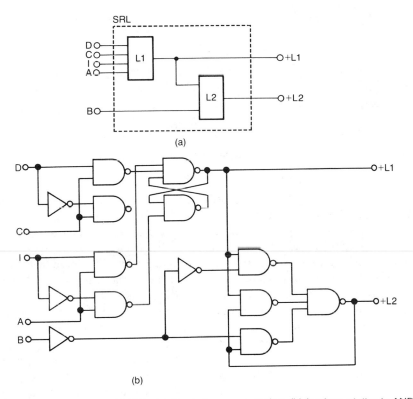

Figure 14-10.   Polarity-hold SRL. (a) Symbolic representation. (b) Implementation in AND-INVERT gates.

points, C, A and B, are primary inputs and can be controlled directly. Furthermore, let's assume that all clocks are at the "off" value, which is all zeros. In a system operation, the C clock will be turned on and off—that is, C will go zero to one, stay at the up level for greater than the minimum latch-up time for the latch, and then return to zero. It is assumed that the data input will be level during this period. This will be discussed in greater detail later on in the chapter.

The data value is then captured in the L1 latch. Its value is now observable on the +L1 output of the SRL. That +L1 is also available to the second latch, the L2 latch. Once the C clock is off, the B clock input can be turned on and off. Thus, the value that was loaded into the L1 can now be loaded into the L2 latch. This would be the order of operation if one were using a master/slave kind of design: the L1 latch would be the master; the L2 latch would be the slave. This is system function.

Let's now take a look at the shift or scanning function. The scanning function works in the following way: The I value would come from a primary input that would be a scan-in primary input. The A clock would be turned on and off,

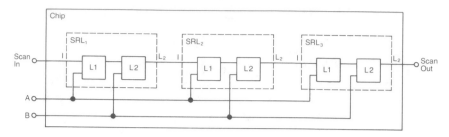

Figure 14-11.   Interconnection of SRLs at the chip level.

again staying on long enough to latch up the value—that is, greater than the minimum latch-up time in the L1 latch. Once the A clock was off and the value was suitably latched up in the L1 latch, the B clock could then be turned on and off. So whatever value was loaded into the L1 latch with the A clock would now be loaded into the L1 latch, as well as the L2 latch. The value that was on the scan-in primary input would now be available on the scan-out or L2 output of the latch. If there were more than one SRL in the system design, then that L2 would be connected to the next I input of the next latch. The choice of order is arbitrary; the closest one on the physical design would be adequate.

An example of establishing a scan path for three SRLs at the chip level is given in Figure 14-11. The figure shows a scan input on the left, a scan output on the right, and A and B clocks controlling the L1 and L2 latches, respectively. The designer is responsible for connecting the scan input to the I input of the L1 and then the L2 of the first SRL to the I input of the second SRL, and so on, to complete this scan path. Again, the ordering of the SRLs is arbitrary. At this packaging level, only four pins are necessary—three primary inputs and one primary output—the object being to use as few pins as possible and clearly use some silicon area for the gates to form the shift register latches.

There are two main aspects of the LSSD design system: the level sensitive design and the scan capability. The following rules are necessary to support this structure. This, however, is not the only set of rules; there probably could be another set or other structures.

Rules for Level Sensitive Design.

1. All internal storage elements must be polarity hold, hazard-free latches.
2. Latches can be controlled by two or more nonoverlapping clocks so that:
   I) A Latch X may feed the data port of another Latch Y if and only if the clock that sets the data into Latch Y is not clocked by Latch X.
   II) A Latch X may gate a Clock C_I to produce a gated Clock C_{IG} that drives another Latch Y if and only if the Clock C_{IG} is not clocked to Latch X, where C_{IG} is any clock produced from C_I.

3. It must be possible to identify a set of clocked primary inputs from which the clocked input SRLs are controlled, either through a simple powering tree or through logic that is gated by SRLs and/or nonclocked primary inputs. In addition, the following rules must hold:

    I)   All clocked inputs to all SRLs must be held at their "off" state when all clocked primary inputs are held to their "off" state.

    II)   The clocked signal that appears at any clocked input of an SRL must be controlled from one or more clocked PIs so that it is possible to set the clocked input or the SRL to the "on" state by turning any one of the corresponding PIs to its "on" state and by setting the required gating conditions from SRLs and/or nonclocked PIs.

    III)   No clock can be ANDed with either the true value or the complement value of another clock.

4. Clocked primary inputs may not feed the data inputs to latches, either directly or through combinational logic, but may feed only the clocked inputs to the latches or the primary output.

Any network that follows these first four rules will be considered to be level sensitive. The following rules are necessary in order to implement the scan capability, that is, the ability to shift values into all the latches and shift them out.

5. All system latches are implemented as part of an SRL. All SRLs must be interconnected to form one or more shift registers, each of which has an input, an output, and shift clocks available at the terminals of the package.

6. Some primary input sensitizing condition, referred to as a scan state, must exist so that:

    I)   Each SRL or scan-out PO is a function of only the single receiving SRL or scan-in PI and its shift register during the shifting operation.

    II)   All clocks, except the shift clocks, are kept "off" at the SRL inputs.

    III)   Any shift clock to the SRL may be turned on or off by changing the corresponding clocked primary input for each clock.

A network that follows all the rules stated above will be an LSSD network that is level sensitive and has scan capability. Given these two attributes, the network will then be able to have tests generated for it so that there will be no races in the test patterns. Thus, the test patterns can be delivered to the network via the shifting mechanism and be observed by shifting out.

Let's look at the general structure, and then a specific example of a 3-bit binary counter designed in LSSD, something that we call double-latch design. Figure 14-12 shows the general form of a double-latch design, so named because the output of the L2 latch is used for the system output. ("Double-latch" implies going through the L1, then the L2, and then back out.) A more common term in general, unstructured design is "master/slave" design.

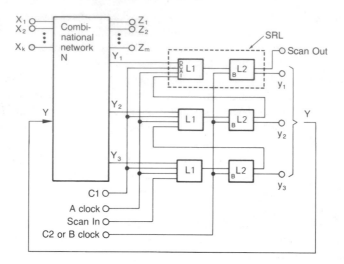

Figure 14-12. LSSD double-latch design.

The scan or shift capability is there, and there is an A clock, a B clock, and a C2 clock, which is a system clock. This C2 clock shares the B clock input, and is used as a system clock during the system portion of the operation and as a shift clock during the shifting operation. A scan path is established from a scan-in primary input to the scan-out primary output going through all the SRLs.

Y's are the next state, and y's are the present state. X_1 through X_k are the primary inputs, and Z_1 through Z_m are the primary outputs of the network.

Let's take the example of a 3-bit counter (which will involve more than a 3-bit counter, as will be seen shortly). Figure 14-13 has three parts: Figure 14-13(a) is the count that we would like and three states to go through—that is, it

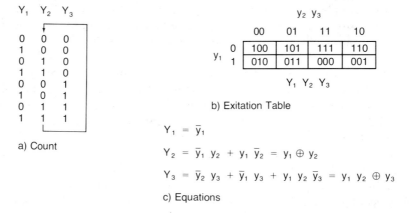

Figure 14-13. Equations for 3-bit binary counter. (a) Count. (b) Exitation table. (c) Equations.

is a binary counter that counts from one to eight. The exitation table for the binary counter is shown in Figure 14-13(b). Remember that y is the present state, and Y is the next state we want to go to.

For example, if we were at Y_1, Y_2, and $Y_3 = 001$, we would then like to go to 101, if we notice that the table entry for the point 001 is 101. This tells you what the next value should be.

From the exitation table, one can drive the equations given in Figure 14-13(c). This is not to imply that, when designing in LSSD, Scan Path, or any of these other techniques, such a formal technique has to be employed. This is just a

$$Y_1 = \bar{y}_1 X_1$$
$$Y_2 = (y_1 \oplus y_2) X_1$$
$$Y_3 = (y_1 y_2 \oplus y_3) X_2$$
$$Z_1 = y_1 y_2 y_3 X_2$$

(a) Functional Equations

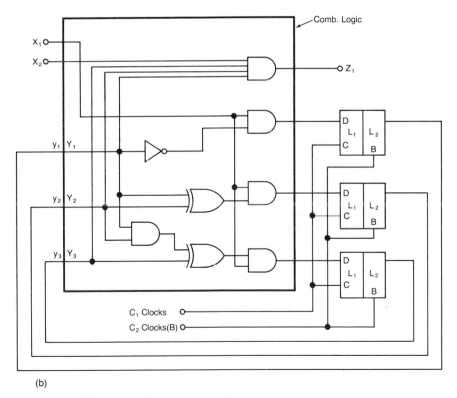

(b)

Figure 14-14.   LSSD 3-bit binary counter. (a) Functional equations. (b) Network implementation.

mechanism to help one through it for the first time; then, normal types of sequential design techniques can be employed, as they usually are.

The equations for Y_1, Y_2, and Y_3 are given in Figure 14-13(c).

At this point, let's add to the network to make it more credible. We would like to find a way to start the counter at the 000 position; thus, we will use a primary input called X_1. When X_1 is equal to zero, we get the C1 clock followed by the C2 clock. The system latches will be at the zero state and will be in the system mode of operation. Note that the A clock is off.

The output required for the network, Z_1, will equal one for the eighth cycle after the latches have been reset. This output will be gated with X_2 of the primary input so that during the reset process or prior to it, no spurious pulses will appear on the Z_1 output. Figure 14-14(b) shows the network in the same structure as Figure 14-12: the combinational logic with primary inputs X_1, X_2; primary output Z_1; latch inputs Y_1, Y_2 and Y_3; SRLs sitting on the right-hand side; and the combinational logic function implementing the functions shown in Figure 14-14(a).

Figure 14-15 shows the timing diagrams. Note that C1 and C2 are the system clocks. These reoccur periodically and in the same relationship to one another. Furthermore, they are not active at the same time. This is to eliminate any races between the latches. X1, X2, both equal to zero when the System Clock C1, C2 comes along, will initialize the latches, and the output C1 will still be at zero.

After C2 goes back to zero, the X1, X2 lines can then be changed to one. Once they are changed to one, which is sufficient delay time, a C1, C2 clocks can occur. With this clock cycle, the system latches could go to state 100 and will then continue the count, as represented in Figure 14-13(a). Once the count 111 is obtained, the output Z1 will go to one for one cycle.

The output Z1 will change to one sometime after the C2 clock goes active for that particular cycle. That is why there is a shaded area, or an X region or unknown value. The output Z1 will stay one until the counter goes back to a different state. If it is allowed to continue to count without any intervention of X1, it will then immediately go back to 000, and the count will start again. This

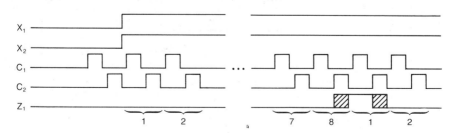

Figure 14-15. Timing diagram for LSSD 3-bit binary counter.

is an example of an LSSD implementation of a simple binary counter using double-latch design.

Note that if one wanted to test the output of Stuck At Zero, one essentially would have to functionally count through a count of eight until the output achieved the proper value.

Looking at the combinational network whose primary output is Z1 and whose other three outputs are Y_1, Y_2, and Y_3, which is the next state for the latches, assume that we would like to test Z1, the output Stuck At Zero. In order to test this, we need y_1, y_2, and y_3 all equal to one and X2 equal to one also. Given that state and that primary input, Z1 in the good machine will go to one.

Furthermore, if we hold X1 equal to zero (X1 being the primary input) and we sample with the System Clock C1 into the L1 latch, we can tell whether the value 000 gets latched into the three L1 latches. This will test the three outputs of the combinational network for Stuck At 1 values.

How would this test be applied in an LSSD environment? Well, to begin, the values of y_1, y_2, and y_3 of 111 will be loaded into the three SRLs via the shift mechanism. Essentially, the scan-in primary input will be held to a one, the C1 clock will be off, the A clock will go on and off, and the B clock (which is the C2 input as well) will be turned on and off. This will load the lower SRL to a one; the upper two will still be at unknown values. We then use another AB clock here, and the result will be that both the lower two SRLs will now be loaded to one, and if we use a third AB clock, all the SRLs will be loaded to one. This is precisely the value that we need.

Now, X2 is held to a one, and X1 is held to a zero. That gives all the inputs on the left-hand side of the combinational network the values that are required for the test.

The output Z1 will now be observed. Z1 is now equal to a one in the good machine. Furthermore, the Y_1, Y_2, Y_3 will be equal to 000 in the good machine. This value cannot be observed directly on a primary output. The way these three values are observed is by now turning the System Clock C1 on and off. The result is that all three zeros are now loaded into the L1 latches of the three SRLs. This is followed by a B clock. Both the L1 and L2 contain these three values of zero.

To observe them, three pairs of AB clocks are applied with the C1 clock held off. Thus, if we use AB/AB/AB, the end result is that all three values have been observable on the scan-out primary output. Thus, in essence, we have been able to observe the values that were sitting on the inputs of the SRLs via the shifting mechanism, and, hence, we have completed one test pattern cycle for an LSSD design.

The network shown in Figure 14-14, the 3-bit binary counter, is a double-latch design. By that, it is meant that the data goes into the L1 latch, then into the L2 latch, and then is fed back around into the combinational logic. This can also be seen in Figure 14-12, for LSSD double-latch design. Again, the key point is that the data passes first through the L1 latches and then through the L2 latches. The L1 latches are all clocked by C1; the L2 latches are all clocked by C2. Therefore, there is no violation of Rule 2—in particular, Rule 2I. What that rule is basically saying is that a latch cannot feed another latch that is clocked by exactly the same clock because this would cause a race condition. Similarly, Rule 2I would also cause a race condition with a latch gating a clock, and if that clock is the same that clocks the original latch, another race can ensue, which is what 2I is trying to prevent.

Another way of designing that does not violate Rule 2I uses only the L1 latch—that is, a single-latch design. In this kind of design structure, the data, as shown in Figure 14-16, passes through an L1 latch clocked by C1, and then goes from the output of the L1 latch back into another combinational logic network, into the data port of another L1 latch—but these L1 latches are now clocked by C2, a different system clock. The reason why this is a different system clock is so that the race condition will again be prevented, as specified by Rule 2I. The output of the L1 latches clocked by C2 can then feed back around into the combinational logic. Thus, every closed loop in this network will have two latches in it, each of which is clocked by a different clock. In this situation, a race does not exist as long as enough time is allowed for the data to propagate through the combinational logic network, and the clocks are active for the minimum latch-up time for the L1 latches.

The shift path is shown for this design as well. However, we must note that all the L2 latches do not participate in any system function in this kind of design. Therefore, there is no system use in terms of function for these L2 latches. Hence, single-latch design is the most expensive way to design an LSSD network.

Some nonfunctional system use for reliability or serviceability can be made

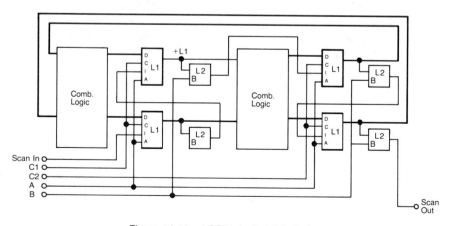

Figure 14-16. LSSD single-latch design.

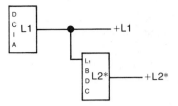

Figure 14-17. L2* SRL.

of the L2 latches. This is called the VLSI oscilloscope because if the B clock is pulsed at any time, a snapshot of the logic values contained in all four of the L1 latches will be obtained. Then, the system clocks (i.e., C1/C2) can be stopped at one's leisure, and the values in the L2 latches can be scanned out by beginning the shifting operation with an A clock rather than a B clock. Thus, the pattern would then be A/B-A/B-A/B-A/B.

This feature could also be used for a checkpoint restart. That is, the scan output could be connected back to the scan input. If the shifting operation started with an A clock, then the four values that are loaded into the L2 latches could be reestablished and loaded back into the L1 latches. In this way, the prior state of the machine could then be established.

Except for these types of operations, the L2 latch does not participate in any logical function within the two combinational logic networks, as shown in Figure 14-16; so this is a rather expensive way of designing. However, some people feel very comfortable in designing a single-latch operation, and prefer this type of design to a double-latch design, even though the performance characteristics of a double-latch design are as good as the performance characteristics of a single-latch design.

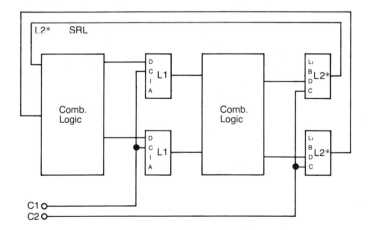

Figure 14-18. Single-latch design with L2* (system function).

As a result of this desire by some designers to design with a single-latch structure, an extension to the original LSSD concept is called the L2* technique. An L2* SRL is shown in Figure 14-17. In essence, it differs from a standard SRL only by having a system clock and a system data port that goes into the L2 latch directly. To distinguish it from a normal L2, it is called an L2* (Das Gupta et al. 1982).

The L1 and L2* latches can be used as independent latches within the design. Figures 14-18 and 14-19 show an example of the single-latch design, as given in Figure 14-16, when implemented with L2* latches.

Figure 14-18 shows the system clocks in the system data paths, and Figure 14-19 shows the scan path. Clearly, the scan path connection is more complicated in the L2* design, but a one-for-one trade-off must be made—one trade-off for wiring and one for the silicon area.

The only rule that must be observed very carefully in an L2* star design is that an L1 and an L2* of the same SRL may not feed the same combinational logic network. The definition of "same SRL" is that L1 is associated with an L2* if the L1 feeds the scan-in port of that L2*; thus, those two latches constitute an SRL.

If they were to feed the same combinational logic network, there is a high likelihood that a pattern would be required where the L1 has a value of one, and the L2 is required to have a value of zero. If scanning is ended with a B clock, this cannot be obtained. Even if scanning stopped with an A clock, a number of counter-examples can be shown that would lead to untestable logic as well.

As a result, it is important to observe the rule stating that the L1 and L2* of the same SRL should not feed the same combinational logic network. If this rule is violated, a lack of fault coverage may occur. It should also be clear that this same rule needs to be followed in a design that mixes both single- and

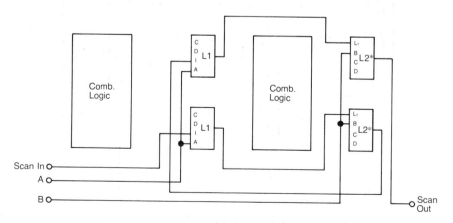

Figure 14-19.   Single-latch design with L2* (scan function).

double-latch designs. With a little imagination, this can be done, and one can see how L1 outputs are used in some places and L2 outputs in others. However, this still does not violate Rule 2I or Rule 2II. In this type of design, it is also important to ensure that the L1 and L2 of the same SRL do not feed the same combinational logic network; if they do, a lack of fault coverage may again occur.

What is the advantage of using this L2* system? From looking at Figures 14-18 and 14-19, it should be clear that there is a significant savings by virtue of the fact that the L2 latch is now put into the system function. If you compare the design in Figure 14-16 with the design in Figure 14-18, there are eight latches in Figure 14-16, but four in Figure 14-18. Essentially, all the overhead associated with the L2's in Figure 14-16 has been eliminated. This design reduces the overhead by a factor of four to one. That is, if the overhead for the double-latch design were 30%, the overhead for the design of Figures 14-18/14-19 would be 7.5%. This is clearly a significant savings that should and will receive considerable attention in the future.

A number of other design techniques are very similar in concept to the LSSD approach. One is the Nippon Electronics Corporation Ltd. approach called Scan Path, another is the Sperry Computers approach called Scan Set, and there are a myriad of others. Whether the approach is LSSD or Scan Path, the basic concept is really important—the fact that controllability and observability for system latches are being used via a shift register function.

The next approach we will discuss is to control and observe the system latches as if they were memory elements rather than a shifting operation. This technique is called Random Access Scan.

Random Access Scan

The Random Access Scan approach differs from LSSD, as mentioned above, by virtue of the fact that it treats the system latches as if they were memory elements. Each memory element is uniquely addressable so that it can be observed and controlled. Figure 14-20 shows the general structure of such an implementation. The latches are configured in an array in the box of Addressable Storage Elements, which are addressed by some Y-decoders and X-decoders. There is a Scan Data In, and there is a single point that can be fanned out to all the latches. All these latches will then funnel down into a single SDO (Scan Data Out). Clock lines are also used, which will be explained shortly.

The basic idea is to address each latch, put the proper value on the Scan Data In and clock it, and that latch will be loaded. If one wants to observe the values in the latches, one must address the latch, and then that value will be observed on the SDO or Scan Data Out line. If some data is to be captured from the combinational logic, a system clock is applied, and turned on and off; then a particular latch is addressed, and its value is on the Scan Data Out line. If a

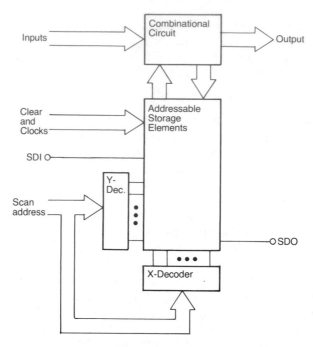

Figure 14-20.   Random Access Scan in/out network.

particular test setup is required, this is again done by addressing the latch, by putting the proper value on the SDI line, hitting a system clock, and thus initializing that latch for the proper value. From there, when testing the combinational logic, we will have the same scan design as before.

Figure 14-21 shows the general latch; this is a single latch, and there are no double latches, except that the latch has two ports into it. One is the system data, which is clocked by CK, and the other is the Scan Data In point, which

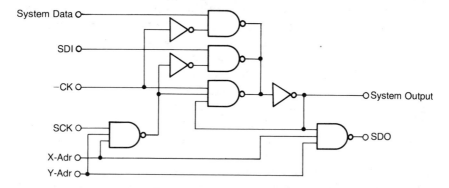

Figure 14-21.   Polarity-hold type addressable latch.

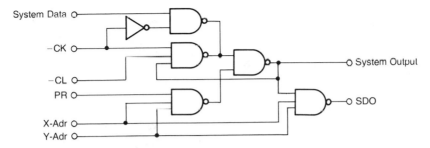

Figure 14-22.   Set/reset type addressable latch.

is clocked by SCK. Where SCK is the Scan In clock, that clock can only affect this particular latch if the proper X and Y addresses are sensitized using the addressing mechanism. Once this is sensitized and the SCK clock is hit, the SDI value is loaded into that particular latch. If one wants to observe that latch, it must be addressed with the proper X/Y address so that it can be observable on SDO. Essentially, all the latches for SDOs are dotted together in one huge dot so that only the net that is being addressed has control.

If one wished to reduce the time taken to set up all these latches for the proper value, a clear line could be used. This line could be forced to zero so that all latches would go to zero. Then only those latches that need to be changed to one must be addressed. This is shown in Figure 14-22.

The disadvantage with this type of design is that the number of I/Os needed at any packaging level is quite significant; you need the address decode lines, the X/Y lines (as shown in Figure 14-20), and the number of clock lines. Wirability could also be a problem because address lines must also go to each of the particular latches. Since this design technique needs more I/Os than would be required for the Scan Design, it is not very attractive. Inputs and outputs (I/Os) are of premium value in the VLSI designs.

This concludes the discussion on structured design approaches. Again, the basic concept of these approaches is to reduce the sequential test generation problem to one of test generation and fault simulation of combinational logic networks. This is achieved by having controllability and observability of all the latch elements in the system.

Because the size of the combinational logic networks that must have test generation is getting larger, interest in self-testing is increasing. This topic will be covered in the next section.

SELF-TESTING

There are three basic forms of self-testing in the area of design for testability: the microprocessor self-stimulated test, in-situ self-testing, and ex-situ self-testing. We will first discuss the microprocessor self-stimulated test at the board level.

Microprocessor Self-Stimulated Test

The microprocessor self-stimulated test basically consists of a microprocessor executing a section of code that may be stored in a ROM or a RAM on the board (See Figure 14-2), or a RAM on a test hardware fixture that is part of the tester. The result is that the microprocessor gives a test of the board logic, usually at or near machine speeds. At the termination of a segment of the test, the tester probes to measure some register in the network via a bus. If the register contains the correct value, the segment of the test has passed, and the board is ready for a new segment to continue the test. The completeness of this test is unknown, as there is usually no logic model to measure the quality of the test against. As a result, the testing of boards is a very iterative process. When a high fallout occurs at the next level of packaging, the reason has to be determined and a test added at the board level to eliminate the bad boards. This can be a very labor-intensive solution, but it may be the only successful one.

In-Situ Self-Testing

The second type of self-test is the in-situ self-test. Intrinsic to this approach is the structured design approach, mentioned above. This form, further, has the system latches participating in both generation of pseudorandom patterns and compressing of the results. The normal system latches are reconfigured into linear feedback shift registers, LFSR, as shown in Figure 14-23, usually comprised of eight or more latch positions. This LFSR can generate a pseudorandom pattern every shift clock cycle. It can also compress test results by "exclusive

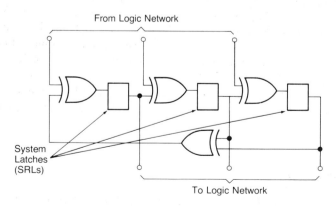

Figure 14-23. System latches for in-situ self-testing.

ORing" the output of the test network, bit by bit, with the contents of the LFSR. This is very similar to the concept used in cyclic reducing codes to compress large amounts of data, and, of course, the signature analysis approach.

The effectiveness of random patterns in an LSSD type environment was demonstrated by Williams and Eichelberger (1977), who concluded that random patterns are very effective at testing combinational logic in an LSSD environment as long as there are not some anomalies in the networks such as PLAs, which usually have high fan-in. An example of results of these types of patterns applied to an LSSD network is shown in Figure 14-24. This concept was also used in the STUMPS approach (discussed below).

A technique recently presented, takes the scan path concept and integrates it with the signature analysis concept, which is an in-situ self-testing approach. It is in situ by virtue of the fact that the system latches are used to generate the random patterns and also used to compress the results of the test. The result is a technique called Built-In Logic Block Observation (BILBO) (Koenemann et al., 1977). Figure 14-25 gives the form of a 3-bit BILBO register. The block labeled L_i ($i = 1, 2, 3$) is the system latches. B_1 and B_2 are control values for controlling the different functions that the BILBO register can perform. S_{IN} is the Scan In input to the 3-bit register, and S_{OUT} is the Scan Out output for the 3-bit register. Q_i ($i = 1, 2, 3$) are the output values for the three system latches. Z_i ($i = 1, 2, 3$) are the inputs from the combinational logic. The structure that this network will be embedded into will be discussed shortly.

There are three primary modes of operation for this register, as well as one

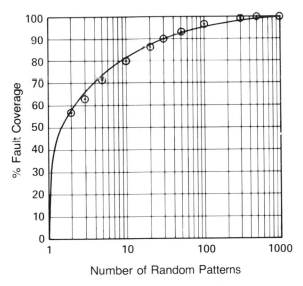

Figure 14-24. Combination logic in LSSD network, fault susceptibility with random patterns.

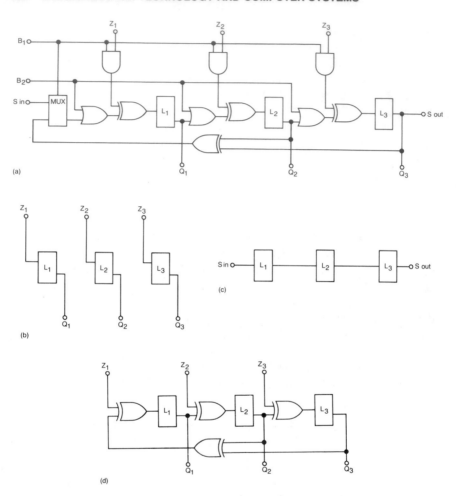

Figure 14-25.   BIBLO and its different modes. (a) General Form of BIBLO register. (b) $B_1B_2 = 11$, system orientation mode. (c) $B_1B_2 = 00$, linear shift-register mode. (d) $B_1B_2 = 10$, signature analysis register with multiple inputs (Z_1, Z_2, Z_3).

secondary mode of operation for it. The first is shown in Figure 14-25(b), that is, with B_1 and B_2 equal to 11. This is a basic system operation mode, in which the Z_i values are loaded into the L_i, and the outputs are available on Q_i for system operation. This would be the normal register function.

When B_1B_2 equals 00, the BILBO register takes on the form of a linear shift register, as shown in Figure 14-25(c). This is similar to Scan Path and LSSD.

The third mode occurs when B_1B_2 equals 10 (Figure 14-25d). In this mode, the BILBO register takes on the attributes of a linear feedback shift register of maximal length with multiple linear inputs. This is very similar to a signature analysis register, except that there is more than one input. In this situation, there are three unique inputs. Thus, after a certain number of shift clocks, say 10, there would be a unique signature left in the BILBO register for the good machine. This good machine signature could be off-loaded from the register by changing from mode $B_1B_2 = 10$ to mode $B_1B_2 = 00$, in which case a shift register operation would exist, and the signature then could be observed from the Scan Out primary output and compared to expected good machine results.

The fourth function that the BILBO register can perform is B_1B_2 equal to 01, which would force a reset on the register. (This is not depicted in Figure 14-25.)

The BILBO registers are used in the system operation, as shown in Figure 14-19. Basically, a BILBO register with combinational logic and another BILBO register with combinational logic as well as the output of the second combinational logic network can feed back into the input of the first BILBO register. The BILBO approach takes one other fact into account: in general, combinational logic is highly susceptible to random patterns. Thus, if the inputs to the BILBO register Z_1, Z_2, Z_3, can be controlled to fixed values, and the BILBO register is in the maximal length linear feedback shift register mode (signature analysis), it will output a sequence of patterns that are very close to random patterns. Thus, random patterns can be generated quite readily from this register. These sequences are called pseudorandom patterns (PN).

If, in the first operation, the BILBO register on the left in Figure 14-26 is used as the PN generator, the output of that BILBO register will be random patterns. This will then give a reasonable test, if sufficient numbers of patterns are applied, of the Combinational Logic Network 1. The results of this test can be stored in a signature analysis register approach, with multiple inputs to the BILBO register on the right. After a fixed number of patterns have been applied, the signature is scanned out of the BILBO register on the right for good machine compliance. If that is successfully completed, then the roles are reversed, and the BILBO register on the right will be used as a PN sequence generator; the BILBO register on the left will then be used as a signature analysis register with multiple inputs from Combinational Logic Network 2 (see Figure 14-27).

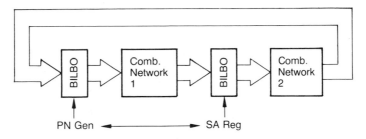

Figure 14-26.   Use of BILBO registers to test Combinational Network 1.

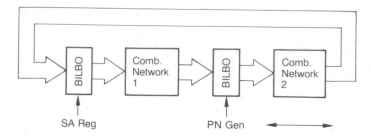

Figure 14-27.   Use of BILBO registers to test Combinational Network 2.

In this mode, the Combinational Logic Network 2 will have random patterns applied to its inputs and its outputs in the BILBO register on the far left. Thus, the testing of the combinational logic networks 1 and 2 can be completed at very high speeds by only applying the shift clocks, while the two BILBO registers are in the signature analysis mode. At the conclusion of the tests, off-loading of patterns can occur, and determination of good machine operation can be made.

This technique solves the problem of test generation and fault simulation if the combinational networks are susceptible to random patterns. There are some known networks that are not susceptible to random patterns, namely, programm-

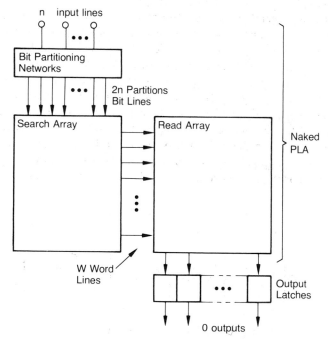

Figure 14-28.   PLA model.

able logic arrays (PLAs) (see Figure 14-28). The reason for this is that the fan-in in PLAs may be too large for a given number of random patterns. (This is only an example of problem networks.) If an AND gate in the search array had 20 inputs, then each random pattern would have $1/2^{20}$ probability of coming up with the correct input pattern to test the third input S-A-1. (An example of fault susceptibility for a PLA is shown in Figure 14-29.) On the other hand, random combinational logic networks with maximum fan-in of 4 can, in general, do quite well with random patterns.

The BILBO technique solves another problem, that of test data volume. In LSSD, Scan Path, Scan/Set, or Random Access Scan, a considerable amount of test data volume is involved with the shifting in and out. With LSSD, or the others, if P patterns are to be applied with maximum SRL length L, the time to load these patterns is proportional, $P \cdot L$. Furthermore, the scanning of patterns in and out is not done at machine speed because measurements must be taken at the scan primary output. Assume that the ratio of speed that the shift register could be shifted by the tester to the speed at which the tester must go in order to do all the forces and experts is K. The value of K is typically about 100–1000 to 1. If P patterns (pseudorandom) are to be applied to a BILBO type network, then after the BILBO registers are loaded, these patterns can be applied in time that is proportional to P/K. Thus, there is a difference of $K \cdot L$ for an LSSD pattern to be applied versus a pseudorandom pattern to be applied. With $L = 100$ and $K = 1000$, $K \cdot L$ would imply 10^5 difference in speed per pattern. Clearly, more pattern must be applied in the BILBO case, but it is believed that there is

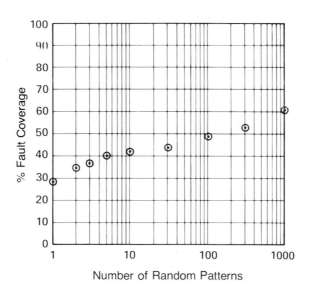

Figure 14-29. Testability vs. number of random patterns for PLA.

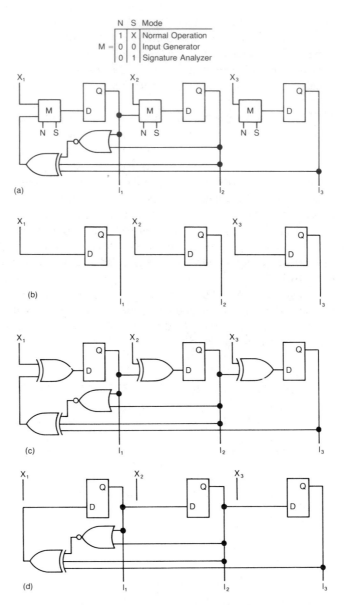

Figure 14-30. Reconfigurable 3-bit LFSR module. (a) General network. (b) N = 1: normal operation. (c) N = 0, S = 1: signature analyzer. (d) N = 0, S = 0: Input generator.

not five orders of magnitude of difference. This difference in speed directly relates to the tester time and thus the cost of the test. Consequently, the test application is much less for BILBO, and the test data volume is also significantly smaller than for the other techniques.

The overhead for BILBO is higher than for LSSD or the other scan techniques because at least one exclusive-or must be used per shift register latch position. There could also be some delay in the system data path. If VLSI has a huge number of logic gates available, this may be a very effective way to use them.

Built-in verification testing, a technique that applies all possible patterns to a combinational network, has been proposed (McCluskey and Bozorgui-Nesbat 1981; McCluskey 1982). This technique compresses the results of each pattern into a linear feedback shift register. After a *complete* set of test patterns has been applied, the results stored in the linear feedback shift register are observed to be compared to the good machine response. The result is that irrespective of the fault model, verification testing will detect the defects (assuming the defective machine does not turn into a sequential machine from a combinational machine). In order to help the network apply its own patterns and accumulate the results of the tests rather than observing every pattern for 2^n input patterns, a structure similar to a BILBO register is used. This register, which has some unique attributes, is shown in Figure 14-30. If a combinational network has 100 inputs, the network must be modified so that the subnetwork can be verified, and thus the whole network will be tested.

Two approaches to partitioning are presented by McCluskey and Bozorgui-Nesbat (1981). The first is to use multiplexers to separate the network, and the second is a sensitized partitioning to separate the network. Figure 14-31 shows the general network with multiplexers, with the network in a mode to test subnetwork G_1. This approach could involve a significant gate overhead to

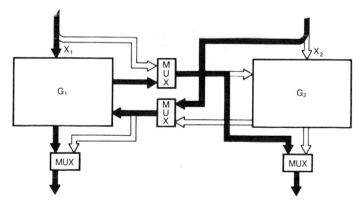

Figure 14-31. Autonomous testing—general network. The solid lines represent the sensitized path for testing network G_1.

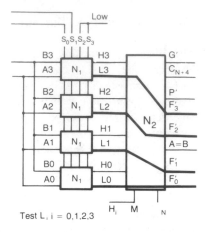

Figure 14-32. Autonomous testing with sensitized partitioning: low.

implement in some networks. Thus, the sensitized partitioning approach is put forth. For example, the 74181 ALU/function generator is partitioned using sensitized partitioning.

By inspecting the network, two types of subnetworks can be partitioned out, four subnetworks N_1 and one subnetwork N_2 (Figures 14-32 and 14-33). By further inspection all the L_i outputs of network N can be tested by holding $S_2 = S_3 = $ low and M = high. Further, all the H_i outputs of network N_1 can be tested by holding $S_0 = S_1 = $ high and M = high because sensitized paths exist through the subnetwork N_2. The output of these networks, N_1, can be tested

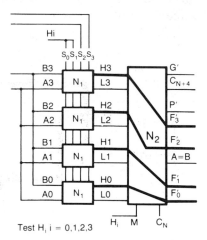

Figure 14-33. Autonomous testing with sensitized partitioning: high.

with the A_i's all tied together and all the B_i's tied together. Further inspection shows a way to test N_2 as well, exhaustively. Thus, far fewer than 2^n input patterns can be applied to the network to test it. In further work (McCluskey 1982; Tang and Chen 1984) the thrust of the research is to establish a mathematical base to determine the minimal number of patterns required to exhaustively test the network as concurrently as possible. This approach makes use of the fact that no output is fed by all the inputs. (If so, no reduction in number of test patterns can be achieved.) It appears that this can be done with some significant savings. (This structure is also an in-situ approach because all the system latches are used to generate and subsume the test results.)

Ex-Situ Self-Testing

The third form of self-testing is ex situ. This structure uses pseudorandom patterns supplied to the system latches by other latches, which exist solely to supply patterns (see Figure 14-34). Furthermore, nonsystem latches are used to compress test results in linear feedback shift registers. The basic testing application initializes all latches, applies pseudorandom patterns, applies system clocks, and then compresses the results of the test.

An example of this technique is the STUMPS approach, shown in Figure 14-35 (Bardell and McAnney 1982). This technique assumes that all chips are designed with LSSD. The nonfunctional latches generate pseudorandom patterns that shift down the scan path. Once the scan path is loaded, a system clock is pulsed, and then the scan path is unloaded. All these operations can run at, or near, system speed; hence, many pseudorandom patterns can be applied. The approach is a compromise for the BILBO approach, in that it is lower in overhead than BILBO; however, it requires a factor L times longer to apply the same number of patterns to the combinational network, where L is the length of the longest scan path. Hence, the speed advantage is K as defined in the prior section.

Another approach using an ex-situ self-testing structure is the on-chip maintenance system proposed by CCD (Resnick 1983).

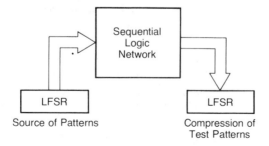

Figure 14-34.   Ex-situ self-testing structure.

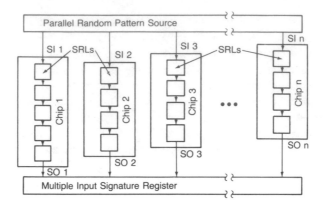

Figure 14-35. Ex-situ STUMPS approach.

IMPACT ON SYSTEM TESTING

A number of these new design techniques have had an impact on system testing. For example, in the IBM System 38 the LSSD structure was used in the formation of system diagnostics (Stolte and Berglund 1979). These diagnostics were generated with the same CAD tools that were used for board testing. The LSSD structure was also integral in the implementation of system diagnostics. The Scan Path structure appears to have been used by NEC in their system testing as well.

In the area of self-testing, microprocessor-based boards have already formed a core of system diagnostics. The new thrust seems clearly directed at using either in-situ or ex-situ self-testing as a key part of system diagnostics in VLSI system designs.

FUTURE DIRECTIONS

The main thrust for structured approaches to design for testability will be in the gate array area of noncaptive IC facilities. This has been slow in coming and will continue at a slow pace. In the area of self-testing, research will continue to assess the techniques mentioned and to look at variations of them. With respect to CAD tools needed to support the self-testing structures, analysis tools will play an even more integral role (Savir et al. 1984; Eichelberger and Lindbloom 1983). The cost of testing represents an increasingly significant percentage of total cost; so researchers are interested in developing more efficient and cost-effective ways of testing.

As we are rapidly approaching ULSI (Ultra Large Scale Integration), new factors apparently will come into play. That is, 10^6 transistors will not be designed from the transistor level up. It is highly likely that there will be integration of subfunctions such as microprocessors, memories of all sorts, interfacing logic, gate arrays, and so on, on the same wafer. If some portion is proprietary (such as the microprocessor), the method of testing each of these subfunctions needs to be communicated. The exact evolution of the technology will be very interesting to observe.

In conclusion, the entire area of testing has been actively explored for a number of years, and will continue to be a fertile field of investigation for both universities and industry in the foreseeable future.

REFERENCES

Ando, H., "Testing VLSI with Random Access Scan," in *Dig. Papers Compcon 80,* IEEE Pub. 80CH1491-OC, pp. 50–52.

Bardell, P. H., and McAnney, W. H., "Self-Testing of Multichip Logic Modules," in *Dig. Papers 1982 International Test Conference,* IEEE Pub. 82CH1808-5, pp. 200–204, Oct. 1982.

Chang, H. Y., Manning, E. G., and Metze, G., *Fault Diagnosis of Digital Systems.* Wiley Interscience, New York, 1970.

Das Gupta, S., Goel, P., Walther, R. G., and Williams, T. W., "A Variation of LSSD and Its Implication on Design and Test-Patterns Generation in VLSI," *Proc.* 1982 International Test Conference, Philadelphia, PA, Nov. 1982, pp. 63–66.

Eichelberger, E. B., and Williams, T. W., "A Logic Design Structure for LSI Testing," *Proc. 14th Design Automation Conference,* IEEE Pub. 77H1216-1C, pp. 462–468, June 1977.

Eichelberger, E. B., and Lindbloom, E., "Random-Pattern Coverage Enhancement and Diagnosis for LSSD Logic Self-Test," *IBM Journal of Res. and Dev.,* 27 (3), 265–272 (May 1983).

Eichelberger, E. B., Muehldorf, E. J., Walther, R. G., and Williams, T. W., "A Logic Design Structure for Testing Internal Arrays," in *Proc.* 3rd USA–Japan Computer Conference, San Francisco, CA, Oct. 1978, pp. 266–272.

Eldred, R. D., "Test Routines Based on Symbolic Logic Statements," *J. Assoc. Comput. Mach.,* 6, (1), 33–36 (1959).

Friedman, A. D., and Menon, P. R., *Fault Detection in Digital Circuits,* Prentice-Hall, Englewood Cliffs, NJ, 1971.

Funatsu, S., Wakatsuki, N., and Arima, T., "Test Generation Systems in Japan," in *Proc.* 12th Design Automation Symp., pp. 114–122, June 1975.

Koenemann, B., Mucha, J., and Zwiehoff, G., "Built-in Logic Block Observation Techniques," in *Dig. Papers, 1979 Test Conference,* IEEE Pub. 79CH1509-9C, pp. 37–41, Oct. 1979.

McCluskey, E. J., "Verification Testing," in *Dig. Papers, 1982 International Test Conference,* IEEE Pub. 82CH1808-5, pp. 183–190, Oct. 1982.

McCluskey, E. J., and Bozorgui-Nesbat, S., "Design for Autonomous Test," *IEEE Trans. Computers, C-30* (11), 866–875 (Nov. 1981).

Resnick, D. R., "Testability and Maintainability with a New 6K Gate Array," *VLSI Design* (Mar./Apr. 1983).

Savir, J., Ditlow, G. S., and Bardell, P. H., "Random Pattern Testability," *IEEE Trans. Computers, C-33* (1), 79–90 (Jan. 1984).

Stewart, J. H., "Future Testing of Large LSI Circuit Cards," in *Dig. Papers 1977, Semiconductor Test Symp.,* IEEE Pub. 77CH1261-7C, pp. 6–17, Oct. 1977.

Stolte, L. A., and Berglund, N. C., "Design for Testability of the IBM System/38," *Dig. of Papers, 1977 Test Conference,* IEEE Pub. 79CH1509-9C, pp. 29–36, Oct. 1979.

Tang, D. T., and Chen, C. L., "Interactive Exhaustive Pattern Generation for Logic Testing," *IBM Journal of Res. and Dev.,* 28 (2), 212–219 (Mar. 1984).

Wadsack, R. L., "Fault Modeling and Logic Simulation of CMOS and MOS Integrated Circuits," *B.S.T.J.*, pp. 1449–1488 (May–June 1978).

Williams, T. W., "The History and Theory of Stuck-At-Faults," *Proc.* 6th European Conference on Circuit Theory and Design, Stuttgart, W. Germany, Sept. 6–8, 1983, pp. 88–81.

Williams, M. J. Y., and Angell, J. B., "Enhancing Testability of Large Scale Integrated Circuits via Testpoints and Additional Logic," *IEEE Trans. Computers*, *C-22*, 46–60 (Jan. 1973).

Williams, T. W., and Eichelberger, E. B., "Random Patterns within a Structured Sequential Logic Design," in *Dig. Papers, 1977 Semiconductor Test Symp.*, IEEE Pub. 77CH1261-7C, pp. 19–27, Oct. 1977.

Williams, T. W., and Parker, K. P., "Design for Testability—A Survey," *IEEE Proceedings of the IEEE, 71* (1), 98–112 (Jan. 1983).

Yamada, A., Wakatsuki, N., Fukui, T., and Funatsu, S., "Automatic System Level Test Generation and Fault Location for Large Digital System," *Proc. 15th Design Automation Conference,* IEEE Pub. 78CH1363-1C, pp. 347–352.

15. Silicon Design Methods and Computer-Aided Design Tools

Ian L. McWalter
Bell-Northern Research

TERMINOLOGY

As with any technical subject, the field of silicon design methods and CAD tools is laden with jargon and conflicting and sometimes confusing terminology. Since it is the intent of this chapter to start at a very basic level—first defining a transistor and then, moving at a fast gallop, giving some insight into how one might usefully put a million of them on one chip—some definitions are in order. While not claiming to give a complete guide to negotiating this particular minefield, I shall attempt to define each term as I use it, if this forms a natural part of the flow. Otherwise, perhaps the chapter-end glossary will serve to cover any items I miss, and it may also be useful as a ready reference guide to readers unfamiliar with the topic. Because we are dealing essentially with silicon, it is likely that any designers attempting to design their own chips will come across terms that are applicable to silicon materials and devices. Although these terms are not essential for an understanding of the chapter, I have included them in the hope that they will be of some use (see also Muroga 1982).

INTRODUCTION

What are large scale integrated circuits (LSI) and very large scale integrated circuits (VLSI)? Why is it necessary to have a uniquely silicon-based design methodology? Why are CAD tools necessary in the design process?

To those of us who have been involved in the explosive growth of silicon technology over the last two decades, these questions seem too obvious to ask. LSI and VLSI are merely natural stages in the evolution of a technology that started with one transistor. We progressed to the logic gate level (four transistors); to registers and counters (50 to 100 transistors); to a multitude of standard products such as memory, microprocessors, and widely applicable random logic functions (1000 to 10,000 transistors); and finally to the current state-of-the-art memory chips, which may have over a million transistors. Whether you call this awe-inspiring progress LSI, VLSI, ULSI (ultra), or ALSI (amazingly) seems to be a marketing issue. The central problem for designers remains that of complexity management, so that it is natural to try to develop increasingly sophisticated computer tools. Along with this development comes an increasing reliance on CAD tools, to the point where although not all tools are essential, a silicon design cannot be implemented without at least some of them.

These musings lead us to an even more fundamental question. What is CAD? We have no problem with the first letter. the C stands for computer, whether it be a mainframe, a minicomputer, or a microprocessor-based workstation. The A is somewhat more problematic. It should mean aided or assisted, but all too often it is interpreted as automated. Indeed, in many cases a considerable degree of automation has been achieved, but only in the sense that data is automatically converted from one form to another. In no sense has the design itself become automated. The original design data is still solely the responsibility of the designer. This confusion has perhaps arisen because the D in CAD stands for design, and the term CAD refers generically to a disparate set of tools performing a wide variety of functions. For a given set of design tools it is usually by no means apparent which portions of the design cycle can be relied upon, and which must be treated with extreme caution. To illustrate this, consider the logic simulator that is one of the basic CAD tools. Most good logic simulators will confirm that a logic design complies with the original intent. However, they usually will not confirm that good timing margins exist, that no race or hazard conditions exist, that the basic modeling information is accurate enough, or that any asynchronous operation is guaranteed. With help from other tools, some of these problems can be addressed, but the basic message is that, as many designers have found at their expense, logic simulation alone does not guarantee satisfactory device operation. This point will be addressed at greater length later in this chapter.

The conclusion here, then, is that perhaps the D in CAD should be replaced by V for verification. The design system should thus be regarded as a series of checks and double checks, to ensure that the final implemented design conforms to the original intent of the designer.

A further reason for reexamining the purpose of CAD is the widespread move in recent years to make silicon available as a design medium for "non–silicon expert" designers. To elaborate on this point a little, CAD tools were largely developed by and for silicon experts—those designers who had sufficient knowledge of silicon as a material, and wafer fabrication as a manufacturing process, to be able to manipulate designs at the transistor level, which in practice meant direct polygon manipulation of the mask information.

As tools became more and more sophisticated in verifying that such designs were correct, it became apparent that the design system could be used to "hide" all silicon-related knowledge from the designer. Therefore non–silicon experts such as system designers could operate at a higher design level, such as the gate or register level, and still have the design translated reliably into a working device. A gate-array system is an obvious example of a system that allows any designer to input design data as a series of predefined functions, but does not require any knowledge of how those functions are constructed. Thus, a gate-array design system, while using much the same tools as any other silicon design system, has a fundamentally different purpose. It aims to restrict a designer to a predefined function set, and disallows making changes to it, as opposed to the original intent of making it much easier for designers to verify their changes. Both approaches have a place in the spectrum of silicon design techniques, but it is evident that the CAD system used for a particular design is intimately related to the design method adopted. It would be safe to say, therefore, that the study of design methods should result in the specification of a CAD system that is very closely coupled to the types of devices that one wishes to design.

No one design system can satisfy all needs, so a careful study of the intended applications is necessary before the correct balance between the desire for first-time success and the need for an efficient, cost-effective chip can be established.

In closing this introduction, let me briefly return to the central issue of complexity management. It seems obvious that single devices containing a million transistors would benefit greatly from computer assistance, and that a design style based on regular arrays of devices would be most suitable. Yet many large systems, such as computers themselves, or large digital telephone switches, contain at least as many devices, if not more, but have been designed and manufactured without the extensive use of CAD. The key here is that large systems consist of many accessible printed circuit boards (PCBs) or hybrid substrates, and go through extensive breadboard and debug cycles. The detailed design is done on manageable chunks, each of which can be isolated and checked. For all practical purposes this is not possible for a system on a chip. Consider the trivial but common example of an incorrectly wired signal. On an accessible board this can be quickly debugged by probing onto large tracks, and fixed with a "red wire." On a piece of silicon, however, the track is only, say 3 microns wide, and it may be buried in an underlying layer. Therefore the act of probing may disturb the operation of the device. It also requires extensive and expensive equipment, and an expert to operate it. Once you have been through all this, the "fix" requires changing mask data and a new, complete fabrication cycle, which is both expensive and time-consuming. In short, mistakes are generally expensive, but mistakes on a piece of silicon are very expensive. We can thus conclude that computer aided design will be beneficial for system design, but that it is essential for very large scale integrated circuit design.

PROCESS TECHNOLOGIES

The basic device on which all semiconductor technology rests is the transistor. Many different types of transistors are possible, but all are based on the principle of conductivity modulation. That is, the characteristics of the device can be controlled by a signal, the effect of which is to determine the degree to which it will conduct. Thus, in the simplest case, we can arrange for the device to be "on" or "off," which is of obvious application in digital logic circuits, or we may have a continuously varying output, which depends on a continuously varying input. The latter arrangement will produce an analog circuit.

The two most used types of transistors are the MOSFET (metal-oxide-semiconductor field effect transistor) and the BJT (bipolar junction transistor). The former device—which was first patented in the late 1920s although the means to fabricate it did not exist until much later (1960s)—is a voltage-controlled device. This means that the controlling signal is a voltage applied to the "gate" that modulates the conductivity between source and drain. Further, the MOSFET is a "surface" device, which means that its behavior is dominated by the properties of the semiconductor material in a thin layer at the surface. This point will be discussed further in the section on "The Rise of MOS." The BJT, by contrast, is a current-controlled device, so that the conductivity between collector and emitter is controlled by a current applied to the base of the device. This device was invented in 1948, and formed the basis of the first integrated circuit some 12 years later. The properties of the BJT are governed mostly by the bulk properties of silicon crystals.

Having contrasted the behavior of these two types of transistors, we should also note that they are essentially similar. Both are characterized by a high input impedance (low input current) and a low output impedance (high output current).

Within these two major classifications further subdivisions occur. MOSFETs may be n-channel or p-channel—referring to the type of carrier (N = electrons, P = holes), and BJTs may be NPN or PNP. MOSFETs are unipolar devices in that the vast majority of the current is carried by a single carrier type, whereas, as the name implies, in bipolar devices both types of carriers are involved. Further classifications involve enhancement and depletion mode devices for MOSFETs and vertical and lateral structures for BJTs, but discussion of these topics is beyond the scope of this chapter.

Having discussed the major types of device that we wish to fabricate, we will now describe the current manufacturing processes and the types of devices they support. In the bipolar field, many different processes exist, but all will provide both of the basic transistor structures, NPN and PNP. The differences are mainly in the dimensions of the structures, which will provide a wide range of speeds and supported voltages. In general, bipolar processes will always provide devices that offer greater speed than MOS processes of comparable dimensions, but will also usually consume more power.

MOS processes fall into three major categories: PMOS, which provides only P-channel devices; NMOS, which provides N-channel devices; and CMOS (complementary MOS), which will provide both. PMOS is now obsolete as a design technology, although the existing product is still manufactured, a point that will be discussed later. NMOS is still an active technology, but CMOS is rapidly becoming the major silicon design medium. NMOS technologies may provide enhancement devices, or both enhancement and depletion devices, while CMOS usually provides N-channel and P-channel enhancement devices, plus a parasitic bipolar device, usually NPN in a P-well process, which is useful in some circumstances.

There also exist a plethora of mixed and special-purpose processes ranging from the relatively simple provision of linear capacitors for mixed analog/digital devices to esoteric high-voltage mixed CMOS and bipolar processes, and silicon on sapphire. Also mention should be made of gallium arsenide, which is a completely different type of semiconductor material from silicon, and which is rapidly gaining a market for very high-speed applications.

The rest of this chapter will concentrate mainly on design methods for MOS circuits, although some mention will be made of bipolar processes. Special-purpose processes beyond what is necessary for mixed analog/digital chips will not be discussed further.

THE RISE OF MOS

The major reason for the delayed introduction of production MOS processes compared to bipolar processes was the difficulty encountered in controlling the surface properties of silicon and, in particular, the fabrication of a high-quality oxide layer to act as the gate insulator. To control the transistor threshold voltage sufficiently, this oxide must contain no contaminants or defects that would lead to an electric charge being incorporated in the oxide, and the interface between the insulator and the silicon should be as nearly perfect as possible. The latter condition is achieved by thermally growing silicon dioxide on the silicon wafer. The wafer is heated to a high temperature while passing oxygen over it, and the resulting oxide is then carefully annealed. However, the former problem took somewhat longer to overcome.

In silicon, electron mobility is approximately two to three times greater than hole mobility, which, in practical terms, means that an N-channel MOS device will be inherently faster than a P-channel device. Despite this advantage, PMOS was the first available MOS process because the contaminants present in the fabrication process, mainly sodium ions, led to the formation of a net positive charge in the gate oxide. This was acceptable for P-channel devices because it merely raised the threshold voltage, and could be overcome by application of a larger negative gate voltage. However, for NMOS it gave rise to serious difficulties in switching the device off because "normal" processing variations in threshold voltage could result in either an enhancement or depletion mode device.

This condition held until the early 1970s, at which time more sophisticated processing techniques eliminated the oxide charge problem and sounded the death knell for PMOS. Thus, while PMOS still exists as a production process on a relatively small scale, manufacturing a few existing standard products, it has not been a design medium for several years.

The seventies can be regarded as the decade of NMOS. Fueled by an explosive increase in demand for semiconductor memory from the computer industry, the NMOS dynamic RAM became the engine of growth in the semiconductor industry. NMOS was (and still is) a simple high-yielding process; and with periodic improvements in processing techniques to reduce defect densities and transistor dimensions, and in circuit techniques to further enhance performance, the steps from 4 kilobits, to 16K, to 64K and 256K per chip were rapidly accomplished. A spin-off of these process improvements was the implementation of NMOS microprocessors of ever increasing power; and it was also realized that provision of a linear capacitor element, which formed the storage capacitor in a DRAM, could be used in analog circuits. By utilizing this capacitor along with a suitable switching arrangement, analog filters (known as switched-capacitor filters), whose performance depended only on the ratio of capacitors, not on their absolute value, could be implemented. Because a silicon process controls these ratios to a very high degree of accuracy, it became possible to integrate high-quality filters on the same chip as digital functions. This development led to a boom in the use of MOS in the telecommunications industry, most notably through a device known as a Filter-CODEC, which converts voice signals to a digital format for transmission. For the first time it became cost-effective to treat the Filter-CODEC as a "per line" function in a telephone switch. It can thus be argued that the current preeminence of digital switching came about through the emergence of NMOS technology, and in particular from its use of memory.

Meanwhile, the huge leaps forward in processing technology were also being reflected as dramatic improvements in the CMOS world. CMOS had long been regarded as a niche technology because of its complexity, which made it costly, and its low speed; but it began to emerge as the technology of choice. This happened because CMOS is amenable to the highest levels of integration, as it consumes the lowest power—and large-scale integration, the larger the better, is what the silicon design world is all about. When this capability was coupled with the fact that power consumption rapidly becomes a key cost factor in system design, it became evident that in spite of a small (and becoming smaller) premium for a CMOS device over its NMOS counterpart, a cheaper system might still result if component packing density could be improved and expensive cooling equipment dispensed with. This trend became most apparent with the conversion of many NMOS microprocessors to CMOS.

The caveat here is that it is never wise to take low power consumption for granted. The potential saving in power through the use of CMOS is achieved by adhering to appropriate design techniques. CMOS power consumption increases linearly with frequency of operation, and with the number of circuits switched at any given time, and due regard should be given to this characteristic during initial design.

The second major influence in promoting CMOS was the growth in the CAD field. It had long been realized that CMOS was a simpler design medium than either NMOS or bipolar design, and as will be discussed in a later section, the concept of gate-array and standard cell design had existed for many years. The coupling of CAD improvements with CMOS process improvements has led to the emergence of many excellent, narrowly focused design systems. These design systems mainly address the lower-volume end of the device market, which can be characterized as "semicustom." The CAD tools obviate the need for silicon expertise on the part of the designer; and since CMOS is a relatively simple structure to model if approached correctly, CAD will lead to a high probability of first-pass success. Thus the type of customer whose primary needs are a low-volume, custom design solution, coupled with a rapid design cycle, will be well served by a semicustom, CMOS, CAD system. In the last few years the number of such customers has been legion, leaving little doubt that semicustom CMOS will become the primary hardware design medium over the next few years.

DESIGN TECHNIQUES FOR DIFFERENT TECHNOLOGIES

The basic principles of design of electronic circuits are well established, and fall generally into two categories. Digital circuit design concepts are usually made concrete through the use of Boolean algebra and timing diagrams, while analog design, which is much more varied, proceeds through the use of basic filter theory, for example, and the use of frequency domain analysis. Such concepts are very general in nature and thus are independent of a particular process technology. However, the creation of real structures that implement the design is highly dependent on the nature of the available devices. In practice, this means that in order to produce an efficient, optimized design, the technology will influence not only the basic building blocks used, but also some of the higher-level design decisions. In short, all design will be based on a series of trade-offs, and these trade-offs will be strongly influenced by the technology.

In order to make this idea more concrete, let us consider the different trade-offs involved in implementing a random logic design, using CMOS as opposed to NMOS. Both processes are perfectly capable of implementing our basic design as represented by Boolean equations, state diagrams, timing diagrams, and basic logic schematics. That is, the first state of the design process can be technology-independent. However, the results we achieve will be markedly different in the two technologies. The issues discussed will be logic style, packing density, yield, design margin, and the necessity for special process-related design techniques.

The first major difference between NMOS and CMOS becomes apparent when we are deciding whether to use static or dynamic logic. Static logic is inherently easier to deal with from a design standpoint because outputs remain valid, independent of the clock. In dynamic logic, however, outputs are only valid for the duration of a gating clock and must be read within a specified time period. In a large, complex, random logic design, this gives rise to a higher probability of errors due to timing margins and clock skews. In CMOS the choice is quite natural because the basic complementary structure is inherently static and consumes no power unless the input is changing. However, in the NMOS technology, a basic inverter will consume power in one state unless the output is gated by a clock, so that static logic must be used sparingly in situations that require it. The advantage here goes to CMOS.

In order to consider differences in packing density between the two technologies, we must consider the photolithographic design rules governing the layout of transistors. Technologies are usually characterized by their minimum feature size, so that a 3 micron technology implies that the minimum drawn dimension for a transistor length is 3 microns. However, this does not imply that all dimensions can be drawn at 3 microns. Also, rules such as overlaps between polygons on different layers, and spacings between conductors and active device regions, must be considered. Without going into further detail, the upshot is that, because the P-channel pullup device in a CMOS inverter is always bigger than both the small equivalent pullup plus the gating transistor in an NMOS inverter, greater packing density can be achieved for dynamic NMOS logic than for static CMOS logic if all common design rules are equal.

NMOS is inherently a simpler process than CMOS; it requires fewer masking steps (6 to 8 as opposed to 11 to 13), and therefore from a defect density viewpoint will always yield more dies per wafer for a given die size. Coupled with the potential for higher packing density for the same function, NMOS will always give a lower-cost final product, provided that fabrication cost is the dominant factor, which historically has been the case. Thus NMOS has the advantage still, although as complexities increase, and test and package costs become a greater proportion of the overall cost, this advantage is rapidly decreasing, and for a growing class of devices may already have disappeared.

A key aspect of good integrated circuit design is design margin, for both internal timing of the logic functions and parametric performance, that is switching points and propagation delays as seen at the pins of the device. A requirement for a good, robust, manufacturable design is that good margin be maintained over a wide range of temperature and voltage conditions. For both NMOS and

CMOS, the major factor influencing variations in performance due to normal spreads in processing is the polysilicon gate line width, which will determine to a large degree the speed of minimum-sized devices. Although this holds true for both processes, the major consideration here is that—from the point of view of both design simplicity and ease of checking by CAD tools—static CMOS logic is less error-prone than is dynamic, "set of the pants" NMOS logic. Errors are always costly, so the advantage here goes to CMOS.

Another major difference between CMOS and NMOS technologies is the output voltage swing obtained from the basic logic structures. A CMOS inverter output will swing from VDD to VSS, and, if correctly designed, will have a switching point close to the mid-rail voltage (usually 2.5 V for modern 5 V processes). However, for enhancement mode NMOS processes, as the output of an inverter rises to the "1" level, the pullup device starts to turn off and thus to slow down. The basic reason for this is that for the transistor to be on, the gate voltage must be held one threshold voltage above the source. If this gate voltage is at VDD, then the maximum output of the inverter will be one threshold below this, and it will take quite a while to get there because the transistor is switching off as it rises. This problem is further exacerbated by the fact that in order to get comparable noise margins to CMOS, substrate bias is introduced, so that the threshold of the transistors is raised from the normal (0.7 V approximately) to about 2 V. In many cases this can be accommodated, but for those cases where a fast, full rail output is desired, a technique known as bootstrapping is used. This technique involves connecting a capacitor between the output node and an additional transistor, which precharges the capacitor when the output is low, and is then disconnected as the output rises. This voltage is then applied to the gate of the device, so that the gate voltage is "booted" by the rising source voltage, and enables the transistor to stay on, so that the output reaches full rail. This technique is expensive because an extra transistor and a capacitor must be added, and is error-prone because the performance is now determined by a rather complex "bootstrap ratio," which is very difficult to model in a logic simulator. The advantage here clearly goes to CMOS.

Much has been said in the literature about the advantages of synchronous design. First we must define what we mean by this. Although the world is an inherently asynchronous place—the chip designer has no knowledge of, say, when the high priority interrupt will take place, or when the telephone will go off-hook—most of the activities inside a chip occur at specific clock edges. In the simplest case, all functions are performed at one edge of a single input master clock. Any external signal that has no fixed relationship to this clock must wait until at least the next appropriate clock edge before action is initiated on chip. This situation cannot be handled by any simulator, or even by any automated tester. Problems due to asynchronous events are usually of a statistical nature. That is, every so often when a particular combination of events takes place, a particular timing relationship occurs between input and clock that causes a system error. Synchronous design at the chip level does not address this class of problem. However, at the chip level synchronous design is, in principle, highly advantageous, in terms of both design simplicity and ease of implementation of CAD tools to check the device, as well as ease of manufacturing testing.

Finally, let us now discuss, briefly, what synchronicity means within a single chip. Since the very act of dividing and distributing an incoming clock signal on chip will necessarily result in skews in the arrival time of clocks at different circuits, synchronous is definitely *not* synonymous with simultaneous. However, if we have adequate techniques, in terms of equalizing loads and verifying that limits on skews have not been violated, then we can ensure that all relevant outputs become valid at the same time, in the sense that all set-up, hold, and pulse-width requirements have been met.

In general then, as far as the necessity for synchronous design is concerned, the considerations are no different for NMOS than for CMOS, but in most cases the design problem is easier and more accurately modelable in CMOS than in NMOS, which may use special techniques such as bootstrapping and back bias.

The subject of design for testability has been rapidly gaining prominence in recent years, mainly because of the increasing complexity of silicon chips, and hence the much greater difficulty in testing internal functions. We will discuss this at greater length later, in the sections concerning individual CAD tools. Here, it is sufficient to note that synchronous, static logic is easier both to test and to simulate than a dynamic or asynchronous structure, and that CMOS is a more suitable technology, in that it can represent an easier problem to solve with current testing and CAD tool technology.

A design style that calls for the availability of predefined, standard building blocks is obviously an easier one to manage for most designers, who wish to be primarily concerned with the function of the chip. Once again CMOS emerges as the winner because it lends itself more readily to the creation of the basic logic elements in standard format. All gates can be defined to have the same drive capability and the same switching points, and to be serviced by a small group of buffer drivers for special circumstances. This can be achieved with very little loss in terms of efficiency of implementation. NMOS, by contrast, loses efficiency very rapidly if general-purpose cells are used, for the same reasons that make the special-purpose techniques outlined above necessary. The advantage here is very firmly with CMOS.

The final major factor we shall discuss is latch-up, the phenomenon of extremely high current consumption, perhaps leading to destruction of the chip, caused by the triggering of parasitic npnp structures in CMOS. These structures in fact form the class of devices known as thyristors, or silicon controlled rectifiers (SCRs), and are widely used for controlling large currents. However, in CMOS they constitute a problem that results from fabricating both types of MOS device on the same substrate. This does not occur in NMOS, so this technology clearly has the advantage. But, considerable design effort has gone into designing protection structures at the periphery of the chip, and into careful design of cells, to minimize this problem. There have also been somewhat extravagant claims, of "latch-up–free" processes, leading to a belief in some quarters that the problem has been solved.

However, this is not completely true; all CMOS processes will always have the potential to latch up, and indeed the problem tends to get worse as feature sizes shrink. However, by designing the process carefully and using rigorous design techniques for peripheral structures, one can alleviate the problem so that latch-up will occur only under extremely unusual circumstances. One can also use system level techniques, such as power supply sequencing, to ensure that the external conditions necessary for latch-up do not occur.

Thus, despite some specific advantages in final product cost and in the area of latch-up, NMOS has lost the battle, and CMOS should, and indeed has, become the technology of choice for most digital functions.

Finally, let us consider bipolar design techniques briefly. As stated earlier, bipolar devices are inherently of higher speed and higher power consumption, as well as more complex to model than MOS devices. Bipolar chips have tended to be less highly integrated because of the higher power consumed, and CAD has been used much less because the problems are much more difficult to solve. Thus bipolar design methods have remained in the province of the silicon expert, and there are probably as many techniques as there are designers! However, in recent years, as with MOS, considerable advances have been made, and bipolar (TTL and ECL) gate arrays have emerged in the marketplace, which will do much to extend the use of bipolar integrated circuits in the high-speed and analog areas to which they are best suited.

DESIGN CHOICES

The design choices to be discussed here refer not only to the technology used, but also to the design style employed. The first and obvious step is to decide whether or not an application-specific integrated circuit (ASIC) is required. Although one of the main points stressed in this chapter is that CAD tools are making silicon design progressively easier, it is still by no means trivial. Even the fastest and simplest of the silicon design techniques still may take significantly longer than a PCB design using off-the-shelf components, and specification changes late in the system design cycle are much less easy to accommodate. Thus it is usually space, power, and cost savings that result in the decision to design customer silicon.

Once this decision has been made, the choice of technology and style are intimately related to the type of device to be designed, and to the perennial design issues of final product cost, design cycle time, product volume, packaging,

R & D expenditures, and performance. The purpose of this section is not to give hard and fast rules for making such choices, as none exists, but rather to give an indication of the different approaches possible.

Let us first define, somewhat arbitrarily, three types of devices and three design styles. The types of device are: all digital, less than 5000 gates where one gate is equivalent to four transistors; all digital (greater than 5000 gates); and mixed analog/digital. The three styles will be full-custom, semicustom, and full-custom for mixed analog/digital.

The distinction between digital chips less than 5000 gates and those that are greater represents a real difference in design techniques. At some level, chosen here as 5000 gates, a chip designed as a single entity as random logic would become so complex as to consume prohibitively large amounts of time and money at the initial design stage, and also be so error-prone as to require many iterations to achieve the desired chip function. Thus, at some point it is necessary to move to more structured, hierarchical techniques, which may also be array-based in that much of the function is achieved through memories, programmable logic arrays, or even arrays of small random logic blocks. Thus, although in some circumstances it may be advantageous, for reasons of cost and performance, to design a large circuit using "small chip" techniques, let us assume that there is a cutoff point for the design of random logic chips. Of course, the contrary argument could be made, that smaller chips would also benefit from the use of hierarchical techniques. Practically, however, most CAD tools operate in a "flat" fashion. Taking as examples a digital logic simulator and a process design rule checker, these tools operate on gate level primitives in the former case, and on polygon information in the latter. Therefore, even if the input presented to these tools is hierarchical in nature, it must be flattened, usually be an automated preprocessing step, before the tool is run. The outputs obtained will then have lost much of the original design information, and could make understanding of results and the consequent debugging phases of design more complex. In summary, hierarchy injects a layer of complexity into the design process, and should only be introduced if the trade-off between complexity of tool use and complexity of chip size is well understood. The rule should be to "keep it flat" for as long as possible, and then to introduce a minimum number of levels of hierarchical abstraction as required.

Now let us define our design styles. "Full-custom" implies that the designer has complete freedom to design at the transistor level, functional I/O points can be placed where they are most convenient, cells performing the same functions may be altered according to local drive requirements and load conditions, and all placements of cells and routing between them is done by hand. In semicustom design much less freedom is allowed to the designer. No transistor-level design is done, all functions are predefined, and the I/O points of cells are fixed, independent of individual routing requirements. All cells will also have similar drive capabilities, producing excessive overdrive capabilities in some instances and the requirement for buffering in others. Both of these cases result in an area penalty when compared to the optimal design. Finally, and most important, for silicon design the term "semicustom" has become associated with automatic placement and routing of the cells and wiring on the chip. This will obviously lead to a reduced design time to first silicon, although again at the expense of the chip area.

For mixed analog/digital chips, a fusion of two very different design skills is necessary. By its nature, analog circuitry is very sensitive to issues such as noise, crosstalk, and distortion, to which digital circuitry is immune (in most cases); so the layout of the analog section is very critical. However, in addition to the internal sources of disturbance within an analog section, interference may also arise from neighboring digital sections. Thus, in general, even if the digital function is such that it may be laid out using automatic techniques, usually its interconnections to the analog section must be done by hand, with due regard to parasitic coupling between signals. A further complication in the mixed analog/digital world is that very few, if any, simulators are capable of handling both analog and digital functions at the same time, so that any signals that pass between these two different worlds are difficult to check using CAD tools.

Like the distinction between different types of chips, the distinction between design styles is blurred; individual designers may use a combination of many techniques. Thus a given chip may be composed of both full-custom and semicustom digital blocks, depending on the requirements of the functions into which the chip is partitioned. In a mixed analog/digital chip, the analog section will almost always use full-custom techniques although programmable switched capacitor filter chips have emerged; but the digital logic that accompanies it may be full-custom or semicustom.

Let us now consider each type of chip in the light of the benefits and drawbacks of each design style. For all digital chips less than 5000 gates, the full-custom approach produces the most efficient, lowest cost, highest performance piece of silicon. It will also have the longest design cycle time, require the most expertise in terms of designer skills, and be least likely to be right the first time. This approach is for silicon experts only. A point worth noting here is the comparison between custom product and standard product. A custom chip, or ASIC, can be designed in either of the custom styles, but standard products, by virtue of their high volume, are very sensitive to issues of yield, hence cost, and performance. They are always designed using the full-custom techniques outlined above, and tend to utilize "every trick in the book."

Moving on to the semicustom approach to the smaller logic chips, there are significant differences between the most common semicustom approaches of gate array and standard cell, which will be discussed at greater length in the next section. Here we will compare only the broad definition of the semicustom design style with the full-custom approach. The design cycle time will be much shorter, and the chances of achieving the intended function the first time will be higher.

However, the cost per function will be higher; the opportunity to use the full capabilities of the selected technology is much reduced; and, because we are using only predefined elements, it is likely that there will be "gate wastage" in the sense that one cannot provide as much useful function with the gates available as with the full-custom approach.

For designs greater than 5000 gates, it is advisable to use a structured hierarchical design approach. Thus, the design will be partitioned into functional blocks, which may themselves be composed of subblocks, so that each section may be specified, designed, and checked separately. As stated earlier, the number of hierarchical levels should be carefully controlled, and due regard should be given to the available CAD tools when one is functionally partitioning the chip. A primary consequence of taking a hierarchical approach is to further blur the distinction between the full-custom and semicustom design approaches. For very large "system on a chip" design, a complete, full-custom approach becomes unmanageable because a very large design team would be necessary to complete the design in a reasonable period of time. One can envisage a situation where individual blocks could be tailored, or where interconnect between blocks involving analog or critical timing signals might be hand-crafted. However, in general, the sheer size of the undertaking will require more regularity in design, so that the use of predefined functions, on-chip memories, and programmable logic arrays for many random logic functions will become the norm. Thus, it becomes prohibitively expensive to achieve the benefits of full-custom design; so as product lifetimes become shorter, the advantages of a shorter design cycle time and a reduced manufacturing "learning curve" conferred by the semicustom design will cause the latter to predominate.

Finally, let us briefly discuss mixed analog/digital chips and design styles. No currently available simulator can handle the operations necessary to model the complete chip with a reasonable degree of accuracy, so it becomes necessary to introduce a level of hierarchy immediately. Hence the "full chip" can never be simulated fully as a single entity, and design will proceed as two hierarchies that remain separate below the full chip level (McWalter 1983). This point will be discussed at greater length in the section on CAD tool requirements.

GATE ARRAYS AND STANDARD CELLS

So far the term "semicustom" have been used with no regard to the considerable differences between the two most popular techniques, standard cells and gate arrays. It could be argued that only the gate-array approach is truly semicustom because it involves the use of preprocessed wafers, with only the last few mask steps, usually contact placements and metal interconnect, being used to "personalize" the design. By contrast, standard cell techniques require the customization of all mask steps, so that wafers must be manufactured from scratch for each new design. Let us now examine this situation.

The concept of the gate array dates back over 15 years, and is based on a single cell that can be reconfigured and joined with its neighbors to perform any logic function. This cell may be as simple as a two-transistor structure, which leads to efficient implementation of inverters, AND and OR structures, and so on, but less efficient latches and flip-flops; or it can be as complex as a 16-transistor cell, which is good for flip-flops but leads to gate wastage for more simple structures. Thus, modeling becomes easier, as all structures are identical, and it is possible to construct a highly focused CAD system, ensuring a high probability of success. This last point is the key reason for the recent surge in popularity of semicustom silicon. After many years of trying, the CAD tools finally arrived that made silicon accessible to the "non-expert" designer.

In the equally venerable standard cell approach, by contrast, each function is specifically built, using "full-custom" techniques, and is then frozen so that the chip designer cannot alter it. This means that each cell contains only as many transistors as required. Usually constraints are applied in the design of the cells, so that, for example, they all have the same height. Power is routed through them at the same points, so that they can be butted together, and inputs and outputs are available at both the top and bottom of the cell, to ease any interconnect routing problems.

In considering the advantages and disadvantages of each technique, note that these techniques may be considered complementary in the armory of design styles, although it is also true that there is considerable overlap in the market being addressed.

For the design of purely random logic chips, both approaches are very similar in the initial design phases. The design capture stage uses the same tools, the types of function available are similar, the same simulation tools are necessary, and the mechanics of generating test stimuli and test programs are identical. However at the back-end of the process, involving layout of the chip and fabrication, considerable differences emerge. Gate arrays are usually automatically placed and routed, with each gate array having a fixed number of gates, arranged in rows, interconnect channels of fixed width, and a fixed number of inputs, outputs, and power pads. If more gates are needed, or the interconnect requires larger routing channels, then the designer must move up to the next fixed size of array, with consequent potential wastage of gates. However, once layout is complete, only two or three mask layers need to be made, preprocessed wafers will be available, and the prototyping turn-around time will be, in principle, only a few days.

For standard cells only the cells specified by the designer will be placed on the chip. Automatic placement and routing will still be used, but now the length of the cell rows and the width of the interconnect channels will be dynamic, depending on the chip. The number of pads and their placement will vary from design to design. All mask layers must be created, as there is no prior knowledge of where the individual cells will be placed on the chip; so manufacturing turn-around will always be longer than for gate arrays.

The next level of integration for both approaches is the addition of special-purpose blocks, memory arrays in particular. Much the same considerations apply here; gate arrays will provide memory in fixed chunks, similar to standard product, whereas with standard cells only as much memory as is desired can be added. In both cases, improvements to automatic placement and routing must be made; for now, rather than a channel router, a block router must be used, in

Figure 15-1. Hand-packed mixed analog/digital NMOS chip.

recognition that fixed areas of the chip contain different types of structures. More sophisticated gate arrays may offer the opportunity to reconfigure certain portions of the chip into either memory or logic gates, improving efficiency.

In conclusion, the initial design time is the same for both approaches. Gate arrays will have a shorter fabrication cycle and will have the advantage if this is a significant proportion of total design time; whereas standard cells will have a lower cost on a per-function basis, all other parameters being equal. Another factor in reducing total design cycle time is that, because gate-array design systems are less flexible than standard cell systems, they allow less room for error, which will be reflected in a higher first-pass success rate. There will always be the possibility of a specification error in both systems, but implementation errors are less likely in gate arrays.

To make the preceding ideas more concrete, Figures 15-1 through 15-4 show

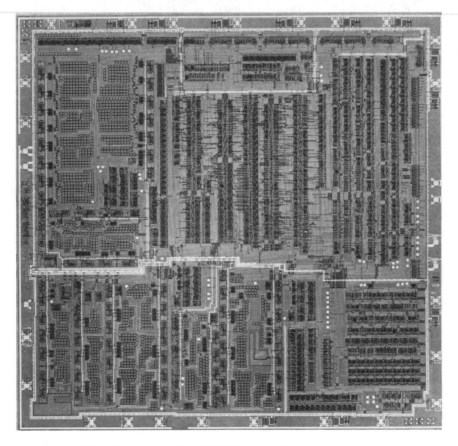

Figure 15-2.   Mixed analog/digital CMOS chip, using standard cell techniques.

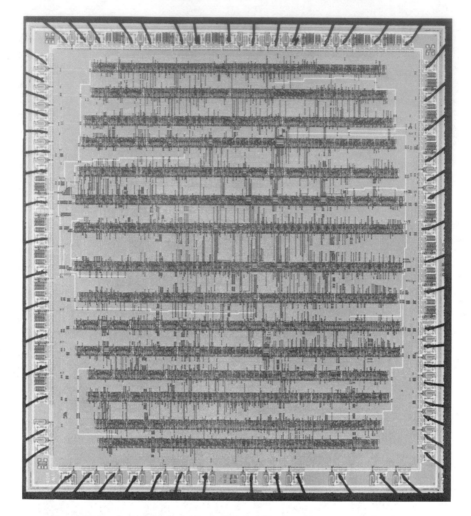

Figure 15-3.   Automatically laid out digital chip using standard cell techniques.

chips designed using different styles. The first is a NMOS mixed analog/digital chip designed using full-custom techniques. All the cells were individually designed by the chip designers, and placement and routing were also manual. Although structure is evident in each section of the chip, notice that cells can be of different height, and can be oriented either horizontally or vertically, or "popped in" at the edges of the chip. The chip is densely packed, with a ratio of active area to interconnect of about 1:1.

Figure 15-2 shows a mixed analog/digital CMOS chip, with the digital section using mainly standard cells. As can be seen, there is more regularity; cells are

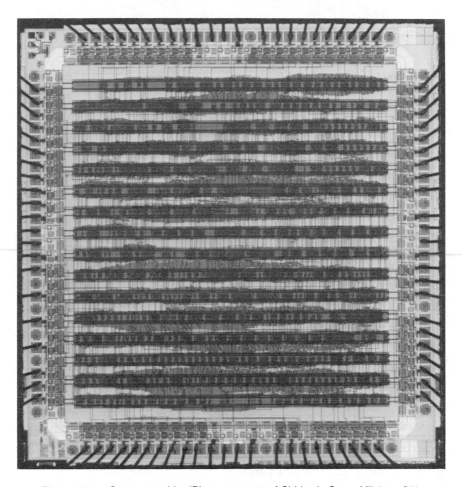

Figure 15-4. Gate array chip. (Photo courtesy of SLI Logic Corp., Milpitas, CA)

arranged in rows with routing channels, or are placed back to back. Different orientations now occur at the block level rather than the cell row level, although once again placement and routing was done by hand.

Figure 15-3 shows a standard cell chip that was automatically placed and routed. The high degree of regularity is evident here, with all cells placed in horizontal rows, and routing taking place in variable-width channels. This is once again a CMOS chip, and it should be noted that internal circuitry is placed as far as possible from the periphery of the chip. This is done to minimize the change of latch-up. The ratio of active area to interconnect is about 1:3.

The final chip, shown in Figure 15-4, is a commercial gate array, and shows the features described earlier, with fixed routing channels, fixed cell rows, and fixed pads.

REQUIREMENTS FOR A SILICON CAD SYSTEM

In order to define the requirements of a CAD system, we must first examine the steps necessary to implement a silicon design. As stated previously, some CAD tools, such as a mask pattern generator,, can now be considered essential; but other tools can be classified as so desirable that they have become requirements. It is also important to define the purpose of the CAD system in terms of what types of design will be attempted, and the degree of flexibility allowed.

Many different combinations are possible; so in the interests of clarity, after defining the general steps involved, we will discuss the requirements for two types of design systems. The first will be a relatively inflexible standard cell design system, for digital random logic design, whereas the second will be a much more sophisticated system for mixed analog/digital design, which could also incorporate on-chip memory.

The basic design flow is shown in Figure 15-5. Although it is shown as a series of sequential steps, activities will overlap in practice, and there will also be iterations in loops around certain sections. The objective of this system is to produce two outputs, one of which is the "PG Tape," used to generate the masks necessary to fabricate the device. The second is a test program, which will enable sorting of good devices from bad when fabrication is complete.

Of these steps, we may say that CAD tools are essential for the layout and PG phases; a simulation tool is highly desirable (which would make design capture and test stimuli generation tools essential); and an aid for test program generation is desirable if the target tester is automated (which is mostly the case), but not necessary if stand-alone bench testers will be used. Aids for post-layout simulation, which means mainly extraction of parasitic loading effects introduced at the layout stage, are highly desirable but not essential if such effects are taken into account in the initial design. Design rule checking is always essential at some point in the system, though not necessarily as a one-shot effort at the end.

Before going on to describe, in general terms, the kind of aids necessary for each of these design steps, let us first examine the nature of simulation as it relates to silicon. The "why" of simulation has been addressed previously. Basically it replaces the traditional breadboarding step in design, and its purpose is to demonstrate that the design will behave as required. However, a distinction must be drawn between the simulator itself and the modeling information it requires in order to correctly determine the behavior of the circuit. The simulator itself is basically an algorithm, with a variety of features that enable the designer to set global values for simulation, and is usually to a large extent technology-independent. Thus the same logic simulator can be used for NMOS, CMOS, bipolar, or even PCB simulation. The "model library," however, is technology-

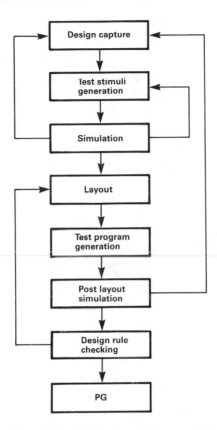

Figure 15-5. Basic silicon design process.

specific, and incorporates the detailed behavior and sometimes also the timing information of the building blocks for that technology. Usually this behavior is described in terms of primitive elements such as gates (inverters, NANDs, latches, etc.), or transistor elements (N-channel, P-channel, NPN, PNP, etc.). These are embedded in the simulator; so all models must be described in terms of the primitives that the simulator understands.

The second aspect of modeling the device behavior, the generation of input stimuli for, and expected responses from, the device, is always completely under the designer's control. The algorithm implemented by the simulator is fixed (to a large degree, model libraries may be available and fixed), although the design system may give the designer the flexibility to create his or her own higher-level models. But the input stimuli, which represent the questions the designer wishes to ask the device, and the responses, which represent the expected answers, are the designer's alone. The quality of any given simulation will depend on the extent to which the designer's input asks "all the right questions." Simulators will only respond to the questions asked, and the accuracy of the answers obtained will depend on the accuracy of the modeling information provided.

To underscore this point, a simulation is very much an abstraction of real behavior, and, when compared to breadboarding techniques, is unlikely to uncover any surprises in the design due to an unexpected set of input conditions. Of course, the same point can apply to prototype breadboards if they are not given a thorough workout under system conditions.

Nevertheless, simulation is still an excellent way of debugging a design, and of making sure that basic concepts are correct (i.e., expected input do indeed give rise to expected outputs), and that system timing requirements are achievable with the chosen technology.

Referring again to Figure 15-5, we shall now describe the aids required for each step. Design capture is necessary in order to describe the device in the language of the simulator and to provide the necessary information for chip layout. The minimum requirement here is a simple text editor, which allows the designer to input a set of control commands, followed by a description of the device network, its cells, and interconnections, which is known as a net list. A different, although not necessarily better, technique is to employ a graphical schematic capture tool. This popular aid enables the designer to capture his or her original schematic, using predefined symbols and boxes and connecting them up by using a graphics editor. A compilation is then performed to produce a net list in the format of the simulator. The designer usually must still input control commands as text. This approach offers no increase in functionality over the text method because the graphics tool is limited by what the simulator can understand, but many designers feel more comfortable at the schematic level, so the capture may proceed faster and have fewer input errors. For analog and transistor-level circuit simulation, it is necessary for the schematic capture tool to allow the assignment of values to a symbol, such as a resistor value, the width-to-length ratio of an MOS transistor, or the emitter area of a bipolar device. This feature is not necessary for logic schematic capture.

Test stimulus generation requires no aids for circuit and/or analog filter simulation. Input signals such as sinusoids or voltage ramps can usually be generated by control commands within the simulator. However, for digital logic simulation, it is necessary to generate test vectors. The essential aid is, again, some form of text editor to format the basic "1's" and "0's" of the pattern, although this form of entry is tedious and error-prone if the pattern gets too lengthy. A better aid would be some form of test pattern specification language, which would allow a higher-level description of repeated patterns and subroutine capabilities.

The simulators are the heart of silicon CAD systems. In general, there are four classes of simulator. The circuit simulator uses highly detailed transistor level models and can usually also accommodate components such as resistors, capacitors, and diodes, and even in some cases inductors and transformers for board-level simulations. This type of simulator is the most accurate but usually cannot cope with circuit sizes greater than a few hundred elements. It will also enable both time and frequency domain analyses to be performed. The next class of simulator is known as a switch level simulator. This is basically a logic simulator, but the primitive elements are transistors modeled as switches, rather than as higher-level gates. This simulation technique is less accurate than circuit simulation, but consumes much less central processing unit (CPU) and can handle larger circuits.

The switch level simulator can be regarded as sitting halfway between the circuit simulator and the gate level logic simulator. The latter is the third class of simulator and can handle extremely large circuits with relatively little CPU expenditure, by using models such as inverters, NAND and NOR gates, and flip-flops as its primitive elements. It will only allow inputs and outputs to take on a very limited set of values—0, 1, X (unknown), and Z (high impedance)—although it may also allow strengths such as "pulled," "driven," or "charged" to be associated with these conditions. Thus modeling takes place at a much higher level than the continuously variable inputs and outputs of a circuit simulator. The logic simulator is the least accurate of the three classes of "structural" simulators, as its information depends completely on the logic models, which do not explicitly contain any basic device information. The logic simulator may be of the unit delay type, which means that no timing information is contained within the gate models. In this case a second timing simulator is also required in order to model circuit behavior.

The final class under discussion is the behavioral simulator. Unlike the first three types, this kind of simulation makes no pretense of accuracy, or "real-world" modeling. It is intended as a high-level modeling aid, which allows a designer to try out ideas at the "black-box" level, with no knowledge of the internal structures of the functions. As a simple example, such a tool allows a designer to specify an addition as $(A + B)$, rather than a series of gates or transistors that form a full adder. The emphasis is on inputs and outputs rather than on implementation, so that this type of simulation is capable of giving results for highly complex systems with very little expenditure of processing power. The emphasis here, then, is on establishing the right functions, rather than on getting the functions right.

From the designer's perspective, chip layout tools may be completely invisible, in that the designer provides a net list to a support group, which may be a silicon vendor, who then run an automatic layout tool. However, some systems provides for designer interaction, so that critical signals may be given high priority, or different layout algorithms used, depending on the needs of the particular function. Non–silicon-expert designers vary in their desire to be involved at the layout phase of design, although most probably would rather forgo the experience. However, for standard cell design, considerably better layout results can be achieved with some intelligent designer input than if the program is allowed to run freely. This point can also apply to gate-array designs, although they are less flexible.

In order to complete a layout, a graphics editor that allows for manipulation at the individual mask level is also essential, at least for the silicon vendor although not for the chip designer. Even if automatic layout tools are to be used, it is usually necessary to add common manufacturing features, such as alignment aids, scribe channels and chip names, and vintage codes. It is also common for a few manual "tidy-up" operations to be performed on the device before it is submitted for manufacturing, in order, for example, to cope with failures of the automatic layout software, to reroute clock signals identified as critical, or to accommodate special power and ground routing requirements.

The original building blocks, or cells, on which the design is based are also first implemented using the graphics layout editor, although again this activity is invisible to the chip designer.

Test program generation is firmly the province of the chip designer, since it is unlikely that anyone would be able to afford the luxury of an assigned test engineer for every chip. In this regard, the CAD industry has generally had a very difficult time because a multitude of automatic test equipment exists, each with its own unique syntax, its own method of generating timing signals, and its own pin electronics for interfacing to the device under test (DUT). Also, most designers prefer to concentrate on traditional issues such as cost, performance, and correctness of design, as opposed to making life easier for manufacturing; so it is easy to see why this essential function has been neglected when compared to the effort spent on general-purpose simulators, layout tools, or design rule checkers.

Fortunately, the problems inherent in devising a general-purpose test program generator are analogous to those encountered in simulators or rule checkers, in that each of these needs a technology-specific input in the form of cell libraries for a simulator and a specific design rule set for checkers. These inputs are normally provided for each specific application, either by users themselves or under contract by the CAD tool vendors. For a test program generator, the problem that needs to be addressed is the combining of test patterns with parametric tests, such as propagation delays, set-up and hold checks, input leakage, and output drive and switching levels. In addition, such common features are required as continuity checking, data-logging capability for characterization, and general control commands for setting up timing generators and defining pins. Each of these items can be addressed in a general way by the test program generator, and then a post processor can format the design information in the syntax of the target tester. As with design rule checkers, these post processors may be written by the end user, or the vendor may provide them. The central message here is that all the information necessary to formulate a complete test of the device should be contained in the design file used for simulation, and therefore should also be used for generating the test program. As a significant side benefit, such an approach should also ensure that chip designers take responsibility for testing a device rather than handing it over to a test engineer at the "end" of the design cycle.

Post-layout simulation was described earlier in this section as being "not absolutely essential" if all the loading effects arising from the physical layout of the chip are correctly accounted for. Except in very limited circumstances, this "correct by construction" approach cannot be completely relied upon. It is thus advisable to have the facility to extract interconnect loading effects from the chip layout, and to apply these real numbers to the original simulation, thereby replacing the estimated loads. It should be the aim of every designer to ensure that post-layout simulation merely confirms a good design margin, but, in reality, sometimes layout change will be necessary.

So far we have considered only full chip extraction of interconnect loads for digital chips. However, for analog functions and for the original design of the standard digital building blocks, the need for extraction and post-layout simulation is unambiguous. The problem is also greater because here we are concerned not only with interconnect effects but also with extraction of transistor sizes, which will rarely be exactly as originally designed. Resistance effects of junctions and gates must also be included because much of the interconnect in cells is not in metal. One must also account for the existence of parasitic effects that take the form of interactions between unrelated signals. The type of tool used here is known as a circuit extractor, which effectively produces a description of the cell or analog block in the language of the circuit simulator to be used.

Design rule checking is another essential phase of the integrated circuit design process. We have already hinted that this procedure is not necessarily best performed as a "full chip" check at the end of the initial design phase. In fact, as chip sizes become larger, such checks are extremely expensive, and consume inordinate amounts of processing power. The layout designer who never makes mistakes is extremely rare, so it is likely that several passes through the checker will be necessary before all rule violations have been corrected. Thus rule checking is best done in a hierarchical fashion, with all cells being verified with a complete rule set. A second level of edge checks can then be run, to make sure that rules will not be violated when cells are butted together; and because most chips are composed of macro blocks, which contain both cells and interconnect, a third level of checks is run to ensure that the interconnect does not come too close to cell borders.

Rule checking clearly should not be in the domain of the chip designer but is the responsibility of the cell designers. In many cases a single complete chip rule check is run, as a double check, because so many different combinations of cell placements and interconnect routing are possible that it is very difficult to produce a complete set of abstracted "edge rules."

The last phase of the design process, the production of a PG tape, is essential, but usually is completely beyond the ken of the chip designer. The PG tool basically takes the polygon mask data of the finished chip and "fractures" the data into smaller polygons. These data are used to drive the PG or E-beam mask-making machine. The total number of polygons produced on each mask layer by such a program translates into the flash count for the PG machine, which basically represents the number of exposures necessary to complete the mask. To put this in perspective, fairly large chips, say 0.75 cm on a side, would require several hundred thousand flashes to complete the metal layer in a 3 micron process. At this level many hours are required on a mechanical system, so that electron-beam scanning techniques are preferred.

Before we describe the particular requirements for two specific types of design systems, it should be noted that we have discussed only those CAD tools that are central, and in varying degrees, visible to the chip designer. In most complete CAD systems, many more tools may exist, both for purposes of internal data translation between different formats and to provide special functions and utilities in support of the tools themselves. There are also many different approaches to simulation, timing verification, production of cell libraries and array structures, and automatic layout. However, space limitations prevent further discussion of this subject here.

Standard Cell Logic Design System

From the chip designer's point of view, a design system for random logic design using standard cells should look like Figure 15-6.

We have discussed each of these tools in general; now let us describe some of the specific features that would make such tools attractive. This is by no means an exhaustive list; many variations are possible.

For graphical schematic capture, a context-dependent menu scheme is probably the most effective form of designer interface. In this scheme, the designer would be shown the symbols that are available to be placed, along with their names, which should give some idea of the function represented. By simply pointing to the symbol on the menu, and then pointing to the position in which it should be placed, the designer can build up the schematic. The next step in the process is to connect the boxes together; so pin positions and designations should be clearly identified on the symbols, and the act of drawing line segments should be as simple as possible. In addition, commands for moving symbols, deleting symbols, moving and deleting lines, and sundry other graphics editing commands are necessary. It should also be noted that this initial phase of the design in-corporates the inherent inflexibility of the standard cell system. The designer can only use those functions that have been preprogrammed into the capture tool, and for which logic models and cell layouts exist. This will lead to non-optimal design in many circumstances. The output of the schematic capture tool is a net list, written in the format of the logic simulator, and perhaps also

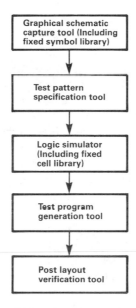

Figure 15-6.   CAD tool set for standard cell design.

containing some key universal control commands. In a good system this same file will also be used for automatic layout, without manual intervention; so all design change should be made at the schematic level and not in the text of the output file.

A final point on design style, sometimes forgotten in the excitement of using a "pretty" color graphics machine, is that individual design blocks are limited by the "page size" of the graphics terminal screen. It is necessary to have a good scheme for naming off-page connectors, and the design must be structured to be easily understood functionally, so that debugging can proceed quickly. As many designers have found from bitter experience, leafing through tens, or even hundreds, of pages of schematics can be extremely frustrating.

The minimum requirement for test stimuli generation is a text editor, to input the "1's" and "0's" for each pin at each clock cycle, which is a format required by both simulators and test machines. Usually this takes the form of rows and columns of numbers. Each column represents a pin, and each row a clock tick. Additional information is also required to specify which pins are inputs, which are outputs, and which are bidirectional. In the simulator, one may also specify "don't care" and high impedance. The designer should beware of this, as in the real world a "don't care" is either a "1" or a "0," and a high impedance will be pulled either high or low. It is thus necessary to ensure that "don't cares" are eventually specified arbitrarily, and unless special arrangements are made, compare strobes are disabled appropriately.

A much better aid for test vector generation (each row of the pattern is a "vector") is a pattern specification language. This form of tool reduces the burden of writing a vector for each clock tick by allowing the designer to conveniently specify repeated patterns. Thus, a long period of time in which there is no input or output activity, but with the clock operating, could be specified as a clock burst, so that a thousand clock ticks could be represented by one line of code. Similarly, large repeating sections of data could be specified in one subroutine called many times, and patterns that are algorithmic in nature could be specified in a high-level language, with automatic conversion to test vector format.

This will make the generation of large numbers of test vectors much easier because we are still at the stage where the limiting factor is designer time, rather than the time it takes to apply test vectors.

A final word of caution on this most critical stage of any integrated circuit design: Most simulators can readily handle test vectors and timing simulations that are not implementable on currently available testers. Even though these simulator features may be useful in the early phases of design, it is essential that the criteria for generation of test vectors be finally limited by the capabilities of the test equipment, and not those of the simulator. This should lead to the creation of only one test pattern for all purposes.

Logic simulation is the point in the design cycle where the designer actually receives answers to the questions posed by the circuit schematic and test vectors. In addition to the net list and test vectors already produced, the simulator also requires control commands and display commands in order to produce a reasonable output. The control commands take the form of global or local commands to perform such operations as: setting the clock cycle time; performing race and hazard checks; increasing delays to account for temperature, power supply, and process variations; and deciding whether or not to compare outputs to expected values. Subsidiary commands may allow the user to perform diagnostics on the circuit, preset values to a given level, or save a simulation so that it can be later restarted from a given point. Display commands allow users to format their output so that they may specify, for instance, whether the output is to be printed as "1's" and "0's", or plotted as timing waveforms; which signals are to be displayed and over what time period; and whether output is displayed at specified times, or only when signals change state. The central output should be whether any compare errors were found, and whether all nodes were exercised by the test patterns.

Thus starts the fairly lengthy process of debugging, fixing the circuit schematic, and resimulating. The designer can proceed to the next stage only when fully satisfied that the correct functions are being achieved, that no compare errors are flagged, that all race and hazard warnings are adequately explained, and that the design has good timing margins.

In creating the manufacturing test program, the designer's job is to provide a complete test pattern, a set of parametric specification limits, and the definition of each pin on the device. The test pattern should be identical to the simulation pattern, but the designer may also have to add some information with regard to clock rates, timing generators, and output strobes. The parametric specifications usually consist of switching points, propagation delays, current consumption, input and output leakage tests, and so on. The pin definitions tell the tester which pins are inputs, which are outputs, and which are special, such as power and ground and bidirectionals.

It is important to realize that there is strong interaction between the pattern and the parametric tests in many circumstances. The pattern must be constructed so that pins are in the correct condition when measurements are made. It may be as simple as setting all outputs high for leakage measurements, or as complex as running a subroutine during dynamic current tests. It also may be necessary to reinitialize the device after the pattern has been stopped for a particular test. In this case it helps to have an initialization subroutine, which can be called as required.

A final point, which usually escapes the attention of most designers, is that more than one test program is required. A minimum set would normally consist of three programs: one for wafer probe test, one for final package test, and a third for quality assurance. The first program is intended to weed out devices on the wafer that have manufacturing defects. Because of the nature of the probing equipment, this program will normally run at a much lower speed than is normal for the device. A complete pattern should be run, however, so that only those dies that are functionally good will go through the expensive packaging procedure. Once the dies are packaged, they are tested again, this time at a much higher rate, although it may not be possible to run at the full operating speed. In this case the speed performance of the circuits must be inferred from selected propagation delay measurements. This can be done; although most commercially available testers cannot run patterns faster than 10 to 20 MHz, they are capable of timing measurements that are accurate to less than a nanosecond. It is also usual to test only at room temperature and nominal voltage in manufacturing; so due allowance must be made for shifts in performance over the full operating range. This "guardband" in specification is intended to ensure that all lots are of acceptable quality, when put through their paces (usually on a sample basis) by the quality assurance program. This program contains the hard limits of the device specification, and usually also contains extra sophistication, such as the means for data-logging results for later analysis. In a sensible scheme of things this program would also be used for initial characterization of the device, although such a program may also contain special tests that do not need to be performed in production.

In the system we are describing here, the designer will not be involved in layout, except perhaps to provide information on critical signal paths and special power routing requirements. However, once layout has been completed by the automatic tool, the designer must verify the device. This verification consists of rerunning the simulation and analyzing in detail the real internal delays on the chip. Obviously, if the simulation fails this step, then the circuit must be debugged and changed, throwing the designer back to the schematic capture phase. Even if the simulation passes, it is advisable to check the results of timing analysis for the most heavily loaded signals to ensure that the chip has adequate margin.

Full-Custom Mixed Analog/Digital Design System

As the title of this section implies, mixed analog/digital designs are still largely in the domain of the silicon expert, although some semicustom techniques for simpler analog functions are emerging. The essential difference for the digital section of the chip is that the designer, or more usually the design team, has control of the chip layout. Many of the same tools are used, however, so we will only address the major differences in this section. This is best done by reference to Figure 15-7, which is a general representation of the layout hierarchy for a mixed A/D chip (McWalter 1983). Alongside each box the major functions to be performed are described.

Taking the digital section first, and proceeding upward, if all digital primitive elements are to be designed from the transistor level, the tools required will be a circuit extractor, a circuit simulator, and a design rule checker, all of which have been previously described. In addition, logic models for each primitive cell must be constructed, for later use by the logic simulator. The final item, I/O generation, applies to both the symbolic representation of the cell for schematic capture and an abstracted version of the layout that contains the real locations of all I/O points. The latter is advisable to avoid complications during hand layout.

A more common technique for CMOS chips is to mix the standard cell approach with a few customized elements. In this case the contraint is that all custom cells follow the same format as the standard library, in order to avoid problems when the two are merged. Once all the digital primitives necessary for the chip have been designed and verified, a macro block approach is adopted. These blocks may consist of as many as a thousand gates or more, so that circuit simulation is no longer possible. Once again, a mix and match approach is possible, with customized macros being joined by standard ones. In addition to gate level and functional or behavioral simulation, it is also expedient to perform timing simulation and interconnect loading extraction at the macro block level. This will minimize any surprises caused by extremely long signal lines traversing several macro blocks. Block I/O generation involves the same steps as for primitive cells, in that schematic symbols and layout representations should be constructed.

The final item for macro blocks, interconnect expansion, refers to a technique

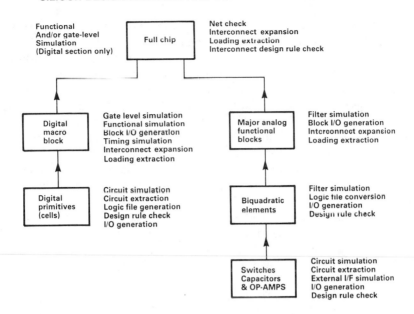

Figure 15-7. Layout heirarchy for mixed analog/digital chip.

whereby all interconnect, both within the block and between blocks, is done by hand on a graphics editor. This interconnect takes the form of centerlines, so that the designer draws only a single line between each I/O point that represents the center of the track. This technique requires the availability of a CAD tool to perform expansions of the centerlines to the desired width and to ensure that the spacing between them conforms to design rules. Contacts between layers are placed automatically, depending on the intersections between interconnect layers, or they may be manually forced by the designer.

Another type of tool that would increase the flexibility of the digital design system is a structural array builder, which would allow the simple construction of regular structures such as memories or programmable logic arrays (PLAs). Such a tool would require only high-level input from the designer, such as Boolean equations for a PLA or type and size for a memory. The array builder would then construct both the physical layout of the block and the logic model for that block.

Moving on to the analog side of the hierarchy, the boxes in Figure 15-7 refer largely to the design of switched capacitor filters, although many of the steps are applicable to other types of analog function. Starting again at the bottom, it can be seen that many of the same functions must be performed to create operational amplifiers, switches, and capacitor elements as were employed in the creation of digital primitives. The key difference is that because many analog sections "talk" to the outside world, it is advisable to simulate the performance of the main elements with the external components that will surround the chip in practice. This usually requires a circuit simulator that is more sophisticated than is necessary simply for the integrated circuit, particularly if inductors and transformers are involved. Proceeding to the next level, many filters are composed of biquadratic elements, which consist of particular arrangements of switches, op-amps, and capacitor banks. Here there are two marked differences from the digital world. First, a filter simulator is necessary, which will perform frequency domain analyses of the biquadratic structures. This may be a feature within the circuit simulator, but more often is a special-purpose tool, built around the mathematics used to describe analog filters. A second point is that most, if not all, connectivity checkers are built around digital simulators, and operate on a format that is different from that used to describe filters. In most practical cases it is more convenient to construct a CAD tool that converts the analog format to digital for net checking purposes, as opposed to creating a whole new checker. The biquad level of hierarchy can obviously be omitted if the analog structures are not filters.

The next level of hierarchy is the functional block, which may be an nth-order filter, and A/D or D/A converter, or simply a sample and hold or a comparator. Once again, a circuit simulation, usually in the frequency domain, must be performed, and here a special program may be necessary if the number of elements in the block exceeds the capability of the circuit simulator.

Now we get to the really difficult stage, the merging together of digital and analog blocks into a single chip. No one simulator is capable of completely simulating a complex mixed chip. Although there has been much talk of a "mixed-mode" simulator that accommodates both, a useful practical tool has not yet emerged. Thus, all checking of signals that cross between the analog and digital sections, such as switch clocks and control signals, must be done by abstracted means. In practice, this tends to mean the manual creation of files that are intended solely for checking and are not simulated. The issues of noise interactions between digital and analog sections, and crosstalk and distortion mechanisms that occur as a result of signals crossing each other, are not addressed by simulators at all. The result usually is that several iterations of the chip are necessary before all parametric specifications are well within limits. However, design systems are capable of ensuring that all signals are connected correctly, enabling the designer to concentrate on the more subtle problems associated with analog circuitry, rather than crossed wires.

DESIGN FOR TESTABILITY

This topic has been addressed at several points in the preceding sections, but is so important that it merits further discussion. The design flows already presented clearly place the responsibility for device testing on the chip designer; for, as semicustom designs proliferate, the expense of hiring a test engineer for every chip will become prohibitive. Indeed, the "throw it over the wall to manufacturing" approach to design has always been expensive, and most of the ideas in this section are also applicable to PCB design (Bennetts 1984).

To begin with, there are two distinct purposes in simulation and testing of custom integrated circuits. The first of these is functional verification of device behavior—is the device suitable for its intended applications? The second purpose is related to the nature of the silicon manufacturing process—has the device been manufactured in accordance with the specification? To elucidate further, in the world of silicon it is much more difficult to "see" defects than on a PCB. Thus, although die visual inspections take place, it is much less likely that all defects will be uncovered. Moreover, with the much greater interconnect density and the inaccessibility of internal nodes, it becomes clear that the onus is on the test program to find manufacturing defects on the chip. Thus there are two types of test: functional and structural. The first type is difficult to perform directly via CAD tools because it is impossible to establish whether the designer has simulated all the functions. The second type of test is more amenable to help from CAD tools because they can establish whether all nodes have been toggled. However, such testing in itself is not sufficient.

Before we elaborate on testability techniques and test pattern verification, let us dispose of the issue of CAD tools for functional testing. By and large, they consist of pattern generation aids. The basic philosophy is to make it easy to generate functional patterns so that the designer does not run out of steam and miss a few. This will probably suffice for the first few million patterns, although the volume of data will cause problems because test rates are usually in the several megahertz region. However, for devices such as microprocessors, which have a dauntingly, if not infinitely, large series of functions, such pattern generation is not enough. The problem then moves into the realm of methods and architectures. To date, ad hoc testing techniques have evidently sufficed in many cases, although the testing of microprocessors has not been without problems. As we move to much higher levels or integration, with complex processors embedded deep within customized structures, new techniques will become necessary. I do not know what these will be. Rather than a general solution, there will probably be a series of application-specific solutions, serviced by a series of CAD tools.

Structural testing, on the other hand, is a more readily soluble problem than functional testing, and has received much more attention from CAD developers. The basic role of CAD in verifying testability is to calculate the degree of fault coverage of a given test pattern applied to a specific chip. Design techniques such as scan designer can also be implemented, allowing automatic generation of test patterns. Further techniques may involve "built-in" signature analysis circuits, in which the input pattern is generated by the chip itself, and the results are compared with a fault-free signature. There are also the "exhortation" techniques, which take the form of requests to designers to avoid logical redundancy, use synchronous circuits, use combinational rather than sequential logic, and maximize controllability and observability. All these techniques are obviously sensible in themselves, but, equally obviously, they cannot be applied to all design circumstances.

The first question we have when trying to determine the fault coverage of a given pattern concerns the types of fault we are trying to model. In practice, many different types of manufacturing defect may occur, which will vary from technology to technology. The effects of such defects will also vary, but it has been found that most situations can be covered by the so-called single stuck-at model. This basically involves the modeling of each node as "stuck at 1" (SA1) or "stuck at 0" (SA0). The test pattern is then applied to the device in order to discover whether the fault has been detected at the output of the chip. In the serial method of fault simulation, this procedure is repeated for each node. For a circuit with 5000 nodes, which is not unusual, 10,000 simulations would have to be performed with a complete pattern; so an extra layer of sophistication is necessary to make fault simulation more cost-effective. This could take the form of parallel or concurrent fault modeling, where multiple versions of the circuit are modeled in each run, or, more simply, a statistical form of serial modeling could be implemented. The latter is the more common technique, and involves the selection of a random sample of faults, say 10% of the total. If we get 90% fault coverage on 10% of the faults, then it is statistically likely that we will get 90% coverage on them all.

The second question in verifying patterns for fault coverage is: how much coverage is enough? In a perfect world the answer would be 100%. However, we already know that we cannot simulate all possible faults, so it would appear sensible to try to completely cover the ones we can model. For reasons related to the "exhortation" techniques mentioned earlier, it is common for a circuit to contain structures in which certain failures are inherently undetectable. The prime example of this is logical redundancy, which may have been introduced accidentally, or may serve to mask static hazards or increase drive capability. It is also possible that, although a node is toggled internally, it is not observable on a device output, or alternatively that a node cannot be toggled without a lengthy and complex sequence of input patterns. The latter is a controllability problem. While every effort should be made to maximize controllability and observability, the results of fault simulation often show coverage to be in the 80 to 90% region. Techniques are available, such as scan design and some forms of combinational logic, whereby patterns can be generated automatically that guarantee 100% stuck-at fault coverage, but they may not be readily applicable in many designs.

The designer should take heart, however, because certain statistical factors are favorable. The fact that fault coverage is at 90% dos not mean that 10% of devices shipped will malfunction. In fact, many devices on a wafer will be fault-free, and will function whether you test them with your 90% pattern or not.

There is also nothing to say that only one defect will occur on bad dies, and in fact "clustering" of defects is often observed in silicon manufacturing processes. This obviously increases the chances of fault detection, as it is unlikely that two or three defects will occur only at those nodes that have been deemed undetectable by the CAD tool. The key point here is that we are trying to achieve the best possible quality level, in terms of shipped product, at the budgeted cost. Perfection is not possible, but an acceptable quality level (AQL) of well under 1% certainly is. The problem then is to relate the test pattern fault coverage to AQL for a particular class of device and a particular manufacturing technology. The answer received will depend on many factors, not the least of which will be design judgment.

Finally, let us consider analog testing. The parametric test problems associated with linear devices are well known, and relate mainly to parametric measurements such as noise, distortion, and crosstalk. These measurements are difficult to perform, and tend to be time-consuming if sufficient accuracy is to be guaranteed. However, from the point of view of fault coverage, things become simpler than with digital circuits. This is so because the simple existence of, say, a sine wave at the output, demonstrates that the op-amps, switches, clocks, and capacitors are free of manufacturing defects. Thus, if a complete set of functional parametric measurements is made, there is no need to take special precautions for fault coverage. Nevertheless, most designers of mixed analog/digital chips also include test points within the circuit for debugging purposes, and it is usually wise also to bring them to the outside world if possible.

SILICON COMPILERS

The concept of silicon compilers arose by analogy with the world of computer software. Thus programmers started with machine language programs (equivalent to transistor level design), and then progressed to ever higher-level descriptions that could be automatically compiled into machine language. The analog of this in hardware design is direct translation of behavioral-level descriptions into customized silicon structures, whose base elements are built into the compiler, but whose final form is programmable by the designer. Thus, the ideal silicon compiler would take this behavioral description and compile it into a silicon chip automatically. The chip would be "correct by construction," and would not need further verification.

Although there are many points of similarity between software development and silicon hardware development, there are also some significant differences, which must be taken into account, and which will influence the development of this exciting and fast-moving field.

Let us first examine the roots of silicon compilation (Mead and Conway 1980; Ayres 1983) and see how it has progressed to its current level. If we consider a regular device, such as a random access memory (RAM), it is easy to see that, once we have the base element (the 1-bit RAM cell), a CAD tool can be constructed to place as many of them as we desire in an array of m bits by n bits. A second tool could place the address decoders and sense amplifiers in the appropriate places; add buffers and device pins, and we have a device. Whether it works or not is a function of the original design of the building blocks and the manufacturing technology used. Thus, the first silicon compilers were simply array placement tools, and were in fact the same tools that were being used in more general-purpose design systems.

The next step was to add further basic building blocks to the design system, so that now we could easily construct arrays of several different types of memory, and also such regular structures as PLAs and perhaps ALUs (arithmetic and logic units).

The problem with this is that we have not defined what is correct functionally, only what is correct physically. Thus, the next phase of compilers, which is the current stage of development, added the ability to either input or extract electrical and functional information. We can now specify the function of a PLA via Boolean equations, and extract the performance characteristics. Similarly, for a RAM we may input the number of words required and the length of each word in bits, and again extract the performance and size information. A solution to the random logic problem can also come from a gate array, where we input a gate-level description in text format in a language that provides capabilities for do loops and subroutines, and then compiles the results, using a predetermined library of basic elements. The state of the art in silicon compilation then is "the structural compiler" rather than a functional or behavioral compiler. Thus, the designer must input information that relates the physical structure of the chip.

The compiler also is still restricted by the library of components to which it has access, each of which must be carefully constructed by silicon cell designers, although recently a number of "cell compilers" have emerged, easing this task. The disadvantage of the cell compiler approach, from the chip designer's point of view, is that it takes the designer back to the stage of transistor level design. This is a retrograde step because the purpose of silicon compilers is to ease the complexity management problem by moving to ever higher levels of abstraction. However, cell compilers will become a very useful tool for technology providers, as opposed to technology users. Here we also see the basic difference between software and hardware compilers. The output of a software compiler is basically "1's" and "0's", which are hurled at a piece of hardware known as a computer. The output of a hardware compiler is at a much higher level, and consists of predetermined cells. It is the availability of such cells that is the limiting factor. Until the hardware compiler can work directly at the basic silicon level, which consists of polygons on different mask layers, the analogy will remain incomplete.

A further, and central, problem in the development of silicon compilers is the language to be used to describe the chip. It is clear that the output of a silicon compiler should be the test tape and PG tape that current design systems provide. It is far from clear what is the most suitable means of initially describing the chip. Current methods include traditional gate-level schematic capture; menu-driven inputs, which present the designer with choices depending on the type of array selected; and hardware description languages. These vary from simple net lists to sophisticated "programming" languages, which allow features such as loops, conditionals, typing, and concurrency. It remains to be seen which, if any, becomes a standard, but it seems certain that silicon compilation will become the basis of future design systems.

CONCLUSION

The concept of silicon compilation arose by analogy with software, and also from the coming together of many CAD tools. These tools had largely been developed as a result of particular problems in the integrated circuit design process. Recent dramatic improvements in the ease of use of such tools, and their integration into recognizable, self-consistent design systems, are likely to make custom silicon the primary hardware design medium of the future. As with definitions of LSI and VLSI, whether or not such systems are called silicon compilers seems to be largely a marketing issue. The point that emerges clearly, however, is that as the use of silicon technology expands, and its capabilities increase, custom integrated circuits will become a primary factor in system design decisions.

GLOSSARY

BJT: bipolar junction transistor.
Bipolar process: a silicon process that produces both types of bipolar junction transistor.
Emitter, base, collector: the three terminals of a BJT device. When the base/emitter junction is forward biased, a small current will be established in the base input. When the collector base junction is reverse biased, a much larger current will be established between collector and emitter.
MOSFET: metal oxide semiconductor field effect transistor.
Gate, source, drain, substrate: the four terminals of an MOS device. A voltage set up on the gate of the device establishes an electric field across the nonconductive gate insulator (hence, field effect transistor), creating a conducting path, known as the inversion region (carrier type is the opposite of that in the substrate material) between the source and drain. When a voltage is applied to the drain and the source is grounded, a current will be established. The minimum gate voltage necessary to support this current is known as the threshold voltage, depends on the characteristics of the material and the potential difference between the source and the substrate.
CMOS process: a silicon process that produces both N-channel and P-channel MOS devices.

N-type: a semiconductor material that has been doped so that the majority carriers are electrons. The materials that produce this effect in silicon are elements such as phosphorous, arsenic, and antimony.

P-type: a semiconductor material that has been doped so that the majority carriers are holes. The materials that produce this effect in silicon are elements such as boron and aluminum.

N-well: a form of CMOS process in which the N-channel transistor is formed in the P-type substrate material, and the complementary P-channel transistor is formed in an N-well.

P-well: a form of CMOS process in which the P-channel transistor is formed in the N-type substrate material, and the complementary N-channel transistor is formed in a P-well.

Twin tub: a form of CMOS process in which both types of well are produced during the manufacturing cycle, although one well is a modification of the substrate doping for threshold control.

NMOS process: a silicon process that produces only N-channel devices, although both enhancement and depletion mode devices are usually available.

PMOS process: a silicon process that produces only P-channel devices; now obsolete as a design medium.

Metal gate: an MOS device whose gate material is formed from metal. Such devices were the earliest form of MOS technology.

Silicon gate: an MOS device whose gate material is formed from polycrystalline silicon that has been heavily doped. Most current MOS processes employ this technique because it allows "self-aligned gates"; that is, the gate itself forms a mask for correct formation of the source and drain regions. The advantage of this is that one process step is eliminated, and parasitic capacitance is reduced, giving rise to faster transistors. As an additional feature, the silicon gate device is very stable under stress, and the polysilicon provides an excellent barrier against contaminants.

Enhancement mode: an enhancement mode MOS device is OFF if the gate–source voltage is zero; that is, a gate voltage must be applied to turn it ON.

Depletion mode: a depletion mode MOS device is normally ON; that is, a gate voltage must be applied to turn it OFF.

VDD, VSS: the normal nomenclature for the power supply rails in an MOS circuit. VDD is the voltage applied to the drain, and VSS is the voltage applied to the source.

Epitaxial layer: a layer of single crystal material grown on a substrate of single crystal material, usually of different doping type or radically doping density.

TTL: transistor-transistor logic; a form of bipolar logic, usually in SSI and MSI form, that has been the standard for most PCB logic design for many years.

ECL: emitter coupled logic; another form of bipolar logic, characterized by very high speeds and equally high power consumption.

Gallium arsenide: a semiconductor material that has much higher mobility than silicon, resulting in higher-speed operation. It competes with ECL for speed but at lower levels of power dissipation. Difficulties with higher levels of integration and cost make it unlikely that gallium arsenide will compete directly with silicon in the general-purpose market.

Integrated circuit: a circuit consisting of a number of transistors ranging currently from two (an SSI inverter) to more than one million (a 1-megabit RAM), fabricated on the same substrate. The substrate is usually silicon, but could also be gallium arsenide, or an insulating material, as in the case of silicon on sapphire (SOS).

Wafer: a silicon disc, usually from 3 inches to 6 inches in diameter and about one-fortieth of an inch thick, which forms the substrate material for integrated circuits.

Chip, die: synonyms for integrated circuit. These terms refer to the finished entities that result after a wafer is scribed and broken.

Process: usually consisting of as many as several hundred discrete steps, the mysterious means by which a raw silicon wafer is turned into many integrated circuits containing the required circuit elements. The process consists of such steps as oxidation, photoresist deposition, mask patterning, ion implantation, and polysilicon and metal deposition. The patterns are established by means of a series of masks.

Photolithography: a process by which a light-sensitive material is deposited on a substrate, selectively exposed in certain areas using masks, and developed. The underlying material is then operated upon, e.g., etched or implanted. The light-sensitive material is then removed, leaving behind the desired pattern.

Reticle: a glass plate that contains the mask information for one layer of one die etched in chromium. The pattern is etched by photolithographic techniques.

Mask: a glass plate that contains the information for one layer of all the dies that are to be fabricated on a wafer. The mask is made by transferring the information from the reticle, photolithographically, by a "step and repeat" process.

PG: pattern generation; a computer-controlled opto-mechanical system that takes the design information from the designer's "PG tape" and creates the reticles. Exposure of the reticles is controlled by blades that form polygons according to the data provided. These machines are capable of forming shapes as small as 3 microns square (or less), with accuracies of a tenth of a micron or so, on the underlying reticle.

E-beam: the even higher-technology version of the PG machine, which replaces the mechanical blades with an electron beam that scans the reticle, exposing it only in the desired places. This is now the mask-making technique of choice. It may also be used for "direct write" onto the wafer itself, thus obviating the need for masks. X-ray systems are also being developed for the latter application.

SSI: small scale integration; a few transistors on a chip, forming basic logic gates, NOT, AND, OR, etc.

MSI: medium scale integration; a few tens of transistors on a chip, providing functions such as shift registers, counters, multiplexors, etc.

LSI: large scale integration; hundreds or thousands of transistors on a chip, providing high-level functions such as microprocessors and their peripherals.

VLSI: very large scale integration; many thousands or hundreds of thousands of transistors on a chip, forming "system level functions" or large memories. The break point between LSI and VLSI is a matter of taste and marketing.

CAD: computer aided design; usually refers to a series of computer programs that assist a designer in verifying the function of a device. Complete CAD systems currently consist of many separate programs.

CAM: computer aided manufacturing; programs written to drive computer-controlled production machinery.

CAT: computer aided testing; a branch of CAD whose aim is to ease the production of test programs.

Schematic capture: a CAD tool that enables a chip designer to "draw" a circuit schematic in a natural way, i.e., as one would draw it on a piece of paper. The output of the tool is a net list, usually suitable for both simulation and automatic layout.

Simulation: the computer equivalent of breadboarding. The simulator allows the designer to confirm the operation of a circuit, provided that the appropriate model libraries are available.

Simulator primitives: the predefined elements within a simulator, which the simulator knows how to deal with, and which form the basis of the required model libraries.

HDL: hardware description language; a text language that may be as simple as a net list, or many contain complex features such as typing, concurrency, looping, and conditional statements. It may also support functional and behavioral simulation.

Mixed mode simulator. Several definitions are abroad! It could be a mixed gate-level and functional simulator, or mixed gate-level, functional, and behavioral. More normally it is a mixed analog/digital simulator.

Silicon compiler. Again many definitions are possible. Perhaps the simplest would be: a single tool that allows the designer to describe a chip and then produces both a model and a layout. By this definition silicon compilers do not currently exist because all design systems require the designer to use more than one tool. However, array compilers exist that are coming close to this definition.

DFT: design for testability; a series of techniques and CAD tools whose aim is to ensure that all devices are testable in manufacture.

Test vectors: a series of "1's" and "0's" that represent the inputs and outputs of the chip. Each vector has a length equal to the number of pins on a device, and represents one clock tick. A series of test vectors is known as a test pattern.

Test program: a complete program for an automatic tester that constitutes the specification to which the device will be shipped. The test program consists of the test pattern and a series of parametric measurements.

Fault coverage: the degree to which a test pattern detects all modeled faults in a circuit, known as the fault coverage of that pattern. It is usually expressed as a percentage.

ATE: automatic test equipment; a computer-controlled test machine used for production testing of silicon chips and PCBs.

Layout editor: a graphical CAD tool that allows for manipulation of cell or chip layouts at the polygon level.

Automatic layout: a method by which the net list representing the chip design is fed into a CAD tool that automatically places the cells and routes the interconnect.

DRC: design rule check; a tool that checks that all photolithographic design rules have been obeyed, so that the chip can be fabricated. It is necessary to write a set of rules for each technology.

ERC: electrical rule check; a CAD tool that checks that no basic electrical rules have been violated; i.e., it checks for power and ground shorts and opens, floating inputs, and unconnected outputs. It may also be programmed with particular device structures so that it can be used to indicate unusual structures that should be checked manually.

Connectivity checker: a CAD tool that verifies that a chip layout exactly corresponds, from an interconnect viewpoint, with the original net list. In principle this is unnecessary with a trustworthy automatic layout tool, but it is essential if the layout was touched by human hand!

Switched capacitor filter: a type of active filter suitable for direct integration on a silicon chip. The resistor element of a conventional filter is replaced by a switch and a capacitor, which when suitably clocked emulates resistor behavior. In this way the performance of the filter is affected only by capacitor ratios, which are well controlled in a silicon process, as opposed to absolute values of capacitors or resistor ratios, which are not.

Biquad: a second-order filter element. Many switched capacitor filters are constructed from cascaded biquadratic elements.

Workstation: the hardware and software available to the chip designer. The workstation may consist of a stand-alone processor, a processor that is also linked to a host machine for larger tasks, or a terminal configuration talking directly to the host.

Gate-array: a semicustom design technique in which the basic building block is a single cell configuration. The final chip is personalized by contact and metal masks only.

Standard cell: a semicustom technique in which the basic building blocks are a series of cells forming basic logic functions. Each chip requires a complete mask set for manufacture.

Macrocell: a higher-level function composed of the basic building blocks employed by the design system. This concept is applicable to both gate arrays and standard cell designs.

Cell library: a fixed library of elements available to the designer, which consists of both basic blocks and macrocells. Each element in the library will have an associated schematic capture symbol, logic model, and layout.

REFERENCES

Ayres, Ronald F., *VLSI Silicon Compilation and the Art of Automatic Microchip Design,* Prentice-Hall, Englewood-Cliffs, NJ, 1983.

Bennetts, R. G., *Design of Testable Logic Circuits,* Addison-Wesley, Reading, MA, 1984.

McWalter, I. L., "Design Techniques for Mixed Analog/Digital Custom VLSI Circuits," presented at the 2nd International Conference on the Impact of High Speed and VLSI Technology on Communication Systems, London, Nov. 1983.

Mead, C., and Conway, L., *Introduction to VLSI systems,* Addison-Wesley, Reading, MA, 1980.

Muroga, S., *VLSI System Design,* John Wiley, and Sons, New York, 1982.

16. Design Verification of VLSI Circuits

Kaoru Okazaki

LSI Research and Development Laboratory

and

Isao Ohkura

Kita-Itami Works

The rapid progress of VLSI design and fabrication technologies is enabling designers to produce denser and more complex integrated circuits. With increasing circuit integration, however, logic design error-discovered after fabrication of VLSI chips are becoming more time-consuming and expensive to correct. Thus, design verification—which aims to ensure before fabrication that the VLSI circuit will function exactly as intended—is becoming vitally important. The main activities in developing a VLSI circuit are functional (logic) design, physical (layout) design, and test generation. In each design activity, sophisticated verification tools are used to confirm the correctness of the design.

This chapter addresses design verification at the logic design stage. The design verification at this stage consists mainly of verification of functional correctness, estimation of performance, and design rule check if rules exist. This chapter first surveys VLSI design flow and verification activities. Then, logic simulators, which have been widely used and will continue to be used mainly as a design verification tool, are briefly reviewed, with emphasis on delay models used therein. Also, timing verification techniques are surveyed. A final section describes a new exact delay logic simulator and a new simulation-based timing verifier, adopting a concept of "skew and logic error." This section has been written as a self-contained entity so that the interested reader may consult it directly, without reading other parts of the chapter.

VLSI DESIGN FLOW AND DESIGN VERIFICATION

The delay time of each logic element constructing a VLSI circuit is affected much more by the loading conditions than an equivalent logic circuit constructed of standard SSI and MSI parts. Therefore, individual logic elements of the VLSI circuit have no fixed delay properties until the final physical layout is established. A hierarchical layout design method, that is, a block-oriented design, minimizes the fluctuation of delay properties in a functional logic block at the physical layout design. This allows rigid timing design and is effective in attaining high performance in a VLSI circuit. Figure 16-1 shows a typical design flow for custom logic VLSI circuits, especially in the case of a cell-based approach. The design proceeds in general by the top-down process for logic design and the bottom-up process for physical implementation. In a more severe design example, some feedback design loops would exist, and iterative designs or specification refinements would be needed.

The design process is roughly divided into five steps, as shown in the figure. In each design step the correctness of the design is confirmed by various kinds of verification tools.

In system design, the system architecture is determined to meet the specification. The architecture is examined and verified by various levels of system simulations (Tokoro et al. 1978; Efron and Gordon 1964; Barbacci 1981) in the progress of the design—the processor-memory-switch (PMS) level, the programming or instruction level, the micro-instruction level, and the register transfer level. If the logic system is very large, it is divided into plural VLSI chips according to delay time, power dissipation, chip size, pin numbers, and so on.

Logic design consists of clarifying the logic diagram on the basis of architecture. Some other circuits may be added to improve testability of the VLSI chip, if necessary. The logic elements used in the design are standard cells that contain primitive logic gates and functional blocks, both of which are already prepared or will be designed in the block design step. Logic simulation or timing

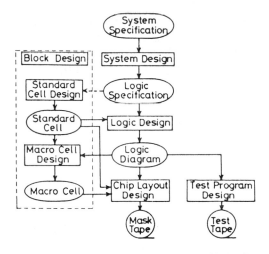

Figure 16-1. A typical design flow for VLSI circuits.

simulation is used for functional verification and performance estimation. A function level logic simulation (Chappel et al. 1976; Tasaki et al. 1980) is practical for large macro blocks or the whole VLSI chip, and a gate level logic simulation (Syzenda and Thompson 1975a; Wilcox and Rombeek 1976) is available for relatively small functional blocks. The delay time specification of each block or gate is also determined in this design stage. The load drivability is selected by considering the delay specification and the loading conditions estimated from the fan-outs of the gate. A timing verifier (MacWilliams 1980) or a path delay analysis tool (Hitchcock et al. 1982; Wold 1978) is also available for estimating the performance (speed and timing margin) of the circuit.

Besides the function and timing verification, a design rule check is also performed in this design step, if the circuit is implemented in compliance with a particular circuit design method such as level sensitive design, scan design, synchronous design, and so on (Eichelberger and Williams 1977). The testability of the circuit is also checked by a testability analysis tool (Goldstein 1979), and if it is not satisfactory, the design is modified.

Block design consists of constructing logic elements that will be used in logic design and chip layout design. The logic elements may be classified into two types of logic blocks: the standard cell and the macro cell. The standard cell consists of one or a few primitive logic gates, with some geometric restrictions on pin position and height. The macro cell is a functional logic block that has a few dozen or a few hundred gates like MSIs, and is constructed of standard cells or generated by manual design, a symbolic design (Ozaki et al. 1984; Williams 1978), or a cell generator (Chu and Sharma 1984). The macro cell group includes high-level macro blocks that are hierarchically constructed by lower-level blocks. There are no restrictions on pin position and size of the cell, but the shape usually must be a rectangle. According to the design level, various kinds of verification tools are executed for function and performance validation. At the device level, the fluctuation of electrical characteristics due to the fabrication condition is sometimes evaluated by a two- or three-dimensional device simulation (Barron 1969; Toyabe et al. 1978). At the circuit level, the dc, ac characteristics are estimated by a circuit simulation (Nagel May 1975; ASTAP Manual) or a timing simulation (Rabbat et al. 1975; Chawla et al. 1975). At the logic level, the functional operations of macro blocks are simulated by a logic simulation, and the timing relations are verified by using an exact delay logic simulation (Ohkura et al. 1980; Okazaki et al. 1983), a timing simulation, or a path delay analysis.

For layout design verification, a connectivity check program is used that compares the reference (intended) interconnections of logic elements with the interconnection data extracted from mask artworks (real implementation).

Chip layout design consists of building the VLSI chip physically by placement of the logic blocks prepared in the block design and routing between them. An intercell connectivity check will be performed if manual modifications are executed. The functional operation and the circuit performance (or timing) are verified again after layout design. Hazardous conditions are detected by a delay simulation based on the physical layout information (real wire lengths). In the case of a VLSI logic circuit, an exact delay simulation based on the real loading conditions is more practical than a circuit simulation or a timing simulation because the circuit complexity is very high. According to the simulation results, the drivability, mainly of interblock drivers, that was set in the block design will be modified to attain the system specification. As in block design, path delay analysis may be used for performance validation.

Test program design consists of preparing a program to test the VLSI chip. The program usually includes dc, ac and logic function test items. The logic function test sequence is the most difficult to prepare because a sequence is desired that will detect all the assumed faults perfectly. The input test sequence is evaluated by a fault simulation (Syzenda and Thompson 1975b). The output test sequence for a given input sequence is generated by a logic simulator. Automatic test generation in a practical computer time is available for a particular circuit, such as a circuit based on the level sensitive and scan design rule (Eichelberger and Williams 1977).

DESIGN VERIFICATION TECHNOLOGY IN LOGIC DESIGN

Function verification and timing verification are the two major tasks in logic design. Up to now, logic simulation has been the main technology for both function and timing verification. Recently, for a particular type of logic circuit to be described later, several techniques have been developed that verify the timing property separately from the logic function. To make each design verification reliable, it is essential that the gap between the circuit model constructed on a computing machine and the physical circuit be reduced as much as possible. In this section, the delay models of logic elements are described first; then logic simulation and timing verification technologies are briefly reviewed.

Delay Modeling of Logic Element

Delay models have been investigated during development and refinement of logic simulators, and are currently used also in timing verification tools.

The first delay model implemented in event-driven simulators is a nominal delay model in which a single delay value t_p is assigned to each kind of logic type, such as a 2-input NAND, 3-input NAND, and so on. Some devices, however, have different delay times for rise and fall transitions due to various electrical parameters and device structures. Such devices can be modeled by assigning two delays, t_{pHL} for a transition from 1 to 0 and t_{pLH} for a transition from 0 to 1. The model is referred to as a rise/fall delay model, which closely approximates the timing properties for many device technologies. Pulse width modulation due to different delay times for rise and fall transitions can be simulated by the model.

A more precise modeling of the delay is called a delay ambiguity model, or a min/max delay model. Commercially available logic circuits, SSIs and MSIs, operate with a propagation delay somewhere between a minimum value t_{pdm} and a maximum value t_{pdM}. These delays define an ambiguity region of duration t_{pdM}-t_{pdm}, and the signal changes its value at some time within this region. In logic simulation, the ambiguity region propagates through the elements in additive fashion. That is, each element contributes its own ambiguity in the switching times as the signal change propagates through the element. Thus, the region at the output node becomes wider than that at the input, as shown in Figure 16-2. Therefore, the model gives an overly pessimistic or worst-case behavior. The delay ambiguity model can be incorporated into the nominal delay model and the rise/fall delay model. Other methods proposed for delay ambiguity modeling are the Monte Carlo simulation (Breuer and Friedman 1976), where all combinations of the delays can be considered, and the probability model (Magnhagen 1977), where the delay distribution is approximated a Gaussian curve. However, as for the elements in VLSI circuits, these delay ambiguity models may not be essential because the elements in the VLSI will be fabricated simultaneously under the same process.

Another circuit delay that should be considered is an inertial delay that models some minimum energy required to switch a logic gate. In most of the present simulators, the inertial delay is usually handled by a high frequency rejection procedure that filters out any output pulse with width less than the inertial delay. Delay modeling of complex functional elements, such as flip-flops and MSI devices, is more complicated. A macroscopic modeling (Evance 1978) has been proposed, which includes setup and hold times, minimum pulse width of the clock signal, set signal and reset signal, variable path delay times depending on logic operations, and other factors.

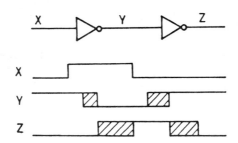

Figure 16-2.   Propagation of ambiguity region in a logic element. Shaded regions show the ambiguity region.

Logic Simulation

The functional correctness of a designed VLSI circuit is verified by using logic simulators. Most of the existing logic simulators are classified into three categories, as abstractions of the circuit descriptions: function level, gate level, and switch level.

At the function level, a circuit is described in terms of functional blocks such as registers, counters, ALUs, processors, and so on. Function level simulators can handle the circuit described at this level. Some function simulators (Syzenda and Thompson 1975a; Wilcox and Rombeek 1976) provide such functional blocks as built-in simulation primitives, which are already verified by simulator developers. Some other function simulators (Chappel et al. 1976; Sasaki et al. 1980) provide specially designed function description languages (or behavior description languages) so logic designers can prepare their own functional blocks in the function simulation. Function simulation tends to run much faster than gate level simulation (to be described); so the whole VLSI circuit can be simulated. However, function simulation generally tends to be less accurate in its timing. At the gate level, a circuit is described in terms of primitive logic gates and is simulated by a gate level logic simulator. Because the gate level simulator can incorporate accurate delay models (described previously), the simulator is suitable for performance analysis of the circuit, especially taking account of real loadings after layout design is finished. At the switch level, particularly in MOS technology, a circuit is described in terms of MOS transistors, which are bidirectional. A switch level simulator (Bryant 1981) can handle such a bidirectional element, but it is difficult to treat delays accurately because of the bidirectionality.

In a top-down design approach, simulation may start at the function level and then proceed toward the gate level and finally the transistor level, progressing partially and gradually. A mixed-level logic simulator, which can handle several levels of circuit descriptions simultaneously and adjust the details of delay modeling, is useful for this design procedure. Nevertheless, a single-level simulation is most desirable for accurate evaluation of timing property, and that level is, for logic simulation, the gate level or a mixture of gate and switch levels if the circuit includes bidirectional MOS elements.

Timing Verification

Timing verification consists of validating the ability of a VLSI circuit design to meet the intended performance criteria, and checking the circuit to determine that no potential hazardous condition exists. Any hazardous conditions produced will be due to spikes, races, violations of timing constraints on memory elements such as setup and hold times, minimum pulse width of clock, and set and reset signals of a flip-flop.

Most of the existing logic circuits are sequential, categorized into two types: synchronous and asynchronous. An asynchronous sequential circuit will respond every time to the state changes of inputs and/or internal memories, so that a sequence of the signal changes; thus the delay time of each logic element making up the circuit plays an essential role in guaranteeing that the circuit operates correctly. It is impossible to separate timing verification from function verification. The timing verification of the asynchronous sequential circuit is primarily a detection of the possibility of critical conditions such as hazards, spikes, races, and so on. The asynchronous circuit is used mainly to achieve high performance (speed) or low operating power. In contrast, the synchronous sequential circuit works by synchronizing state transitions of internal memory elements to a clock provided externally; so most signals in the circuit can change only during particular parts of the clock period. Timing verification of the synchronous sequential circuits is rather simple and can be performed separately from function verification. Timing verification of this type of circuit consists mainly of performance validation and detection of hazardous conditions along the clock lines. Performance is verified by examining whether or not the longest delay time of the combinational circuits between internal memory elements is less than the clock period desired (Hitchcock et al. 1982; Wold 1978).

There are a number of approaches to the timing verification of a VLSI circuit, which can be classified into two main categories: the dynamic method, based on simulation, and the static method, based on path delay calculation. The dynamic method detects potentially hazardous conditions by simulating a circuit under a certain input pattern, and analyzing the simulation results during the simulation or afterward. There are two types of simulation technique. One is a conventional logic simulation and the other a change–stable simulation. The approach using the conventional logic simulation is applicable to any style of logic circuit, either synchronous or asynchronous sequential circuits; but it has the disadvantage of requiring a great computational effort, by running through a large number of patterns to test all the worst-case timing paths. The approach using a change–stable simulation adopts a concept of signal state in the time domain—changing state and stable state—in addition to the logical values used in the conventional logic simulators. It is computationally efficient and can verify almost all the worst-case timing paths because the input patterns can be remarkably reduced by using changing and stable states for data inputs instead of combinations of 1 and 0 (Wilcox and Rombeek 1976). However, this approach can work only on the synchronous sequential circuits. In change–stable simulation, most of the signals in an asynchronous circuit tend substantially to be in a changing state, so the change–stable simulation is not applicable to timing verification of asynchronous circuits.

The static method consists of computing delay times for all the paths of a combinational logic circuit between memory elements, and validating circuit performance. There are two methods: path enumeration and worst-case path

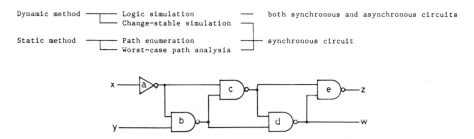

Figure 16-3. Paths in a combinational circuit. There are 13 paths. For example, the path
x—a—b—c—d—e—z— cannot be sensitized.

analysis. The path enumeration method enumerates all the paths in the circuit
and computes the delay time of each path (Wold 1978). The method tends to
have a long processing time because the number of paths through a graph representing a circuit grows exponentially with the size of the graph. The worst-
case path analysis computes the greatest (worst-case) path delay of the circuit
(Hitchcock et al. 1982; Bening et al. 1982). This method computes the latest
time at which the signal changes reach each gate, starting from inputs of the
circuit. The worst-case path analysis method can run much faster than the path
enumeration method. The static method does not need input patterns but requires
some effort to eliminate paths that cannot be sensitized by any input pattern (see
Figure 16-3). Although a specific path can easily be ignored by user's instruction
in the path enumeration method, the worst-case path analysis method cannot
accept such a function without cutting all other paths at the same time, and thus
gives a pessimistic result.

Timing verification techniques are summarized below.

A NEW EXACT DELAY LOGIC SIMULATOR AND A NEW SIMULATION-BASED TIMING VERIFIER

For design verification, it is indispensable to construct an accurate model of a
logic circuit and predict behaviors of the circuit accurately and precisely. Although circuit simulators and timing simulators are able to analyze not only delay
time but also transition time, the circuit scale is practically limited. Circuit
simulators cannot provide for the entire VLSI circuitry, and timing simulators
cannot take care of the circuit through a large number of patterns in a practically
acceptable run time. Mixed-level simulators, which aim for accuracy of circuit
simulation and efficiency of logic simulation, still have a problem, in that timing
accuracy would be lost in the interface between timing and logic levels. Logic
simulators, on the other hand, are available for a circuit up to tens of thousands
of gates, and the single-level feature can avoid the above problem of mixed-
level simulators. Although logic simulators can handle several kinds of delay

times, the previous delay models, described above, are still insufficient to predict behaviors of the actual circuit for lack of the transition time effect (slope of the waveform). Therefore, precise modeling of the delay, including the effect of transition time, is an important problem for design verification.

Another aspect of design verification is timing verification, used to detect the occurence of hazardous conditions that would cause incorrect circuit operation, as well as to analyze circuit performance. Several techniques have been developed to compute the maximum operating speed of a circuit separately from its logical behaviors, but their applicability is limited. These techniques apply only a particular group of circuits—synchronous circuits—and would give rather pessimistic results.

There are essentially many demands for asynchronous circuits, and for rigid timing design even in synchronous circuits. The only way to verify these designs would be by a precise and exact simulation. The timing verification discussed in this section is used for detection of hazardous conditions, in order to confirm that a circuit can operate correctly within the tolerance of fabrication parameters and input timings (input skew).

In this section, an exact delay model introducing transition time effects [precise device analyses are not described here; see the literature (Tokuda et al. 1983)], and a simulator employing this model, will be described. Some experimental results are also given. A new simulation-based timing verifier based on the concept "skew and logic error" (defined later) will be presented, which is very helpful in correcting designs if errors do exist.

Characterization of Gate Delay

Before we describe delay time of a gate, we must have a delay time definition. The waveforms of the input and output nodes of a gate are illustrated in Figure 16-4. The delay times t_{pLH} and t_{pHL} are defined by the intervals between two times at which the input and the output voltages become equal to the threshold voltage V_T, respectively. The rise time t_r and the fall time t_f are defined by the

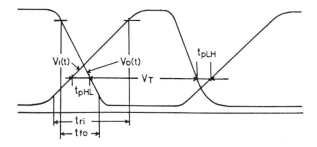

Figure 16-4.   Definition of delay times and transition times.

time duration in which the input or the output node voltage swings between 0.1 VPP and 0.9 VPP. With this definition, the delay times are always positive for any values of the load capacitance and the input transition time t_{rI} or t_{fI}.

It is well known that the delay time of a gate, especially an MOS device, is largely affected not only by the loading capacitance of the gate but also by the slope of the input waveform (transition time) (Tokuda et al. 1983; Bening 1969; Koehler 1978). Generally, the greater the transition time is, the greater the delay time becomes. The transition time of a signal is determined by the drivability of a gate that drives the signal and the slope of the input signal of the gate. This implies that the delay time of a gate may be affected by all of the predecessor-gates through which a signal change is propagated. However, it has been shown by our theoretical analyses and by numerical analyses of circuit simulations that the transition time of an input signal of a gate can be quite approximately given by using only the drivability and the loading capacitance of the gate driving the signal (Tokuda et al. 1983). Thus, the delay time of a gate is given by a function of the load capacitance and the drivability of the gate and the load capacitance and the drivability of the gate that feeds that input. Consequently, the delay times of a gate whose logical function is inversion (not), such as an inverter, are given by the following equations:

$$t_{pLH} = K_{tv} C_{Lr} [K_{Or} + K_{lr}(C_{If}/C_{Lr})^{Kr}]$$
$$t_{pHL} = K_{tv} C_{Lf} [K_{Of} + K_{lf}(C_{Ir}/C_{Lf})^{Kf}]$$

where C_{Ir}, C_{If}, C_{Lr}, and C_{Lf} are the capacitances normalized by the drivability of the gate that drives the capacitances, and K_{Or}, K_{lr}, K_r, K_{Of}, K_{lf}, and K_f are the coefficients determined empirically by a numerical analysis or a simulation. K_{tv} is a coefficient introducing the operating voltage and temperature effects. The normalized capacitance is given by dividing the actual capacitance by the drivability, that is:

$$C = (C_w + < \text{sum of input capacitances of the fan-out gates} >)/K_{dr}$$

where C_w is an interconnection wire capacitance, and K_{dr} is the drivability of a gate. Note that a value of K_{dr} is generally different for rise and fall transitions, and thus the normalized capacitances are different for rise and fall transitions. For an inverted-output gate, a rising input causes a falling output, and thus C_{If} is used in t_{pLH} and C_{Ir} in t_{pHL}. A comparison of the delay times given by the circuit simulations and those calculated by the above delay time equations is shown in Figure 16-5.

For a multiple input gate, such as a three-input NOR gate, each input node has a different capacitance, and thus the delay time of the gate varies, depending upon which input causes the output transition. The above delay time equations

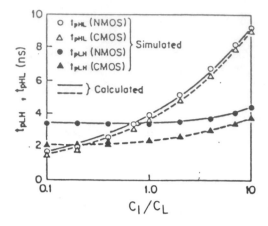

Figure 16-5. Simulated and calculated delay times of NMOS and CMOS inverters.

can be also applied to a calculation of delay times of a complex gate, such as an AND-OR-INV, in which the transistors are connected in series and parallel fashion, having the appropriate drivability values.

Multiple Rise/Fall Delay Model

As discussed above, a gate model has to have different delay values corresponding to each input node. For this requirement, the multiple rise/fall delay model has been developed. A gate model is normally evaluated by two basic operations: a delay operation causing signal delay and a functional operation to give the output value. By using the arrangement of two sequential operations, there are two possible ways to set up the gate model: the input side delay model and the output side delay model (see Figure 16-6).

The input side delay model applies the delay operation first and then uses the functional operation to yield the output:

$$z = F(D(x_1, x_2, \ldots, x_n)) = F(D_1(x_1), D_2(x_2), \ldots, D_n(x_n))$$

where $(x_1, x_2, \ldots, x_n)$ are inputs, and z is the output. F is the functional operation, and D is the delay operation. The delay operation for the usual NOR gate is as follows:

$$D_j = t_{pHLj} \text{ for } x_j = 0/1$$
$$= t_{pLHj} \text{ for } x_j = 1/0$$

Here, the notations 0/1 and 1/0 mean rising transition and falling transition, respectively.

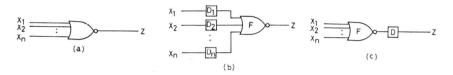

Figure 16-6. Gate model for a logic gate. (a) n-input NOR gate. (b) Input side delay model. (c) Output side delay model.

The latter approach performs two operations in reverse order, which is referred to as the output side delay model:

$$z = D(F(x_1, x_2, \ldots , x_n))$$

In the delay operation of this model, the delay value is chosen in accordance with the logically active input causing the output transition. Furthermore, when the overlapping input transition causes an output transition, the delay is calculated by a procedure specific to each element type. For example, consider the overlapping input transition of a NOR gate. Because only one of the switching transistors (enhancement transistors in enhancement-depletion configuration) becomes ON to make the output 0, the delay time is determined by the most rapidly rising input. On the other hand, all the inputs must become 0 to make the output 1, and thus the delay time is determined by the most slowly falling input. Consequently, the delay operation of the NOR gate is given as follows:

$$D = \text{MINIMUM } (t_{pHLj}(x_j = 0/1), \ldots) \text{ for } z = 1/0$$
$$= \text{MAXIMUM } (t_{pLHj}(x_j - 1/0), \ldots) \text{ for } z = 0/1$$

Although either model may be applied in the logic simulation, there are performance differences in efficiency and timing analysis. The input side delay model needs a computationally complex procedure to handle detailed timing. A simplified input side delay model in which the delay operation is replaced by a delay element can be easily implemented in a conventional event-driven logic simulator. However, following this model, the element count increases greatly in number, and thus the simulation run time will increase. Also the model would exhibit overly pessimistic results for the timing property. Thus the output side delay model is adopted in our new logic simulator.

In summarizing the gate model, the output side delay model, referred to as the multiple rise/fall delay model in this chapter, is characterized by the following features:

1. Output side delay.
2. Delay table that stores the multiple rise/fall delay times from inputs to output.

3. Input selection process to find logically active input.
4. Additional delay calculation for the overlapping input transition.

Consideration of Resistive Wire

In the above discussion, only capacitive load is considered. This is accurate for a metal wire. If the wire is polysilicon, however, the resistivity cannot be ignored. Figure 16-7 shows an example in which a signal fans out through metal and polysilicon wires, and the node voltage responses obtained by the circuit simulator SPICE2. The resistivity of the polysilicon wire slightly reduces its capacitive loading, but the pure delay (called diffusion delay) (Mead and Conway 1980) of the polysilicon wire and the larger transition time of the signal wave are introduced. If the resistivity is high, the transition time of the signal at the input node of the gate G_p is approximately proportional to the time constant of an equivalent RC network of the polysilicon wire. Thus a polysilicon wire is modeled on a pure delay element that has a delay time proportional to resistance times capacitance. The delay times of the driving and driven gates G_s, G_p, and G_m are calculated by the equations given previously, with the following capacitance values:

$$G_s: C_L = C_m + K_1 C_P + K_2 C_g, C_I = C_1$$

$$G_p: C_L = C_2, C_I = C_m + K_3 C_p + K_4 C_g$$

$$G_m: C_L = C_3, C_I = C_m + K_1 C_P + K_2 C_g$$

where C_m, C_p, and C_g are the capacitance of the metal wire, the capacitance of the polysilicon wire, and the input capacitance of the gate G_p, respectively. A final network model in the logic simulation is shown in Figure 16-7(c). The reason why the delay element is inserted between G_s and G_p without adding the wire delay to the gate delay is that this delay is the pure delay, and is not related to the internal timing behavior of the gate.

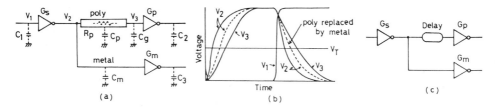

Figure 16-7. An example of resistive wire connection. (a) A circuit simulation. (b) Node voltages as a function of time obtained by SPICE2. The dotted line shows the node voltage V_2 in case polysilicon wire is replaced by metal wire. (c) a circuit model in logic simulation.

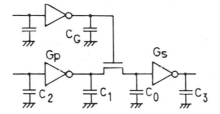

Figure 16-8. Configuration for transmission gate.

Consideration of Transmission Gate

Modeling of a circuit including transmission gates is more complex. The delay time of a transmission gate is not determined just by its input capacitance and loading capacitance because a transmission gate is a nonactive device and itself cannot drive a load capacitance. There are two operation modes of a transmission gate: synchronous and asynchronous. In the synchronous mode, a transmission gate becomes conductive after the data input becomes stable. On the other hand, in the asynchronous mode, the input changes in the conductive state. Under the synchronous mode, the delay time is affected by the capacitance of the gate input and a ratio of capacitances at output and data input nodes. Under the asynchronous modes, the delay time of the transmission gate is not affected by the data input waveform. However, the delay time of the preceding (driving) gate becomes larger than that in the synchronous mode because the gate has to drive the larger load capacitance. An analogous argument is true for the succeeding (driven) gate. In this case, the load capacitance at the input node of the driven gate varies.

Consequently, an accurate delay simulation can be achieved if the operation mode of each transmission gate in a circuit can be limited to one of these modes, individually. This limitation may not be impractical. Violation of the operation mode imposed on each transmission gate can be easily checked during simulation. The loading conditions in delay time calculations of the driving gate G_p and the driven gate G_s in Figure 16-8 are listed below:

transmission gate	G_s	G_p
synchronous mode	$C_L = C_3, C_I = C_1 + K_1C_0$	$C_L = C_1, C_I = C_2$
asynchronous mode	$C_L = C_3, C_I = C_1 + K_2C_0$	$C_L = C_1 + K_3C_0, C_I = C_2$

New Exact Delay Logic Simulator

The newly developed logic simulator, which is called the multiple media delay simulator, is table-driven and event-oriented and uses a time wheel method to schedule circuit activities. The circuit to be simulated is internally expressed with tables that store element type properties, fan-in lists, fan-out lists, and delay values. Here, the element type means logical function such as NAND, NOR, and so on. The delay table stores rise/fall delay values from each input to output and timing constraints such as criteria of narrow pulse filtering, setup and hold times for flip-flops, and so forth. The element type properties include an input selection rule for delay determination (described later) and a delay calculation code (max, min, average, etc.) for overlapping input transition. These codes are specified in advance for each element type, but a designer can modify the codes within the limit of codes provided.

A basic simulation cycle for the multiple media delay simulator is shown in Figure 16-9. A unique feature of the simulator is a procedure that determines the delay value of a gate, changing its logic state (step 4 in Figure 16-9). As in conventional logic simulators, simulation is continued by evaluating elements whose inputs change and scheduling elements whose output will change in the future. In the process of delay value determination, because an element has multiple sets of rise/fall delay times from inputs to output, the multiple media delay simulator requires a procedure to find inputs that cause the output transition. To do this, it is necessary to define for each element type what type of input transition should be selected according to the output transition, as well as to know inputs that are changing in the present. The latter can be easily processed because the signal changes in the present are held on the time wheel in the linked list structure in the event-oriented simulator. The definition of input transitions,

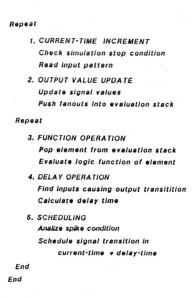

Figure 16-9. Basic simulation cycle of the multiple media delay simulator.

an input selection rule, is done specifically on an element-type-by-element-type basis. For example, the rising (falling) output of a NAND gate is a result of falling (rising) inputs, so inputs with transition opposite to that of the output are chosen. The multiple media delay simulator applies to each element type one of the following selection rules or a combination of them:

1. Select inputs with transition opposite to that of the output.
2. Select inputs with the same transition as that of the output.
3. Select inputs with any transition.
4. Select control (clock, direct set, or reset) input.

If the element has two types of inputs, control and data inputs, as in a clocked-NAND gate, the selection rules are applied separately to each type of input.

As an example that demonstrates the accuracy of delay modeling, Figure 16-10 shows the typical response of a two-input NOR gate with appropriate loadings in the multiple rise/fall delay model. The response obtained with the multiple rise/fall delay model is in close agreement with the response from the SPICE2 circuit simulation, which is also shown in the figure. For comparison, responses of the simplified input side delay model and the conventional single rise/fall delay model are also shown in the figure. The response of the single rise/fall delay model gives a greater delay time, and the response of the input side delay model exhibits a spike that is not found in the response of SPICE2. As another example, Figure 16-11 shows the simulated and measured delay times on some paths in a fabricated LSI chip and demonstrates the accuracy of the multiple rise/fall delay simulation in practical applications.

Table 16-1 shows processing times observed for a circuit with approximately 1000 gates. The multiple media delay simulation employing the multiple rise/ fall delay model is not so slow considering its accuracy (compare model-1 with model-3), and is far superior to the simplified input side delay model, that is, a

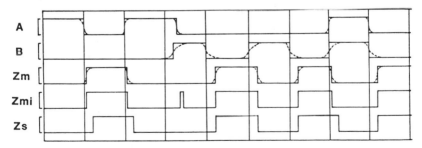

Figure 16-10. Responses of two-input NOR gate ($z = \overline{A + B}$) with appropriate capacitive loadings. Node *B* has a larger capacitance than that of node *A*. *Zm* is obtained by the multiple rise/fall delay model, *Zmi* by the simplified input side delay model, and *Zs* by the single rise/fall delay model. The dotted lines are obtained by SPICE2.

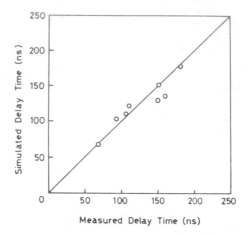

Figure 16-11.   Comparison of simulated and measured delay times in a fabricated LSI.

method that realizes the multiple rise/fall delay model in a conventional logic simulator by inserting a delay element just in front of the corresponding input of a gate (compare model-1 with model-2).

DELAY MODEL	ELEMENT COUNT	RUN TIME, MIN.
model-1	1045	5.1
model-2	2523	9.1
model-3	1045	4.2

model-1: multiple rise/fall delay model.
model-2: simplified input side delay model.
model-3: single rise/fall delay model.

New Simulation-Based Timing Verifier

The actual fabricated chips are different in their delay and loading conditions from those assumed in the simulation because the fabrication process parameters fluctuate. Moreover, owing to the input timing skew that occur in testing or actual usage on a printed circuit assembly, the chips are exposed to certain input timings that are different from the inputs to the logic simulator. The new simulation-based timing verifier proposed in this section validates during the simulation that the chips can tolerate these fluctuations.

It is very beneficial for the designer to know the origin of a timing error, if one is detected. For this purpose, a new classification for timing errors has been introduced: skew error and logic error. Skew error is a timing error caused by input timing ambiguity; it can be avoided primarily by modifying the timing of input patterns. Of course, the circuit can be modified if it is more desirable to do so. Logic error is a timing error caused by improper logic circuit design; it can be avoided only by redesigning the circuit in its logical or physical aspects. That being so, the "skew and logic error" concept greatly helps the designer, as a timing check based on the concept not only reveals the occurrence of timing errors, but also gives a good idea of what to do to correct the circuit.

Figure 16-12 shows examples of skew and logic errors for the setup time constraint of the flip-flop, where t_{su} and t_{sk} are the setup time and the skew of the input timing, respectively, and t_{dc} is the interval between the transitions of the data signal and the clock signal. In Figure 16-12(a), t_{dc} is smaller than t_{su}, and t_{dc} is determined only by the difference of the delay times between two combinational logic circuits. Thus, this case is a "logic" setup time error. In Figure 16-12(b), t_{dc} is smaller than $t_{su} + 2t_{sk}$. Here, $2t_{sk}$ means the maximum skew that might occur between two input signals. Since t_{dc} is determined by not only the difference of the combinational circuit delay times but also the timing of two input signals, it is possible to make t_{dc} larger than $t_{su} + 2t_{sk}$ by modifying the timing of two inputs. Thus, this case is a "skew" setup time error. In Figure 16-12(c), where the circuit configuration is the same as in Figure 16-12(a), t_{dc} is also smaller than $t_{su} + 2t_{sk}$. However, there is no possibility of input skew, so the condition $t_{dc} < t_{su} + 2t_{sk}$ is ignored. Because t_{dc} is greater than t_{su}, this case is not an error.

The timing errors treated in this timing verifier are spike, race, hazard, and timing constraints of memory elements. "Spike" means a narrow pulse to which a circuit might not respond correctly. "Race" indicates a timing relationship between two signals that would lead the circuit into another logic "state" if the sequence of the signal changes were reversed. "Hazard" refers to a situation that does not produce a spike in the logic simulation but can produce a spike with skew or process fluctuations. It is not abnormal for spikes or hazards to occur in the combinational circuits, but if they propagate to the control pins (or lines) of the memory elements or the memory circuits (looped circuits), the circuit

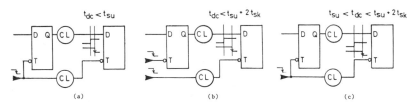

Figure 16-12.   Examples of setup time violations. (a) "Logic" setup time error. (b) "Skew" setup time error. (c) No error. The quantity t_{su} is the setup time check width, and t_{sk} is the skew of input timing.

might operate incorrectly. Timing constraints include setup and hold times, and minimum pulse width of clock, set, and reset signals. Basically, a timing error is declared to have occurred when one of the following conditions is detected:

1. A race between clock and set or reset or between set and reset at a flip-flop or latch is occurring.
2. One of the timing constraints imposed upon the memory elements is violated.
3. A spike appears at the clock, set, or reset pin (or signal) of a flip-flop.
4. A hazard that occurred in any part of the circuit propagates and reaches the clock, set, or reset pin (or signal) of a flip-flop.
5. A race is occurring at the "control" signals of a looped circuit.
6. A hazard or a spike appears on the "control" signals of a looped circuit.

An example of a race at a looped circuit is shown in Figure 16-13. In this case, if the sequence of signal transitions of the "control" signals c and d were in reverse order, the circuit could hold the value 1.

These timing clocks are performed during the simulation, but there are two problems with processing techniques. One involves hazard propagation, and the other a mechanism to decide whether an error is a skew error or a logic error. Although a spike is simulated explicitly in the logic simulators, a hazard usually is not treated explicitly; so a particular mechanism of hazard propagation is required for the timing check on hazard. The new simulation-based timing verifier has introduced "hazard event," which is a flag of hazard existence and holds the gate name at which the hazard has occurred (see Figure 16-14b). When a hazard event reaches a gate, the event is examined to determine whether it propagates through the gate or not. For the example of a NAND gate, if every input except the input at which the hazard event propagates has logical 1, then the hazard event appears at the output. If one of the other inputs has logical 0, the hazard event disappears. For the latter problem, the concept of the "first-ancestral signal" of an event (signal change including hazard event) has been introduced. This is an original event that successively produces some events and also the event of interest. The first-ancestral signal is always a primary input. For example, the

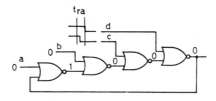

Figure 16-13. Example of a race at a looped circuit. The quantity t_{ra} is the logic error check width for the race check.

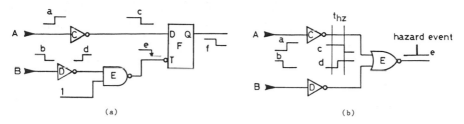

Figure 16-14. The first-ancestral signal transition propagation on the event. (a) An ordinary event. Here, the first-ancestral signal of the ordinary event f is B. (b) A hazard event. Here, the first-ancestral signals of the hazard event e are A and B. The quantity t_{hz} is the logic error check width for hazard check.

first-ancestral signal of the event f in Figure 16-14(a) is B. An ordinary event has one first-ancestral signal, and a hazard event has two, as shown in Figure 16-14. The first-ancestral signal can easily be carried on the events in our simulation-based timing verifier; the verifier is based on the multiple media delay simulator, and the simulator has a function to find the input that causes the output transition of each gate, as described previously in the delay determination mechanism. The algorithm for judging whether an error is a skew error or a logic error is described below in the case of the setup time constraint, without generality. Assume that the logic error width of a setup time is t_{su}, and the skew of input timing is t_{sk}. The quantity t_{dc} is again the interval between the transitions of the data signal and the clock signal.

Step 1. Find a situation $t_{dc} < t_{su} + 2t_{sk}$ by checking every time that the data signal change or the active clock signal change occurs. (Of course, the set and the reset are nonactive.)
Step 2. Judgment: If the first-ancestral signal of the data signal transition is not the same as that of the clock signal transition, then the situation is declared to be a skew error. If they are same, check $t_{dc} < t_{su}$. If it is true, the situation is declared to be a logic error. If it is not true, the situation is not an error.

CONCLUSION

An accurate modeling method of delay, time-dependent on not only the loading but also the slope of the input waveform, has been described. An accurate delay logic simulator that employs the delay model has been presented. The simulator can accurately simulate multiple input to output delays, overlapping input transition and the circuit with an MOS transmission gate. Some experimental results have shown that the simulator provides a fast and accurate design verification for VLSI design.
A new simulation-based timing verifier has been proposed. The primary func-

tion of the timing verifier is detection of potentially hazardous conditions in order to confirm that a circuit design can operate correctly within the tolerance of fabrication parameters and input timing skew. A new concept for the timing problem, "skew and logic error," has been introduced, which the designer will find very helpful in circuit correction.

ACKNOWLEDGEMENT

The authors wish to thank Drs. H. Oka, Y. Gamoh, K. Sato, and H. Matsushita and Mr. T. Tokuda for their encouragement and helpful discussions.

REFERENCES

ASTAP General Information Manual, IBM Corp., PA.

Barbacci, M. R., "Instruction Set Processor Specifications (ISPS): The Notation and Its Applications," *IEEE Trans. Computers, C-30,* 24–40, 1981.

Barron, M. B., "Computer Aided Analysis of Insulated Gate Field-Effect Transistors," Stanford Electronic Laboratories, Rep. No. 5501-1, Stanford University, 1969.

Bening, L. C., "Simulation of High Speed Computer Logic," *Proc. 1969 Design Automation Workshop,* pp. 103–112, 1969.

Bening, L. C., Lane, T. A., and Smith, J. E., "Development in Logic Network Path Delay Analysis," *Proc. 19th Design Automation Conference,* pp. 605–615, 1982.

Breuer, M. A., and Friedman, A. D., *Diagnosis and Reliable Design of Digital Systems,* Computer Science Press, Woodland Hills, CA 1976, pp. 174–254.

Bryant, R. E., "MOSSIM: A Switch-Level Simulator for MOS LSI," *Proc. 18th Design Automation Conference,* pp. 786–790, 1981.

Chappel, S. G., et al., "Functional Simulation in the Lamp System," *Proc. 13th Design Automation Conference,* pp. 42–47, 1976.

Chawla, B. R., et al., "MOTIS—An MOS Timing Simulator," *IEEE Trans. Circuits and Systems, CAS-22,* 901–910 (1975).

Chu, K. C., and Sharma, R., "A Technology Independent MOS Multiplier Generator," *Proc. 21st Design Automation Conference,* pp. 90–97, 1984.

Efron, R., and Gordon, G., "General Purpose Digital Simulation and Example of Its Applications," *IBM System Journal, 3,* 22–34 (1964).

Eichelberger, E. B., and Williams, T. W., "A Logic Design Structure for LSI Testability," *Proc. 14th Design Automation Conference,* pp. 462–468, 1977.

Evance, D. J., "Accurate Simulation of Flip-Flop Timing Characteristics," *Proc. 15th Design Automation Conference,* pp. 398–404, 1978.

Goldstein, L. H., "Controllability/Observability Analysis of Digital Circuits," *IEEE Trans. Circuits and Systems, CAS-26,* 685–693 (1979).

Hitchcock, R. B., Sr., Smith, G. L., and Cheng, D. D., "Timing Analysis of Computer Hardware," *IBM J. Res. Dev. 26,* 100–105 (1982).

Koehler, D., "Computer Modeling of Logic Modules under Consideration of Delay and Wave-shaping," *Proc. IEEE,* 1294–1296 (1978).

MacWilliams, T. M., "Verification of Timing Constraints on Large Digital Systems," *Proc. 17th Design Automation Conference,* pp. 139–147, 1980.

Magnhagen, B., "Practical Experiences from Signal Probability Simulation of Digital Designs," *Proc. 14th Design Automation Conference,* pp. 216–219, 1977.

Mead, C., and Conway, L., *Introduction to VLSI Systems,* Addison Wesley, Reading, MA, 1980.

Nagel, L. W., "SPICE2: A Computer Program to Simulate Semiconductor Circuits," University of California, Berkeley, ERL Memo REL-M520 (May 1975).

Ohkura, I., Okazaki, K., and Horiba, Y., "A New Exact Delay Logic Simulation for ED MOS LSI," *IEEE ICCC'80,* pp. 953–956, 1980.

Okazaki, K., Moriya, T., and Yahara, T., "A Multiple Media Delay Simulator for MOS LSI Circuits," *Proc. 20th Design Automation Conference,* pp. 279–285, 1983.

Ozaki, M., Watanabe, M., Kakinuma, M., Ikeda, M., and Sato, K., "MGX—An Integrated Symbolic Layout System for VLSI," *Proc. 21st Design Automation Conference,* pp. 572–579, 1984.

Rabbat, N. B., et al., "A Computer Modeling Approach for LSI Digital Structures," *IEEE Trans. Electron Devices, ED-22,* (1975).

Sasaki, T., et al., "MIXS: A Mixed Level Simulator for Large Digital System Logic Verification," *Proc. 17th Design Automation Conference,* pp. 626–633, 1980.

Syzenda, S. A., and Thompson, E. W., "Digital Logic Simulation in a Time-Based, Table-Driven Environment—Part 1, Design Verification," *Computer, 8,* 24–36 (1975a).

Syzenda, S. A., and Thompson, E. W., "Digital Logic Simulation in a Time-Based Table-Driven Environment—Part 2, Parallel Fault Simulation," *Computer, 6,* 38–49 (1975b).

Tokoro, M., et al., "A Module Level Simulation Technique for Systems Composed of LSI's and MSI's," *Proc. 15th Design Automation Conference,* pp. 418–427, 1978.

Tokuda, T., Okazaki, K., Sakashita, K., Ohkura, I., and Enomoto, T., "Delay-Time Modeling for ED MOS Logic LSI," *IEEE Trans Computer-Aided Design, CAD-2,* 129–134 (1983).

Toyabe, T., et al., "A Numerical Model of Avalanche Breakdown in MOSFET's," *IEEE Trans. Electron Dev., ED-25,* 825–832 (1978).

Wilcox, P., and Rombeek, H., "F/Logic—An Interactive Fault and Logic Simulation for Digital Circuits," *Proc. 13th Design Automation Conference,* pp. 68–73, 1976.

Williams, J. D., "STICS—A Graphical Compiler for High Level LSI Design," *Proc. 1978 National Computer Conference,* pp. 289–295, 1978.

Wold, M. A., "Design Verification and Performance Analysis," *Proc. 15th Design Automation Conference,* pp. 264–270, 1978.

17. The Complexity of Design Automation Problems

Sartaj Sahni*[+]
University of Minnesota

Atul Bhatt[++]
Sperry Corp.

Raghunath Raghavan*[+++]
Mentor Graphics

Over the past 20 years, the complexity of the computer design process has increased tremendously. Traditionally, much of the design has been done manually, with computers used mainly for design entry and verification, and for a few menial design chores. Such labor-intensive design methods cannot survive very much longer, primarily for two reasons.

The first reason is the rapid evolution of semiconductor technology. Increases in the levels of integration possible have opened the path for more complex and varied designs. LSI technology has already taxed traditional design methods to the limit. With the advent of VLSI, such methods will prove inadequate. As a case in point, the design of the Z8000 microprocessor, which qualifies as a VLSI device, took 13,000 man-hours and many years to complete. In fact, it has been noted (Noyce 1977) that, industry-wide, the design time (in man-hours per month) has been increasing exponentially with increasing levels of integration. Clearly, design methods will have to go from labor-intensive to computer-intensive.

Secondly, labor-intensive methods do not adequately accommodate the increasingly stringent requirements for an acceptable design. Even within a given

*The work of these authors was supported in part by the National Science Foundation under grants 78-15455 and MCS 80-005856.

[+]Address: Department of Computer Science, University of Minnesota, Minneapolis, MN 55455.

[++]Address: Sperry Corporation, P.O. Box 43942, St. Paul, MN 55164.

[+++]Address: Mentor Graphics, Beaverton, OR 97005.

technology, improvements are constantly sought in performance, cost, flexibility, reliability, maintainability, and so on. This increases the number of iterations in the design cycle, and thus requires smaller design times for each design step.

Industry-wide, the need for sophisticated design automation (DA) tools is widely recognized. To date, most of the effort in DA has concentrated on the following stages of the design process: physical implementation of the logic design and testing. DA for the early stages of the design process, involving system specification, system architecture, and system design, is virtually non-existent. Though DA does not pervade the entire design process at this time, there are a number of tools that aid, rather than automate, certain design steps. Such computer-aided design tools can dramatically cut design times by boosting designer productivity. We shall restrict our attention to problems encountered in developing tools that automate, rather than aid, certain design steps.

In the light of the need for more advanced and sophisticated DA tools, it becomes necessary to reexamine the problems tackled in developing such tools They must be thoroughly analyzed and their complexity understood. (The term "complexity" will be defined more precisely, using concepts from mathematics and computer science, later in this chapter.) A better understanding of the inherent difficulty of a problem can help shape and guide the search for better solutions to that problem. In this chapter, several problems commonly encountered in DA are investigated, and their complexities analyzed. Emphasis is on problems involving the physical implementation and testing stages of the design process.

We shall first present a brief description, in general terms, of the DA problems considered. Then, the concepts of complexity and nondeterminism are introduced and elaborated upon in the next section, which also includes other background material.

A section on "Complexity of Design Automation Problems" analyzes the problems discussed earlier. Each problem is mathematically formulated and described in terms of its complexity. Most problems under discussion are shown to be NP-hard. Finally, a brief section explores ways of attacking these problems via heuristics and what are called "usually good" algorithms.

The book edited by Breuer (1972a) and his survey paper (1972b) provide a good account of DA problems, techniques for solutions, and their applications to digital system design. Though these efforts are over 15 years old, the problems formulated therein are still representative of the kinds of problems encountered in designing large, fast systems using MSI/LSI technology and a hierarchy of physical packaging. The book by Mead and Conway (1980) describes a design methodology that appears to be appropriate for VLSI. This methodology gives rise to a number of design problems, some of which are similar to problems encountered earlier, and some that have no counterpart in MSI/LSI-based design styles. W. M. van Cleemput (1976) has compiled a detailed bibliography on DA-related disciplines. The computational aspects of VLSI design have been studied in the book by Ullman (1984). David Johnson's ongoing column "The

NP-Completeness Column" in the *Journal of Algorithms*, Academic Press, is a good source for current research on NP-completeness. For example, the December 1982 column was devoted to routing problems (Johnson 1982).

SOME DESIGN AUTOMATION PROBLEMS

There are numerous steps in the process of designing complex digital hardware. It is generally recognized that the following classes of design activities occur:

1. *System Design*. This is a very high-level architectural design of the system. It also defines the circuit and packaging technologies to be utilized in realizing the system. (While it might appear that this is a bit too early to define the circuit and packaging technologies, such is not the case. It is necessary if one is to obtain cost/performance and physical sizing estimates for the system. These estimates help confirm that the system will be well-suited, cost- and performance-wise, to its intended application.) Clearly, system design defines the nature and the scope of the design activities to follow.

2. *Logical Design*. This is the process by which block diagrams (produced following the system design) are converted to logic diagrams, which are basically interconnected sets of logic gates. The building blocks of the logic diagrams (e.g., AND, OR, and NOT gates) are not necessarily representative of the actual circuitry to be used in implementing the logic. (For example, programmable logic arrays, or PLAs, may be used to implement chunks of combinational logic.) Rather, these building blocks are primitives that are "understood" by the simulation tools used to verity the functional correctness of the logic design.

3. *Physical Design*. This is the process by which the logical design is partitioned and mapped into the physical packaging hierarchy. The design of a package, or module, at any level of the physical packaging hierarchy includes the following activities: (a) further partitioning of the logic "chunk" being realized between the submodules (which are the modules at the next level of the hierarchy) contained within the given module; (b) placement of these submodules; and (c) interconnection routing.

The design process is considered to be essentially complete following physical design. However, another important premanufacturing design step is prototype verification and checkout, wherein a full-scale prototype is fabricated as per the design rules, and thoroughly checked. The engineers may make some small changes and fixups to the design at this point ("engineering changes").

These design steps exist both in "conventional" hardware design (i.e., using MSI/LSI parts) and in VLSI design. In conventional design, the design steps mentioned above occur more or less sequentially, whereas in some VLSI design methodologies, there is much overlap, with system, logical, and physical design decisions occurring, in varying degrees, in parallel.

The design step that has proved to be the most amenable to automation is physical design. This step, which contains the most tedious and time-consuming detail, was also the one that received the most attention from researchers early on. Our discussion on DA problems will concentrate on the class of physical design problems and, to a lesser extent, on testing problems.

In discussing physical design automation problems, we shall further classify them into various subclasses of problems. Although these subclasses are intimately related (in that they are all parts of a single problem), it is preferable to treat them separately because of the inherent computational complexity of the total problem. Actually, each of these problems represents a general class of problems whose precise definition is strongly influenced by factors such as the level (in the wiring hierarchy of IC chip, circuit card, backplane, etc.) of design, the particular technology being employed, the electrical constraints of the circuitry, and the tools available for attacking the problems. The specific problem, in turn, influences the size of the problem, the selection of parameters for constraints and optimization, and the methodology for designing the solution techniques.

Some Classes of Design Automation Problems

Implementation Problems. For lack of a better term, we shall classify as "implementation problems" all those problems encountered in the process of mapping the logic design onto a set of physical modules. These implementation problems include the following types of problems:

Synthesis. This problem deals with the translation of one logical representation of a digital system into another, with the constraint that the two representations be functionally equivalent.

This problem arises because the building blocks of the logic design are determined by the functional primitives understood by the logic simulation system, and not by the functionalities of the circuits most conveniently implemented in the given semiconductor technology. (So, though the choice of the underlying technology influences logic designers, insofar as they take advantage of its strengths and compensate for its weaknesses, the primitives in which the design eventually gets expressed are not altered.) Consequently, there is a need to rewrite the design in terms of the circuit families supported by the given technology.

Synthesis is a major bottleneck in designing computer hardware. Manual synthesis, apart from being slow, is quite error-prone. Designers often find themselves spending half their time correcting synthesis errors. Unfortunately, automated synthesis is still far from viable, and much work needs to be done in this area.

Partitioning. The partitioning problem is encountered at various levels of the system packaging hierarchy. In very general terms, the problem may be described as follows. Given a description of the design to be implemented within a given physical package, the problem is to subdivide the logic among the subassemblies (i.e., packages at the next level of the hierarchy) contained within the given package, in a way that optimizes certain predetermined norms.

Quantities of interest in the partitioning process are:

1. The number of partition elements (i.e., distinct subassemblies) (Kodres 1969).
2. The size of each partition element. This is an indication of the amount of space needed to physically implement the chunk of logic within that partition element.
3. The number of external connections required by each subassembly.
4. The total number of electrical connections between subassemblies (Habayeb 1968; Lawler 1962).
5. The (estimated) system delay (Lawler 1969). This points to the fact that proper partitioning is an extremely key element in optimizing the system performance. In fact, in many design methodologies, partitioning, at least at the early (and critical) stages of the design process, is still done manually by extremely skilled designers, in order to extract the maximum performance from the logic design.
6. The reliability and testability of the resulting system implementation.

Construction of a Standard Library of Modules. The library is a set of fully-designed modules that can be utilized in creating larger designs. The problem in creating libraries is deciding the functionalities of the various modules that are to be placed in the library. The process involves balancing the richness of functionality provided against the need to keep within reasonable bounds the number of distinct modules (which is related to the total cost of creating the library). Notz et al. (1967) proposed measures that aid in the periodic update of a standard library of modules.

The problem of library construction is intimately related to the partitioning problem. The library should be constructed with a good idea of what the partitioning will be like in the various logic designs that use that library (though parts of a library may be based on earlier successful subdesigns.) On the other hand, partitioning is often done based on a good understanding of the library's contents.

In SSI terms, the library is the 7400 parts catalog. In LSI terms, the parts in the library are far more complex functionally. Library construction is usually quite expensive. The logical and physical designs of each part in the library have to be totally optimized to extract the maximum performance while requiring the least space and power, given a specific semiconductor technology.

Selection of Modules from a Standard Library. Given a partition of a circuit along with a standard library of modules, the selection problem deals with finding a set of modules with either minimal total cost or minimal module count to implement the logic in the partition.

Placement Problems. In the most general terms the placement problem may be viewed as a combinational problem of optimally assigning interrelated entities to fixed locations (cells) contained in a given area. The precise definition of the interrelated entities and the location is strongly dependent on the particular level of the backplane being considered and the particular technology being employed. For instance, we can talk of the placement of logic circuits within a chip, of chips on a chip carrier, of chip carriers on a board, or of boards on a backplane. As stated earlier, the particular level influences the size of the problem, the choice of norms to be optimized, the constraints, and even the solution techniques to be considered.

The optimization criterion in placement problems is generally some norm defined on the interconnections, and in practice a number of goals must be satisfied. The main goal is to enhance the wirability of the resulting assembly while ensuring that wiring rules are not violated. Some of the norms used are listed below:

1. Minimizing the expected wiring congestions (Clark 1969).
2. Avoidance of wire crossovers (Kodres 1962).
3. Minimizing the total number of wire bends (in rectilinear technologies) (Pomentale 1965).
4. Elimination of inductive cross-talk by minimum shielding techniques.
5. Elimination/suppression of signal echoes.
6. Control of heat dissipation levels.

It can be seen that satisfying all the above-mentioned goals is virtually impossible. In most practical applications the norm minimized is the total weighted wire length.

In the context of VLSI, the placement problem is concerned almost exclusively with enhancing wirability. A difference is that the cell shapes and locations are not fixed, and the relative (or topological) placement of the cells is the important thing. Before absolute placement on silicon occurs, the cell shapes and the amount of space to be allowed for wiring have to be determined. This gives placement a different flavor from the MSI/LSI context. Furthermore, the term "placement" does not have a standard usage in VLSI. For instance, it has been used to describe the problem of determining the relative ordering of the terminals emanating from a cell.

Wiring Problems.   Also referred to as interconnection or routing problems, wiring problems involve the process of formally defining the precise conductor paths necessary to properly interconnect the elements of the system. The constraints imposed on an acceptable solution generally involve one or more of the following:

1. Number of layers (planes in which paths may exist).
2. Number and location of via holes (feedthrough pins) or paths between layers.

3. Number of crossovers.
4. Noise levels.
5. Amount of shielding required for cross-talk suppression.
6. Signal reflection elimination.
7. Line width (density).
8. Path directions (e.g., horizontal and/or vertical only).
9. Interconnection path length.
10. Total area or volume for interconnection.

Because of the intimate relationship that exists between the placement and wiring phases, many of the norms considered for optimization are common to both phases.

Approaches to wiring differ according to the nature of the wiring surface. There are two broad approaches: one-at-a-time wiring and a two-stage, coarse–fine approach.

One-at-a-time wiring is most appropriate in situations where the wiring surface is wide open, as is typically the case with PC cards and back-panels. In practice, one-at-a-time wiring is generally viewed as entailing the following subproblems:

1. Wire list determination.
2. Layering.
3. Ordering.
4. Wire layout.

Wire list determination involves making a list of the set of wires to be laid out. Given a set of points to be made electrically common, there are a number of alternative interconnecting wire sets possible. Layering assigns the wires to different layers. The layering problem involves minimizing the number of layers, such that there exists an assignment of each wire to one of the layers that results in no wire intersections anywhere. The ordering problem decides when each wire assigned to a layer is to be laid out. Since optimal wire layout appears to be a computationally intractable problem, all wire-layout algorithms currently in use are heuristic in nature. Therefore, the sequence or ordering not only affects the total interconnection distance, but also may lead to the success or failure of the entire wiring process. Last but not least, the wire layout problem, which seems to have attracted more interest than the others, deals with how each wire is to be routed, that is, specifying the precise interconnection path between two elements.

A criticism of one-at-a-time wiring has been that, when a wire is laid out, it is done without any prescience. Thus, when a wire is routed, it might end up blocking a lot of wires that have yet to be considered for wire routing. To alleviate this problem, a two-stage approach, based on Hashimoto and Stevens (1971), is often used. Here, the wiring surface is usually divided into rectangular areas called channels. In the first stage, often referred to as *global routing*, an algorithm is used to determine the sequence of channels to be used in routing each connection. In the second stage, usually called *channel routing*, all connections within each channel are completed, with the various channels being considered in order.

This two-stage approach, which is generically referred to as the *channel routing approach*, is very appropriate for wire routing inside ICs, where the wiring surface is naturally divided into channels. There are, however, situations where the channel routing approach is not appropriate. In particular, it is not very appropriate for wide open wiring surfaces, where all channel definition becomes artificial. With wire terminals located all over the wiring surface, it becomes difficult to define nontrivial channels while ensuring that all terminals are on the sides of (and not within) channels. Even when channel definition is possible, applying the channel routing approach to relatively uncluttered wiring surfaces results in many instances of the notorious channel intersection (or switchbox) problem. The effects of the coupling of constraints between channels at channel intersections are usually undesirable, and can destroy the effectiveness of the channel routing approach when the number of channel intersections is large in relation to the problem size.

For very large systems, where the number of interconnections may be in the tens of thousands, an interesting approach to the general wiring problem, given a wide open wiring surface, is the *single row routing approach*. It was initially developed by So (1974) as a method to estimate very roughly the inherent routability of the given problem. However, it produces very regular layouts (which facilitate automated fabrication), so it has been adopted as a viable approach to the wiring problem.

Single row routing consists of a systemic decomposition of the general multilayer wiring problem into a number of independent single layer, single row routing problems. There are four phases in this decomposition (Ting and Kuh 1978):

1. Via assignment: In this phase, vias are assigned to the different nets such that the interconnection becomes feasible. Note that after the via assignment is complete, wires to be routed are either horizontal or vertical. Design objectives in this phase include: (a) Minimizing the number of vias used. (b) Minimizing the number of columns of vias that result.
2. Linear placement of via columns: In this phase, an optimal permutation of via columns is sought that minimizes the maximum number of horizontal tracks required on the board.
3. Layering: The objective in this phase is to evenly distribute the edges on a number of layers. The edges are partitioned between the various layers in such a way that all edges on a layer are either horizontal or vertical.
4. Single row routing: In this final phase, one is presented with a number of

single row connection patterns on each layer. The problem here is to find a physical layout for these patterns, subject to the usual wiring constraints.

If one of the objectives during the via assignment phase is minimizing the projected wiring density, the single row wiring approach in fact becomes an effective application of a two-stage channel routing-like approach to situations in which the wiring surface is wide open.

The apparent complexity of the general wiring problem has sparked investigations into topologically restricted classes of wiring problems. One such class of problems involves the wiring of connections between the terminals of a single rectangular component, with wiring allowed only outside the periphery of the component. A norm to be minimized is the area required for wiring.

Another restricted wiring problem is *river routing*. The basic problem is as follows. Two ordered sets of terminals $(a_1, a_2 \ldots, a_n)$ and $(b_1, b_2 \ldots, b_n)$ are to be connected by wires across a rectangular channel, with wire i connecting terminal a_i to terminal b_i, $1 \leq i \leq n$. The objective is to make the connections without any wires crossing, while attempting to minimize the separation between the two sets of terminals (i.e., the channel width).

River routing has found applications in many VLSI design methodologies. When a top-down design style is followed, it is possible to ensure that, by and large, the terminals are so ordered on the perimeter of each block that, in the channel between any adjacent pair of blocks, the terminals to be connected are in the correct relative order on the opposite sides of the channel (i.e., a river routing situation exists).

The requirement concerning the proper ordering of the terminals of each block is admittedly quite difficult to always meet. However, it is a less severe requirement to meet than the one imposed in design systems such as Bristle Blocks (Johannsen 1979). In the latter system, wiring is conspicuously avoided by forcing the designer to design modules in a plug-together fashion; the blocks must all fit together snugly, and all desired connections between blocks are made to occur by actually having the associated terminals touch each other. In such a design environment, all channel widths are zero, and thus there can be no wiring.

Testing Problems.   Over the years, fault diagnosis has grown to be one of the more active, albeit less mature, areas of design automation development. Fault diagnosis compromises: (a) fault detection, and (b) fault location and identification.

The unit to be diagnosed can range from an individual IC chip, to a board-level assembly comprising several chip carriers, to an entire system containing many boards. For proper diagnosis, the unit's behavior and hardware organization must be thoroughly understood. Also essential is a detailed analysis of the faults for which the unit is being diagnosed. This, in turn, involves concepts such as fault modeling, fault equivalence, fault collapsing, fault propagation, coverage analysis, and fault enumeration.

Fault diagnosis is normally effected by the process of *testing*. That is, the unit's behavior is monitored in the presence of certain predetermined stimuli known as tests. Testing is a general term, and its goal is to discover the presence of different types of faults. However, over the years, it has come to mean testing for physical faults, particularly those introduced during the manufacturing phase. Testing for nonphysical faults (e.g., design faults) has come to be known as *design verification*, and it is just emerging as an area of active interest. In keeping with the industry trend and to avoid confusion, we shall use the terms "testing" and "design verification" in the sense described above.

Testing. The central problem in testing is *test generation*. The majority of the effort in testing has been directed toward designing automatic test generation methods. Test generation is often followed by test verification, which deals with evaluating the effectiveness of a test set in diagnosing faults. Both test generation and test verification are extremely complex and time-consuming tasks. As a result, their development has been rather slow. Most of the techniques developed are of a heuristic nature. The development process itself has consistently lagged behind the rapidly changing IC technologies. Hence, at any given time, the testing methods have always been inadequate for handling designs that use concurrent technologies. Most of the automatic test generation methods existing today were developed in the 1970s. They mainly addressed fault detection, and were based on one of the following: path sensitizing, D-algorithm (Roth 1966), or Boolean difference (Yan and Tang 1971). They basically handled logic networks implemented with SSI/MSI level gates. Also, most of the techniques considered only combinational networks, and almost all of them assumed the simplified single stuck-at fault model. As regards test verification, formal proof to date has been almost impossible in practice. Most verification is done by fault simulation and fault injection.

The advent of LSI and VLSI, while improving cost and performance, has further complicated the testing problem. The different architectures and processing complexities of the new building blocks (e.g., the microprocessor) have rendered most of the existing test methods quite incapable. As a result, it has become necessary to reinvestigate some of the aspects of the testing problem. Take, for instance, the single stuck-at fault model. For years, the industry has clung to this assumption. While being adequate for prior technologies, it does not adequately cover other fault mechanisms, such as bridging shorts or pattern sensitivities. Furthermore, for testing microprocessors, PLAs, RAMs, ROMs, and complex gate arrays, testing at a level higher than the gate level appears to make more sense. This involves testing at the functional and algorithmic or behavioral levels. Also, more work needs to be done in the area of fault location and identification. The method presented in Abramovici and Breuer (1980) attempts to achieve both fault detection and location without requiring explicit fault enumeration.

Finally, there is the prudent approach of designing for testability in order to simplify the testing problem. Design for testability first attracted attention with the coming of LSI. Today, with VLSI, its need has become all the more critical. One of the main problems in this area is deriving a quantitative measure of testability. One way is to analyze a unit for its controllability and observability (Goldstein 1979), quantities that represent the difficulty of controlling and observing the logical values of internal nodes from the inputs and outputs, respectively. Most existing testability measures, however, have been found to be either crude or difficult to determine. The next problem is deriving techniques for testability design. A comprehensive survey of these techniques is given in Grason and Nagle (1980) and Williams and Parker (1982). Most of them are of an ad hoc nature, presented either as general guidelines or hard and fast rules. A summary of these techniques appears in Figure 17-1.

Design Verification. The central issue here is proving the correctness of a design. Does a design do what it is supposed to do? In other words, we are dealing with testing for design faults. The purpose of design verification is quite clear. Design faults must be eliminated, as far as possible, before the hardware is constructed, and before prototype tests begin. The increasing cost of implementing engineering changes, given LSI/VLSI hardware, has enhanced the need for design verification.

```
                    DESIGN FOR TESTABILITY--TECHNIQUES


           AD HOC TECHNIQUES
                 -"Bed of Nails"
                 -Isolation
                 -Signature Analysis
                 -Partitioning

           STRUCTURED TECHNIQUES
                 -Address Multiplexing
                 -Scan Path
                 -Level Sensitive Scan Design(LSSD)
                 -Scan-Set
                 -Random Access Scan Techniques

           SELF-TEST/BUILT-IN TECHNIQUES
                 -Built-In Logic Block Observation(BILBO)
                 -Testing with Walsh Coefficients
                 --Syndrome Testing
                 -Autonomous Testing
```

Figure 17-1. Summary of techniques for design for testability.

Compared to physical faults, design faults are more subtle and serious and can be extremely difficult to handle. Hence, the techniques developed for physical faults cannot be effectively extended to design faults. To date, very little effort has been devoted to formalizing design verification. Designs are still mostly checked by ad hoc means such as fault simulation and prototype checkout.

Like other disciplines, design verification has not been spared the impact of LSI/VLSI. Some of these influences are listed below.

Ratification. Matching the design specification, known as ratification, was accomplished in the pre-LSI era by gate level simulation This may no longer be sufficient. The simulations need to be more detailed, and they need to be done at higher levels, such as the functional and behavioral levels. Also, techniques are required for determining the following:

1. Stopping rules for simulation.
2. The extent of design faults removed.
3. A quantitative measure for the correctness or quality of the final design.

Validation. In the pre-LSI days, this was restricted to the testing of hardware on the test floor. Moreover, the testing process was not formalized or systematic, and hence lacked thoroughness and rigor. Today, validation mainly involves testing the equivalence of two design descriptions. The descriptions may be at different levels. Thus, before being compared, they need to be translated to a common level. For example, one can construct symbolic execution tree models of the design descriptions to be compared.

Timing Analysis. In the past, it was sufficient to analyze only single "critical" paths. The technology rules of LSI/VLSI are so complex that identification of these critical paths has become extremely difficult. Statistical timing analysis methods need to be investigated, in order to cope with the tremendous densities and wide range of tolerances imposed by LSI/VLSI.

Finally, research has also started in developing design techniques to alleviate the need for, and/or facilitate, the design verification process.

COMPLEXITY AND NONDETERMINISM

Complexity

By the *complexity of an algorithm*, we mean the amount of computer time and memory space needed to run it. These two quantities will be referred to, respectively, as the time and space complexities of the algorithm. To illustrate, consider the procedure MADD (Algorithm 17-1), an algorithm to add two m $\times$ n matrices together.

line procedure MADD(A,B,C,m,n)
    {compute C = A + B}
1 **declare** A(m,n),B(m,n),C(m,n)
2 **for** i ← 1 **to** m **do**
3   **for** j ← 1 **to** n **do**
4   C(i,j) ← A(i,j) + B(i,j)
5   **endfor**
6 **endfor**
7 **end** MADD

Algorithm 17-1. Matrix addition.

The time needed to run this algorithm on a computer comprises two components: the time to compile the algorithm and the time to execute it. The first of these two components depends on the compiler and the computer being used. This time is, however, independent of the actual values of n and m. The execution time, in addition to being dependent on the compiler and the computer used, depends on the values of m and n. It takes more time to add larger matrices.

Since the actual time requirements of an algorithm are very machine-dependent, the theoretical analysis of the time complexity of an algorithm is restricted to determining the number of steps needed by the algorithm. This step count is obtained as a function of certain parameters that characterize the input and output. Some examples of often-used parameters are: number of inputs; number of outputs; magnitude of inputs and outputs; and so forth.

In the case of our matrix addition example, the number of rows m and the number of columns n are reasonable parameters to use. If instruction 4 of procedure MADD is assigned a step count of 1 per execution, then its total contribution to the step count of the algorithm is mn, as this instruction is executed mn times. Sahni (1981, Chapter 6) discusses step count analysis in greater detail. Since the notion of a step is somewhat inexact, one often does not strive to obtain an exact step count for an algorithm. Rather, asymptotic bounds on the step count are obtained. Asymptotic analysis uses the notation; O, Ω Θ, and o. These are defined below.

Definition: [Asymptotic Notation] $f(n) = O(g(n))$ (read as "f of n is big oh of g of n") iff there exist positive constants c and n_0 such that $f(n) \leq cg(n)$ for all n, $n \geq n_0$. $f(n) = \Omega(g(n))$ (read as "f of n is omega of g of n") iff there exist positive constants c and n_0 such that $f(n) \geq cg(n)$ for all n, $n \geq n_0$. $f(n)$ is $\Theta(g(n))$ (read as "f of n is theta of g of n") iff there exist positive constants c_1 c_2, and n_0 such that $c_1 g(n) \leq f(n) \leq c_2 g(n)$ for all n, $n \geq n_0$. $f(n) = o(g(n))$ (read as "f of n is little o of g of n") iff $\lim_{n \to \infty} n/g(n) = 1$.   □

The definitions of O, Ω, Θ, and o are easily extended to include functions of more than one variable. For example, $f(n,m) = O(g(n,m))$ iff there exist positive constants c, n_0, and m_0 such that $f(n,m) \leqslant cg(n,m)$ for all $n \geqslant n_o$ and all $m \geqslant m_0$.

Example 17-1: $3n + 2 = O(n)$ as $3n + 2 \leqslant 4n$ for all n, $n \geqslant 2$. $3n + 2 = \Omega(n)$ and $3n + 2 = \Theta(n)$. $6*2^n + n^2 = O(2^n)$. $3n = O(n^2)$. $3n = o(3n)$ and $3n = O(n^3)$. □

As illustrated by the previous example, the statement $f(n) = O(g(n))$ only states that $g(n)$ is an upper bound on the value of $f(n)$ for all n, $n \geqslant n_0$. It does not say anything about how good this bound is. Notice that if $n = O(n)$, $n = O(n^2)$, $n = O(n^{2.5})$, etc. In order to be informative, $g(n)$ should be as small as a function of n as one can come up with such that $f(n) = O(g(n))$. So, while we shall often say that $3n + 3 = O(n)$, we shall almost never say that $3n + 3 = O(n^2)$.

As in the case of the "big oh" notation, there are several functions $g(n)$ for which $f(n) = \Omega(g(n))$. $g(n)$ is only a lower bound on $f(n)$. The theta notation is more precise than both the "big oh" and omega notations. The following theorem obtains a very useful result about the order of $f(n)$ when $f(n)$ is a polynomial in n.

Theorem 17-1: Let $f(n) = a_m n^m + a_{m-1} n^{m-1} + \cdots + a_0$, $a_m \neq 0$.
(a) $f(n) = O(n^m)$
(b) $f(n) = \Omega(n^m)$
(c) $f(n) = \Theta(n^m)$
(d) $f(n) = o(a_m n^m)$ □

Asymptotic analysis may also be used for space complexity.

Although asymptotic analysis does not tell us how many seconds an algorithm will run for or how many words of memory it will require, it does characterize the growth rate of the complexity (see Figure 17-2). So, if procedure MADD takes 2 milliseconds (ms) on a problem with $m = 100$ and $n = 20$, then we expect it to take about 16 ms when $mn = 16,000$ (the complexity of MADD is $\Theta(mn)$). For sufficiently large values of n, a $\Theta(n^2)$ algorithm will be faster than a $\Theta(n^3)$ algorithm.

We have seen that the time complexity of an algorithm is generally some function of the instance characterisitics. This function is very useful in determining how the time requirements vary as the instance characteristics change. The complexity function may also be used to compare two algorithms A and B that perform the same task. Assume that algorithm A has complexity $\Theta(n)$ and algorithm B is of complexity $\Theta(n^2)$. We can assert that algorithm A is faster than algorithm B for "sufficiently large" n. To see the validity of this assertion, observe that the actual computing time of A is bounded from above by n for some constant c and for all n, $n \geqslant n_1$ while that of B is bounded from below by dn^2 for some constant d and all n, $n \geqslant n_2$. Since $cn \leqslant dn^2$ for $n \geqslant c/d$,

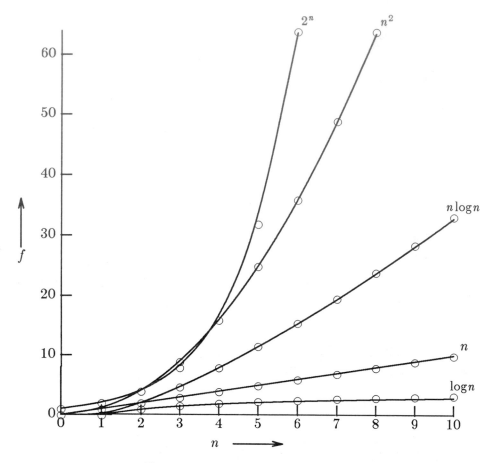

Figure 17-2. The growth rate of complexity.

algorithm A is faster than algorithm B whenever $n \geqslant \max\{n_1, n_2, c/d\}$. One should always be cautiously aware of the presence of the phrase "sufficiently large" in the assertion of the preceding discussion. When deciding which of the two algorithms to use, we must know whether the n we are dealing with is in fact "sufficiently large." If algorithm A actually runs in $10^6 n$ milliseconds while algorithm B runs in n^2 milliseconds and if we always have $n \leqslant 10^6$, then algorithm B is the one to use. To get a feel for how the various functions grow with n, you are advised to study Figure 17-2 very closely. As is evident from the figure, the function 2^n grows very rapidly with n. In fact, if an algorithm needs 2^n steps for execution, then when n = 40 the number of steps needed is approximately $1.1*10^{12}$. On a computer performing one billion steps per second, this would require about 18.3 minutes. If n = 50, the same algorithm would run for about 13 days on this computer. When n = 60, about 36.56 years will be required to execute the algorithm, and when n = 100, about $4*10^{13}$ years

Table 17-1 Time for f(n) instructions on a 10^9 instr/sec computer

n	$f(n) = n$	$f(n) = n\log_2 n$	$f(n) = n^2$	$f(n) = n^3$	$f(n) = n^4$	$f(n) = n^{10}$	$f(n) = 2^n$ [a]
10	.01 µs	.03 µs	.1 µs	1 µs	10 µs	10 sec	1 µs
20	.02 µs	.09 µs	.4 µs	8 µs	160 µs	2.84 hr	1 ms
30	.03 µs	.15 µs	.9 µs	27 µs	810 µs	6.83 d	1 sec
40	.04 µs	.21 µs	1.6 µs	64 µs	2.56 ms	121.36 d	18.3 min
50	.05 µs	.28 µs	2.5 µs	125 µs	6.25 ms	3.1 yr	13 d
100	.10 µs	.66 µs	10 µs	1 ms	100 ms	3171 yr	$4*10^{13}$ yr
1,000	1.00 µs	9.96 µs	1 ms	1 sec	16.67 min	$3.17*10^{13}$ yr	$32*10^{283}$ yr
10,000	10.00 µs	130.3 µs	100 ms	16.67 min	115.7 d	$3.17*10^{23}$ yr	
100,000	100.00 µs	1.66 ms	10 sec	11.57 d	3171 yr	$3.17*10^{33}$ yr	
1,000,000	1.00 ms	19.92 ms	16.67 min	31.71 yr	$3.17*10^2$ yr	$3.17*10^{44}$ yr	

Abbreviations: µs = microsecond = 10^{-6} seconds; ms = millisecond = 10^{-3} seconds; sec = seconds; min = minutes; hr = hours; d = days; yr = years.

will be needed. So, we may conclude that the utility of algorithms with exponential complexity is limited to small n (typically $n \leq 40$).

Algorithms that have a complexity that is a polynomial of high degree are also of limited utility. For example, if an algorithm needs n^{10} steps, then using our one billion step per second computer we will need 10 seconds when n = 10; 3,171 years when n = 100; and $3.17*10^{13}$ years when n = 1000. If the algorithms complexity had been n^3 steps instead, then we would need one second when n = 1000, 16.67 minutes when n = 10,000; and 11.57 days when n = 100,000.

Table 17-1 gives the time needed by a one billion instruction per second computer to execute a program of complexity f(n) instructions. One should note that currently only the fastest computers can execute about one billion instructions per second. From a practical standpoint, it is evident that for reasonably large n (say n > 100) only algorithms of small complexity (such as n, nlogn, n^2, n^3, etc.) are feasible. Further, this is the case even if one could build a computer capable of executing 10^{12} instructions per second. In this case, the computing times of Table 17-1 would decrease by a factor of 1000. Now, when n = 100 it would take 3.17 years to execute n^{10} instructions, and $4*10^{10}$ years to execute 2^n instructions.

Another point to note is that the complexity of an algorithm cannot always be characterized by the size or number of inputs and outputs. The time taken is often very data dependent. As an example, consider Algorithm 17-2. This is a very primitive backtracking algorithm to determine if there is a subset of $\{W(1),W(2), \ldots W(n)\}$ with sum equal to M. This problem is called the *sum of subsets* problem. Procedure SS is initially invoked by the statement:

$$X \leftarrow SS(1,M)$$

```
procedure SS(i,P)
 {determine if W(i:n) has a}
 {subset that sums to P}
 global W(1:n),n
 if i>n then return(false)
 case
  :W(i) = P: return(true)
  :W(i)<P: if SS(i + 1,P − W(i))
        then return(true)
        else return(SS(i + 1,P))
      endif
  :else: return (SS(i + 1,P))
 endcase
end SS
```

Algorithm 17-2.

Procedure SS returns the value *true* iff a subset of W(1:n) sums to M. Observe that if $\sum_{i=1}^{a} W(i) = M$, then SS runs in $\Theta(n)$ time. If no subset of W(1:n) sums to M, then SS takes $\Theta(2^n)$ time to terminate. For other cases, the time needed is anywhere between $\Theta(n)$ and $\Theta(2^n)$. So, the complexity of SS is $\Omega(n)$ and (2^n). In cases like SS where the complexity is quite data-dependent, one talks of the best case, worst case, and average (or expected) complexity. A precise definition of these terms can be found in Sahni (1981). Here, we shall rely on our intuitive understanding of these terms.

Most of the complexity results obtained to date have been concerned with the worst-case complexity of algorithms. The value of such analyses may be debated. Procedure SS has a worst-case time complexity of $\Theta(2^n)$. This just tells us that there exist inputs on which this much time will be spent. However, it might well be that for the inputs of "interest," the algorithm is far more efficient, perhaps even of complexity $O(n^2)$.

As another example, consider the much publicized Khachian algorithm for the linear programming problem. This algorithm has a worst-case complexity that is a polynomial function of the number of variables, equations, and size (i.e., number of bits) of the coefficients. The Simplex method is known to have a worst-case complexity that is exponential in the number of equations. So, as far as the worst-case complexity is concerned, Khachian's algorithm is superior to the Simplex method. Despite this, Khachian's algorithm is impractical, whereas the Simplex method has been successfully used for years to solve reasonably large instances of the linear programming problem. The Simplex method works well on those instances that are of "interest" to people.

It has been shown (Dantzig 1980) that under reasonable assumptions, the expected (or average) complexity of the Simplex method is in fact $O(m^3)$, where m is the number of equations [the expected number of pivot steps in 3.5 m, and each such step takes $O(m^2)$ time].

Thus, the notion of average complexity seems to better capture the complexity one might observe when actually using the algorithm. Average complexity analysis is, however, far more difficult than worst-case analysis and has been carried out successfully for only a limited number of algorithms. As a result, this chapter is concerned mainly with the worst-case complexity of design automation problems. One should keep in mind that while many of the design automation problems will be shown to be "probably" intractable in terms of worst-case complexity, these results do not rule out the possibility of very efficient expected behavior algorithms.

Nondeterminism

The commonly used notion of an algorithm has the property that the result of every step is uniquely defined. Algorithms with this property are called deterministic algorithms. From a theoretical framework, we can remove this restriction on the outcome of every operation. We can allow algorithms to contain operations whose outcome is not uniquely defined but is limited to a specific set of possibilities. The machine executing such operations is allowed to choose any one of these outcomes. This leads to the concept of a *nondeterministic algorithm*. To specify such algorithms we introduce three new functions:

(a) **choice**(S) . . . arbitrarily choose one of the elements of set S.
(b) **failure** . . . signals an unsuccessful completion.
(c) **success** . . . signals a successful completion.

Thus the assignment statement $X \leftarrow$ **choice**$(1:n)$ could result in X being assigned any one of the integers in the range $[1,n]$. There is no rule specifying how this choice is to be made. The **failure** and **success** signals are used to define a computation of the algorithm. One way to view this computation is to say that whenever there is a set of choices that leads to a successful computation, then one such set of choices is made, and the algorithm terminates successfully. A nondeterministic algorithm terminates unsuccessfully iff there exists no set of choices leading to a success signal. A machine capable of executing a nondeterministic algorithm in this way is called a *nondeterministic machine*.

Example 17-2. Consider the problem of searching for an element x in a given set of elements $A(1)$ to $A(n)$, $n \geq 1$. We are required to determine an index j such that $A(j) = x$ or $j = 0$ if x is not in A. A nondeterministic algorithm for this is:

$$j \leftarrow \textbf{choice}(1:n)$$
if $A(j) = x$ **then** print (j); **success endif**
print ("0") **failure.**

From the way a nondeterministic computation is defined, it follows that the number "0" can be output iff there is no j such that $A(j) = x$. The computing times for **choice, success,** and **failure** are taken to be $O(1)$. Thus the above algorithm is of nondeterministic complexity $O(1)$. Note that since A is not ordered, every deterministic search algorithm is of complexity at least $O(n)$. $\square$

Because many choice sequences lead to a successful termination of a nondeterministic algorithm, the output of such an algorithm working on a given data set may not be uniquely defined. To overcome this difficulty, one normally considers only decision problems, that is, problems with answer 0 or 1 (or true or false). A successful termination always yields the output 1, whereas unsuccessful terminations always yield the output 0.

In measuring the complexity of a nondeterministic algorithm, the cost assignable to the **choice**(S) function is $O(\log k)$ where k is the size of S. So, strictly speaking, the complexity of the search algorithm of Example 17-2 is O(log n). The time required by a nondeterministic algorithm performing on any given input depends upon whether or not there exists a sequence of choices that leads to a successful completion. If such a sequence exists, then the time required is the minimum number of steps leading to such a completion. If no choice sequence leads to a successful completion, then the algorithm takes O(1) time to make a failure termination.

Nondetermination appears to be a powerful tool. Algorithm 17-3 is a nondeterministic algorithm for the sum of subsets problem. Its complexity is O(n). The best deterministic algorithm for this problem has complexity $O(2^{n/2})$ (see Horowitz and Sahni 1974).

procedure NSS(W,n,M)
　　declare X(1:n),W(1:n),N,M
　　for i ← 1 **to** n **do**
　　　X(i) ← **choice**({0,1})
　　end

　　if $\sum_{i=1}^{n} W(i)$ = M **then success**

　　　　　　　　　　else failure
　　endif
　end NSS

Algorithm 17-3. Nondeterministic sum of subsets algorithm.

NP-Hard and NP-Complete Problems

The *size* of a problem instance is the number of digits needed to represent that instance. An instance of the sum of subsets problem is given by (W(1),W(2), . . . , W(n),M). If each of these numbers is nonnegative and integer, then the instance size is $\sum_{i=1}^{a} \log_2 W(i) + \log_2 M$ if binary digits are used. An algorithm is of *polynomial* time complexity iff its computing time is O(p(m)) for every input size m and some fixed polynomial p( ).

Let P be the set of all decision problems that can be solved in deterministic polynomial time. Let NP be the set of decision problems solvable in polynomial time by nondeterministic algorithms. Clearly, P ⊆ NP. It is not known whether P = NP or P ≠ NP. The P = NP problem is important because it is related to the complexity of many interesting problems (including certain design automation problems). There exist many problems that cannot be solved in polynomial time unless P = NP. Because intuitively, one expects that P ≠ NP, these problems are in "all probability" not solvable in polynomial time. The first problem that was shown to be related to the P = NP problem, in this way, was the problem of determining whether or not a propositional formula is satisfiable. This problem is referred to as the *Satisfiability problem*.

Theorem 17-2: Satisfiability is in P iff P = NP.

Proof: See Horowitz and Sahni (1978) or Garey (1979).   □

Let A and B be two problems. Problem A is *polynomially reducible* to problem B (abbreviated A reduces to B, and written as A α B) iff the existence of a deterministic polynomial time algorithm for B implies the existence of a deterministic polynomial time algorithm for A. Thus, if A α B and B is polynomially solvable, then so also is A. A problem A is *NP-hard* iff Satisfiability α A. An NP-hard problem A is *NP-complete* iff A ϵ NP.

Observe that the relation α is transitive (i.e., if A α B and B α C, then A α C). Consequently, if A α B and Satisfiability α A, then B is NP-hard. So, to show that any problem B is NP-hard, we need merely show that A α B where A is any known NP-hard problem. Some of the known NP-hard problems are:

NP1: Euclidean Steiner Tree (Garey et al. 1977)
Input: A set X = $\{(x_i,y_i)|1 \leq i \leq n\}$ of points.
Output: A fine set Y = $\{(a_i, b_i)|1 \leq i \leq m\}$ of points such that the minimum spanning tree for X∪Y is of minimum total length over all choices for Y. The distance between two points (t,u) and (v,W) is $[(t - v)^2 + (u - w)^2]$.

NP2: Manhattan Steiner Tree (Garey et al. 1977)
Input: Same as in NP1.
Output: Same as in NP1, except that the distance between two points is taken to be $|t - v| + |u - w|$.

NP3: Euclidean Traveling Salesman (Garey and Johnson 1976b).
Input: Same as in NP1.
Output: A minimum-length tour going through each point in X. The Euclidean distance measure is used.

NP4: Euclidean Path Traveling Salesman (Papadimitriou 1977)
     (also called Euclidean Hamiltonian Path)
Input: Same is in NP1.
Output: A minimum length path that visits all points in X exactly once. The Euclidean distance measure is used.

NP5: Manhattan Traveling Salesman (Garey and Johnson 1976b)
Input: Same as in NP3.
Output: Same as in NP3, except that the Manhattan distance measure is used.

NP6: Manhattan Path Traveling Salesman (Papadimitriou 1977)
   (also called Manhattan Hamiltonian Path)
Input: Same as in NP1.
Output: Same as in NP4, except that the Manhattan distance measure is used.

NP7: Chromatic Number I (Ehrlich et al. 1976).
Input: A graph G that is the intersection graph for straight line segments in the plane.
Output: The minimum number of colors needed to color G.

NP8: Chromatic Number II (Ehrlich et al. 1976)
Input: Same as in NP5.
Output: "Yes" if G is 3-colorable and "No" otherwise.

NP9. Partition (Karp 1972)
Input: A Multiset $A = \{a_i | 1 \leq i \leq n\}$ of natural numbers.
Output: "Yes" if there is a subset $B \subseteq \{1, 2, \ldots, n\}$ such that $\sum_{i \varepsilon \beta} a_i = \sum_{i \varepsilon \beta} a_i$.

"No" otherwise.

NP10: 3-Partition (Garey and Johnson 1975)
Input: A multiset $A = \{a_i | 1 \leq 1 \leq 3m\}$ of natural numbers, and a bound B, such that ε (i) $\sum_{a_i \varepsilon A} a_i = mB$ (ii) $B/4 < a_i < B/2$ for $1 \leq i \leq 3m$
Output: "Yes" if A can be partitioned into m disjoint sets $A_1, A_2 \ldots A_m$ such that, for $1 \leq i \leq m$, $\sum_{a_j c A_i} = B$; "No" otherwise.

NP11: Knapsack (maximization) (Karp 1972)
Input Multisets $P = \{p_i | 1 \leq i \leq n\}$ and $W = \{w_i | 1 \leq i \leq n\}$ of natural numbers and another natural number M.
Output: $x_i \varepsilon \{0,1\}$ such that $\sum_i p_i x_i$ is maximized and $\sum_i w_i x_i \leq M$.

NP12: Knapsack (minimization)
Input: Same as in NP11, except replace set P by $K = \{k_i | 1 \leq i \leq n\}$.
Output: $x_i \varepsilon \{0,1\}$ such that $\sum_i k_i x_i$ is minimized and $\sum_i w_i x_i \geq M$.

NP13: Integer Knapsack (Lucker 197)
Input: Multiset $W = \{w_i | 1 \leq i \leq n\}$ of nonnegative integers and two additional nonnegative integers M and K.
Output: "Yes" if there exist nonnegative integers x_i, $1 \leq i \leq n$ such that $\sum w_i x_i \leq M$ and $\sum w_i x_i \leq K$. "No" otherwise.

NP14: Quadratic Assignment Problem (Sahni and Gonzales 1976).

$Input:$ $c_{i,j}$, $1 \leqslant i \leqslant n$, $1 \leqslant j \leqslant n$.

$d_{k,q}$, $1 \leqslant k \leqslant m$, $1 \leqslant q \leqslant m$.

$Output:$ $x_{i,k}$ $\varepsilon\{0,1\}$, $1 \leqslant i \leqslant n$, $1 \leqslant k \leqslant m$, such that

$$(a) \quad \sum_{t=1}^{m} x_{i,k} \leqslant 1, 1 \leqslant i \leqslant n$$

$$(b) \quad \sum_{k=1}^{n} x_{i,k} = 1, 1 \leqslant k \leqslant m$$

$$\text{and} \quad \sum_{\substack{i,j=1 \\ t \neq j}}^{n} \{ \sum_{\substack{k,q=1 \\ k \neq q}}^{m} c_{i,j} \, d_{k,q} \, x_{i,k} \, x_{j,q}\} \text{ is minimized.}$$

A listing of over 200 known NP-hard problems can be found in Garey (1979). The importance of showing that a problem A is NP-hard lies in the P = NP problem. Because we do not expect that P = NP, we do not expect NP-hard problems to be solvable by algorithms with a worst-case complexity that is polynomial in the size of the problem instance. From Figure 17-2 and Table 17-1 it is apparent that if a problem cannot be solved in polynomial time, then it is intractable, for all practical purposes. If A is NP-complete and if it does turn out that P = NP, then A will be polynomially solvable.

COMPLEXITY OF DESIGN AUTOMATION PROBLEMS

In the following section, we illustrate how one goes about showing that a problem is NP-Hard or NP-complete. We consider three examples from the design automation area. Over 30 design automation problems are then described. With each problem, a discussion of its complexity is included.

Showing Problems NP-Hard and NP-Complete

Circuit Realization. In this problem, we are given a set of r modules. Associated with module i is a cost c_i, $1 \leqslant i \leqslant r$. Module i contains m_{ij} gates of type j, $1 \leqslant j \leqslant n$. We are required to realize a circuit C with gate requirements $(b_1 b_2, \ldots, b_n)$; i.e., circuit C consists of b_j gates of type j. $(x_1, \ldots, x_r)$ realizes circuit C iff

$$\sum_{i=1}^{r} m_{ij} x_i \geqslant b_j, 1 \leqslant j \leqslant n$$

and x_i is a natural number, $1 \leqslant i \leqslant r$.

The cost of the realization $(x_i, \ldots, x_r)$ is $\sum_{i=1}^{r} c_i x_i$. We are interested in obtaining a minimum cost realization of C.

Theorem 17-3: The circuit realization problem is NP-hard.

Proof: From the preceeding section, we see that it is sufficient to show that Q α circuit realization, where Q is any known NP-hard problem. We shall use Q = NP13 = integer knapsack (see preceding section).

Let $(w_1 w_2, \ldots, w_p)$, M, and K be any instance of the integer knapsack problem. Construct the following circuit realization instance:

$$n = 1; b_1 = K; r = p; m_{i,1} - w_i, 1 \leqslant i \leqslant p; c_i = w_i, 1 \leqslant i \leqslant p$$

Clearly, the least-cost realization of the above circuit instance has a cost at most M iff the corresponding integer knapsack instance has answer "yes." So if the circuit realization problem is polynomially solvable, then so also is NP13. But NP13 is NP-hard. So, circuit realization is also NP-hard. □

In order to show that an NP-hard problem Q is NP-complete, we need to show that it is in NP. Only decision problems (i.e., problems for which the output is "yes" or "no") can be NP-complete. So, the circuit realization problem cannot be NP-complete. However, we may formulate a decision version of the circuit realization problem: Is there a realization with cost no more than S? The proof provided in Theorem 17-3 is valid for this version of the problem too. Also, there is a nondeterministic polynominal time algorithm for this decision problem (Algorithm 17-4). So, the decision version of the circuit realization problem is NP-complete.

```
procedure CKT(b,S,m,r,n,c)
    {Bs there a circuit realization with cost ≤ S?}
    declare r, n, c(r), m(r,n), b(n), S, x(r)
    q < max {b(j)}
    for i ← 1 to r do {obtain χᵢs nondeterministically}
        x(i) ← choice(0:q)
    endfor
    for i ← 1 to n do {check feasibility}
```
$$\text{if } \sum_{j=1}^{r} m(j,i)*x(i) < b(i) \text{ then failure endif}$$
```
    endfor
```
$$\text{if } \sum_{t=1}^{r} c(i)x(i) > S \text{ then failure}$$
```
                    end CKT
```

Algorithm 17-4.

Euclidean Layering Problem.   A wire to be laid out may be defined by the two end points (x,y) and (u,v) of the wire. (x,y) and (u,v) are the coordinates of the two end points. In a Euclidean layout, the wire runs along a straight line from (x,y) to (u,v). Figure 17-3(a) shows some wires laid out in a Euclidean manner. Let $W = \{ [(u_i,v_i),(x_i,y_i)] \mid 1 \leq i \leq n\}$ be a set of n wires. In the Euclidean layering problem, we are required to partition W into a minimum number of disjoint sets $W_1, W_2, \ldots, W_k$ such that no two wires in any set W_i cross. Figure 17-3(b) gives a partitioning of the wires of Figure 17-3(a) that satisfies this requirement. The wires in W_1 and W_2 can now be routed in separate layers.

Theorem 17-4: The Euclidean layering problem is NP-hard.

Proof: We shall show that the known NP-hard problem NP7 (Chromatic Number I) reduces to the Euclidean layering problem. Let $G = (V,E)$ be any intersection graph for straight line segments in the plane. Let W be the corresponding set of straight line segments. Note that $|W| = |V|$ as V has one vertex for each line segment in W. Also, (i,j) is an edge of G iff the line segments corresponding to vertices i and j intersect in Euclidean space. From any partitioning $W_1,W_2 \ldots$ of W such that no two line segments of any partition intersect, we can obtain a coloring of G. Vertex i is assigned the color j iff the line segment corresponding to vertex i is in the partition W_j. No adjacent vertices in G will be assigned the

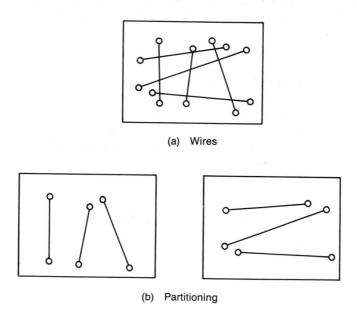

(a)   Wires

(b)   Partitioning

Figure 17-3.   Partitioning techniques.

same color, as the line segments corresponding to adjacent vertices intersect and so must be in different partitions. Furthermore, if G can be colored with k colors, then W can be partitioned into $W_1, \ldots, W_k$.

Hence, G can be colored with k colors iff W can be partitioned into k disjoint sets, no set containing two intersecting segments. So, if we could solve the Euclidean layering problem in polynomial time, then we could solve the chromatic number problem NP7 in polynomial time by first obtaining W as above and then using the polynomial time algorithm to minimally partition W. From the partition, a coloring of G can be obtained. Since NP7 is NP-hard, it follows that the Euclidean layering problem is also NP-hard. □

The above equivalence between NP7 and the Euclidean layering problem was pointed out by Akers (Breuer 1972).

A decision version of the Euclidean layering problem would take the form: Can W be partitioned into ≤ K partitions such that no partition contains two wires that intersect? The proof of Theorem 17-4 shows this decision problem is NP-hard. Procedure ELP (Algorithm 17-5) is a nondeterministic polynomial time algorithm for this problem. Hence, the decision version of the Euclidean layering problem is NP-complete.

```
procedure ELP(W,n,k)
  {n = |W|}
  wire set W, integer n,k;
  L(i) ← 0, 1≤i≤k
  for i←1 to n do {assign wires to layers}
     j ← choice(1:k)
     L(j) ← L(j) ∪ {[(u_i, v_i), x_i,y_i)]}
  endif
  success
end ELP
```

Algorithm 17-5.

Rectilinear Layering Problem. This problem is similar to the Euclidean layering problem (see above). Once again, we are given a set $W = \{[(u_i,v_i),(x_i,y_i)] \mid 1 \le i \le n\}$ of wire end points. In addition, we are given a p × q grid with horizontal and vertical lines at unit intervals. We may assume that each wire end point is a grid point. Each pair of wire end points is to be joined by a wire that is routed along grid lines alone. No two wires are permitted to cross or share the same grid line segment. We wish to find a partition W_1, $W_2, \ldots W_k$ of wires such that k is minimum and the end point pairs in each partition can be wired as described above. The end point pairs in each partition can be connected in a rectilinear manner in a single layer. The complete wiring will use k layers.

Theorem 17-5. The rectilinear problem is NP-hard.

Proof: We shall show that if the rectilinear layering problem can be solved in polynomial time, then the known NP-hard problem NP10 (3-Partition) can also be solved in polynomial time. Hence, the rectilinear layering problem is NP-hard.

Let $A = \{a_1, a_2, \ldots, a_{3m}\}$; B; $\sum a_i = mB$; $B/4 < a_i < B/2$ be any instance of the 3-Partition problem. For each a_i, we construct a size subassembly and enforcer subassembly ensemble as shown in Figure 17-4(a). Figure 17-4(b) shows how the ensembles for the n a_is are put together to obtain the complete wiring problem. The grid has dimensions $(B + 1) \times (i_1 + i_2 + 1)$ here

$$i_1 = \sum a_i + m = mB + m$$

and

$$i_2 = \sum [a_i + 2(m - 1)] + m = mB + 6m^2 - 6m + m$$

Note that all wire end points are along the bottom edge of the grid.

The valve assembly shown in Figure 17-4(b) is similar to the enforcer subassembly except that it contains m wires instead of m − 1. As is evident, no two wires of the valve assembly can be routed in the same layer. Hence, at least m layers are needed to wire the valve.
An examination of the ensemble for each a_i reveals that:

(i)   No two wires in the enforcer subassembly can be routed on the same layer, obviously.
(ii)   A wire from the size subassembly cannot be routed on the same layer with a wire from the enforcer subassembly.
(iii) All wires in a size subassembly can be routed on the same layer.

Therefore, at least m layers are required to route each ensemble. Hence, the rectilinear layering problem defined by Figure 17-4(b) needs at least m layers.
If the 3-Partition instance has a 3-Partition $A_1, A_2, \ldots, A_m$, then only m layers are needed by Figure 17-4(b). In layer i, we wire the size subassemblies

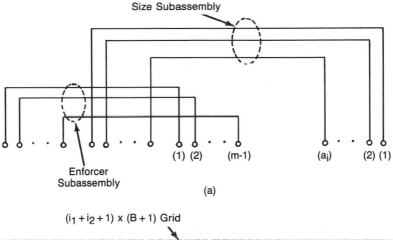

Size Subassembly

Enforcer
Subassembly

(1) (2)   (m-1)   (a$_i$)   (2) (1)

(a)

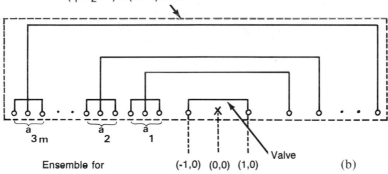

(i_1 + i_2 + 1) x (B + 1) Grid

a
3 m

a
2

a
1

X

Valve

Ensemble for   (-1,0) (0,0) (1,0)   (b)

Figure 17-4. Rectilinear layering problem.

for the three a_js in A_i as well as one wire of the valve and one wire from each of the 3m − 3 enforcer subassemblies corresponding to the 3m − 3 a_js not in A_i.

On the other hand, if Figure 17-4(b) can be wired in m layers, then there is a 3-Partition of the a_is. Because no layer may contain more than one wire from the valve, each layer contains exactly B wires from the size ensembles. If a wire from the size ensemble for a_i is in layer j, then all a_i wires from this ensemble must be in this layer. To see this, observe that the remaining m − 1 layers must each contain exactly one wire from a_i's enforcer subassembly and so can contain no wires from the size subassembly. Hence, each layer must contain exactly three size ensembles. The 3-Partition is therefore A_i = {j | the size subassembly for j is in layer i}.

So, Figure 17-4(b) can be wired in m layers if the 3-Partition instance has answer "yes." Hence, the rectilinear layering problem is NP-hard. □

As in the case of the problems considered above under "Circuit Realization" and "Euclidean Layering Problem," we may define a decision version of the rectilinear layering problem and show that this version is NP-complete.

Mathematical Formulation and Complexity of Design Automation Problems

Implementation Problems.

IP1: Function Realization
Input: A Boolean function B and a set of component types $C_1, C_2, \ldots, C_k$. C_i realizes the Boolean function F_i.
Output: A circuit made up of component types $C_1, C_2, \ldots, C_k$ realizing B and using the minimum total number of components.
Complexity: NP-hard. The proof can be found in Ibarra and Sahni (1975), where IP1 is called P6.

IP2: Circuit Correctness
Input: A Boolean function B and a circuit C.
Output: "yes" if C realizes B and "no" otherwise.
Complexity: NP-hard. The proof can be found in Ibarra and Sahni (1975), where it is called P5. It is shown that tautology reduces to P5 (IP2).

IP3: Circuit Realization
Input: Circuit requirements $(b_1, b_2, \ldots b_n)$ with the interpretation that b_i gates of type i are needed to realize the circuit; modules 1, 2, . . . , r with composition m_{ij}, where module i has m_{ij} gates of type j; module costs c_i, where c_i is the cost of one unit of module i.
Output: Nonnegative integers, $x_1, x_2, \ldots, x_n$ such that

$$\sum_i m_{ij} x_i \geq b_j, \ i \leq j \leq n$$

and $\sum_i c_i x_i$ is minimized.

Complexity: NP-hard. See above, section on "circuit Realization."

IP4: Construction of a Minimum-Cost Standard Library of Replaceable Modules
Input: A set $\{C_1, C_2, \ldots, C_n\}$ of logic circuits such that circuit C_i contains y_{ij} circuits of type j, $1 \leq j \leq r$; and a limit, p, on the number of circuits that can be put into a module.

Output: A set $M = \{m_1, m_2, \ldots, m_k\}$ of module types, with module m_i containing a_{ij} circuits of type j, such that:

(i) $\quad \sum_j a_{ij} \leq p, 1 \leq i \leq k$

(ii) $\quad \sum x_{ij} a_{jq} \geq y_{iq}, 1 \leq i \leq n$ and $1 \leq q \leq r$

$\qquad x_{ij} =$ smallest number of modules m_j needed in implementing C_i.

(iii) $\sum_i \sum_j x_{ij}$ is minimum over all choices of M.

Complexity: NP-hard. Partition (NP9) reduces to IP4 as follows. Let $A = \{a_1, a_2, \ldots, a_n\}$ be an arbitrary instance of NP9. Construct the following instance of IP4. The set $\{C_1, C_2, \ldots, C_n, C_{n+1}\}$ has the composition

$$y_{ii} = a_i, 1 \leq i \leq n;$$
$$y_{ij} = 0, 1 \leq i,j \leq n \text{ and } i \neq j;$$
$$y_{i,n+1} = a_i, 1 \leq i \leq n;$$
$$p = (\sum_i a_i)/2.$$

Clearly, there exists a set M such that $\sum_i \sum_j x_{ij} = n + 2$ iff the corresponding partition problem has answer "yes."

IP5: Construction of a Standard Library of a Minimum Number of Replaceable Module Types

Input: Same as in IP4. In addition, a cost bound C is specified.

Output: A minimum cardinality set $M = \{m_1, m_2, \ldots m_k\}$ with $a_{ij}, 1 \leq i \leq k$, $1 \leq j \leq r$ as in IP4 for which there exist natural numbers x_{ij} such that

$$\sum_j x_{ij} a_{jm} \geq y_{im}, 1 \leq m \leq 4, 1 \leq i \leq n$$

and $\sum x_{ij} \leq C$.

Complexity: NP-hard. Partition (NP9) reduces to IP5, as follows. Given an arbitrary instance of partition, the equivalent instance of IP 5 is constructed exactly as described for IP4. In addition, let $C = n + 2$. Clearly, $k = 2$ can be achieved iff the corresponding partition problem has answer "yes."

IP6: Minimum Cardinality Partition

Input: A set $V = \{1, 2, \ldots n\}$ of circuit nodes; a symmetric weighting function w(i,j) such that w(i,j) is the number of connections between nodes i and j; a size function s(i) such that s(i) is the space needed by node i; and constants E and S that are, respectively, bounds on the number of external connections and the space per module.

Output: Partition $P = \{P_1, P_2, \ldots, P_k\}$ of V of minimum cardinality such that:

(a) $\sum_{i \in P_j} s(i) \leqslant S, \ 1 \leqslant j \leqslant k$

(b) $\sum_{i \in P_j \text{ and } q \in P_j} w(i,q) \leqslant E, \ 1 \leqslant j \leqslant k$

Complexity: NP-hard. Partition (NP9) reduces to this problem, as follows. Let $A = \{a_1, a_2, \ldots, a_n\}$ be an arbitrary instance of partition. Equivalent instance of IP6:

$$s(i) = a_i, \ 1 \leqslant i \leqslant n;$$
$$S = (\sum_i a_i)/2;$$
$$w(i,j) = 0, \ 1 \leqslant i,j \leqslant n;$$
$$E = 0.$$

There is a minimum partition of size 2 iff the partition instance has a subset that sums to S.

IP7: Minimum External Connections Partition I
Input: V, w, s, and S as in IP6.
Output: A partition P of V such that:

(a) $\sum_{i \in P_j} s(i) \leqslant S, \ 1 \leqslant j \leqslant k$

(b) $\sum_j \{ \sum_{i \in P_j \text{ and } q \in P_j} w(i,q) \}$ is minimized.

Observe that the summation of (b) actually gives us twice the total number of interpartition connections.

Complexity: NP-hard. This problem is identical to the graph partitioning problem (ND14) in the list of NP-complete problems in Garey (1979).

IP8: Minimum External Connections Partition II
Input:1 V, w, s, and S as in IP6. A constant r.
Output: A partition $P = \{P_1, \ldots, P_k\}$ of V such that:

(a) $k \leq r$

(b) $\sum\limits_{i \in P_j} s(i) \leq S, \ 1 \leq j \leq k$

(c) $\max\limits_{j}\{ \sum\limits_{i \in P_j \text{ and } q \in P_j} w(i,q)\}$ is minimized.

Complexity: NP-hard. Partition (NP9) can be reduced to IP8 as described above for IP6.

IP9: Minimum Space Partition
Input: V, w, s, and E as in IP6. In addition, a constant r.
Output: A partition $P = \{P_1, P_2, \ldots P_k\}$ of V such that:

(a) $k \leq r$

(b) $\sum\limits_{i \in P_j \text{ and } q \in P_j} w(i,q) \leq E, \ 1 \leq j \leq k$

(c) $\max\limits_{j}\{ \sum\limits_{i \in P_j} S(i)\}$ is minimized.

Complexity: NP-hard. Partition (NP9) can be reduced to IP9 in a manner very similar to that described for IP6.

IP10: Module Selection Problem
Input: A partition element A' (as in the output of IP6) containing y_i circuits of type i, $1 \leq i \leq r$; a set $M = \{m_j \mid 1 \leq j \leq n\}$ of module types, with z_j copies of each module type m_j. Each m_j has a cost h_j and contains a_{ij} circuits of type i, $1 \leq i \leq r$, $1 \leq j \leq n$.
Output: An assignment of non-negative integers, $x_1, x_2, \ldots, x_n$, $0 \leq x_j \leq z_j$

to minimize the total cost $\sum\limits_{j=1}^{n} x_j h_j$ and subject to the constraint that all circuits

in A' are implemented, i.e., $\sum\limits_{j=1}^{n} a_{ij} x_j \geq y_i$, $1 \leq i \leq r$.

Complexity: NP-hard. IP10 contains the 0/1 Knapsack problem (NP12) as a special case. Given an arbitrary instance w, M, K of the 0/1 Knapsack problem, the equivalent instance of IP10 has $r = 1$; $Y_1 = M$; and $z_j = 1$, $a_{ij} = w_j$, $h_j = k_j$; $i = 1$, $1 \leq j \leq n$.

Placement Problems.

PP1: Module Placement Problem
Input: m; p; s; $N = \{N_1, N_2, \ldots, N_s\}$, $N_i \subseteq \{1, \ldots, m\}$; $D(p \times p) = [d_{ij}]$; and $W(1 : s) = [w_i]$. m is the number of modules. p is the number of available positions (or slots, or locations); s is the number of signals; N_i, $1 \leqslant i \leqslant s$ are signal nets; d_{ij} is the distance between positions i and j; and w_i is the weight of net N_i, $1 \leqslant i \leqslant s$.
Output: $X(m \times p) = [x_{ij}]$ such that $x_{ij} \in \{0,1\}$ and

(a) $\displaystyle\sum_{j=1}^{p} x_{ij} =$

(b) $\displaystyle\sum_{i=1}^{m} x_{ij} \leqslant 1$

(c) $\displaystyle\sum_{t=1}^{s} w_i f(i,X)$ is minimized.

x_{ij} is 1 iff module i is to be assigned to position j. Constraints (a) and (b), respectively, ensure that each module is assigned to a slot and that no slot is assigned more than one module. f(i,X) measures the cost of net N_i under this assignemnt. This cost could, for example, be the cost of a minimum spanning tree; the length of the shortest Hamiltonian path connecting all modules in the net; the cost of a minimum Stiner tree; etc. In general, the cost is a function of the d_{ij}s.
Complexity: NP-hard. The quadratic assignment problem (NP14) is readily seen to be a special case of the placement problem PP1. To see this, just observe that every instance of NP14 can be transformed into an equivalent instance of PP1 in which $|N_i| = 2$ for ever net and f(i,X) is simply the distance between the positions of the two modules in N_i. So, PP1 is NP-hard.

PP2: One-Dimensional Placement Problem
Input: A set of components $B = \}b_1, b_2, \ldots, b_n\}$; a list $L = \{N_1, N_2, \ldots, N_m\}$ of nets on B such that:

$N_i \leqslant B$, $1 \leqslant i \leqslant m$;

$\cup N_i = B$; $N_i \cap N_j = \emptyset$, $i \neq j$.

Output: An ordering σ of B such that the ordering

$$B_\sigma = \{b_{\sigma(1)}, b_{\sigma(2)}, \ldots, b_{\sigma(n)}\}$$

minimizes max {number of wires crossing the interval between $b_{\sigma(i)}$ and $b_{\sigma(i+1)}$) $| 1 \leq i \leq n - 1$}.

Complexity: NP-hard. The problem is considered in Goto et al. (1977).

Wiring Problems.

WP1: Net Wiring with Manhattan Distance
Input: A set P of pin locations, $P = \{(x_i, y_i) \mid 1 \leq i \leq n\}$, set F of feedthrough locations, $F = \{(a_i, b_i) \mid 1 \leq i \leq m\}$, and a set $E = \{E_k \mid 1(< = i(< = r\}$ of equivalence classes. Each equivalence class defines a set of pins that are to be made electrically common.
Output: Wire sets $W_i = \{[(t_j{}^i, u_j{}^i), (v_j{}^i, w_j{}^i)] \mid 1 \leq j \leq q_i\}$ such that all pins in F_i are made electrically common. Each $(t_j{}^i, u_j{}^i)$ and $(v_j{}^i, w_j{}^i)$ is either a pin location in E_i or is a feedthrough pin. No feedthrough pin may appear as a wire end point in more than one W_i. The wire set, $\cup W_i$ is such that $\sum_{i,j}(| t_j{}^i -$

$v_j{}^i | + u_j{}^i - w_j{}^i |$) is minimum.

Complexity: NP-hard. The Manhattan Steiner Tree problem (NP2) is a special case of WP 1. To see this, let

$$E = \{P\} \text{ and } F = \{(a_i, b_i) \mid (a_i, b_k) \notin P\} = P$$

Hence WP1 is NP-hard.

WP2: Net Wiring with Euclidean Distance
Input: P, F, and E as in WP1.
Output: Wire sets as in WP1 but $\sum_{i,j}[t_j{}^i - v_j{}^i)^2 + (u_j{}^i - w_j{}^i)^2]$ is minimized.

Complexity: NP-hard.
The Euclidean Steiner tree problem (NP1) is a special case of WP2, in exactly the same way that NP2 was a special case of WP1. Hence, it is NP-hard.

WP3: Euclidean Spanning Tree
Input: A set P of pin locations, $P = \{(x_i, y_i) \mid 1 \leq i \leq n\}$;
Output: A spanning tree for P of minimum total length. The distance between two points (a,b) and (c,d) is the Euclidean metric $[(a - c)^2 + (b - d)^2]$.
Complexity: Polynomial. An $O(n \log n)$ algorithm to find the minimum spanning tree is presented in Shamos and Hoey (1975).

WP4: Manhattan Spanning Tree
Input: P, as in WP3.
Output: A spanning tree for P of minimum total length. The distance between two points (a,b) and (c,d) is the Manhattan metric $| a - c | + | b - d |$.
Complexity: Polynomial. An O(n log n) algorithm that finds the minimum spanning tree is presented in Hwang (1979a).

WP5: Degree Constrained Wiring with Manhattan Distance
Input: P, E, and F as in WP1 and a constant d.
Output: W_is as in WP1 but with the added restriction that at most d wires may be incident on any pin or feedthrough location.
Complexity: NP-hard. WP5 contains the Manhattan Hamiltonian path problem (NP6) as a special case. To see this, let $E = \{P\}$, $d = 2$, and $F = \emptyset$. Hence, WP5 is NP-hard.

WP6: Degree Constrained Wiring with Euclidean Distance
Input: P, F, and E as in WP1 and a constant d.
Output: W_is as in WP2 but with the added restriction that at most d wires may be incident on any pin.
Complexity: NP-hard. WP6 contains the Euclidean Hamiltonian path problem (NP4) as a special case. The argument is analogous to that presented for WP5.

WP7: Length Constrained Wiring (Manhattan Distance)
Input: P, F, and E as in WP1, and a constant L.
Output: W_is as in WP1 with the added requirement that

$$| t_j{}^i - v_j{}^i | + | u_j{}^i - w_j{}^i | \leq L$$

Complexity: NP-hard.
WP7 contains WP1 as a special case, when $L = \infty$. Since WP1 is NP-hard, so is WP7.

WP8: Length Constrained Wiring (Euclidean Distance)
Input: P, F, and E as in WP1, and a constant L.
Output: W_is as in WP2 with the added requirement that

$$[(t_j{}^i - v_j{}^i)^2 + (u_j{}^i - w_j{}^i)^2] \leq L$$

Complexity: NP-hard. WP8 contains WP2, which is itself NP-hard, as a special case, when $L = \infty$.

WP9: Euclidean Layering Problem I
Input: A set W of wires, $W = \{[(u_i, v_i), (x_i, y_i)] \mid 1 \leq i \leq n\}$. (u_i, v_i) and (x_i, y_i) are the coordinates of the end points of wire i.

Output: A partitioning W_1, W_2, . . . , W_k of W such that $W_i \cap W_j = \emptyset$, $i \neq j$; $\cup W_i = W$ and no two wires in any W_i intersect. The end points of wires are connected by straight wires (i.e., in the Euclidean manner). k is to be minimum.
Complexity: NP-hard. See above, section on "Euclidean Layering Problem."

WP10: Euclidean Layering Problem II
Input: W as in WP9 and a constant r.
Output: A partitioning of W into sets W_1, W_2, . . . , W_r, and X. No two wires in any W_i intersect when end points are connected by a straight wire. $| X |$ is minimum.
Complexity: NP-hard when r = 3. The corresponding intersection graph is 3-colorable iff x $-$ $\emptyset$. Since deciding 3-colorability of intersection graphs is NP-hard (NP8), WP10 is also NP-hard.

WP11: Manhattan Layering Problem I
Input: Same as in WP9.
Output: Same as in WP9, except that the end points of each wire are connected in a Manhattan manner (i.e., a straight run along the x-axis and a straight run along the y-axis).
WP12: Manhattan Layering Problem II
Input: Same as in WP10.
Output: Same as in WP10, except that wire end points are connected in a Manhattan manner.
Complexity: Status unknown.

WP13: Rectilinear Layering Problem
Input: A p $\times$ q grid; a wire set W $-$ $\{[(u_i, v_i) (x_i, y_i)] \mid 1 \leqslant i \leqslant n\}$. (u_i, v_i) and (x_i, y_i) are grid points that are the end points of wire i. All the wires are constrained to be routed along the grid lines only.
Output: A partition of W into W_1, W_2, . . . , W_r such that

(i) $W_i \cap W_j = \emptyset$, $i \neq j$; $\cup W_i = W$
(ii) All wires $\in W_i$ can be routed along the grid lines without intersections.
(iii) r is a minimum.
Complexity: NP-hard. See above, section on "Rectilinear Layering Problem."

WP14: Grid Routing.
Input: Set W of wires as in WP9 and a rectangular m $<$s n grid. The end points of wires correspond to grid points.
Output: A routine for each wire such that no two wires intersect and all wire segments are on grid segments.
Complexity: NP-hard (Kramer and van Leeuwen 1982; Richards 1984).

WP15: Single Bend Grid Routing
Input: Same as in WP14.
Output: Maximum number of wires that can be routed on the grid using at most one bend per wire.
Complexity: NP-hard (Raghavan et al. 1981).

WP16: Minimum Layer Single Bend Grid Routing
Input: Same as in WP14.
Output: Minimum number of layers needed to route all the wires in W using at most one bend per wire.
Complexity: NP-hard (Raghavan et al. 1981).

WP17: Single Row Layering Problem
Input: A set of vertices $V = \{1, 2, \ldots, n\}$ evenly spaced along a line; a list of nets $L = \{N_1, N_2, \ldots, N_m\}$ such that $N_i \subseteq V$, $1 \leq i \leq m$; $\cup N_i = V$; $N_i \cap N_j = \emptyset$ $i \neq j$; integers c_u and c_l: the respective upper and lower street capacities.
Output: A decomposition of L into $L_1, L_2, \ldots, L_r$ such that
(i) $L_i \cap L_j = \emptyset$, $i \neq j$; $\cup L_i = L$
(ii) All nets $\in L_i$ have single layer single row realizations that require no more than c_u and c_l tracks in the upper and lower streets, respectively.
(iii) r is minimum.
Complexity: NP-hard. By setting $c_u = 0$ and $c_l = B + 1$,3-Partition (NP10) can be reduced to WP15 in a manner very similar to that described for WP13.

WP18: Single Row Routine with Nonuniform Conductor Widths
Input: V and L as in WP17; in addition a natural number valued function t, where t_i is the width of the conductor used to route net N_i.
Output: A layout for the nets that minimizes
    max {total width required in the upper street, total
            width required in the lower street}
Complexity: NP-hard.
Partition (NP9) reduces to this problem, as shown. Given an arbitrary instance of partition $A = \{a_1, a_2, \ldots, a_n\}$ the equivalent instance of WP16 is:

$$V = \{1, 2, \ldots, 2n\}$$
$$N_i = \{i, 2n + 1 - i\}, 1 \leq i \leq n$$
$$t_i = a_i, 1 \leq i \leq n$$

Clearly, there exists a realization with upper street width = lower street width = $(\sum a_i)/2$ iff the corresponding partition instance has answer "yes."

WP19: Single Row Routing Problem
Input: V and L as in WP17.

Output: A layout for L that minimizes
    max }number of tracks needed in upper street, number of   tracks
needed in lower street}
Complexity: NP-hard. See Arnold (1982).

WP20: Minimum Width Single Row Routing
Input: V and L, as in WP17.
Output: A layout for L that minimizes (number of tracks needed in upper
street + number of tracks needed in lower street).
Complexity: Status unknown.

WP21: Single Row Routing with Fewest Bends I
Input: V and L, as in WP17.
Output: A layout for L that minimizes the total number of bends in the wiring
paths.
Complexity: NP-hard (Raghavan and Sahni 1984).

WP22: Single Row Routing with Fewest Bends II
Input: V and L, as in WP17.
Output: A layout for L that minimizes the maximum number of bends in any
one wire.
Complexity: NP-hard (Raghavan and Sahni 1984).

WP23: Single Row Routing With Fewest Interstreet Crossings I
Input: V and L, as in WP17.
Output: A layout for L that minimizes the total number of conductor crossings
between the upper and lower streets.
Complexity: NP-hard (Raghavan and Sahni 1984).

WP24: Single Row Routing With Fewest Interstreet Crossings II
Input: V and L, as in WP17.
Output: A layout for L that minimizes the maximum number of conductors
between an adjacent pair of nodes.
Complexity: NP-hard (Raghavan and Sahni 1984).

WP25: One Component Routing
Input: A rectangular component of length l and height h having n pins along its
periphery and a set of two point nets defined on these pins.
Output: A two-layer wiring of the nets such that all vertical runs are on one
layer, and all horizontal runs are on the other. Wires can run only around the
component. The area of the smallest rectangle that circumscribes the component
and all routing paths is minimized.

Complexity: $O(n^3)$ (La Paugh 1980a). When an arbitrary number of rectangular components are present and each net may consist of several pins, the routing problem is NP-hard (Szymanski and Yannanakis 1982).

WP26: River Routing

Input: Two ordered sets $X = (x_1, x_2, \ldots, x_n)$ and $Y = (y_1, y_2, \ldots, y_n)$ of pins separated by a wiring channel. Each set is divided into blocks of consecutive pins. While the relative ordering of blocks if fixed, their relative positioning is not.

Output: A one-layer wiring pattern connecting x_i to y_i, $1 \le i \le n$. The channel dimensions necessary to accomplish this wiring are given by the vertical distance (separation) needed between the two rows of pins and the horizontal length (spread) of the channel. The channel area is the product of spread and separation. The output wiring should optimize these channel dimensions.

Complexity: If the wiring channel is assumed to be a single layer grid (hence wires must be rectilinear) a placement of the blocks that minimizes the channel separation can be found in O(n log n) time; for a given separation a placement with minimal spread can be determined in O(n) time; and a placement minimizing channel area may be obtained in $O(n^2)$ time (Leiserson and Pinter 1981). When the position of each pin is not fixed and wires are not constrained to run along grid lines (but must still consist of horizontal and vertical runs with some minimum separation), the channel separation can be minimized in $O(n^2)$ time (Dolev 1981). When the wires can take on any shape and the pin positions are fixed, minimum length wiring can ba done in $O(n^2)$ time (Tompa 1980). If two layers are allowed with one devoted to horizontal runs and the other to vertical runs, then the minimum separation can be found in O(n) time (Siegel and Dolev 1982), provided up to two vertical wires are permitted to overlap. The offset that minimizes the separation for both the single layer and restricted two layer case can be found in O(n log n) time (Siegel and Dolev 1982). Furthermore, the offset that leads to the minimum area circumscribing rectangle may be found in $O(n^3)$ time for both cases.

WP27: Channel Routing

Input: A set of nets. Each net is a pair of pins on opposite sides of a rectangular channel.

Output: A two-layer wiring of the nets such that no wire has more than one horizontal segment. Horizontal segments are to be laid out in one layer and vertical segments in the other. The number of horizontal tracks used is minimized.

Complexity: NP-hard (La Paugh 1980b). The problem remains NP-hard if doglegs are allowed, and nets are permitted to contain any number of pins from both sides of the channel (Szymanski 1982). Several good heuristics for two layer channel routing exist (Deutsch 1976; Yoshimura and Kuh 1982; Marek-Sadowska and Kuh 1982; Rivest et al. 1981; Fidducia and Rivest 1982). All of these allow doglegs and those of Marek-Sadowska and Kuh (1982) and Rivest et al. (1981)

permit horizontal and vertical segments to share layers. Lower bounds on the number of tracks needed are developed in Brown and Rivest (1981). Routine in the T-shaped and X-shaped junctions that result from the intersection of rectangular channels is considered in Pinter (1981).

Fault Detection Problems. Let C be an n-input 1-output combinational circuit. Let Z be the set of all possible single stuck-at-0 (s-a-0) and stuck-at-1 (s-a-1) faults. The tuple $(i_1, i_2, \ldots, i_n, j, F(0), F(1))$ is a fault detection test for C iff each of the following is satisfied:

(a) $i_k \varepsilon \{0,0\}$, $1 \leqslant k \leqslant n$; $j \varepsilon \{0,1\}$

(b) $F(0) \subseteq Z$; $F(1) \subseteq Z$; $F(0) \cap F(1) = \emptyset$; $F(0) \cup F(1) \neq \emptyset$

(c) (i) There is a s-a-O fault at one of the locations in F(O) iff C with inputs $i_1, i_2, \ldots, i_n$ has output j.
     (ii) There is a s-a-1 fault at one of the locations in F(1) iff C with inputs $i_1, i_2, \ldots, i_n$ has output j.

A *test set,* T, is a set of fault detection tests. T is a test set for $L \subseteq Z$ iff (i) the union of all the F(0)s for the tests in T is L and (ii) the union of all the F(1)s for the tests in T is L. If $L = Z$ then T is a test set for C. Circuit C is irredundant iff it has a test set.

FDP1: Irredundancy
Input: A combinatorial circuit C.
Output: "yes" iff the circuit is irredundant (i.e., all s-a-0 and s-a-1 faults can be detected by I/O experiments; "no" otherwise.
Complexity: NP-hard. See problem P1 in Ibarra and Sahni (1975).

FDP2: Line Fault Detection
Input: Same as for FDP1.
Output: "yes" iff a fault in a particular input line can be detected by I/O experiments; "no" otherwise.
Complexity: NP-hard. See problem P2 in Ibarra and Sahni (1975).

FDP3: All Faults Detection
Input: Same as for DP1.
Output: "yes" iff all single input faults can be detected by I/O experiments; "no" otherwise.
Complexity: NP-hard. See problem P3 in Ibarra and Sahni (1975).

FDP4: Output Fault Detection
Input: Same as for FDP1.

Output: "yes" iff faults in the output line can be detected by I/O experiments; "no" otherwise.
Complexity: NP-hard. See problem P4 in Ibarra and Sahni (1975).

FDP5: Minimal Test Set
Input: Same as for FDP1.
Output: If C is irredundant, a minimal test set for C.
Complexity: NP-hard. See Ibarra and Sahni (1975.

HEURISTICS AND USUALLY GOOD ALGORITHMS

Having discovered that many of the interesting problems that arise in design automation are computationally difficult (in the sense that they are probably not solvable by a polynomial time algorithm), we are left with the issue of alternate paths one might take in solving these problems. The three most commonly tried paths are:

(a) Obtain a heuristic algorithm that both is computationally feasible and obtains "reasonably" good solutions. Algorithms with this latter property are called *approximation algorithms*. We are interested in good, fast [i.e., low-order polynomial; say O(n), O(n log n), O(n^2), etc.] approximation algorithms.

(b) Arrive at an algorithm that always finds optimal solutions. The complexity of this algorithm is such that it is computationally feasible for "most" of the instances people want to solve. Such an algorithm will be called a *usually good algorithm*. The Simplex algorithm for linear programming is a good example of a usually good algorithm. Its worst case complexity is exponential. However, it is able to solve most of the instances given it in a "reasonable" amount of time (much less than the worst-case time).

(c) Obtain a computationally feasible algorithm that "almost" always finds optimal solutions. An algorithm with this property is called a *probabilistically good* algorithm.

Approximation Algorithms

When evaluating an approximation algorithm, one considers two measures: algorithm complexity and the quality of the answer (i.e., how close it is to being optimal). As in the case of complexity, the second measure may refer to the worst case of the average case.

There exist several categories of approximation algorithms. Let A be an algorithm that generates a feasible solution to every instance I of a problem P. Let $F^*(I)$ be the value of an optimal solution, and let F'(I) be the value of the solution generated by A.

Definition: A is an *absolute approximation* algorithm for P iff $| F^*(I) - F'(I) | \leq k$ for all I, with k a constant. A is an $f(n)$-*approximate* algorithm for P iff $| F^*(I) - F'(I) | / F^*(I) | \leq f(n)$ for all I. n is the size of I, and we assume that $| F^*(I) | > 0$. An $f(n)$-approximate algorithm with $f(n) \leq \varepsilon$ for all n and some constant ε is an ε-*approximate* algorithm.

Definition: Let $A(\varepsilon)$ be a family of algorithms that obtain a feasible solution for every instance I of P. Let n be the size of I.

$A(\varepsilon)$ is an *approximation scheme* for P iff for every $\varepsilon > 0$ and every instance I, $| F^*(I) - F'(I) | / F^*(I) \leq \varepsilon$. An approximation scheme whose time complexity is polynomial in n is a *polynomial time approximation* scheme. A *fully polynomial time approximation* scheme is an approximation scheme whose time complexity is polynomial in n and $1/\varepsilon$. For most of the heuristic algorithms in use in the design automation area, little or no effort has been devoted to determining how good or bad (relative to the optimal solution values) they are.

In what follows, we briefly review some results that concern design automation. The reader is referred to Horowitz and Sahni (1978, Chapter 12) for a more complete discussion of heuristics for NP-hard problems. For most NP-hard problems, the problem of finding absolute approximations is also NP-hard. As an example, consider problem IP3 (circuit realization). Let:

$$\min \sum_i c_i x_i$$

(1) subject to

$$\sum_i m_{ij} x_i \geq b_j, \ 1 \leq j \leq n$$

and

$$x_i \geq 0 \text{ and integer}$$

be an instance of IP3. Consider the instance:

$$\min \sum_i d_i x_i$$

(2) subject to

$$\sum_i m_{ij} x_i \geq b_j, \ 1 \leq j \leq n$$

and

$$x_i \geq 0 \text{ and integer}$$

where $d_i = (k + 1)c_i$. Since the values of feasible solutions to (2) are at least $k + 1$ apart, every absolute approximation algorithm for IP3 must produce optimal solutions for (2). These solutions are, in turn, optimal for (1). Hence, finding absolute approximate solutions for any fixed k is no easier than finding optimal solutions. Horowitz and Sahni (1978, Chapter 12) provide examples of NP-hard problems for which there do exist polynomial time absolute approximation algorithms. It has long been conjectured (Gilbert and Pollak 1968) that, under the Euclidean metric,

$$\frac{\text{length of minimum spanning tree}}{\text{length of optimum Steiner tree}} = \frac{F'}{F*} = 2/3$$

Hence,

$$\frac{|F' - F*|}{F*} \leq \frac{2 - 3}{3} \leq 0.155$$

Hence, the O(n log n) minimum spanning tree algorithm in Shamos and Hoey (1979) can be used as a 0.155-approximate algorithm for the Euclidean Steiner tree problem.

For the rectilinear Steiner tree problem, it is known (Hwang 1979b; Lee et al. 1976) that

$$\frac{\text{length of minimum spanning tree}}{\text{length of optimum Steiner tree}} = \frac{F'}{F*} \leq 3/2$$

Hence,

$$\frac{|F' - F*|}{F*} \leq 1/2$$

The O(n log n) spanning tree algorithm in Hwang (1979a) can be used as a 0.5-approximate algorithm for the Steiner tree problem.

Because both the Euclidean and rectilinear Steiner tree problems are strongly NP-hard, they can be solved by a fully polynomial time approximation scheme iff P = NP. (See Horowitz and Sahni 1978, Chapter 12] for a definition of strong NP-hardness and its implications).

Shamos and Hoey (1975) suggest an O(n log n) approximation algorithm that finds a traveling salesman tour that is not longer than twice the length of an optimal tour, using the Euclidean minimum spanning tree. This is a 1-approximate algorithm, and it is possible to do better. Christofedes (1976) contains a 0.5-approximate algorithm for this problem. Sahni and Gonzalez (1976) have shown that there exists a polynomial time ε-approximation algorithm for the quadratic assignment problem iff $P = NP$.

Usually Good Algorithms

Classifying an algorithm as "usually good" is a difficult process. From the practical standpoint, this can be done only after extensive experimentation with the algorithm. The Simplex method is regarded as good only because it has proved to be so over years of usage on a variety of instances. An analytical approach to obtain such a classification comes from probabilistic analysis. Karp (1975, 1976) has carried out such an analysis for several NP-hard problems. Such analysis is not limited to algorithms that guarantee optimal solutions. Karp (1977) analyzes an approximation algorithm for the Euclidean traveling salesman problem. The net result is a fast algorithm that is expected to produce near optimal salesman tours. Dantzig (1980) analyzes the expected behavior of the Simplex method.

Kirkpatrick et al. (1983; Vecchi and Kirkpatrick 1983) have proposed the use of simulated annealing to obtain good solutions to combinatorially difficult design automation problems. Experimental results presented in these papers as well as in Nahar et al. (1985) Golden and Skiscim (1984), and Romeo et al. (1984) indicate that simulated annealing does not perform as well as other heuristics when the problem being studied has a well-defined mathematical model. However, for problems with multiple constraints that are hard to model, simulated annealing can be used to obtain solutions that are superior to those obtainable by other methods. Even in the case of easily modeled problems, simulated annealing may be used to improve the solutions obtained by other methods.

CONCLUSIONS

Under the worst-case complexity measure, most design automation problems are intractable. This conclusion remains true even if we are interested only in obtaining solutions with values guaranteed to be "close" to the value of optimal solutions. The most promising approaches to certifying the value of algorithms for these intractable problems appear to be probabilistic analysis and experimentation. Another avenue of research that may prove fruitful is the design of highly parallel algorithms (and associated hardware) for some of the computationally more difficult problems.

REFERENCES

Abramovici, M., and Breuer, M. A., "Fault Diagnosis Based on Effect–Cause Analysis: An Introduction," *Proc. 17th Design Automation Conference*, pp. 69–76, 1980.

Arnold, P. B., "Complexity Results for Circuit Layout on Double-Sided Printed Circuit Boards," bachelor's thesis, Harvard University, 1982.

Breuer, M. A., "The application of Integer Programming in Design Automation," *Proc.* SHARE Design Automation Workshop, 1966.

Breuer, M. A. Editor, *Design Automation of Digital Systems,* Vol. 1, *Theory and Techniques,* Prentice-Hall, Englewood Cliffs, NJ 1972a.

Breuer, M. A., "Recent Developments in the Automated Design and Analysis of Digital Systems," *Proc. IEEE, 60* (1), 12–27 (1972b).

Brown, D. and Rivest, R., "New Lower Bounds for Channel Width," in *VLSI Systems and Computations,* Kung et al., Editors, Computer Science Press, Potomac, MD, 1981, pp. 178–185.

Christofedes, N., "Worst-Case Analysis of New Heuristic for a Traveling Salesman Problem," Mgmt. Science Research Report, Carnegie Mellon University, 1976.

Clark, R. L., "A Technique for Improving Wirability in Automated Circuit Card Placement," Rand Corp. Report R-4049, Aug. 1969.

van Cleemput, W. M., *Computer Aided Design of Digital Systems,* 3 vols., Digital Systems Lab., Stanford University, Computer Science Press, Potomac, MD, 1976.

Dantzig, G., "Khachian's Algorithm: A Comment," *SIAM News, 13* (5) (Oct. 1980).

Dejka, W. J., "Measure of Testability in Device and System Design," *Proc.* 20th Midwest Symposium on Circuits and Systems, pp. 39–52, Aug. 1977.

Deutsch, D., "A Dogleg Channel Router," *Proc. 13th Design Automation Conference,* pp. 425–433, 1976.

Dolev, D., et al., "Optimal Wiring between Rectangles," *Proc.* 13th Annual Symposium on Theory of Computing, pp. 312–317, 1981.

Ehrlich, G. S., Even, S., and Tarjan, R. E., "Intersection graphs of Curves in the Plane," *J. Combin. Theol., Ser. B, 21,* 8–20 (1976).

Eichelberger, E. B., and William, T. W., "A Logic Design Structure for LSI Testability," *Proc. 14th Design Automation Conference,* pp. 462–468, 1977.

Fidducia, C., and Rivest, R., "A Greedy Channel Router," *Proc. 19th Design Automation Conference,* pp. 418–424, 1982.

Garey, M. R., *A Guide to the Theory of NP-Completeness,* W. H. Freeman and Co., San Francisco, CA, 1979.

Garey, M. R., and Johnson, D. S., "Complexity Results for Multiprocessor Scheduling under Resource Constraints," *SIAM J. Comput., 4,* 397–411 (1975).

Garey, M. R., and Johnson, D. S., "Near-Optimal Graph Coloring," *JACM, 23,* 43–49, (1976a).

Garey, M. R., and Johnson, D. S., "Geometric Problems," *Proc.* 8th Annual ACM Symposium on Theory of Computing, ACM, New York, 1976b, pp. 10–22.

Garey, M. R., Graham, R. L., and Johnson, D. S., "The Complexity of Computing Steiner Minimal Trees," *SIAM J. Appl. Math., 32,* 835–859 (1977).

Gilbert, E. N., and Pollak, H. O., "Steiner Minimal Trees," *SIAM J. Appl. Math.,* 1–29 (Jan. 1968).

Golden, B., and Skiscim, C., "Using Simulated Annealing to Solve Routing and Location Problems," University of Maryland, College of Business Administration, Technical Report, Jan. 1984.

Goldstein, L. H., "Controllability/Observability Analysis of Digital Circuits." *IEEE Trans. Circuits and Systems, CAS-26,* (9) 685–693 (Sept. 1979).

Goto, S., Cederbaum, I., and Ting, B. S., "Suboptimum Solution of the Backboard Ordering with Channel Capacity Constraint," *IEEE Trans. Circuits and Systems, CAS-24,* 645–652 (Nov. 1977).

Grason, J., and Nagle, A. W., "Digital Test Generation and Design for Testability," *Proc. 17th Design Automation Conference,* pp. 175–189, 1980.

Habayeb, A. R., "System Decomposition, Partitioning, and Integration for Microelectronics," *IEEE Trans. System Science and Cybernetics, SSC-4,* (2), 164–172 (July 1968).

Hashimoto, A., and Stevens, J., "Wire Routing by Optimizing Channel Assignment within Large Apertures," *Proc. 8th Design Automation Conference,* pp. 155–169, 1971.

Horowitz, E., and Sahni, S., "Computing Partitions with Applications to the Knapsack Problem," *JACM, 21,* 277–292 (1974).

Horowitz, E., and Sahni, S., *Fundamentals of Computer Algorithms,* Computer Science Press, Potomac, MD, 1978.

Hwang, F. K., "On Steiner Minimal Trees with Rectilinear Distance," *SIAM J. Appl. Match,* 104   114 (Jan. 1976).

Hwang, F. K., "Rectilinear Minimal Spanning Trees," *JACM, 26,* 177–182 (1979a).

Hwang, R. K., "Rectilinear Steiner Trees," *IEEE Trans. Circuits and Systems, CAS-26,* 75–77 (1979b).

Ibaraki, T., Kameda, T., and Toida, S., "Generation of Minimal Test Sets for System Diagnosis," University of Waterloo, 1977.

Ibarra, O. H., and Kim, C. E., "Fast Approximation Algorithms for the Knapsack and Sum of Subset Problems," *JACM, 22,* 463–468 (1975).

Ibarra, O. H., and Sahni, S., "Polynomially Complete Fault Detection Problems," *IEEE Trans. Computers, Vol. C-24,* 242–249 (Mar. 1975).

Johannsen, D., "Bristle Blocks: A Silicon Compiler," *Proc. 16th Design Automation Conference,* 1979.

Johnson, D., "The NP-Completeness Column: An Ongoing Guide," *Algorithms, 3* (4), 381–395 (Dec. 1982).

Karp, R., "On the Reducibility of Combinatorial Problems," in R. E. Miller and J. W. Thatcher, Editors, *Complexity of Computer Computations,* Plenum Press, New York, 1972, pp. 85–103.

Karp, R., "The Fast Approximate Solution of Hard Combinatorial Problems," *Proc.* 6th Southeastern Conf. on Combinatorics, Graph Theory, and Computing, Winnipeg, 1975.

Karp, R., "The Probabilistic Analysis of Some Combinatorial Search Algorithms," University of California, Berkeley, Memo No. ERL-M581, Apr. 1976.

Karp, R., "Probabilistic Analysis of Partitioning Algorithms for the Traveling Salesman Problem in the Plane," *Math. of Oper. Res., 2* (3), 209–224 (1977).

Kirkpatrick, S., Gelatt, C., Jr., and Vecchi, M., "Optimization by Simulated Annealing," *Science, 220* (4598), 671–680 (May 1983).

Kodres, U. R., "Formulation and Solution of Circuit Card Design Problems through Use of Graph Methods," in G. A. Walker, Editor, *Advances in Electronic Circuit Packaging,* Vol. 2, Plenum Press, New York, 1961, pp. 121–142.

Kodres, U. R., "Logic Circuit Layout," *Digest Record* of the 1969 Joint Conference of Mathematical and Computer Aids to Design, Oct. 1969.

Kramer, M. R., and van Leeuwen, "Wire-Routing Is NP-Complete," Technical Report, Computer Science Dept., University of Utrecht, The Netherlands, 1982.

La Paugh, A., "A Polynomial Time Algorithm for Routing around a Rectangle," *Proc.* 21st Annual IEEE Symposium on Foundations of Computer Science, pp. 282–293, 1980a.

La Paugh, A., "Algorithms for Integrated Circuit Layout: An Analytic Approach," MIT-LCS-TR-248, doctoral dissertation, MIT, 1980b.

Lawler, E. L., "Electrical Assemblies with a Minimum Number of Interconnections," *IEEE Trans. Electronic Computers* (Correspondence), *EC-11,* 86–88 (Feb. 1962).

Lawler, E. L., Levitt, K. N., and Turner, J., "Module Clustering to Minimize Delay in Digital Networks," *IEEE Trans. Computers, C-18,* 47–57 (Jan. 1969).

Lawler, E. L., "Fast Approximation Algorithms for Knapsack Problems," *Proc.* 18th Annual IEEE Symposium on Foundations of Computer Science, pp. 206–213, 1977.

Lee, J. H., Bose, N. K., and Hwang, F. K., "Use of Steiner's Problem in Suboptimal Routing in Rectilinear Metric," *IEEE Trans. Circuits and Systems, CAS-23,* 470–476 (July 1976).

Leiserson, C., and Pinter, R., "Optimal Placement for River Routing," in *VLSI Systems and Computations,* Kung et al., Editors, Computer Science Press, Potomac, MD, 1981, pp. 126–142.

Lueker, G. S., "Two NP-Complete Problems in Nonnegative Integer Programming," Report No. 178, Computer Science Lab., Princeton University, Princeton, NJ, 1975.

Marek-Sadowska, M., and Kuh, E., "A New Approach to Channel Routing," *Proc.* 1982 ISCAS Symposium, IEEE, pp. 764–767, 1982.

Mead, C., and Conway, L., *Introduction to VLSI Systems,* Addison-Wesley, Reading, MA, 1980.

Nahar, S., Sahni, S., and Shragowitz, E., "Experiments with Simulated Annealing," Design Automation Conference, 1985.

Notz, W. A., Schischa, E., Smith, J. L., and Smith, M. G., "Large Scale Integration; Benefitting the Systems Designer," *Electronics,* 130–141 (Feb. 20, 1967).

Noyce, R. N., "Microelectronics," *Scientific American,* 62–69 (Sept. 1977).

Padadimitriou, C. H., "The Euclidean Traveling Salesman Problem Is NP-Complete," *Theoretical Computer Science, 4,* 237–244 (1977).

Pinter, R., "Optimal Routing in Rectilinear Channels," in *VLSI Systems and Computations,* Kung et al., Editors, Computer Science Press, Potomac, MD, 1981, pp. 153–159.

Pomentale, T., "An Algorithm for Minimizing Backboard Wiring Functions," *Comm. ACM, 8* (11), 699–703 (Nov. 1965).

Raghavan, R., and Sahni, S., "The Complexity of Single Row Routing," *IEEE Trans. Circuits and Systems, CAS-31* (5), 462–472 (May 1984).

Raghavan, R., Cohoon, J., and Sahni, S., "Manhattan and Rectilinear Routing," Technical Report 81-5, University of Minnesota, 1981.

Richards, D., "Complexity of Single-Layer Routing," *IEEE Trans. Computers,* 286–288 (Mar. 1984).

Rivest, R., Baratz, A., and Miller, G., "Provably Good Channel Routing Algorithms," in *VLSI Systems and Computations,* Kung et al., Editors, Computer Science Press, Potomac, MD, 1981, pp. 153–159.

Romeo, F., Vincentelli, A., and Sechen, C., "Research on Simulated Annealing at Berkeley, *Proc. ICCD,* 652–657 (Oct. 1984).

Roth, J. P., "Diagnosis of Automatic Failures: A Calculus and a Method,", *IBM J. Syst. Dev.,* No. 10, 278–291 (1966).

Sahni, S., *Concepts in Discrete Mathematics,* Camelot Publishing Co., Fridley, MN, 1981.

Sahni, S., and Gonzalez, T., "P-Complete Approximation Problems," *JACM, 23,* 555–565 (1976).

Shamos, M. I., and Hoey, D., "Closest Point Problems," *Proc.* 16th Annual IEEE Symposium on Found. of Comp. Sc., pp. 151–163, 1975.

Siegel, A., and Dolev, D., "The Separation for General Single Layer Wiring Barriers," in *VLSI Systems and Computations,* Kung et al., Editors, Computer Science Press, Potomac, MD, 1982, pp. 143–152.

So, H. C., "Some Theoretical Results on the Routing of Multilayer Printed Wiring Boards," *Proc.* IEEE Symposium on Circuits and Systems, pp. 296–303, 1974.

Szymanski, T., "Dogleg Channel Routing Is NP-Complete," unpublished manuscript, Bell Labs, 1982.

Szymanski, T., and Yannanakis, M., unpublished manuscript, 1982.

Ting, B. S., and Kuh, E. S., "An Approach to the Routing of Multilayer Printed Circuit Boards," *Proc.* IEEE Symposium on Circuits and Systems, pp. 902–911, 1978.

Tompa, M., "An Optimal Solution to a Wiring Routing Problem," *Proc.* 12th Annual ACM Symposium on Theory of Computing, pp. 161–176, 1980.

Ullman, J., *Computational Aspects of VLSI,* Computer Science Press, Potomac, MD, 1984.

Vecchi, M., and Kirkpatrick, S., "Global Wiring by Simulated Annealing," *IEEE Trans. Computer Aided Design, CAD-2* (4), 215–222 (Oct. 1983).

Williams, T. W., and Parker, K. P., "Design for Testability: A Survey," *IEEE Trans. Computers, C-31,* (1), 2–15 (19820.

Wojtkowiak, H., "Deterministic Systems Design from Functional Specifications," *Proc. 18th Design Automation Conference,* pp. 98–104, 1981.

Yan, S. S., and Tang, Y. S., "An Efficient Algorithm for Generating Complete Test Sets for Combinatorial Logic Circuits," *IEEE Trans. Computers,* 1971.

Yao, A., "An O(| E | loglog | V |) Algorithm for Minimum Spanning Trees," *Information Processing Letters, 4* (1), 21–23 (1975).

Yoshimura, T., and Kuh, E., "Efficient Algorithms for Channel Routing," *IEEE Trans. DA,* 1–15 (1982).

18. Hierarchical Automatic Layout Algorithms for Custom Logic LSI

Hidekazu Terai

Central Research Laboratory Hitachi, Ltd.

As the number and variety of VLSIs that must be designed have increased dramatically with rising applications in electronic equipment, master-slice LSIs have proved effective in meeting these design needs. The method enables lower-cost design of these LSIs, which is important because of their comparatively low production volume. For this reason, automatic placement and routing programs have become essential design automation (DA) tools for this type of LSI design (Kamikawai et al. 1976; Ueda et al. 1978; Tanaka et al. 1981a,b; Shiraishi and Hirose 1980; Yoshizawa et al. 1979; Goto 1981; Matsuda et al. 1982; Terai et al. 1983).

The master-slice approach is superior to the custom LSI approach in terms of reducing design cost and period, but a shortcoming is that area utilization in the master-slice LSI is less than that of a customized LSI. For example, the number of gates in a CPU for a small/medium-scale computer reaches several tens of thousands, so constructing the CPU by using customized LSIs is much more effective than assembling several master-slice LSIs (Kasai et al. 1984; Erdelyi et al. 1984). Therefore, the need for an automated design system for custom VLSIs (up to 100k gates) based on the standard cell approach has greatly increased (Ohno et al. 1982; Terai et al. 1982).

The following concepts must be incorporated into the layout of a DA system: (1) a unified logic input language and design data base, so that both standard cell approach VLSIs and master-slice VLSIs can be designed using the same interface to the designer; (2) object partitioning design and hierarchical design approaches—the object partitioning design to divide objects into arbitrary sizes, and the hierarchical design to lay out VLSIs of up to several tens of thousands of gates in a short period and with reduced effort; (3) a new floor planning method, to permit attainment of the smallest possible areas; (4) a wiring method interconnecting different levels of the hierarchy, to achieve a high wiring density; and (5) a wide variety of verification functions in addition to automatic placement and routing, to ensure that new VLSIs can be fabricated with no errors.

This chapter describes layout design methodology for custom logic VLSIs, and placement and routing algorithms. This discussion utilizes CMOS chips that are currently used in major electronic equipment. First, we briefly overview a design automation configuration that is now in use. Then we present a layout model and a hierarchical structure that forms the basis for following discussions. Placement and routing algorithms and their evaluations are covered in the next three sections, with floor planning algorithms covered first, followed by cell-based placement and hierarchical routing procedures, respectively. These three topics are closely interrelated in hierarchical layout design methodology.

A final section gives an example of actual VLSI chips designed by using the DA system. Here, procedures described in earlier sections are incorporated.

DESIGN AUTOMATION SYSTEM CONFIGURATION

A design automation system for practical use in VLSI design is shown in Figure 18-1. The system consists of five subsystems sharing a common design data base. The first is the logic design subsystem, which includes a logic input, logic diagram output, and Logic stimulator. The second is the layout subsystem, which is composed of two major parts: automatic layout and manual (interactive) layout. Third, the verification subsystem includes a logical/physical matching check, a

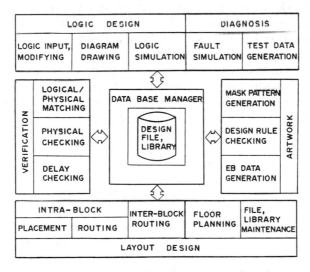

Figure 18-1. Design automation system configuration.

physical check, and a delay check. The logical/physical matching check verifies the correspondence of logic design to physical design. Wiring rule violations, such as short circuits, are check out by the physical check. The delay check calculates the signal propagation time along the gates and wiring paths. The fourth subsystem generates test data sequences for testing chips; this subsystem consists of a test pattern generator and a fault simulator. Finally, the artwork subsystem converts symbolic layout data to physical mask pattern data. A design rule check program checks fabricating process rules, such as minimum spacing between wiring mask patterns. Finally, electron beam or photolithography data is generated.

LAYOUT HIERARCHY AND DESIGN METHODOLOGY

In this section, basic concepts of hierarchical layout design and layout models for each level of the hierarchy are described.

HIERARCHICAL STRUCTURES OF VLSI CHIPS

Hierarchical design is a well-known, efficient approach to the layout of huge numbers of transistors. A hierarchical structure for a custom VLSI chip is shown in Figure 18-2. A complete layout for VLSIs with gates numbering several tens of thousands probably would cause problems with design management. To overcome these difficulties, the concept of partitioning the design object was introduced (Terai et al. 1982). This concept keeps the design within reasonable limits. The idea is to partition a chip into any number of subchips with predefined shapes before interblock routing is processed. These subchips are designated block assemblies (or simply assemblies). The general concept behind hierarchical division is to partition design objects between levels, whereas this block assembly concept divides objects on the same level.

There are two main reasons for introducing the block assembly concept. First, it eases the memory bottleneck in computers. In placing blocks on a chip, the whole chip should be considered in order to optimize construction. However, once the position of each block is specifically determined, it is not always

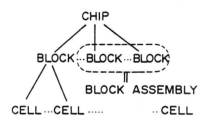

Figure 18-2. Hierarchical structure of custom VLSI chip.

necessary to deal with the entire chip for interblock routing. In fact, certain problems occur because of the increased number of wiring tracks. Lee's algorithm (maze running algorithm) is a well-known, efficient routing procedure for complicated wiring models. A fundamental problem in this algorithm is that the memory requirement sometimes exceeds available memory capacity limits, even in large main frame computers. For example, if a 10 mm × 10 mm chip is wired using 3 micro intervals, then 3000 by 3000 entries are required as a mesh table. This means that 18M bytes are necessary for the main memory (for Lee's router, 2 bytes/entry are generally assigned). Therefore, the chip must be partitioned for this routine. The second reason for using this concept is to improve design work efficiency. Layout work efficiency will increase with partitioning because each XY drawing is fairly small, making it easy to check and modify. Moreover, each partitioned unit can be independently processed. This enables concurrent design, thus reducing design periods.

Layout Model

The relationship between wiring layers and the corresponding layout hierarchy is shown in Fig. 18-3. It is assumed that three wiring layers, Poly Si and two levels of metal, are available. This figure shows that Poly Si and the first metal layer (All) are used in cells, and Poly Si and two metal layers (A11 and A12) are used for intrablock wiring. Poly Si cannot be used for interblock wiring, so only two metal layers are then available. This is so because the wiring resistance value of Poly Si is so high that interblock wiring, which is longer than itnrablock wiring is likely to cause undesirable signal propagation delays.

The layout models for each hierarchy are shown in Figure 18-4. A cell is a minimum logical unit consisting of 1 to about 50 gates. The height and width vary according to the logic function. A block is a basic unit of logic design, and has about 500 to 1000 gates. A chip consists of blocks (inner logic parts) and I/O buffer arrays (peripheral logic parts). The chip can be partitioned into block assemblies after all block locations have been determined. By creating virtual edge pins on each partitioning boundary, each block assembly can be routed

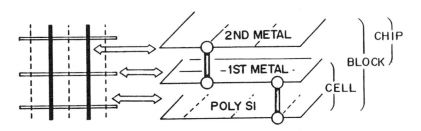

Figure 18-3. Wiring layer model.

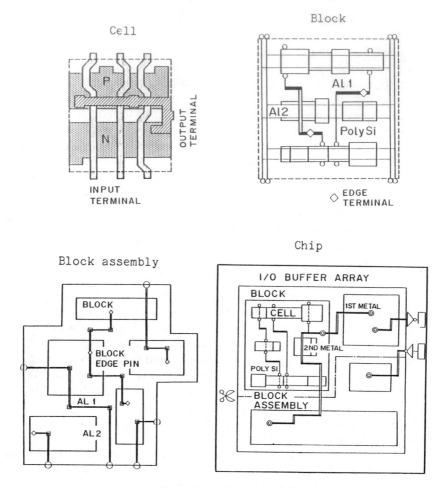

Figure 18-4.   Layout model.

concurrently. Consequently, the entire chip layout is completed by joining the assemblies together like a jigsaw puzzle.

AUTOMATIC FLOOR PLANNING

This section considers floor planning to be block placement problems on a chip (Leblond 1983; Otten 1982; Ueda et al. 1985; Kozawa et al. 1984). The "Combine and TOP down placement (CTOP)" algorithm has been developed to achieve block placement by minimizing both interblock routing and dead spaces (Kozawa et al. 1984). This algorithm consists of a bottom-up combining process and a top-down placement process.

The concept of "combine operation" is first explained by examining several definitions:

Hyperblock: A block or set of blocks having a rectangular shape.

Combine: A new hyperblock, K, can be constructed by joining two hyperblocks, I and J, either horizontally or vertically, and obtaining their rectangular enclosure. This is defined as a "combine operation" and is written $'K = I \oplus J'$, where $\oplus$ is the direction of the combine. If I and J are horizontally combined, $\oplus$ represents X, and if vertically combined, $\oplus$ represents Y.

Combine value: If the combine $'I\ J'$ is given, the combine value, $P(I \oplus J)$, is defined by:

$$p(I \oplus J) = S(I \oplus J) + C(I, J)*S(I \oplus J)**a \qquad [1]$$

$$S(I \oplus J) = \frac{A(I) + A(J)}{A(I \oplus J)} \ (\leq 1) \qquad [2]$$

$$C(I, J) = \frac{2*Tc(I, J)}{2*Tc(I, J) + Td(I, J)} \ (\leq 1) \qquad [3]$$

$$Tc(I, J) = \sum_{n1 \in N1} \frac{1}{|n1|\#} \qquad [4]$$

$$Td(I, J) = \sum_{n2 \in N2} \frac{1}{|n2|} \qquad [5]$$

where:

I,J : hyperblock

$\oplus$: direction of combine

A(I) : area of I

N1 : set of common nets between I and J

N2 : set of nets that connect to either I or J

$|n|\ \#$: number of block terminals in a net n

a : adjustable parameter (> 1)

In the above formula, $S(I \oplus J)$, which is the first term of $P(I \oplus J)$, represents the ratio of the I and J area to the area of $I \oplus J$. If $S(I \oplus J)$ is large, the dead space is relatively small, and placing I and J adjacently causes dead space to decrease. On the other hand, $C(I,J)$ presents connectivity between I and J. It consists of a conjunction between I and J ($=2*Tc(I,J)$) and a disjunction ($=td(I,J)$), which is considered a multipoint net. If $C(I,J)$ is large, placing I adjacent to J causes decreasing wiring space. To evaluate wiring space, however, dead space must be thought to have a negative effect on wiring space, however, dead space must be thought to have a negative effect on wiring space. Typically, if

A(I1) = A(I2), A(J1) = A(J2), C(I1, J1) = C(I2, J2), and A(I1 ⓓ J1) > A (I2 ⓓ J2), the wiring space in I1 ⓓ J1 is larger than in I2 ⓓ J2. Considering this fact, C(I,J)*S(I ⓓ J)**2 (the second term of P(I ⓓ J)) can be used to evaluate wiring space. The parameter "a" is determined experimentally.

The CTOP algorithm is composed of four steps: Step 1: Initial Combine (Figure 18-5a). If p(I ⓓ J) is large for two hyperblocks, I and J, adjacent placement is desirable for decreasing wiring and dead space. Thus, the following process is repeated in this step until there remains only one hyperblock:

1. Calculate P(I ⓓ J) for any pairs of hyperblocks, I and J, that have common nets.
2. Select Im, Jm, Dm such that P(Im Dm Jm) is maximum, and a new hyperblock is created Im Dm Jm.

However, performing this process without constraints allows considerable hyperblock growth in the same direction (X or Y). This leads to a large amount

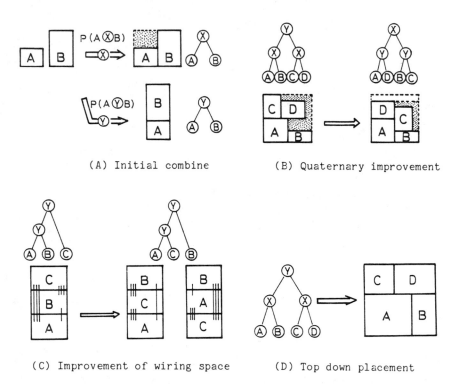

(A) Initial combine   (B) Quaternary improvement

(C) Improvement of wiring space   (D) Top down placement

Figure 18-5.   Combine and TOP-down Placement (CTOP) flow.

of dead space in the final stage. In order to avoid this tendency, an intermediate upper bound of the hyperblock is designated, and combining is performed step by step. This combine process is represented by a labeled binary tree, called a "combine tree."

Step 2: Quaternary Improvement of Dead Space (Figure 18-5b). In this step, the dead space is decreased by changing the combine tree constructed during Step 1. Here, minor changes in it are necessary because major changes for decreasing dead space may cause an increase of wiring space. From the root of the combine tree to its leaves, the following process is repeated:

1. For each hyperblock K, select four hyperblocks whose ancestor is K, and which have no ancestor-descendant relations each other.
2. Recombine them so that the dead space is minimum under the condition that both the horizontal and vertical sizes of hyperblock K are not larger than its initial ones.

Step 3: Improvement of Wiring Space (Figure 18-5c). In this step, the size of any hyperblock does not change, and only the wiring space is improved. The following process is repeated in the bottom-up manner:

1. First, select a subtree such that all nodes have the same label (X or Y).
2. Next, as in Step 1, recombine the leaves in the same direction according to connectivity $C(I,J)$.

Step 4: Top-Down Placement (Figure 18-5d). The following process is again repeated from the root of the combine tree to the leaves.

1. For each hyperblock having "sons," position them relative to one another in accordance with the combine direction so as to decrease the pseudo-wire length.

This concludes the procedure.

A result of automatic floor planning is shown in Figure 18-6.

CELL PLACEMENT PROCEDURE

This section presents some algorithms to generate the block shown above in Figure 18-4(b). The placement procedure is composed of two major processes: initial placement and an improvement phase. Various kinds of algorithms have been developed and reported (Hanan et al. 1972; Goto 1980; Sato et al. 1981; Kambe et al. 1982).

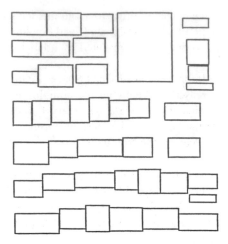

Figure 18-6.   Example of automatic floor planning results.

Initial Placement

Initial placement is a process whereby each cell is arranged in a two-dimensional array. Two typical methods are explained below.

Zigzag Placement Algorithm.   In this type of placement, a two-dimensional structure is generated by "folding" the one-dimensional placement results, as shown in Figure 18-7. Strongly connected cells tend to be placed in a row, but are not arranged facing each other across a channel, as the cell row made by a simple folding is rather long. Therefore, the simple zigzag algorithm has a defect in that the total horizontal wire length cannot be significantly reduced. (The effectiveness of considering total horizontal wire length is discussed later.) To overcome this defect, a multi-zigzag placement algorithm has been developed (see Figure 18-7).

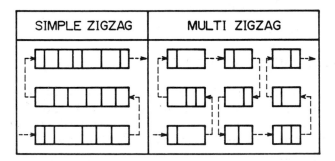

Figure 18-7.   Zigzag Placement Algorithm.

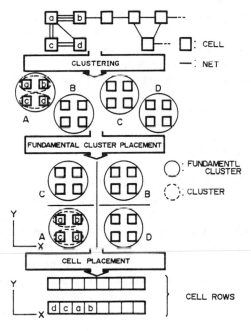

Figure 18-8.   Two-dimensional Clustering Placement algorithm.

Clustering Placement Algorithm.   The One-dimensional Clustering Placement (OCP) algorithm, consisting of a clustering process and a top-down linear
placement process, was developed in the early seventies (Schuler and Ulrich
1972). If simple zigzag placement is performed after OCP, almost all pairs of
strongly-connected cells may be placed in the X-direction. However, to obtain
good placement results (to reduce total horizontal wire length), it is necessary
for pairs to be placed not only in the X direction but also in the Y-direction. As
a result, the Two-dimensional Clustering Placement (TCP) algorithm was developed (Kozawa et al. 1983). This algorithm consists of the following three
steps (see Figure 18-8):

Step 1: Clustering. Various sizes of cells are grouped into clusters of approximately the same size according to their degree of logical connection. Such
clusters are called "fundamental clusters."

Step 2: Fundamental Cluster Placement. In this step, the two-dimensional
relative placement of fundamental clusters is determined by top-down partitioning. This process is performed by repeating initial partitioning and partitioning
improvement.

The initial partitioning procedure consists of two processes. The first process
makes a binary tree that represents the degree of connection between fundamental

clusters. The second process, controlled by the size adjustment function, repeats decomposition of a tree node that crosses the cut line into two nodes. It then places them in relation to each other from the top to the bottom of the tree. This process continues until no crossing node exists.

Step 3. Cell Placement. Placement of cells in each fundamental cluster is then determined by the OCP algorithm.

This concludes the TCP algorithm.

Placement Improvement

Calculation of Pseudo-Wire Length. Generally, the sum of the pseudo-wiring length is used as an evaluation function in a placement improvement procedure. Calculation methods (A) through (D) shown in Figure 18-9 have already been suggested. The method selected depends on the particular object's characteristics. Methods (A) through (D) were first adopted for printed circuit boards and master-slice LSIs having fixed placement locations. However, a variable wiring region, which is a characteristic especially important to custom logic LSIs, should be designed for 100% wiring in a minimal space. Therefore, trunk length (method (E) in Figure 18-9) is considered to be the most suitable evaluation function.

With the block model shown in Figure 18-4, the block area is divided into

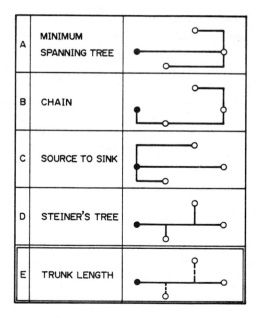

Figure 18-9. Calculation methods for pseudo-wire lengths.

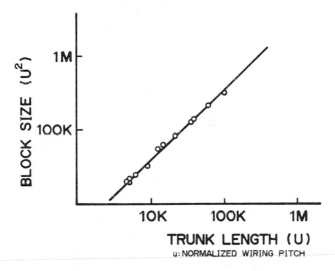

Figure 18-10. Relationship between total trunk length and block area.

the cell area and the wiring region. Because the cell area is fixed, only the wiring region is variable. As a result, the length of cell sequences should be uniform, and the number of tracks in the wiring region minimized. Figure 18-4 shows that branch length (Poly Si and Al2) does not affect the block area, but the number of tracks (Al1) significantly influences it. Therefore, minimization of total trunk length is effective for minimizing the number of tracks.

The relationship between total trunk length and block area for several blocks of different gate number is shown in Figure 18-10. The vertical and horizontal axes in the figure represent block size and total trunk length normalized by the wiring pitch, respectively. These experimental values all lie on a straight line. Therefore, it is shown experimentally that the total trunk length dominates the block size, and a procedure minimizing total trunk length is suitable for LSI layout.

Net Balance Algorithm. One-dimensional iterative placement improvement is summarized in the following procedure:

"Repeat Steps 1 to 3 until all cells in the block remain at their own location forever."

Step 1: Select a cell row.

Step 2: Compute the optimal location for a specified cell.

Step 3: Move the cell to its new location only if this causes a decrease in the pseudo-wire length sum. This concludes the procedure.

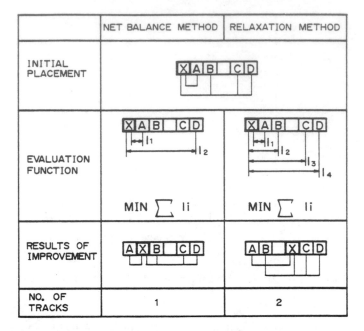

	NET BALANCE METHOD	RELAXATION METHOD
INITIAL PLACEMENT		
EVALUATION FUNCTION	MIN $\sum$ li	MIN $\sum$ li
RESULTS OF IMPROVEMENT		
NO. OF TRACKS	1	2

Figure 18-11.   Net balance placement algorithm.

The objective function used to compute the optimal location must be set carefully. To solve this problem, an algorithm known as "Net balance placement" (NB) has been developed (Terai et al. 1982).

This algorithm incorporates placement improvement based on a placement that approximates the total trunk length minimization. this improvement is used mainly with cell sequences placed in a line.

The NB method is compared with the earlier relaxation method (Kozawa et al. 1972) in Figure 18-11. The evaluation function of the relaxation method counts one net three times (12, 13, 14 indicated in Figure 18-11). As a result, this placement would require two tracks, as shown in the figure. On the other hand, the NB method employs an evaluation function that regards one net as one trunk.

Problem examples are now given to illustrate the above discussion. For the relaxation method, the formulation is as follows:

Problem 1

Let $\{ a_i \mid i = 1,N \}$ be a given set whose elements consist of x number of pin coordinate values of the pins of all nets collected to cell "X." Let:

$$g(x) = \sum | a_i - x |$$ [1]

be a distance function when cell "X" is located at point x. Then, derive solution xO, which will minimize g(x).

Solution xO is:

$$xO = a \frac{N + 1}{1} \text{ (N is odd)} \qquad [2]$$

or

$$xO = a \frac{N}{2} \text{ (N is even)} \qquad [3]$$

where { a li } is sorted in ascending order.

It is evident that xO provides an optimal location for cell "X" is the routing path is radial, as shown in Figure 18-12. However, the channel router used in the intrablock routing procedure makes a Steiner tree, as shown in Figure 18-12. Therefore, g(x) in Eq. (1) is not always a suitable function for the channel router. Now, the Net balance method is formulated as follows:

Problem 2

Let $a_1, a_2, \ldots {}_n : a_i > 0$ be N points (pins) on the x-coordinate, and let these points be partitioned into M groups, $S_1, S_2, \ldots S_M$ (nets). In addition, let

$$1(S) - \max(S) - \min(S) \qquad [4]$$

represent a mapping function for set S, where max(S) and min(S) are maximum and minimum values for set S elements. Then, arrive at a solution xO such that

$$f(x) = \sum 1(S_j \cup \{x\}) \qquad [5]$$

is minimum.

The solution is: "xO is a median point of sequence $p_1, p_2, \ldots p_{2M}$, which is sorted in ascending order. Here, p is an element of set

$$p = \{ \max(S_j), \min(S_j) \mid j = 1,M \}." \qquad [6]$$

Block areas are compared in Table 18-1. Column (A) represents values using the function f(x)—Net balance placement—and column (B) represents that for g(x) as described above.

As shown in Table 18-1, net balance placement results in more effective reductions of block size.

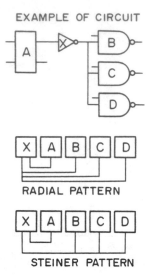

RADIAL PATTERN

STEINER PATTERN

Figure 18-12.   Wiring patterns.

Evaluation of Placement Procedures

The area qualities are now compared for three types of placement procedures, A, B, and C. In procedure A, a two-dimensional cell array is produced through simple zigzag folding (see Figure 18-7). This occurs after applying NB improvement to a random one-dimensional arrangement. In Procedure B, an OCP algorithm produces one-dimensional cell placement, and multi-zigzag folding (see Figure 18-7) generates a two-dimensional array. Procedure C makes a two-dimensional cell array directly by using the TCP algorithm, and no zigzag folding is applied. In all cases, NB improvement is applied to the two-dimensional cell array. The area comparison for the above three procedures is shown in Figure 18-13. The figure clearly shows that Procedure C produces the optimum result

Table 18-1. Comparison of block areas.

	(A) mm^2	mm^2	(B) mm^2
BLK1		1.86	1.87
BLK2		1.13	1.16
BLK3		1.19	1.22
BLK4		1.08	1.09
BLK5		1.56	1.56

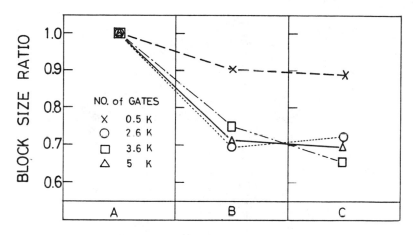

Figure 18-13.   Area comparison of placement procedures.

in total. The area of Procedure C is about 30% smaller than that of Procedure
A. Layout drawings of Procedures A and C for the same 5K gate block logic
are shown in Figure 18-14. From the figure, the advantages of Procedure C are
evident.

ROUTING PROCEDURE

This section discusses routing problem relating to hierarchical layout and multi-
levels of wiring. In addition, two kinds of routine algorithms suitable for each
hierarchy are presented.

Routine Problem for Hierarchical Design

In many hierarchical design cases, the lower level of the hierarchy is processed
as a "black box." Consequently, no attention is paid to it while the upper level
is being designed. However, this idea is inadequate for multilayer wiring problems (up to three layers). It was shown in Figure 18-3 that each wiring layer is
assigned to the design hierarchy. Here, the first and second metal layers are used
in common for intra- and interblock routing. This means that interblock routing
is allowed to cross previously laid-out block regions. To do this, automatic
wiring should be able to make the remaining portions of a track left in a lower-
level design available in the upper level.

   In order to satisfy the above requirements, two kinds of algorithms should be
adopted: channel router (CR) (Hashimoto and Stevens 1971; Horino et al. 1971;
Kozawa et al. 1974; Yoshimura and Kuh 1982) and maze router (MR) Lee 1961;
Hoel 1976; Hightower 1983). Generally, channel router can perform routing in

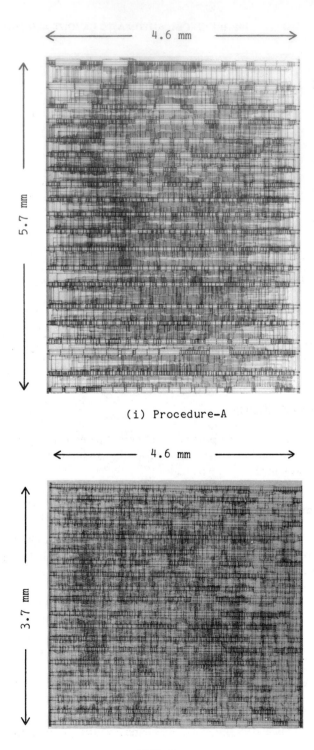

4.6 mm

5.7 mm

(i) Procedure-A

4.6 mm

3.7 mm

(ii) Procedure-C

less CPU time than maze router. On the other hand, maze router has superior wiring potential. As shown in Figure 18-4, the block is considerably restricted within certain limits in its layout model, but the block assembly has fewer restrictions than the block. By considering the differences between models and the features of each algorithm, the channel and maze routers are preferably applied to intrablock routing and intra-assembly (interblock) routing, respectively.

Intrablock Routing

The general flow of intrablock routing is summarized in Figure 18-15. The general procedural flow consists of two kinds of procedures: global routing and detail routing. Global routing covers Step 1 through 3. Detail routing is represented by Step 4.

Step 1: Pin Assignment. As shown in the cell portion of Figure 18-4, a cell has dual equipotential terminals on its upper and lower sides. As a result, it is necessary to decide on the terminal to be wired. A terminal is selected that narrows the wiring space (the number of the first metal channels between cell rows).

Step 2: Layer Assignment. Two kinds of layers (Poly-Si and second metal) are available for y-direction wiring. It is well-known that the resistance value of Poly-Si is about 1000 times that of metal. With respect to a critical net, such as a clock signal, therefore, the layer used is assigned according to the net length (y direction) in this step.

Step 3: Feed-through Assignment. Feed-through points in a cell row are determined for connecting cells located at a distance that is farther away than two rows.

Step 4: Track Assignment. Step 3 has transformed all vertical wiring to the boundaries of each wiring space between two cell rows. This is performed so that the wiring problem is reduced to routing a channel with terminals on each edge. The algorithm is a track assignment method employing trunk-division for cyclic constraint or "constraint loops" (Kozaka et al. 1974).

Interblock Routing

Each block and block assembly terminal is wired by using a maze-running algorithm. Some improvements for increasing wirability have been added to the classic algorithm, such as equipotential wiring. This wiring improvement is used to route not only the block/assembly edge pins but also the equipotential paths

Figure 18-14.   Layout examples of two kinds of placement procedures.

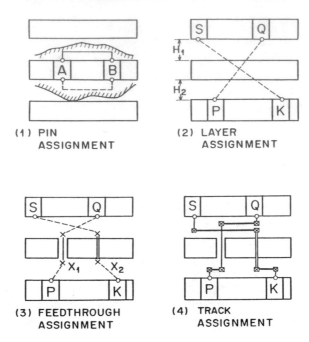

(1) PIN
ASSIGNMENT

(2) LAYER
ASSIGNMENT

(3) FEEDTHROUGH
ASSIGNMENT

(4) TRACK
ASSIGNMENT

Figure 18-15.   Intrablock routing flow.

existing in the previously routed blocks on the lower level of hierarchy. Equipotential path wiring has proved effective for increasing wirability. In designing a chip (described in 18), the number of wiring failures was reduced to 9 (99.7% at wirability) by using this technique. Before using this technique, the failure number was 25 (99.1% at wirability).

EXAMPLE OF AUTOMATIC LAYOUT DESIGN

An example of a VLSI including 74K transistors designed by the hierarchical approach is shown in Figure 18-16. This VLSI is part of the CPU for small/medium-scale computers, and is composed mainly of a 32-bit ALU, registers, and some control units. Design features of the chip are:

Chip size: 12 mm × 12 mm
No. of transistors: 74,000
No. of random gates: 17,000
No. of cells: 5700
No. of blocks: 40
RAM: 2k bits
No. of package pins: 200
Technology: 2 micron CMOS

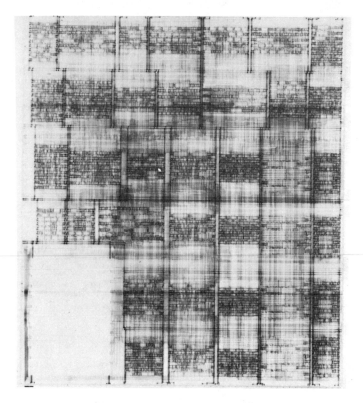

Figure 18-16. Example of custom VLSI layout.

The total CPU time for routine blocks on this chip (partitioned into four block assemblies) was 256 minutes on a three MIPS computer. In addition, the number of unconnected pins-to-pins was 9, where the sum of pin-to-pins to be connected is 2726. The channel utility was 49.1% at the first metal and 47.0% at the second metal. The total wiring length in the chip was 10,892 mm. The average wiring length was 0.46 mm for the intrablock net, and 7.6 mm for the interblock net. The layout design effort amounted to less than 10 man-months, and the chip was fabricated without error on the first design.

CONCLUSIONS

This chapter has presented the following subjects concerning a hierarchical method for designing a VLSI chip layout:

1. A concept of partitioned layout design and related layout models.
2. An automatic floor planning algorithm with block placement function.

3. Several cell placement procedures and their evaluation results with respect to block area performance.
4. A routing method suitable for a hierarchical physical layout.

Through practical chip design, these layout methods and algorithms have proved effective for designing logic VLSIs with gates numbering several tens of thousands.

REFERENCES

Erdelyi, C. K., et al., "A Comparison of Mixed Gate Array and Custom IC Design Method," *ISSCC Digest of Technical Papers,* 14–15 (1984).

Goto, S., "An Efficient Algorithm for Two-Dimensional Placement Problem in Electrical circuit Layout," *IEEE Trans. Circuits and Systems, CAS-28* (1), 12–18 (1980).

Goto, S., "An Automated Layout Design System for Masterslice LSI Chip," *Proc.* Europ. Conf. on Circuit Theory and Design, pp. 631–635, 1981.

Hanan, M., et al., "Placement Techniques," Chapter 5 in *Design Automation of Digital Systems,* Vol. 1, Prentice-Hall, Englewood Cliffs, NJ, 1972, pp. 213–282.

Hashimoto, A., and Stevens, S., "Wire Routing by Optimizing Channel Assignment within Large Apertures," *Proc.* 8th Design Automation Workshop, pp. 155–169, 1971.

Hightower, D., "The Lee Router Revisited," *Proc.* IEEE Int. Conf. on Computer Design: VLSI in Computers, pp. 136–139, 1983.

Hoel, J. H., "Some Variations of Lee's Algorithm," *IEEE Trans. Computers, C-25* (1), 19–24 (1976).

Horino, H., Kitazume, Y., and Hirano, C., "Routing Algorithm between Two Blocks or Elements," Report of Technical Group on Semiconductors and Semiconductor Devices, IECE Japan SSD 70-64, 1971 (in Japanese).

Kambe, T., Chiba, T., Kimura, S., Inufushi, T., Okuda, N., and Nishioka, I., "A Placement Algorithm for Polycell LSI and Its Evaluation," *Proc. 19th Design Automation Conference,* pp. 655–662, 1982.

Kamikawai, R., Kishida, K., Osawa, A., Yasuda, I., and Chiba, T., "Placement and Routing Program for Masterslice LSIs," *Proc. 13th Design Automation Conference,* pp. 245–250, 1976.

Kasai, R., et al., "An Integrated Modular and Standard Cell IC Design Method," *ISSCC Digest of Technical Papers,* pp. 12–13, 1984.

Kozawa, T., Horino, H., Watanabe, K., Nagata, M., and Fukuda, H., "Block and Track Method for Automated Layout Generation of MOS-LSI Arrays," *ISSCC Digest of Technical Papers,* pp. 62–63, 1972.

Kozawa, T., Horino, H., Ishiga, T., Sakemi, J., and Sato, S., "Advanced LILAC—An Automated Layout Generation System for MOS/LSI," *Proc.* 11th Design Automation Workshop, pp. 26–46, 1974.

Kozawa, T., Terai, H., Ishii, T., Hayase, M., Miura, C., Ogawa, Y., Kishida, K., Yamada, N., and Ohno, Y., "Automatic Placement Algorithms for High Packing Density VLSI," *Proc. 20th Design Automation Conference,* pp. 175–181, 1983.

Kozawa, T., Miura, C., and Terai, H., "Combine and Top Down Placement Algorithm for Hierarchical Logic VLSI Layout," *Proc. 21st Design Automation Conference,* pp. 667–669, 1984.

Leblond, A., "CAF: A Computer-Assisted Floorplanning Tool," *Proc. 20th Design Automation Conference,* pp. 747–753, 1983.

Lee, C. Y., "An Algorithm for Path Connections and Its Applications," *IRE Trans. Electronic Computers, EC-10* (3), 346–365 (1961).

Matsuda, T., Fujita, T., Takamizawa, K., Mizuhara, H., Nakamura, H., Kitajima, F., and Goto, S., "LAMBDA: A quick, low cost layout design system for masterslice LSIs," *Proc. 19th Design Automation Conference,* pp. 802–808, 1982.

Ohno, Y., Yamada, N., Sato, K., Sakataya, Y., Endo, M., Horikoshi, H., and Oka, Y., "Integrated Design Automation System for Custom and Gate Array VLSI Design," *Proc.* IEEE Conf. on Circuits and Computers, pp. 512–515, 1982.

Otten, R., "Automatic Floorplan Design," *Proc. 19th Design Automation Conference,* pp. 261–267, 1982.

Sato, K., et al., "MILD—A Cell Based Layout System for MOS LSI," *Proc. 18th Design Automation Conference,* pp. 828–836, 1981.

Schuler, D. M., and Ulrich, E. G., "Clustering and Linear Placement," *Proc.* 9th Design Automation Workshop, pp. 50–56, 1972.

Shiraishi, H., and Hirose, F., "Efficient Placement and Routine Techniques for Master Slice LSI," *Proc. 17th Design Automation Conference,* pp. 458–464, 1980.

Tanaka, C., Murai, S., Nakamkura, S., Ogihara, T., Terai, M., and Kinoshita, K., "An Integrated Computer Aided Design System for Gate Array Masterslices: Part 1 Logic Reorganization System LORES-2," *Proc. 18th Design Automation Conference,* pp. 59–65, 1981.

Tanaka, C., Murai, S., Tsuji, H., Yahara, T., Ozaki, K., Terai, M., Katoh, R., and Tachibana, M., "An Integrated Computer Aided Design System for Gate Array Masterslices: Part 2 the Layout Design System MARS-M3," *Proc. 18th Design Automation Conference,* pp. 812–819, 1981b.

Terai, H., Hayase, M., Ishii, T., Miura, C., Kozawa, T., Kishida, K., and Nagao, Y., "Automatic Placement and Routing Program for Logic VLSI Design Based on Hierarchical Layout Method," *Proc.* IEEE Int. Conf. on Circuits and Computers, pp. 415–418, 1982.

Terai, H., Hayase, M., Ishii, T., Miura, C., Ogawa, Y., Kozawa, T., Sato, Y., Sasaki, H., and Hiyama, T., "Performance Analysis of Automatic Placement and Routing for Large-Scale CMOS Masterslices," *Proc.* IEEE Int. Conf. on Computer Design, pp. 536–539, 1983.

Ueda, K., Sugiyama, Y., and Wada, K., "An Automated Layout System for Masterslice LSI: MARC," *IEEE J. Solid State Circuits, SC-13* (5), 716–721 (1978).

Ueda, K., Kitazawa, H., and Harada, I., "CHAMP; Chip Floor Plan for Hierarchical VLSI Layout Designing," *IEEE Trans. Computer-Aided Design, CAD-4* (1), 12–22 (1985).

Yoshimura, T., and Kuh, E. S., "Efficient Algorithms for Channel Routing," *IEEE Trans. Computer Aided Design, CAD-1* (1), pp. 25–35 (1982).

Yoshizawa, H., Kawanishi, H., Goto, S., Kishimoto, A., Fujinami, Y., and Kani, K., "Automatic Layout Algorithms for Masterslice LSI," *Proc.* IEEE Int. Conf. on Circuits and Systems, pp. 470–473, 1979.

19. Layout Compiler-Annealing Applied to Floorplan Design

Ralph H. J. M. Otten

IBM Corporation Thomas J. Watson Research Center and
Eindhoven University of Technology Department of Electrical Engineering

Annealing was announced as a general method for combinatorial optimization, and its first applications were very successful. The mathemeaticl model of annealing is a time-inhomogeneous, irreducible Markov chain with a score associated with each state, and the variations in the transition probabilities over time controlled by a single control parameter. The goal is to find a state with a low score. Ever since the first reports of its impressive results, obtained in placement of modules on gate arrays, researchers have tried to use the method on other problems, and have worked on rigorous theoretical statements about its performance. Soon it became clear that annealing was not suitable for finding high-quality results for every combinatorial problem with reasonable efficiency, and that, if it was suitable for a certain problem, the efficiency was not obtained for every formulation of that problem and for every schedule of control parameter values. This raises two questions. What are the essential properties the state space must have for an efficient application of the annealing method? And what is the best schedule, that is the best sequence of values t_i for the control parameter, given a suitable state space?

Both questions are presently unanswered. Some promising research by Greg Sorkin and others at IBM's Watson Research Center, aimed at answering the first question, introduced the concept of ultrameticity in the annealing context. Early experience indicates that state spaces with a high degree of ultrametricity [measured, for example, by the correlation between the longest two distances (minimum number of intermediate states between two states) in any set of three states] usually have better anneraling results than spaces with a low degree of ultrametricity. Also, it seems, but this is less surprising, that the distance,, defined above, and the barrier distance, defined as the smallest maximum score over any set of intermediate states, have to be correlated. However, none of these results has enough rigor to be useful in state space analysis at present.

The strongest result on the convergence of annealing is due to Bruce Hajek. A necessary and sufficient condition for convergence with probability one to a global minimum, if t_i is a monotonously nonincreasing sequence with limit zero, and the barrier distance is symmetric, is:

$$\sum_{i=1}^{\infty} \exp\left(-\frac{d}{t_i}\right) = \infty$$

where d is maximum barrier distance from any state to a state with a strictly lower score. This implies, not surprisingly, that no schedule can guarantee a convergence to a state with the lowest score in the space because any such schedule has to become infinitely slow in the end.

In the absence of useful answers to the two basic questions, applications of annealing have to be guided by intuition and experiment, beside analyses based on the theory of Markov chains. Such an approach is presented in the second part of this chapter. The analysis is more or less along the lines presented by Michele Lundy at a workshop in Yorktown Heights in April 1984, where she tried to formulate a general framework for annealing applications to well-known NP-hard problems. One of her important observations is that under certain mild restrictions the equilibrium density of the chain for a given value of the control parameter does not depend on the topology of the state space, but the probability of a good result after a given number of steps heavily depends on the chosen topology.

In the second part of this chapter an almost self-contained presentation of the annealing algorithm is presented. The goal of the analysis given there is to come to an algorithm that can be used in an automatic way; that is, the program automatically adapts its schedule to the size, topology, and score function of the problem, depending on some external parameters to control the time spent by the algorithm. This is necessary in the context where the algorithm is going to be used, in a layout compiler, that is, a system of programs that produce the mask specification of an integrated circuit from its functional specification without any user interaction. The first part of this chapter extensively describes the basic ideas, and some of their implementations, in such a compiler. The task to which the annealing algorithm is to be applied is carefully specified. The last part of this chapter formulates this task in a form suitable for the annealing algorithm.

Acknowledgment

My recent research on the topics discussed in this chapter has been done together with Lukas van Ginneken. He implemented the net assignment part of the floorplan design phase, and invented the new heuristic for finding short interconnection trees, discussed at the end of the section on "Net Assignment." Also, I greatly appreciate his contributions to the annealing research and implementation.

Thanks also to Marian Mack, who runs many of the implementations of the ideas in this chapter, and shares her experiences with me. Her suggestions led to several improvements in the original implementations.

LAYOUT COMPILATION

The environment is discussed in terms of input, output, and data bases, and the outline of a complete layout design process is proposed. The process is interpreted as an application of the principles of stepwise refinement to the design of complex integrated circuits. We emphasize the rationale behind the chosen system decomposition, the accepted restraints, and the methods for cell assembling.

Stepwise Refinement

The technique of stepwise refinement has been shown to be effective in the development of computer programs. It was explicitly formulated in a famous paper by Niklaus Wirth (1971), who viewed the design of a structured program as a sequence of refinement steps. Starting with a clear problem statement that specifies the relation between the input and the output data, the task is progressively refined, by decomposition into subtasks, each having an equally clear specification. The sequence of refinement steps terminates when all tasks are specified in a chosen programming language. The constructs of that language should be a direct translation of tasks resulting from the final refinement steps. To be effective, they have to form a small but powerful enough repertoire. This method thus entails a hierarchical structure. (A hierarchy is either a set of hierarchies or an atom. In this case each hierarchy represents a task, and each task translatable in a construct is an atom.)

Stepwise refinement may also be viewed as postponing implementation decisions, to avoid committing a program prematurely to a specific implementation. Each decision should leave enough freedom to following stages to satisfy the constraints it created, and at the same time rearrange the available data so that further meaningful decisions are possible in the next step. So, concurrent with a gradual stiffening of the design, information is progressively organized so that more and more detailed decisions can be derived.

The principles of stepwise refinement obviously apply to any complex design task based on a top-down strategy rather than to a process of combining independently developed subdesigns. Completely specified subdesigns, in general, are difficult to handle because the flexibility and the information needed to adapt them to their environment often are not available when they are designed. On the other hand, the application of stepwise refinement in layout design raises a number of questions. First, what information is available in the initial stage of layout design? One difficulty in answering this question involves separating layout design from the other tasks in a silicon compiler. This will be another question that immediately arises is, what relevant information can one derive at the intermediate stages before fixing the geometrical details in the final stage? This will be the subject of the section on "Floorplan Design." A section on "Cell Assemblers" will consider the translation of the results of the last refinement steps.

Layout Design in a Silicon Compiler

This section describes the environment of layout design in a silicon compiler. The various relations of the layout design part with its environment are indicated in Figure 19-1. They suggest the division of this section into four subsections. The output of a silicon compiler is the first subject; the primitive constructs used to translate the result of the final refinement steps into a layout are discussed. There are some strict constraints on these constructs, the so-called design rules, whose impact on the compiler is outlined next. In discussing the information generated by the preceding parts of the silicon compiler, and therefore available to the layout design part as input, the question of the initial data is considered. Finally, the interface with other parts of the silicon compiler is discussed.

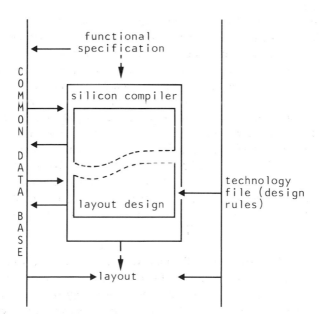

Figure 19-1. The environment of layout design.

The Ouput: A Layout. The ultimate task of a layout design system is to produce a layout, a set of data that uniquely and completely specifies the geometry of the circuit. Usually this data is an encoding of patterns, two-block partitions of the plane. The term "mask" will be used for each plane with a pattern, even if that plane is not exactly one of the real masks used in the fabrication. A layout is then translated into a sequence of processes that selectively change the characteristics of the silicon according to those patterns, thus realizing the functional specification available as input to the layout design procedures. Whereas the layout design system has considerable freedom in deriving geometry from functional specifications, the result of that translation procedure is fixed. Up to 40 different patterns are sometimes used in the sequence of selective exposures of the wafer surface. However, many of these are implied by other patterns in the sequence. For present-day technologies the geometrical specification of 8 to 15 planes suffices to specify the layout.

From device theory, general restrictions on the shapes of the regions seldom can be derived. Lithography techniques, however, do sometimes have their limitations. Quite often only orthogonal artwork is acceptable. This leads to regions that are unions of iso-oriented rectangles. There are examples of circuits that indicate that other layout primitives are more efficient, such as hexagons in some systolic arrays. Rarely is a restriction to rectangles and combinations thereof detrimental, whereas the cell design algorithms and layout data bases profit from such a restraint. The rectangle is therefore accepted as the basic construct in the present system.

The rectangle constraint is also accepted for the compounds of layout primitives that form the atoms and even the hierarchies entailed by the method of stepwise refinement. Consequently, each hierarchy will be a rectangle dissection in the final layout, that is, a rectangle subdivided into nonoverlapping rectangles. The restriction to rectangles might seem rather arbitrary. However, in a truly top-down design it is very difficult to be particular about these shapes because the shapes of its constituent parts still have to be determined. A good estimate for the shape of the enclosing region is of great value in determining the shape and positions of the constituent parts, and hardly a constraint if these parts have a high degree of flexibility. The earlier estimates become in some sense even self-fulfilling because the parts mostly can be fit nicely in the environment by using their flexibility. Besides, choosing rectangles as the only constructs in the repertoire simplifies the formulation of design decisions, and lowers the complexity of deriving these decisions, as will be seen later.

Technological Constraints. To improve chances for successful integration of the circuit, and increase yield when the circuit goes into production, patterns are required to satisfy certain rules, the so-called *design rules*. A first classification distinguishes roughly two classes of rules: *numeric rules,* quantifying extensions of, and spacings between, patterns in a plane and in combinations of planes, and *structural rules,* enforcing and prohibiting certain combinations.

There usually are a large number of numeric rules. Very few, however, are critical in a layout. For example, the spacing between two separate pieces of metal in the same layer is bounded below by different numbers, depending on whether or not there are contacts to other layers in one or both of these metal pieces. In a wiring algorithm there rarely is a good reason for trying to use all these different minima. Instead, the maximum of all the rules that might apply is taken as the pitch for the metal in that layer. The reason for specifying the different rules for all special cases is mostly for the (manual) optimization of small pieces of layout that are used repetitively, such as memory cells. The numeric rules are almost exclusively specifications of lower bounds. This does not imply that the concerned extensions and spacings can be arbitrarily large. Making them arbitrarily large might impair the functioning of the devices in the circuit and increase delays, and it decreases the yield. The rules are formulated as minimum rules only because it is assumed that the layout design techniques will try to keep the total chip small.

Good algorithms working with these rules produce valid layouts in a wide range of values for these lower bounds. Of course, the algorithms do not produce optimal layouts for all combinations of values in these rules, but they should produce acceptable solutions for all practical value sets. The latter requirement is much more difficult to maintain under changes in structural rules because these changes often require completely different decisions. The rules usually increase the dependence between different masks. This is particularly problematic if the metal layers are involved. Rules that forbid or enforce certain overlaps between patterns in the metal layer masks and other masks affect the wiring routines, which often are generic algorithms solving some cleverly isolated interconnection problem. Introducing structural constraints often invalidates the assumptions made during the isolation.

The Input: The Functional Hierarchy.   Isolating tasks of an integrated approach is dangerous because of their mutual dependence. Taking this dependence into account by iterations over several design tasks is highly undesirable because of the time complexities involved and convergence properties. Clearly, since the final result has to be a complete specification of the masks, the later steps are mainly based on layout considerations. And, since the complete functional specification is the input to the compiler, the early decisions or refinement steps have to be predominantly based on function and testing arguments. In between, many steps, such as logic decomposition and data path definition, have a significant influence on the final layout, and many functionally almost equivalent decisions may have completely different consequences for the layout and its design process. The boundary between layout design and other tasks of a silicon compiler is therefore quite fuzzy.

The early refinement steps in a silicon compiler will lead to a functional hierarchy, probably reflecting the functional interdependence of the hierarchies (modules). Functional interdependence and connectivity are often highly correlated, and the latter is an important basis for decisions in layout design. The functional hierarchy, therefore, represents useful data for the layout design part. Besides, it is also expedient to keep this hierarchy easily recoverable, as will be argued in the following section on data integrity.

The considerations that lead to a functional hierarchy mostly ignore other important aspects of layout design. For example, in the design of digital systems the isolation and implementation of execution units is often established quite early. The remainder, control and interrupt, is left logically completely specified, but mainly unstructured. A layout with a decomposed, or even partly duplicated, control unit might be more efficient than a layout with a decomposed, or even partly duplicated, control unit might be more efficient than a layout in which this part has been kept together. Several sections of the control can be placed closer to specific execution units to which they are heavily connected. If the connectivity with the rest of the control is relatively low, this might save wiring area. It is also possible that the decomposition goes beyond what is useful for layout decisions. The layout design part may, therefore, choose to ignore parts of the decomposition, initially or throughout. Nevertheless it is assumed that, possibly after some clustering around seeds and some pruning, the design data is completely hierarchically structured. That structure is considered part of the initial data for layout design. The hierarchies and atoms are called *modules* in this context. The formal definition of a module implies such a hierarchy.

A module $\mathcal{M}$ is defined to be a collection of modules $\{m_1, m_2 \ldots, m_m\}$ where $m = |\mathcal{M}|$, and an incidence structure $IS = (\mathcal{M} \cup \{\mathcal{M}\}, \mathcal{P}, \mathcal{N})$. The modules m_i are the submodules of $\mathcal{M}$, and comodules of each other. $\mathcal{M}$ is their unique supermodule. There is exactly one module without a supermodule. This module represents the entire system to be integrated. Cells are modules with an empty set of submodules. The others are called *compounds*. The hierarchy can be represented as a rooted tree: the modules are represented by the nodes; the root represents the system; the leaves represent the cells; the internal nodes represent the compounds. Each node representing a submodule is the end of an arc that started in the node representing its supermodule.

With regard to the incidence structure IS, the module and its submodules are considered to be subsets of the set of pins $\mathcal{P} = \{p_1, p_2, \ldots, p_p\}$. Also the *signal nets*, forming $n = \{n_1, n_2, \ldots n_n\}$, are considered to be subsets of $\mathcal{P}$. *Pins* are for the moment merely a mechanism for relating modules and their supermodule with signal nets. The incidence structure can be represented by a bipartite graph $(\{\mathcal{M}\} \cup \mathcal{M} \cup \mathcal{N}, \mathcal{P})$, the *potential graph*, (See Figure 19-2.)

Cells can be of various types. The type determines the cell's flexibility and how to obtain the layout of a cell. The most rigid cell type is the inset cell. Its configuration and pin positions are fixed and stored in a master or user library or completely implied by the algorithm generating the cell. The layout design system can only assign a location and an orientation to such a cell. Other cells

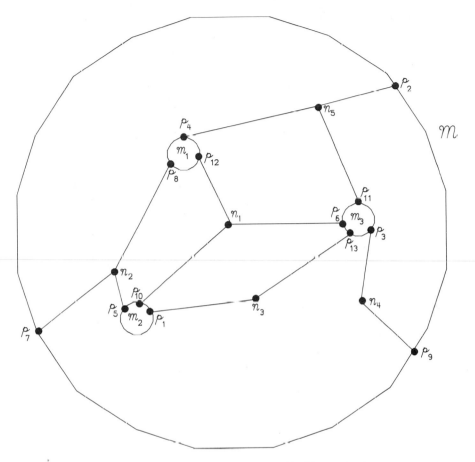

Figure 19-2.   Illustration of the concept of incidence structure.

have pointers to algorithms that can determine their layouts. These cells usually have a higher degree of flexibility. The algorithms are such that estimates about the environment can be taken into account. This certainly is true for general-purpose cells such as *macros*. These cells have a decomposition of their own into circuits of a particular family. Another, less flexible, example of a general-purpose cell is a programmable logic array. Its potential for adapting to its environment is sometimes further diminished by optimizations, such as row and column folding, that impose stricter constraints on the sequences in which nets enter the cell region. There are also cells, such as cells generated by algorithms, that, depending on a few parameters, construct special-purpose subcircuits such as arithmetic logic units, rotators, and adders. These cells have limited flexibility, such as permitting stretching in one direction. Stretchability is often important in avoiding pitch adjustments in data buses.

Data Integrity.   The single most important consideration in designing complex systems is conceptual integrity. An important aspect of this integrity is storage of the data of a design between the various stages. As pointed out in the previous subsection, the design has a hierarchical structure when entering the layout design stage. The modules in that hierarchy may have specific meaning for certain parts of the system. For example, they may have a functional model associated with them. Such a model makes possible simulation of that module in its environment. Extensive circuit simulation will have been performed on the system before the layout is considered. Yet, certain important performance aspects heavily depend on parasitic elements and final device parameters, which are not known until the layout is determined. Therefore simulation is also important during and after establishment of the layout. This requires that the simulation part be able to find the modules for which a model is known, and assign values to parameters that represent the influence of device realizations and parasitics. It is expedient that the results of the layout design process be stored in a way compatible with the data representation delivered by previous design procedures. In the preceding subsection it was established that that data is hierarchically structured. A hierarchy is mostly represented by an unordered tree. It would be convenient if the layout design procedure could preserve that structure, possibly refined and ordered. Refinement here means that leaves can be replaced by hierarchies, of which the root takes the place of the replaced leaf; and subtrees consisting of a module with all its submodules can be replaced by any tree with the same root and the same leaves, but a number of additional internal nodes.

Floorplan Design

The question of what meaningful information can be derived at intermediate stages, that is, by refinement steps in the layout phase of the design, is addressed in this section. First, some definitions that are useful for specifying flexibility in module shape are introduced. Then, the need for floorplan design is discussed. A useful restraint in generating floorplan structures is derived, and some compatible algorithms are indicated, in the next two subsections. Under the restraint of the subsection on "Floorplan Topologies," it is possible to determine the geometry of a floorplan under a variety of constraints imposed on its contour, without violating the shape constraints of the individual modules. The algorithm for this task is presented in the subsection on "Floorplan Optimization." Global wiring, that is, the assignment of nets to specific regions in a rectangle dissection, is discussed in the last subsection.

Shape Constraints.   For every cell in the hierarchy there is an algorithm that tries to adapt the cell to its estimated environment, while generating its detailed layout. This preliminary environment has to be created on the basis of estimates concerning the area needed by each cell, feasible (rectangular) shapes for it, and the external interconnection structure. The size and the shape of a cell are constrained by the amount and type of circuitry that has to be accommodated in that cell. It is reasonable to expect one dimension of the enclosing rectangle not to increase if the other dimension is allowed to increase. Constraints satisfying that requirement are called *shape constraints*. The precise definition follows.

A *bounding function* is a right-continuous, nonincreasing, positive function of one variable defined for all real values not smaller than a given positive constant.

The *bounded area* of a bounding function f is the set of pairs of real numbers (x,y) such that $f(x)$ is defined, and $y \geq f(x)$.

The *inverse* f^{-1} of a bounding function f is a bounding function defining a bounded area with exactly those (y,x) for which (x,y) is in the bounded area of f.

The *shape constraint* of a rectangle is a bounding function of one of its dimensions. The bounded area of that function is the set of all permissible pairs of dimensions. (Some examples are given in Figure 19-3.)

Useful shape constraints of compounds in the functional hierarchy can be derived in a straightforward manner from the shape constraints of cells.

Placement vs. Floorplan Design. The estimation of the rectangle in which the module is going to be realized is controlled through the shape constraints. Guidelines for the position of such a rectangle among all the other rectangles are contained in the functional hierarchy, which gives some indication about which modules belong together functionally, and in the incidence structures associated with the modules. In the context of layout, these incidence structures are often called *net lists*.

Utilizing the data (shape constraints, net lists, and functional hierarchy), the cells have to be arranged in a rectangle. This enclosing rectangle is often desired to be as small as possible, and sometimes it is constrained in aspect ratio or completely specified. If the cells were fixed objects, this would be the classical placement problem. However, in this context the cells are allowed to take any

Figure 19-3. Examples of shape constraints. The bounded area is indicated by shading. The examples are an inset cell with a fixed orientation, an inset cell with free orientation, and a cell with a minimum area and minimum dimensions.

shape not excluded by the shape constraint. This generalization of placement is called *floorplan design*.

Both floorplan design and placement are guided by a number of objectives, which are not easy to formulate in a single object function. This can be illustrated by the following typical combination of objectives. The first is primarily concerned with the realization of the interconnections. A common figure of merit for it is total wire length, often estimated by summing the perimeters of the rectangles that enclose all module centers connected to the same net. At the same time it is desirable to give the cells rectangular regions in which they can be efficiently allocated. The first objective is of a rather topological nature, working with concepts such as "close," "neighbor," and "connectivity." The latter is more a geometrical objective. Major concepts for it are "deformation," "dead area," "aspect ratio," and "wiring space." To relate the two objectives, an additional refinement step, using an intermediate structure capturing much of the data affecting one of the objectives, might be helpful.

Floorplan Topologies (Otten 1982a). It has already been observed that in the final floorplan the modules will be rectangle dissections in which each submodule is either a rectangle dissection itself or, in the case of cells, a rectangle. Creating a preliminary environment for the cells is essentially generating certain aspects of the rectangle dissection in which each cell is an undivided rectangle. Because the shape of the cells is not yet known in that stage, the geometrical details of the rectangle dissection cannot be determined. Less restrictive aspects of a rectangle dissection are its neighbor relations, that is, which cells share a particular line segment in the rectangle dissection. The set of neighbor relations is called the *topology* of the rectangle dissection. This topology is useful information that can be generated at an intermediate stage of the refinement process. Usually, enough freedom is left for the cell assembling procedures after fixing the topology, and such a topology provides useful information about the environment of the cells. Therefore, the first task is designing a floorplan is to determine its topology. A reasonable decomposition of that task, certainly in light of the discussion of the subsection on "Data Integrity," is to take one module at a time, starting with the root of the functional hierarchy, and progressing downward such that no module is treated before its supermodule is. This translates the functional hierarchies into nested rectangle dissections.

```
module list: = (root);
while module list is not empty do
    select .// from module list;
        if .// is compound
            then FLOORPLAN( .// )
            else ASSEMBLE( .// ) end;
        add submodules of .// to module list;
        delete .// from module list
    end
```

In spite of the constraints accepted so far, the floorplan design problem is still complex. For example, given its topology and the shape constraints of its cells, finding the smallest floorplan is an NP-hard problem. There also is no pseudo-polynomial algorithm for it, since the corresponding decision problem is strongly NP-complete (Stockmeyer 1983). At this point one may ask whether the class of topologies for which the previous problem, and hopefully several other problems, can be solved in polynomial time is still large enough to include an efficient floorplan topology for all practical cases. To answer that question that class has to be identified.

A concise way of representing the topology of a rectangle dissection is by its *polar graph,* a plane, directed graph without cycles. There are three bijective relations between elements of this graph and its associated dissection: edges correspond one-to-one with undivided rectangles, vertices with the elements of one set of iso-oriented line segments, and inner faces with the line segments in the other set (Figure 19-4). Many floorplans designed in practice, and all the floorplans of the successful, more special layout styles, have polar graphs that are two-teriminal series-parallel digraphs. Such a digraph is one of the following three types:

1. A digraph consisting of two vertices joined by a single arc.
2. A digraph obtained by identifying the sources and identifying the sinks of at least two two-terminal series-parallel digraphs.
3. A digraph obtained from a sequence of at least two two-terminal series-parallel digraphs by identifying the source of each (except the first one) with the sink of the preceding one.

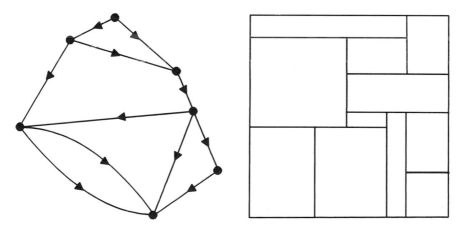

Figure 19-4.   A rectangle dissection and one of its polar graphs.

It is exactly for the class of floorplans that have such a digraph as polar graph that the above optimization problem is efficiently solvable.

A first observation is that, as with any two-terminal series-parallel graph, such a topology can be represented by a much easier to handle data structure, namely, an ordered tree. By restricting floorplans to such topologies, the procedure given earlier in this section thus automatically achieves what was found convenient in the discussion at the end of the subsection on "Data Integrity." It orders and refines the given functional hierarchy according to one of the rules there described, and the layout structure can be stored consistently in a natural way.

The tree replacing a two-terminal series-parallel graph is called the *decomposition tree* of that graph. Its leaves correspond with the arcs, and its internal nodes correspond with the two-terminal series-parallel subgraphs of the original graph. Consequently, each leaf represents an undivided rectangle, and each internal node represents a rectangle dissection, also with a two-terminal series-parallel graph. The rectangle dissections represented by the endpoints of tree-arcs starting from the same tree-vertex are placed next to each other in the same order, either from the left to the right or from the top to the bottom, depending on whether the corresponding two-terminal series-parallel graphs are connected in parallel or in series. A rectangle dissection with a two-terminal series-parallel graph as polar graph is a rectangle dissected by a number of parallel lines into smaller rectangles that might be dissected in the perpendicular direction. Such structures are called *slicing structures* (Figure 19-5), and the associated tree is a slicing tree. Each vertex represents a slice. It either contains only one cell, or is a juxtaposition of its *child slices*. In the latter case, that slice is said to be the *parent slice* of its child slices, and these child slices are the *sibling slices* of each other. The sibling slices are ordered according to their position in the parent slice (for example, left to right and top to bottom).

When a (preliminary) geometry is associated with the slicing tree, the *longitudinal dimension* of a slice is the dimension it inherits from its parent slice. The other dimension is its *latitudinal dimension*. The latitudinal dimension of the common ancestor slice is the length of the side it transmits to its child slices. The other dimension is its longitudinal dimension. The *shape constraint* of a slice is a bounding function of its longitudinal dimension. The bounded area of that function is the set of all permissible pairs of longitudinal and latitudinal dimensions.

Implementations.   There are several ways of obtaining slicing structures. A well-known method is the mincut algorithm (Lauther 1980). If applied in its pure form it leads to binary slicing trees, but one clearly can extend it to produce general slicing trees. The procedure "mincut" assigns to a set of modules a decomposition type and an ordered two-block partition. Calling the procedure with the set of all modules $\mathscr{M}$ and a given aspect ratio A yields a slicing tree T. Seed selection is fixing the positions of modules in a non-empty subset of

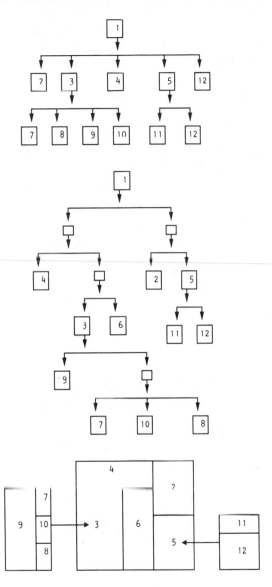

Figure 19-5.   A hierarchy, a slicing tree, and the corresponding rectangle dissection.

$\mathscr{P}$, seed1, relative to the positions of modules in another subset, seed2, disjoint from seed1. The selection can be made on the basis of previous decompositions and the type of decomposition just selected. The partition ($\mathscr{P}_1$, $\mathscr{P}_2$) must be such that $\mathscr{P}_1$ contains seed1, but not seed2, and the difference between the total area of modules in $\mathscr{P}_1$ and the total area of modules in $\mathscr{P}_2$ does not exceed a value depending on the total area of modules in $\mathscr{L}$ and the aspect ratio A. Among the partitions satisfying these requirments one with the minimum (or a small) number of nets common to both blocks is selected.

```
procedure mincut( . //: set of modules;A : real);
    var q : queue;
        seed1,seed2, ℒ, ℒ₁, ℒ₂ : set of modules;
        typ : 1,2,3; exp : 1, – 1;
    begin enqu (( . //,A),q); T : =  . //
        while not empty (q) do
            ( ℒ,B) d: = dqu(q) ;
            if | ℒ | > 1 then
            if B ⩾ 1 then
                C : = B; typ : = 2; exp : = 1
                else C : = B⁻¹; type : = 3; exp : = – 1 end;
                (seed1,seed2) : = seedselection( ℒ ,T,typ);
                ( ℒ ₁, ℒ ₂ : partition ( ℒ,C,seed1,seed2);
                T : = update-tree(T, ℒ ₁, ℒ , ℒ₂);
                enqu(( ℒ ₁,(C Σ( ℒ ₁)/Σ( ℒ ))ᵉˣᵖ),q);
                enqu(( ℒ ₂,(C Σ( ℒ ₂)/Σ( ℒ ))ᵉˣᵖ),q);
            end
        end
    end mincut;
```

The mincut method does not use an intermediate structure that captures globally a large part of the topological aspects of the input, as suggested above in "Placement vs. Floorplan Design." Each dissection divides the problem into smaller problems, but it is difficult to take into account decisions in one part when handling the other parts.

Methods using an intermediate structure are also known. One such structure is a point configuration in which the topological properties of the input are somehow translated into a closely related geometrical concept, namely, distance—and, because the configuration will be embedded in a plane, more particularly distance in the two-dimensional euclidean space. High connectivity is reflected in relatively short distances. A useful metric in this context is the so-called *dutch metric*. It assigns to each pair of modules a distance equal to:

$$d_d(m_1, m_2) = \sqrt{1 - \frac{\Sigma\{w(n)|\, n \cup m_1 \neq \phi \wedge n \cup m_2 \neq \phi\}}{\Sigma\{w(n)|\, n \cup m_1 \neq \phi \vee n \cup m_2 \neq \phi\}}}$$

(where w is a function assigning weights to the nets). It has the nice property that the ensueing distance space is always embeddable in a euclidean space, although seldom in the euclidean plane. It is, however, straightforward to construct a two-dimensional point configuration in which the distances d_p are such that $\sum (d_a^2 - d_p^2)$ is minimum. It comes down to determining the eigenvectors of the *schoenberg matrix* with the two largest associated eigenvalues (Otten 1982b). Also by calculating a partial eigensolution the weighted sum of distances can be minimized, under the constraint that the configuration must have a certain spread (Hall 1970). As discussed before, it is important in a top-down approach to take the environment into account, and the preferred distance metric in layout design is often minkowski-1, because of the orthogonal artwork required by many lithography techniques. Both eigensolution methods, however, are based on euclidean distances, and cannot take the pin positions on the perimeter into account. In the section "Annealing Applied to Floorplan Design" (below), a method for obtaining a point configuration will be described that does not have these two disadvantages.

Properties of the final rectangle dissection have to be derived from the point configuration and the shape constraints. The topological considerations should be taken into account by preserving relative positions in the point configuration and keeping modules close together if they are represented by points with short distances between each other. The geometrical aspects should be taken care of by keeping track of, for example, deformation implied by the dissections. Shrinking is a method that efficiently achieves these goals for slicing structures. For reasons of clarity its discussion will be limited to designs with exclusively flexible modules. Mostly, shapes close to square are preferred for flexible modules because square shapes in general require less wiring space. Now, assume that a given slice $\mathcal{B}$ with $\mathcal{B} \geq 2$ modules is to be sliced in a number of child slices. $\mathcal{B}$ is separated from its sibling slices by lines parallel to one axis, and the slicing must now be tried parallel to the other axis. Supported by Figure 19-6, a visualization follows of a process for determining the shrink factor of possible slicing lines with a given orientation.

Think of a module as a square around the point representing it, with its sides parallel to the axes. The sizes of these squares are such that each pair of squares has overlap, and the ratio of their areas is equal to the ratio of the area estimates given to the corresponding modules. The squares are simultaneously shrunk, leaving their centers fixed and preserving the area ratios. At some point during the shrinking process there will be a line with a given orientation that divides the modules into two blocks without intersecting any of the squares. The amount of shrinking necessary to reach that point is the shrink factor to be assigned to the corresponding line. Shrinking is continued, and another line with the same orientation and separating other blocks will be intersection-free. Again, the corresponding factor will be assigned to that line. The process can be continued until all $| \mathcal{B} | - 1$ factors are assigned. The lines with relatively high shrink factors are candidates for slicing lines. The accepted lines partition the set of modules into blocks, each of which is treated in the same way. The selection of the lines usually takes other aspects into account. One of these, *deformation,*

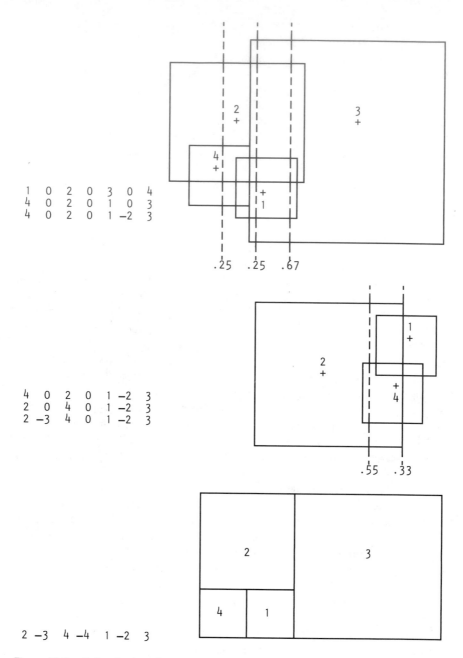

Figure 19-6. Determination of shrink factors. In this small example the squares around the module centers are drawn in their position before the shrinking process. The amounts of shrinking necessary to have the vertical dashed lines intersection-free are given for each line. After acceptance of the line with highest shrink factor, the three modules at the left are treated in the same way. The consequences for the shorthand tree are given for each step. The final tree and configuration complete the picture.

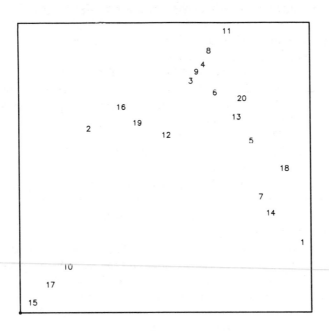

2 -4 16 -5 19 -3 12 -3 15 -4 17 -4 10 -2 8 -5 4 -5 9 -5 3 -4 11 -5 6 -5 13 -6 20 -3 5 -4 18 -3 7 -5 14 -4 1

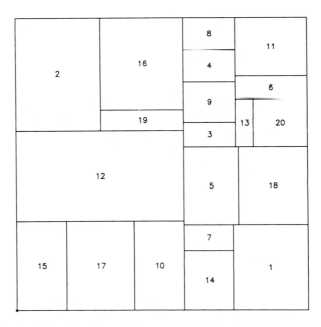

Figure 19-7.   A point configuration with a consistent shorthand tree and rectangle dissection.

will be discussed after introduction of the *shorthand tree,* a data structure that keeps track of the lines that are accepted.

The shorthand tree is useful for storing partial structure trees. It is a one-dimensional array with $2|\mathcal{M}| - 1$ integers. The modules are represented as positive integers. After completion of the slicing structure, these positive integers will be separated from each other by $|\mathcal{M}| - 1$ negative integers representing the slicing lines. The values of these negative integers indicate the level of slicing. Initially all $|\mathcal{M}| - 1$ positions for the slicing indicators are filled with zeros. The positive integers are ordered according to the coordinates of the corresponding points on the axis perpendicular to the first set of slicing lines. At the corresponding positions in the array, the zeros are replaced by a -2 (0 and -1 are reserved for separating eventual bonding pad macros). In each of the subsequent steps the groups of (two or more) positive integers separated by negative integers are treated in a similar way. They are ordered according to the coordinates not used in the preceding step, and the positions corresponding to the accepted slicing lines are filled with an integer 1 lower than the indicator of the preceding step. The algorithm continues until all zeros in the shorthand tree are replaced by negative integers. The concept is illustrated in Figure 19-6, and Figure 19-7 shows a point configuration, the shorthand tree obtained from that configuration, and a corresponding rectangle dissection.

The preferred shape for flexible modules is a square. It is, however, in general not possible to accommodate the modules of a set $\mathcal{B}$ as squares in a slice with area

$$\Sigma(\mathcal{B}) = \sum_{m \in B} \alpha(m),$$

$\alpha(m)$ being the area estimate for module m, to say nothing of such a slice with given dimensions, l_1 and l_2. To assess the deviation from square form of the modules in $\mathcal{B}$, *deformation* is defined as follows:

$$\mathrm{def}(\mathcal{B}, l_1, l_2) = \sum_{m \in A} \left(\frac{\sqrt{\alpha(m)}}{\min(l_1, l_2)} + \frac{\min(l_1, l_2)}{\sqrt{\alpha(m)}} - 2 \right)$$

$$+ \ \mathrm{def}(\mathcal{B} \setminus \mathcal{A}, \max(l_1, l_2) - \frac{\Sigma(\mathcal{A})}{\min(l_1, l_2)}, \min(l_1), l_2))$$

where $\mathcal{A} = \{ m \in \mathcal{B} | \alpha(m) > (\min(l_1, l_2))^2 \}$, can be used. When $\mathcal{A} = 0$ the deformation, as defined, will be zero. This does not mean that all modules of $\mathcal{B}$ fit into an $l_1 \times l_2$ slice as nonoverlapping squares; it only is an indication that the modules are small in comparison with that slice, and a reasonable packing can be expected.

```
procedure lot( . //: set of modules;T:slicing tree;A : real);
   var q :queue; s,ss : slice;
   begin enqu (ancestor,q);
      dim[ancestor].long : = √ A * Σ( . //) ;
      dim[ancestor].latt : = √ Σ( . //)/A ;
      while not empty(q) do
         s : = dqu(q) ;
         for each child ss of s do
            dim[ss].long : eq dim[s].latt;
            dim[ss].latt : = dim[s].long * Σ(ss)/ Σ(s) ;
            if not leaf(ss)
            then enqu(ss,q) end;
         end
      end
   end lot;
```

To use this definition of deformation it must be possible to calculate the dimensions of the slices in a partial structure tree, given that tree, the area estimates of the modules, and the preferred aspect ratio of the supermodule. This is quite easy when all modules are flexible. The calculation starts with deriving the outer dimensions from the total area and the aspect ratio. One of these dimensions is inherited by the child slices of the ancester. The other dimensions of these child slices are obtained by dividing the sum of the areas of the modules in each of them by the inherited dimension. Proceeding in this way, each slice inheriting the dimension calculated for its father slice will yield the dimensions of all slices.

As the computation is performed top-down, it can also be applied to partial slicing trees. So, for every decision, proximity information (shrink factors) and shape information (deformation) are available. For floorplans with a large number of modules not too different in size, the first slicing lines are mostly determined on the basis of proximity information. Only later, when few modules have to be allocated in a slice, will deformation usually dominate the selection. In the final stage, one may even consider determining the deformation incurred by all possible slicing trees (because the ordering does not affect the deformation, one should generate only one tree out of every isomorphism class) that can be formed with the few modules that go into such a low-level slice.

Generalization of the process presented here for flexible modules with a given area estimate is not very difficult, but a procedure calculating the dimensions of each slice in a partial slicing tree is needed. Such a procedure will be presented in the next section.

Floorplan Optimization. Given the slicing tree, and the shape constraints of the modules (represented by the leaves of that tree), what is the best rectangle dissection corresponding to that tree? A procedure will now be described that produces such a geometry if "best" can be specified in terms of a contour score. A *contour score* c is a function of two variables, defined for a convex subset Γ of the pairs of positive real numbers, that is quasi-concave and monotonously nondecreasing in its two arguments; that is,

$$\forall_{x_0,x_1,y_0,y_1\in\Gamma}[x_1 \geqslant x_0 \wedge y_1 \geqslant y_0 \rightarrow c(x_1,y_1) \geqslant c(x_0,y_0)]$$

and:

$$\forall_{x_1,x_2\in\Lambda} \forall_{0\leqslant\lambda\leqslant1} [c(x_1) \leqslant c(x_2) \rightarrow c(x_1) \leqslant c((1 - \lambda)x_1 + \lambda x_2)]$$

Area and perimeter are examples of contour scores. Therefore, it is possible, using the procedure of this section, to construct the smallest rectangle dissection with a given slicing topology and given shape constraints for its undivided rectangles. Also, the smallest with a given aspect ratio, or with a lower and upper bound on the aspect ratio, can be produced. The complexity of the algorithm given is polynomial. As mentioned under "Floorplan Topologies," this problem is NP-hard for more general dissections.

Inset cells have piecewise linear shape constraints. Such constraints can be conveniently represented by a sequential list of their breakpoints. This is not the case for flexible cells, and possibly other cell types occurring in practice. Of course, any shape constraint can be approximated by a piecewise linear bounding function with arbitrary accuracy. From the discussion of flexible cells in the preceding paragraph, it is clear that a piecewise linear approximation with three breakpoints suffices, considering the limited accuracy of any area estimation for the given examples (Figure 19-8).

The shape constraint of a compound slice can be derived from the shape constraints of its child slices. In the final configuration, these child slices have to have the same longitudinal dimension, which is the latitudinal dimension of their parent. The inverse of the compound's shape constraint is defined only on the intersection of the intervals on which the shape constraints of its children are defined. Its smallest possible longitudinal dimension for a given feasible latitudinal dimension x is the sum of the values of the shape constraints of the children at x. So, the shape constraint of a compound is obtained by addition of the shape constraints of its children in the interval where they are all defined, and determination of the inverse of the resulting bounding function. These operations are easy for piecewise linear shape constraints, represented by a list of their breakpoints ordered according to the respective longitudinal dimensions. For each breakpoint of any child of which the first coordinate x is in the mentioned intersection, the shape constraints of all the children have to be evaluated and added. If the result is y, then (y,x) is a breakpoint of the parent's shape constraint. Ordering all these new breakpoints according to the y-values yields a consistent representation for the shape constraint of the corresponding compound slice.

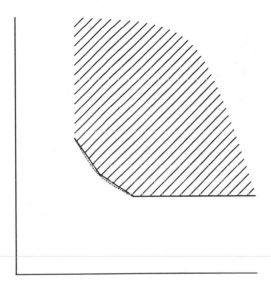

Figure 19-8.   Piecewise linear approximation of the shape constraint of a module with given minima on area and dimensions.

The ability to obtain the shape constraints of a slice by adding the shape constraints of the child slices and inverting the result enables the shape constraint of the enveloping rectangle. The bounded area of that shape constraint is the set of all possible outer dimensions of the total configuration. A contour score always assumes its minimum value over the bounded area at the boundary determined by associated shape constraints. This is a consequence of the monotonicity of shape constraints and contour scores. For piecewise linear shape constraints, that minimum will be assumed at at least one of its breakpoints because of the quasi-concavity of the contour score. So, to find an optimum pair of dimensions for the common ancestor slice, the contour score only has to be evaluated at the breakpoints of its shape constraint in the convex set of permissible pairs.

Given the longitudinal dimension of a slice and its shape constraint, its lati-tudinal dimension can be found by evaluating its shape constraint for the given longitudinal dimension. After derivation of the shape constraint for the common ancestor and determination of a dimension pair for which the contour score assumes a minimum, the longitudinal dimensions of its children are known. So, for each of them the latitudinal dimension, which in turn is the longitudinal dimension of its children, can be calculated. Continuing in this way will finally yield the dimensions of all slices in the configuration. If the shape constraint has a zero right derivative at the point where it has to be evaluated, some slack area will be included; that is, the slice can be realized in a smaller rectangle without affecting its environment. In order to have the wiring channels connect

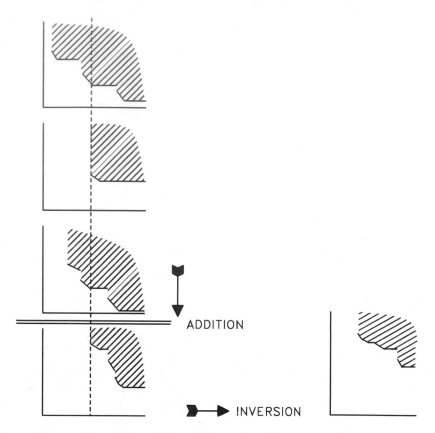

Figure 19-9. Addition and inversion of shape constraints.

to other wiring channels at both ends, this slack should be taken up by slices containing only one cell.

The algorithm consists of three parts:

1. Visit the nodes of the topology in depth-first order, and just before returning to the parent determine the shape constraint by adding the shape constraints of its children and inverting the result.
2. Evaluate the contour score for each of the breakpoints of the shape constraint of the common ancestor, and select a dimension pair for which the smallest value of the contour score has been found.
3. Visit the nodes of the topology in depth-first order, and before going to any of its children determine the latitudinal dimension by evaluating its shape constraint for the inherited longitudinal dimension.

Clearly, when in the first step the same procedure has been applied to all children of a certain slice, the shape constraints of these child slices are known, and combining them in the way described yields the shape constraint of their parent. The process will end with determining the shape constraint of the common ancestor slice, and then the shape constraints of all slices are known. As explained earlier, the contour score will be evaluated at each of its breakpoints. The dimensions associated with the minimum value will become the dimensions of the enveloping rectangle. Thus, after completion of the second step the longitudinal dimension of the primogenitive (and all the other children) of the common ancestor is known. Together with the shape constraints, this is enough information to begin the process of the third step. At the beginning of a visit to a node in the structure tree, representing a certain slice, the latitudinal dimension of that slice can be determined by evaluating its shape constraint at the value of the dimension that it inherits from its parent slice.

So, completing all three steps yields the dimensions of all slices in an optimum configuration for the given topology and cell shape constraints. Determination of the position coordinates of the slices from these dimensions and the topology is straightforward. Also it is easy to determine what orientation the inset cells can have in this optimum configuration.

Net Assignment (van Ginneken and Otten 1985).   The more complex the circuit to be integrated, the more dominant the wiring is in the final layout, as can easily be learned by examining existing integrated circuits. Though part of the wiring can be realized on top of active devices, particularly when there is more than one metal layer, a considerable portion of the chip, called the wiring space, is used exclusively for the realization of the incidence structures of the modules. If the cells are realized in iso-oriented rectangular regions, the wiring space can be seen as the union of nonoverlapping rectangles. The selection of the rectangles that together form the wiring space affects the efficiency of the wiring procedures. It constrains the sequence in which the wiring can be generated, and determines the algorithms that can be used for the wiring, as well as the number of different algorithms needed to perform that task.

```
procedure dfs1(s : slice);
    var : ss : slice; q : queue;
begin for each ss child of s do
    enqu(ss,q);
        if ss ≠ leaf then dfs1(ss) end
    end;
    ss : = dqu(q);
    F[s] : = F[ss];
    while not empty(q) do
        ss := dqu(q);
            F[s] : = addition(F[s],F[ss])
        end;
    F[s] : = inversion(F[s])
end dfs1;
```

```
procedure dfs2(s : slice; x : dimension);
   var ss : slice; y : dimension;
begin y : =  value(F[s],x);
   dim[s].long : =  x;
   dim[s].latt : =  y;
   for each ss child of s do dfs2(ss,y) end;
end dfs2;

procedure value(FF : breakpointlist; x : dimension) : dimension;
      {evaluates FF at x};

procedure minloc(FF : breakpointlist) : dimension;
         {returns the abscissa of the breakpoint
            with the optimum contour score}
begin dfs1(ancestor);
      chipl : =  minloc(F[ancestor]);
      dfs2(ancestor,chipl)
end.
```

Here also, slicing structures offer considerable advantages over general rectangle dissections. In the first place, they imply a decomposition of the wiring space into the minimum number of rectangles, which are in a one-to-one correspondence with the slicing lines. To distinguish these undivided rectangles from the ones that correspond with the cells in the functional hierarchy (in the slicing tree both kinds are represented by leaves), they are called *junction cells*. Second, feasible sequences for generating the wiring can easily be derived from the slicing tree. A possible rule here is to prepare the junction cells in a sequence based on the length of the path from the leaf representing the junction cell to the root of the slicing tree, such that the longer this path, the earlier the wiring in that cell has to be generated. And third, all these rectangles can be wired by using the same kind of algorithm, usually called a channel router, though not necessarily one in the strictest sense (Otten 1982a).

As the wiring consumes quite a high percentage of the total area, it is useful to have early estimates for this space, so that the sequence of floorplan calls can take these estimates into account in the design of the nested floorplans. Several objectives may be important in realizing the interconnections, many of which are directly related to the size of these nets. Two problems arise immediately. The first one is a consequence of the interpretation of a floorplan as a topology rather than a geometrical configuration. Yet, in order to measure the size of a net, some metric is necessary. The second problem is the need for an ambience in which the wide diversity of objectives can be formulated and optimized.

An often used structure for approaching these problems is the plane graph determined by the rectangle boundaries in the rectangle dissection. Each rectangle corner is a vertex, and the line segments between them are the edges. This graph depends on the geometry of the rectangle dissection, and this geometry is not known in the floorplan design stage. A closely related graph can be defined for slicing structures, for there is a one-to-one correspondence between the junction cells of slicing structures with the same topology. Also the relation τ generated by the T-intersections between junction cells is invariant over the rectangle dissections with the same slicing tree (Otten 1982a). The expression $jc_1 \tau jc_2$ indicates that a junction cell jc_2 intersects with a higer level junction cell jc_1. τ can be stored in a matrix J. This matrix is useful for finding the identity of a T-junction between two junction cells. It has rows for the horizontal junction cells and columns for the vertical junction cells. If $jc_1 \tau jc_2$, then the identifying number of this T-junction is stored in the corresponding matrix position; otherwise there is a 0. The rows and columns of J can be ordered according to the estimated coordinates of the associated junctions cells. (The pattern of nonzeros in J then gives a picture of a rectangle dissection having the correct floorplan.) This ordering, of course, is not independent of geometry.

A graph known as the *basic graph* also depends on the estimated geometry. The quantity τ is its set of vertices, and there is an edge between two vertices if the corresponding T-junctions are on the same junction cell, and there is no T-junction in between. Each edge has a length equal to the difference between the coordinates of the two junction cells it connects. A distance matrix D can be derived from this graph. It is a square matrix with a row and a column for each vertex in τ. Each entry is a number equal to the length of the shortest path between the associated vertices. The lengths of the shortest paths can be conmputed from the single edge distances by the Yamada–McNaughton algorithm, slightly modified for speed-ups possible because of the properties of slicing structures. All paths from vertices in one slice to vertices in an abutting sibling slice have to cross vertices on the slicing line that separates them, so the distances within a compound slice can be computed recursively in a bottom-up fashion. Note that the triangle inequality holds for the distance space thus created.

The above data structures are used in the following net assignment procedure. It proceeds on a net-by-net basis, in an arbitrary sequence. Each time that a net is routed, extra vertices, representing the modules this net has to connect, are added to the graph. Extra edges connect each new vertex with the T-junctions that surround its module. Pins that can occur only at one side of the module only get the edges that lead to the vertices on that side. The length of the extra edges is used to represent the possibility of duplicating a pin. For a porous module the length of the extra edges is short, to make it possible for the net to pass through the module. If this is not allowed, then the edges are made long enough that passing through the module will be avoided. Also extra vertices are added that represent the pins of the environment. Modules might have multiple pins on their periphery that connect to the same net. In that case, a vertex is added to the graph for every pin. The problem to be solved by the net router for each net is to determine a connected subgraph of the extended basic graph. This subgraph has to contain all vertices representing modules to be connected by the net, and the sum of the lengths of its edges has to be small. This subgraph is a Steiner tree if that sum if minimum. After the net has been routed, the extra vertices and edges are deleted.

This effectively isolates the net routing problem for each net, for an environment is created in which any Steiner tree heuristic can run. The input of the net router is the distance matrix D of the extended basic graph, and the subset of vertices (modules) to be connected. It returns a number of vertex pairs. The edges on a shortest path between vertices of each pair are edges of the connecting subgraph. Each edge corresponds with an interval in the associated junction cell. It is possible that there are several shortest paths. This freedom can be used for optimizing another criterion, for example, controlling the density in the junction cell.

Any net that passes through a module is split into several nets. Thus every new net has only one pin to a module. To make this possible, a new netlist has to be constructed during the net assignment. The new nets will inherit properties and identification from the net from which they originated.

The net assignment is stored as a number of intervals assigned to the junction cells that contain them. If the segment ends at a module vertex, then the module is recorded as the terminal. If it ends at a T-junction, then the orthogonal junction cell is the terminal. The exact positions of the terminals are determined by the algorithms that generate the modules and route the junction cells. A net in a junction cell is represented by a number of intervals. If those intervals belong to the same net, and are to be connected by the channel cell router, then they have an endpoint in common. In this way the net can never become unconnected because of updates in the sizes of the modules afterwards. However, it can happen that a loop is created if the sizes of modules change considerably. This can occur with two segments in the same junction cell that were originally connected outside the junction cell. If the sizes change very much, the segments can shift until they overlap.

When an interval of a net connects to a module, the side on which the pin should preferably be placed is indicated. This information is used by the algorithm that generates the modules to adapt the positions of the pins to external wiring. The pins are given a position on one of the sides. The net enters the module in the graph via a vertex on its boundary, often one of the vertices on the corners of the module. If the entrance vertex is on one of the sides, then that side is chosen. If it is on one of the corners, we choose the side facing the highest-level junction cell of the entrance vertex. The only exception occurs when a two-terminal path runs from one module to another module, and they are abutting. In this case, the side facing the junction cell that separates them is always chosen.

```
procedure spantree (in: r, D, C; out: T(V,B,r) );
var p, q, s, S;
begin
    V : = {r}
    S : = C \ {r};
    B : = { };
```

```
while not (S = { }) do
    (p,q):→(p,q)ε((V∩C) × S)∧
        ∀{(x,y)|(x,y)ε((V∩C) × S)}[D[p,q]≤D[x,y]]
    S : = S \ {q};
    V : = V U {q};
    if B = { } then B : = (p, q)}
        else newvertex (s); V : = V U {s};
        B: = (B\{(p,pred(p))})U{(pred(p),s),(s,p),(s,q)};
    end
end
end spantree;
```

If minimization of the total wire length is the important objective, a Steiner tree on the above structure is wanted. Though slicing yields a considerable saving in computation time, this problem still requires exponential worse-case time (if $P \neq NP$). Therefore, one has to restort to procedures yielding a small connecting subgraph. The edge set of this subgraph is denoted by W in the pseudo-programs given at the end of this section. The set of vertices that have to be in that subgraph is denoted by C. Most existing methods for finding Steiner trees, exact and heuristic, are based on either a shortest spanning tree algorithm, or a three-point Steiner tree algorithm. The heuristic presented here uses both algorithms in two stages. First it determines the topology class of trees, and then it selects the optimum tree in that class.

The topology class is represented by a binary tree $T(V,B,r)$, with vertices V, branches B, and root $r \in V$. The binary tree itself is a member of the topology class. The other topologies are obtained by contraction of edges of the tree T. The leaves and the root of the bonary tree correspond to the vertices in C. The remaining internal vertices are candidates for Steiner points. In the final result some of these vertices can coincide, yielding a tree with fewer Steiner points.

An algorithm that finds the shortest spanning tree in a complete graph is used to determine the topology class. The vertices of that graph are the vertices in C. The length of an edge is the length of the path in the extended basic graph as given by the matrix D. The shortest spanning tree can be found in $O(|C|^2)$ time. The shortest spanning tree cannot be longer than twice the length of the Steiner tree. While the algorithm determines the spanning tree, it constructs a "matching" binary tree. Matching means that the spanning tree has a topology that can be derived from the topology of the binary tree by contraction of some edges. This is done by adding three edge "stars" instead of edges. The shortest edge connecting a vertex p already in the tree with a vertex q not yet in the tree is found. The edge between the p and the vertex one edge closer to the root, $pred(p)$, is removed, and the new vertex q is connected by means of a star that takes the place of the removed edge.

For the second stage, it is irrelevant how the binary tree was found. The algorithm finds the optimum tree in the topology class determined by the binary tree in $O(n^2|C|)$ time, where n is the number of vertices in the extended basic graph. If there exists a Steiner tree with a topology in the determined class, the result will be a Steiner tree. If it is not, then still the shortest tree with this topology is found, which is usually not much longer than the optimum.

This stage consists of two recursive depth-first search visits to the tree. The first one determines the length of all subtrees connecting to any vertex. From the information provided by this procedure, the length of the tree is determined. The second procedure traces back the shortest result in this tree, while choosing the Steiner points.

After assigning all nets to the junction cells, we can determine densities in the junction cells, and on that basis a fairly accurate estimation of the wiring space is possible. This can be used in the floorplan optimization by treating the junction cell as an inset cell with fixed orientation and minimum longitudinal dimension equal to zero. The latitudinal dimension must be greater than the width estimated on the basis of the cell's density. In addition to obtaining accurate estimates for the shapes of the rectangles, the net assignment also yields information about the location of external nets for the floorplans and cells to be designed later.

Cell Assemblers

The refinement steps described in the previous section determine a topology for every compound of the functional hierarchy. If these topologies are restricted to slicing structures, floorplan design replaces each subtree of which the vertices represent a certain module and all its submodules by another tree of which the root represents the selected module, and the leaves represent its submodules. This is in accordance with one of the rules suggested at the end of the section on "Data Integrity." The other type of refinement considered there was the replacement of a leaf in the functional hierarchy (a *function cell*) by a tree decomposition. The reasons for not having this decomposition in the initial tree can be quite diverse. For example, the decomposition suitable for the functional design may be be far from optimal for layout design. In that case, such a hierarchy is pruned. Clearly, a data base problem has to be resolved when this happens. It might also happen that the decomposition is suitable for using in the layout program, but more specialized programs are needed than the general floorplanning scheme described. Most often, however, there is no need for further decomposition from the functional design point of view, but flexibility is increased if the layout design procedures use some inherent decomposition.

Algorithms for designing cells, possibly using such a decomposition, are called *cell assemblers*. The task of a cell assembler is to determine the internal layout of its cell (with respect to a reference point in that cells' region) on the basis of a suitable specification and data bout its environment. There may be quite a diversity of cell assemblers in a silicon compiler system. The application range of the silicon compiler is highly dependent on the set of implemented cell assemblers. Whereas most of the decisions during floorplan design are to a high degree technology-independent, cell design is dominated by the possibilities and limitations of the target technology. The numeric design rules are stored as numbers whose value is assigned to certain variables in the cell assembler. The structural rules are to be incorporated in the algorithms, if possible in the form of case statements, so that a variety of rules can be satisfied.

```
procedure select (in: D[1..n,1..n], T(V,B,r); out: W );

procedure up (t); var i;
    if leaf(t) then
        for i: = 1 to n do t.v[i] : = D[t,i] end
    else up(left(t)); up(right(t));
        for i: = 1 to n do
            t.v.[i] : = left(t).v[i] + right(t).v[i] end;
        for i: = 1 to n do
            t.v[i] : = min{s|1≤m≤n∧s = t.v[m]+D[i,m]} end
    end
end up;

procedure down(t,k); var i;
    if not leaf (t) then
        i :→ t.v[i] = min {t.v[j]|1 ≤ j ≤ n};
        W : = W U { (i,k) };
        down(left(t),i);
        down(right(t),i);
    end
end down;

W : = { }
up(ι);
s :→ (r,s)εB
down(s,r);
end select;
```

The layout of a slice is obtained by first obtaining the layout of all its child slices except junction cells, and then calling the appropriate assemblers for the junction cells. Visiting the slicing tree in depth first order and performing the above operations when returning to the parent slice enables the program to determine the "chip" coordinates (coordinates with respect to a unique point on the chip) of all layout elements in the parent slice before leaving the corresponding vertex. The translation of that result into the rectangles of the various masks is also performed at this point. This translation is straightforward.

In the following two subsections, some types of cell assemblers are described. The first is devoted to function cells, with special attention to their possible decomposition. The second deals with realizing given interconnection lists in a junction cell.

Function Cells. The task of a cell assembler is extremely simple for inset cells whose internal layout is stored in a master or user library. From the topology determined in the floorplan design process and the shape constraints of all cells, including the junction cells, an estimate can be derived for the rectangle that is going to accommodate that module. Reasonably accurate data about the position of the nets to be connected to that cell is generated by the net assignment process. On the basis of that data, the assembler has to decide which orientation must be given to the inset cell, and how it must be aligned with its sibling slices.

For function cells whose internal layout is not stored in a library, it may still be implied by the specification. For example, if a cell is going to be realized as a programmed logic array, the specification is either a personality matrix or a set of Boolean expressions. In the former case the assembler does nothing else than perform a straightforward translation and handle the result in the same way as the stored inset cells. If the array is specified by a set of Boolean expressions, the assembler must have a so-called pla-generator. The shape of the resulting array is still difficult to control, but the pin positions can be adapted to the results of the net assignment, performed during the floorplan design stage. Some sophisticated pla-generators use techniques such as row and column folding to make the area of the array smaller. This constrains the choice in pin positions considerably, and might lead to a higher area consumption because of the complex wiring around that array. A pla-generator in a cell assembler should at least be able to take the results of the net assignment into account (Demicheli and Sangiovanni-Vincentelli 1983).

Regular arrays such as programmed logic arrays and memories have an obvious decomposition into array cells with no or only slight variations in their dimensions. Their positions in the array are heavily constrained, so this decomposition cannot be used to manipulate the shape of the array. There also are cells that have a natural or given decomposition that can be used for that purpose. These cells are called *macros*. They are decomposed into circuits that either are selected from a prespecified catalogue or can be designed with a simple algorithm from a function specification. The reason for keeping such a macro from the floorplan design stage is that the circuits have certain properties that make special layouts very efficient. For example, the catalogue, or the simple algorithm, may have a constraint that gives all cells in the macro the same width and the same positions for general supply pins, such as power supply and clock pins. In that case a polycell layout style is suitable for the macro. It forces the cells to be distributed over columns, but the number of columns can be chosen freely. Therefore, the aspect ratio of the macro can be influenced. Also the pin distribution around the periphery can be prescribed on the basis of data about the environment (Brayton et al. 1984).

Decompositions like those in macros occur very often, but for special cells that are frequently used, it sometimes is worthwhile to implement a special algorithm producing a highly optimized layout. In word-organized digital systems these special cells often process a number of bit vectors. The layout as a whole may benefit from aligning these cells so that the buses carrying these bit vectors do not have to be matched to the pitch of each individual cell. Also buses that pass over such a cell without making any contact have to be accommodated. These requirements imply a certain kind of flexibility, such as stretchability and variable bus pitches. If possible, such an algorithm must be able to produce these highly optimized cells for several bus dimensions and a range of performance requirements.

Junction Cells. Junction cell assemblers are closely related to channel routers (Burstein 1983) becuase of the way they are isolated and the moments on which they are called. The junction cell is a rectangular area of which the latitudinal sides are parts of the longitudinal sides of junction cells that are represented in the slicing tree by vertices closer to the root. When the assembler is called for a certain junction cell, the longitudinal coordinates of the entry points of the nets are known. There are several ways a net may enter the junction cell: from the longitudinal sides of that cell, from a higher metal layer, from the latitudinal sides, and perhaps in still other ways. The task of the assembler is to realize all the required interconnections in a rectangle with an as small as possible latitudinal dimension. The longitudinal dimension has to be commensurate with the latitudinal dimension of the parent slice, so increasing the longitudinal dimension of the channel should be avoided if possible.

THE ANNEALING ALGORITHM

Simulated Annealing* was introduced into layout design by Kirkpatrick et al. (1983). Their description, however, leaves several gaps that have to filled in order to use the ideas of annealing in an automated environment. One of these is the schedule, that is, the number of configurations evaluated at the subsequent values of the control parameter, and the decrements between those values. The schedule has a big influence on the performance of the algorithm. The second section of this discussion describes a way of automatically adapting the schedule to individual problems. In the first section the behavior for constant values of the control parameter is described, and some quantities that are used to control the annealing are introduced.

Annealing Chains

Annealing chains are Markov chains with fixed transition probabilities; that is, the transition matrix of the chain does not vary in time. The probabilities satisfy a number of requirements, indicated in the text by a Greek prefix. From these requirements the equilibrium properties of the chain are derived. Three aggregate

---

*C. D. Gelatt, Jr, e.a., "Optimization of an Organization of Many Discrete Elements," patent issued.

functions are defined for such a chain. The influence of the control parameter is chosen in such a way that simple relations between the aggregate functions are enforced.

Definitions. An annealing algorithm works on a state space, that is, a finite set $\mathscr{L}$ of s states, on which a topology is defined. Each state represents an encoding of a configuration. A *score function* ε: $\mathscr{L} \to R^+$ assigns a positive real number to each state. This number is interpreted as a quality indicator, in the sense that the lower this number, the better is the configuration encoded in that state. By defining a set μ of neighbor relations over $\mathscr{L}$ (i.e., $\mu \subseteq \mathscr{L} \times \mathscr{L}$), a topology is endowed to the state set $\mathscr{L}$. The elements of μ are called *moves,* and the states in $(s, s') \in \mu^k$ are said to be connected via a set of k moves. The relation μ is required to be symmetric and antireflexive, mainly because it makes analysis of annealing chains and controlling the annealing algorithm easier.

$$\forall_{s \in \mathscr{L}} \forall_{s' \in \mathscr{L}} [(s, s') \in \mu \to (s', s) \in \mu] \tag{μ1}$$

$$\forall_{s \in \mathscr{L}} [(s, s) \notin \mu] \tag{μ2}$$

The transitive closure of μ has to be the universal relation over $\mathscr{L}$.

$$\bigcup_{k=1}^{\infty} \mu^k = \mathscr{L} \times \mathscr{L} \tag{μ3}$$

This is obviously necessary if each pair of states has to be connected via a number of moves. The *diameter* $\phi(\mathscr{L})$ of a space $\mathscr{L}$ in which μ is satisfied is defined as the smallest integer such that:

$$\forall_{(s, s) \in \mathscr{L} \times \mathscr{L}} \exists_{k \leq \phi(\mathscr{L})} [(s, s') \in \mu^k]$$

A *selection probability* β: $\mu \to (0,1)] \subset R$ is assigned to each move. Of course,

$$\forall_{s \in \mathscr{L}} \left[\sum_{s' \in s\mu} \beta((s, s')) = 1 \right] \tag{β1}$$

but β is also required to satisfy:

$$\forall_{(s, s') \in \mu} [\beta((s, s')) = \beta((s, s))] \tag{β2}$$

for reasons similar to those for introducing (μ1) and (μ2).

Another probability function, called the *acceptance probability,* assigns a number to a pair of scores. The number not only depends on the values of these scores, but also on a positive real number, the *control parameter*. This function $\alpha: R \times R \times R^+ \rightarrow (0,1] \subset R$ has the following properties:

$$\forall_{(\varepsilon,\varepsilon')\in R^2} \forall_{t\in R^+}[\varepsilon \geq \varepsilon' \rightarrow \alpha(\varepsilon,\varepsilon',t) = 1] \tag{α1}$$

$$\forall_{(\varepsilon,\varepsilon')\in R^2} \forall_{(\varepsilon',\varepsilon'')\mathbf{R}^2} \forall_{t\in R^+} [\varepsilon \leq \varepsilon'$$
$$\leq \varepsilon'' \rightarrow \alpha(\varepsilon,\varepsilon',t)\alpha(\varepsilon',\varepsilon'',t) = \alpha(\varepsilon,\varepsilon'',t)] \tag{α2}$$

(α1) is adopted more for historical reasons than out of necessity; (α2) has the same background as (μ.1–2) and (β2).

The following (infinite) Markov chain will be called the *annealing chain*. It uses the above probability functions to move from one state to another.

```
select presentΔstate
    from ℒ
    at random;

loop
        select nextΔstate
            from presentΔstateμ
            according to β;

        if random([0,1]) < α(ε(presentΔstate), ε(nextΔstate), t)
        then presentΔstate: = nextΔstate end

    end;
```

The Equilibrium Density.   the relative frequency of steps in the annealing chain with presentΔstate = s is denoted by $\delta(s,t)$. The function δ: $\mathcal{P} \times R^+ \rightarrow [0,1]$ is called the *equilibrium density*. An explicit expression can be obtained for δ under the requirements imposed on the move set and the selection and acceptance function. The formalism using a nonnegative matrix P_{ss}, the so-called *transition matrix*, to describe the Markov chain is convenient in deriving that expression. If the states are numbered in such a way that the score function is monotonously nondecreasing with the state index, this matrix has the following entires off the diagonal:

$$i < j: p_{ij} = a_{ij} b_{ij}$$
$$i > j: p_{ij} = b_{ij}$$

where $a_{ij} = \alpha(\varepsilon(s_i),\varepsilon(s_j),t)$ and $b_{ij} = \beta((s_i, s_j))$ if $(s_i, s_j) \in \mu$, and $b_{ij} = 0$ if $(s_i, s_j) \notin \mu$. The entries on the diagonal are chosen such that the matrix becomes a (row-)stochastic matrix; that is, all row sums are equal to 1.

Moreover, P is primitive, because the transitive closure of μ is required to be $\mathscr{L} \times \mathscr{L}(\mu 3)$, and $p_{ij} > 0$, whenever $(s_i, s_j) \in \mu$. Consequently, the famous theorem of Perron-Frobenius applies (Wielandt 1950). The equilibrium vector d_{1s} is, therefore, the unique solution of:

$$d_{1s}P_{ss} = d_{1s}$$
$$d_{1s}1_{s1} = 1$$

Now, $\displaystyle\sum_{i=1}^{s} a_{1i}p_{ij} = \sum_{i=1}^{j-1} a_{1i}a_{ij}b_{ij} + a_{1j} - \sum_{k=1}^{j-1} a_{1j}b_{jk}$

$$- \sum_{k=j+1}^{s} a_{1j}a_{jk}b_{jk} + \sum_{i=j+1}^{s} a_{1i}b_{ij} = a_{1j},$$

because $a_{1i}a_{ij} = a_{1j}$ for $1 \leq i \leq j$ ($\alpha 2$), $a_{1j}a_{jk} = a_{1k}$ for $1 \leq j \leq k$ ($\alpha 2$), and $b_{jk} = b_{kj}$ ($\beta 2$). This proves that:

$$d = \frac{1}{\displaystyle\sum_{i=1}^{s} a_{1i}} (a_{11}, a_{12}, \ldots, a_{1s})$$

A number of important properties of the chain have thus become apparent from this derivation:

P0. The equilibrium density of an annealing chain, with $\varepsilon(s_0) = \varepsilon_0 = \min_{s \in \mathscr{L}} \varepsilon(s)$, is

$$\delta(s,t) = \frac{\alpha(\varepsilon(s_0), \varepsilon(s), t)}{\displaystyle\sum_{s' \in \mathscr{L}} \alpha(\varepsilon(s_0), \varepsilon(s'), t)} \tag{1.1}$$

P1. The equilibrium density is independent of β! Consequently, β can be changed (within the limits given by ($\beta 1$–$\beta 2$)), without disturbing the equilibrium.

P2. The chain in equilibrium is an ergodic process, because P is primitive. This means that in estimating the expectation of the scores, only a single realization of the annealing chain has to be observed for a sufficient number of steps. In general, a large number of separate realizations would be necessary to estimate the expectation for $\varepsilon(s)$. The number of realizations needed, however, will be considerably lower than the required number of steps when observing a single realization.

These properties depend crucially on the assumptions made in the previous section. Relaxing these requirements a little may still yield similar, but more complicated, properties. However, because these requirements hardly limit the applicability of the algorithm, they have been adopted throughout in this chapter.

From Eq. (1.1) an explicit expression for the score density can be derived:

$$\hat{\delta}(\varepsilon,t) = \frac{|\{s|\varepsilon(s) = \varepsilon\}| \, \alpha(\varepsilon_0,\varepsilon,t)}{\sum_{\{\varepsilon'\}} |\{s|\varepsilon(s) = \varepsilon'\}| \, \alpha(\varepsilon_0,\varepsilon',t)} = \frac{\hat{\delta}(\varepsilon,\infty)\alpha(\varepsilon_0,\varepsilon,t)}{\sum_{\{\varepsilon'\}} \hat{\delta}(\varepsilon',\infty)\alpha(\varepsilon_0,\varepsilon',t)} \qquad (1.2)$$

Chain Statistics.   Some aggregate functions are useful in characterizing the chain. The first two are the average value of ε:

$$E(t) = (\varepsilon) = \sum_{s \in \mathscr{S}} \delta(s,t) \, \varepsilon(s) = \sum_{\{\varepsilon\}} \varepsilon \, \delta(\varepsilon,t) \qquad (1.3)$$

and its variance:

$$\sigma^2(t) = |(\varepsilon^2) - E^2(t)| \qquad (1.4)$$

Another function, also associating a number with each chain in equilibrium, will be useful. The number is to be reasonable measure of the accessibility of the state space. This *accessibility function H* is determined by the density δ, and therefore, $H: R^s \to R$. H should be a continuous and symmetric function of the components of δ. There are a number of properties such an accessibility function is expected to have. First, H should take its maximum value when all states are equally accessible:

$$\forall_\delta \left[H(\delta) \leq H\left(\frac{1}{s}1_s\right) \right] \qquad (h1)$$

Further, the value of H should not depend on the description of the chain. Suppose the state space $\mathscr{S}$ is partitioned into p state spaces forming the partition $\mathscr{S} = \{\mathscr{S}_1, \mathscr{S}_2, \ldots \mathscr{S}_p\}$. The Markov chain can then be viewed as a composition of several chains, one indicating in which block the presentΔstate is, and p chains indicating which state in the selected block is the presentΔstate. The equilibrium densities of these chains are, respectively:

$$\delta\mathscr{S}(\mathscr{S}_i,t) = \sum_{s \in \mathscr{S}i} \delta(s,t) \text{ and } \delta\mathscr{S}_i(s,t) = \frac{\delta(s,t)}{\sum_{s' \in \mathscr{S}i} \delta(s',t)}$$

The accessibility measure should not depend on the chosen viewpoint, and therefore:

$$H(\delta) = H(\delta\mathscr{S}) + \sum_{i=1}^{p} \left(\left(\sum_{s \in \mathscr{S}i} \delta(s,t)H(\delta\mathscr{S}_i)\right)\right) \qquad (h2)$$

Finally, adding an impossible state to the state set should not change the accessibility function:

$$H(d_1, d_2, \ldots, d_s) = H(d_1, d_2, \ldots, d_s, 0) \tag{h3}$$

The only function, but for an arbitrary positive factor, satisfying these requirements is:

$$H(\delta) = H(t) = - \sum_{s \in \mathscr{P}} \delta(s, t) \ln(\delta(s, t)) \tag{1.5}$$

with $x \ln(x)$ defined to be 0 for $x = 0$ (Khinchin 1957). Because δ only depends on t for a given chain in equilibrium, H can be seen as a function of t.*

Acceptance. Two very important properties of annealing chains follow by adding the following requirements:

$$\forall_{(\varepsilon, \varepsilon') \in R^2} [\lim_{t \to \infty} \alpha(\varepsilon, \varepsilon', t) \to 1] \tag{α3}$$

$$\forall_{(\varepsilon, \varepsilon') \in R^2} [\varepsilon \leq \varepsilon' \to \lim_{t \to 0} \alpha(\varepsilon, \varepsilon', t) \to 0] \tag{α4}$$

P3. For a sufficiently high value of the control parameter t, the relative frequency is the same for all states.

Consequently, the accessibility for high values of t will be close to

$$H_\infty = \ln |\mathscr{P}| \tag{1.6}$$

P4. For a sufficiently low value of the control parameter t, presentΔstate is almost exclusively a state with $\varepsilon \approx \varepsilon_0$. If s_0 is the only state with $\varepsilon = \varepsilon_0$ (i.e., ε has a unique global optimum over $\mathscr{P}$), presentΔstate will be s_0 for an arbitrarily large proportion of the time for t low enough.

If the system has only one global minimum the accessibility will come arbitrarily close to 0 by choosing a t that is low enough. If there are several minima, the accessibility will approach

$$H_0 = \ln |\{s \in \mathscr{P} | \varepsilon(s) = \varepsilon(s_0)\}| \tag{1.7}$$

---

*H is often called the entropy function. However, to avoid confusion with the entropy of a probability density, used later in this chapter, and with the usual definition of the entropy of a Markov chain, H will be referred to as the accessibility.

(α1–4) do not uniquely determine the acceptance function. However, adding a few mild conditions restricts the choice for α considerably. The first two are almost implied by calling t a "control parameter." They require $\alpha(\varepsilon,\varepsilon',t)$ to be continuous and monotonous in t.

$$\forall_{(\varepsilon,\varepsilon')\in R^2} \ \forall_{t_0\in R^+} \ [\lim_{t\to t_0} \alpha(\varepsilon,\varepsilon',t) = \alpha(\varepsilon,\varepsilon',t_0)] \qquad (\alpha 5)$$

$$\forall_{(\varepsilon,\varepsilon')\in R^2} \ \forall_{t\in R} + \ \forall'_t{}_{\in R} + \ [t > t' \to \alpha(\varepsilon,\varepsilon',t) \geqslant \alpha(\varepsilon,\varepsilon',t')] \qquad (\alpha 6)$$

The following restriction requires α only to depend on the difference of the scores and the control parameter t:

$$\forall_{(\varepsilon,\varepsilon')\in R^2} \ \forall_{\Delta\varepsilon\in R^+} \ \forall_{t\in R^+} \ [\alpha(\varepsilon,\varepsilon + \Delta\varepsilon,t) = \alpha(\varepsilon',\varepsilon' + \Delta\varepsilon,t)] \qquad (\alpha 7)$$

Finally, $\alpha(\varepsilon,\varepsilon',t)$ should be differentiable with respect to ε' for $\varepsilon' > \varepsilon$.

$$\forall_{(\varepsilon,\varepsilon')\in R^2} \ \forall_{t\in R^+} \ \left[\lim_{\Delta\varepsilon\to 0} \frac{\alpha(\varepsilon,\varepsilon' + \Delta\varepsilon,t) - \alpha(\varepsilon,\varepsilon',t)}{\Delta\varepsilon} \right.$$
$$\left. = \lim_{\Delta\varepsilon\to 0} \frac{\alpha(\varepsilon,\varepsilon' - \Delta\varepsilon,t) - \alpha(\varepsilon,\varepsilon',t)}{-\Delta\varepsilon} \right] \qquad (\alpha 8)$$

Now, because of (α2) and (α7):

$$\alpha(\varepsilon,\varepsilon + \Delta\varepsilon,t) = g_t(\Delta\varepsilon) = \frac{\alpha(\varepsilon_0,\varepsilon + \Delta\varepsilon,t)}{\alpha(\varepsilon_0,\varepsilon,t)}$$

Therefore:

$$\frac{\partial}{\partial\varepsilon}\alpha(\varepsilon_0,\varepsilon,t) = \lim_{\Delta\varepsilon\to 0} \frac{\alpha(\varepsilon_0,\varepsilon + \Delta\varepsilon,y) - \alpha(\varepsilon_0,\varepsilon,t)}{\Delta\varepsilon}$$
$$= \alpha(\varepsilon_0,\varepsilon,t) \lim_{\Delta\varepsilon\to 0} \frac{g_t(\Delta\varepsilon) - g_t(0)}{\Delta\varepsilon} = \alpha(\varepsilon_0,\varepsilon,t) \ c(t)$$

where $c(t)$ is the value of the right-hand derivative of g_t at 0. This simple differential equation gives $\alpha(\varepsilon_0,\varepsilon,t) = k \exp(\varepsilon \ c(t))$.

P5. For $\varepsilon' > \varepsilon$ the acceptance function satisfying (α1–8) can only be of the form

$$\alpha(\varepsilon,\varepsilon',t) = \exp((\varepsilon' - \varepsilon) \ c(t))$$

with $c(t)$ continuous and monotonous, $\lim_{t\to 0} c(t) = -\infty$ and $\lim_{t\to\infty} c(t) = 0$.

The commonly adopted choice for $c(t)$ is $-t^{-1}$, so that:

$$\alpha(\varepsilon,\varepsilon',t) = \min\left(\left\{\,,\exp\left(-\frac{\varepsilon'-\varepsilon}{t}\right)\right\}\right) \qquad (1.8)$$

This function has a number of properties that can be used advantageously in controlling and analyzing annealing chains. For example, it enforces a simple relation between E and H, for:

$$\frac{dE}{dt} - t\frac{dH}{dt} = \sum_{s\in\mathscr{S}} (\varepsilon(s)\dot\delta(s,t) + t\dot\delta(s,t)(1 + \ln(\delta(s,t))))$$

$$= \left(t - t\ln\sum_{s\in\mathscr{S}}\exp\left(-\frac{\varepsilon(s')}{t}\right)\right)\sum_{s\in\mathscr{S}}\dot\delta(s,t)$$

Putting $\delta(s_i,t) = \dfrac{d_i}{\sum\limits_j d_j}$, where because of Eq. (1.8) $\dot d_k = \dfrac{\varepsilon(s_k)}{t^2}d_k$, makes it clear that:

$$\sum_{s\in\mathscr{S}}\dot\delta(s,t) = \frac{\displaystyle\sum_i\sum_j\frac{\varepsilon(s_i)}{t^2}d_id_j - \sum_i\sum_j\frac{\varepsilon(s_j)}{t^2}d_id_j}{\left(\displaystyle\sum_j d_j\right)^2} = 0$$

and therefore $\dfrac{dE}{dt} = t\dfrac{dH}{dt}$. Also $\dfrac{dE}{dt} = \sum\limits_{s\in\mathscr{S}}\varepsilon(s)\dot\delta(s,t)$, and using the above notation once more yields:

$$\frac{dE}{dt} = \frac{1}{t^2}\left(\frac{\displaystyle\sum_i\varepsilon^2(s_i)d_i}{\displaystyle\sum_k d_k} - \frac{\displaystyle\sum_i\varepsilon(s_i)d_i\sum_j\varepsilon(s_j)d_j}{\left(\displaystyle\sum_k d_k\right)^2}\right) = \frac{\langle\varepsilon^2\rangle - \langle\varepsilon\rangle^2}{t^2}$$

P6. If the adopted acceptance probability is according to Eq. (1.8), then:

$$\frac{dE}{dt} = t\frac{dH}{dt} = \frac{\sigma^2}{t^2} \qquad (1.9)$$

This makes it possible to monitor the accessibility during the annealing if a quasi-equilibrium is maintained during the whole process.

The Schedule

The goal of the annealing algorithm is to find a state s with $\varepsilon(s)$ close to $\varepsilon(s_0)$. For a very low t the chain would be almost all the time in such a state after reaching equilibrium (P4), but it would take a huge number of steps to reach and detect that situation. For very high values of t the chain will be almost immediately in equilibrium (P3). Also reestablishing equilibrium from an equilibrium situation at a slightly different t can be done in relatively few steps ($\alpha 5$). The algorithm therefore performs the above chain for several, generally decreasing, values of t, each time with a limited number of steps, thereby reducing the space accessibility of confining the chain more and more to states close to—hopefully—a global minimum. The decrements in t and the number of steps per value of t are the parameters that characterize the schedule.

Before starting with gradually decreasing t, the algorithm goes through an initialization phase to obtain initial values for t, the aggregate functions, among others, an estimated value for the minimum accessibility, and an estimate for the number of steps necessary, and to set up the data structures that are maintained during the schedule. Among these data structures there is at least one storing the current configuration (state). Preferably, this data structure should uniquely represent one solution to the problem. Having several representations for the same solution makes the state space unnecessarily large, and consequently makes the algorithm slow. Also, invalid configurations (i.e., configurations that do not correspond to solutions to the real problem) should not be introduced carelessly. They not only enlarge the state space, but might isolate good valid configurations. Another aspect of state encoding is ease of implementing the moves and evaluating the scores. These operations have to be performed many times, and should be implemented very efficiently.

Move Set Control.   The topology of the state space is determined by its move set. Although the topology does not influence the equilibrium density, it has a considerable effect on the convergence properties of the annealing algorithm. A small space diameter is desirable for having all states easily reachable. Yet the difference between the scores of two by move-connected states has to be relatively small. The choice of the move set should be a compromise between a small space diameter and a smooth score variation. Fortunately, because of property P1, this compromise can be adjusted (within the limits of $\beta 1$ and $\beta 2$) to the current acceptance function. When t is low, moves with a big increase in the score have a low probability of getting accepted, and moves with a big decrease in the score most likely do not exist. A considerable speed-up can be obtained by preventing the selection of these moves, if that class of moves can be efficiently identified. Often it is possible to extract without much difficulty a parameter l from each move, such that the average change in the score $\Delta\varepsilon$ is almost or completely monotonously related to l. The value of l for a particular move is called its movelength. In that case it often is possible to keep moves with low acceptance probability from getting selected by limiting the movelength on the basis of their average score change, for example, $/ < L$, where:

$$L = \max \left(\{ /_{min} \} \cup \{ / | \overline{\Delta \varepsilon}(/) \leqslant -t \ln (\zeta) \} \right) \tag{2.1}$$

The table representing the function $\overline{\Delta \varepsilon}(/$ can be assembled during the initialization, and updated while running the annealing algorithm with the moves selected and evaluated. Of course, the adaptive selection function must be implemented such that $\beta 1$ and $\beta 2$ are satisfied at all times.

Control. The schedule must be controlled in such a way that the process stays in quasi-equilibrium and yet converges quickly to a global optimum. This has to be achieved by determining the decrements in t, and the number of moves per value for t. The decrements must be chosen such that the steps do not disturb the equilibrium density too much. For example:

$$\forall_{s \in \{ s | E(t) - \kappa\sigma < \varepsilon(s) < E(t) + \kappa\sigma \}} \left[1 - \gamma^* < \frac{\delta(s, t - \Delta t)}{\delta(s, t)} < 1 + \gamma^* \right] \tag{2.2}$$

Because $\dfrac{\delta(s, t - \Delta t)}{\delta(s, t)} \approx \dfrac{\delta(s, t) - \Delta t \, \dot{\delta}(s, t)}{\delta(s, t)} = 1 - \dfrac{\varepsilon(s) - E(t)}{t^2} \Delta t$, this requirement is equivalent to:

$$\Delta t < \frac{\gamma}{\sigma} t^2$$

where γ replaces the quotient $\dfrac{\gamma^*}{\kappa}$. The number of moves for every selected value of t should be big enough to obtain useful information about $E(t)$ and $\sigma^2(t)$. This is the only requirement when Δt is chosen conservatively.

A Stop Criterion. If the decrements in t are small enough to keep the process in quasi-equilibrium, a fairly general stop criterion can be used. It derives from the observation that the improvement possible by lowering t further must be much smaller than the improvement obtained by decreasing t stepwise from its initial value. The latter equals $E_\infty - E(t)_2$. An upper bound for the improvement still possible, if t is in the interval containing $t = 0$ and with $\dfrac{d^2E}{dt^2} > 0$, is t $\dfrac{dE}{dt}$. Using Eq. (1.9) this gives:

$$\frac{\sigma^2}{t(E_\infty - E(t))} < \theta \tag{2.3}$$

θ being a very small positive number. The left-hand side of Eq. (2.3) does not represent the ratio between improvement possible and improvement achieved outside the given interval, but its value is close to 1 for high values of t, and drops slowly until it enters that interval. So it is harmless outside that interval in the sense that it cannot cauase an untimely stop.

The above criterion is only reliable if quasi-equilibrium has been maintained during the annealing process. An indication for this condition is the value of H, the accessibility. The value of H at high values of t is close to H_∞, which can be calculated with Eq. (1.6). The decrements in H can be calculated by using the relation in Eq. (1.9). The value of H when the step criterion is satisfied should be close to H_0. If the schedule is too fast, the process is likely to get trapped in a local minimum, for a while or forever. In the latter case H will stay much too high. In the former case H will drop at too low a value for t, and consequently drops quickly and, finally, below H_0. Of course, both effects can occur in the same process, thereby canceling each other's influence, and H may be close to H_0 when the stop criterion is satisfied, without convergence to a global minimum.

Initialization. The purpose of the initialization is to obtain values for E_τ, σ_∞, and the value of t with which to start the annealing process. The function $\Delta\varepsilon(\,/\,)$ has to be tabulated in the initialization if the move length is to be adapted according to Eq. (2.1).

One way of doing this, is setting t at a value such that the initial probability of accepting the move with the biggest change in ε is reasonably high. By approximating this maximum score change, an initial t can be calculated for a given probability ξ of accepting such a big change in ε:

$$t_b = \frac{\max(\Delta e)}{-\ln\xi} \tag{2.4}$$

Usually it is not difficult to approximate $\max(\Delta\varepsilon)$. After simulating the chain with $t = t_b$ for a sufficient number of steps, an estimate for σ^2 is available. If $t_b \gg \sqrt{\sigma^2}$, then t_b can be accepted as initial value for the control parameter. This follows from the requirement that $E(t)$ must initially be much closer to E_∞ than the standard deviation in order to be able to reach all states easily, and the fact that for high values of t, assuming a normal density for the ε over $\mathscr{P}$, E depends on t according to $E_\infty - \sigma^2/t$, with σ^2 independent of t. During the simulation (ε) can be updated together with σ^2. So, if the value of t_b calculated with Eq. (2.4) does exceed the standard deviation $\sqrt{\sigma^2}$ by a considerable amount, t_b is accepted as the initial control parameter, E_∞ is set to (ε), and σ_∞ is made equal to the measured standard deviation. If this is not the case, the procedure is repeated with a higher value for t_b.

A problem with that approach is the the number of steps has to be determined. This number should be high enough to yield reliable values for E and σ^2. This number might be quite high, because the scores observed are not mutually independent. Another way of obtaining reliable estimates for E_∞ and σ_∞ is to generate a number of mutually independent random states $s^{(1)}, s^{(2)}, \ldots, s^{(r)}$ and evaluate ε for these states. Because both the expectation and the variance of these s exist, the central limit theorem can be applied.

$$\text{probability} \left\{ \frac{1}{r} \sum_{i=1}^{r} (\varepsilon(s^{(i)}) - E_\infty) < \frac{\tau \sigma_\infty}{\sqrt{r}} \right\}$$

$$= \frac{1}{\sqrt{2\pi}} \int_{-\tau}^{\tau} \exp\left(-\frac{1}{2}x^2\right) dx = \Phi(\tau)$$

If this probability is required to be higher than p_m for a relative error smaller than π, the number of random states required is:

$$r > \left(\frac{\Phi^{-1}(p_m)}{\rho}\right)^2 \left(\frac{\sigma_\infty}{E_\infty}\right)^2 \tag{2.5}$$

Further, if σ^2 is estimated by:

$$\frac{1}{r-1} \sum_{i=1}^{r} \left(\varepsilon(s_i) - \frac{1}{r} \sum_{j=1}^{r} \varepsilon(s_j)\right)^2 \tag{2.6}$$

the expected relative error, assuming that the scores are normally distributed over the states, is (Hoel 1956):

$$\sqrt{\frac{2}{r-1}} \tag{2.7}$$

An estimate for σ_∞ having been obtained, t_b can be set such that $1 < \dfrac{t_b}{\sigma_\infty} \ll \dfrac{1}{\gamma}$.

ANNEALING APPLIED TO FLOORPLAN DESIGN

To apply annealing to the task of floorplan design, the relation between states and floorplans must be specified. Also, a score function has to be defined in such a way that low score configurations will be translated into high-quality floorplans. This is the subject of this part of the chapter. First, a more general problem, the p-dimensional box problem, will be introduced. The annealing algorithm for this problem will be completely specified; and some remarkable observations, made on the basis of many runs of the annealing algorithm on this type of problems, will be summarized. Then the relation of the box problem to floorplan design in the context described in the first part of the chapter will be discussed.

The p-Dimensional Box Problem

Box problems are rather abstract constructions that are closely related to typical layout problems. For example, the linear placement problem, where modules have to be arranged side by side in such a way that their mutual interconnections have a short total length, is one of them. Here the multi-dimensional generalization is formulated first. Then the annealing algorithm used for this problem is specified. Experience with this algorithm is reported to conclude this discussion.

Problem Formulation.   The data for the p-dimensional box problem consist of a positive integer p, an incidence structure $(\mathcal{M}, \mathcal{L}, \mathcal{I})$, a set $\mathcal{A}$ of p axes, and four sets of p functions. In the usual terminology for incidence structures, the elements of $\mathcal{M}$ are called *points*, the elements of $\mathcal{L}$ are called *blocks*, and the elements of $\mathcal{P}$ are called *flags*. Points and blocks are both subsets of $\mathcal{P}$. Whenever a point and a block have an element in common, they are said to be *incident* with each other. An *axis* is a permutation over $\mathcal{M}$. Each configuration s consists of p axes, and can therefore be interpreted as a function s: $\mathcal{M}^p \to \{1 \ldots |\mathcal{M}|\}^p$. The quantity $s_{\text{axis}}(m)$ is the position of the point m in the given axis.

The sets of functions play an important role in evaluating the scores. There is a set of p functions v_{axis}: $\mathcal{M} \to R^+$, called *size functions*, a set of p functions w_{axis}: $\mathcal{L} \to R^+$, called *weight functions*, and two sets of p functions $u_{\text{axis}}^{\leftarrow}$ and $u_{\text{axis}}^{\rightarrow}$: $\mathcal{L} \to !B = \{0,1\}$, called *exit functions*. The size functions are used to space the points in an axis; that is, each point will get a position in the p-dimensional euclidean unit cube according to:

$$x_{\text{axis}}(m) = \frac{\frac{1}{2}v_{\text{axis}}(m) + \sum\limits_{i < s_{\text{axis}}(m)} v_{\text{axis}}(s_{\text{axis}}^{-1}(i))}{\sum\limits_{m \in M} v_{\text{axis}}(m)} \tag{1.1}$$

All coordinates are positive and smaller than 1.

A p-dimensional *box* is a set of p intervals in R. The size of a box is the sum of the lengths of its intervals. The box of a block is the smallest box that encloses all points to which that block is incident, where when $u_{\text{axis}}^{\leftarrow} = 1$ the number 0 must be included in the interval of that axis, and when $u_{\text{axis}}^{\rightarrow} = 1$ the number 1 should be included in the interval of that axis. The weight of a box associated

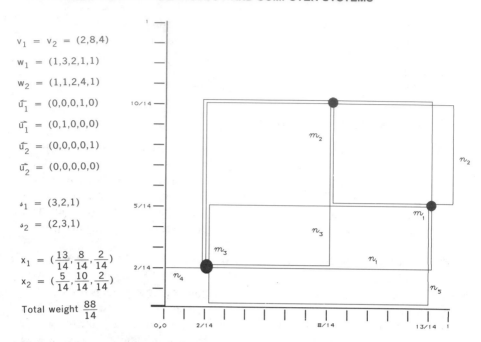

$v_1 = v_2 = (2,8,4)$

$w_1 = (1,3,2,1,1)$

$w_2 = (1,1,2,4,1)$

$\vec{u}_1^- = (0,0,0,1,0)$

$\vec{u}_1^+ = (0,1,0,0,0)$

$\vec{u}_2^- = (0,0,0,0,1)$

$\vec{u}_2^+ = (0,0,0,0,0)$

$\delta_1 = (3,2,1)$

$\delta_2 = (2,3,1)$

$x_1 = (\dfrac{13}{14}, \dfrac{8}{14}, \dfrac{2}{14})$

$x_2 = (\dfrac{5}{14}, \dfrac{10}{14}, \dfrac{2}{14})$

Total weight $\dfrac{88}{14}$

Figure 19-10. A two-dimensional box configuration with the incidence structure of Figure 19-2. Lines are slightly displaced to show the boxes.

with a block is the sum of the lengths of its intervals, each multiplied by the weight of its axis for that block. The weight of a configuration is the sum of the weights of the boxes of all blocks in $\mathcal{A}$.

The formulation is most easily understood by considering a two-dimensional example (Figure 19-10).

The Annealing Formulation. In order to use the annealing algorithm for finding how weight configurations, the state space and the score function have to be specified. The state set is easy. Each state corresponds one to one with a configuration. Therefore, they can be represented as p one-dimensional arrays, each containing all the numbers between 1 and $|\mathcal{U}|$ exactly once. The total number of states is $(|\mathcal{U}|!)^p$. Using Stirling's formula, this yields for the maximum accessibility:

$$H_\infty \approx |\mathcal{A}||\mathcal{U}| \ln (|\mathcal{U}|) + \frac{|\mathcal{A}|}{2} \ln (2\pi|\mathcal{U}|) - |\mathcal{A}||\mathcal{U}| \quad (1.2)$$

Each move is a transposition in an axis; that is, for two distinct points the integers assigned to them by s_{axis} are interchanged. Therefore, $(\mu 2)$ is obviously satisfied, because no transposition leaves the state unchanged. For every transposition there has to be another transposition that can restore the previous state, namely, the one interchanging the integers assigned to the same two points in the same axis. This is required by (μ1). A sufficient condition for satisfying (μ3) is that all transpositions involving points where the integers differ by only one be in the move set. However, it might take up to $p| \mathcal{M}|(| \mathcal{M}| - 1)$ of these moves to reach a certain state from a given present state. The minimum number of moves needed to get from a state to any other state, if all possible transpositions are in the move set, is at most $p(| \mathcal{M}| - 1)$. Consequently, $p(| \mathcal{M}| - 1) \leqslant (\mathcal{P}) \leqslant p| \mathcal{M}|(| \mathcal{M}| - 1)$ if the moves of move length 1 are in the move set. The move length l is defined to be the modulus of the difference between the integers involved. Because this number is exactly one plus the number of points in the axis between the two points involved, it is a rather faithful indication for the changes in the weight of the boxes of the incident blocks, whereas changes in weights of boxes are relatively small. It also is quite simple to extract this parameter for every proposed move. These properties make l suitable for controlling the topology of the move set according to Eq. (2.1) in the second part of this chapter. To summarize:

$$\mu = \left\{ (s, s') \, | \, \exists !_{x \in \mathcal{A}} \, [(y \neq x \to s_y \equiv s'_y) \wedge (|\{ m| \ s_x(m) \neq s'_x(m)\}| \right.$$
$$\left. = 2) \wedge (\sum_{m \in \mathcal{M}} | s_x(m) - s'_x(m)| \leqslant 2L_x \} \right\} \quad (1.3)$$

The selection function is such that each axis is chosen with equal probability, and each transposition in the selected axis with move length less than L_{axis} is equally probable.

$$L_{\text{axis}} \leqslant \frac{1}{2}(| \mathcal{M}| - 1): \beta = \cfrac{1}{p(L_{\text{axis}}| \mathcal{M}| - \frac{1}{2}L_{\text{axis}}(L_{\text{axis}} + 1))}$$

$$L_{\text{axis}} > \frac{1}{2}(| \mathcal{M}| - 1): \beta = \frac{1}{p(L_{\text{axis}}} | \mathcal{M}| - \frac{1}{2}(| \mathcal{M}| - L_{\text{axis}} - 1)(| \mathcal{M}| - L_{\text{axis}}))$$

$$(1.4)$$

It is quite easy to see that the requirements (β1–2) imposed upon selection functions are satisfied by the settings of Eqs. (1.4).

```
Begin  s: = ŝ ; k: = −1;
    H∞ : = |𝒜| |𝓜| ln (|𝓜|) + |𝒜|/2 ln (2π|𝓜|) − |𝒜| |𝓜| ;
    H₀ : = ln(#global minima) ;

    repeat t: = t_b; k: = k + 1; E : = E∞; H : = H∞;
        for each axis ε 𝒜 do L_axix : = |𝓜| = 1 end;
```

repeat e: $= 0$, esq: $= 0$;
  for h: $= 1$ to $|\mathcal{M}|$ do
    axis $: =$ random $(\mathcal{A})$;
    modl $: =$ random $(\mathcal{M})$;

    repeat mod2 $: =$ random $(\mathcal{M})$
    until $(\text{modl} \neq \text{mod2})$ # $(|s_{\text{axis}}[\text{modl}] - s_{\text{axis}}[\text{mod2}]| \leq L_{\text{axis}})$;

    $\ell: =$ swap$(s_{\text{axis}}[\text{modl}], s_{\text{axis}}[\text{mod2}])$
    if random$([0, 1)) < \exp \left(-\dfrac{\varepsilon(\ell) - \varepsilon(s)}{t} \right)$

    then $s : = \ell$;
      if $\varepsilon(s) < \varepsilon(\hat{s})$ the $\hat{s} : = s$ end
    end;

    e: $= e + \dfrac{\varepsilon(\phi)}{|\mathcal{M}|}$; esq: $=$ esq $+ \dfrac{\varepsilon(\phi)^2}{|\mathcal{M}|}$
  end;

  H: $= H + \dfrac{e - E}{t}$; E: $= e$; $\sigma^2: = |$ esq $- E^2 |$;

  t $: = t - \dfrac{1}{2^k} \dfrac{\gamma}{\sigma} t^2$;

  for each axis $\varepsilon \mathcal{A}$ do
    $L_{\text{axis}} : = \max(\{1\} \cup \{ \ell \mid \overline{\Delta \varepsilon} (\ell) < -t \ln (\zeta) \})$
  end;

until $\dfrac{\sigma^2}{t(E_x - E(t))} < \theta$;
 until $\dfrac{H - H_0}{H_x - H_0} < \eta$
end;

The discussion in the second part of the chapter ("The Annealing Algorithm") and the adaptation for p-dimensional box problems in this section lead to a module body as given. Repeatedly, an annealing chain is activated and run for a number of steps equal to the number of points. It is the loop controlled by the variable h. Inside the loop, data is created for obtaining the average score and the average squares of the score over the states visited. After completion of that loop, the values for the chain statistics are recorded, the value of the control parameter for the next chain is determined, and the upper bounds for its move lengths are evaluated. This is repeated until the stop criterion is satisfied. Then the postmortem check on the accessibility is applied. If the result is negative, the whole annealing process is repeated, but twice as slowly. After some experience, the setting of the constants γ, ζ, and θ should be such that restarting the annealing seldom occurs. To get an idea about their values, set $\gamma = .01$, $\zeta = .5$, and $\theta = .0001$.

Usually, the score of a configuration is simply its weight. Sometimes, however, there is a problem when the exit functions are trivial (for example, all $u \equiv 0$), or the same for all axes. In that case, a good solution for any axis is a good solution for all axes. Selecting the configuration with all axes having the same good point sequence (or reverse if $\overleftarrow{u} \equiv \overrightarrow{u}$) yields a low score, but a configuration with all its points on a diagonal of the unit cube. This mostly is not a useful solution. To prevent it, the correlation between the axes should be kept low. The weight of a configuration is then multiplied with a correlation measure to obtain its score. The correlation between two axes is defined as follows:

$$\rho(x_i, x_j) = \frac{3| . \mathscr{U} |(| . \mathscr{U} | + 1)^2 - 12 \sum_{m \in . \mathscr{U}} (s_{x_i}(m) \, s_{y_j}(m))}{2(| . \mathscr{U} | \quad | . \mathscr{U} |^3)} \tag{1.5}$$

To use the preferred method of initialization, the second method under "Initialization" (above), a method for generating random configurations is necessary. This is not difficult for p-dimensional box problems, for it takes $m - 1$ transpositions to generate a random permutation of m elements (de Balbine 1967). Though it might be more efficient to evaluate ε from scratch for each random configuration thus obtained, than to use the updating procedure for each transposition, the latter produces data for the $\Delta\varepsilon$ table.

Chain Statistics. Evidence has been presented (Otten and van Ginneken 1984) that for p-dimensional box problems and probably a much larger class of problems, the chain statistics E, H, and σ^2 as functions of the control parameter obey the following relations quite closely.

$$t \geq T: E(t) - E_\infty = -\frac{\sigma_\infty^2}{t}, \qquad \sigma(t) = \sigma_\infty, \qquad H(t) - H_\infty = -\frac{\sigma_\infty^2}{2t^2}$$

$$t_e < t \leq T: E(t) - E_\infty = \frac{\sigma_\infty^2}{T}\left(\frac{t}{T} - 2\right), \quad \sigma(t) = \sigma_\infty \frac{t}{T}, \qquad H(t) - H_\infty$$

$$= \frac{\sigma_\infty^2}{T^2}\left(\ln\left(\frac{t}{T}\right) - \frac{1}{2}\right) \tag{1.6}$$

where t_e is very small, namely,

$$t_e \approx T \exp\left(\frac{T^2(H_0 - H_\infty)}{\sigma_\infty^2} + \frac{1}{2}\right)$$

That means that, after the initialization, all parameters that determine almost completely the variation with t in the chain statistics are known, except for T.

This parameter separates two regions of control. One region, $t > T$, shows a behavior that is consistent with the assumption that the scores are normally distributed over the states. This normal density is maintained with constant standard deviation, and with hyperbolic decrease in the average score for decreasing values of t. This, of course, cannot continue because that would imply an unbounded decrease in score. At some point, $t \approx T$, the lower bound on the score becomes noticeable. Remarkably, the score average starts to decrease linearly with t after that point, and shows this behavior until some very low value $t \gtrsim t_e$ is reached. In (Otten and van Ginneken 1984) it is shown that the most likely score density in that range is a gamma density with $\left(\dfrac{\sigma_\infty}{T}\right)^2$ degrees of freedom. The value of T depends on the size and the connectivity of the given incidence structure. A precise functional relation between T and the characteristics of the input has not been determined yet.

For $t < t_e$ the individual characteristics of the problem dominate the behavior. The annealing, therefore, has to continue for a while after reaching the point where $t = t_e$. Therefore, $\theta \ll \dfrac{\dfrac{1}{2T}}{t_e} - 1$.

Box Problems and Floorplan Design

There are some clear similarities between the input data for the floorplan design task as described in the first part of this chapter, and the formulation of box problems in the previous section. Moreover, the result of annealing applied to the two-dimensional box problem is a point configuration in the euclidean plane, and this was recognized as a suitable intermediate structure in floorplan design. So, if it were possible to set the weights such that a good solution for the two-dimensional box problem corresponded with a good intermediate structure in the floorplan design procedure, the annealing algorithm could be used for generating such a structure.

Interpretation of Box Configurations. Both problems have an incidence structure as a starting point. Points in the incidence structure of the box problem obviously correspond with the modules in floorplan design. Their sizes, assigned to them by the v functions, can be the (estimated) areas of the modules; that is, the image of a module $m \in \mathcal{M}$ is its (estimated areas under the function v_{axis} for all axes. The blocks of the box problem are replaced by nets in the floorplan design task. The exit functions indicate whether a net is connected with the outside world (if any $u_{\text{axis}}(n)$ is 1, then n is a global net), and at which side or corner it is made available by the (preliminary) environment.

A good configuration for a two-dimensional box problem, interpreting the floorplan design data as described in the preceding paragraph, and with all weight functions as constant functions with 1 as their image value, corresponds, after appropriate slicing, to a floorplan in which the total wiring length, measured center to center in the manhattan way, is not far from the shortest possible, if the modules have approximately the same size, the floorplan has to fit in a square region, and each net has not more than three pins. Of course, these restrictions are not acceptable for floorplan design, and have to be abolished or diminished by manipulating the weight functions.

This method, and no global method in existence, efficiently handles floorplan design in which the modules differ by orders of magnitude in size. Some preanalyses of the data, and some decisions concerning the large modules, are necessary in those cases. Which range of sizes can be handled also depends on how many modules there are, and how large the total area is. The more modules, and the larger the total area, the bigger the range in sizes usually can be. The positioning with Eq. (1.1) then is sufficiently accurate for one to estimate the wire length, and to make decisions on that basis. The subsequent slicing process uses the relative positions, while taking other factors, such as the shape of the modules, into account. If the modules are of exactly size or shape, more efficient annealing formulations are possible.

Weight Functions for Floorplan Design.   A frequently used estimate for the length of a net is half the perimeter of the smallest rectangle (with the same orientation as the axes) enclosing all the points representing modules connected to that net. This is a tight lower bound for the length of the Steiner tree in the plane with minkowski-1 (manhattan) distances. For two- or three-pin nets, the Steiner tree can always be realized with this length. For nets with more pins, it is possible that the Steiner tree is longer. The following argument gives a basis for correcting the estimates. Consider a rectangular area with a large number of points connected by a Steiner tree. If the number of points is large enough, then the effects of the sides of the rectangle can be ignored. For random point configurations in this rectangle, the Steiner tree will have a certain average length. If there are x times as many points in x times as large an area, it is easy to see that the total expected length of the Steiner tree for this case will be also be x times as long. The estimate for the length, however, grows only with the length of the perimeter, leaving a factor in the order of $\sqrt{x}$ to be corrected. The correction factor should have a value of 1 for $|n| = 2$ or 3. This leads to a correction factor for the length of the Steiner tree of:

$$sf(n = \sqrt{\max\{1, |n| - 2\}} \tag{2.1}$$

In general, the contour of a floorplan as estimated while establishing a preliminary environment, or given by the fabrication guidelines, cannot be guaranteed to be a square. The configurations in a two-dimensional box problem, however, are without exception contained in a unit square. This is not important for most slicing procedures that use point configurations because they mostly depend only on the relatively positions of the points on the axes. However, if wire length is the objective function to be minimized, the deviation from square has to be taken into account in the score evaluation. This is not difficult, for different weight functions can be defined for distinct axes. Assuming a factor equal to 1 for the first axis, the weight function for the second axis can be multiplied by a factor equal to the aspect ratio:

$$ar_x = 1 \text{ and } ar_y = \text{aspect ratio} \tag{2.2}$$

The fact that each axis can have its own weight function can also be used to distinguish the two wiring directions. This may be important if wires parallel to one axis have electrical properties different from the wires parallel to the other axis, for example, because they are allocated in different layers. It may also be important to keep the modules more porous in one direction than in the other. So, each axis may get a directional factor df_{axis} measuring the importance of preferring one direction over the other for the wiring.

Finally, it might be important to keep certain nets small while other nets are not that critical. For example, nets carrying signals whose delay immediately affects the overall speed of a digital system should be kept short because net size and propagation delay are monotonously related in many present-day technologies. This and other consideration led to a factor cf assigned to each net measuring the relative importance of its size.

The discussions of this section lead to the following weight function:

$$w_{axis}(n) = (1 + b|\rho_{xy}|) \cdot cf(n) \cdot df_{axis} \cdot ar_{axis} \tag{2.3}$$

where b has the value 1 when the exit functions do not effectively differentiate the axes, and otherwise the value 0.

Adaptation to Special Cases.   The method has also been used for placement problems. Sometimes modifications in the basic method were desirable to increase the quality of the results. For example, several layout styles require modules to be arranged in columns. A more accurate estimation of the wiring length is possible by changing the v function for one axis. Consecutive points on that axis get the same coordinate. The corresponding modules are considered to be in the same column. How many modules are assigned to each column depends on their sizes. The distance between points with different coordinates on that axis is proportional to one plus the number of columns between the two columns containing the corresponding two modules.

Note, however, that the number of configurations equivalent to a given configuration may be quite large now because transpositions in the special axis involving points with the same coordinate do not affect the score. The same is true for the configurations with the lowest score. In order for the postmortem check on the accessibility to be usable, the number of equivalent configurations has to be estimated, or rather the logarithm of this number. This can be done by generating an arbitrary configuration, and determining the number of modules in each column c. Using Stirling's formula again yields:

$$H_0 \approx \sum_{c \in C} \left(|c| \ln(|c|) + \frac{1}{2} \ln (2\pi |c|) - |c| \right) \qquad (2.4)$$

With modifications like the one described here for layout styles with columns of modules, the basic annealing formulation can be adapted to a variety of problems. A difficult question remains, of whether the annealing algorithm does as well on those variations as on the pure box problem. Certainly, formulations in which moves are connected with wild score variations should be avoided. Studies to obtain more precise characterizations of well-formulated annealing tasks are presently being conducted.

REFERENCES

de Balbine, G., "Note on Random Permutations," *Mathematics of Computation, 21* (100), 710–712 (1967).

Brayton, R. K., Chen, C. L., McMullen, C. T., Otten, R. H. J., and Yamour, Y. J., "Automated Implementation of Switching Functions as Dynamic CMOS Circuits," *Proceedings* of the Custom Integrated Circuits Conference, 1984, pp. 346–350.

Burstein, M., "Channel Routing," to be published.

Demicheli, G., and Sangiovanni-Vincentelli, A., "Multiple Constrained Folding of Programmable Logic Arrays: Theory and Applications," *IEEE Transactions on Computer-Aided Design, CAD-2* (3), 151–167 (1983).

van Ginneken, L. P. P. P., and Otten, R. H. J. M., "Global Routing," IEEE International Symposium on Circuits and Systems, Kyoto, 1985.

Hall, K. M., "An r-dimensional quadratic placement algorithm," *Management Science,* No. 17, 219–229 (1970).

Hoel, P. G., *Introduction to Mathematical Statistics,* Wiley, New York, 1956.

Khinchin, A. I., *Mathematical Foundations of Information Theory,* translated by R. A. Silverman and M. D. Friedman, Dover Publications, New York, 1957.

Kirkpatrick, S., Gelatt, C. D., Jr., and Vecchi, M. P., "Optimization by Simulated Annealing," *Science, 220* (4598), 671–380 (May 13, 1983).

Lauther, U., "A Min-cut Placement Algorithm for General Cell Assemblies Based on a Graph Representation," *Journal of Digital Systems, 4,* 21–34 (1980).

Otten, Ralph H. J. M., "Layout Structures," IBM Research Report RC9657, Thomas J. Watson Research Center, Yorktown Heights, NY, 1982a (Preliminary version in *Proceedings* of IEEE Large Scale Systems Symposium, Virginia Beach, VA, 1982, pp. 349–353.

Otten, R. H. J. M., "Automatic Floorplan Design," IBM Research Report, RC 9656, Yorktown Heights, NY, 1982b: (Also, in *VLSI Technologies in Graphics,* H. Fuchs Editor, IEEE Computer Society, 1983; Preliminary version in *Proceedings* of the 19th Design Automation Conference, Las Vegas, 1982, pp. 261–267.)

Otten, R. H. J. M., "Efficient Floorplan Optimization," *Proceedings* of IEEE International Conference on Computer Design, Port Chester, NY, Oct. 1983.

Otten, R. H. J. M., and van Ginneken, L. P. P. P., "Annealing: The Algorithm," IBM Research Report, RC 10861, Dec. 1984.

Stockmeyer, L. J., "Optimal Orientations of Cells in Slicing Floorplan Design," IBM Research Report, RJ 3731, IBM Research Laboratory, San Jose, CA, 1983.

Wielandt, H., "Unzerlegbare, nicht-negative Matrizen," *Mathematische Zeitschrift, 52,* 642–648, 1950.

Wirth, N., "Program Development by Step-wise Refinement," *Communications of the ACM, 14,* 221–227 (1971).

20. Automatic Placement

Bryan T. Preas
Xerox Palo Alto Research Center

Patrick G. Karger
Tektronix/CAE Systems Division

Physical design of digital circuits consists of transforming a circuit design specification into a representation that can be used in the manufacture of the physical circuit. The speed with which this transformation takes place is greatly enhanced by the use of automatic layout techniques. Automatic layout is a subset of the physical design process that maps a structural description of the circuit into a physical description consisting of geometric coordinates for all of the circuit components and the interconnection wiring. The structural description that is an input to the layout process consists of a list of circuit primitives, or *components,* that are to be included in the layout, and a list of *signal sets* indicating the terminals or *pins* on the components that are to be made electrically common by the layout process. The *interconnection nets,* or simply *nets,* are the connections among the signal sets. Soukup (1981) and Breuer (1972) provide good overviews of the physical design process.

Automatic layout consists of two primary functions: the positioning of the components onto the layout surface, called *placement,* and then interconnecting them by wiring, called *routing,* according to a set of *design rules.* Figure 20-1 shows an example of a circuit that was produced by automatic placement and routing. It is a standard cell design. Although the placement and routing functions are intimately related and are interdependent, historically they have been separated because of computational complexity. [There has been at least one attempt at performing automatic placement and routing in parallel (Loosemore 1979), but this approach does not have a large following.] Automatic placement, the subject of this chapter, is responsible for determining the location of circuit components that comprise the circuit being designed, subject to the *constraints* imposed by the designer and the design rules.

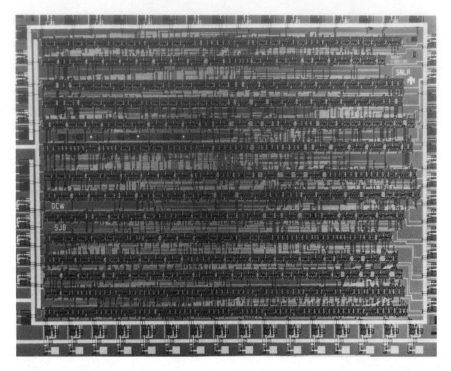

Figure 20-1.   A standard cell layout. (Photo courtesy of Sandia National Laboratories, Alburquerque, NM)

Good placement is a key aspect of automatic layout, but it sometimes receives insufficient attention. A poor placement can leave the router with a difficult or impossible task, whereas a good placement can make a router's job easy. Also, placement directly determines the minimum length of the interconnection wiring; and because wiring delay is often the dominant part of the response time of electrical signals on the nets, placement often determines the performance of the physical circuit.

The design of a digital system and the associated placement subproblem are typically defined hierarchically. For example, modern computers are usually packaged as boards carrying very large scale integrated (VLSI) circuits. VLSI circuit definitions have varying numbers of hierarchy levels, but a typical division might be functional blocks, registers, gates, and transistors (Preas 1979). Placement algorithms normally operate on one level of hierarchy at a time; the placement of the (sub-) components within a single (higher level) component is normally considered as a separate problem with boundary conditions defined by the other components within the overall circuit design.

Placement methods fall into two groups: *constructive* and *iterative*. Constructive methods produce a *complete placement* (where all circuit components have assigned positions) based on a *partial placement* (in which some or all components do not have assigned positions) as input, whereas iterative methods seek to improve a complete placement by producing a new, better, complete placement.

This chapter concentrates on the actual placement algorithms and says little about the nature of the components being placed or the *placement surface* on which the components reside. As a result of this *technology-independent* presentation, the placement concepts are applicable to digital circuit design for a wide range of circuit manufacturing technologies, including *printed circuit boards* and *hybrid integrated circuits,* as well as silicon integrated circuits (*gate arrays, standard cell* designs, and *structured custom* designs).

This chapter defines the placement problem, and categorizes and reviews the placement techniques that are currently available. It concentrates on placement of components in the upper and intermediate levels of the design hierarchy. Specifically excluded from discussion is the generation of the lowest or *leaf* cells of the hierarchical structure tree. The leaf cells are considered to be atomic objects for automatic placement. In a typical standard cell or gate array design, they would be the small scale integration (SSI) gates found in the cell library. We do not address the positioning problems associated with direct computation of cells, or *silicon compilation.* This chapter is intended as an introduction and guide to placement literature, with references that were selected to illustrate placement concepts rather than to constitute a catalog of placement references.

The following section defines the placement problem and introduces some useful terminology. Later sections discuss the models, or abstractions, that are used by placement algorithms; constructive and iterative placement methods; and applications of placement algorithms.

DESCRIPTION OF THE PLACEMENT PROBLEM

The Placement Problem

The placement problem consists of mapping the components in the design specification onto *positions* on the placement surface. Pins on the components define locations at which the circuitry within the component connects to interconnection routing among the circuitry. *External pins* form the interface between the circuitry inside and outside of the component being laid out; they are the pins of the component at the next higher level of hierarchy. The subsets of pins, termed *signal sets,* that are to be connected by wiring to form electrically common interconnection nets are also a placement input.

The actual goal of placement is to determine positions for components that permit completely automatic routing of the interconnection wiring while honoring any number of other (possibly conflicting) goals, such as minimizing cross-talk among the signals, equalizing heat dissipation across the placement surface, and maximizing circuit performance. Because such goals are difficult to cast into an objective function that can be evaluated by a computer, a more restricted objective must be substituted. When a placement function derives a good placement as measured by the restricted objective, it is hoped that the placement is also good as measured against the actual goal. The restricted objective function, then, is defined on the interconnection nets and the components, and is used as a metric to compare different placements. Ideally, it should reflect as accurately as possible (or at least be correlated with) the actual goal defined earlier, as well as be fast to compute. These conflicting goals have lead to a plethora of objective functions, some of which are discussed in the next section.

It is possible to define an objective function for small placement problems in such a way that an optimum solution can be determined. For example, Uehara and van Cleemput (1981) and Nair et al. (1985) present methods for determining the optimum placement of transistors within a restricted domain. Chandrasekhar and Breur (1982) illustrate the complexity of finding the optimum placement for two rectangular components. However, in general, placement problems are much more complicated; practical gate array and standard cell designs can contain a few hundred to several thousand components, and it has been shown that the general placement problem is nondeterministic polynomial (NP)-complete. Sahni and Bhatt (1980) and Donath (1980) present good discussions of the complexity of placement as well as other design automation problems. Because the general placement problem is NP-complete, and large numbers of components are involved, an optimum solution cannot be guaranteed. Hence methods based on heuristics must be used.

Gate Assignment and Pin Assignment

Although our topic is placement, gate assignment and pin assignment are closely related to placement, and, conceptually, they share many attributes. A gate is a subelement of a component. An example is the component shown in

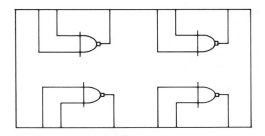

Figure 20-2.   A component containing four functionally equivalent NAND gates.

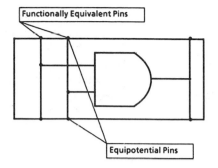

Figure 20-3. Pin assignment nomenclature.

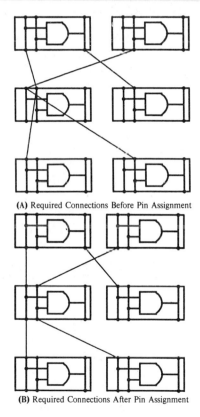

Figure 20-4. An example illustrating the effectiveness of pin assignment. (a) Required connections before pin assignment. (b) Required connections after pin assignment.

Figure 20-2, which consists of four NAND gates; these gates are functionally equivalent and can be used interchangeably. Gate assignment consists of mapping the logical gates in the design specification onto the *functionally equivalent gates* on the components (Nishioka et al. 1978). This is analogous to placement where the gates are now the placeable objects, and the gate sites specify positions where the gates are to be placed.

As illustrated in Figure 20-3, the pin assignment is to assign net connections to *functionally equivalent pins* or *equipotential pins* (Koren 1972; Mory-Rauch 1978; Brady 1984). Here the design specification pins can be considered the placeable objects, and the components can be considered the placement surface which contains positions for the (functionally equivalent or equipotential) physical pins. The importance of pin assignment in improving routability is shown in Figure 20-4; such changes in pin assignment can have a significant impact in difficult problems.

ABSTRACT MODELS OF PLACEMENT

Placement algorithms operate on abstractions, or models of physical circuits. The important aspects of electronic circuits that must be modeled are the logic elements, the interconnections, and the placement surface. The models of these aspects (the devices and components, the interconnection nets, and the carrier, respectively) and their relation to the physical circuit are discussed in this section.

Devices and Components

Automatic placement systems typically use an *object/instance* paradigm to describe the logic elements used in circuits. The physical definition of a *device* (the object) describes how a logic element is constructed from transistors or other primitives; these elements may be leaf cells which are obtained from a library, or they may have been constructed previously by the layout system.

Two views of a device consisting of one NAND gate are shown in Figures 20-5 and 20-6. Figure 20-5 shows the mask layers required to fabricate the NAND gate in a standard cell design, while Figure 20-6 illustrates the properties of the device that are important for automatic placement. Many of the details of the device definition are abstracted away by a placement function; only those properties that are important to the electrical and physical interface with the rest of the circuit and the placement surface are considered. These properties include the physical size and shape of the object, and the location in three dimensions (geometric coordinates and the physical interconnection layer) of the pins and power supply connections. The device definition must also include *restrictions* (constraints or hints) on component placement that are associated with the device. For example, a bonding pad is restricted to lie only on the periphery of an

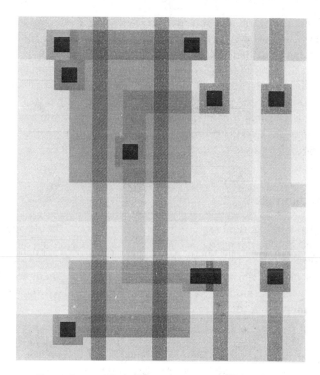

Figure 20-5. Mask layer definition of a NAND gate.

integrated circuit, and some gate arrays have input/output drivers that must reside in special positions close to the bonding pads.

Depending on the accuracy of representation required by the layout system, other details of the device may be included in the definition. A placer, which is used in a layout system to exploit the existing connections among equipotential pins, should include these interconnections in the object definition. Other information that may be included is power dissipation, electrical strength or *drive* of the output pins, capacitance of the input pins, or the locations of material within the gate that may influence routing outside (or over) the device.

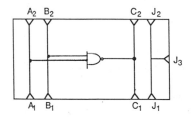

Figure 20-6. Device definition of a NAND gate.

Whereas devices carry the information pertaining to the definition of logic elements, the components carry the information pertaining to their application; so a component is an instance of a device. Because the components are those elements that are placed, they must carry the placement information: current location and orientation (the rotation and reflection). A component should also carry application-oriented placement restrictions; for example, performance constraints may dictate that a component be near a second component with which it shares a signal set. Akers (1981) provides a discussion of positional constraints and methods to determine if the constraints can be satisfied.

Interconnection Nets

The logic elements, as represented by their abstractions, the devices and components, must be interconnected according to the signal set specifications. It is the responsibility of the placement function to position the components so that the interconnection wiring can be effectively routed. Thus, an important element within a placer is the modeling of the interconnection. Two important models are the interconnection topology and the evaluation metrics.

Interconnection Topology. Assume the pins of a signal set to be vertices of an undirected graph; then the connections among them form a tree. The automatic layout system, the devices used by the circuit, the router, and the manufacturing technology combine to impose restrictions on the form of the connection tree. Figure 20-7 shows examples of the interconnection forms. The most general form, called a Steiner tree (Chang 1972), permits vertices of the connection graph at locations other than the pins, and places no restrictions on the *degree* (the number of incident connections) of the vertices. This is typical of connections for integrated circuits. A more restrictive interconnection method is the spanning tree, in which the vertices are restricted to the pin locations (Prim 1957). Other restrictions may apply. For example, *wire-wrap* fabrication imposes constraints on the degree of the vertices because the posts that implement the pins can have a limited number of wraps (typically three). An even more restricted method of interconnection is the chain, where no branching of the interconnection wiring (i.e., degree of vertices less than or equal to two) is permitted (Lin 1965). Placement functions may model the interconnection tree as a complete graph for computational simplicity, although the interconnection net will be routed more simply. Various placement systems may model these interconnection topologies directly, or they may use the abstractions described next.

Interconnection Metrics. While the quality of a placement may depend on the interaction of components' shapes (see below, under "Component Shapes"), it is a strong function of the interconnection nets. In this section, metrics are

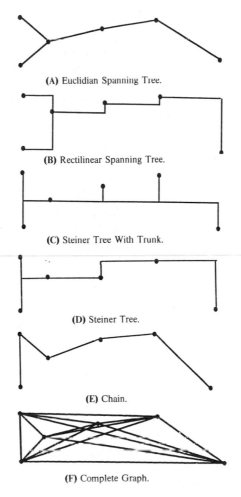

(A) Euclidian Spanning Tree.

(B) Rectilinear Spanning Tree.

(C) Steiner Tree With Trunk.

(D) Steiner Tree.

(E) Chain.

(F) Complete Graph.

Figure 20-7. Forms of interconnection models. (a) Euclidian-spanning tree. (b) Rectilinear spanning tree. (c) Steiner tree with trunk. (d) Steiner tree. (e) Chain. (f) Complete graph.

discussed to compare alternative placements that are a function of the nets. A very large number of metrics, ranging from simple to complex, are divided into two classes: those that assume that the net routings do not interact, and those that account for the net interaction, or *congestion*.

The simplest interconnection metrics assume that the nets can be routed without interfering with each other; naturally this does not account for interaction among the nets. One widely used metric is total *wire length:* the sum over all the nets of the lengths of the interconnection tree, where the length of a tree is the sum of the individual connections within the tree (Hanan and Kurtzberg 1972a). An approximation to wire length of an individual net is the half perimeter of the smallest rectangle enclosing the pins of the net (Schuler and Ulrich 1972). This approximation is the same as the rectilinear, minimal Steiner tree length for two or three pin nets, while for four and five pin nets it is within twice the width of the enclosing rectangle of the Steiner length (Hanan 1966). A more accurate, but more time-consuming, approximation to the Steiner length is the minimal spanning tree length (Hwang 1976).

Another technique used to measure placement quality is to model the connections as *springs* that exert forces on components (Fisk et al. 1967; Quinn 1975). This leads to a force metric where a good placement is one that minimizes the sum of forces on the components. A related method exploits the convexity of this metric to perform quadratic minimization (Blanks 1984, 1985b).

The above measures of placement quality quantify only the amount of wiring; they do not keep track of where the wiring resources are required. This can lead to heavy wiring buildup or congestion, and routing failures as shown in Figure 20-8.

The second major class of interconnection metrics incorporates the interaction among the nets in the measure of placement quality. The congestion may be a global measure such as the number of nets that cross a *cut line,* as described in Breuer (1977) and Lauther (1979), or a local measure such as *track density* within a routing channel (Persky 1976; Kozawa et al. 1983). An even more complex congestion measure is *routability* (Persky 1976). In this case, the placement metric measures cyclic routing constraints because extra routing area is required to resolve constraint cycles. Karger and Malek (1984) propose a metric that uses wire length as a metric and models routing resource congestion as constraints.

An intuitive measure of placement quality that accounts for congestion is illustrated in Figure 20-9. This force vector drawing, showing the complete graph of some of the signal sets, provides a visual indication of placement quality; a good placement looks uncongested. Congestion metrics often correlate better

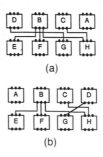

(a)

(b)

Figure 20-8.   Wire length reduction can cause routing failures. (a) Two tracks required; all connections are routed. (b) Shorter wire length, with three tracks required; failure occurs.

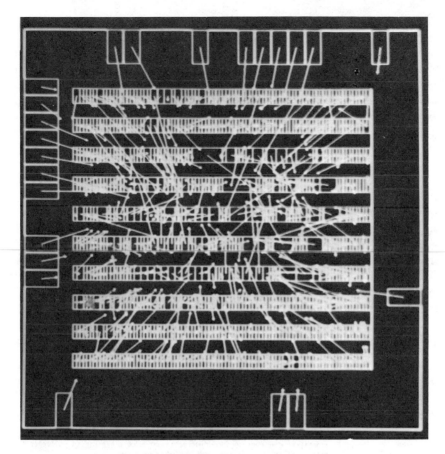

Figure 20-9.   A "force vector" diagram showing a gate array placement. (Photo courtesy of Tektronix/CAE Systems Division, Austin, TX)

with routability compared to wire length metrics because the locations where routing resources are needed are included in the metric.

Carrier

The characteristics of the physical surface onto which the logical elements are placed must be modeled by the placement system. The abstraction of the physical surface is called the *carrier*. Carrier models divide into two categories: fixed coordinate and relative coordinate. The fixed coordinate category, characterized by a rigid placement surface, is typical of printed circuit boards and gate arrays; the overall size as well as the positions at which the components can be placed, called *slots,* have geometric coordinates that are fixed throughout the placement

function execution. This gives rise to a placement model in which placement
assigns components to slots that carry the geometric coordinates. Figure 20-10
shows the base (unprogrammed) die of a gate array. Close inspection of the
figure reveals the rows of slots in which the components will be placed. Between
each pair of rows are the routing areas.

The second category includes the relative coordinate placement surface models.
Standard cell and structured custom integrated circuit design styles permit automatic layout systems to adjust the positions of the components so that all of
the required wiring can be accommodated in the routing areas among the components; so the component positions are defined relative to the routing areas.
For example, standard cell row locations are located relative to the routing
channels, and the component positions within the row are defined relative to the
row origin. Structured custom placement surfaces with components modeled as
arbitrary rectangles have a more complex topology than the standard cell designs,
but are still based on relative coordinates. In this design style, component positions are defined as nodes or arcs on a directed graph (Kani et al. 1976; Preas
1979; Preas and van Cleemput 1979). A placement surface model with the ability
to position more complex component shapes (Preas and Chow 1985) is also
available. The necessary background for general graph models can be found in
Deo (1974).

This section has discussed the abstract models that placement algorithms must

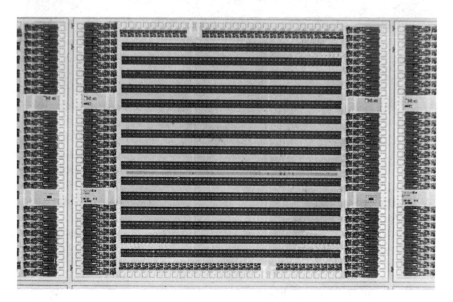

Figure 20-10.   A base (unprogrammed) gate array. (Photo courtesy of National Semiconductor Corporation, Santa Clara, CA)

manipulate in order to solve the placement problem. These models provide the foundation for the placement algorithms described in the next two sections.

CONSTRUCTIVE PLACEMENT ALGORITHMS

An overview of current placement techniques is provided in this and the following section. The techniques reviewed do not form an exhaustive list, but rather demonstrate the major approaches. These techniques are divided into constructive placement and iterative placement. The constructive placement algorithms, discussed first, share the characteristic that their input is a partial placement and their output a complete placement. Although some constructive placement algorithms permit a *seed* placement as an initial condition, the ability to operate on unplaced components sets these algorithms apart from the iterative algorithms that are described in the next section. Constructive placement algorithms are generally used for initial placement, and are normally followed by one or more (iterative) placement improvement algorithms.

Constructive placement techniques are discussed in Hanan et al. (1976a, 1978); in these discussions, constructive placement, as represented by cluster growth (called "constructive initial placement"), is compared with various placement improvement algorithms. Richard (1984) and Palczewski (1984) discuss constructive placement algorithms as a class, although both use a different taxonomy than is used here. Palczewski (1984) algorithms are divided into the following classes: cluster growth, partitioning-based placement, global techniques, and branch and bound techniques. The basic algorithms have been combined and extended in an attempt to improve placement quality. These classes, as well as the combinations, are discussed below.

Cluster Growth

Cluster growth constructive placement is a bottom-up implementation method that operates by selecting components and adding them to a partial, or incomplete, placement. These methods are differentiated from other placement methods in that cluster growth chooses and places the components independently. Cluster growth is analogous to crystal growth because subsequent components are placed around the seed components, causing the placement to grow like a crystal; it has also been called epitaxial growth (Soukup 1981).

The generic cluster growth algorithm is provided in Figure 20-11, while its operation is illustrated in Figure 20-12. In Figure 20-12, the dots represent the slots in which the rectangular components may be placed. The first step is to determine a seed placement. The components in the seed placement and their positions may be chosen by the user in order to guide the placement process,

```
Determine component(s) for seed
PLACE the seed component(s)
UNTIL all components are placed DO
    SELECT component(s) from the unplaced set
    PLACE the selected component(s)
ENDLOOP
```

Figure 20-11.   Generic cluster growth algorithm.

or may be determined algorithmically (Schweikert 1976). Next, unplaced components are sequentially selected and placed in relation to those components already placed. This process continues until all components are placed.

The selection function of the cluster growth algorithms determines the order in which unplaced components are included in the placement. The order is determined by how "strongly" the unplaced components are connected to the placed components. The placement function determines the best positions for the selected components. Typically the positions adjacent to previously placed components, the *candidate* positions, are investigated by calculating the metric, or *score*, resulting from placing the selected component at the candidate positions. The component is then placed at the candidate position that results in the best score. By necessity the score must be based on incomplete information because there are unplaced components that cannot contribute to the scoring. It is often necessary to choose among candidate positions that produce the same score. These *ties* can be resolved by a secondary metric such as the candidate position closest to the "center-of-gravity" of the component directly connected to the candidate component. Hanan and Kurtzberg (1972a) and Hanan et al. (1976a)

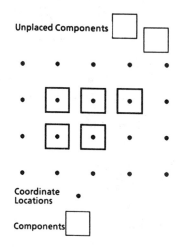

Figure 20-12.   Cluster growth placement.

provide an extensive discussion of selection and place functions for cluster growth. The definition of the selection and positioning functions differentiates the various cluster growth techniques. The more important selection and positioning algorithms are described next.

Random Placement.   Random constructive placement is a degenerate form of cluster growth. For this method the selection function reduces to randomly choosing an unplaced component and placing it in a randomly chosen position. While this circuit is fast and easy to program, it utilizes no circuit-specific information. Consequently, poor results may be expected (Magnuson 1977; Hanan et al. 1976a,b). A closely allied method is arbitrary constructive placement in which unplaced components are selected arbitrarily (e.g., in sequential order) and are placed in arbitrary (e.g., the next available) positions.

Pair Linking.   Pair linking (Kurtzberg 1965; Hanan and Kurtzberg 1972a) is a constructive placement algorithm in which the selection function chooses the unplaced component that is the most strongly connected to any single placed component. The chosen component is then positioned as closely to its (strongly connected) pair as possible.

Cluster Development.   The selection function of the cluster development algorithm (Kurtzberg 1965; Hanan and Kurtzberg 1972a) chooses the unplaced component that is most strongly connected to all of the placed components, and the position function chooses a place for that component with respect to the other components to which that component is connected. A closely related cluster growth method is described in Schweikert (1976). This method uses an "inside–outside" selection function, and the position function uses an exhaustive search of the available positions that account for placement restrictions. Other cluster development algorithm descriptions appear in Schuler and Ulrich (1972), Cox and Carroll (1980), and Kang (1983); these algorithms share the common feature that they position the components in a line and then "fold" the line to form a two-dimensional layout. A knowledge-based approach to cluster development is reported in Odawara et al. (1985).

Cluster Growth Algorithms.   The performance of the cluster growth algorithms is dependent on the number of interconnections and the number of pins per interconnection, but in general the dominant factor is the number of components. Kurtzberg (1965) shows that the cluster growth and pair linking algorithms have a computational complexity of n^2 where n is the number of components. Their performance is discussed in Hanan and Kurtzberg (1972a) and Palczewski (1984). These algorithms are easy to implement, but modern systems tend to favor the bipartitioning or global methods that are described next.

Partitioning-Based Placement

Placement algorithms based on partitioning that separates components into two or more parts while reserving space for the components during the partitioning process are widely used in modern layout systems. This "top-down" design approach tends to avoid heavy wiring congestion typically found in the center of the carrier. These algorithms differ from the cluster growth algorithms in that the partitioning-based algorithms consider all interconnections in parallel and then move the components in steps by partitioning them into specific areas of the placement surface. The *bipartitioning* or min-cut algorithms, illustrated in Figure 20-13, divide the components into two sets such that the number of weighted connections is minimized between the sets, and the area of the components is approximately equal. This process is repeated recursively until each partition contains only one component. Partitioning heuristics relating to this process are described in Kernighan and Lin (1970) and Schweikert and Kernighan (1972). Breuer (1977) describes the concepts used in this approach and introduces the placement problem. Corrigan (1979) describes a bipartitioning placement system with more sophisticated partitioning sequences.

Another partitioning-based placement approach, particularly well suited for standard cell and gate array placement problems, is described in Wada (1982), Kozawa et al. (1983), and Kambe et al. (1982). This approach first partitions the components into rows and then uses a bipartitioning approach to place the components within the rows.

These partitioning-based algorithms make use of the interconnection infor-

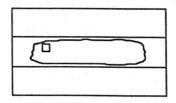

(A) Component Has Freedom Within Area.

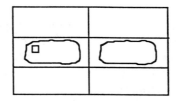

(B) Component Has Freedom Within Reduced Area.

Figure 20-13.   Min-cut placement. (a) Component has freedom within area. (b) Component has freedom within reduced area.

mation at a global level and defer local considerations until late in the placement process (Lauther 1979). Although these procedures produce good results, they tend to be computationally expensive (Murai et al. 1980). Also, these methods tend to be difficult to use when components are constrained to fixed positions, and when natural partitions of the circuit do not correspond to the specified partitioning sequences.

Global Placement

While all placement techniques keep global notions of better or worse placements, the global placement techniques are distinguished by the way in which they move components. Global methods move all of the components in parallel along an n dimensional gradient. This contrasts to clustering methods, which consider each component (or at most a small number of components) sequentially, and to partitioning methods, which first divide the components by partitioning and then deal with the components individually. It seems that superior results would be obtained by considering all of the interconnections simultaneously and moving all the components in parallel to find a global solution; however, there is still insufficient data to support this conclusion. The global methods discussed are quadratic assignment and convex optimization.

Quadratic Assignment. Quadratic assignment methods solve a special case of the placement problem. In general, placement algorithms must deal with signal sets containing two or more pins, while the quadratic assignment can only deal with pairs of pins. Usually the quadratic assignment problem is formulated as follows: given a cost matrix $C = [c_{ij}]$, where c_{ij} is the strength of the connection (number and weight of connections) between two components i and j, and a distance matrix $D = [d_{kl}]$, where d_{kl} is the distance between positions k and l, find the minimum of

$$G = \Sigma \, c_{ij}{}^* d_{p(j)p(i)}$$

over all permutations, p, of the components' positions. Hanan and Kurtzberg (1972a) describe the quadratic assignment problem and how a placement problem can be transformed into an associated quadratic assignment problem. However, an optimum solution to the associated quadratic assignment problem does not guarantee an optimum solution to the original placement problem. An efficient technique for solving large assignment problems has been reported in Hung and Rom (1980). The linear assignment problem may also be applied in placement algorithms; Akers (1981) provides a good description of the linear assignment problem and its application to placement. Although the assignment methods may use quadratic metrics, they do not exploit the convexity of the metric; techniques that do are described next.

Convex Function Optimization. Through appropriate abstractions that produce a convex domain (primarily the absence of slots and absence of components with finite size) and the use of a quadratic metric, it is possible to map a placement problem into one that may be solved using convex optimization techniques. The advantage of convex optimization defined on a convex domain is that there is only one minimum for the objective function. Although these methods suffer when mapping the abstract problem formulation back into the physical domain (actual components must be mapped into slots on the placement surface), good results have been reported. A general optimization reference is Walsh (1975), while a discussion of quadratic optimization applied to placement may be found in Blanks (1985b).

Quinn (1975) uses a force analogy to derive a system of equations in quadratic form, modeling the interconnection nets with a complete graph with weighted connections. Repulsive forces serve to incorporate the realism of the placement surface into the carrier, but this destroys the convexity of the resulting metric. Cheng and Kuh (1984) model the components and their interconnections as a resistive electrical circuit and then use relaxation to solve the resulting equations. This approach requires some fixed-position components—otherwise, all of the components would collapse to a single point. A justification for quadratic metrics is provided in Fukunaga et al. (1983). This reference uses a state space approach to derive a global equation matrix, with the eigenvectors corresponding to the two smallest eigenvalues determining the placement. Assignment by relaxation (Hung and Rom 1980) is used to map the global solution onto the carrier.

These techniques in general suffer from an inability to deal with practical constraints such as as finite component area and constraints imposed on component positions. These drawbacks are addressed in Blanks (1985a,b) through the use of quadratic constraints and a two-step procedure to map the global (ideal) placement onto the carrier without violating any physical constraints. Sha and Dutton (1985) attempt to alleviate the need for the second step by encoding all geometric information into constraints placed on the scoring function. Because few comparisons have been performed, it is difficult to evaluate the relative merits of these global methods versus other placement methods.

Branch and Bound

Branch and bound techniques can be used to find an optimum solution to small placement problems (Hanan and Kurtzberg 1972a; Gilmore 1962). Lawler and Wood (1966) and Hillier and Lieberman (1980) provide introductions to general branch and bound techniques. The basic idea of the technique is as follows: Assume that an upper bound on the score for a placement problem is known (normally this is obtained from a constructive placement algorithm). The first step is to partition the partial placements into subsets. Lower bounds of the score for the placements within the subsets are computed. Those subsets that have lower bounds greater than the current upper bound are excluded from further consideration. A remaining subset, perhaps the one with the smallest lower bound, is further partitioned, and the process continues until the best solution is found. The lower bound, then, allows the decision tree to be pruned, and potentially results in reductions in the computational requirements. The more accurately the lower bound is calculated, the sharper the pruning. However, there is a trade-off; it takes longer to calculate a more accurate lower bound. Because of the need either to calculate accurate lower bounds or to explore a large number of branches, this technique is usually quite time-consuming. Many heuristics have been suggested to quickly prune the tree (Hanan and Kurtzberg 1972a) and thus reduce the number of branches explored. Most of these heuristics have a computational complexity of n^3 or n^4. Branch and bound algorithms do not appear in modern layout systems; practical problem sizes have grown so large that run times are excessive even with the heuristic bounds computation.

Combinations of Constructive Algorithms

Although each of the constructive placement algorithms surveyed in this section possesses desirable characteristics, each also exhibits some undesirable features. As a result, enhancements or combinations of approaches have been proposed. Wipfler et al. (1982) report a combined force and min-cut approach in which the min-cut function is used to account for actual component geometries that the basic force method is unable to handle. Richard (1984) uses a combination of partitioning and cluster growth, while the use of linear assignment combined with cluster growth is reported in Akers (1981).

ITERATIVE PLACEMENT

The goal of iterative placement is to manipulate a complete placement to produce an improved complete placement. This process is performed within an iteration loop that continues until a stopping criterion is met. For example, the stopping criterion might be relative or absolute improvement in the placement metric, or perhaps the time expended in the iterative process. Within one iteration loop, components are selected and moved to alternate locations. If the resulting configuration is better than the old, the new configuration is retained; otherwise, the previous configuration is returned. The improvement process and the more important iterative improvement algorithms are discussed in this section.

Three Phases of Iterative Placement

Many different iterative placement techniques exist. Although they differ substantially, they share the same underlying structure and have three main phases: selection, movement, and scoring. The generic form for an iterative improvement

```
SCORE current placement
UNTIL stopping criterion is satisfied DO
    SELECT component(s) to move
    MOVE selected component(s) to trial positions
    SCORE trial placement
    IF trial_score < current_score THEN
        current_score ← trial_score
    ELSE
        MOVE selected component(s) to previous positions
ENDLOOP
```

Figure 20-14. Generic iterative placement algorithm.

placement algorithm is shown in Figure 20-14. These three phases are discussed next.

Selection.   The selection function chooses the components to participate in movement. This mechanism reduces the set of all possible combinations of components to move simultaneously to a computationally feasible subset of those combinations. The selection process may simply select components to be interchanged in a predefined sequence (such as trying all possible pairs), or may involve intelligence to select those components that are placed poorly. Incorporating intelligence into the selection process typically allows placement to converge more quickly, but may not improve the quality of the solution.

One possible selection technique is to sequence through all possible s-tuples of component combinations. It is questionable whether using values of $s > 2$ achieves significant improvement over simply trying all pairs ($s = 2$). Gilmore (1962) and Reiter and Sherman (1965) report improvement by using $s > 2$, while Garside and Nicholson (1968) report little improvement at the expense of a significant increase in computation time. For placement algorithms that incorporate computationally intensive scoring metrics, it may be impractical to score all of the combinations generated by an s-tuple selection scheme. It is possible to extend the selection scheme by including a predictor that is computationally simpler than the scoring metric but has a sufficient correlation with the scoring metric to filter a large number of the combinations that are "obviously" going to fail. Such predictor mechanisms can substantially reduce the number of combinations that must be scored, but may overlook potentially successful combinations.

Random pair selection chooses a pair of components at random from all of the available components. At first this selection technique may seem to have little merit; however, one class of placement algorithms, simulated annealing (discussed later), requires that the placement space be explored at random.

Another selection technique uses force vectors as an aid to decide the components that should be moved (Fisk et al. 1967; Hanan et al. 1976a). The interconnections are modeled as springs (as discussed in the section on abstractions). The components selected are the ones exhibiting strong forces, which indicates they are far from their equilibrium positions.

Movement. Once the components are selected for trial interchange, the movement function determines the new locations for the selected components. If a pair of components is selected, then the components' positions are interchanged. It must, of course, be physically possible for each component to fit at the location of the other component.

Multiple way exchange involves moving more than two components (Cote and Patel 1980). In this case, each component has multiple positions to which it can be relocated. One movement scheme orders the components from worse-placed to best-placed and then chooses new positions for the components in this order. Even though multiple way exchange allows combinations of component placements to be found that pairwise exchange might not find, experience shows multiple way exchange leads to only slightly better results at a large increase in computation time. The force methods discussed previously also lead to movement heuristics because the position where the force on a component is minimum can be computed.

Sometimes iterative placers are limited to moving components among positions on the carrier that are slots for certain types or sizes of components. These

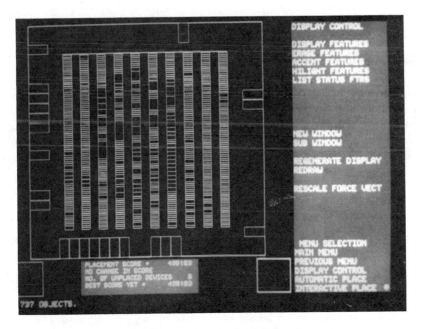

Figure 20-15. An example of static slot definition for a gate array.

positions are based on the initial positions of the components being considered. An example of this static slot definition for a gate array carrier is shown in Figure 20-15. This restriction eliminates the need to check dynamically for component overlap, and thus results in much faster placement techniques.

Scoring. After the selected components have been moved to new positions, the evaluation function is invoked to measure the quality of the new arrangement. Both the interconnection models and the scoring metrics were discussed in the section on abstract models of placement. At this stage in the placement process, the representation of interconnections should be as accurate as possible consistent with the time available for iterative improvement. This means that a connection tree should match the actual routes that will be generated. The scoring metric may be the same as that used by the initial placer or may incorporate more detail about the routing resources. The iterative placement scoring metric should work or may incorporate more detail about the routing resources. The iterative placement scoring metric should work in concert with that used by the initial placer; otherwise, any intelligence in the initial placement is wasted.

Iterative Improvement Techniques

Now that the three main functions within the iterative placement loop have been described, some of the iterative algorithms will be reviewed in terms of these functions.

Pairwise Interchange. In pairwise interchange each component is selected in turn to be the *primary* component and is trial-interchanged with each other component. If a trial interchange results in an improvement in the placement, the interchange is accepted; otherwise, the scoring metric can be used as the basis for acceptance of an exchange. This technique results in $n*(n - 1)/2$ trial interchanges, making the computational complexity (On^2), where n is the number of components. Some placement systems using pairwise exchange have been reported in Hanan et al. (1976a), Schweikert (1976), Iosvpovici et al. (1983), and Khokhani and Patel (1977).

Neighborhood Interchange. The neighborhood interchange technique is similar to the pairwise interchange technique; however, the primary component is interchanged only with components in its vicinity. This vicinity (distance or the number of components included) can be set by the user. If the number of components included in the vicinity is m, the run time then becomes proportional to $(n*m)/2$ if there are n components. Placement systems using neighborhood interchange are described in Hanan et al. (1976a).

Force Directed Interchange.   Force directed interchange uses a force analog, as discussed previously, to select components to move, as well as to determine the positions to which the components should be moved. However, the position to which the selected component should be moved may already be occupied by another component. It is possible that the location of the selected component is not a good position for the component occupying its favored positions. In simple force directed interchange this problem is avoided by trial-interchanging the selected component with the three adjacent components in the direction of the desired location. That is, the primary component is interchanged with its nearest horizontal, vertical, and diagonal neighbors in the direction of the desired location. If the desired location is collinear with the current location, then only one interchange takes place, with the component immediately adjacent in the desired direction. Placement techniques utilizing force directed interchange are described in Hanan et al. (1976a).

Force Directed Relaxation.   The force directed relaxation technique is similar to the force directed exchange method in the calculation of force vectors for each component. In this technique, however, the primary component is positioned in the compatible slot nearest to the desired zero force point (Hanan et al. 1976a). The component that was occupying that slot is chosen as the next primary component. The process results in a series of components to be relaxed. This series terminates when the primary component is moved to an empty slot (i.e., a slot that did not have an occupant component). When a series terminates, the new score of the placement is calculated and compared to the previous score. If the placement improved, the series is accepted, and all components remain in their current positions; otherwise, the series is rejected, and all components are returned to their previous positions. The process terminates when all components have either initiated or are contained in an accepted series. The primary components to initiate series is based on the choice of descending order of the number of signal sets to which each component belongs. Fisk et al. (1967) describe a similar technique but one without slot definitions.

Force Directed Pairwise Relaxation.   The force directed pairwise relaxation method also uses force vectors to find the zero force target locations for each component. In this method, however, the primary component is not allowed to initiate a series as in force directed relaxation. Instead the primary component A is trial-interchanged with a component B in the vicinity of the target location only if the target location of B is in the vicinity of component A. The interchange is accepted only if the placement score improves. The components are chosen to be primary components in descending order based on the number of signal sets to which they belong. Once a component has participated in an accepted interchange, it can no longer be a primary component. The placement process ceases when all components have been chosen for relaxation. Placement systems using force directed pairwise relaxation have been reported by Hanan et al. (1976a) and Kozawa et al. (1983).

Unconnected Sets. Steinberg (1961) describes a technique that selects components by subdividing them into subsets of components that have no signal sets in common, or *unconnected sets* of components. Each component in such a subset can be trial-placed without considering the other components because they are unconnected. Each component in the subset can be moved to each location occupied by components in the subset and the score computed, or the placement of these components can be formed by solving the resultant assignment problem. After each unconnected subset has been processed in this manner, the cycle is complete. The procedure can be either terminated or repeated in an iterative process. The size of the unconnected subsets, the number of unconnected subsets generated, and the number of cycles performed are parameters of the technique. Experience with this technique (Ciampi 1975; Hanan et al., 1976a,b) indicates limited success because the number of components moved by any step is small, and nets with large numbers of pins (e.g., clocks) prevent component movement.

Simulated Annealing. The previous techniques share the characteristic that trials are accepted only if the score does not increase. This constraint can cause the placement to reach a local score minimum but miss the global optimum. An example of this is shown in Figure 20-16. In this example, a placer starts at placement state A, accepts only decreasing scores, and gets stuck at a local minimum at state B. However, if it accepted score increases, it might climb out

$$P(i) = e^{-E(i)/K_b t}$$

where: $P(i)$ = probability of configuration i
$\quad\quad$ $E(i)$ = energy of configuration i
$\quad\quad$ K_b = Boltzman's constant
$\quad\quad$ t = temperature

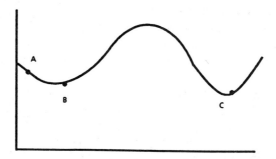

Figure 20-16. A monotonically decreasing algorithm may get stuck in a local minimum.

of the local minimum and might find the global minimum at state C. Kirkpatrick et al. (1983) describe an optimization technique, simulated annealing, that addresses this problem. This technique utilizes an algorithm (Metropolis et al. 1953) developed to find the lowest energy configuration of a collection of confined molecules, where the probability of any configuration i is given by the Boltzman distribution:

$$P(i) = e^{-E(i)/K_b t}$$

where $P(i)$ = probability of configuration i, $E(i)$ = energy of configuration i, K_b = Boltzman's constant, and t = temperature.

The optimization of a circuit placement with a very large number of components is analogous to the process of annealing, in which a material is melted and cooled slowly so that it will crystallize into a highly ordered state. The energy corresponds to the placement score, and an "annealing schedule" specifies the beginning and ending scores as well as the rate at which the score is lowered (melting and freezing temperatures and the rate at which the temperature is lowered). The method begins with a random initial placement. An altered placement is generated, and the resulting change in score ΔE is calculated. If $\Delta E < 0$ (the system went to a lower energy level), then the move is accepted. If $\Delta E > 0$ then the move is accepted with probability $e^{-\Delta E/K_b t}$. This method has been used extensively as an iterative improvement technique. The generic iterative placement algorithm is rearranged as shown in Figure 20-17 to illustrate the simulated annealing concepts.

Simulated annealing can climb out of local minima and can find the global optimum if an infinite number of iterations are performed at each temperature (Romeo and Sangiovanni-Vincentelli 1985). Because it is impossible to perform an infinite number of iterations, heuristics have been developed for determining the number of iterations at each temperature. Some of these heuristics are simple (Kirkpatrick et al. 1983; Romeo et al. 1984), while others are quite sophisticated (Romeo and Sangiovanni-Vincentelli 1985).

One problem with simulated annealing is determination of the annealing schedule. Heuristics for determining the "melting" and "freezing" points may be simple (Kirkpatrick et al. 1983; Romeo et al. 1984) or complex (White 1984). Knowing how fast to "cool" the system is also problematic. It is desirable to cool the placement rapidly to reduce the computation time. However, if the system is cooled too quickly, it can become stuck in a local minimum. Most approaches use the conservative technique of updating T as $T \leftarrow alpha*T$ where alpha is greater than 0.95, but this produces very long run times.

In order to decrease the run time, a "rejectionless" method has been devised in which the probability of selecting a move is based on its probability of being accepted (Greene and Supowit 1984). The technique requires keeping information about all possible moves, which is both space- and time-consuming. An accep-

```
T ← T0
SCORE current placement
UNTIL freezing point reached DO
    UNTIL equilibrium at current temperature reached DO
        SELECT and MOVE components randomly
        SCORE trial placement
        ΔS ← trial_score - current score
        IF ΔS < 0 THEN current score ← trial score
        ELSE BEGIN
            r ← uniform random number {0 <= r <= 1}
            IF r < e^{-ΔS/T} THEN
                current_score ← trial score
            ELSE
                MOVE selected components
                to previous positions
        END
    ENDLOOP
    T ← alpha * T {0 < alpha < 1}
ENDLOOP
```

Figure 20-17.   Simulated annealing algorithm.

tance rate crossover point exists that is problem-dependent. If the acceptance rate is above the crossover point, then standard simulated annealing is faster. Once the acceptance rate falls below the crossover point, the rejectionless method is faster. This suggests a two-step method where standard simulated annealing is used above the crossover point, and the rejectionless method is used below.

Regardless of the method used, simulated annealing is computationally intensive compared to other techniques. The question is whether it achieves better results in a reasonable amount of time. Unfortunately, few comparative results exist. Experiments reported in Nahar et al. (1985) compared simulated annealing to other heuristics and concluded that several other heuristics performed as well as or better than simulated annealing for a fixed amount of computation time. It would be more meaningful to allow simulated annealing heuristics to run to completion and give the final results and the amount of computation time required to achieve those results.

Summary of Placement Techniques

While finding the optimum solution to the general placement problem is NP-complete, the placement methods discussed provide heuristic techniques that attempt to provide a "good" placement. These techniques are divided into two categories: constructive placement and iterative placement. The computational complexity of these heuristics normally ranges from n^2 to n^4. Many other heuristics are possible with varying complexities, but the methods discussed represent the most popular placement techniques.

APPLICATIONS OF PLACEMENT ALGORITHMS

So far, discussion has been limited to general models and algorithms for component placement, but the ways in which the algorithms are applied vary significantly, depending on various problem and design environment attributes. Some of these attributes are explored in this section: problem size, component shapes, and floor planning.

Problem Size

Currently most placement algorithms operate on a *flat* representation, meaning that the placement algorithms do not make use of information at higher or lower levels of the design hierarchy. However, because placement algorithms are typically $O(n^2)$ or higher, it may be possible to consider all components to be placed simultaneously in a very large circuit. At this point there are three choices: partitioning, hierarchical placement, or hardware accelerators.

Partitioning.   In this approach, flat placement problems are divided into separate but interdependent subproblems that are solved sequentially. The subproblems are not independent, as the decisions made in one subproblem affect the others; decisions that improve the placement within one partition might degrade the overall placement. However, subdividing the problem reduces its complexity. For an algorithm with complexity of $O(n^2)$ (n is the number of components), a problem subdivided into p parts has a complexity of $O(p(n/p)^2)$ or $O(n^2/p)$. Khokhani et al. (1981) describe a method that partitions a circuit into "super nodes," and that uses constructive initial placement and iterative placement to improve the assignment of super nodes to "super locations." Next the super nodes are decomposed into primitives and placed. Szepieniec and Otten (1980) present a partitioning approach that "slices" a layout recursively to perform the partitioning.

Hierarchy.   In hierarchical placement, problem complexity is reduced by subdividing the circuit into multiple levels. The top level divides the circuit into first-level components, which are themselves partitioned into lower-level components. This subdivision continues recursively until (for purposes of placement) primitive placement components are defined. The hierarchical subdivision may be the same as that defined in the design specification (Preas and Gwyn 1978; Preas 1979), or the layout system may provide automatic hierarchical partitioning (Payne and van Cleemput 1982) before invoking placement algorithms separately on the resultant components.

Hardware Accelerators.   It is possible to reduce the execution times through the use of special-purpose hardware optimized for placement tasks, or multiple processors that allow parallel processing. Utilization of hardware accelerators normally requires algorithmic modifications; hardware accelerators can be applied in placement problems by partitioning the problems into "independent" subproblems and assigning each subproblem to a separate processor.

A discussion of a hardware accelerator implementation with comparative results is given in (Spira and Hage (1985). An implementation for a placement accelerator with applications to force directed interchange, force directed pairwise relaxation, and neighborhood interchange is described in Iosupovici et al. (1983). Other placement accelerators involve extensive concurrency on an array of processors (Chyan and Breuer 1983); in this case, care must be taken to avoid the oscillation that may occur when independent problems are being optimized concurrently. Discussion of hardware accelerators for general computer aided design is provided in Blank (1984).

Component Shapes

The basic placement algorithms discussed in the two preceding sections do not consider the shapes of the components or the interaction of a component's sizes with the wiring requirements. Classical placement algorithms assume that placement quality is a function only of the interconnections and is not affected by the shapes of components (Hanan and Kurtzberg 1972b; Hanan et al. 1976a). Such assumptions model the components as "point objects."

An important class of standard cell and gate array placement surfaces arranges the components in rows (columns can be converted to rows by rotation); in effect, the components have only one important dimension. In this case, an accurate placement metric must include the width of the rows, as well as the heights of the routing channels, so simple wire length no longer provides a sufficient metric (Persky 1976). A more accurate placement metric is chip area (Preas and Gwyn 1978) for standard cell designs or completion rate for gate array designs.

Structured custom layout problems must model the components with two dimensions because both their height and width can affect the layout (Preas 1979; Preas and van Cleemput 1979; Lauther 1979). In this case, routing area is the only appropriate metric, but it is very difficult to compute.

Floor Planning

Floor planning is traditionally treated as a separate problem, but it is actually a placement problem that has an extra degree of freedom because the shapes of the components can change (Chow 1985; Otten 1982; Heller et al. 1982; Slutz

1980). Early in the design process of a typical structured custom integrated circuit, when the subcomponents have not yet been designed, their shapes may be altered by the layout system or the designer. At this time, a "floor planning placer" may vary the shapes of (some of) the components (subject to some constraints), as well as their positions, in order to optimize the placement.

Choice of Algorithms

This chapter reviews and discusses the major placement algorithms that are currently available. A large number of choices are available because digital circuits have a wide variety of attributes that lead to different placement algorithm implementations. The primary attributes to take into account when choosing a placement approach are the following:

- Quality of results required
- Problem characteristics
- Computer time available
- Programming time

An early choice is that of the data structure to implement the abstract models of the placement problem. Fixed coordinate designs methodologies (printed circuits and gate arrays) use absolute coordinates, while standard cell and structured custom methodologies must be built on a layer that supports relative coordinate manipulation. Of course, much more care must be accorded implementation for large placement problems and those that require very good placement results.

SUMMARY

This chapter defines the placement problem, and categories and reviews the currently available placement techniques. It concentrates on placement of components in the upper and intermediate levels of the design hierarchy. The actual placement algorithms are emphasized, and little is said about the nature of the components being placed or the placement surface on which the components reside. As a result of this technology-independent presentation, the placement concepts are applicable to digital circuit design for a wide range of circuit manufacturing technologies. Placement methods fall into two groups: constructive and iterative. Constructive methods produce a complete placement (where all circuit components have assigned positions) based on a partial placement (in which some or all components do not have assigned positions) as input, while iterative methods seek to improve a complete placement by producing anew, better, complete placement.

Automated placement is extremely important because it has a great influence on the amount and location of wiring required to interconnect the components. Therefore, the placement phase impacts the router's ability to complete the required interconnections, as well as affecting the performance of the digital circuit. A good placement must be generated where the metric that is optimized closely models the important factors in routing. The models and algorithms chosen for a particular layout system depend on many factors. When one is considering speed versus performance trade-offs, it is difficult, if not impossible, to find a single model and algorithm that work best for all circuits encountered. Several models and algorithms that address this problem have been presented.

Good placement is critical in order to generate high-quality layouts. Inclusion of superior placement techniques is therefore essential to the success of a design automation system.

ACKNOWLEDGMENT

The authors express their appreciation to K.A. Preas for her care and patience in editing this chapter as well as insuring the readability of the resultant work.

REFERENCES

Akers, S. B., "On the Use of the Linear Assignment Algorithm in Module Placement," *Proceedings of the 18th Design Automation Conference,* 1981, pp. 137–144.

Blank, T., "a Survey of Hardware Accelerators Used in Computer Aided Design," *IEEE Design and Test of Computers, 1* (3), 21–39 (Aug. 1984).

Blanks, J. P., "Initial Placement of Gate Arrays Using Least-Squares Methods," *Proceedings of the 21st Design Automation Conference,* 1984, pp. 670–671.

Blanks, J. P., "Near-Optimal Placement Using a Quadratic Objective Function," *Proceedings of the 22nd Design Automation Conference,* 1985a, pp. 609–615.

Blanks, J. P., "Use of a Quadratic Objective Function for the Placement Problem in VLSI Design," doctoral dissertation, University of Texas at Austin, 1985b.

Brady, H. N., "An Approach to Topological Pin Assignment," *IEEE Transactions on Computer-Aided Design of Integrated Circuits and Systems, CAD-3* (3), 250–255 (July 1984).

Breuer, M. A., Editors, *Design Automation of Digital Systems,* Volume One, *Theory and Techniques,* Prentice-Hall, Englewood Cliffs, NJ, 1972.

Breuer, M. A., "A Class of Min-Cut Placement Algorithms," *Proceedings of the 14th Design Automation Conference,* 1977, pp. 284–290.

Chandrasekhar, M. S., and Breuer, M. A., "Optimum Placement of Two Rectangular Blocks," *Proceedings of the 19th Design Automation Conference,* 1982, pp. 879–886.

Chang, S., "The Generation of Minimal Trees with a Steiner Topology," *Journal of the Association for Computing Machinery, 19* (4), 699–711 (Oct. 1972).

Cheng, C. K., and Kuh, E. S., "Module Placement Based on Resistive Network Optimization," *IEEE Transactions on Computer-Aided Design of Integrated Circuits and Systems, CAD-3* (3) 218–225 (July 1984).

Chow, C. S., "Phoenix: An Interactive Hierarchical Topological Floorplanning Placer," master's thesis, Department of Electrical Engineering and Computer Science, Massachusetts Institute of Technology, 1985.

Chyan, D., and Breuer, M. A., "A Placement Algorithm for Array Processors," *Proceedings of the 20th Design Automation Conference*, 1983, pp. 182–188.

Ciampi, P. L., "A System for Solution of the Placement Problem," *Proceedings of the 12th Design Automation Conference*, 1975, pp. 317–323.

Corrigan, L. I., "A Placement Capability Based on Partitioning," *Proceedings of the 16th Design Automation Conference*, 1979, pp. 406–413.

Cote, L. C., and Patel, A. M., "The Interchange Algorithms for Circuit Placement Problems," *Proceedings of the 17th Design Automation Conference*, 1980, pp. 528–534.

Cox, G. W., and Carroll, B. D., "The Standard Transistor Array (STAR), Part II: Automatic Cell Placement Techniques," *Proceedings of the 17th Design Automation Conference*, 1980, pp. 451–457.

Deo, N., *Graph Theory with Applications to Engineering and Computer Science*, Prentice Hall, Englewood Cliffs, NJ, 1974.

Donath, W. E., "Complexity Theory and Design Automation," *Proceedings of the 17th Design Automation Conference*, 1980, pp. 412–419.

Fisk, C. J., Caskey, D. L., and West, L. E., "ACCEL: Automated Circuit Card Etching Layout," *Proceedings of the IEEE*, 55 (11), 1971–1982 (Nov. 1967).

Fukunaga, K., Yamada, S., Stone, H. S., and Kasai, T., "Placement of Circuit Modules Using a Graph Space Approach," *Proceedings of the 20th Design Automation Conference*, 1983, pp. 465–471.

Garside, R. G., and Nicholson, T. A. J., "Permutation Procedure for the Backboard-Wiring Problem," *Proceedings of the IEEE*, 115 (1), 27–30 (Jan. 1968).

Gilmore, P. C., "Optimal and Suboptimal Algorithms for the Quadratic Assignment Problem," *Journal of the Society for Industrial and Applied Mathematics*, 10 (2), 305–313 (June 1962).

Goto, S., Cederbaum, I., and Ting, B. S., "Suboptimum Solution of the Back-board Ordering with Channel Capacity Constraint," *IEEE Transactions on Circuits and Systems*, CAS-24, no 11, November 1977, pp. 645–652.

Greene, J. W., and Supowit, K. J., "Simulated Annealing without Rejected Moves," *Proceedings of the International Conference on Computer Design*, 1984, pp. 658–663.

Hanan, M., "On Steiner's Problem with Rectilinear Distance," *SIAM Journal on Applied Mathematics*, 14 (2), 255–265 (Mar. 1966).

Hanan, M., and Kurtzberg, J. M., "Placement Techniques," in M. A. Breuer, Editors, *Design Automation of Digital Systems. Volume One, Theory and Techniques*, Prentice-Hall, Englewood Cliffs, NJ, 1972a, Chap. 5, pp. 213–282.

Hanan, M., and Kurtzberg, J. M., "A Review of the Placement and Quadratic Assignment Problems," *SIAM Review*, 14 (2), 324–342 (Apr. 1972b).

Hanan, M., Wolff, P. K., Sr., and Agule, B. J., "Some Experimental Results on Placement Techniques," *Proceedings of the 13th Design Automation Conference*, 1976a, pp. 214–224.

Hanan, M., Wolff, P. K., Sr., and Agule, B. J., "A Study of Placement Techniques," *Journal of Design Automation and Fault-Tolerant Computing*, 1 (1), 28–61 (Oct. 1976b).

Hanan, M., Wolff, P. K., Sr., and Agule, B. J., "Some Experimental Results on Placement Techniques," *Journal of Design Automation and Fault-Tolerant Computing*, 2, 145–168 (May 1978).

Heller, W. R., Sorkin, G., and Maling, K., "The Planar Package Planner for System Designers," *Proceedings of the 19th Design Automation Conference*, 1982, pp. 253–260.

Hillier, F. S., and Lieberman, G. J., *Introduction to Operations Research*, Holden-Day, San Francisco, 1980.

Hung, M. S., and Rom, W. O., "Solving the Assignment Problem by Relaxation," *Operations Research*, 28 (4), 969–982 (July–Aug. 1980).

Hwang, F. K., "On Steiner Minimal Trees with Rectilinear Distance," *SIAM Journal on Applied Mathematics*, 30 (1), 104–114 (Jan. 1976).

Iosupovici, A., King, C., and Breuer, M. A., "A Model Interchange Placement Machine," *Proceedings of the 20th Design Automation Conference*, 1983, pp. 171–174.

Kambe, T., Chiba, T., Kimura, S., Inufushi, T., Okuda, N., and Nishioka, I., "A Placement Algorithm for Polycell LSI and Its Evaluation," *Proceedings of the 19th Design Automation Conference*, 1982, pp. 655–662.

Kang, S., "Linear Ordering and Application to Placement," *Proceedings of the 20th Design Automation Conference*, 1983, pp. 457–464.

Kani, K., Kawanishi, H., and Kishimota, A., "ROBIN: A Building Block LSI Routing Problem," *Proceedings of the International Symposium on Circuits and Systems*, 1976, pp. 658–661.

Karger, P. G., and Malek, M., "Formulation of Component Placement as a Constrained Optimization Problem," *Proceedings of the International Conference on Computer Design*, 1984, pp. 814–819.

Kernighan, B. W., and Lin, S., "An Efficient Heuristic Procedure for Partitioning Graphs," *Bell System Technical Journal*, 49 (2), 291–307 (Feb. 1970).

Khokhani, K. H., and Patel, A. M., "The Chip Layout Problem: A Placement Procedure for LSI," *Proceedings of the 14th Design Automation Conference*, 1977, pp. 291–297.

Khokhani, K. H., Patel, A. M., Ferguson, W., Sessa, J., and Hatton, D., "Placement of Variable Size Circuits on LSI Masterslices," *Proceedings of the 18th Design Automation Conference*, 1981, pp. 426–434.

Kirkpatrick, S., Gelatt, C. D., and Vecchi, M. P., "Optmization by Simulated Annealing," *Science*, 220 (4598), 671–680 (May 1983).

Koren, N. L., "Pin Assignment in Automated Printed Circuit Board Design," *Proceedings on the 9th Design Automation Workshop*, 1972, pp. 72–79.

Kozawa, T., Terai, H., Ishii, T., Hayase, M., Miura, C., Ogawa, Y., Kishida, K., Yamada, N., and Ohno, Y., "Automatic Placement Algorithms for High Packing Density VLSI," *Proceedings of the 20th Design Automation Conference*, 1983, pp. 175–181.

Kurtzberg, J. M., "Algorithms for Backplane Formation," in *Microelectronics in Large Systems*, Spartan Books, Washington, D. C., 1965, pp. 51–76.

Lauther, U., "A Min-Cut Placement Algorithm for General Cell Assemblies Based on a Graph Representation," *Proceedings 16th Design Automation Conference*, 1979, pp. 1–10.

Lawler, E. L., and Wood, D. E., "Branch-and-Bound Methods: A Survey" in *Operations Research*, 14 (4), 699–719 (July–Aug. 1966).

Lin, S., "Computer Solutions of the Traveling Salesman Problem," *Bell System Technical Journal*, 44 (10) 2245–2269 (Dec. 1965).

Loosemore, K. J., "Automated Layout of Integrated Circuits," *Proceedings of the IEEE International Symposium on Circuits and Systems*, 1979, pp. 655–668.

Magnuson, W. G., "A Comparison of Constructive Placement Algorithms," in *IEEE Region 6 Conference Record*, 1977, pp. 28–32.

Metropolis, N., Rosenbluth, A. W., Rosenbluth, M. N., Teller, A. H., and Teller, E., "Equation of State Calculations by Fast Computing Machines," *Journal of Chemical Physics, 21* (6), 1087–1092 June 1953.

Mory-Rauch, L., "Pin Assignment on a Printed Circuit Board," *Proceedings of the 15th Design Automation Conference*, 1978, pp. 70–73.

Murai, S., Kakinuma, M., Imai, M., and Tsuji, H., "The Effects of the Initial Placement Techniques on the Final Placement Results," *Proceedings of the IEEE Conference on Circuits and Computers*, 1980, pp. 80–82.

Nahar, S., Sahni, S., and Shragowitz, E., "Experiments with Simulated Annealing," *Proceedings of the 22nd Design Automation Conference*, 1985, pp. 748–752.

Nair, R., Bruss, A., and Reif, J., "Linear Time Algorithms for Optimal CMOS Layout," in *VLSI: Algorithms and Architecture*, P. Bertolazzi and F. Luccio, Editors, North-Holland: Elsevier Science Publishers B. V., 1985, pp. 327–338.

Nishioka, I., Kurimoto, T., Yamamoto, S., Shirakawa, I., and Ozaki, H., "An Approach to Gate Assignment and Module Placement for Printed Wiring Boards," *Proceedings of the 15th Design Automation Conference*, 1978, pp. 60–69.

Odawara, G., Iijima, K., and Wakabayashi, K., "Knowledge-Based Placement Technique for Printed Wiring Boards," *Proceedings of the 22nd Design Automation Conference*, 1985, pp. 616–622.

Otten, R. H. J. M., "Automatic Floorplan Design," *Proceedings of the 19th Design Automation Conference*, 1982, pp. 261–267.

Palczewski, M., "Performance of Algorithms for Initial Placement," *Proceedings of the 21st Design Automation Conference*, 1984, pp. 399–404.

Payne, T. S., and van Cleemput, W. M., "Automated Partitioning of Hierarchically Specified Digital Systems," *Proceedings of the 19th Design Automation Conference*, 1982, pp. 182–193.

Persky, G., "PRO—An Automatic String Placement Program for Polycell Layout," *Proceedings of the 13th Design Automation Conference*, 1976, pp. 417–424.

Preas, B. T., "Placement and Routing Algorithms for Hierarchical Integrated Circuit Layout," doctoral dissertation, Department of Electrical Engineering, Stanford University, 1979.

Preas, B. T., and Chow, C. S., "Placement and Routing Algorithms for Topological Integrated Circuit Layout," *Proceedings of the International Symposium on Circuits and Systems*, 1985, pp. 17–20.

Preas, B. T., and Gwyn, C. W., "Methods for Hierarchical Automation Layout of Custom LSI Circuit Masks," *Proceedings of the 15th Design Automation Conference*, 1978, pp. 206–212.

Preas, B. T., and van Cleemput, W. M., "Placement Algorithms for Arbitrarily Shaped Blocks," *Proceedings of the 16th Design Automation Conference*, 1979, pp. 474–480.

Prim, R. C., "Shortest Connection Networks and Some Generalizations," *Bell System Technical Journal*, 36 (6), 1389–1401 (Nov. 1957).

Quinn, N. R., "The Placement Problem as Viewed from the Physics of Classical Mechanics," *Proceedings of the 12th Design Automation Conference*, 1975, pp. 173–178.

Reiter, S., and Sherman, G., "Discrete Optimizing," *Journal of the Society for Industrial and Applied Mathematics*, 13, (3), 864–889 (Sept. 1965).

Richard, B. D., "A Standard Cell Initial Placement Strategy," *Proceedings of the 21st Design Automation Conference*, 1984, pp. 392–398.

Romeo, F., Sangiovanni-Vincentelli, A., and Sechen, C., "Research on Simulated Annealing at Berkeley," *Proceedings of the International Conference on Computer Design*, 1984, pp. 652–657.

Romeo, F., and Sangiovanni-Vincentelli, A., "Probablistic Hill Climbing Algorithms: Properties and Applications," *Proceedings of the 1985 Chapel Hill Conference on VLSI*, 1985, pp. 393–417.

Sahni, S., and Bhatt, A., "The Complexity of Design Automation Problems," *Proceedings of the 17th Design Automation Conference*, 1980, pp. 402–411.

Schuler, D. M., and Ulrich, E. G., "Clustering and Linear Placement," in *Proceedings of the 9th Design Automation Workshop*, 1972, pp. 50–56.

Schweikert, D. G., "A 2-Dimensional Placement Algorithm for the Layout of Electrical Circuits," *Proceedings of the 13th Design Automation Conference*, 1976, pp. 408–416.

Schweikert, D. G., and Kernighan, B. W., "A Proper Model for the Partitioning of Electrical Circuits," *Proceedings of the 9th Design Automation Workshop*, 1972, pp. 57–62.

Sha, L., and Dutton, R. W., "An Analytical Algorithm for Placement of Arbitrarily Sized Rectangular Blocks," *Proceedings of the 22nd Design Automation Conference*, 1985, pp. 602–608.

Slutz, E. A., "Shape Determination and Placement Algorithms for Hierarchical Integrated Circuit Layout," doctoral dissertation, Department of Electrical Engineering and Computer Science, Stanford University, 1980.

Soukup, J., "Circuit Layout," *Proceedings of the IEEE*, 69 (10), 1281–1304 (Oct. 1981).

Spira, P. M., and Hage, C., "Hardware Acceleration of Gate Array Layout," *Proceedings of the 22nd Design Automation Conference*, 1985, pp. 359–366.

Steinberg, L., "The Backboard Wiring Problem: A Placement Algorithm," *SIAM Review, 3,* (1), 37–30 (Jan. 1961).

Szepieniec, A. A., and Otten, R. H. J. M., "The Genealogical Approach to the Layout Problem," *Proceedings of the 17th Design Automation Conference,* 1980, pp. 535–542.

Uehara, T., and van Cleemput, W. M., "Optimal Layout of CMOS Functional Arrays," *IEEE Transactions on Computing, C-30,* 305–312 (May 1981).

Wada, M. M., "SLOOP: Sandia Standard Cell Layout Optimization Program: A User's Manual," Sandia National Laboratories, SAND82-0331, March 1982.

Walsh, G. R., *"Methods of Optimization,"* Wiley, New York, 1975.

White, S. R., "Concepts of Scale in Simulated Annealing," *Proceedings of the International Conference on Computer Design,* 1984, pp. 646–651.

Wipfler, G. J., Wiesel, M., and Mlynski, D. A., "A Combined Force and Cut Algorithm for Hierarchical VLSI Layout," *Proceedings of the 19th Design Automation Conference,* 1982, pp. 671–677.

21. High Density Routing for Printed Wiring Board

Isao Shirakawa
Osaka University

No matter how high the integration density of a VLSI becomes, and no matter how large a VLSI system grows, *PWBs* (printed wiring boards) are still the dominant means of achieving system level interconnections. Thus development is continuing on more sophisticated algorithms to raise layout performance for high density PWBs. Recent advances in the technology of microelectronics, in conjunction with trends toward growing system complexity, have changed design environments on PWBs so that: (1) the *between-pins capacity* [i.e., the number of wiring tracks permitted to run between two consecutive pins of a standard *DIP* (dual in-line package)] is specified to be two or more; (2) demands for PWBs of four or more signal layers are growing exponentially; and (3) with the proliferation of different types of VLSI packages apart from standard DIPs (e.g., flat packages, chip carrier packages, and plug-in packages), interconnection complexities are going to avalanche.

When layout specifications for PWBs undergo such changes, conventional routers, such as *maze routers* (Lee 1961; Hadlock 1977) and *line-search routers* (Mikami and Tabuchi 1968; Hightower 1969), are beset with various difficulties. As pointed out by Doreau and Abel (1978), the deficiency common to these conventional routers is that they lack "topological fluidity," that is, the capability to defer detailed wire patterns until all interconnections have been considered. Especially for line-search routers [which are widely used as the most powerful automatic routing tools to achieve a greater part (ordinarily more than 85 percent) of interconnections on a PWB], this deficiency is fatal when the between-pins capacity is specified to be two or more, and hole positions are to be fixed. Thus, for the purpose of high density routing on PWBs, a number of layout concepts designed to embody "topological fluidity" must be incorporated into an automatic routine scheme. Moreover, the greater the interconnection complexities on PWBs are, the more the time and energy wasted in manual or semiautomatic layout design to complete failures of automatic routine. Thus, as interconnection complexities of PWBs multiply, the development of high quality rerouting facilities becomes increasingly urgent.

This chapter investigates a number of routing schemes to cope with these trends toward growing interconnection complexities on PWBs. First, two major conventional routers—maze routers and line-search routers—are surveyed, and then a single-row router (Tsukiyama et al. 1980) is outlined. A new routing scheme (Asahara et al. 1981) is also surveyed, with emphasis on algorithm implementation along with routing performance. The goal is to realize topological fluidity by means of a single-row router combined with a line-search router such that the search for available segments in the line-search router is implemented channel by channel, with all interconnections within each channel completed later by the single-row router. Finally, a rerouting scheme (Shirakawa and Futagami 1983) is reviewed, that has been applied to single-layer PWBs but can be modified to be of practical use for multilayer PWBs.

CONVENTIONAL ROUTERS

Most existing routing systems for PWBs are constructed of several distinctive routers so that a merit of one may compensate for a defect of another. Usually these routers are combined so that one is executed first, followed by the second working on the failures, followed by the third working on the failures of the first two, and so on. Routers conventionally used in this manner are ordinarily classified into two categories. One category consists of those that are intended for finding a route whenever one exists. Maze routers (Lee 1961; Hadlock 1977) are typical of this type. A main drawback of these routers is that they are time-consuming and require much memory space. The other category consists of routers that have originated from those in the first category and try to compensate for defects in them by being less general and more efficient. Line-search routers (Mikami and Tabuchi 1968; Hightower 1969) are typical of this category. A main drawback of these routers is that they cannot always find a route even if one exists.

To discuss how these routers are implemented, we first consider a routing model on PWBs. Unless otherwise specified, we henceforth restrict our attention only to two-layer PWBs, as discussion can be easily extended to the multilayer case. Given a PWB, a *layered grid graph* $G = (V,E)$ where V and E denote sets of vertices and edges, respectively, can be defined according to routing specifications concerning between-pins capacity and hole positions, as illustrated in Figure 21-1 (Aoshima 1985). In this modeling, each edge drawn by a broken line represents the available position of either a drilled-through hole for a pin of a component to be mounted or a plated-through hole for interconnections

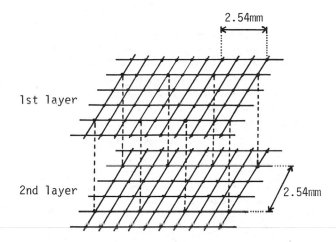

(a)   Case of between-pins capacity two.

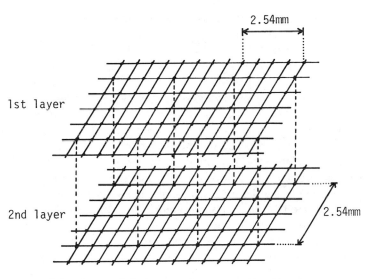

(b)   Case of between-pins capacity three.

Figure 21-1.   Examples of layered grid graphs. (a) Case of between-pins capacity two. (b) Case of between pins capacity three.

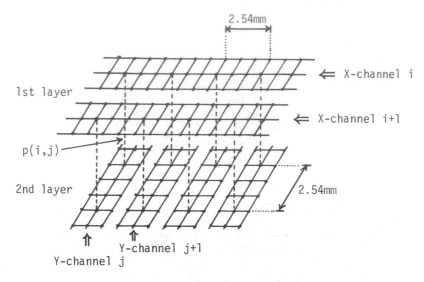

Figure 21-2. A layered grid graph for the use of the *XY* technique.

between layers, and each edge drawn by a solid line represents a unit segment of a track on which a conductor path is to be allocated.

Generally the so-called *XY technique* (Mikami and Tabuchi 1968) is adopted to achieve high density routing, which indicates that all conductor paths are confined to run in the *X* direction on one layer and in the *Y* direction on the other layer. In this case, we have only to modify a layered grid graph *G* as illustrated in Figure 21-2 (note that this figure shows a case of between-pins capacity two). Thus, whether the between-pins capacity of a given PWB is specified to be two or more, and whether or not the *XY* technique is adopted, a PWB routing problem can be formulated by means of a layered grid graph constructed accordingly, as follows: Let a *net* denote a set of pins to be made electrically common, and let a *net list* represent a set of all nets to be realized on a PWB. Assume that a layered grid graph $G = (V,E)$ has been constructed according to routing specifications for a given PWB, and the placement of components has been determined on it. Then, given a net list $L = \{N1, N, \ldots, Nm\}$, each net Ni specifies a set Ei of edges in G. Thus, a routing problem of our interest is to seek a set of connected subgraphs Gi $(i = 1, 2, \ldots, m)$ of G, each Gi containing all edges of Ei, such that any Gi and Gj $(i \neq j)$ are pairwise vertex-disjoint.

Maze Router

Maze routers consist essentially in finding a path for a *from-to*, which represents a set of any two connected portions for a net remaining to be interconnected, by means of labeling expansion around one until the other is reached. There are two different approaches to maze routine; one is the *BFS* (breadth-first search) approach of Lee (1961), and the other the *DFS* (depth-first search) approach of Hadlock (1977). For simplicity, these approaches henceforth will be dedicated to path finding for a from-to consisting only of two specified pins s and t.

BFS Maze Router.

Step 1: Let $S = \{s1, s2\}$ ($T = \{t1, t2\}$) be a set of end vertices of the edge in a given layered grid graph $G = (V,E)$ that corresponds to pin s (t). Let all vertices in V be unlabeled and unchecked. Put $i \leftarrow 0$, and then attach label i to $s1$ and $s2$.

Step 2: Attach label $i + 1$ to each unlabeled vertex that is adjacent to any vertex of label i and has not yet been used for any other net. If there is no vertex labeled $i + 1$, then halt (*comment:* at this stage it turns out that there is no path between S and T).

Step 3: If $t1$ or $t2$ is labeled, then check it, and go to Step 4. Otherwise put $i \leftarrow i + 1$, and go to Step 2.

Step 4: Chose arbitrarily an unchecked vertex of label i that is adjacent to any checked vertex, and check it. If $i = 0$, then halt (*comment:* at this stage, a sequence of checked vertices specifies a path between S and T). Otherwise, put $i \leftarrow i - 1$, and repeat Step 4.

We next describe a DFS maze router. In this approach two stacks, P-STACK and N-STACK, are provided. At each stage of labeling expansion around a vertex u, for each vertex v that is nearer to (farther from) T than u, an ordered pair (v, u) is pushed on P-STACK (N-STACK); thus not only the information of tracing backward from T to S can be retrieved, but also the next labeling should be expanded at the first vertex in the ordered pair on the top of P-STACK prior to that on the top of N-STACK.

DFS Maze Router.

Step 1: Let $S = \{s1,s2\}$ ($T = \{t1,t2\}$) be a set of end vertices of the edge in a given layered grid graph $G = (V,E)$ that corresponds to pin s (t). Initialize both P-STACK and N-STACK to empty. Let all vertices in V be unlabeled and fatherless. Put $u \leftarrow s1$ and push $s2$ on P-STACK.

Step 2: Label u. If u is $t1$ or $t2$, then halt (*comment:* at this stage a path is found by tracing fathers, one at a time, from T to S). Otherwise, for each unlabeled vertex v that is adjacent to u and has not yet been used for any other net, execute the following:

1. If v is on the same layer as u and is nearer to $t1$ or $t2$ than u, then push (v, u) on P-STACK.

2. Otherwise, push (v, u) on N-STACK.

Step 3: If P-STACK is empty, then go to Step 4. Otherwise, pop the topmost element, say (y, x), from P-STACK. If y is labeled, then repeat Step 3. If y is unlabeled, then put $u \leftarrow y$ and father $(y) \leftarrow x$ (*comment: x* is set to be the father of y), and go to Step 2.

Step 4: In N-STACK is empty, then halt (*comment:* at this stage it turns out that there is no path between S and T). Otherwise, exchange the names between P-STACK and N-STACK (*comment:* that is, let P-STACK $\leftarrow$ N-STACK and make N-STACK empty), and then go to Step 3.

It can be readily verified that the processing times of these BFS and DFS maze routers are bounded both by $O(|V + E|)$ per from-to, where $|\,.\,|$ denotes the number of elements of a set. Given a PWB, let $G(k) = (V(k), E(k))$ be a layered grid graph when the between-pins capacity is specified to be k. Then it may be readily verified that $|V(1)| : |V(2)| : |V(3)| (|E(1)| : |E(2)| : |E(3)|) = 4 : 9 : 16$. Considering that for a PWB of medium size with routing area of 30 cm $\times$ 23 cm (12 in. $\times$ 9 in.) we have $|V(1)| \cong 2 \times 240 \times 180 = 8.6 \times 10^4$ and $|E(1)| \cong 2|V(1)| + 120 \times 90 = 1.8 \times 10^5$, and that on such a scale of a PWB the number of from-tos is ordinarily around 1000 or more, we can see how much time is necessary if maze routers attempt to interconnect all from-tos on a PWB. Thus, to make the best use of maze routers, they are to be implemented at the final stage of the whole routing process, working on failures of other routers so far executed.

In terms of routine performance, there is a great difference between the BFS and DFS maze routers: As Hadlock (1977) insists, as long as there exists a path for a from-to, the number of labeled vertices in the former is generally larger than that in the latter, and hence the latter finds a path in shorter time than the former. On the other hand, when no path can be available for a from-to, a set of vertices labeled in the BFS maze router is equal to that in the DFS maze router, but because of stacking and unstacking operations the latter spends more time than the former. According to our implementation results (Asahara et al. 1982), at the stage of about 1 70% success rate (that is, about 70% of the from-tos attempted in maze routing succeed in interconnection), the average processing time per successful from-to in the BFS routing is almost twice as great as in the DFS routing; while at the stage of about a 30% success rate, the average processing time per successful from-to in the BFS routing is almost the same as that in DFS routing. Thus, it may be concluded that while the success rate is comparatively high, such as 70% or more, the DFS maze router is superior to the BFS maze router; but when it drops to a low level, such as 30%, the BFS maze router is preferred to the DFS maze router.

It should be added that one effective way to cope with the inefficiency of maze routers is to introduce the XY technique into the routing process (Asahara et al. 1981), which implies that mae routing is to be implemented in a modified graph such as that of Figure 21-2. This modified version of maze routing yields the constraint of preventing X- or Y-directional segments from being generated on the layer of Y- or X-directional use, respectively, so that the routing performance can be improved (Asahara et al. 1981, 1982). Thus this version may well be executed prior to the conventional BFS and DFS mae routers stated above.

Line-Search Router

Line-search routers are intended to compensate for the inefficiency of maze routers. With the use of the XY technique, path finding in line-search routers is implemented *line by line,* that is, track by track or channel by channel. There are two methods of line-search routing; one is that of Mikami and Tabuchi (1968), and the other that of Hightower (1969). A primary difference between them lies in the expansion mechanisms for seeking line segments reachable from vertices of a from-to. Given a from-to $\{ S, T\}$, the former approach tries to expand line segments reachable from S and T in a BFS fashion so that a path between S and T can be constructed of as few line segments as possible, while the latter attempts to expand line segments reachable from S and T in a DFS fashion such that each each stage a line segment is chosen to escape from the obstacle with which the last segment meets. In what follows, a routing procedure based on the former approach is outlined, which will be combined with a single-row router later.

First, several terminologies are defined: For the sake of convenience, the first (second) layer of a layered grid graph G is for X-directional (Y-directional) use, and X-directional (Y-directional) channels, abbreviated simply to X-*channels* (Y-*channels*), are defined as exemplified in Figure 21-2, where X-channels (Y-channels) are numbered from top to bottom (from left to right). For simplicity, a from-to $\{ S,T\}$ is confined to such a case of $S = \{s1, s2 \}$ and $T = \{t1,t\}$, where $s1$ and $s2$ ($t1$ and $t2$) are end vertices of the edge corresponding to a specified pin s (t). By a *block* $B = (i,j)$ we denote a subgraph composed of vertices and edges at the intersection of X-channel i and Y-channel j. In each block $B = (i,j)$ there is only one edge connecting the first and second layers (i.e., an edge corresponding to a hole position), which is denoted by $p(i,j)$. By an X-*channel segment* between blocks $(i1,j)$ and $(i2,j)$ [Y-*channel segment* between blocks $(i,j1)$ and $(i,j2)$] we mean that a subgraph composed of vertices and edges on the first layer between blocks $(i1,j)$ and $(i2,j)$ [Y-channel segment between $(i,j1)$ and $(i,j2)$] is said to be *feasible* if neither $p(i1,j)$ nor $p(i2,j)$ [neither $(i1,j)$ nor $p(i2,j)$] has been occupied by any other net, and the number of nets allocated to each block on the channel segment is less than its routing capacity. Given a block $B = (i,j)$, let a maximal feasible X-channel (Y-channel) segment passing $B = (i,j)$ be designated as an X-channel (Y-channel) segment of *level 0 with respect to* $B = (i,j)$, denoted by XO-SEG(B) (YO-SEG(B)).

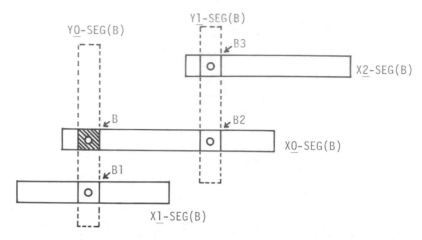

Figure 21-3.   Example of channel segments of level 0, 1, 2 with respect to block B.

Recursively, for each $k = 1, 2, \ldots$ let a maximal feasible X-channel (Y-channel) segment that passes a block i',j') on $Yk - 1\text{-SEG}(B)$ $[Xk - 1 - \text{SEG}(B)]$, with $p(i',j')$ not occupied by any other net, and that has not yet been expanded with respect to B, be referred to as an X-channel (Y-channel) segment of *level* k *with respect to* B, denoted by $Xk - \text{SEG}(B)$ $[Yk - \text{SEG}(B)]$, as illustrated in Figure 21-3.

Line-Search Router.

Step 1: Let $S = \{s1,s2\}$ ($T = \{t1,t2\}$) be a set of vertices of the edge in a layered grid graph G that corresponds to pin $s(t)$. Denote by $Bs = (is,js)$ and $Bt = (it,jt)$ the blocks that contain edges $(s1,s2)$ and $(t1,t2)$, respectively. Let K indicate the upper bound to the sum of the largest levels of X- and Y-channel segments to be used for interconnecting S and T. Put $h \leftarrow 0$ and $k \leftarrow 0$.

Step 2: For each $Xh - \text{SEG}(Bs)$, search for any $Xk\text{-SEG}(Bt)$ that intersects with $Xh\text{-SEG}(Bs)$ at a block $B = (i,j)$ with $p(i,j)$ not occupied by any other net. If such $Xk\text{-SEG}(Bt)$ is found, then go to Step 7.

Step 3: For each $Yh\text{-SEG}(Bs)$, search for any $Yk\text{-SEG}(Bt)$ that intersects with $Yh\text{-SEG}(Bs)$ at a block $B = (i,j)$ with $p(i,j)$ not occupied by any other net. If such $Yk\text{-SEG}(Bt)$ is found, then go to Step 7.

Step 4: For each $Xh\text{-SEG}(Bs)$, search for any $Yk\text{-SEG}(Bt)$ that intersects with $Xh\text{-SEG}(Bs)$ at a block $B = (i,j)$ with $p(i,j)$ not occupied by any other net. If such $Yk\text{-SEG}(Bt)$ is found, then go to Step 7.

Step 5: For each $Yh\text{-SEG}(Bs)$, search for any $Xk\text{-SEG}(Bt)$ that intersects with $Yh\text{-SEG}(Bs)$ at a block $B = (i,j)$ with $p(i,j)$ not occupied by any other net. If such $Xk\text{-SEG}(Bt$ is found, then go to Step 7.

Step 6: If $k + h = K$, then halt (*comment:* at this stage this router proves to fail in path finding). Otherwise, if $h < k$, then $h \leftarrow h + 1$, and go to Step 2; else if $h = k$, then $k \leftarrow k + 1$, and go to Step 2.

Step 7: By tracing back the blocks at the intersections of X- and Y-channel segments expanded from S and T, find a set of feasible channel segments that interconnect S and T. Halt.

SINGLE-ROW ROUTER

The single-row routing approach, first introduced by So (1974), has attracted a great deal of interest in terms of realizing "topological fluidity" (Tsukiyama et al. 1980; Asahara et al. 1981; Doreau and Koziol 1981). The basic concept of this approach is to find a path for each net by upward and downward zigzagging in a single channel of a layered grid graph, as outlined below. For example, consider a channel consisting of three tracks on which a net list $L = \{N1, N2, N3\}$ is defined, as shown in Figure 21-4, where it should be noticed that each net is defined as a set of vertices that corresponds to a set of hole positions to be made electrically common. The interconnection of each net Ni is realized by means of a set of paths consisting of horizontal and vertical line segments. This means of realization for a given net list in a channel is called *single-row routing* (Ting et al. 1976, where upward and downward zigzagging is allowed, but not forward and backward zigzagging. In this routing, let R denote a track on which vertices corresponding to hole positions are located, and let the routing space above (below) R be designated as the *upper* (*lower*) *street,* where the number of horizontal tracks in the upper (lower) street is referred to as the *upper* (*lower*) *street capacity.* Thus, in the above example, net list L is realizable with upper and lower street capacities two and one, respectively. Henceforth, without loss of generality, single-row routing is discussed under the following assumptions:

1. Every net of a given list consists of at least two vertices.
2. Every vertex belongs to a net.
3. Any net does not contain a pair of consecutive vertices.

For assumption (3), some explanation is necessary: If a net Nk contains a pair of consecutive vertices i and $i + 1$, then delete vertex $i + 1$ and seek a reali-

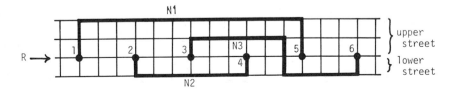

Figure 21-4.   Example of single-row routing.

zation. After that we can complete the realization by connecting vertices i and $i + 1$ by means of a path on track R, without changing the upper and lower street capacities.

Given a net list $L = \{N1, N2, \ldots, Nm\}$, an *interval graphical representation* (Kuh et al. 1979) is defined associated with an ordered sequence of L, as follows. For example, given a net list $L = \{N1, N2, N3, N4, N5, N6, N7\}$, with $N1 = \{1,5\}$, $N2 = \{2,4\}$, $N3 = \{3,8\}$, $N4 = \{7,10\}$, $N5 = \{6,14\}$, $N6 = \{9,11,13\}$, and $N7 = \{12,15\}$, consider a sequence of $s = (N1, N5, N2, N3, N4, N7, N6)$. Then the interval graphical representation of s is depicted as in Figure 21-5(a), where each horizontal line segment represents the interval covered by a net, and they are arranged according to the order in s. Obviously, there are $m!$ ordered sequences for a net list of m nets, and hence a total of $m!$ interval graphical representations. In an interval graphical representation, let the *reference line* (Kuh et al. 1979) be defined as the continuous line segments connecting all vertices of nets in succession from left to right. For example, the reference line of the interval graphical representation of Figure 21-5(a) is drawn by broken lines as shown. Now, stretch out the reference line and map it into track R. In accordance with this topological mapping, let each interval line be transformed into a path composed of horizontal and vertical line segments, as illustrated in Figure 21-5(b), where the portions above and below the reference line are mapped into paths in the upper and lower streets, respectively. As readily seen, this topological mapping yields a realization of a given net list. Thus, for each interval graphical representation there is a corresponding unique realization.

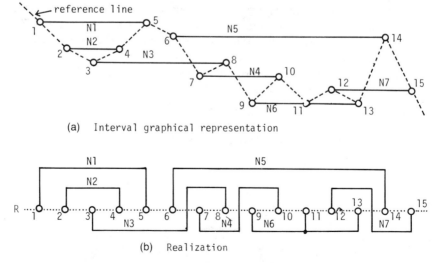

(a)   Interval graphical representation

(b)   Realization

Figure 21-5.   Interval graphical representation for single-row routing. (a) Interval graphical representation. (b) Realization.

Given an interval graphical representation associated with an ordered sequence s of net list L, draw a vertical line at each vertex i belonging to any net, and define the *cut number* $c(i)$ as the number of interval lines cut by the vertical line at vertex i, ignoring the one to which vertex i belongs. Let us also define the *upper* and *lower cut numbers* $cu(i)$ and $cw(i)$ as the numbers of interval lines cut by the vertical line above and below vertex i, respectively (Ting et al. 1976). Obviously, we have $c(i) = cu(i) + cw(i)$. Let C represent the maximum number of $c(i)$, and let Cu and Cw denote the maximum numbers of $cu(i)$ and $cw(i)$, respectively. The quantities Cu and Cw denote the maximum numbers of $cu(i)$ and $cw(i)$, respectively. The quantities Cu and Cw represent the necessary numbers of tracks in the upper and lower streets, respectively, in the realization corresponding to the interval graphical representation associated with sequence s. Thus, the problem of single-row routing that interests us is formulated as follows: Given a net list L and a pair of upper and lower street capacities Ku and Kw, respectively, find an ordered sequence s of net list L such that $Cu \leqq Ku$ and $Cw \leqq Kw$.

Necessary and Sufficient Conditions for Realizability

Let $\{1, 2, \ldots, r\}$ bet a set of vertices on R with which a given net list L is defined. Let us denote by $[i,j]$ $(i \leqq j)$ a closed interval on R between i and j. Associated with net list L, let an interval $I = [i,j]$, such that $c(k) \geqq h$ for all k on I and $c(i - 1) = c(j + 1) = h - 1$, be referred to as an *h-interval*. For any interval $I = [i,j]$ on R, let $\overline{L}(I)$ denote a set of nets that have no vertex on I, but have two vertices $(a(< i)$ and $b(>j)$; and let $L(I)$ be the union of $\overline{L}(I)$ and a set of nets containing any vertex on I. For example, for the net list L of Figure 21-6(a), there are three 2-intervals $I1 = [3,4]$, $I2 = [7,10]$, and $I3 = [12,13]$, as illustrated in the figure. With the use of these terminologies, we obtain a set of necessary and sufficient conditions that a given net list L be realizable by single-row routing with prescribed upper and lower street capacities Ku and Kw, respectively (Tsukiyama et al. 1980). In general, the generation of an ordered sequence s satisfying the realizability condition is very hard. Fortunately, however, for the special case of $ku + Kw \leqq 4$ and $| Ku - Kw | \leqq 1$, which is of great use in PWB routing with a specified between-pins capacity (note here that single-row routing with upper and lower street capacities Ku and Kw, respectively, accords with the routing specification of between-pins capacity $Ku + Kw$), the realizability condition can be described so that the generation of an ordered sequence s can be efficiently executed as follows.

Realizability Condition for Single-Row Routing.
As far as the the upper and lower street capacities Ku and Kw, respectively, are restricted to the case of $Ku + Kw \leqq 4$ and $| Ku - Kw | \leqq 1$, a given net list L is realizable with

prescribed upper and lower street capacities Ku and Kw, respectively, if and only if:

> *CASE 1 ($Ku = 1$ and $Kw = 0$; or $Ku = 0$ and $Kw = 1$):*
> Step A1: $C = 9$.
> *CASE 1 ($Ku = Kw = 1$):*
> Step B1: $C \leq 1$.
> *CASE 3 ($Ku = 2$ and $Kw = 1$; or $Ku = 1$ and $Kw = 2$):*
> Step C1: $C \leq 2$,
> Step C2: for any interval I, $|\overline{L}(I)| = 1$, and
> Step C3: there exists a sequence s such that for any 2-interval I, the single net of $\overline{L}(I)$ precedes the other nets of $L(I)$ in s.
> *CASE 4: ($Ku = Kw = 2$):*
> Step D1: $C \leq 3$,
> Step D2: for any 3-interval I, $|\overline{L}(I| = 2$, and
> Step D3: there exists a sequence s such that for any 3-interval I, one of the $\overline{L}(I)$ precedes the other nets of $L(I)$, and the other of the $\overline{L}(I)$ follows the other nets of $L(I)$ in s.

Thus, the main problem at this stage is to construct an efficient algorithm to generate an ordered sequence s of a given net list L so as to satisfy the realizability condition. Because in Case 1 and Case 2 the generation of s is trivial, we will restrict ourselves only to Case 3 and Case 4.

Special Case of Ku $= 2$ *and* Kw $= 1$. In order to generate a sequence s satisfying condition C3, let us introduce a directed graph GN constructed such that each vertex Ni corresponds to a net Ni in a given net list, and there exists an edge (Ni, Nj) incident from vertex Ni into vertex Nj if and only if net Ni is required to precede net Nj in s [specifically, for any 2-interval I, Ni is the single net of $\overline{L}(I)$ and Nj is contained in $L(I) - \overline{L}(I)$]. Because we can construct a sequence s of vertices of any *acyclic* graph (a graph containing no cycle is said to be acyclic) such that for any edge (Ni, Nj), Ni precedes Nj in s, it follows that if graph GN is acyclic, there exists a sequence s satisfying condition C3. The algorithm will explore every 2-interval I from left to right, to check whether graph GN, which is obtained by adding edges incident from the vertex corresponding to the single net of $\overline{L}(I)$ into the vertices corresponding to the nets of $L(I) - \overline{L}(I)$, is acyclic or not. It terminates with a solution of single-row routing as long as the realizability conditions C1, C2, and C3 are satisfied.

Single-Row Router for the Case of Ku $= 2$ *and* Kw $= 1$.

Step 1: Let L be given a net list. Let $GN = (V, E)$ be a graph with $V = L$ (*comment:* that is, each vertex Ni corresponds to a net Ni) and $E = \phi$ (empty).

Step 2: If condition C1 is not satisfied, then halt (*comment:* at this state L proves to be unrealizable).

Step 3: Seek all 2-intervals of net list L. If there is no 2-interval, then construct arbitrarily a sequence s of L (*comment:* this s satisfies C3), and go to Step 6. Otherwise, let $I1, I2, \ldots, I\ell$ be 2-intervals located from left to right in this order. Put $k \leftarrow 1$.

Step 4: If $k = \ell + 1$, then go to Step 5. Otherwise, execute the following:
1.
If condition C2 is not satisfied for 2-interval Ik, then halt (*comment:* at this stage L proves to be unrealizable).
2. For each ordered pair (Ni, Nj) of nets with $Ni \in \overline{L}(Ik)$ and $Nj \in L(Ik) - \overline{L}(Ik)$, add an edge (Ni, Nj) to graph GN.
3. If GN is not acyclic, then halt (*comment:* at this stage condition C3 proves to be violated).
4. Put $k \leftarrow k + 1$, and repeat Step 4. Step 5:
Construct a sequence s of the vertices of GN such that for every edge (Ni, Nj) of GN, Ni precedes Nj in sequence s. Step 6:
According to an interval graphical representation associated with sequence s, realize L. Halt.

For example, consider a net list L of Figure 21-6(a), for which there are three 2-intervals $I1$, $I2$, and $I3$. Apply the above procedure to this L: At Step 4 we can obtain an acyclic graph GN as shown in Figure 21-6(b), and at Step 5, we

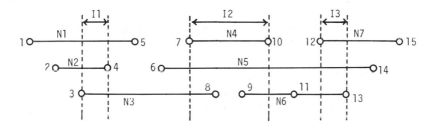

(a) A given net list and 2-intervals I1, I2, and I3.

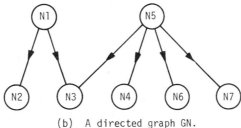

(b) A directed graph GN.

Figure 21-6. A given net list L and acyclic graph GN for L. (a) A given net list and 2-intervals $I1$, $I2$, and $I3$. (b) A directed graph GN.

can construct a sequence s, say $s = (N1, N5, N2, N3, N4, N7, N6)$. Associated with this x, we have an interval graphical representation as illustrated in Figure 21-5(a) for which L is realized as shown in Figure 21-5(b).

Special Case of Ku = Kw = 2. In order to generate a sequence s satisfying condition D3, let us introduce a directed graph GN constructed in a similar way so that vertex Ni corresponds to a net Ni in L, and each edge (Ni,Nj) indicates that Ni should precede Nj in s. This time, however, for any 3-interval I with $\overline{L}(I(= \{Na,Nb\}$ and $L(I) - \overline{L}(I = \{Nx,Ny\}$, a set of four edges of the form either $\{(Na,Nx), (Nz,Ny), (Nx,Nb), (Ny,Nb)\}$ or $\{(Nb,Nx), (Nb,Ny), (Nx,Na), (Ny,Na)\}$ is to be added to GN. The algorithm will explore every 3-interval from left to right, to check whether graph GN, which is obtained by a set of four edges stated above, is acyclic or not. It terminates with a solution of single-row routing as long as the realizability conditions D1, D2, and D3 are satisfied.

Single-Row Router for the Case of Ku = Kw = .

Step 1: Let L be a given net list. Let $GN = (V,E)$ be a graph with $V = L$ and $E = \phi$.

Step 2: If condition D1 is not satisfied, then halt (*comment:* at this stage L proves to be unrealizable).

Step 3: Seek all 3-intervals of net list L. If there is no 3-interval, then construct arbitrarily a sequence s of L (*comment:* this s satisfied D3), and go to Step 6. Otherwise, let $I1, I2, \ldots, I\ell$ be 3-intervals located from left to right in this order. Put $k \leftarrow 1$.

Step 4: If $k = \ell + 1$, then go to Step 5. Otherwise, execute the following:
1. If condition D2 is not satisfied, then halt (*comment:* L proves to be unrealizable).
2. Let $\overline{L}(Ik) = \{Na,Nb\}$ and $\overline{L}(Ik - L(Ik) = \{Nx,Ny\}$. Add to GN the following four edges: either (a) $(Na,Nx), (Na,Ny), (Nx,Nb), (Ny,Nb)$ or (b) $(Nb,Nx), (Nb,Ny), (Nx,Na), (Ny,Na)$ so that GN should be acyclic.
3. If GN cannot be acyclic, then halt (*comment:* at this stage conditions D3 proves to be violated).
4. Put $k \leftarrow k + 1$, and repeat Step 4.

Step 5: Construct a sequence s of the vertices of graph GN such that for every edge (Ni,Nj) of GN, Ni precedes Nj in sequence s.

Step 6: According to an interval graphical representation associated with sequence s, realize L. Halt.

For example, let us pay attention to a net list shown in Figure 21-7(a), for which there are four 3-intervals $I1$ through $I4$. Apply the above procedure to this L: At Step 4 we can obtain an acyclic graph GN as shown in Figure 21-7(b), and at Step 5, we can construct a sequence s, say $s = (N8, N5, N1, N7,$

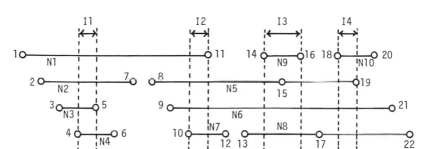

(a)   A given net list and 3-intervals I1 − I4.

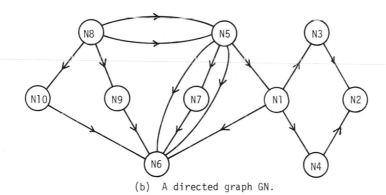

(b)   A directed graph GN.

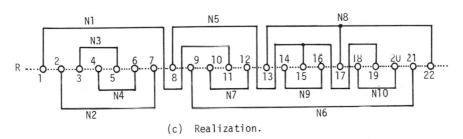

(c)   Realization.

Figure 21-7.   Example of graph *GN* and realization of a net list *L*. (a) A given net list and 3-intervals *I1–I4*. (b) A realized graph *GN*. (c) Realization.

N9, N10, N6, N3, N4, N2). Associated with this *s, L* is realized as illustrated in Figure 21-7(c).

We now touch on time complexities of these two procedures. Suppose that a data structure is appropriately provided, and that a given net list is composed of *m* nets that are defined with the use of *r* vertices on *R*. Then the single-row routers described here for the cases of $(Ku,Kw) = (2,1)$ and $(Ku,Kw) = (2,2)$ are both of time complexity $O(m \times r)$. Recently a fast $O(r)$-time procedure has been devised for the case of Ku 3 and Kw 3 (Han and Sahni 1984). Considering that usually the number of nets to be interconnected within a channel on a PWB is bounded by a small number, the single-row routers considered above are still of practical use.

A ROUTING SCHEME BY MEANS OF LINE-SEARCH ROUTER COMBINED WITH SINGLE-ROW ROUTER

According to So's (1974) approach, the general multilayer wiring problem can be decomposed into a number of phases, such as layering, via assignment, and single-row routing. This approach was introduced originally for large scale blackboard wiring, where a multilayer PWB has fixed geometries; that is, each layer has fixed plated-through hole positions, uniformly spaced on a rectangular grid, such that positions for pins (drilled-through holes to reach all layers) and vias (plated-through holes for interconnection between layers) alternate on each row, with each column of pins corresponding to that of terminal pins of a circuit card, that is, a PWB to be mounted on a backboard. Apart from such a multilayer PWB, in circuit card wiring, interconnection is implemented for each net defined on terminal pins of different components, such as discrete elements, SSIs, MSIs, LSIs, and VLSIs. Thus, a circuit card cannot have such fixed geometries, and this approach is not of direct use for such PWB wiring. However, because the single-row routing can be regarded as one of the most practical means to realize topological fluidity, we cannot leave this routing scheme unused for circuit card wiring. In terms of fruitful application of the single-row routing to circuit card wiring, there remains the key problem of how to assign nets to each channel, or, in other words, how to define a net list in each X- or Y-channel. To solve this problem, a number of routing concepts have been proposed and incorporated into routing systems (Asahara et al. 1981; Doreau and Koziol 1981). In what follows, a routing scheme (Asahara et al. 1981) based on a line-search router combined with a single-row router is outlined, together with implementation results.

For simplicity, henceforth let a from-to $\{S,T\}$ be of the form $S = \{s1,s2\}$ and $T = \{t1,t2\}$, with S and T corresponding to pins s and t, respectively, to be interconnected. To combine the line-search router with the single-row routers, both described above, a number of additional terminologies are necessary: According to a given PWB of between-pins capacity $Ku + Kw$ with $Ku + Kw \leq 4$ and $0 \leq Ku - Kw \leq 1$, we construct a layered grid graph G similar to the one as illustrated in Figure 21-2, such that in each X- and Y-channel, there are Ku and Kw tracks in the upper and lower streets, respectively. By setting the routing capacity of each block to be $Ku + Kw$, *feasible* X- and Y-channel segments are defined just as before. However, an X-channel (Y-channel) segment of level k, with respect to block $B = (i,j)$, is defined in a little different way, as follows: An X-channel (Y-channel) segment of *level 0 with respect to B*, denoted by $X\underline{0}$-SEG(B) [$Y\underline{0}$-SEG(B)], is defined in just the sam way as before; that is, it represents a maximal feasible X-channel (Y-channel) segment passing B. On the other hand, for each $k = 1, 2, \ldots$, an X-channel (Y-channel) segment of *level k with respect to B*, denoted by $X\underline{k}$-SEG(B) [$Y\underline{k}$-SEG(B)], is defined recursively as a maximal feasible X-channel (Y-channel) segment that, as illustrated in Figure 21-8, (a) passes a block $Bk = (ik,jk)$ on $Y\underline{k-1}$-SEG(B) [$X\underline{k-1}$-SEG(B)], with $p(ik,jk)$ not occupied by any other net, such that new net $\{p(ik-1,jk-1), p(ik,jk)\}$ (where put $Bo = B$, that is, $i0 = i$ and $j0 = j$) is realizable by single-row routing with prescribed upper and lower street capacities Ku and Kw, respectively, together with all nets in the net list so far allocated to the same channel, and (b) has not yet been expanded with respect to B.

Now, let us modify the line-search router described before, as follows:

1. If two channel segments intersect with each other at Steps 2 or 3, then let $B' = (i',j')$ and $B'' = (i'',j'')$ be the end blocks of the channel segment newly found to interconnect blocks Bs and Bt, and check whether new net $\{p(i',j'), p(i'',j'')\}$ is realizable by single-row routing with prescribed capacities Ku and Kw, together with all nets in the net list so far allocated to the same channel. If it is realizable, then let the procedure go to Step 7.

2. If two channel segments intersect with each other at Steps 4 or 5, then let two channel segments newly found to interconnect blocks Bs and Bt be between $B = (i,j)$ and $B' = (i',j')$ and between $B = (i,j)$ and $B'' = (i'',j'')$, and check whether each of the new nets $\{p(i,j), p(i',j')\}$ and $\{p(i,j), p(i'',j'')\}$ is realizable by single-row routing with prescribed capacities Ku and Kw, together with all nets in the net list so far allocated to the same channel. If both of them are realizable, then let the procedure go to Step 7.

3. Suppose that at Step 7 a channel path just found between Bs and Bt starts at block $Bs = (is,js)$, turns at blocks $Bu = (iu,ju)$, $Bv = (iv,jv)$, $\ldots$, $Bz = (iz,jz)$ in this order, and arrives at $Bt = (it,jt)$. Then each of new nets $\{p(is,js), p(iu,ju)\}$, $\{p(iu,ju), p(iv,jv)\}$, $\ldots$, $\{p(iz,jz), p(it,jt)\}$ is to be added to the corresponding net list.

At the termination of the line-search router with this modification, the net list constructed for each channel is guaranteed to be realizable by the single-row router with prescribed capacities Ku and Kw. Thus, the interconnection of the net list in each channel can be completed independently later on. This means that the line search router combined with the single-row router attains topological fluidity.

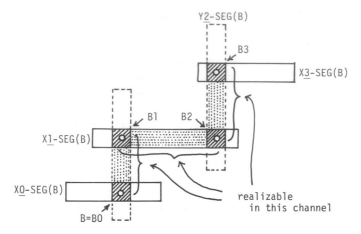

Figure 21-8. Channel segments of level k = 0, 1, 2, 3.

Routing Mode

By using this routing scheme in conjunction with maze routers, a routing system is constructed to deal with PWBs of different between-pin capacities according to a user's specifications. Several variations are introduced in executing this routing scheme and the maze routers, as outlined below:

A. Line Search Router + Single-Row Router
   MODE 1: Ku = 1 and Kw = .
   MODE 2: Ku = Kw = 1.
   MODE 3: Ku = 2 and Kw = 1.
   MODE 4: Ku = Kw = 2.

Thus, MODE k is supposed to be applied to PWBs with specification of between-pins capacity k. It should be remarked here that there are a number of ways to apply the line-search router, which may affect the wirability of the system. The wiring performance is observed to be fairly sensitive to the following factors:

(a) The number to which the parameter K of the line-search router may be raised.
(b) Whether or not a restriction is imposed on wire pattern generation such that the router is executed for all from-tos first by setting K = 0, then for the failures by setting K = 1, and so on.

According to benchmark tests on (a), the ratio of from-tos realized with K = 4 or more to all those realized by the line-search router is less than 3%; moreover, there are a number of PWBs for which the total wiring performance obtained by setting $K = 4$ is less than that obtained by setting $K = 3$. Thus, the line-search router with $K = 3$ is expected to attain the highest wiring performance. As to (b), the total wiring performance with such a restriction imposed on wire pattern generation has shown a downward tendency against the one without it. Thus, the line-search router is to be executed without such a restriction.

B. Maze Router

MODE A: In this mode, BFS and DFS maze routers are executed with the use of the *XY* technique; that is, these routers are applied to layered grid graphs as exemplified in Figure 21-2.

MODE B: In this mode, BFS and DFS maze routers are executed with no restriction imposed on generating wire patterns; that is, maze routers are applied to layered grid graphs as illustrated in Figure 21-1.

C. Net Ordering

According to benchmark tests on several net ordering schemes, a policy to select a from-to to the shortest distance among the remaining from-tos to any routing stage seems best in both line-search routers and maze routers.

Implementation Results

The routers described have been programmed and run on an NEC ACOS 77/900 computer. Some of the implementation results are shown in Tables 21-1 and 21-2, where $T1$, $T2$, and $T3$ are processing times such that:

- $T1$ is spent by line-search routers.
- $T2$ is spent by single row routers.
- $T3$ is spent by maze routers.

The abbreviation W.P. indicates the wiring performance, defined as (# inter-connected from-tos)/(# from-tos to be interconnected) $\times$ 100 (%). It is observed from these tables that wirability to the routing system tends to rise rapidly as the between-pins capacity increases, and that there is no much difference among processing times of Modes 1 through 4, although the numbers of vertices in layered grid graphs in Modes 2, 3, and 4 are 2.25, 4, and 6.25 times as many as that in Mode 1, respectively.

ROUTING AND REROUTING SCHEMES FOR SINGLE-LAYER PWB

Many routing procedures for multilayer PWBs have been introduced and extensively developed, as described. However, there has not been much specific

Table 21-1 Routing results for a PWB of size 280mm × 203 mm with 517 nets and 1142 from-tos on 3668 pins.

LINE-SEARCH ROUTER + SINGLE-ROW ROUTER			MAZE ROUTER					
MODE	T1/ T2(sec.)	W.P.(%)	MODE	T3(sec.)	W.P.(%)	TOTAL W.P.(%)	T1 + T2 + T3 (sec.)	# VIAS
1	362.7/ 33.5	77.5	A/ B	297.3	5.8	83.3	693.9	1,285
2	408.3/ 41.0	94.3	A/ B	544.2	2.5	96.8	993.5	1,468
3	261.1/ 43.7	99.5	A/ B	29.6	0	99.5	334.4	1,185
4	251.1/ 41.2	100				100	293.2	1,094

development reported for automatic routing on those single-layer PWBs and hybrid ICs that are common for numerous analog appliances. Difficulties of routing on these single layer boards or substrates are summarized as follows:

1. A variety of wire patterns must be generated with the aid of not only horizontal and vertical wire segments but also oblique ones, to raise wiring performance.
2. In multilayer PWBs, wire patterns usually are generated by using he *XY*-technique. However, in single-layer PWBs, once a wire pattern is generated, any horizontal and vertical wire segment of this pattern will prevent all vertical or horizontal wire segments crossing it from being laid out for other nets thereafter. Thus, in the single-layer case, once generated any wire pattern becomes by a great obstruction to path finding thereafter; so wiring performance is much more affected by net ordering in single-layer PWBs than in the multilayer case.

In what follows, routing and rerouting schemes for single-layer PWBs are outlined (Shirakawa and Futagami 1983; Futagami et al. 1982).

To cope with problem (1) above, another grid graph $G = (V,E)$, as illustrated in Figure 21-9, is defined such that (a) each vertex corresponds to a position for a pin of a component to be mounted or a snapping point of a wire pattern to be generated, and (b) each edge corresponds to a unit segment of track on which a wire segment is to be allocated.

If any diagonal edge is used for a wire pattern, then another diagonal edge that crosses over it is prohibited from use in order to preserve the planarity of the wire patterns.

Table 21-2 Routing results for a PWB of size 183 mm × 274 mm with 481 nets and 911 from-tos on 2179 pins.

	LINE-SEARCH ROUTER + SINGLE-ROW ROUTER			MAZE ROUTER				
MODE	T1/ T2(sec.)	W.P.(%)	MODE	T3(sec.)	W.P.(%)	TOTAL W.P.(%)	T1+T2+T3 (sec.)	# VIAS
1	336.6/ 21.1	66.4	A/ B	484.5	6.5	72.9	839.2	1003
2	436.2/ 29.0	85.2	A/ B	1175.9	3.4	88.6	1641.1	1326
3	446.8/ 33.6	94.6	A/ B	599.1	3.2	97.8	1079.5	1379
4	161.1/ 34.8	100				100	195.9	1080

To cope with problem (2) above, a sort of dynamic ordering scheme is introduced, where the routing process is implemented through a number of stages in such a way that, at each stage: (a) an interference relation for unconnected from-tos is sought; (b) according to the interface information, paths are generated for them so that each may make a detour to keep away from the other; and (c) a maximal set of paths that do not intersect with each other is chosen for wire patterns, as summarized below.

Routing Scheme

The routing scheme is executed by stage: At any stage, let L be a set of incomplete from-tos, and let W be a set of paths so far generated for interconnection.

Step 1: Apply the maze router *independently* to all from-tos in L, one at a time, such that the shortest path $P(v, v')$ for each $\{v, v'\} \in L$ should not intersect any path in W, but paths $P(u, u')$ and $P(w, w')$ for any pair of from-tos $\{u, u'\}$ and $\{w, w'\}$ in L may be permitted to intersect with each other. Let the shortest path $P(v, v')$ generated for each $\{v, v'\}$ in L be called a *trial routing*.

Step 2: For each pair of trial routings $P(u, u')$ and $P(w, w')$ that intersect with each other, compute the length of the shortest detour to be made for $P(w, w')$ to avoid $P(u, u')$, as follows:

Apply the maze router to $\{w, w'\}$ under the constraint that any vertex on trial routine $P(u, u')$ or any path in W should not be visited, but those on trial routine $P(v, v')$ for any $\{v, v'\} \neq \{u, u'\}$ in L may be visited. If such a shortest path $P'(w,w)$ is found, then the difference of length between $P(w,w')$ and $P'(w,w')$ is designated as the *detour length* for $P(w,w')$ to avoid $P(u,u')$.

Step 3: Based on detour lengths for all pairs of intersected trial routings, find a maximal set of paths that do not intersect with each other, one at a time, by repeated application of path modification to trial routings so that the detour length to be extended should be minimal.

Step 4: Add a set of paths obtained in Step 3 to W, and delete a set of the corresponding from-tos from L.

Rerouting Scheme

At the termination of the routing process, let $L*$ be a set of all interconnected from-tos, and let $W*$ be a set of paths determined for from-tos in $L*$. Let L be a set of all incomplete from-tos, and we may assume that for any net of L a path cannot be found that does not visit any vertex on a path in $W*$. Given an incomplete from-to $\{u,u'\} \in L$, define a set $OB(u,u')$ of *obstacle from-tos* for $\{u,u'\}$ constructively, as follows:

Step 1: $OB(u,u') \leftarrow \phi$.

Step 2: By means of a breadth-first search, attach labels A and B to vertices that are reachable from u and u', respectively, not visiting any vertex on a path in $W*$.

Step 3: If there exists a path $P(z,z')$ in $W*$ between z and z' with $\{z,z'\} \in L* - OB(u,u')$ such that $P(z,z')$ contains a pair of vertices x $(\neq z,z')$ and y $(\neq z,z')$ that are adjacent to vertices labeled A and B, respectively, then add $\{z,z'\}$ to $OB(u,u')$, and repeat Step 3. Otherwise, halt.

Thus, if we strip any path $P(z,z')$ with $\{z,z'\}$ $OB(u,u')$, then from-to $\{u,u'\}$ proves to be interconnected.

We now describe a rerouting scheme that consists of a main routine REROU-

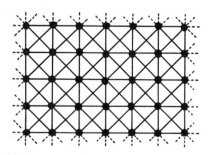

Figure 21-9. A grid graph for a single-layer PWB.

TER and a subroutine PATHFINDING. In this scheme, Depth denotes the depth of recursive to PATHFINDING, and Max represents an upper limit to Depth (according to benchmark tests, setting Max = 3 or 4 is sufficient in practice).

—**procedure** REROUTER;
—**begin** let L be a set of incomplete from-tos; sort L in increasing order of distance in grid graph G;
   —**for** each $\{u,u'\}$ on L do
   —**begin** apply the maze router to $\{u,u'\}$ under the constraint that any path in $W*$ should not be visited;
   —**if** $\}u,u'\}$ is interconnected *then* add the path between u and u' to $W*$
   —**else begin**
      Depth ← 1;
      initialize STACK to empty;
      PATHFINDING($\{u,u'\}$, DEPTH)
—**end**
—**end**
—**end** REROUTER;

—**procedure** PATHFINDING($\{u,u'\}$, Depth);
—**begin** find a set Z of from-tos in $OB(u,u')$ that are not on STACK;
—**for** each from-to $\{w,w'\} \in Z$ *do* calculate the length of a path $P'(u,u')$ between u and u' that is obtained by applying the maze router under the constraint that any path in $W* - \{P(w,w') \in W*\}$ should not be visited; sort Z in increasing order of length of $P'(.)$;
—**for** each from-to $\{w,w'\}$ on Z *do*
—**begin** delete $P(w,w')$ from $W*$;
—**if** there are a pair of vertex disjoint paths $R(u,u')$ between u and u' and $Q(w,w')$ between w and w', any of which does not visit a path in $W*$ (*comment:* checking whether these two paths exist or not is executed by means of the so-called *two paths algorithm* (Siloach 1980; Ohtsuki 1980) *then*
—**begin** add these two paths $R(u,u')$ and $Q(w,w')$ to $W*$; delete $\{u,u'\}$ from L; halt this procedure (*comment:* rerouting for interconnection of from-to $\{u,u'$ results in success)
—**end**
   —**else if** Depth $\neq$ Max *then*—**begin** Depth ← Depth + 1; push $\{u,u'\}$ on
                                    STACK; PATHFINDING($\{w,w'\}$,Depth)
—**end**
   —**else** add $P(w,w')$ to $W*$ (*comment:* $P(w,w')$ is registered again as a path in $W*$
—**end;** Depth ← Depth − 1;
—**if** Depth = 0 *then* halt this procedure (*comment:* rerouting for interconnection of $\{u,u'\}$ results in failure)   —**else** pop a from-to from top of STACK
—**end** PATHFINDING;

The rerouter described above mains mainly at the interconnection of a maximal set of incomplete from-tos. Hence, it is possible that some paths making long detours can be improved to shortcut through open spaces generated by shifting other conductor paths. Such path improvement may enable blockages against some incomplete from-tos to be released. In these circumstances, the rerouter is modified to carry out such improvements, a procedure that is omitted here because of space limitations (see Shirakawa and Futagami 1983).

CONCLUDING REMARKS

This chapter has described a number of routing concepts and schemes for high density PWBs. Because of the different types of VLSI packages that can now be mounted on a PWB, as well as the rapid advances in system-level scale for PWBs, routing complexity and wiring performance impact each other in a vicious circle. Therefore, development is continuing on novel routing methodologies to cope with the proliferation of routing complexities. One of the most promising approaches is to introduce parallel processing schemes into routing procedures. To do this, a few parallel routing concepts have been proposed, mainly associated with maze routers. However, such parallel implementations of maze routers are used only to speed up the routing process. More drastic parallelism should be devised for every routing phase to improve routing performance.

REFERENCES

Aoshima, K., "Implementations and Evaluations of Routing Procedures," *Memo. Computer Graphics Tutorial Forum*, Feb. 21, 1985 (in Japanese).

Asahara, S., Odani, M., Ogura, Y., Shirakawa, I., and Ozaki, H., "A Routing System Based on a Single-Row Router and Its Wirability," *Internat'l J. Circuit Theory and Applications*, 9, 286–296 (1981).

Asahara, S., Odani, M., Ogura, Y., Shirakawa, I., and Ozaki, H., "An Analysis of Wiring Performances of a Routing System for High Density Printed Wiring Boards," *Trans. IECE*, J65-A, 159–166 (1982) (in Japanese).

Doreau, M. T., and Abel, L. C., "A Topological Based Non-Minimum Distance Routing Algorithm," *Proc. 15th Design Auto. Conf.*, pp. 92–99, 1978.

Doreau, M. T., and Koziol, P., "TWIGY: A Topological Algorithm Based Routing System," *Proc. Design Auto. Conf.*, pp. 746–755, 1981.

Futagami, S., Shirakawa, I., and Ozaki, H., "An Automatic Routing System for Single-Layer Printed Wiring Boards," *IEEE Trans. CAS-29*, 46–51 (1982).

Hadlock, F. O., "A Shortest Path Algorithm for Grid Graphs," *Networks*, 7, 323–334 (1977).

Han, S., and Sahni, S., "Single-Row Routing in Narrow Streets," *IEEE Trans.*, CAD-3, 235–241 (1984).

Hightower, D. W., "A Solution to Line Routing Problems on the Continuous Plane," *Proc. 6th Design Auto. Workshop*, pp. 1–24, 1969.

Kuh, E. S., Kashiwabara, T., and Fujisawa, T., "On Optimum Single-Row Routing," *IEEE Trans.*, CAS-26, 361–368 (1979).

Lee, C. Y., "An Algorithm for Path Connections and Its Applications," *IRE Trans.*, EC-10, 346–365 (1961).

Mikami, T., and Tabuchi, K., "A Computer Program for Optimal Routing of Printed Circuit Connection," *Proc. IFIP Congress*, pp. 1474–1478, 1968.

Ohtsuki, T., "The Two Disjoint Path Problem and Wire Routing Design," *Proc. 17th Symp. Res. Inst. Electrical Comm.*, Tohoku Univ., pp. 257–267, 1980.

Shirakawa, I., and Futagami, S., "A Rerouting Scheme for Single-Layer Printed Wiring Boards," *IEEE Trans., CAD-2*, 267–271 (1983).

Siloach, Y., "A Polynomial Solution to the Undirected Two Paths Problems," *J. Assoc. Comput. Mach.*, *27*, 445–456 (1980).

So, H. C., "Some Theoretical Results on the Routing of Multilayer Printed Wiring Boards," *Proc. IEEE ISCAS*, pp. 296–303, 1974.

Ting, B. S., Kuh, E. S., and Shirakawa, I., "The Multilayer Routing Problem: Algorithms and Necessary and Sufficient Conditions for the Single-Row Single-Layer Case," *IEEE Trans., CAS-27*, 768–778 (1976).

Tsukiyama, S., Kuh, E. S., and Shirakawa, I., "An Algorithm for Single-Row Routine with Prescribed Street Congestions," *IEEE Trans., CAS-27*, 765–771 (1980).

22. Cell Based Physical Design for VLSI

Ulrich Lauther
Siemens AG, ZTI DES

With the continuing decrease in size of the features that can be reliably fabricated on silicon, and the growing complexity of systems to be integrated on a single chip, nonmanufacturing problems are becoming more prevalent. Among them are the physical design of very large scale integrated circuits, verification of the designs, and testing of the resulting chips in production. In this chapter we look into the problem of physical design (or layout generation) and discuss some approaches to its solution.

With the complexity of VLSI chips growing exponentially, traditional layout methods employing a low level representation of mask geometry (rectangles and polygons on various mask levels describing transistors and connections) have become increasingly obsolete, just as computer programming at the machine code level became impractical decades ago.

The same principles that are successfully used in the design of large software systems can be applied to the design of hardware: the principles of modularity, abstraction, hiding, and hierarchy. A module in software systems is a piece of code with a well-defined function, a concise interface to the calling environment, and implementation details hidden from the user. It is the building block from which large systems can be put together in a hierarchical way. The same holds for the building blocks or "cells" in a cell based layout style (or layout system). Predefined layout blocks with well-defined function, guaranteed behavior, and a clear interface can be used as black boxes to be combined hierarchically into larger and larger functional units that finally make up the whole system to be implemented.

Technological and circuit level electrical details are hidden in low level prede-signed blocks (leaf cells), so they can be assembled even by designers unfamiliar with details of the target technology; for example with today's automated design systems (see Koch and Nett 1984 for an advanced specimen of this group) and predesigned cell libraries, former PCB designers can develop CMOS chips without detailed knowledge of that technology. Expert knowledge is needed, of course, for those who are responsible for the development and maintenance of cell libraries. We can view a library as a bulk of coagulated work and knowledge freely accessible to those who use it.

Another parallel to software design is obvious: we know that in large software systems a programmer can produce about 10 lines of code per day (a figure that does not depend on the programming language used (Walston and Felix 1977), and that a layout designer draws about 10 features per day (Lattin 1979). In the same way that 10 lines of a high level language will express much more than 10 lines of assembler code, it is much more efficient to place and connect 10 building blocks than to draw 10 transistors. Just as an experienced programmer may produce more efficient code at machine level than a compiler can do, the experienced IC designer may produce a more dense layout designing at transistor level than her/his colleague dealing with less flexible layout blocks. But in both cases the complexity of the task no longer permits low level manual optimization.

There is another reason that compels us to cost-efficient design of large systems: as integrated circuits become larger, their function becomes more specialized, and production volume per chip type decreases. This would increase the share of development cost per chip if cost-efficient design methods were not employed. Design at the cell level not only makes design more transparent; because of the reduction of the number of layout pieces one has to deal with at each level of the hierarchy, these methods lend themselves to automated design; placement of cells and routing of connections can be done with a high degree of automation.

The main topic of this chapter will be a characterization of the problems associated with automated design at the cell level for various cell based design styles, and methods for their solution.

DEVELOPMENT OF LEAF CELLS

As described above, a cell based design uses a hierarchy of cells to implement the desired function. At the lowest level of this hierarchy we find predefined building blocks from a cell library, often called leaf cells because they form the leaves in the hierarchy tree. Before we discuss our main topic—how to assemble and connect cells—we will briefly discuss methods for layout development of leaf cells.

The traditional and most frequently used method for designing leaf cells is to draw the layout at the CRT of a turnkey graphics system (see VLSI Design Staff 1984 for an overview). These systems (CALMA, COMPUTER VISION, and APPLICON are the most popular of the commercially available ones) are layout editors that behave like an electronic pencil and razor. They allow one to draw rectangles and more complex polygons, to associate these figures with mast layers (usually indicated by colors), and to store these entities in a hierarchically organized database.

These systems typically are "dumb" in the sense that they do not "know" about the role an inserted figure plays in the design—whether it is intended to make a connection, form a transistor, or do something else. The systems are optimized for fast response—for instance, when the current window is changed— and adapt very easily to changing technology because they do not know the target technology at all. The drawback of this philosophy is that no rules referring to the technology can be checked; usually extensive design rule checking (Tsukizo et al. 1983; Baird 1978; Lauther 1981) must be done when the design is considered completed.

The same holds for electrical verification: quite sophisticated feature extraction programs (Hofmann and Lauther 1983) are needed to extract the semantics (transistors, connections), form the geometry, and feed these elements into a circuit simulator.

Another drawback of these systems is that all of the data entered into the internal database has the character of constants. If design rules change, there is no economical way to adapt the geometric data to the new rules keeping the topology of the circuit intact.

A more modern way to enter a design into a design system is symbolic design (see Taylor 1984) for an overview and bibliography). Here the designer does not explicitly handle basic layout elements such as rectangles and polygons without a priori meaning, but (s)he designs in terms of symbols ("sticks") representing wires, transistors, contacts, and so on. The designer does not have to take care of the final dimensions of these features and of the spacing between them. This task is done automatically by a compaction (better: spacing) program that transforms the symbolic representation to final geometry and adjusts the spacing between features so that design rules are observed, and the area of the design is minimized. Because topology of the design (expressed by the so-called sticks diagrams) and the metric are conceptually separated in this method, a design can easily be adapted to changing design rules.

For this purpose, only the compactor has to be run again using the current set of design rules. The design is "technology updateable" (Lee et al. 1981). This feature is a must, especially for a cell library to be used in many projects over a long period of time. Otherwise, each minor change of technological parameters would render the total library obsolete. Electrical verification by circuit simulation also becomes simpler with this concept. Because the designer indicates her/his intention (by choosing appropriate symbols), no complicated feature extraction is needed: a relatively simple interface program between the symbolic layout systems and the circuit simulators will do the job.

A more recent approach to resizing of layouts attempts to extract the intended topology directly from mask data instead of from sticks. In this way, traditionally handcrafted layouts can be processed (Schiele 1984).

Sometimes not only design rules but also behavioral characteristics of a cell should be adaptable to actual needs—the driving power of an output stage, the width (number of bits) of a counter, and so on. For these purposes it is most convenient to describe the layout with a procedural language. This could be a special layout language or a set of procedures embedded into any high-level general-purpose programming language such as C, PASCAL, or LISP (Rowson 1980; Rivest 1980). This approach surely gives the largest degree of flexibility; on the other hand, the immediate feedback of interactive graphics is missing, and verification under various settings of the parameters is virtually impossible. The design process becomes a programming task with all its known virtues and difficulties; therefore, this method is not widely accepted at this time.

CELL BASED LAYOUT STYLES

The three principal methods used for cell based design are the gate array, the standard cell, and the general cell approach. These cell design concepts impose different restrictions on placement and routing, calling for different mathematical models and solution methods. Therefore, we will first characterize these concepts and then discuss related placement and routing problems before going on to describe algorithms for automated or computer aided layout.

Gate Arrays

Currently the most popular cell based approach to layout is the gate array method (sometimes also referred to as the uncommitted logic array (ULA) approach). A gate array is a partially predefined and prefabricated chip (master) that carries an array of transistor groups (basic cells). Various logic functions can be realized by superimposing specific interconnection patterns on the transistors of basic cells. These patterns (intracell wiring) are stored in a library, each defining a specific function (e.g., 2-input-NAND gate) to be represented on the chip. From the algorithm point of view, placing such an interconnection pattern is equivalent to placing a functional block, similar to the way components are assigned to predefined slots on a PC board. When all the functional blocks required for the implementation of the intended function have been placed, they have to be interconnected according to a net list. Generation of these interconnections (intercell wiring) is the routing problem in gate arrays.

Because all active devices of a gate array are prefabricated, only intra- and intercell wiring is individual to a specific integrated circuit. Only mask and processing costs connected with wiring have to be charged to the usually low

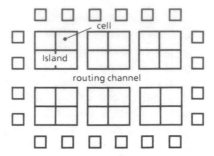

Figure 22-1. Scheme of an island type gate array.

production volume of the specific design, whereas design costs for the master can be charged to the high production volume of the master.

More advantages of the gate array approach stem from the predefinition of the master; not only active devices, but also power supply lines are predefined. Layouters or layout programs do not have to deal with related problems. The same holds for the design of the chip periphery, with regard to specialist knowledge concerned with alignment parks, guard structures, the cutting frame, and so on.

All of these characteristics make gate arrays attractive if a large family of similar ICs, each with a low production volume, must be designed and fabricated within a short time—a situation that typically arises in the design of mainframe computers. Therefore, system houses were the first to adopt this design strategy.

If the production volume per chip increases, drawbacks of the approach become more pronounced. Usually not all of the predefined gates can be used in a specific design, and unused gates represent wasted silicon real estate. A similar problem arises with the routing area: in designing the master, sufficient routing area must be allocated for the most complex design the master is intended for. Again the result will be wasted area in the bulk of applications.

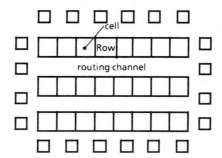

Figure 22-2. Scheme of a row type gate array.

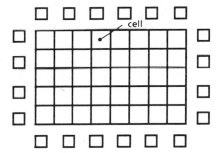

Figure 22-3. Scheme of a matrix type gate array.

The geometric structure of a master can greatly vary. We see three principal architectures:

1. *The island type:* Cells are arranged in blocks, surrounded by channels that are used for intercell wiring (see Figure 22-1).
2. *The row type·* Cells are arranged in rows, separated by row parallel routing channels for intercell wiring (see Figure 22-2). If a net has terminals in more than one of these channels, cell rows have to be crossed using predefined "feed-through cells," electrically equivalent (i.e., internally connected) terminals of a logical cell, or just over-the-cell routing if it is permitted in the specific technology used.
3. *The matrix type:* Here cells are closely packed, and no distinguishable routing channels exist. All routing is done over the cells. Figure 22-3 shows the principle, and in Holzapfel et al. (unpublished) a specific master of this type is discussed. An interesting point of this architecture is that the row type can be seen as an offshoot of this more general type—if only ever other row is used for placement of active cells, the remaining rows of unused cells can be utilized as routing channels.

These different master types call for different approaches to placement and especially to routing; to a certain degree a specialized automatic placement and routing package has to be provided for the master type, if it is called for, in order to achieve satisfactory results.

Standard Cells

The main drawback of gate arrays—poor area utilization—can be avoided if we tailor the selection of library cells and the width of routing channels to actual needs, sacrificing the concept of prefabrication. Proceeding from the row type gate array, we arrive at the standard cell concept by this generalization.

Standard cells are cells with—within a cell family—equal or nearly equal cell heights, whereas the width of cells varies according to the complexity of the implemented function. Cell terminals (inputs and outputs) are located at the top and/or bottom edge of cells, with power supply (and sometimes clock) terminals placed along the two sides of the cells. Cells are arranged in rows, connecting power (and clock) terminals by abutment. Each pair of rows includes a row parallel routing channel that accommodates most of the wiring, and whose width depends on the amount of wiring it has to accommodate. Nets that must cross a row use feed-through cells, pairs of electrically equivalent cell terminals, or over-the-cell routing, just as in row type gate arrays.

Advantages of the standard cell concept as compared to gate arrays are obvious. Because only cells that will be used are placed on the chip, and the width of routing channels is adaptable to needs, the chip area is usually smaller than that of a functionally equivalent gate array chip. Mapping of functional blocks to library cells is easier because of the freely variable cell width. Finally, automatic layout generation—especially fully automatic routine—is easier to achieve because of the "soft" routing channels.

There are drawbacks as well. With individual adjustment of routing channels and free placement of cells within rows, prefabrication of active devices is no longer possible. The geometry of all masks defining a cell must be stored in the cell library, and all the processing steps have to be done for each individual chip type. Therefore, a larger production volume is needed to achieve costs per chip comparable to gate arrays. Automatic layout generation must handle not only cell placement and routing within the chip's core, but also fabrication-dependent preparation of the chip periphery. Figure 22-4 shows the principle of a standard cell chip.

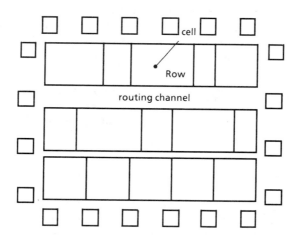

Figure 22-4.   Scheme of a standard cell based chip.

General Cells

Not all circuit designs can be mapped to a gate array or to a chip composed exclusively of standard cells, for the following reasons:

- Large functional units cannot be efficiently implemented as one cell within the rigid framework of a unique cell height and row structure.
- The number of standard cells that can be used on one chip is limited; if cell rows become too long, power supply lines in the cell layout become too narrow to carry the current needed to supply the cells within the row.
- Automatic layout programs tend to become slow if the number of cells they must deal with exceeds a certain limit.

The general cell concept solves these problems.

General cells are cells of rectangular but otherwise unrestricted shape, with terminals located at all four edges. Because of greatly varying cell sizes, matrix or row structures make no sense for these cells; rather, they must be placed in an unstructured mode to minimize unused silicon real estate, maximize wirability, and minimize signal delays.

The main advantage of general cells is that they lend themselves to the concept of hierarchy: a set of standard cells can be used as one general cell. Other typical cells are PLAs, RAMs and ROMs, which can be generated from a functional description. Figure 22-5 shows the principle of a hierarchically structured general cell layout.

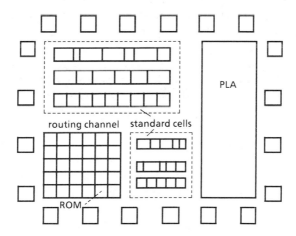

Figure 22-5.   Scheme of hierarchically structured general cell layout.

AUTOMATIC PLACEMENT AND ROUTING

The functional/logical design of a circuit produces a list of cells and pads to be placed, and a net list specifying connections between their terminals. The cells must be placed in a way that will minimize chip area and/or signal delays, and the connections must be realized on a small set of layers (currently one to three, according to the technology).

Although placement and routing interact heavily—placement more or less determines the topology of wires, and the area needed for wiring determines the spacing between cells—these problems usually are handled as individual steps to achieve algorithmic tractability of the whole problem. As we will see, even further decomposition into substeps is needed in most cases.

The character of the resulting subproblems of placement and routing can be very different for the various cell concepts. In the following discussion, we will characterize these subproblems and describe associated solution methods. In a final section, we will elaborate on solution methods successfully applied in a specific general-cell automatic placement and routing program.

Partitioning of the layout problem into subproblems calls for a mechanism to control the solution of a subproblem in such a way that subsequent subproblems can be solved, and the overall design goal met. For this purpose—as the "glue" between subtasks—objective functions are defined that express the design goal of a subtask. An objective function should reflect the desired goal as precisely as possible, but should be evaluated fast.

Placement Techniques

The goal of placement is to position and orient all the cells and pads of a circuit in such a way that restrictions specific to the cell concept, to the technology, and to the circuit at hand are handled, and subsequent steps (i.e., routing) can be accomplished successfully.

The natural objective function to be minimized in placement would be the difficulty of routing the required connections between cells. Routability ensured, the total chip area, which is closely related to manufacturing cost, should be minimized (if not fixed a priori, as in gate arrays).

This objective function is far too complex to be used directly in placement algorithms. Rather, simple functions are employed for optimization, which can be evaluated rapidly, and which hopefully reflect the desired goal. Typical functions used for this purpose are the estimated total wire length, the wiring density at sample locations on the chip, and the potential energy of an associated physical system (see below).

To evaluate wire length during placement we need some model for the topology of connections because the actual embedding of the wires representing a net is unknown during the placement phase. Two point connections usually are modeled

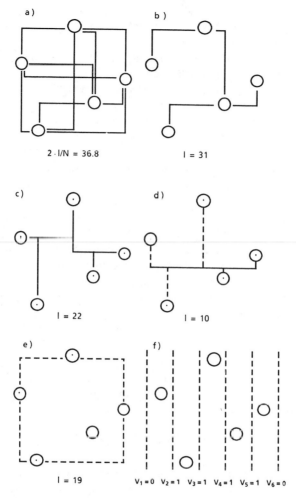

Figure 22-6.   Various models for the topology of one net and associated cost measures. (a) Complete graph. (b) Minimal spanning tree. (c) Steiner tree. (d) Trunk length. (e) Enclosing rectangle. (f) Cut line values.

as a "Manhattan path," that is, as a sequence of wire segments running parallel to one of the two coordinate axes. The associated wire length is referred to as the "Manhattan distance,"

If a net has to connect more than two terminals, several alternative models are available (see Figure 22-6):

- *Complete graph:* Two point connections are assumed between all pairs of the N terminals in the net. Because this results in $N(N-1)/2$ connections in the model, whereas only $(N - 1)$ connections are needed in the layout, the model connections have to be weighted with the factor $1/N$ in all evaluations of the objective function.

- *Minimum spanning tree:* Here, from the complete graph those $(N - 1)$ two point connections are chosen that connect the N terminals with minimal total length. The well-known minimum spanning tree algorithm of Prim (1951) can be used for this purpose. (Careful implementation yields $(N*N)$ time and $O(N)$ space complexity.) Minimal spanning trees can be calculated only if some placement already exists; otherwise, terminal locations would not be known.
- *"Optimal" spanning tree:* This technique was used in the NOMAD system for PCB layout (Shupe 1975). It does not presume a placement to exist. The subset of two point connections between pairs of modules. If, for instance, net 1 connects modules A and B, and net 2 connects modules A, B, C, and D, then the connection (A, B) would be preferred over (C, D) for net 2 because A and B should be placed near each other anyway.
- *Steiner tree:* A Steiner tree is the shortest tree connecting a given set of points in a plane, possibly using additional branching points. It gives the best approximation to the final layout of wires, but being expensive to calculate it is seldom employed.
- *Trunk length:* Trunks are the channel parallel parts of nets in a standard cell layout. Their length is usually adopted as an objective function in standard cell placement (Kozawa et al. 1983).
- *Minimum enclosing rectangle:* The half periphery of the smallest rectangle enclosing all terminals of a net usually correlates well with actual wire length and can easily be calculated.

Density-oriented objective functions can be defined using the notion of cut lines (Breuer 1977) (see Figure 22-6f). Cut lines have an associated value, that is, the number of nets with terminals on both sides of the cut line (nets that must cross the line at least once). Given a set of cut lines, we can try to minimize the sum of cut line values, their maximum value, or the value of cut lines in some given sequence.

Even with most of these simplified objective functions, the resulting mathematical problems cannot be solved exactly because they belong to the class of NP-complete problems (Sahni and Bhatt 1980). Therefore, we resort to heuristic approaches to find good solutions using a reasonable amount of computer time.

The known approaches to placement can be classified as constructive or iterative improvement methods. For both types there are sequential and simultaneous variants. Constructive methods build a placement from scratch, whereas iterative improvement tries to modify a given placement so that the objective function is decreased.

A typical sequential-constructive approach would be to place cell after cell until all cells are laid down. In each step we choose the next cell to be placed according to a selection criterion (usually the one with the most connections to cells already placed) and search for the best location for this cell (the one that causes the smallest increase of the objective function). A comprehensive study of sequential algorithms for initial placement as well as for iterative improvement can be found in Hanan et al. (1976).

Two typical simultaneous-constructive techniques map the placement problem to a mechanical or electrical system, for instance, cells to masspoints and connections to springs (Quinn 1975; Antreich et al. 1982), or cells to resistors and locations to nodes (Cheng and Kuh 1984), and then solve the resulting equality system simultaneously.

Another constructive approach, which has become popular in the last half decade, is a min-cut placement (Breuer 1977; Guenther 1969; Lauther 1979). The set of cells to be placed is partitioned into subsets so that the number of nets cut by the partitioning is minimal, and the size of subsets is approximately equal. Applying this procedure recursively to resulting subsets yields a top-down placement procedure that tends to minimize wiring density. This technique apparently belongs to the class of placement algorithms that are controlled by a density-oriented objective function. Partitioning is in itself a difficulty task. Usually variants (Fiduccia and Mattheyses 1982; Krishnamurthy 1984) of the famous Kernigan–Linn heuristic (Kernighan and Linn 1979; Schweikert and Kernighan 1972) are used as approximations.

Placement of Gate Arrays. All of the techniques outlined above are well suited to placement of gate arrays. Because of prefabrication of the master, only specific locations (slots) can be used to place a cell, and some slots may be legal only for some cell types. Placement algorithms that are not based on the notion of slots need an additional step to move cells to legal positions. Linear assignment techniques (Akers 1981; Burkhard and Zimmerman 1980) are well suited for this purpose. Iterative improvement techniques are easily implemented because the swapping of a pair of cells does not affect the location of other cells.

Placement of Standard Cells. Placement of standard cells is slightly more complex than gate array placement because no slots are available, and the goal to be achieved is twofold: routability within a small area and minimization of dead space.

Standard cell placement naturally can be broken down into two major steps: assignment of cells to rows and determination of the best permutation of cells within a row.

In the first step, it is crucial to achieve rows of nearly equal length; otherwise, dead space would be introduced. If cells cannot be crossed by vertical wires, additional feed-through cells usually are inserted to bridge the rows. The space used by these cells must be taken into account when row lengths are calculated.

If a general two-dimensional placement procedure is used, a special step is needed to balance row lengths.

Final determination of the permutation of cells within rows cannot be done until the global topology of nets (which channels to use, where to cross the rows) has been fixed. (This is done in loose or global routing.)

To minimize chip area, the permutation of cells within rows would not be controlled by estimated wire length, but by the density of wiring in the two adjacent channels, that is, by the maximal number of wires running in parallel in these two channels. As standard cells usually have different widths, swapping of two cells within a row usually results in a translation of all of the cells within the interval defined by these two cells; this renders the evaluation of such an operation more expensive than the swapping of equally sized cells on a gate array. Typical standard cell systems have been described in Kozawa et al. (1983) Killer and Lauther (1977), Persky et al. (1977), and Dunlop and Kernighan (1985).

Placement of General Cells.   In general-cell placement, consideration of dead space becomes even more important. For any placement modifications a data structure that provides for fast recalculation of cell positions is crucial. One way to accomplish this is not (only) to store cell locations but relations between adjacent cells (Brooks et al. 1940; Zibert 1974; Kani et al. 1976; Preas and Gwyn 1978). Using this technique, cell positions are defined implicitly by a system of inequalities that be solved rapidly by longest-path calculations on an associated directed acyclic graph.

In Lauther (1979) we introduced an algorithm that combines the min-cut principle with this graph representation technique in a powerful and efficient placement tool for general cells. We will come back to this method in another section of this chapter.

Routing

Placement accomplished, we are faced with the general routing problem: given a set of points in the plane (cell terminals) and a set of obstacles (cell boundaries, predefined routing, etc.), subsets of these points (nets) must be connected using wires on a specified number of layers and contacts at points where a connection changes layers. The width of wires usually is layer-specific, is sometimes net-specific, and for special nets (power supply) even changes within a net. Spacing of wires is also layer-dependent and often dominated by spacing of contacts, which tend to be significantly wider than wires. To make the problem even more complex, cell terminals may be extended, and cells may be movable to allow for adjustment of wiring space to actual needs.

The naive approach—to tackle this general routing problem directly, routing wire by wire and applying one of the well-known search strategies (Lee 1961; Hightower 1969)—is doomed to failure for problems of any significant size, for the following reasons:

- Connections routed successfully in early stages tend to block connections attempted later.
- Routing connecting by connection may lead to overcrowded areas on the chip.
- There is no way to adapt the placement of cells to wiring space needs recognized during routing.
- The potential space for a connective to be laid down is very large, so we can expect prohibitive consumption of resources and storage, as well as runtime.

The solution to all these problems again embodies the principle of "divide and conquer": we divide the routing task into subtasks and at the same time divide the routing plane into smaller regions that can be processed independently of each other, using specialized routing algorithms for the resulting special routing problems. The two major subtasks generated in this way are global routing and detailed routing of subregions.

Global Routing.   The task of global routing (sometimes called loose routing) is threefold. The first task is to determine the topology of nets in such a way that congestion of routing regions and resulting conflicts in detailed routing are avoided. This is most important, especially for gate arrays where routing space is fixed. Concentrating on topology and largely ignoring metrics helps one to consider nets concurrently and to speed up this process. The second task is to partition the overall routing problem into special, local routing problems and to partition the global netlist accordingly into local subnetlists. Finally, global routing yields an estimate of the wiring space needed for final routing. This estimate can be use to adjust placement to provide the space required (if such adjustments are possible within the framework of the underlying cell concept).

Crucial for global routing is an appropriate model for the routing space. In standard cell and general cell layout systems, the routing regions usually are modeled by a channel graph. One way to define this graph is as follows: edges of the graph represent routing channels, and nodes represent channel intersections. Additional edges are introduced where cells may be crossed. Associated with the edges is the geometric length of the channel and its capacity, that is, the number of wires it can accommodate. The global routing problem in this setting is to generate a tree on the channel graph for each net is so that the total length of wiring and the overflow of channel capacity is minimized.

A heuristic often used for this problem is to define a cost function on channel edges comprising geometric length and actual wiring density within the channel, and then to generate for each net a minimal tree on the graph that connects its terminals. A drawback of this sequential approach is that the nets are not truly considered concurrently; nets processed early do not "see" the final channel densities. Iterative deletion and reinsertion of nets can be applied to evade this problem. Final channel densities may be used to adjust the placement of cells.

When all the nets have been assigned to channels, we can break down the global netlist into the channel-specific subnetlists. In these netlists, cell terminals should belong to one (local) net, if they belong to the same global net, and if they are to be connected within the channel. To generate this piece of information, we assign local net numbers to cell terminals doing a depth-first search (DFS) (Tarjan 1972) on the tree generated for the net. In this search a current net number is maintained and assigned to cell terminals of the net as they are encountered. Whenever the DFS leaves a channel, the current net number is pushed onto a stack, and a new net number is generated. Coming back to a channel visited before, the old net number is retrieved from the stack, and marking of terminals is resumed with that number.

To describe the interface between two channels, it is convenient to create pseudo-pins for wires that leave or enter a channel. A local net is then defined as a set of cell terminals and pseudo-pins carrying the same local net number.

For gate arrays that do not expose a rigid separation between cell area and routing space but allow extensive over-the-cell routing, another way to model routing space is more appropriate: a coarse grid is defined, corresponding to the array of basic cells on the master. A capacity (in terms of wires and contacts) is associated with cells of the grid and/or with boundaries between cells. Classical maze routers (Lee 1961; Soukup and Royle 1981) could then be applied to the coarse grid to find the rough topology of each net sequentially.

Burstein and Pelavin (1983a) recently showed that linear programming can be used in this model to consider interacting nets concurrently and to achieve good solutions to the global routing problem efficiently. The basic idea is to reduce the wiring problem on an $n \times m$ grid to a hierarchically defined sequence of 2×2 problems. Dependencies between nets to be routed on a 2×2 grid can be stated and solved as a linear program of constant size, independent of the number of nets involved.

Application of simulating annealing techniques to the global routing problem is discussed in Vecchi and Kirkpatrick (1983), and more work related to gate array global routing can be found in Ting and Tien (1983) and Tsukiyami et al. (1983).

Detailed Routing.   Routing problems resulting from global routing are not only smaller in size than the original routing problem (smaller number of terminals and fewer connections to be made), but also show a special structure that can be exploited to devise specialized, efficient routing algorithms. Below we will list and characterize some of these special problems.

Single-Row Routing. In a single-row routing problem all terminals to be connected are located on a straight line. Connections are made within the upper and lower street along this line using one layer only. The capacity of these two streets is limited, as is the number of wires that may be fed from one street into the other between two consecutive terminals. This problem originally came up in routing of PCBs with regularly arranged dual in line packages, but it may be relevant also for some gate array types. Kuh et al. (1979) found necessary and sufficient conditions for single-row problems to be routable. (Ranghawan and Sahni (1982) report on an enumerative but efficient algorithm to find an optimal solution for such problems. Optimality, in this context, means that the maximal number of horizontal wires running in parallel (or the number of "tracks" used) is minimal. Further work on this problem is reported in Han and Sahni (1984), Tarng et al. (1984), and Carter and Breuer (1983).

Channel Routing. In a channel routing problem the terminals to be connected are located along two parallel straight lines (the channel border lines) enclosing a (w.l.o.g. horizontal) routing channel.

Additional "free" terminals are used to describe connections leaving or entering the channel at the two open ends. All connections have to be made within the channel, which itself is free of any obstacles. The channel width (the distance between the two borderlines) is not specified as part of the problem, but is determined by the router. The same holds for the vertical position of the free terminals. The main goal of channel routing is to minimize channel width (or equivalently the number of tracks used to embed wires). As a generalization of the problem we may permit channels with rigged borderlines forming bays, and the router should take advantages of these bays.

A standard approach to channel routing is to embed horizontal parts of a net ("trunks") in one layer and vertical parts ("branches") predominantly in a second layer (Hashimoto and Stevens 1971). Thus, the routing problem is reduced to the combinatorial problem of how to assign trunks to tracks such that no two

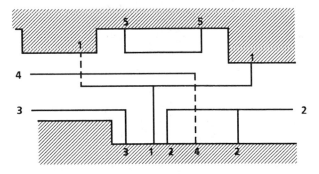

Figure 22-7.   A channel with rugged borderlines. Net 2 connects two fixed terminals and one free terminal.

trunks or branches of different nets touch or overlap each other. Apparently, a trunk to be connected to a terminal at the top shore of the channel has to be embedded on a higher track than one to be connected to a bottom terminal at the same horizontal position (or within a short horizontal distance); otherwise, a conflict would arise between the two associated branches. We can compile a list of all such conditions (to be conveniently stored as a direct graph) before actually embedding any wires.

Typically, channel routing algorithms are algorithms that work on the "vertical constraint graph" derived in this way, considering all potential conflicts concurrently. This technique results in much better performance (in terms of computer runtime as well as completion rate) than more traditional routing algorithms, which try to embed wire by wire, employing breadth-first (Lee 1961) or depth-first (Hightower 1969) search techniques in the routing space. Although the channel routing problem has been shown to be NP-complete (Szymanski 1985), algorithms exist that are fast and achieve optimal or near-optimal solutions in most practical cases (Kernighan et al. 1973); Yoshimura and Kuh 1982; Burstein and Pelavin 1983b; Rivest and Fiduccia 1982; Yoshimura 1984). A lower bound for the channel width is the maximum number of wires that must cross any cross section of the channel (the channel density). Channel density is easy to calculate and can be used to assess the quality of solutions.

If the vertical constraints graph is acyclic, we can always find a solution; otherwise, cycles have to be broken in a preprocessing pass that introduces jogs in the trunks involved in the cycle (Asano et al. 1977). Open connections will result only if sufficient space for jog insertion is not available. More general routing methods such as the exhaustive search strategy of a maze router may be used to solve these problems. Alternatively, a slight modification of placement will achieve routability in most cases.

If an overall routing problem is broken down into channel routing problems, pseudo-pins will be used to describe the interface of perpendicular, adjacent channels; wires leaving one channel (free terminals) play the role of fixed terminals in the other channel.

As the position of free terminals is determined by the router itself, the adjacent channel can be processed only after the primary channel has been finished. Thus, the form of channel crossings imposes an order on the sequence in which channels are to be processed: for each T-shaped channel crossing, the base channel has to be processed before the crosspiece channel (Kawanishi et al. 1973). These order restriction may form a loop, where no "safe" routing can be found. It has been shown (Szepienec and Otten 1980; Supowit and Slutz 1983) that have a safe routing order exists if the chip topology corresponds to a slice structure, and that no safe order exists if the chip contains the infamous "pin-wheel" (see Figure 22-8). A chip has slice structure if it can be divided by a routing channel into two subchips, both of which consist of exactly one cell or are slice-structured themselves.

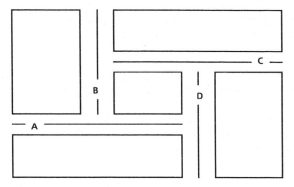

Figure 22-8. The infamous "pin-wheel." Channel *A* must be routed before *D*, *D* before *C*, *C* before *B*, *B* before *A*.

Channel routing can be applied to gate arrays, to standard cells, and to general cell layouts, and it should be used as the routing workhorse in these systems. More details on channel routing systems can be found in Lauther 1984); application in a general cell environment is discussed in Lauther (unpublished) and in the final section of this chapter.

River Routing. If we further restrict the class of routing problems to be considered, we arrive at the river routing problem, in which nets are required to contain exactly two terminals, one at each of the two channel borders (i.e., each net crosses the channel). A second condition is that the order in which nets appear as terminals be the same along both shores. If these conditions are met, the wires can all be embedded in one layer without crossing each other. The objective in river routing is to find an embedding that minimizes channel width and (with lower priority) the number of bends needed in wires. Exact and efficient solutions are known for this class of routing problems (Pinter 1982).

Switch Box Routing. If, on the other hand, we lessen problem restrictions and permit fixed terminals at all four sides of the "channel" (this implies fixed dimensions of the routing region), we get a much more difficult problem, the switch box routing problem (Soukup 1981; Hsu 1982). Not only is the solvability of a given instance of the problem in question, but if a solution exists, we have no algorithm that will guarantee that we will find the solution. Again, a slight restriction of the problem class changes the situation: if only two terminal nets are allowed, then the problem can gracefully be mapped to a network flow problem, and solution methods derived from this view will guarantee that we will find the solution if one exists, and will do so efficiently (Mehlhorn and Preparata 1983).

Clearly, the key to planning a routing system is to break the overall routing problem down into special problems that can be solved efficiently. In the following section we will discuss a general cell placement and routing system in which the routing problem has been totally broken down into channel and routing problems.

PLACEMENT AND ROUTING IN THE CALCOS SYSTEM

In this section we will discuss in more detail placement and routing in a specific system to demonstrate how the techniques presented above can be combined to yield a powerful, flexible, and safe tool for physical design.

The intention in designing the CALCOS system was to provide a system for hierarchical design of custom or semicustom integrated circuits. In the typical application we have a mix of general cells laid out by automatic module generators such as PLA, ROM, and RAM; handcrafted special-purpose cells; and blocks of random logic composed of standard cells assembled by a specialized tool (not to be discussed here). Knowing the shortcomings of purely automatic approaches, we decided to integrate flexible interactivity into the system to make full use of the superior pattern recognition ability and expert knowledge of the human designer. To prevent introduction of errors during manual interaction, all manipulations are checked online for consistency or controlled in such a way that inconsistencies are not introduced in the first place. In an interactive session the designer should be able to explore a variety of alternative solutions. To this end, not only placement and placement modification but also routing must be fast, because only after complete routing can the quality of placement be properly assessed.

Placement

The main objective in general cell placement is minimization of silicon real estate, which can be divided into three components: cell area, which is constant; wiring area, which more or less corresponds to total wire length or average wiring density; and unused, wasted space, which results from poor relative placement of blocks with differing dimensions.

For manual interactions, which are used mainly to reduce dead space, as well as for automatic steps, we need a method to manipulate a set of rectangular objects in the plane efficiently and in such a way that a legal arrangement of these objects is guaranteed at all times.

To provided fast routing, we decided to use channel routing exclusively. Therefore, we must break down the routing space totally into channel routing regions, and must find a safe routing order for these channels.

All these requirements (minimization of wiring density, ease of consistent manipulation, and natural definition of channels and their routing order) are met by a combination of two concepts: top-down min-cut placement and polar graph

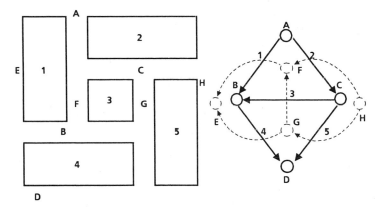

Figure 22-9. Correspondence between a set of blocks and channels and a pair of mutually dual planar graphs.

representation. Here, the relative placement of cells and routing channels is represented by a pair of mutually dual graphs. There is a one-to-one correspondence between edge pairs (primal/dual) and cells, between nodes in the primal (dual) graph and horizontal (vertical) channels, and finally between path lengths and coordinates (see Figure 22-9).

Initial Placement. To construct an initial placement in this setting we proceed as follows: At the outset, the set of all cells to be placed is represented by a square with area equal to the sum of cell areas plus estimated wiring area. Concurrently, the set of cells is represented by a pair of mutually dual edges whose length is equal to the dimensions of the initial square. The set of cells is then partitioned into two subsets of approximately equal area in such a way that the number of nets incident to cells in different subsets (i.e., subsets that are cut by the partitioning) is minimal. This partitioning divides (or slices) the initial square into two rectangles representing the two subsets; in the concurrent graph representation the primal (dual) edge is split into two serial (parallel) edges (see Figure 22-10a).

The same algorithm is then applied recursively to the resulting subsets (optionally changing the cut direction for best fitting of cells into available space); it terminates when each subset contains exactly one cell (see Figure 22-10g).

The initial floorplan of cells generated by this method represents the cells with their true area (plus estimated share of wiring space) but not with their true shape. To take cell shapes into account, the edges in the primal (dual) graph are weighted with the height (width) of the corresponding cell. Calculation of the longest paths in both graphs from any node to the sink then yields the cell (or channel) coordinates (see Figure 22-11).

This calculation is a very fast operation, taking O(N) steps for N cells. The

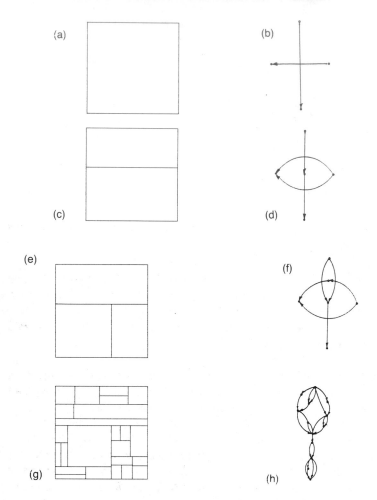

Figure 22-10. Sequence of states during initial placement. (a, b) Initial square and corresponding pair of edges. (c, d) First segmentation. (e, f) First step recursion. (g, h) Final slice structure.

initial placement outlined here is described in more detail in Lauther (1979). The algorithm attempts to minimize wiring density in each step, allowing very global rearrangements in the first steps, whereas final steps concentrate on local optimization.

The need to create a sliced chip plan (to permit exclusive use of channel routing) is met in a very natural way. The sequence in which channels (cutlines) are generated also defines a safe routing order; the channel created first must be routed last.

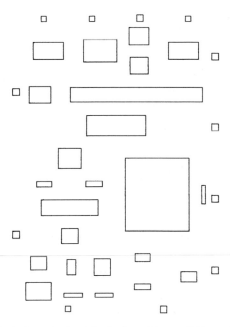

Figure 22-11.   Initial placement after introduction of true cell dimensions.

Placement Compaction.   Introduction of true cell shapes may result in a lot of dead area. The underlying graphical representation allows the use of relatively simple algorithms that reduce the amount of this wasted space.

The simplest of these operations is rotational optimization. Each cell may be rotated by 90° (in the graph representation the two weights of a pair of dual edges are swapped), resulting in 2^{+N} different placements, all with the same slice structure. Stockmeyer (undated) devised an $O(N \log N)$ algorithm that finds the optimal arrangement for a series-parallel graph or (equivalently) a slice structure. For a more general graph structures the problem can be solved by implicit enumeration (branch and bound) (Zibert 1974) or (approximately) by relatively simple greedy algorithms.

A more sophisticated device is two-dimensional compaction. (One-dimensional compaction is pointless because, in our representation, the layout is compact in each direction separately anyhow.) The basic idea here is to cut the longest (critical) path in one of the two graphs by splitting a node into two subnodes (subchannels) joined by a dummy edge to keep the planar graph structure intact (see Figure 22-12). In the geometric representation, splitting of a node results in a broken cutline. Figure 22-13 shows the arrangement of cells after rotational optimization and after two-dimensional compaction.

Unfortunately, this compaction technique destroys the slice structure of the

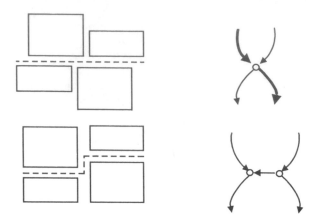

Figure 22-12.   The critical path is broken to achieve a more compact layout.

layout (or the series-parallel structure of the two graphs representing the place-ment). Therefore, if this operation has been applied, or if we start from a user-specified preplacement of cells, a new slice structure must be generated before we proceed to routing. For this purpose a procedure quite similar to min-cut placement was developed. The only difference is that the amount of geometric distortion inflicted upon the placement is minimized, instead of the number of nets cut in a partitioning step.

Our exclusive employment of channel routing certainly restricts the class of chip structures feasible. The infamous pin-wheel cannot be processed; but, be-

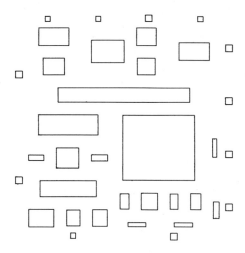

Figure 22-13.   Result of rotational optimization and two-dimensional compaction.

cause of the concept of rugged borderlines for channels, the penalty in terms of chip area should be small.

Interactive Placement Manipulation.   The polar graph representation can also be used for the implementation of powerful interactive manipulation aids. Using a CRT, cells can be rotated, mirrored, or removed from the cell formation and inserted again at another location. Unlike the usual practice in conventional graphic systems, the set of all cells is immediately rearranged when one cell has been moved.

For this purpose, all user requests (ROTAT, INTERCHANGE, MOVE) are transformed into equivalent modifications of the underlying polar graph structure; then the longest paths are recalculated, yielding new cell positions.

In Figure 22-14 we can observe two such operations. First cell Z1 is moved to the site below Z2; then Z3 is moved above Z4 with simultaneous rotation. Next the COMPACT command is executed, resulting in a major rearrangement of cells in the bottom part of the chip plan.

Routing

Routing is done in two passes. In the first pass, global routing, we determine for each net the channels it has to run through, and for each channel, the width

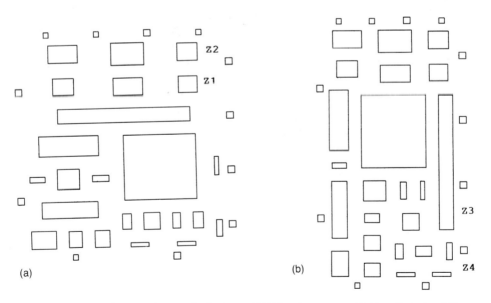

(a)   (b)

Figure 22-14.   (a) Interactive manipuilation; cell Z1 has been moved tot he site below Z2. (b) Further manipulation; Z3 was rotated by 90° and inserted above Z4. New compaction results in a significant rearrangement of cells.

needed to accommodate the pertinent nets. In the second pass, detailed routing, the exact position of path segments and contacts has to be found for each net.

Global Routing. Detailed routing will be done channel by channel using a channel local netlist. It is the task of global or loose routing to break down the global netlist into these channel-specific netlists. We try to control loose routing to minimize the length of connections and the congestion of channels. Knowing loose routes for all nets, we can estimate the required width of each channel and adjust cell placement accordingly.

Another task of global routing in our system (to be discussed in a separate paragraph) is to enforce a special topology of power and ground nets that permits planar embedding of these nets, and to adjust the width of power and ground branches according to the current they have to carry.

Our global router works as follows. First we generate a channel graph describing the topology of channels. Edges of this graph represent channel segments (subchannels), and nodes represent channel intersections. This graph can easily be constructed from the two placement graphs. Each node of the placement graph (representing a main channel) is mapped into a series of channel graph edges (subchannels). Each subchannel separates exactly two cells and can be considered as an additional edge in the placement graph weighted with the width of the subchannel. Inclusion of these edges (dotted in Figure 22-15) in longest path calculation results in a cell placement with sufficiently spaced cells. Exclusive use of these edges for placement adjustment leads to "soft" channels and reduced chip area.

To control global routing, a cost function is defined on channel edges comprising geometric length and crowdedness of the channel. Having built this data structure, we calculate for each net a minimal tree on the channel graph connecting all terminals of the net. The three is constructed by an approximation algorithm based on a series of shortest path calculations, using the cost function

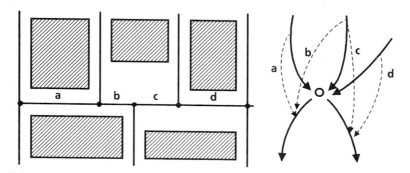

Figure 22-15.   One node of the placement graph is mapped into a main channel consisting of subchannels *a . . . d*.

defined on edges. The width of subchannels used for the trees is updated for each net.

Placement Adjustment during Routing. Until global routing cells are free to move, only sufficient distances to neighbor cells have to be observed. The situation changes when one or more channels have been routed: cells adjacent to a routed channel may still float as a group, but relative to each other and relative to the wiring of that channel their positions are fixed. To accomplish this behavior we have to modify the definition of our placement graphs and the associated procedure for longest path calculations. Until now, edges of the graph represented inequalities between cell or channel coordinates. We now introduce special edges that express equalities and are used to lock together cells adjacent to a channel.

Routing of a channel involves three major substeps. First, a description of the channel to be routed (borderlines, pins, design rules etc.) is fed to the channel router. Second, the router is invoked, returning the actual channel width and a set of wires and vias. The third step is concerned with placement adjustment: the two leftmost cells of the channel are locked to the crosspiece channel left of the current channel; this prevents further lateral shifts between all the cells belonging to the current channel. (Cells that are not leftmost are separated by vertical channels that have already been routed and fixed.) For vertical locking, the lengths of subchannel edges belonging to the current channel are adjusted according to the actual channel width reported by the router, and fixed as well. Finally, longest path calculation is invoked to update all cell positions (positions change if the channel width determine by the router differs from the estimate derived in global routing). Wires and vias returned by the router are described relative to a channel-specific coordinate system; therefore, they automatically move with floating clusters of wired cells. Prerequisite for the technique described is a fast algorithm for longest path calculation in an acyclic directed graph containing special edges expressing fixed distances between end nodes. In Lauther (unpublished) we report such an algorithm that works with $O(n + v)$ time and space complexity for a graph with n edges and v nodes.

Demarcation of Channels. We have described how channels are defined in a natural way from the placement of cells, but have not yet stated exactly where the domain of one channel ends and that of an adjacent channel begins. In our system this demarcation of channels is done by the channel router itself. Whenever a channel is routed, the router calculates and returns hulls at the two channel ends enclosing all the wiring of that channel at tightly as possible.

In routing a channel, adjacent channels are one of two types. Either they are tributary channels to the current one, in which case they have already been processed, and their hull becomes part of the borderline defining the current channel (other border points are derived from the cells situated along the two

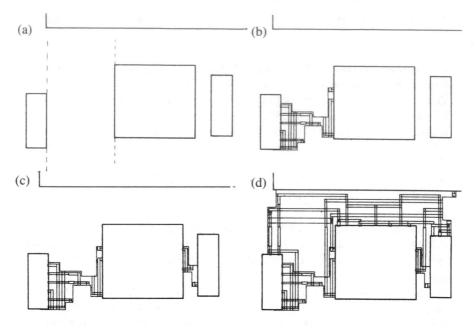

Figure 22-16.   Sequence of channel-routing, demonstrating the process of border generation.

shores of the channel); or they are adjacent to one of the two channel ends, in which case they will be routed later, using the hull just being generated as an input.

The hull polygone generated at the channel end also defines one coordinate of pseudo-pins that describe connections entering at that channel end. The other coordinate is a result of track assignment for these connections inside the channel. Figure 22-16 demonstrates this process of border generation and usage. In Figure 22-16(a), we see two cells defining an elementary routing channel and the two borderlines derived from the cells. In Figure 22-16(b), the channel has been routed. We see the resulting wires, the adjustment of cell positions, and the two hulls with associated pseudo-pins generated at channel ends. In Figure 22-16(c), one of these hulls with pseudo-pins serves as one of the inputs for a perpendicular channel, which is shown routed in Figure 22-16(d). Figure 22-17 shows a complete chip broken down into routing regions by the technique described.

Power and Ground Routing.   Power and ground nets need special processing fro two reasons: first, the width of wires has to be adapted to the current they carry to avoid excessive voltage drops or migration problems; second, in most technologies these nets have to be laid out in one layer to avoid voltage drops in a layer with high resistance (polysilicon) or at via holes.

For planar power and ground routing we implemented the scheme proposed by Syed and El Gamal (1982). One of the two nets (w.l.o.g. ground) is embedded in horizontal channels that are always above the power net, and in vertical channels that are always to the left of power. Starting from cell terminals, power (ground) connections are routed to the left (right) in horizontal channels and upward (downward) in vertical channels, resulting in a tree rooted at the upper left (lower right) corner of the chip. This scheme works under the condition that power (ground) terminals are located at the top or left (bottom or right) edge of cells. If this is not the case, then in addition to the main power and ground trees, local connections are needed to wrap power and ground around to the correct side of the cells.

To achieve the described topology of power and ground nets, we had to modify global routing and local channel routing. In global routing, tree generation (for power and ground) is not controlled by the cost function defined on channel edges, but by traffic rules to be observed at channel intersections. Further, global routing must generate addition pseudo-pins to take of local wrap-around connections.

In local channel routing, a vertical order has to be observed between horizontal wires. To describe this order we assign order values to (local) nets during global routing. The ground nets get a high order value, the power nets a low value,

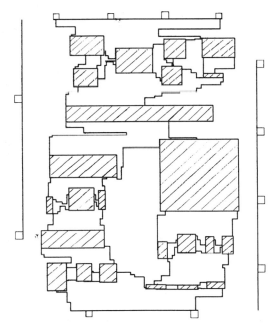

Figure 22-17.   A chip totally broken down into routing channels.

and signal nets the neutral value zero. Now the router has to observe the following rules:

1. Nets with order value $\neq \phi$ have to be arranged vertically according to order values.
2. Nets with order values $= \phi$ whose x-interval includes a bottom (top) terminal of a net order with value $\neq \phi$ have to be routed above (below) that net.

The first rule implements the overall topology of power and ground trees, whereas the second rule ensures that power and ground terminals can be connected to channel-parallel parts of these trees without having to cross other wires. The proper arrangement of wrap-around connections is achieved similarly to appropriate order values assigned during global routing.

Another piece of information gathered in global routing that must be transferred

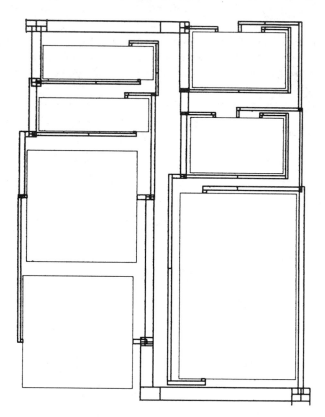

Figure 22-18. Two interdigitating power and ground trees.

to local channel routing is the width of power and ground wire segments. For this purpose, a DFS is made on power and ground trees, starting from the root and accumulating current values during backtracking steps over edges. The leaves of the trees—the power and ground terminals of cells—are labeled with the estimated current consumption of the cells. Figure 22-18 shows an example of two interdigitating power supply trees generated by the algorithms described. The roots of these trees are connected to supply rails inside a peripheral frame (not shown in the figure) that supply peripheral cells (pads) embedded in this frame and serve also to connect the power and ground trees to the corresponding pads.

Local Channel Routing.   Space limitations prevent discussion of our channel router in detail, but we will deal with some of its key features and concepts.

Because of the integrated processing of signal nets and power supply nets, the router must work gridless in the vertical direction; otherwise, too much space would be wasted by the mapping of nonstandard wires to multiples of a basic pitch. Because wires leaving a channel end are used as terminals along the shore of a perpendicular neighbor channel, gridless processing is needed in the horizontal direction as well. The ungridded mode of channel routing is implemented as follows: All processing along the channel is done by a series of plane sweeps (Shamos and Hoey 1976), a technique that does not need the notion of a grid. In the vertical direction we first assign nets to tracks using one of the standard algorithms (which are inherently grid-based), and apply compaction afterward to reduce channel width and to enforce design rules in the vicinity of wires of nonstandard width.

Track assignment—assignment of horizontal wires (trunks) to tracks in such a way that no conflicts arise in the embedding of vertical wires (branches), and that channel width is minimized—has been thoroughly covered in the literature, but it is just one of a variety of steps needed. Among these steps are:

1. Generation of constraints between trunks due to pins at opposite shores of the channel within a short horizontal distance.
2. Generation of constraints between trunks according to order values, to enforce the special topology of power and ground nets.
3. Introduction of jobs in trunks to break directed cycles of constraints generated in steps 1 and 2.
4. Track assignment, taking rugged borders into account.
5. Compaction in the vertical direction, taking spacing of via holes into account (only where they are required).
6. Generation of branches, preferring one of the two layers available (metal in a one-metal technology).

7. Generation of hulls demarcating the domain of the channel at the two channel ends.
8. Calculation of pseudo-pin positions.

Steps 1 through 3 and 5 make heavy use of plane sweep techniques to achieve a good performance. They have been partially described in Lauther (1984). For track assignment we use a variant of the algorithm published by Yoshimura (1984), as it is easily adaptable to our concept of rugged channel borders.

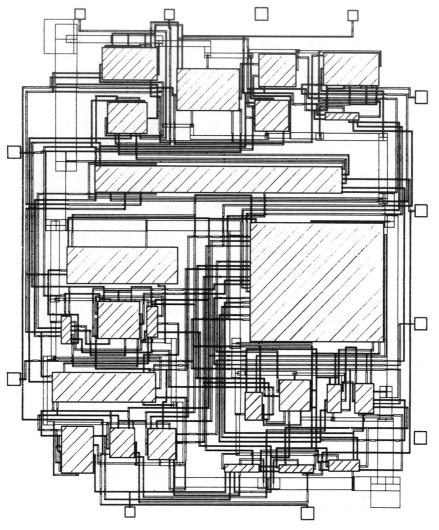

Figure 22-19. A chip with fully automatic routing including planar power and ground.

For compaction, a graph is built whose nodes represent trunks and whose edges describe spacing requirements. The graph is constructed in a way similar to Lengauer's algorithm (1983). Longest path calculation on this graph yields final trunk positions.

The channel router was implemented in PASCAL using approximately 8000 lines of code. It may be of interest that track assignment itself takes only 370 lines.

The total CPU time used for automatic placement and routing of the example in Figure 22-19 is 22 sec on an MIPS machine. This circuit contains 24 cells and 10 pads connected by 55 nets using 259 cell terminals. About 80% of the wiring is in the preferred layer.

CONCLUSIONS

After a general discussion of the advantages of cell based physical design and a look at general principles and methods for its automation, we presented in some detail the actual implementation of a working system. The main key to efficiency, both in quality of results and in usage of resources (computer runtime and memory), is the principle of "divide and conquer." The overall layout problem has been divided into the main steps of placement, loose routing, and detailed routing, with the routing itself carried out channel by channel, that is, broken down into small problems that can be solved efficiently.

REFERENCES

Akers, S. B., "On the Use of Linear Assignment Algorithm in Module Placement," *Proc. 18th Design Automation Conf.*, pp. 137–143, June 1981.

Antreich, R. J., Johannes, F. M., and Kirsch, F. H., "A New Approach for Solving the Placement Problem Using force Models," *Proc.* IEEE ISCAS, Rome, 1982, pp. 481–486.

Asano, T., et al., "A Wire Routing Scheme Based on Trunk-Division Methods," *IEEE Trans. Computers, C-26* (8), 764–772 (Aug. 1977).

Baird, H. S., "Fast Algorithms for LSI Artwork Analysis," *J. Design Automation and Fault-Tolerance Computing, 2* (2), 179–209 (1978).

Breuer, M., "Min-Cut Placement," *Design Automation and Fault-Tolerant Computing, 1* (4), 343–362 (Oct. 1977).

Brooks, R. L., Smith, C. A. B., Stone, A. H., and Tutte, W. T., "The Dissection of Rectangles into Squares," *Duke Math. J.*, No. 7, 312–340 (1940).

Burkhard, R., and Zimmermann, U., *Assignment and Matching Problems: Solutions with FORTRAN-Programms*, Springer, 1980.

Burstein, M., and Pelavin, R., "Hierarchical Wire Routing," *IEEE Trans. CAD, CAD-2* (4), 223–234 (Oct. 1983a).

Burstein, M., and Pelavin, R., "Hierarchical Channel Router," *Integration, 1*, 21–38 (1983b).

Carter, H., and Breuer, M., "Efficient Single-Layer Routing along a Line of Points," *IEEE Trans. CAD, CAD-2* (4), 259–266 (Oct. 1983).

Cheng, C., and Kuh, E., "Module Placement Based on Resistive Network Optimization," *IEEE Trans. CAD, CAD-3* (3), 218–225 (July 1984).

Dunlop, A., and Kernighan, B., "A Procedure for Placement of Standard Cell VLSI Circuits," *IEEE Trans. CAD, CAD-4* (1), 92–98 (Jan. 1985).

Fiduccia, C. M., and Mattheyses, R. M., "A Linear-Time Heuristic for Improving Network Partitions," *Proc. 16th Design Automation Conf.*, pp. 175–181 (June 1982).

Guenther, Th., "Die Raeumliche Anordnung von Einheiten mit Wechselbeziehungen," *Elektronische Datenverarbeitung*, No. 6, 209–212 (1969).

Han, S., and Sahni, S., "Single-Row Routing in Narrow Streets," *IEEE Trans. CAD, CAD-3* (3), 235–241 (July 1984).

Hanan, H., Wolf, P. K., and Agule, B. J., "A Study of Placement Techniques," *J. Design Automation and Fault-Tolerant Computing, 1* (1), 28–61 (Oct. 1976).

Hashimoto, A., and Stevens, J., "Wire Routing by Optimizing Channel Assignment within Large Apertures," *Proc. 8th Design Automation Workshop*, pp. 155–159, 1971.

Hightower, D. W., "A Solution to Line Routing Problems in the Continuous Plane," *Proc. 6th Design Automation Workshop*, pp. 1–24, June 1969.

Hofmann, M., and Lauther, U., "HEX: An Instruction Driven Approach to Feature Extraction," *Proc. 20th Design Automation Conf.*, pp. 331–336, June 1983.

Holzapfel, H. P., Michel, P., Koeppel, K., and Neppl, F., "New Architecture Brings Flexibility to Computer Gate Array Design," to be published.

Hsu, C., "A New Two-Dimensional Routing Algorithm," *Proc. 19th Design Automation Conf.*, pp. 46–40, 1982.

Kani, K., Kawanishi, H., and Kishimoto, A., "ROBIN: A Building Block LSI Routing Program," *Proc.* 1976 IEEE ISCAS, pp. 312–340.

Kawanishi, H., Goto, S., Oyamada, T., and Kani, K., "A Routing Method of Building Block LSI," 7th Asilomar Conf. on Circuits, Systems and Computers, 1973.

Kernighan, B. W., and Linn, S., "An Efficient Heuristic Procedure for Partitioning Graphs," *Bell Syst. Tech. J., 49*, 291–308 (Feb. 1979).

Kernighan, B. W., Schweikert, D. G., and Persky, G., "An Optimizing Channel Routing Algorithm for Polycell Layouts of Integrated Circuits," *Proc. 10th Design Automation Workshop*, pp. 50–59, 1973.

Koch, B., and Nett, M., "Computer-Aided Design of Digital Circuits with the VLSI Design System VENUS," *Siemens Forsch.-u. Entwick.-Ber., 13* (5), 215–220 (1984).

Koller, K., and Lauther, U., "The Siemens-AVESTA-System for Computer Aided Design of MOS-Standard Cell Circuits," *Proc. 14th Design Automation Conf.* pp. 153–157, June 1977.

Kozawa, T., et al., "Automatic Placement Algorithms for High Packing Density VLSI," *Proc. 20th Design Automation Conf.*, pp. 175–181, June 1983.

Krishnamurthy, B., "An Improved Min-Cut Algorithm for Partitioning VLSI Networks," *IEEE Trans. Comp., C-33* (5), 438–446 (May 1984).

Kuh, E., Kashiwabara, T., and Fujisawa, T., "On Optimum Single Row Routing," *IEEE Trans. Circuits and Systems, CAS-26*, 361–368 (1979).

Lattin, B., "VLSI Design Technology: The Problem of the 80's for Microprocessor Design," *Proc.* 1st Caltech. Conf. on VLSI, 1979, pp. 247–252.

Lauther, U., "A Min-Cut Placement Algorithm for General Cell Assemblies Based on a Graph Representations," *Proc. 16th Design Automation Conf.*, pp. 1–10, 1979.

Lauther, U., "An (*N* Log *N*) Algorithm for Boolean Mask Operations," *Proc. 18th Design Automation Conf.*, pp. 555–562, June 1981.

Lauther, U., "Anwendung von Plane-Sweep Verfahren bei der Layouterzeugung intergrierter Schaltungen," *Informatik Fachberichte 89*, Springer, 1984, pp. 69–78.

Lauther, U., "Channel Routing in a General Cell Environment," to be published.

Lee, C. M., Chawla, B. R., and Just, S., "Automatic Generation and Characterization of CMOS Polycells," *Proc. 18th Design Automation Conf.*, pp. 220–224, June 1981.

Lee, C. Y., "An Algorithm for Path Connections and its Applications," *IEEE Trans. Electronic Computers, VEC-10* 346–365 (1961).

Lengauer, T., "Efficient Algorithms for the Constraint Generation for Integrated Circuit Layout Compaction," *Proc.* of the WG'83, International Workshop on Graphitheoretic Concepts in Computer Science, Hrsg. M. Nagl and J. Perl, Editors, pp. 219–230.

Melhorn, K., and Preparata, F., "Routing through a Rectangle," Technical Report, 1983.

Persky, G., Deutsch, D., and Schweikert, D., "LTX—A Minicomputer Based System for Automation LSI Layout," *J. Design Automation and Fault-Tolerant Computing, 1* (3) 217–256 (May 1977).

Pinter, R. Y., "On Routing Two-Point Nets across a Channel," *Proc. 19th Design Automation Conf.*, pp. 894–902, 1982.

Preas, B. T., and Gwyn, C. W., "Methods for Hierarchical Automatic Layout of Custom LSI Circuit Masks," *Proc. 15th Design Automation Conf.*, pp. 206–212, 1978.

Prim, R. C., "Shortest Connection Networks and Some Generalizations," *Bell System Tech. J., 36,* 1389–1401 (Nov. 1957).

Quinn, N. R., "The Placement Problem as Viewed from the Physics of Classical Mechanics," *Proc. 12th Design Automation Conf.*, pp. 173–178, 1975.

Ranghawan, R., and Sahni, S., "Optimal Single Row Router," *Proc. 19th Design Automation Conf.*, pp. 38–45, 1982.

Rivest, R. L., "A Description of a Single-Chip Implementation of the RSA Cipher," *LAMBDA, 1* (3), 14–18 (1980).

Rivest, R. L., and Fiduccia, C. M., "A Greedy Channel Router," *Proc. 19th Design Automation Conf.*, pp. 418–424, 1982.

Rowson, J. A., "Procedural vs. Graphical Design," *LAMBDA, 1* (3), 6–7 (1980).

Sahni, S., and Bhatt, A., "The Complexity of Design Automation Problems," *Proc. 17th Design Automation Conf.*, pp. 402–441, June 1980.

Schiele, W., *Entwurfsregelanspassung der Maskengeometrie Integrierter Schaltungen*, VDI-Verlag, Duesseldorf, 1984.

Schweikert, D. G., and Kernighan, B. W., "A Proper Model for the Partitioning of Electrical Circuits," *Proc. Design Automation Workshop*, pp. 56–62, June 1972.

Shamos, I., and Hoey, D., "geometric Intersection Problems," *Proc.* 17th Ann. Conf. Foundations of Computer Science, Oct. 1976, pp. 208–215.

Shupe, Ch. F., "Automatic Component Placement in the NOMAD System," *Proc. 12th Design Automation Conf.*, pp. 162–172, June 1975.

Soukup, J., "Circuit Layout," *Proc. IEEE, 69,* 1281–1304 (1984).

Soukup, J., and Royle, J., "On Hierarchical Routing," *J. Digital Syst., 5* (3), 265–289 (1981).

Stockmeyer, L., "Optimal Orientations of Cells in Slicing Floorplan Design," Research Report, IBM Research Lab., San Jose, CA.

Supowit, K., and Slutz, E., "Placement Algorithms for Custom VLSI," *Proc. 20th Design Automation Conf.*, pp. 164–170, 1983.

Syed, Z., and El Gamal, A., "Single Layer Routing of Power and Ground Networks in Integrated Circuits," *J. Digital Systems, 6* (1), 53–63 (1982).

Szepienec, A., and Otten, R., "The Genealogical Approach to the Layout Problem," *Proc. 17th Design Automation Conf.*, pp. 535–542, June 1980.

Szymanski, T. G., "Doglet Channel Routing Is NP-complete," *IEEE Trans. CAD, CAD-4* (1) 31–41 (Jan. 1985).

Tarjan, R., "Depth-First Search and Linear Graph Algorithms," *SIAM J. Comput., 1* (2), 146–160 (June 1972).

Tarng, T., Marek-Sadowska, M., and Kuh, E., "An Efficient Single-Row Routing Algorithm," *IEEE Trans. CAD, CAD-3* (3), 178–183 (July 1984).

Taylor, S., "Symbolic Layout," *VLSI Design, 5* (3) 34–42 (Mar. 1984).

Ting, B., and Tien, B., "Routing Techniques for Gate Array," *IEEE Trans. CAD, CAD-2* (4), 301–312 (Oct. 1983).

Tsukiyama, S., Harada, I., Fukui, M., and Shirakawa, I., "A New Global Router for Gate Array LSI," *IEEE Trans. CAD, CAD-2* (4), 313–321 (Oct. 1983).

Tsukizo, A., Sakemi, J., Kozawa, T., and Fukuda, H., "Mach: A High-Hitting Pattern Checker for VLSI Mask Data," *Proc. 20th Design Automation Conf.*, pp. 726–731, June 1983.

Vecchi, M., and Kirkpatrick, S., "Global Wiring by Simulated Annealing,qc *IEEE Trans. CAD, CAD-2* (4), 215–222 (Oct. 1983).

VLSI Design Staff, "A Review of CAD Systems for IC Layout," *VLSI Design, 5* (3), 20–25 (Mar. 1984).

Walston, C. E., and Felix, C. P., "A Method of Programming Measurement and Estimation," *IBM Syst. J., 16* (1), 54–73 (1977).

Yoshimura, T., "An Efficient Channel Router," *Proc. 21th Design Automation Conf.*, pp. 38–44, 1984.

Yoshimura, T., and Kuh, E. S., "Efficient Algorithms for Channel Routing," *IEEE Trans. CAD, CAD-1* (1), 25–35 (Jan. 1982).

Zibert, K., "Ein Beitrag zum rechnergestuetzten topologischen Entworf von Hybrid-Schaltungen," doctoral thesis, Techn. Univ. Munich, 1974.

23. Logic Design Expert System

Shigeru Takagi
Nippon Telegraph and Telephone Corp.

Computer-aided design (CAD) is a collection of programs for assisting the digital system designer during each design phase. Although most CAD programs perform tasks according to the decision-making logic of conventional programs, these programs cannot readily accommodate significant amounts of knowledge and cannot solve problems with the competence of a human expert. As knowledge-based expert system technology has advanced, providing insight into how knowledge is formalized and its application to problems, the concept of expert systems for VLSI design has begun to attract attention in the CAD community.

In this chapter, we will formalize and discuss some issues of knowledge-based expert systems for synthesizing digital system circuits from register transfer level specifications. This chapter first delineates knowledge representation paradigms and inference environments. It next presents an overview of the logic design expert system. It finally describes a new data path synthesis, logic circuit synthesis, and data path verification algorithms in detail.

KNOWLEDGE REPRESENTATION AND INFERENCE ENVIRONMENT

An expert system usually consists of a knowledge base and inference mechanisms. The knowledge base is a collection of elementary knowledge fragments. The inference mechanisms search through the knowledge base, select the most promising knowledge segment for solving a problem, and apply it to the problem. There are several knowledge representation paradigms or styles of knowledge programming for knowledge segments. In the Turing machine sense, all common programming languages are universal. However, different techniques for ex-

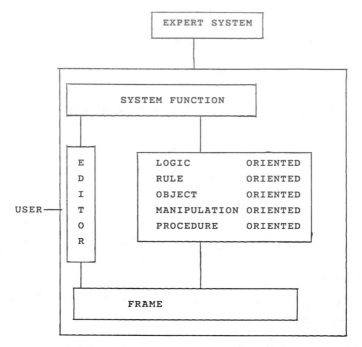

Figure 23-1.   *K*nowledge *r*epresentation and *i*nference environment.

pressing knowledge may be effective, depending on the nature of the problem, and allow a good view of the problem.

Therefore, we have developed the *k*nowledge *r*epresentation and *i*nference *e*nvironment, KRINE (Ogawa et al. 1984), which integrates five knowledge representation paradigms (Figure 23-1).

The main features of each paradigm are now described.

Logic Programming

In the logic programming paradigm, a knowledge segment is represented as a PROLOG (Kowalski 1983) clause. A clause is an expression of the form:

<consequent> - <antecedent1>, <antecedent2>, ---, <antecedentn>.

The symbol "< − " means "implied by" in the clause context. Each consequent and antecedent represent the atomic formula:

$$p(t1, t2, ---, tn)$$

where P is a predicate symbol, and t1 through tn are arguments to the predicate. Letters beginning with * are variables, whereas those not beginning with * are constants. Clauses with both consequents and antecedents are rules. Those without antecedents are facts, and those without consequents arc used as goals. Rules and facts are stored in the knowledge base, while goals represent problems to be solved.

Rule: <consequent> < − <antecedent1>, ---, <antecedentn>.

Fact: <consequent> < −.

Goal: < − antecedent1>, ---, <antecedentn>.

If the goal clause is supplied to the system, the resolution process starts. This process involves matching the antecedents in the goal with consequents in the rule set, and then using those antedecents as subgoals. This procedure continues recursively until either each subgoal is proved by a fact, or a match is unavailable, at which time backtracking occurs, and different choices for the subgoals are made by the inference mechanism. If this backtracking exhaust all possible subgoals, the goal remains unprovable.

Rule-Based Programming

Rule-based programming is very much like logic programming. A rule can be easily translated into a PROLOG clause, and is represented in the form:

If Condition1, Condition2, ---, Conditionn
Then Action1, Action2, ---, Actionm.

If conditions "Condition1" through "Conditionn" are satisfied, then actions "Action1" through "Actionm" are executed. In situations where there are many constraints to be considered, there are no straightforward algorithms for use in reaching the goal. Additionally, there are only collections of empirical or heuristic method for reaching the goal. The problems are best solved by rules or logic programming.

Object-Oriented Programming (Bobrow and Stefik 1981)

In this paradigm, things are formalized as a single type of entity called "object" that combines state and behavior. In general, objects are grouped into classes, all of which have the same structure and the same behavior. Behavior is invoked from an object by sending it a message. Behavior can be described by either PROLOG, rule, or procedure. An object can be implemented with a frame. Digital system components can be naturally described in objects.

Manipulation-Oriented Programming

Manipulation-oriented programming facilitates separation of monitoring program events from the process that can cause those events. In particular, the events of concern here are storage and access to the state of the objects. This is achieved by providing an interface procedure that is always called upon to access or modify the state. Manipulation-oriented programming is effective for simulating or graphic display handling.

Procedure-Oriented Programming

The knowledge of this representation composes instructional sequences. Here, data and procedures are separated, in contrast with object-oriented programming. In KRINE, LISP is used. Well-structured problems whose solutional algorithms are already known, or that involve an extensive numerical calculation, are best described by conventional algorithm (procedural) approaches. These problems include simulation, test pattern generation, and cost evaluation, such as gate-count, and gate-delay.

LOGIC DESIGN EXPERT SYSTEM OVERVIEW

Digital systems such as computers are regarded as machines for executing algorithms. Therefore, digital systems are most concisely and clearly specified with algorithms they execute. The logic design problem is to synthesize digital circuits that execute the specified algorithms. Usually, digital circuits are partitioned into a data path section and a control section, as shown in Figure 23-2. The data path section is structured with data path components such as registers,

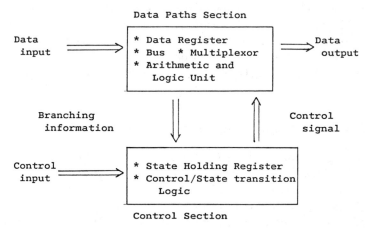

Figure 23.-2. The general model for a digital system.

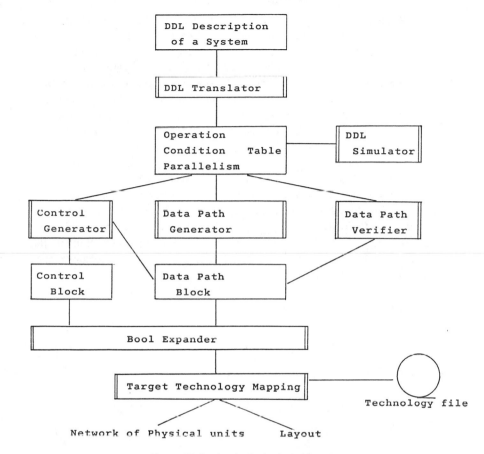

Figure 23-3. Logic design expert system.

arithmetic and logic unit, multiplexors, buses, and data transfer lines. The control circuit is structured with state holding registers and combinational control circuits, and will cause register transfers to take place in the data path section by sending signals on a set of control lines. In some systems the sequencing of control will be influenced by branching information fed back from the data path section. It is our goal to develop a logic design expert system that can synthesize these digital circuits from the algorithmic specifications.

Figure 23-3 shows an overview of the logic design expert system, DE (Takagi 1984), under development. The DE is based on the KRINE system. Its programs are implemented with a mixture of the five knowledge representation paradigms described above.

The DE has two major functions. The first one is synthesis. That is, the algorithm of a digital system is described as a finite state machine in the register transfer level language, DDL (Digital System Description Language) (Duley and Dietmeyer 1968), which is translated into an internal table, and then simulated with the DDL simulator. Based on this internal table, the data path generator synthesizes data path blocks while the control generator synthesizes control blocks. Each block is then expanded in to Boolean expressions by the Bool expander. Each Boolean expression is finally transformed into a network of physical units such as the CMOS standard cell, TTL, and so on. The layout of physical units can also be derived during Bool expansion and physical unit mapping.

The second function is verification. The verifier is intended to be used in two situations. In one, there is a need to change the specifications and to estimate the amount of engineering changes in already designed data paths. In the other, there is a need to collaborate on the DA system with hardware designers (e.g., the hardware skeleton from the designer and the hardware detail from the DA system). The verifier checks the consistency between an implementation and the specifications, and, if an inconsistency exists, it analyzes its cause and makes suggestions on how to compensate for it.

In this chapter, we focus on the problems of data path synthesis, logic circuit synthesis, and data path verification.

HARDWARE DESCRIPTIONS IN DDL

Before presenting the details of the synthesis algorithms, it will be necessary to briefly summarize DDL and look at a specification example.

DDL is a register transfer language with the following features:

1. The system is described be resource declaration statements such as register, register-file, input and output declarations, and an automaton statement.
2. An automaton corresponds to one finite-state machine and is composed of several state statements. An automation takes exactly one of its states in each cycle, which is defined as the period between two adjacent clock pulses.
3. A state statement corresponds to a state and is composed of several execution transition, or condition statements.
4. An execution statement refers to resources, applies a particular function, and sends its result to a destination resource.
5. A transition statement indicates a transition from one state to another.
6. A condition statement is composed of several condition blocks. Each condition block contains several execution, transition, or condition statements. Therefore, in general, a condition exhibits a nesting structure.

Figure 23-4 shows an example of a DDL description of a system (part of a description). For example, at the Stop* state, if the Start-Button is pushed, the

```
<SYSTEM>  SAMPLE;
  <REGISTER>  BR(0 : 15)  SC(0 : 15)  CC  IR(0 : 15);
  <REG-FILE>  GR((0 : 1) (0 : 15));
  <INPUT>     START-BUTTON  INPUT-TERMINAL(0 : 15);
  <OUTPUT>    OUTPUT-TERMINAL(0 : 15);
  <AUTOMATON> CONTROL;
    <STATE>
    STOP*  IF START-BUTTON
              THEN GOTO P1.
           ENDIF.
           END;
    P1     MAR <- SC.
           SC  <- 1+ SC.
           GOTO P2.
           END;
    P2     MEM-READ.
           GOTO P3.
           END;
    P3     IR <- MWR.
           GOTO DEC.
           END;
    DEC    CASE OP
             HJ  SC <- EA.
                 GOTO STOP*.
                 END;
             JNZ IF  GR(GR-F)=0
                     THEN GOTO BRANCH.
                     ELSE GOTO NEXT.
                     ENDIF.
                 END;
             JC  CASE GR-F
                    (0 0) GOTO NEXT. END;
                    (0 1) IF  CC=0
                            THEN GOTO BRANCH.
                          ENDIF.
```

Figure 23-4. Register transfer level behavior description example.

state proceeds to the P1 state. At the P1 state, the content of the sequence counter is transferred to the memory address register, MA, and, at the same time, the content of the sequence counter is incremented and reset to the new value. Then, the state proceeds to the P2 state, and so on. No hardware resources such as ALUs, buses, or multiplexors are declared in these specifications.

Let us not review in the next two sections how data paths and logic circuits are synthesized from these specifications.

DATA PATH SYNTHESIS

If the cost of data paths is not relevant, data paths that satisfy the DDL specifications can be obtained with a straightforward algorithm.

Straightforward Algorithm

1. Translate the specifications into an operation table whose entry consists of an operation and the condition that must be satisfied for the operation to be executed.

    execute condition sink $< -$ operand1 operator operand2

2. Allocate an ALU for each operation and synthesize data transfer paths between it and the resources that are specified as sources or destinations of the operation.
3. Synthesize control circuits based on conditions of the operations, and connect control signals with data paths.

This algorithm, illustrated in Figure 23-5, is adopted by conventional DDL-based logic synthesis systems (Duley and Dietmeyer 1969; Kawato et al. 1979; Endou et al. 1984). However, the synthesized result based on this algorithm

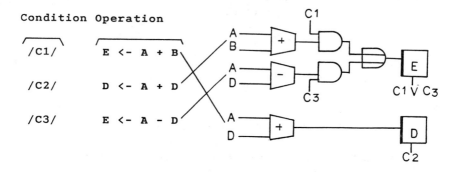

Figure 23-5. Conventional synthesis technique.

contains many redundant circuits, and the resource utilization efficiency is low compared with that of manual design.

Human experts synthesize more optimized data paths. The goal of data path synthesis is to construct a data path implementation scheme that satisfies the DDL specifications, has a reasonable structure in the hardware sense, and has a minimum number of data path components. This goal can be achieved in the following way:

1. Determine the parallelism between operations by analyzing the specifications.
2. Divide the data path synthesis problem into that ALU synthesis subproblem and the data transfer path synthesis subproblem.
3. Follow the philosophy, "Merge resources that do not work in parallel." That is, either (a) merge operations that do not work in parallel into an ALU, or (b) merge data transfer thus which do not work in parallel into a bus or a common multiplexor.

Parallelism Detection between Operations

Let two operations be OPi and OPj. The parallelism between them can be determined using the semantics and syntax of the DDL specifications:

1. Only one state is active in each machine cycle. Thus, if OPi and OPj belong to different states, they can never work simultaneously.
2. If OPi and OPj belong to the same state, however, we analyze the sentence structure of the state and determine the parallelism. The basic idea is that if OPi and OPj belong to different conditional blocks of a condition sentence, they can never work simultaneously. Yet, if OPi and OPj belong to the same conditional block of a condition statement, they may work simultaneously.

ALU Design

Finding the minimum number of ALUs is a problem with non-polynomial hard complexity, and techniques to find a nearly optimal solution have been proposed (Tesng and Siewjorek 1981, 1983). Most of them are theoretical.

However, some practical problems need to be considered in real design. For example, a set of operations that do not work simultaneously is not always merged into an ALU with reasonable hardware, and the DA system cannot expand any kind of ALUs into gate circuits. To cope with these problems, rules that suppress constructing unreasonable ALUs are introduced in the synthesis algorithm.

The data path generator constructs ALUs in the following steps:

1. Let $i = 1$.
2. Pick out an operation from the operation list (abbr. OP-T), and let it be an ALUi.
3. Find an operation that never works simultaneously with an ALUi and does not disturb the reasonableness of an ALUi (details are described in the next paragraph) from OP-T, and merge it with an ALUi. When an operation is merged with an ALU, updating of information occurs, such as ALU functions, ALU source and destination resources, and conditions under which ALU function or data transfer occurs (Figure 23-6). Repeat Step 3 until no OP is found that can be merged with an ALUi.
4. Let $i = i + 1$. Repeat Steps 2, 3, and 4 until OP-T becomes empty.

There are many merge patterns (Tesng and Siewjorek 1983) between an OP and ALUi, depending on their relationship. There are also several factors to take

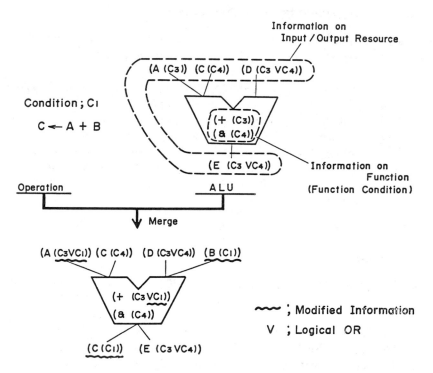

Figure 23-6.   ALU merge concept.

into account to maintain a reasonable ALU structure. For example, a barrel shift function is not usually merged with an ALU. Thus, the procedure for finding an operation and merging it with an ALUi requires the implementation of a collection of ALU synthesis rules (abbr. ALU rule).

An ALU rule example:

IF   1. there is an ALU,
          2. there is an operation,
          3. the ALU and the operation never work simultaneously,
          4. the ALU does not include the function of the operation,
          5. the function of the operation is mergeable with the ALU,
          6. the left input port of the ALU is already connected to the first operand of the operation,
          7. the right input port of the ALU is already connected to the second operand of the operation, and
          8. the output port of the ALU is not connected to the sink of the operation,
THEN   1. add the required function with its condition to the ALU,
          2. update the transfer condition from the first operand,
          3. update the transfer condition from the second operand, and
          4. add a new path from the output port of the ALU to the sink.

An ALU rule summons rules to maintain the reasonableness of an ALU. The rule checks the following three conditions, and, if any of the three conditions is satisfied, the rule judges that the operation is mergeable with an ALUi:

Letting the operator of the operation be Fop, and the operators included in an ALUi be F_1 --- F_n:

(Condition 1) Several ALUs, which are already expanded into gate circuits, are registered in the DA system, such that an ALU can be found that includes the operators Fop, F_1 --- F_n.

(Condition 2) Several gate synthesis procedures, which can expand specific types of ALUs into gate circuits, are registered in the DA system, such that a procedure can be found that can expand an ALU including the operators, Fop, F_1 --- F_n.

(Condition 3) Several sets of operators, which are declared reasonable by designers, are registered in the DA system. A set exists that includes the operators, Fop, F_1 --- F_n.

After synthesis of ALUs, the data transfer paths are synthesized.

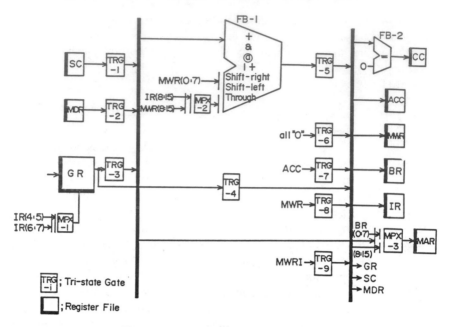

Figure 23-7. Data path synthesis example.

Data Transfer Path Design

Logical information concerning data transfer paths (e.g., source, sink, transfer condition) is derived during ALU design. Data transfer paths, which do not work simultaneously, are merged into a bus or a common multiplexor.

A bus is synthesized (Tesng and Siewjorek 1981) when (a) there are multiple data transfer paths that do not work in parallel or whose sources are the same, and (b) they have multiple destinations.

There are two cases when multiplexors are generated:

1. If there are multiple data transfer paths whose destinations are the same, a resource input multiplexor is generated between the sources and the designation.
2. If there are plural resource input multiplexors that do not work simultaneously, they can be merged in to a common multiplexor.

Figure 23-7 provides a data path synthesis example. Two ALUs and two buses are allocated. The synthesized result is comparable to that of manual design. In this example, a rule is used wherein any operation and any ALU can be merged with each other. If the rule is changed, the result changes. Figure 23-8 presents

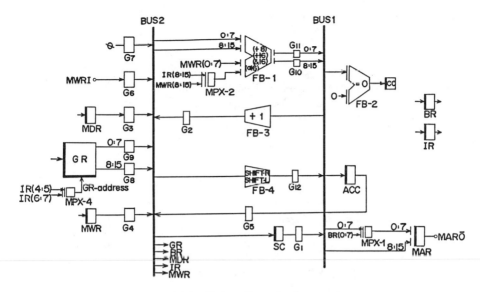

Figure 23-8.   Data path synthesis example.

another example of data path synthesis from the same specification indicated in Figure 23-7. In this example, the following rule is used:

1. SHIFT-RIGHT and SHIFT-LEFT cannot be merged with other functions.
2. SHIFT-RIGHT and SHIFT-LEFT can be merged.
3. INCREMENT cannot be merged with other functions.

If a rule is used wherein no operation can be merged into an ALU, the number of synthesized ALUs is equal to that of a conventional algorithm.

At the data path design level, details of each data path component are not yet fixed. The next step is to synthesize the circuits for each component.

LOGIC CIRCUIT SYNTHESIS

The goal of logic circuit synthesis is to construct a combinational logic circuit that satisfies the functional specifications, satisfies the performance requirements, and has a minimum number of gates or connecting wires. No simple algorithm for achieving this goal is known. Human experts tackle this problem with a variety of knowledge that extends from fundamental knowledge such as Boolean algebra and Boolean expression manipulation to advanced knowledge such as top-down structured design methods and design strategies (Figure 23-9). Let us

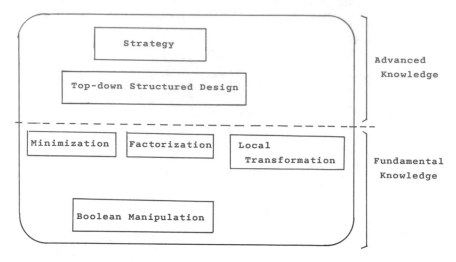

Figure 23-9. Expert's design knowledge.

take a look at these two knowledge extremes and show how the problem is solved.

Fundamental Knowledge

Fundamental knowledge of logic synthesis is composed of the following theoretical or heuristic elements:

Knowledge of Boolean Theorems and Algebraic Properties (Deen et al 1977).

At the detailed logic design level, logic circuits are specified in Boolean expressions. A Boolean expression is composed of logic variables and logic operators. There are only three basic operations: logical product (*), logical sum ($+$), and negation ($-$) operations. The following postulates, laws, and theorems are important in the specification and manipulation of Boolean expressions:

*Boolean theorems:
$$A*0 = 0, A*1 = A, A*A = A, A*\overline{A} = 0, A = A,-$$
$$A+0 = A, A+1 = 1, A+A = A, A+A = 1$$

* De Morgan's theorem:
$$\overline{A*B*C} = \overline{A} + \overline{B} + \overline{C}, \overline{A + B + C} = \overline{A} * \overline{B} * \overline{C}$$

* Absorption theorem:
$$A*(A+B) = A, A*(\overline{A}+B) = A*B, A+A*B = A, A+\overline{A}*B = A+B$$

*Algebraic properties:
A*B = B*A, A+B = B+A, A*(B*C) = A*B*C, A+(B+C) = A+B+C,
A*(B+C) + A*B + B*A

Knowledge of High-Level Functions. Sometimes, circuit functions are specified using high-level operators such as EXCLUSIVE-OR, arithmetic sum, and so on. Knowledge for expanding such high-level functions into basic Boolean operators is required. For example, the arithmetic sum function

$$x(n;1) - A(n;1) \text{ Plus } B(n;1) \text{ Plus Cin}$$

can be expanded into basic Boolean expressions with the following recurrence formula:

$$\text{Carry}(0) = \text{Cin}$$
$$X(i) = A(i) \oplus B(i) \oplus \text{Carry}(i-1)$$
$$\text{Carry}(i) = A(i)*B(i) + A(i)*\text{Carry}(i-1) + B(i)*\text{Carry}(i-1)$$

Heuristic Knowledge for Logic Minimization. Knowledge of Boolean theorems and algebraic properties is very powerful in that it can result in the minimum Boolean expression in time. However, it has the disadvantage of being too time-consuming. With respect to the need for devising a more efficient algorithm, several heuristic algorithms, which are less time-consuming but do not guarantee production of optimal solution, are proposed (Hong et al. 1974; Brown 1981; Brayton et al. 1982).

Local Transformations of Circuits. The heuristic algorithms for logic minimization take a wide view of circuits and optimize circuits globally. However, they are not sufficient for optimizing circuits that are fragments of total circuits and have some specific patterns. Heuristic knowledge to improve circuits locally is proposed (Darringer 1981).

With fundamental knowledge, logic synthesis problems are solved through the following steps:

1. Transform the high-level functional specifications for a function block into Boolean expressions.
2. Minimize the Boolean expressions.
3. Transform the Boolean expressions into circuits.
4. Improve the circuits by local transformations.

Most conventional DA systems employ these steps. However, these steps are designed for solving "shallow problems," whose minimal solution contains at most a few hundred implicants regardless of the number of logic variables. Although control circuits are of this nature, data path components such as ALUs do not fit into this category. The number of implicants increases dramatically as the data path bit-width increases. For example, the n bit parity function,

$$P1 \oplus P2 \text{ ---- } \oplus Pn - 1 \qquad [10.3]$$

has 2^n product terms in the minimal solution. The order of the product term number for an ALU that has an n data bit-width and m control inputs is estimated to be $0(2^{2n+m})$. It is thus quite difficult to solve such complex problems with the procedure described above.

These problems can be solved, however, by integrating the top-down structured design methods and design strategies in the logic design expert system.

Advanced Design Knowledge

Top-Down Structured Design Methods (Takagi 1985a).

A top-down structured design method can be modeled by the problem reduction technique used in the artificial intelligence field. A complex problem is usually decomposed into several smaller problems (Figure 23-10). The original problem X is decomposed into subproblems A, B, and C. Subproblem A is further decomposed into even smaller subproblems AA and AB.

This decomposition is accomplished using a template (circuit architecture) that corresponds to the style of decomposing the original problem into subproblems. Transformation procedures derive the specifications of each subproblem from the original problem. A design method is defined as the combination of a template and transformation procedures.

Function blocks can be classified into one of several functional block types

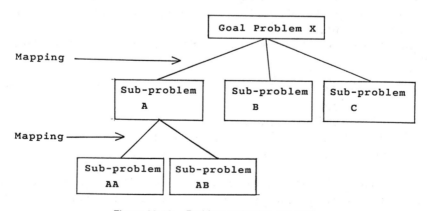

Figure 23-10. Problem reduction concept.

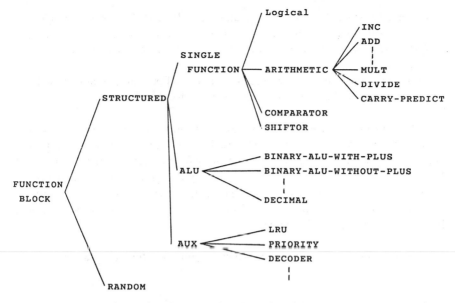

Figure 23-11. Example of function block types.

(Figure 23-11). Each functional block type has its own design methods. For example, the adder block has ripple-adder, carry-predict-adder, and carry-save-adder design methods, and so on. The synthesis of the adder is such a simple problem that it requires no template, but needs only a simple transformation (synthesis) procedure for each design method.

Figure 23-12 shows the transformation procedure for the ripple-adder design method. This procedure can synthesize any bit-width ripple adder. Synthesis

```
Ripple-adder   bit-width

    DO I=0 To Bit-width-1
      Collect  T(I)=A(I) ⊕ B(I),
      Collect  G(I)=A(I) * B(I),
      IF I=0
        THEN  Collect C(I)=Cin
        ELSE  Collect C(I)=G(I) + T(I) * C(I - 1),
        END-IF,
      Collect S(I)=T(I) ⊕ G(I),
    END-DO,


  *Collect ; function to collect expressions
               into the result buffer
```

Figure 23-12. Ripple-adder synthesis procedure.

procedures for single function blocks can usually be coded in a simple programs, as shown in the figure. However, more complex problems, such as ALU logic synthesis, require more sophisticated design methods (see below "Design Method Example").

Design methods can be hierarchically organized. For example, design methods for multipliers should summon design methods for adders.

Problem-Solving Strategy. For the same function specifications, it is possible to synthesize a wide variety of circuits that vary in gate-count and gate-delay. This section shows how a number of different circuits can be synthesized by selecting a specific method or by combining design methods. The algorithm for synthesizing circuits at the target gate-count and gate-delay is also given. *Circuits Synthesizable with Design Methods.* Various circuits can be generated by combining the following three levels.:

1. Template selection: There are several templates for each function block type.

2. Subproblem-solving technique selection: If original problem X is decomposed into subproblems Y1, --, Yn, and there exist M1, --, Mn ways of solving each subproblem, there are then M1x --- xMn possible combinations.

3. Combination of design methods: A subproblem can be solved by combining several design methods. For example, an n bit-width adder can be realized by combining a 1 bit-width adder and an m bit-width adder (n = 1 + m). A simple design method combination mechanism for an adder can be described using predicate calculus, as follows:

Make-adder(*nbit *circuit) < −
PLUS(*nbit *1bit *mbit); *1bit *mbit> 0, *nbit = *1bit PLUS *mbit
An-adder-design-method(*1bit *circuit1),
An-adder-design-method(*mbit *circuit2),
Carry-connect(*circuit1 *circuit2 *circuit).
An-adder-design-method(*nbit *circuit)< − Ripple-adder(* nbit *circuit).
An-adder-design-method(*nbit *circuit)< − Carry-predict-adder(*nbit *circuit).

If several adder designs methods (Ripple-adder design method, Carry-predict-adder design method, and so on) are given, and the Carry-connect method is known, then the Make-adder predicate can generate any circuit that can be realized with the single adder design method or with a combination of two adder design methods. However, a more powerful combination mechanism is possible.

The next example shows how any number of adder design methods can be used in adder synthesis:

> Make-adder1(*bit *circuit)< −
> An-adder-design-method(*bit *circuit).
> Make-adder1(*bit *circuit) < −
> PLUS(*nbit *1bit *mbit),
> Make-adder1(*1bit *circuit1),
> Make-adder1(*mbit *circuit2),
> Carry-connect(*circuit1 *circuit2 *circuit).

The second clause of Make-*adder1 recursively summons itself. With this recursion, any number of adder design methods can be used.

Constraint Satisfaction Technique.
A simple technique to satisfy the required cost-performance is described as follows:

> Make-adder-with-restriction(*nbit *circuit *cost *delay) < −
> Make-adder1(*nbit *circuit),
> Gate-count(*circuit *G),
> = < (*G *cost),
> Delay-count(*circuit *D),
> = < (*D *delay).

The Make-adder-with-restriction predicate generates an adder circuit with Make-adder1 predicate, and checks whether the synthesized circuit satisfies the gate-count and gate-delay. If the requirements cannot be satisfied, the make-adder-with-restriction backtracks and searches another solution. The make-adder-with-restriction is very powerful in that it can find a solution if one exists, although it sometimes has the disadvantage of being too time-consuming. Accordingly, investigation should be made of the following more sophisticated techniques, which prune useless searches:

1. A technique for decomposing original constraints into subconstraints for subgoals.
2. A technique for estimating the gate-count and gate-delay for design methods.
3. A technique for backtracking and selecting a more appropriate design method for reaching a goal effectively.

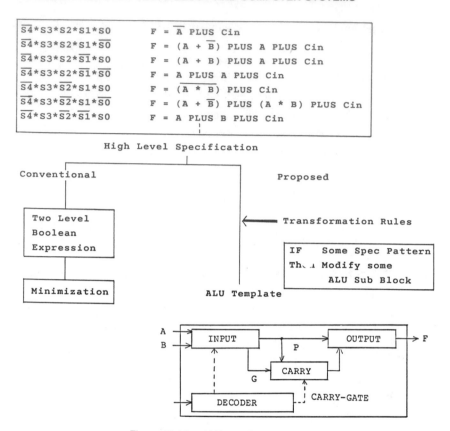

```
S̄4*S3*S2*S1*S0         F =  Ā PLUS Cin
S̄4*S3*S2*S1*S̄0         F = (A + B̄) PLUS A PLUS Cin
S̄4*S3*S2*S̄1*S0         F = (A + B) PLUS A PLUS Cin
S̄4*S3*S2*S̄1*S̄0         F = A PLUS A PLUS Cin
S̄4*S3*S̄2*S1*S0         F = (A * B) PLUS Cin
S̄4*S3*S̄2*S1*S̄0         F = (A + B̄) PLUS (A * B) PLUS Cin
S̄4*S3*S̄2*S̄1*S0         F = A PLUS B PLUS Cin
```

High Level Specification

Conventional   Proposed

Two Level
Boolean
Expression   ←——— Transformation Rules

                        IF   Some Spec Pattern
                        Then Modify some
                                   ALU Sub Block

Minimization   ALU Template

Figure 23-13. ALU template example.

Design Method Example

A design method for a complicated example (ALU) (Takagi 1985b) is shown in Figure 23-13.

High-level Specifications Example.   An ALU has both data input/output terminals and control input terminals. If control signals are supplied to the control input terminals, the ALU will execute operations between input data and send the result to the output terminal. Therefore, ALU specifications can be expressed in terms of data input–output relations and control signals for specifying each input–output relation. For example:

$$\bar{S4}*S3*S2*S1*S0 \qquad F = A \text{ PLUS } Cin$$

means that if the control signals S0 through S3 are "1" and S4 is "0", the ALU executes the arithmetic sum of A and Cin.

Template Example. There are several kinds of ALU templates. No template is superior to all other templates with regard to every function specification. The ALU template in Figure 23-13 is an example devised for medium-speed computers. The philosophy behind this template is as follows:

1. ALU logic is structured as (a) output stage logic is an exclusive-or circuit, (b) carry-predict logic, (c) input logic that generates G and P signals, and (d) decoder logic. Here, the G signal implies nth-bit-carry generation, and the P signal indicates the nth-bit-partial-sum.
2. To perform arithmetic sum operations, the carry is predicted from the G and P signals and is applied along with partial sum P to the output stage exclusive-or circuit.
3. To perform logical operation, the carry-gate signal should be pulled down to "0," and the carry-predict signal should be pulled down to "0."

As an ALU template specifies the skeleton of an ALU, details of ALU template subblocks such as G logic, P logic, and the decoder are not fixed at this level. When an ALU specification is given, ALU template subblocks are synthesized.

Synthesis Procedure Example. The synthesis procedure depends on the ALU templates. The example of synthesis procedure in Figure 23-13 gives procedures (rules) for synthesizing G, P, and carry-gate logic according to the following steps:

1. The G, P, and carry-gate logic is initialized to "0."
2. The function syntax of each function specifications is matched with the "IF" part of the synthesis rules. The matched rule rewrites G, P, or carry-gate as its "THEN" part specifies.
   [Rule example]
   If the function syntax is "control-code F = α PLUSβ PLUS Cin"
   THEN DO P = P Control-code*($\alpha \oplus \beta$)
   DO G = G + Control-code*($\alpha + \beta$)
   DO Carry-gate = Carry-gate + control-code
   Here α and β are expressions that do not contain the PLUS operator.
3. The G, P, and carry-gate logic is minimized.
4. The G, P, and output logic is repeated n (data bit width) times.
5. Carry-predict logic is synthesized.

Synthesized Circuit Example. Figure 23-14 outlines the synthesized circuits for a 32-function ALU.

In LSI design, the total cell area is an important issue. It is known that the total area is approximately proportional to the total fan-in number, which Figure 23-15 presents. Without advanced design knowledge, the fan-in number is fairly

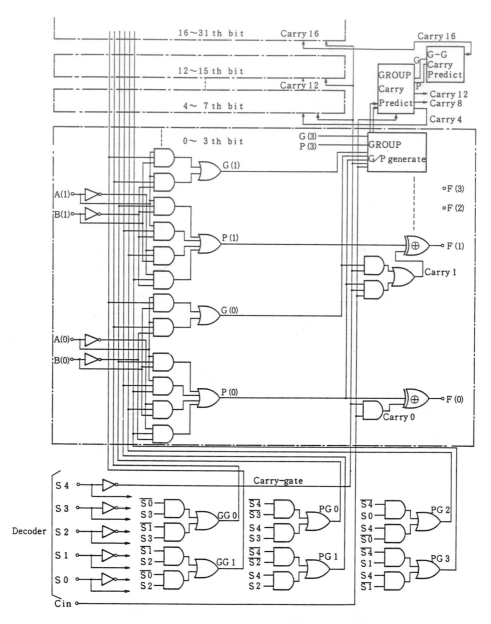

Figure 23-14. ALU synthesis example (32-function).

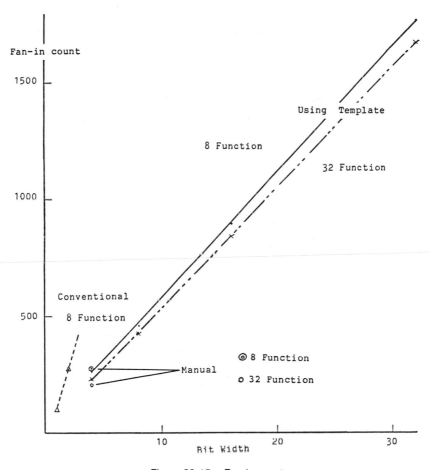

Figure 23-15.   Fan-in count.

large compared with that resulting from manual designs. When advanced design knowledge is adopted, the fan-in number becomes comparable to that of manual design.

Figure 23-16 indicates ALU synthesis time, which without advance knowledge increases dramatically as the ALU data bit-width increases.

The synthesized circuits are technology-independent. Using a technology-specific file, these circuits are transformed into a network of physical units, laid out and routed. The knowledge for mapping logical circuits to a network of physical units and the layout scheme is integrated in the DE. Figure 23-17 shows an ALU mask pattern example.

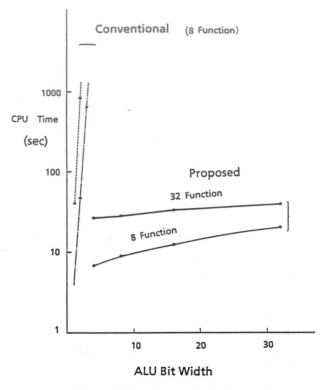

Figure 23.16. ALU synthesis time.

DATA PATH VERIFICATION

There are two approaches to data path verification. The first approach, the conventional one, is simulation. DDL level simulation and gate level simulation are reformed, and the results of both are compared (Kawato et al. 1979).

The second approach, which is proposed in this chapter, is analytical. The main idea is to formalize the designer's data path analysis knowledge into rules and then to utilize them (Takagi 1984, 1985c). The analytical method is advantageous in that it needs no test pattern generation, it can be applied before data path design is completed, and it can be extended to failure analysis and data path revision.

The verification steps are as follows. The DDL specifications are first translated into an operation and a parallelism table. The verifier then performs two tasks: a check to see if each operation can be executed in data paths, and a check to see that no resource conflict exists between operations that act simultaneously. If these conditions are satisfied, verification succeeds. If not, verification fails,

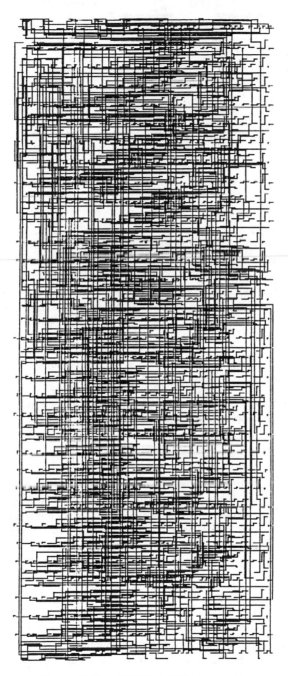

Figure 23-17.   ALU mask pattern example.

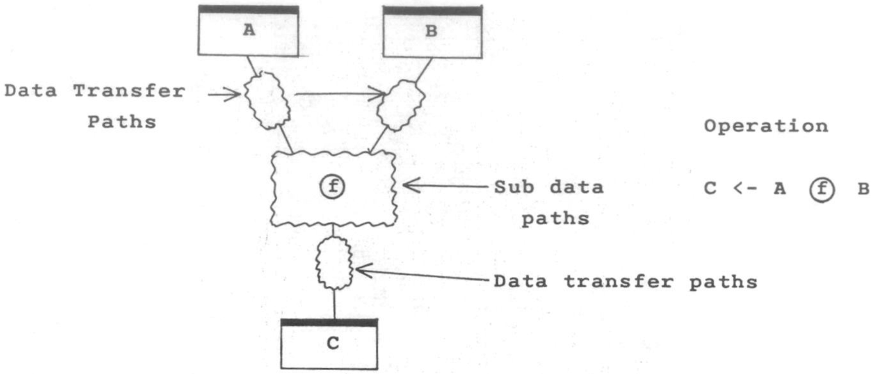

```
Data Transfer →
    Paths                                              Operation

                        (f)  ←——— Sub data        C <- A  (f)  B
                                   paths

                              ←——— Data transfer paths

                        C
```

Top Level Verification Rule

    IF   1) THere are sub data paths which have the
               the required function and
            2) There are data transfer paths between the sub
               data paths and resources which are specified
               as source or sink of the operation
    THEN   The operation can be executed on the data paths

Figure 23-18.   Operation verification concept.

and processing then proceeds to the failure analysis step, which is followed by the data path revision proposal step.

Operation Verification Concept

Figure 23-18 outlines the concept of operation verification. An operation can be executed on data paths if there are sub–data paths that have the required function, and if there are data transfer paths between the sub–data paths and the registers referred to or modified by the operation. These conditions define the top level verification rule. This rule summons the next level verification rules, which are sub–data path and data transfer path verification rules. Both these principles are presented in Figure 23-19.

The sub–data path facility, which can execute the required function, is immediately found if a component exists that has that function. If none exists, the verifier attempts to verify the required function by combination of more basic functions. This can be realized by integrating function synthesis knowledge into rules. For example, the "subtraction" function can be realized with the "add" and the "not" functions.

Data transfer paths are verified in a similar fashion. That is, data transfer paths

Sub Data Paths Analysis

(1)   Simple Match

C <- A ⓕ B

(2)   Analysis by Function Synthesis Knowledge

Knowledge on Function Synthesis

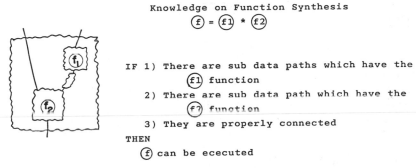

IF 1) There are sub data paths which have the
       ⓕ1 function
    2) There are sub data path which have the
       ⓕ2 function
    3) They are properly connected
  THEN
    ⓕ can be ececuted

Data Transfer Paths Analysis

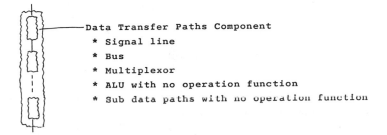
Data Transfer Paths Component
* Signal line
* Bus
* Multiplexor
* ALU with no operation function
* Sub data paths with no operation function

Figure 23-19.   Sub–data path and data transfer path verification concept.

are defined as series of data transfer path components, and a data transfer path component is either a signal line, but, multiplexor, and ALU with a no-operation function or sub–data paths with no-operation function.

Using these rules, the verifier checks the consistency between operations and data paths, and if the verification process succeeds, it reports how each operation is executed in data paths. If the verification process fails, however, the verifier analyzes the cause of the failure.

Failure Analysis

There must be several reasons why the verification process fails. For example, sub–data paths, which can execute the required function, or data transfer paths

Transfer Paths			Function
ALU->C	A->ALU	B->ALU	
X	O	O	O
O	X	O	O
O	O	X	O
O	O	O	X
X	X	O	O
X	O	X	O
X	O	O	X
O	X	X	O
O	X	O	X
O	O	X	X
X	X	X	O
X	X	O	X
X	O	X	X
O	X	X	X
X	X	X	X

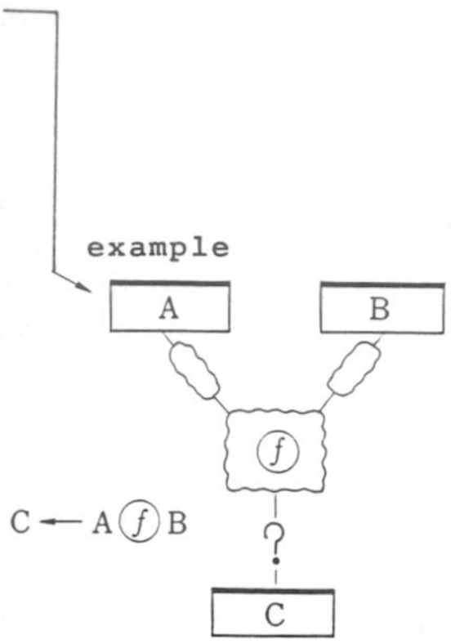

example

A   B

f

$C \leftarrow A \; (f) \; B$

C

O ; Exist

X ; Missing

```
Failure Analysis Rule
If 1) There exist sub data paths which have f function
   2) There exists a data transfer path from left input resource
   3) There exists a data transfer path from right input resource
   4) There exists no data transfer path to destination resource
Then Failure pattern type 1 occurs between     and
```

Figure 23-20.   Failure pattern concept.

might be missing. These defects and the locations where these defects might occur are categorized and formulated into failure patterns (Figure 23-20). A failure pattern, with a failure analysis rule provided for each one points out locations where an inconsistency occurs.

After the failure position is localized, rules to revise data paths are activated. The revision is accomplished by function revision or data transfer path revision rules.

Data Path Revision

Figure 23-21 conceptualizes data path revision. There are thee main principles of the revision. First, if a component does not have the required function, function revision rules suggest adding the required function to the component. If function synthesis is taken into account, patching patterns for the function failure are increased. Second, according to the variety of topological patterns around the source and destination, several patching patterns exists for data transfer paths. Upon inspection of these paths, the topologies around resources are classified, and for each topological pattern, a data transfer path revision rule is provided.

1) Function Revise Concept (example)

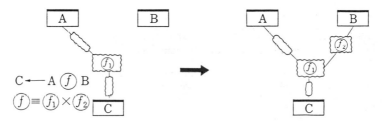

2) Data Transfer Paths Revise Concept (example)

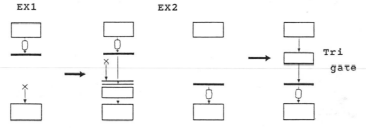

3) Resource Conflict Revise Concept (example)

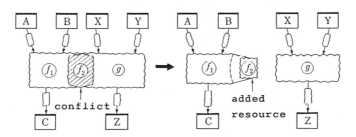

Figure 23-21.   Data path revision concept.

Finally, a resource conflict can be revised by adding new resources to the conflict position and new data transfer path to the new resources.

There are several ways to patch the cause of a failure. The patching rules depend solely on the knowledge of data path design experts. By upgrading the patching rules gradually, the verifier should be able to attain a good data path patching capability.

Data Path Verification Example

Figure 23-22 presents an example of data paths to be verified and their specifications.

The verification results are summarized in Figure 23-23. Here, 1, 2, 3, 4, and

Operations to be Verified on the data paths

OPid	OPERATION	CONDITION
OP1	MEMORY-ADDRESS <- SC	INSTRUCTION-FETCH
OP2	IR <- MEMORY-READ-DATA	MEMORY-READ
OP3	SC <- 1+ SC	MEMORY-READ
OP4	GRO <- GRO + GR1	DECODE-EXECUTE & (IR = ADD)
OP5	GRO <- GRO - GR1	DECODE-EXECUTE & (IR = SUBTRACT)
OP6	GRO <- GRO @ GR1	DECODE-EXECUTE & (IR = EXCLUSIVE-OR)
OP7	BR <- GRO	DECODE-EXECUTE & (IR = LOAD-BR)
OP8	SC <- BR	DECODE-EXECUTE & (IR = JUMP)
OP9	SC <- GRO + BR	DECODE-EXECUTE & (IR = JUMP-AND-LINK)
OP10	GR1 <- SC	DECODE-EXECUTE & (IR - JUMP-AND-LINK)

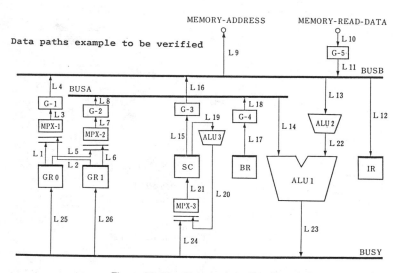

Figure 23-22. Data path verification example.

7 are examples where the operations are successfully verified. Conversely, 5, 6, and 8 are examples where operation verification has failed. In detail:

1. Shows that a simple data transfer operation is verified, reporting on how this operation is executed in the data paths.
2. Shows that two operations, which occur simultaneously, are verified.
3. Shows that the operator of (GRO < − GRO + GR1) is directly matched to the ADD function of ALU1.

```
------------- (1) -----------------------------------------------
-(verifier ((memory-address <- sc)))):
(MEMORY-ADDRESS <- SC IS VERIFIED AS FOLLOWS :
 SC TO MEMORY-ADDRESS :THROUGH: ((SC T) L15 (G_3 ON) L16 (BUSB T) L9))
------------- (2) -----------------------------------------------
-(verifier ((ir <- memory-read-data) (sc <- 1+ sc)))):
(IR <- MEMORY-READ-DATA IS VERIFIED AS FOLLOWS :
 MEMORY-READ-DATA TO IR :THROUGH: (L10 (G_5 ON) L11 (BUSB T) L12 (IR SET)))
(SC <- 1+ SC IS VERIFIED AS FOLLOWS :
 1+ :IN: (ALU3 INCREMENT)
 SC TO (ALU3 IN) :THROUGH: ((SC T) L19)
 (ALU3 OUT) TO SC :THROUGH: (L20 (MPX_3 SEL2) L21 (SC SET)))
------------- (3) -----------------------------------------------
-(verifier ((gr0 <- gr0 + gr1)))):
(GR0 <- GR0 + GR1 IS VERIFIED AS FOLLOWS :
 + :IN: (ALU1 ADD)
 GR0 TO (ALU1 A) :THROUGH: ((GR0 T) L5 (MPX_2 SEL1) L7 (G_2 ON) L8
                                                      (BUSA T) L14)
 GR1 TO (ALU1 B) :THROUGH: ((GR1 T) L2 (MPX_1 SEL2) L3 (C_1 ON) L4
                                     (BUSB T) L13 (ALU2 THROUGH) L22)
 (ALU1 C) TO GR0 :THROUGH: (L23 (BUSY T) L25 (GR0 SET)))
------------- (4) -----------------------------------------------
-(verifier ((gr0 <- gr0 - gr1)))):
(GR0 <- GR0 - GR1 IS VERIFIED AS FOLLOWS :
 - :AS: COLABORATION (ALU2 INVERT) AND (ALU1 ADD-1) THROUGH (L22)
 GR0 TO (ALU1 A) :THROUGH: ((GR0 T) L5 (MPX_2 SEL1) L7 (G_2 ON) L8
                                                      (BUSA T) L14)
 GR1 TO (ALU2 IN) :THROUGH: ((GR1 T) L2 (MPX_1 SEL2) L3 (G_1 ON) L4
                                                      (BUSB T) L13)
 (ALU1 C) TO GR0 :THROUGH: (L23 (BUSY T) L25 (GR0 SET)))
------------- (5) -----------------------------------------------
-(verifier ((gr0 <- gr0 @ gr1)))):
**CAUTION!! : GR0 <- GR0 @ GR1 CAN NOT BE EXCUTED BECAUSE
(FUNCTION @ IS NOT CONTAINED IN ALU1)
    I RECOMMEND TO :
(ADD-FUNCTION (ALU1 C) := (ALU1 A) @ (ALU1 B) WITH (ALU1 ?FUNC-ID))
------------- (6) -----------------------------------------------
-(verifier ((br <- gr0)))):
**CAUTION!! : BR <- CONN GR0 CANNOT BE EXCUTED BECAUSE
(THERE IS NO PATH FROM GR0 TO BR)
  I RECOMMEND TO :
(CONNECT BUSB AND BR BY ?NET-NAME)
------------- (7) -----------------------------------------------
-(verifier ((sc <- br)))):
(SC <- BR IS VERIFIED AS FOLLOWS :
 BR TO SC :THROUGH: ((BR T) L17 (G_4 ON) L18 (BUSA T) L14 (ALU1 SELA)
                          L23 (BUSY T) L24 (MPX_3 SEL1) L21 (SC SET)))
------------- (8) -----------------------------------------------
-(verifier ((sc <- gr0 + br) (gr1 <- sc)))):
(SC <- GR0 + BR IS VERIFIED AS FOLLOWS :
 + :IN: (ALU1 ADD)
 GR0 TO (ALU1 B) :THROUGH: ((GR0 T) L1 (MPX_1 SEL1) L3 (G_1 ON) L4
                                     (BUSB T) L13 (ALU2 THROUGH) L22)
 BR TO (ALU1 A) :THROUGH: ((BR T) L17 (G_4 ON) L18 (BUSA T) L14)
 (ALU1 A) TO SC :THROUGH: (L23 (BUSY T) L24 (MPX_3 SEL1) L21 (SC SET)))
(GR1 <- SC IS VERIFIED AS FOLLOWS :
 SC TO GR1 :THROUGH: ((SC T) L15 (G_3 ON) L16 (BUSB T) L13 (ALU2 THROUGH)
                             L22 (ALU1 SELB) L23 (BUSY T) L26 (GR1 SET)))
**CONFLICTION!!** : (SC <- GR0 + BR) AND (GR1 <- SC) AT :
 (ALU1 L23 BUSY BUSB L13 ALU2 L22)
((INSERT-MPX ?MPX-NAME BETWEEN L26 AND GR1) (CONNECT SC AND (?MPX-NAME 1)
                                                       BY ?NET-NAME))
```

Figure 23.23.   Data path verification example.

4. Shows that the operator - of (GRO $<$ $-$ GRO $-$ GR1) is executed by cooperation between the NOT function of ALU2 and the ADD-1 function of ALU1.

5. Shows an example of failure caused by lack of the required function in data paths, with the verifier suggesting the addition of the exclusive-or function, @, to ALU1.

6. Shows an example of a failure caused by the lack of data transfer paths, with the verifier suggesting the connection of BUSB and BR.

7. Shows that a simple data transfer operation (SC $<$ $-$ BR) is executed with the aid of the through function of ALU1.

8. Shows an example where a resource conflict occurs.

CONCLUSIONS

We have described a new algorithm for the knowledge-based logic design expert system that synthesizes circuits from register transfer level specifications. The proposed algorithm is integrated into the prototype design expert system, DE. The experimental results are encouraging. We therefore feel that the knowledge-based approach to logic design is indeed promising.

The knowledge-based approach will decrease computation time, improve circuit quality, and provide new CAD functions such as verification, patching circuits, and redesign. It is easy to enhance a system's problem-solving ability, as the expert knowledge usually can be segmented into rules and added to the knowledge base. The multiple knowledge representation paradigm adopted in the knowledge representation and inference environment, KRINE, is very useful, because different paradigms allow different problems to be stated concisely and to be executed effectively.

Future research will include expanding both the problem domain and learning aspects of the expert system, as well as incorporating more expert knowledge, and improving the designer interface with several question-answering mechanisms.

REFERENCES

Bobrow, D. G., and Stefik, M., "The LOOPS Manual," XEROX PARC Knowledge-Based VLSI Design Group Memo, KB-VLSI-81-13, 1981.

Brayton, R. K., Hachtel, G. D., Hemachandra, L. A., Newton, A. R., and Sangiovanni-Vincentelli, A. L. M., "A Comparison of Logic Minimization Strategies using ESPRESSO: An APL Program Package for Partitioned Logic Minimization," *Proc.* ISCAS, Rome, 1982, pp. 42–48.

Brown, D. W., "A State-Machine Synthesizer," *Proc. 18th DA Conference*, pp. 301–305, 1981.

Darringer, J. A., "A New Look at Logic Synthesis," *Proc. 17th DA Conference*, pp. 543–549, 1981.

Deen, B., Muchow, K., and Zeppa, A., *Digital Computer Circuits and Concepts*, Prentice-Hall, Englewood-Cliffs, NJ, 1977.

Duley, J. R., and Dietmeyer, D. L., "A Digital System Design Language (DDL)," *IEEE Trans. Comput.*, C-17(9), 850–861 (1968).

Duley, J. R., and Dietmeyer, D. L., "Translation of a DDL Digital System Specifications to Boolean Equation," *IEEE Trans. Comput.*, C-13(4), 305–313 (1969).

Endou, M., Hoshino, T., and Karatsu, O., "Optimization Technique of Logic Synthesis System: ANGEL (in Japanese)," DA 23-5, IPSJ, 1984.

Hong, S.J., Cain, G. R., and Ostapko, D. L., "MINI: A Heuristic Approach for Logic Minimization," *IBM J. Res. Develop.*, 443–458 (1974).

Kawato, N., Saito, T., Maruyama, F., and Uehara, T., "Design and Verification of Large Scale Computer by Using DDL," *Proc. 16th DA Conference*, pp. 360–366, 1979.

Kowalski, R., *Logic for Problem Solving*, North-Holland, Amsterdam, 1983.

Ogawa, U., Shima, K., Sugawara, T., and Takagi, S., "Knowledge Representation and Inference Environment: KRINE," *Proc. FGSC'84*, pp. 643–651, 1984.

Takagi, S., "Rule Based Synthesis, Verification and Compensation of Data Paths," *Proc. ICCD'84*, pp. 133–138, 1984.

Takagi, S., "ALU Logic Synthesis Using Templates," *Proc. 7th ISCAS*, Japan, 1985a.

Takagi, S., "Design Method Based Logic Synthesis," *Proc. 7th CHDL Conference*, 1985b.

Takagi, S., "The Verification and Compensation of Data Paths (in Japanese)," *IPSJ Trans.* 26 1, 1985c.

Tesng, C., and Siewjorek, D. P., "The Modeling and Synthesis of Bus Systems," *Proc. 18th DA Conference*, pp. 471–478, 1981.

Tesng, C., and Siewjorek, D. P., "Facet: A Procedure for the Automated Synthesis of Digital Systems," *Proc. 20th DA Conference*, pp. 490–496, 1983.

24. Fault-Tolerant Multiple Processor Systems—Modeling and Analysis

Kishor Trivedi

Duke University

Robert Geist

Clemson University

In recent years, the demand for computing capacity has sustained a tremendously high rate of growth. Coupled with technological advances, this demand has spurred an increased interest in multiple processor and distributed computing systems. However, many interesting problems in the analysis of such systems must yet be solved in order to provide the designers and users with cost-effective tools for system evaluations. In addition to the need to derive measures of system effectiveness for a given system and its associated parameters, there is a need to provide techniques for selecting the set of parameter values that would optimize system effectiveness in a given setting. The criteria for selection may be based on reliability, availability, performance, cost, or a combination of these measures. As the measures themselves are qualitatively so different, effective combinations, which capture the real interests of the users, are often difficult to formulate.

Many of the tools necessary for these analyses have their foundations in probability theory and statistics. When we consider the analysis of a computer system serving a large number of users, we must account for several types of random phenomena. First, the job arrival patterns are diverse. Second, the resource requirements and the perception of the quality of service will differ from one user to another. Finally, the resources of the computer system are subject to chance failures due to environmental conditions and aging. The theory of stochastic processes plays a crucial role in evaluation of various measures of system effectiveness, such as throughput, response, time, reliability, and availability.

In a larger context, the modeling of modern computer systems provide an important arena within which to view the classic struggle between model accuracy, that is, the extent to which a model of a system faithfully represents the system under study, and model tractability, that is, the extent to which the modeler is able to extract useful information from the model in a cost-effective manner. The difficulty in searching for solutions to this trade-off problem is compounded, within this particular arena, by certain additional complexity constraints, which typically render the classical modeling tools inadequate. One constraint, nearly always present while modeling complex systems, is the huge disparity in state transition rates. A rate ration of 10^{-10} within a single model is not uncommon, leading to the so-called stiff systems of differential, integral, or algebraic equations, for which standard numerical techniques are largely inadequate. Though great progress has been made in recent years on numerical techniques for solving stiff systems (Shampine and Gear 1979), a disparity ratio of 10^{-10} coupled with a system of size 10^5 still appears to be out of reach.

There are three general techniques for system evaluation: testing, simulation, and analytic modeling. Exhaustive testing will likely produce the most accurate results, but will tend to be extremely expensive. Further, if the system being evaluated is not yet available for experimentation, the testing method obviously will not be applicable. In any case, the use of testing techniques implies the need for experimental design and statistical analysis of observed data. When the testing approach is not feasible, the system must be hierarchically decomposed until a level is reached where testing is feasible. Component test data must then be combined into a system evaluation by means of a simulation model, an analytic model, or a hybrid model.

Simulation models can be more realistic than analytic models but are likely to be less realistic than the testing method. Like testing, simulation models also require a careful design of (simulation) experiments and statistical analysis of output data. Simulation models can be driven by either an event trace or distributions of random times to the occurrence of events, in which case procedures to generate random variates and needed (Trivedi 1982). Both simulation and testing often provide an attractive alternative to analytic techniques for modeling computer systems, in that great detail can easily be captured without sacrificing model tractability. On the other hand, unacceptable solution time costs can easily be inherent in this approach. Suppose we wish to estimate the probability p that a system reaches a certain absorbing (model exit) state before some time t by $S(n)/n$, where $S(n)$ is the total number of times we reach the specified state in n simulation or (testing) trials of duration t. Then:

$$\overline{P}(P(X(n) \leqslant k) = \sum_{i=0}^{k} \binom{n}{i} p^i (1 - p)^{n-i} \approx \sum_{i=0}^{k} \frac{(np)^i}{i!} e^{-np} = P(X>1) \quad [10.3]$$

where X is a $k + 1$ stage Erlang random variable with parameter np (Trivedi 1982). If we require at least 95% confidence that we will not underestimate p by more than 10%, then we must have $P(S(n)/n \geqslant .9 *p) \geqslant .95$, that is, $P(X$

$> 1) \leq .05$ where X is a $[.9*np] + 1$ stage Erlang random variable and hence $2npX$ is a chi-square random variable with $2 ([.9*np] + 1)$ degrees of freedom. Thus we require $p (2npX > 2np) \leq .05$, that is, $n > X^{.052}/2p$, where $X^{.052}$ is the high .05 percentile of a chi-square distribution with $2 ([.9*np] + 1)$ degrees of freedom. The problem is just this: a realistic value for p may be 10^{-9}! If so, the inequality above indicates that more than one trillion trials may be necessary! It is difficult to conceive of a nontrivial simulation running on any machine for which a trillion trials would represent a reasonable computation time. Thus testing and simulation models can be both expensive to develop and expensive to run (to obtain statistically significant results), and the alternative of analytic modeling can appear quite attractive, particularly if a wide range of models must be evaluated.

A large number of "standard" analytic models for computer system evaluation are available (Lavenberg 1983; Siewiorek and Swarz 1982; Trivedi 1982). Analytic model types include combinatorial (see, e.g., Chapters 1–5 of Trivedi 1982), Markov (Siewiorek and Swarz 1982; Trivedi 1982), semi-Markov (Ross 1970), renewal (Trivedi 1982), and regression models (Trivedi 1982). Solution techniques include transform methods (Kulkarni et al. 1984), (unpublished), recursive techniques, and numerical solution of differential (Trivedi et al. unpublished), integral (Geist et al. 1983), or algebraic (Lavenberg 1983; Trivedi 1982) equations.

Analytic models of computer systems have traditionally incorporated drastically simplifying assumptions, in order that the models might remain solvable for the desired system measures. The modeler has simply assumed (hoped!) that the system under study would be sufficiently robust that the extracted measures would give at least order-of-magnitude information. In this vein, Markov models of computer system behavior, for which numerous caveats must be issued, have proliferated throughout the literature. However, as more accurate information has become necessary, the drastic simplifications implicit in the assumptions of such models have been found to be unacceptable. The most serious of these is the implicit assumptions that the holding times in the states of the model are exponentially distributed, an assumption often known, in practice, to be highly inaccurate. Such objections might be dismissed with a reference to the classic result of Cox (1955) that any probability density function having a rational Laplace transform can be realized by a ladder network of exponential stages. However, repeated application of the methods of stages leads to a more serious problem, that of state-space explosion. Stiffler argues (Stiffler and Bryant 1982) that models that are sufficiently accurate to represent modern, ultrareliable, fault-tolerant computer systems could easily require 10^5 states. Should we then expand each of these into a series-parallel cascade, we would easily overrun the bounds of tractability.

Nonhomogeneous Markov models effectively remove the exponential holding time assumption without an increase in the size of the state space. Nevertheless, parameter specification for such models can be extremely difficult: the all-important state transition rates, which are, by definition, functions of time, are functions of global time, not time from entry into the given state. In many cases, only the latter information is available to the modeler. Though global time dependence may be available in certain instances and may be entirely appropriate for some processes within the particular computing systems being modeled (e.g., failure processes in a nonrepairable flight control system), it is rarely appropriate and available for an entire system (Trivedi and Geist 1983).

Semi-Markov models provide the luxury of locally time-dependent transition rates, which can be invaluable in allowing accurate specification of parameters for the model. Nevertheless, semi-Markov models of sufficient complexity and size to capture the essential details of the system are extremely difficult to solve for the desired information. Some techniques may be found in Sahner and Trivedi (1985) and Stiffler and Bryant (1982).

Thus our problem domain, the accurate, cost-effective modeling of modern computer systems, is characterized by a number of special constraints that render the classical modeling techniques ineffective when used in isolation. It becomes necessary to combine the use of testing, analytical modeling, simulation modeling, and logical proofs in order to provide satisfactory (cost-effective) methods of system evaluation for the purpose of prediction and validation.

The issues in modeling modern computer systems can be broadly classified into those arising from model construction, model solution (Laprie 1984), and model validation or verification (Sargent 1982). Modeling languages such as fault-trees (Barlow and Lambert 1983), the PMS notation (Siewiorek and Swarz 1982), RESQ (Sauer et al. 1982), and stochastic Petri nets (Dugan et al. 1984; Dugan 1984; Marsan et al. 1984) can be of great value in simplifying the arduous and error-prone task of model construction. The goal of these languages is to provide higher-level constructs to the user and let the modeling package automatically generate the details of the underlying stochastic model. Petri nets appear to be gaining considerable support in this context. A Petri net is a directed bipartite graph whose two vertex sets are called places and transitions. Places contain zero or more tokens, and the state of the net is represented by the number of such tokens in each place. Should all arcs into a transition emanate from places that contain one or more tokens, the transition is said to be enabled. Enabled transitions may fire, that is, remove one token from each input place and add one token to each output place. Extensions include firing time distributions, probabilistic arcs, inhibitor arcs, and logic gates (Meyer et al. 1984). The major advantage of such nets is their facility for the concise specification of inherently concurrent behavior.

When models are used for system evaluation, the modeler may decide to ignore certain features of the system, such as its structure, its workload, its fault-occurrence behavior, or its fault/error-handling behavior. This is often done to simplify the model in order to make it analytically tractable. The modeler assumes that the ignored features do not have a significant influence on system effectiveness. It is then important to validate the model against data collected by testing on the system itself, or, if such is not possible, to analyze the effects of the ignored features on the output of the model. Ignoring important aspects of system behavior can result in true errors in model construction and can cause model predictions to deviate substantially from real system performance. While it may be possible to resort to proof procedures to show that the model is a proper abstraction of the modeled system (Wensley 1978), normally we resort to face validity (Sargent 1982) or depend upon operational (input–output) validation. As part of model validation, it is necessary to determine whether the underlying assumptions are correct by means of mathematical analysis or by the statistical analysis of experimental data. If the assumptions cannot be supported by such analyses, then either the model must be changed, or the effect of the erroneous assumptions on model solution must be bounded.

Typical examples include distributional assumptions and independence assumptions, whose effects may be benign, but certainly must be addressed (McGough et al. 1988; Trivedi et al. 1988). Further, it is possible for systems to start operation in a state other than that which is assumed by the model. Some reliability prediction packages include the analysis of such uncertainty (Geist et al. 1983, 1984; Trivedi et al. 1988). Another class of modeling errors arises from model sensitivity to variations in the input parameters. It is often possible (and necessary!) to analyze the effect of such parametric errors in reliability prediction models (Geist et al. 1984; Trivedi et al. 1988), availability models (Goyal et al. 1986), and queueing models for system performance evaluation (Suri 1983).

Errors in the model solution process (as opposed to those from model construction) represent another major difficulty in system evaluation. These errors can be classified into approximation errors and numerical errors. The former arise in solving complex models through approximation techniques, used because an exact solution is either impossible or computationally expensive. Courtois (1977) has analyzed approximation errors in queueing theoretic models. McGough et al. (1988) have analyzed and bounded approximation errors in certain reliability models. Finally, truncation and round-off are two types of numerical errors, which always must be given careful consideration.

PERFORMANCE EVALUATION

As with general system evaluation, the common techniques for performance evaluation are measurement, simulation, and analytic modeling. The first two techniques require the use of statistical methods to design experiments and to analyze output data. For a treatment of measurement techniques see Ferrari (1983); for a discussion of simulation methods see Lavenberg (1983).

Most current work may be classified as either program-oriented (transaction-oriented) or resource-oriented analysis. In the former, a relatively simple model of system resources is assumed, while the characteristics of transactions are modeled in great detail. In the latter, system resources are modeled in great detail, while a relatively simple model of transaction (program) behavior is assumed.

The general problem of modeling program behavior is quite difficult, but it is often possible to derive bounds on the performance. For several different methods of analyzing sequential programs, the reader is referred to Trivedi (1982). Analytic evaluation of the execution times for programs with internal concurrency can be equated to the analysis of an extended stochastic PERT network (Fix and Neumann 1979). In Sahner and Trivedi (1985), we consider the stochastic analysis of precedence graphs by restricting attention to a class of distributions that are closed under convolution max, and alternation operations. We restrict our attention to a set of graphs that we call special series-parallel; this set is a (proper) subset of series of series-parallel graphs. These graphs, which can be thought of as node activity networks, are assumed to be directed acyclic graphs; and distribution associated with a node is assumed to be a mixture of Erlang distributions. Thus we are able to compute the distribution function of the execution time of a concurrent program in an environment with ample processors. If the program under consideration is not series-parallel, it can often be modeled as a stochastic Petri net. Such nets can often be analyzed by an automatic transformation to a Markov chain (Marsan et al. 1984) or a semi-Markov process (Dugan et al. 1984; Dugan 1984). If neither transformation is possible, the stochastic Petri net can be simulated (Dugan et al. 1984; Dugan 1984).

In general, models for system performance evaluation require three sets of inputs: workload characteristics, hardware configuration, and software configuration. Thus the resource requirements and traffic parameters of individual programs, as well as the mix of programs, must be specified, as must the types of hardware resources, their interconnections, and capacities. Software configuration parameters include scheduling algorithms and file or data base allocations. The outputs of such models include throughput, response times, and queue lengths, both for the system as a whole and for various individual devices within the computing system.

Many different analytical techniques have been used for the solution of performance models. Combinatorial (Goyal and Agerwala 1984) and discrete time Markov models (Trivedi 1982) have been used for the analysis of multiprocessor memory interference. Continuous time Markov models are more popular for modeling CPU–I/0 terminal interactions and for modeling computer communication networks. The fastest techniques are those available for the so-called product-form queueing networks (Lavenberg 1983; Trivedi 1982).

More often, however, attempts to provide exact solutions to analytic models either lead to excessive computation times or make too many assumptions for the results to be useful. As an alternative, various approximation techniques have been used to solve large problems (Lavenberg 1983). These methods include decomposition (Courtois 1977; Lavenberg 1983; Trivedi 1982), iterative (Heidelberger and Trivedi 1982, 1983; Lavenberg 1983), and diffusion approximations (Lavenberg 1983). As noted earlier, it is often necessary to remove the "exponential-distribution" assumption of Markov models. Solution of such non-Markovian models can be obtained by the method of supplementary variables (Baccelli and Trivedi 1983), Markov renewal techniques (Ross 1970), the method of stages (Trivedi 1982), or the method of embedded chains (Lavenberg 1983; Trivedi 1982).

Many software packages are available for the specification and solution of performance models (Sauer et al. 1982). These packages include, for example, the RESQ package developed by the IBM Corporation; CADS and PAWS by the Information Research Associates of Austin, Texas; BEST/1 by BGS Systems; PANACEA and QNA by AT&T Bell Laboratories; and QNAP by INRIA, France (Potier 1986).

Current research in performance modeling includes approximation techniques to solve otherwise intractable models, error analysis of these approximation techniques, extending the range of applicability of tractable analytic models for the evaluation of parallel and distributed systems (Bux 1981; Dugan et al. 1984; Goyal and Agerwala 1984; Goldberg et al. 1983; Heidelberger and Trivedi 1983) and investigation of new, user-based, measures of system performance (Geist 1984; Geist et al. 1988; Geist and Trivedi 1983a).

System performance is often modeled by a network of queue–server pairs (Geist and Trivedi 1982). Resources in computer systems are modeled as servers, resource contention by queues for service, and the job movement from one resource to another by a network flow structure. Under certain simplifying assumptions, these networks have a product-form solution that can be evaluated using elegant algorithms to produce such performance measures as job throughputs, mean response time, mean queue lengths, and so on.

As an example, consider the model, shown in Figure 24-1, for a terminal-oriented multiprocessor system. A collection of users at $M(= = 20)$ terminals will access an $n(= = 3)$ processor system. Service at the processor note is FCFS, and job service demands are exponentially distributed with a mean $E[S]$ $(= = 1)$ sec. Terminal users have an average "think-time" of $E[TT]$ $(= = 15)$ sec, but we make no other particular assumptions concerning the distribution of think-times.

The simple algorithm of Table 24-1 can be used to compute system throughput and response time. The values for the $r[i,k]$, a measure of relative utilization of node i under load k, in this case are given by $r[0,k] = E[TT]^k/k!$, $r[1,k] = E[S]^k/k!$, if $K \leq n$, and $r[1,k] = E[S]^k/n!n^{k-n}$, if $k > n$, After execution of the algorithm of Table 24-1, average system throughput, $E[T] = C[M-1, PREV]/C[M,PREV]$, which, for the parameters given, turns out to be 1.245

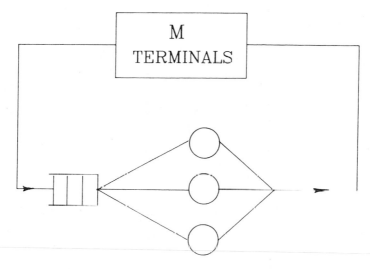

Figure 24-1.   A queueing model of a three-processor system.

jobs/sec. System response time is then easily computed: $E[R] = M/E[T] - E[TT] = = 1.063$, not much more than the expected service time.

In some cases, for instance, when blocking of servers occurs for lack of queue spaces or when a job needs two resources simultaneously, the product-form solution illustrated above no longer holds. Although the queueing network can still be solved by describing the complete Markov chain of the system, the number of states in the Markov chain can become unmangeably large. A well-known approximate solution method called Norton's decomposition (also known as flow equivalent server) is described in Geist and Trivedi (1982) and Trivedi (1982), with applications to a multiprogrammed uniprocessor. During the design process of computer systems, not only the evaluation but also the optimization of certain output measures is needed. Some simple optimization procedures and their limitations are also pointed out in Geist and Trivedi (1982).

RELIABILITY EVALUATION

Testing followed by data analysis is a common technique for reliability/availability estimation of systems. For example, life-testing of a multitude of individual circuit types has been carried out, and the resulting failure rate estimates have been compiled in the *Military Standard Handbook*. Based on this standard, a program called LAMBDA has been written to calculate failure rates of series systems (Siewiorek and Swarz 1982). As another example, logs are kept on outages of telephone trunks, and the observed data has been analyzed to estimate the availability of individual trunks (Spragins et al. 1981). Fault injection testing can also be carried out to estimate fault detection coverage in

Table 24-1.

```
PREV: = 0; CUR: = 1;
C[0,PREV]: = 1; C[0,CUR]: = 1;
for i: = 1 to M do C[i,PREV]: = 0;
for i: = 0 to TOPNODE do

  begin
  for j: = 1 to M do

    begin
    C[j][CUR]: = 0;
    for k: = 0 to j do C[j,CUR]: = C[j,CUR] + C[j-k,PREV] * r[i,k];
    end;

  PREV: = 1-PREV;
  CUR: = 1-CUR;
  end
```

fault-tolerant systems (Lala 1983). Most software reliability efforts calibrate an empirical model based on data collected from tests, and later use the model for predicting future behavior (Goel 1983; Littlewood, 1980; Ramamoorthy and Bastani 1982).

Black box testing, applied to complex fault-tolerant computer systems, may become prohibitively expensive, if not impossible (Trivedi et al. 1980). In such cases, it becomes necessary to hierarchically decompose the system to a level where testing becomes feasible. The results of such tests on subsystems must then be combined using either a simulation or an analytic model.

Simulation models have been used in reliability evaluation in at least two different ways. To assess the effectiveness of various fault/error-handling mechanisms, a common technique is to use the previously mentioned fault injection in a simulation model of the system. Observations of system response to repeated fault injections are then analyzed to obtain estimates of parameters such as fault detectability and duration of detection time (Agrawal 1981). It is also possible to use distribution-driven simulation models for reliability/availability prediction (Dugan et al. 1984; Spragins et al. 1984; Siewiorek and Swarz 1982).

Analytic models or many different types exist for reliability evaluation. Combinatorial methods (Trivedi 1982, Chapters 1–5) have been used in the CARE package (Mathur 1972) and in ADVISER (Siewiorek and Swarz 1982). Markov methods (Trivedi 1982) form the basis of the ARIES package (Makam et al. 1982). The CARE II model uses Markov renewal methods (Stiffler et al. 1975), whereas the CARE III model uses a combination of nonhomogeneous Markov and Markov renewal techniques (Stiffler and Bryand 1982).

In order to construct a reliability model, three sets of inputs are necessary: the system structure, the fault-occurrence behavior, and the fault/error-handling behavior. The first category of inputs provides information regarding the set of resources, their interconnections, and conditions under which the system will fail. The description of the system structure determines the structure of the reliability model. One method of specifying structural information is to use fault-trees (Barlow and Lambert 1975); the CARE III reliability prediction package uses fault-trees for this purpose (Stiffler and Bryant 1982). Another common method of specifying computer system structure is the PMS notation (Siewiorek and Swarz 1982). The reliability prediction package called ADVISER allows the user to specify structural information by means of a PMS diagram plus assertions defining operational states (Siewiorek and Swarz 1982). Yet another powerful tool for specifying system structure is the Petri net. We are exploring this possibility for our joint reliability modeling effort at Duke and Clemson (Dugan et al. 1984; Dugan 19'84; Geist et al. 1948; Trivedi et al. 1988).

The second category of inputs describes the class of faults in the subsystems and their stochastic failure processes. This information is collectively referred to as the fault-occurrence model. Much work remains to be done on this important topic. The required distributions of times-to-failure and the associated parameters can be obtained from standards (Siewiorek and Swarz 1982), from field data (Siewiorek and Swarz 1982), by conducting lifetests of individual subsystems, or by evaluating a reliability model of the subsystem.

Three different types of faults are typically considered in reliability models: permanent, intermittent, and transient. ARIES (Makam et al. 1982), CARE II (Stiffler et al. 1975), CAST (Conn et al. 1977), and SURF (Costes et al. 1981) allow both permanent and transient fault types; CARE III (Stiffler and Bryant 1982) and HARP (Trivedi et al. 1988) allow not only permanent and transient faults but also intermittent faults. To gain tractability, it is commonly assumed that the time-to-failure distributions are exponential. This yields a homogeneous Markov model (e.e., ARIES, HARP Phase I). CARE III and HARP II and III utilize nonhomogeneous Markov models to allow Weinbull-like failure processes. It should be mentioned that this approach restricts us to situations without repair and, in the case of standby redundancy, forces the assumption that the standby unit fails at the same rate as the active unit (Trivedi and Geist 1983). It is possible to sue the Coxian method of stages to allow for nonexponential time-to-failure distributions (Trivedi 1982); the SURF package uses this approach. However, a significant number of stages may be needed to capture interesting distributions; so the state space of the resulting Markov chain can become rather large.

The third category of inputs specifies the fault/error-handling behavior of the system. The most common approach to modeling fault/error-handling behavior is to allow for coverage factors (probabilities) and have the model user specify them directly. However, accurate estimation of coverage factors may be quite elusive. Furthermore, system reliability is known to be extremely sensitive to the coverage factors (Arnold 1973; Bouricius et al. 1969; Trivedi 1984), and thus it is important to provide tools/techniques for the specification of detailed fault/error-handling behavior and for the careful computation of coverage factors. CARE II and CARE III use semi-Markov models to represent the fault/error-handling behavior. Although this technique does represent a major advance beyond the specification of coverage as a single parameter, it is still restrictive, as it tries to capture, in a single state graph, the duration aspects (e.g., duration of a transient fault and holding times of an intermittent fault in its active and benign states) as well as fault-treatment aspects (detection, isolation, reconfiguration) of the fault/error-handling subsystem. Moreover, the most common treatment of transient faults, namely, attempting transient recovery several times before deciding that the fault is permanent, is difficult to represent in a CARE III–like model. It is for these reasons that we have chosen the more flexible modeling technique provided by the Extended Stochastic Petri Net in our Work. In these efforts we have also developed new methods of computing coverage factors that are applicable to a wide variety of fault/error-handling models (Geist et al. 1984; Trivedi et al. 1988; Trivedi 1984a).

In order to solve the problems posed by large state spaces, several approaches have been suggested. A graph pruning technique is applied in ADVISER, structural decomposition is applied in CARE (Mathur 1972 and ARIES, while behavioral decomposition is applied in CAST, CARE II, CARE III, and HARP. The problem of excessively large state spaces is most often addressed through structural decomposition. Computer systems, especially those with some component redundancy, typically exhibit a natural physical division into smaller subsystems, for example, processors, buses, and memory modules. The state space for each subsystem may be sufficiently small that subsystem behavior can then be analyzed in a cost-effective manner using one of the standard approaches mentioned earlier. A tractable model of overall system behavior can then be obtained through a combinatorial aggregation of the subsystem behaviors.

The advantages of this approach in state-space reduction are substantial. We often find that an aggregate state space actually represents a cartesian product of the states in the subsystems, and so the total size of the systems to be analyzed is reduced from a product to a sum of the sizes of the subsystems. On the other hand, the implicit assumption made in the aggregation step, independence of subsystem behavior, is often violated in practice, and, to the extent that it is, the results obtained from the model will be approximations. Further, this approach does not address the issue of widely disparate state transition rates. Should the rate disparities not separate along physical lines, we would still have stiff (albeit smaller) systems to solve.

An alternative to this structural decomposition is behavioral decomposition (Geist and Triendi 1983b; Geist et al. 1984; Trivedi and Geist 1983), in which we attempt to separate the system along temporal lines, according to the relative magnitudes of the state transition rates. To do this, we must first identify a collection of system states that are quasi-absorbing, that is, for which the total rate of exit transitions is, relatively, close to zero. We then form a reduced system consisting of only quasi-absorbing states in the following manner. To each such state, we associate a subsystem consisting of all sequences of states ending in the quasi-absorbing state and having high transition rates throughout the sequence. Of course, a given state in such a sequence may also be a member of similar sequence leading to another quasi-absorbing state. In this case, we must add the sequence and the quasi-absorbing state to the subsystem being formed. Each subsystem thus formed is analyzed in isolation, temporarily ignoring any low-rate exits from the subsystem, thereby avoiding any problems with stiffness. Probabilities of reaching the (quasi-)absorbing states within the subsystem are then obtained. Finally, the overall system is modified by replacing any exit from a quasi-absorbing state into any state belonging to one of the subsystems with direct exits to the (quasi-)absorbing states of that subsystem. The rates are modified, of course, by factors involving the associates subsystem exit probabilities. The reduced system can then be considered in isolation, again without stiffness complications.

Naturally, not all systems are amenable to such decomposition. (For a further characterization of such decompositions and an extension see Bobbio and Trivedi 1984.) The goal, of course, is that all subsystems have identical structure and that the reduced system be sufficiently small to allow cost-effective solution. To the extent that the subsystems differ in structure, we fail to reduce the size of the state space from a product to a sum, and any advantages gained by avoiding stiffness may well be lost to the approximations made in using the decomposition–aggregation approach. Error analysis must account for such approximations (McGough et al. 1988), as well as for those errors arising from any numerical techniques used to solve the subsystems or the reduced version of the overall system.

Nevertheless, certain large collections of systems are susceptible to attack by these methods. Such systems are characterized by long periods of relative inactivity alternating with brief period of extremely high activity. Consider, for example, a Markov model, shown in Figure 24-2, of a three-processor standby-redundant fault-tolerant computer system. The numerically marked states are quasi-absorbing, in that they represent a "system operational, no active faults" state. The units fail at a constant rate λ, which is relatively near zero. Upon failure of an operational unit, the system enters a "fault active" state. From here, if the fault is detected at a constant rate δ, we switch in a stanby unit, take the faulty unit off line, and enter another quasi-absorbing "operational" state. Other exists form the "fault-active" state, which might occur before detection, are the occurrence of another fault, which causes system failure, or the production of an error (at a constant rate p). Should the error be detected (with, say probability p), the presence of the fault will be recognized, and recovery can still occur. Otherwise, the error propogates through the system, causing failure. Realistic values for the ratios λ/p, λ/δ, and λ/ε in modern systems are all in the order of 10^{-8} or less.

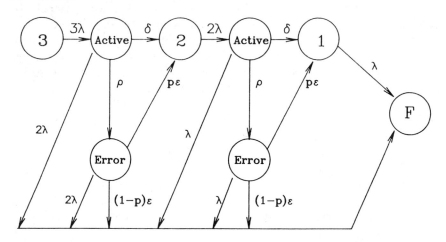

Figure 24-2. A reliability model of the three-processor system.

When we decompose this system along behavioral lines, we obtain two identical "fault/error-handling" subsystems, a copy of which is shown in Figure 24-3. We can easily solve this small Markov model for the limiting values of the state probabilities for quasi-absorbing states F and n:

$$P_{AF}(+\infty) = \lim_{t \to \infty} P_{AF}(t) = (1 - p)\, \rho/(\rho + \delta)$$

$$P_{An}(+\infty) = \lim_{t \to \infty} P_{An}(t) = (\delta + p\, \rho)/(\rho + \delta)$$

We can then form the reduced model, shown in Figure 24-4, with coverage factors $c_n = = P_{An}(+\infty)$, whose solution is relatively easy, when compared to the original model.

However, the error in this approximate solution (which results from ignoring the low-rate transitions from the subsystem) may not be acceptable for all applications. For instance, in life-critical avionics applications, such as flight control systems, demands for accuracy to eight or nine decimal digits are not uncommon in the estimation of systems reliability. Fortunately, we can often incorporate the effect of these low-rate transitions during the integration (formation of the reduced model) step. This technique is best introduced in the context of our present example. System recovery, as modeled with the subsystem of Figure 24-3, is not entirely dependent upon the probability of eventually reaching state n, $p_{An}(+\infty)$. We must reach state n prior to the next fault (a low-transition rate, 2λ or λ, from the subsystem). Thus we must take into account the distribution of recovery times, $F_R(t) = = p_{An}(t)/P_{An}(+\infty)$, and, similarly,

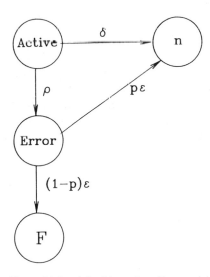

Figure 24-3.   A fault/error-handling model.

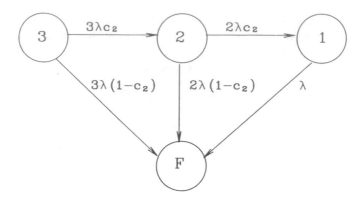

Figure 24-4.   The reduced reliability model.

the distribution of times to failure due to error propagation. If X denotes the time to next fault (premature exit from the subsystem), then we should change the rate modification factor in the reduced model from $c_n = = P_{An} (+ \infty)$ to:

$$c_n = P_{An} (+ \infty)P(R < X)$$

$$= P_{An} (+ \infty) \int_0^{+\infty} \int_r^{+\infty} f_R (r)n\lambda e^{-n\lambda e} \, dx dr$$

$$= P_{An} (+ \infty) \int_0^{+\infty} e^{-n\lambda r} f_R (r) dr$$

$$= P_{An} (+ \infty)L_R (n \lambda)$$

[10.3]

where L_R denotes the Laplace transform of the recovery density function f_R (Trivedi 1984a). In addition, we should reduce the factor $P_{AF}(+ \infty)$ in a like manner, and then add a transition to failure (failure due to two near-coincident faults) with modifying factor $P_{An}(+ \infty)X \leq R) + P_{AF}(+ \infty)P(X \leq F)$. Because we do not distinguish between the two failure modes, we can combine arcs from n to F and obtain the reduced model of Figure 24-4.

As evidence of the accuracy we can obtain with this approach, consider the model of our example with parameter values taken from Table 24-2, which are realistic for modern flight-control systems. Suppose, for purposes of analytical tractability, we attach a fault/error-handling subsystem to only the first fault, from state 3. The exact 100-hour reliability (probability that the system is operational at time $t = 100$ hours) can then be computed, and to ten digits it is .9997303442. The approximate value offered by the approach outlined above is .9997303415.

Table 24-2.

T	100 hr
λ	0.0001/hr
δ	600/hr
ρ	6000/hr
p	.99
ε	1200/hr

More generally, we need to modify reduced model rates by factors that represent the probability of a premature (non-quasi-absorbing) exit from the subsystem. Should all non-quasi-absorbing states of the subsystem that lie on any path to the chosen quasi-absorbing state have identical, constant, low-rate exit transitions (as in our example), the computation of the modifying factor is reduced to a determination of the distributions of times to a quasi-absorbing exit. Moreover, the equation (*) is easily extended to the case in which the low-rate transitions are globally time-dependent, yielding a nonhomogeneous Markov chain for the reduced model (Geist et al. 1984).

Extensions to the reduced model in other directions are analytically more difficult and of questionable value in our applications. Thus we have chosen, instead, to concentrate on what appears to be a more critical issue, subsystem specification and determination of the key information concerning distributions of the times to reach the quasi-absorbing states.

The effect of variations in failure rates and coverage parameters on the predicted reliability can be bounded by analytical methods (Geist et al. 1984; Trivedi et al. 1988). Effects of various other potential errors in the modeling process and in the solution technique need to be classified and analyzed.

We have concentrated here on reliability models of nonrepairable systems. For a survey of reliability/availability models of repairable systems, we refer the reader to Goyal et al. (1986). For a study of decomposition/aggregation for repairable systems see Bobbio and Trivedi (1984).

COMBINED EVALUATION OF PERFORMANCE AND RELIABILITY

The separation of reliability and performance evaluation of a computer system is no longer appropriate for systems with graceful degradation capabilities. Combined evaluation of performance and reliability of such systems is an active area of research (Baccelli and Trivedi 1983; Dugan et al. 1984; Geist et al. 1984; Kulkarni et al. 1948; Meyer 1982; Mitrani and King 1983; Trivedi 1984).

A model for the combined evaluation of performance and reliability clearly must have the three types of inputs (system structure, fault-occurrence model, and fault/error-handling model) needed for a reliability model. In addition, it also needs information on the workload to be processed and the scheduling algorithms to be employed. The information on workload includes the resource requirements and usage patterns of individual programs as well as the mix of programs concurrently using system resources. The common method of specifying resource usage patterns is to use a Markov chain (Sauer et al. 1982; Trivedi 1982), but for programs with internal concurrency, precedence graphs (Sahner and Trivedi 1985) or stochastic Petri nets (Dugan et al. 1984; Dugan 1984; Marsan et al. 1984) are often more appropriate. The mix of programs can be specified by a set of Markov chains or by a stochastic Petri net. Scheduling algorithms refer not only to the procedures for choosing among a set of queued programs when a resource becomes available but also to the procedures for handling a program interruption, either preplanned (e.g., time-slice completion) or unplanned (occurrence of a failure).

Most current models can be classified as either combinatorial, semi-Markov reward processes, or queuing systems subject to failure. These models often differ in the intended measure of system performance. Measures derived using combinatorial methods include the probability of catastrophic failure during the execution of a program and the distribution of elapsed time of a program (Castillo and Siewiorek 1980; Kulkarni et al. 1984; Krishna and Shin 1983). Measures considered by the semi-Markov reward models are the expected reward rate at time t, accumulated reward until time t (Meyer 1982), and accumulated reward until system failure (Beaudry 1978; Trivedi 1984). We have recently shown that task-oriented models (the former) and resource-oriented models (the latter) are duals of each other (Kulkarni et al. 1984). The measures evaluated by queuing models include the program response times, queue lengths, and utilizations of various servers (Baccelli and Trivedi 1983; Gaver 1962; Mitrani and King 1983; Nicola 1983).

Existing models an also be divided along the lines of approximate versus exact, capacity-based versus throughput-based versus response-time based, no resource contention versus resource contention, and transient versus steady-state. Early models for the evaluation of fault-tolerant systems were capacity-based. For instance, Beaudry (1978) assumed that system performance was proportional to the number of available processors, and considered a Markov reward model for which she evaluated the accumulated reward until failure.

Other models have allowed for nonlinearities inherent in system performance measures due to queuing effects. Many of these efforts have taken advantage of the fact that the times-to-failure and times-to-repair are usually several orders of magnitude larger than the time to complete the execution of a program. Therefore, the model can be solved by decomposing system behavior into submodels and later combining the results of the submodels. Such a model can be considered a Markov reward process. For instance, Meyer (1982) considered an $M/M/n$ queue with finite buffer as a performance submodel to compute the steady-state throughputs (or reward rates) in different structure states. He then combined these with exponential times-to-failure distributions for the processors and buffers to compute the distribution of the accumulated reward in the bounded utilization period [0, t]. Such models are further extended in Kulkarni et al. (1988), and computational techniques are also discussed.

Several authors have considered exact models that include resource-contention (queuing) program characteristics, and failure and repair processes. Gaver (1962) considered a single server $M/G/1$ queue with constant failure rate and generally distributed repair times. He derived an expression for steady-state average response time, for different types of failure interruptions. Recently, Nicola (1983) extended Gaver's model to allow the simultaneous presence of different types of failure interruptions. Mitrani and King (1983) studied an $M/M/n$ degradable system with constant failure and repair rates. Because of the exponential service time distribution, the type of preemption due to failure was not relevant. Assuming perfect coverage, they provided a numerical procedure for computing the steady-state average response time. In Baccelli and Trivedi (1983), we considered a two-processor standby-redundant system with a Poisson arrival stream of jobs. The service-time distribution system was general, and the preemptive-repeat-different discipline was used. We also provided a numerical procedure to compute the average response time.

We can continue the analysis of the queueing system introduced earlier, where we now allow the processors to fail (terminals are assumed to be fault-free for this example). Suppose our three processors are subject to failure (and subsequently repair). Failures seldom occur—let's say, on the average, only once in 2 months (independent random arrivals). Repairs can take significant time too— let's say, on the average, 3 weeks—but are exponentially distributed, so most repairs take less time than this. With the very reasonable assumption that the system reaches steady state between failures (repairs), we can quickly obtain a new estimate for $E[R]$. We use the algorithm of Table 24-1 to analyze one-, two-, and three-processor systems in isolation, and then combine these results with the steady-state availability of each such system. For the parameters indicated, we find that the expected throughput, $E[T]$, has dropped to 1.165 sec, and thus expected response time, $E[R]$, has jumped to 2.162 sec, more than double the earlier value.

This last analysis leads to a disturbing question of current interest: in this context, is $E[R]$ meaningful? Are we not merely averaging excellent behavior and horrible behavior, and then giving a performance estimate that will never really be experienced? Several alternative approaches are under investigation, including development of sufficient machinery to provide a complete specification of the response time distribution (Kulkarni et al. 1984), as well as alternative measures of system "goodness" (Geist et al. 1987). Much work remains to be done.

CONCLUSION

We have presented a brief overview of probabilistic models used in the analysis of fault-tolerant multiple processor systems. For a more detailed study, the reader is referred to several books (Courtois 1977; Ferrari et al. 1983; Lavenberg 1983; Siewiorek and Swarz 1982; Trivedi 1982) and survey papers (Bux 1981; Goyal et al. 1986; Geist and Trivedi 1982; Geist and Trivedi 1983b; Laprie 1984; Ramamoorthy and Bastani 1982; Trivedi 1984a,b).

REFERENCES

Agrawal, V. D., "Sampling Techniques for Determining Fault Coverage in LSI Circuits," *J. Digital Systems*, 189–202 (1981).

Arnold, T. F., "The Concept of Coverage and Its Effect on the Reliability Model of a Repairable System," *IEEE Trans. Comp.*, 251–254 (Mar. 1973).

Baccelli, F., and Trivedi, K., "Analysis of M/G/2 Standby Redundant System," *Proc.* IFIP Symposium, PERFORMANCE '83, College park, MD, 1983 (North-Holland).

Barlow, R., and Lambert, H., "Introduction to Fault Tree Analysis," in *Reliability and Fault-Tree Analysis*, SIAM, Philadelphia, 1975, pp. 7–35.

Beaudry, M. D., "Performance Related Reliability for Computing Systems," *IEEE Trans. Comp.*, 540–547 (June 1978).

Bobbio, A., and Trivedi, K., "An Aggregation Technique for the Transient Analysis of Stiff Markov Systems," submitted Nov. 1984.

Bouricius, W., Carter, W., and Schneider, P., "Reliability Modeling Techniques for Self-Repairing Computer Systems," *Proc.* 24th Annual ACM National Conference, pp. 295–309, 1969.

Bux, W., "Local-Area Subnetworks: A Performance Comparison," *IEEE Trans. Communications*, 1465–1473 (Oct, 1981).

Castillo, X., and Siewiorek, D. P., "A Performance-Reliability Model for Computing Systems," *Proc.* 1980 Int. Symp. on Fault-Tolerant Computing, Portland, ME, June 1980, pp. 187–192.

Conn, R., Merryman, P., and Whitelaw, K., "CAST—A Complementary Analytic Simulative Technique for Modeling Fault-Tolerant Computing Systems," *Proc.* AIAA Computers in Aerospace Confeence, Los Angeles, Nov. 1977, pp. 6.1–6.27.

Costes, A., Doucet, J., Landrault, C., and Laprie, J., "SURF—A Program for Dependability Evaluation of Complex Fault-Tolerant Computing Systems," *Proc.* IEEE FTCS-11, 1981, pp. 72–78.

Courtois, P., *Decomposability: Queueing and Computer Science Applications*, Academic Press, New York, 1977.

Cox, D., "The Use of Complex Probabilities in the Theory of Stochastic Processes," *Proc. Cambridge Philosophical Society, 51*, 313–319 (1955).

Dugan, J., Bechta, "Extended Stochastic Petri Nets: Analysis and Applications," Ph.D. dissertation, Duke University, Durham, NC, 1984.

Dugan, J., Bechta, Trivedi, K., Geist, R., and Nicola, V., "Extended Stochastic Petri Nets: Analysis and Applications," *Proc.* IFIP Symposium, PERFORMANCE '84, Paris, France, 1984 (North-Holland).

Ferrari, D., Serazzi, G., and Zeigner, A., *Measurement and Tuning of Computer Systems*, Prentice-Hall, Englewood Cliffs, NJ, 1983.

Fix, W., and Neumann, K., "Project Scheduling by Special GERT Networks," *Computing 23*, 299–308, (1979).

Gaul, W., "On Stochastic Analysis of Project Networks," in M. A. H. Dempter et al. Editors, *Deterministic and Stochastic Scheduling*, D. Reidel Publishing Co., 1982.

Gaver, D. P., "A Waiting Line with Interrupted Service, Including Priorities," *J. Royal Statistical Society*, Series B24, 73–90 (1962).

Geist, R., "Perception-Based Configuration Design of Computer Systems," *Information Processing Letters*, 18 (1984).

Geist, R., and Trivedi, K., "Queueing Network Models in Computer System Design," *Mathematics Magazine*, 67–80 (1982).

Geist, R., and Trivedi, K., "The Integration of User Perception in the Heterogeneous M/M/2 Queue," *Proc.* IFIP Symposium, PERFORMANCE '83, College Park, MD 1983a (North-Holland).

Geist, R., and Trivedi, K., "Ultra-High Reliability Prediction for Fault-Tolerant Computer Systems," *IEEE Trans. Comp.*, 1118–1127 (Dec. 1983b).

Geist, R., Trivedi, K., Dugan, J. B., and Smotherman, M., "Design of the Hybrid Automated Reliability Predictor," *Proc.* 5th IEEE/AIAA Digital Avionics Systems Conf., pp. 16.5.1–16.5.8, Nov. 1983.

Geist, R., Trivedi, K., Dugan, J.B., and Smotherman, M., "Modeling Imperfect Coverage in Fault-Tolerant Systems," *Proc.* 14th Int. Conf. on Fault-Tolerant Computing, Orlando, June 1984.

Geist, R., Stevenson, D., and Allen, R., "The Perceived Effect of Breakdown and Repair on the Performance of Multiprocessor Systems," submitted.

Goel, A., "A Guidebook for Software Reliability Assessment," Syracuse University, Tech. Report, 1983.

Goldberg, A., Popek, G., and Lavenberg, S., *Proc.* 9th Int. Symp. on Comp. Performance Modeling, Measurement, and Evaluation, *PERFORMANCE '83*, College Park, MD, May 1983, pp. 251–268.

Goyal, A., and Agerwala, T., "Performance Analysis of Future Shared Storage Systems," *IBM J. Res. and Development*, 95–108 (Jan. 1984).

Goyal, A., Lavenberg, S., and Trivedi,K., "Probabilistic Modeling of Computer Systems Availability," The Conference on Statistical and Computational Problems in Probability Modeling, *Annals of Operations Research* (1986).

Heidelberger, P., and Trivedi, K. S., "Queueing Network Models for Parallel Processing with Asynchronous Tasks," *IEEE Trans. Comp.* (Nov. 1982).

Heidelberger, P., and Trivedi, K., "Analytic Queueing Models for Programs with Internal Concurrency," *IEEE Trans. Comp.* 73–82 (Jan. 1983).

Krishna, C.M., and Shin, K. G., "Performance Measures for Multiprocessor Controllers," *Proc.* 9th IFIP International Symposium on Computer Performance Modeling, College Park, MD, 1983.

Kulkarni, V., Nicola, V., and Trivedi, K., "On Modeling the Performance and Reliability of Multimode Computer Systems," *Proc.* Int. Workshop on Modeling and Performance Evaluation of Parallel Systems, Grenoble, Dec. 1984 (North-Holland).

Kulkarni, V., Nicola, V., Trivedi, K., and Smith, R., "A Unified Model for the Analysis of Job Completion Time and Performability Measures in Fault-Tolerant Systems," *JACM*.

Lala, J. H., "Fault Detection, Isolation and Reconfiguration in FTMP: Methods and Experimental Results," *Proc.* 5th IEEE/AIAA Digital Avionics Systems Conference, 1983.

Laprie, J., "Trustable Evaluation of Computer Systems Dependability," in G. Iazeolla, P. J. Courtois, and A. Hordigk, Editors, *Mathematical Computer Performance and Reliability*, Elsevier Science Publishers, B. V. (North-Holland), Amsterdam, 1984.

Lavenberg, S. (ed.), *Modeler's Handbook*, Academic Press, New York, 1983.

Littlewood, B., "Theories of Software Reliability: How Good Are They and How Can They be Improved?" *IEEE Trans. Software Engineering*, 489–500 (Sept. 1980).

Makam, S., Avizienis, A., and Grusas, G., "UCLA ARIES 82 User's Guide," Computer Science Department Report No. CSD-820830, Aug. 1982.

Marsan, M. A., Balbo, G., and Conte, G., "A Class of Generalized Stochastic Petri Nets for the Performance Evaluation of Multiprocessor Systems," *ACM Trans. Computer Systems* (May 1984).

Mathur, F. P., "Automation of Reliability Evaluation Procedures through CARE - the Computer-Aided Reliability Estimation Program," *Proc.* AFIPS Fall Joint Computer Conference, Vol. *41*, pp. 65–82, 1972.

McGough, J., Smotherman, M., and Trivedi, K., "The Conservativeness of Reliability Estimates Based on Instanteous Coverage," To appear, *IEEE Trans. Comp.*

Meyer, J., "Closed-form Solutions of Performability," *IEEE Trans. Comp.* 648–657 (July 1982).

Meyer, J., Movagher, A., and Sanders, W., "Stochastic Activity Networks: Structure, Behavior, and Application," Tech. Rept., Industrial Technology Institute, Univ. of Michigan, Dec. 1984.

Mitrani, I., and King, P. J. B., "Multiserver Systems Subject to Breakdowns: An Empirical Study," *IEEE Trans. Comp.*, 96–98 (Jan. 1983).

Nicola, V., "A Single Server Queue with Mixed Types of Interruptions," EUT Report 83-E-138, Eindhaven University of Technology, The Netherlands, 1983.

Potier, D., Editor, *Proc.* International Conference on Modeling Techniques and Tools for Performance Analysis, Paris, May 1986.

Ramamoorthy, C. V., and Bastani, F. B., "Software Reliability—Status and Perspectives," *IEEE Trans. Software Engineering*, 354–371 (July 1982).

Ross, S.M., *Applied Probability Models with Optimization Applications*, Holden-Day, San Francisco, 1970.

Sahner, R., and Trivedi, K., "Performance and Reliability Analysis Using Direct Graphs," submitted, Mar. 1985.

Sargent, R., "Verification and Validation of Simulation Models," in F. E. Celliar, Editor, *Progress in Modeling and Simulation*, Academic Press, London, 1982.

Sauer, C., MacNair, E., and Kurose, J., "The Research Queueing Package: Past, Present and Future, *Proc.* National Computer Conference, 1982.

Shampine, L. F., and Gear, C. W., "A User's View of Solving Stiff Ordinary Differential Equations," *SIAM Review, 21*, 1–17 (1979).

Siewiorek, D., and Swarz, R., *The Theory and Practice of Reliable System Design*, Digital Press, 1982.

Spragins, J., Markov, J., Doss, M., Mitchell, S., and Squire, D., "Communication Network Availability Predictions Based on Measurement Data," *IEEE Trans. Communications*, 1482–1491 (Oct. 1981).

Stiffler, J. et al., an Engineering Treatise of the CARE II Dual mode and Coverage Models," Final Report, NASA Contract L-180844, Nov. 1975.

Stiffler, J., and Bryant, L., "CARE III Phase III Report—Mathematical Description," NASA Contract Report 3566, Nov. 1982.

Suri, R., "Robustness of Queueing Network Formulae," *JACM* (July 1983).

Trivedi, K., *Probability and Statistics with Reliability, Queueing, and Computer Science Applications*, Prentice-Hall, Englewood Cliffs, NJ, 1982.

Trivedi, K., "Reliability Evaluation for Fault-Tolerant Systems," invited paper, *Mathematical Computer Performance and Reliability*, G. Iazeolla, P. J. Courois,and A. Hordijk, Editors, Elsevier Science Publishers B. V. (North-Holland), Amsterdam, 1984a, pp. 403–414.

Trivedi, K., "Modeling and Analysis of Fault-Tolerant Systems," *Proc.* International Conference on Modeling Techniques and Tools for Performance Analysis, Paris, May 1984b.

Trivedi, K., and Geist, R., "Decomposition in the Reliability Analysis of Fault-Tolerant Systems," *IEEE Trans. Reliability* (Dec. 1983).

Trivedi, K., Gault, J., and Clary, J., "A Validation Prototype of System Reliability in Life-Critical Applications," *Proc.* Pathways to System Integrity Symp., Nat. Bureau of Standards, 1980.

Trivedi, K., Geist, R., Smotherman, M., and Dugan, J., "Hybrid Reliability Modeling of Fault-Tolerant Computer Systems," to appear, *Int. J. Computers and Electrical Engineering*.

Wensley, J. et al. "SIFT: The Design and Analysis of a Fault-Tolerant Computer for Aircraft Control," *Proc. IEEE, 66* (Oct. 1978).

25. Parallel Execution of Logic Programs

Doug DeGroot
Texas Instruments

Logic programming (Kowalski 1979) has emerged as a surprisingly successful computational paradigm in the last few years, and it is continuing to gain support from many fields within computer science.

Perhaps one of the most promising aspects of logic programming is its seemingly overabundant capability of being parallelized. Many models of parallel logic program execution have appeared in the last few years. Although all of these possess certain key characteristics that distinguish them from each other, the majority belong to either the class of and-parallel execution models or the or-parallel class, or even to both. These two classes, and-parallel and or-parallel, are certainly the most well known and well studied. Space limitations prevent a thorough discussion of either and certainly of both in this chapter; instead, only an introduction to several major models of and-parallelism is presented. These models are compared and contrasted to each other to give the reader an understanding of some of the main problems in and-parallel execution of logic programs and to describe several promising solutions.

PROLOG PROGRAM STATEMENTS

Prolog is based on Horn-Clause logic. Three types of statements (clauses) are allowed in Prolog:

1. Fact.
2. Conclusion ← Premises.
3. ← Query.

The first two are used to build programs; the third is used to interact with existing programs. Facts are simply that. They are unconditional assertions. For example, typical facts in a program might be:

    father(john, tom).
    isa(fido, dog).
    is(jane, pretty).
    disease(type(leukemia), location(blood)).

Facts can be arbitrarily complex as in:

    female(jane, age(20), children(0),
        home(dallas), occupation(student)).

or:

    family(father(tom), mother(sue),
        children ([joe,jane,judy])).

They can also refer to unknown items, as in:

                    likes (X, honey).

where the variable X refers to all persons. Variables in a clause begin with capital letters and are universally quantified. This last fact says that everyone (actually, "everything") likes honey.

The second type of statement is called a "rule." A rule may be regarded as saying, "the (single) conclusion is true if (all of) the premises are true," or, "to prove (show) that the goal is true, prove (show) that the subgoals are true." Typical rules are written as follows:

    grandparent (X,Y) ← parent (X,Z) & parent (Z,Y).
    grandfather (X,Y) ← grandparent (X,Y) & male (X).
    desirable(X) ← single(X) & rich(X).
    leave(X) ← entry(X,Y) & open(Y) & exit (X,Y).

The first rule says that X is the grandfather of Y if X is the parent of some person Z, and that person Z is the parent of Y.

Queries are written in the same way that rules are written but without a conclusion. For example, the query

            ←flight (X, Paris) & cost (X,Y) & lt (Y,850).

can be interpreted as, "Is there a flight to Paris that costs less than $850.00?"

With these three simple statement types plus a variety of special, "built-in" execution primitives, sophisticated and efficient general-purpose symbol-pro-

```
qsort(X.Unsorted,Sorted) ←
    partition(Unsorted,X,Smaller,Larger) &
    qsort(Smaller,S__smaller) &
    qsort(Larger,S__larger) &
    append(S__smaller,X.S__larger,Sorted).
qsort(nil,nil).

partition(X.Xs,A,Smaller,X.Larger) ←
    lt(A,X) &
    partition(Xs,A,Smaller,Larger).
partition(X.Xs,A,X.Smaller,Larger) ←
    ge(A,X) &
    partition(Xs,A,Smaller,Larger).
partition(nil,nil,nil).

append(nil,X,X).
append(X.L1,L2,X.L3) ← append(L1,L2,L3).
```

Figure 25-1.   The quicksort program.

cessing programs can be quickly developed. As an example, Figure 25-1 illustrates the quicksort program written in Prolog.

It is important to notice that in Prolog clauses, the scopes of all variables and subgoal arguments are strictly local. There are no such things as global variables. Thus an X in one clause is entirely different from all other X's in other clauses or in a recursive-clause call. All Xs within the same clause represent the same X, however, as we would expect. The scopes of all the variables, then, are limited entirely to the single clause within which they appear.

Given a simple query from the user, Prolog considers it a goal and tries to solve it. If it can find a fact that "matches" the query exactly, Prolog considers that it has solved the query. If it can find no fact, but it can find a rule whose head "matches" the query, it replaces the query by the subgoals in the matching rule. These subgoals are placed in a goal stack, and the first subgoal in the stack becomes the current goal. Prolog continues solving subgoals until they have all been solved, if possible; if all subgoals cannot be solved, Prolog reports failure for the original query. If it can solve all subgoals, Prolog reports success. Figure 25-2 illustrates this process, although in a simplified manner. The matching process referred to is called *unification;* it is described in detail below. Notice that the goal stack is indeed managed as a stack, resulting in a depth-first search through the program search space for a solution.

In the example, at each step there is one and only one rule or fact that matches a given query (goal). This is rarely the case in "real" programs. Consider for a

Program

1. a ← b & c.
2. b ← d.
3. c.
4. d ← e & f.
5. e.
6. f.

Query: ← a.

Action	*goal stack*
initial query	a
match a with rule 1	b, c
match b with rule 2	d, c
match d with rule 4	e, f, c
match e with fact 5	f, c
match f with fact 6	c
match c with fact 3	empty
done - goal stack empty	

Figure 25-2. The query reduction process.

moment a bedroom with two entries, a door and a window. We can represent this with two facts:

1. entry(bedroom,door).
2. entry(bedroom,window).

Given the query "leave(bedroom)," the Prolog system will attempt to see if the goal can be solved; in other words, if there is a way to leave the bedroom. The previous rule for "leave" can be used:

leave(X) ← entry(X,Y) & open(Y) & exit(X,Y).

Note that "leave(bedroom)" matches "leave(X)" if we substitute "bedroom" for X. Doing this yields a variant of the original rule:

leave(bedroom) ← entry(bedroom,Y) & open(Y) & exit(bedroom,Y).

The goal-solving process described above is allowed to substitute values for variables in its attempt to solve a goal. If during a match a variable in the head of a clause gets changed to a value, all occurrences of that variable in the rest of the clause also get changed. But if a variable in the current subgoal gets changed to a value, all other occurrences of that variable in the goal stack get changed. If two different variables are matched together, either one can be changed into the other. As a result of the given match, the variable X gets assigned the value "bedroom." Every X in the clause gets changed to "bedroom" as a result of this one matching action. To solve the new, resulting "leave(bedroom)" rule, the three subgoals must now be solved. Prolog attempts to solve the three subgoals in the order in which they appear in the goal statement, sequentially, from left to right. (This ordering is important, as will be shown later.) The first subgoal is "entry(bedroom,Y)." Given the previous two "entry" facts, Prolog can solve the "entry(bedroom,Y)" subgoal in either of two ways, by matching Y with "door" or by matching Y with "window." Prolog considers the facts in the order in which they appear in the program, from top to bottom. The first fact states that there is a door that is an entry to the bedroom; to match "entry(bedroom,Y)" with "entry(bedroom,door)," Prolog sets Y to "door" everywhere Y occurs in the goal stack. Since facts are unconditional, no subgoals have to be solved in order to use this fact, and so the first subgoal of the "leave(bedroom)" goal statement has been solved.

Prolog now attempts to solve the second subgoal, which has now become "open(door)." If this subgoal can also be solved, Prolog will attempt to solve the third subgoal. Finally, if it too can be solved, all subgoals will have been solved, and so therefore the original goal of leaving the bedroom will have been solved—by opening the door and leaving through it. At this point Prolog will report success to the user and consider its job done.

Suppose however, that upon discovering that the door was an entry to the bedroom, the "open(door)" subgoal could not be solved, because of the lack of a key, a broken door knob, or something similar. Instead of failing the whole goal, Prolog *backtracks* to the next previously successful subgoal. In this case, Prolog backtracks to the "entry(bedroom,Y)" subgoal. Here it finds that there is another way to solve this subgoal, by using the "entry(bedroom,window)" fact. It sets Y to "window" and then tries to solve the two remaining subgoals with the new value of Y. If it can now find a way to open the window and exit through it, the original goal will have again been solved, but in an entirely different way.

UNIFICATION

As described above, when Prolog looks for a fact or head of a rule that matches with a subgoal, it may have to assign values to some of the variables in the subgoal, in the clause, or both. For example, to solve the "leave(bedroom)" query, the variable X in the "leave(X)" rule was changed to "bedroom." In this way, the subgoal was made to match the rule head. This process is called *unification*. Before unification can be explained, some terminology is required.

```
function unify(term1,term2):substitutions;
begin
  if variable(term1) or variable(term2)
    then return (term1 = term2)
  else if constant(term1) and constant(term2)
          and term1 = term2
    then return ()
  else if functor__of(term1) =
        functor__of(term2) and
        nbr_.args__of(term1) =
        nbr__args__of(term2)
    then begin
      S__tot:=();
      for n:=1 to nbr__args(term1) do begin
        S:=unify(arg(term1,n),arg(term2,n));
        if S = fail then return fail
        else add S to S__tot;
      end;
      return S__tot;
    end;
  else return fail;
end; {unify}
```

Figure 25-3. A unification algorithm.

Constants, numbers, strings, and variables are all examples of *terms*. A *functor* is any alphanumeric symbol, such as "father," "entry," and so on. If "f" is a functor and t_1, t_2, tn are terms, then $f(t_1, t_2, \ldots ,tn)$ is also a term. Terms can be arbitrarily complex, as can be seen from the recursive definition. Unification in Prolog involves an attempt to find a set of variable substitutions that makes two terms equal. A simple unification algorithm is shown in Figure 25-3. More detailed discussions of the unification process may be found in Robinson (1965).

Whenever the current subgoal matches (unifies) with a fact or rule head, any variables that are assigned values through unification become changed to their new value throughout any remaining subgoals in the goal stack or in the rule body, if one. If backtracking occurs, the previous unifications are undone, thereby restoring any changed variables in the goal stack or clause to their previously unassigned state.

With this brief introduction to Prolog, we now turn to the parallel execution of logic programs.

AN INTRODUCTION TO PARALLELISM IN LOGIC PROGRAMS

In this section, a brief introduction to both and-parallelism and or-parallelism is given. A few differences between the two are discussed.

Given a specific subgoal to solve, there may be several different clauses that unify with this subgoal and which can thus potentially be used in satisfying the subgoal. Of all the unifying clauses, some may eventually lead to failure, while others may lead to success. Sequential Prolog interpreters select the first unifying clause in hopes that it will lead to success. If instead it leads to failure, backtracking will occur, and the second unifying clause will be selected. If this second clause also leads to failure, backtracking will force the third unifying clause to be selected, and so on. Execution continues in this manner until one clause leads to success, or until all the clauses have been tried, and all lead to failure. It is important to note that at any particular time only one clause is being investigated to see if it leads to success.

Or-parallelism involves the simultaneous execution of two or more of the clauses instead of just one. Suppose all the unifying clauses are selected for simultaneous execution; because the first clause or the second or the third, etc., may lead to success, and because these clauses are investigated in parallel, this form of parallelism is called or-parallelism. Note that even though only one clause needs to lead to success in order to satisfy the original query, several selected clauses may lead to success, and thus more than one answer may be produced for a given query when in fact only one is needed. It may frequently be desirable, however, as in database applications or in expert systems, to derive all answers. In such cases, every unifying clause must be investigated for every subgoal to see if it leads to success. Or-parallelism clearly offers the potential for producing all these answers faster than sequential Prolog because the different alternatives are pursued in parallel. If each subgoal has, on the average, only a small number of clauses that unify with it, potential speedups may be small, however. Thus, or-parallelism may be best suited for applications that possess many solutions to typical queries.

An important aspect of or-parallelism is that backtracking is eliminated as a basic operation of the execution model.

Instead of investigating various alternative clauses for a subgoal in parallel, *and-parallelism* is concerned with trying to solve two or more subgoals in parallel. For each subgoal, only one clause is investigated at a time, and thus no or-parallelism is exploited. To solve the original query, a series of sets of subgoals will be produced, and for each set, the first subgoal, the second subgoal, and so on must all be solved in order to solve the set—thus the name "and-parallelism." If at any time a subgoal in a subgoal set cannot be solved, backtracking to a previous subgoal set must occur.

In or-parallelism, all the various parallel activities are working on different problems, each a potential solution to the original query. Because the problems are different, no cooperation of communication between the different activities is required. The advantages for parallel processing are clear. Sophisticated management of a large, distributed data space imposes the greatest problems (Cepielewski 1984; Crammond 1985).

```
gfather(G,C) ← father(G,F) & father(F,C).
father(tom,sue).
father(tom,joe).
father(john,sam).
father(joe,jane).
```

Figure 25-4. A grandfather example.

In contrast, all subgoals in a subgoal set in and-parallelism are working on the same problem. Each must succeed if the subgoal set is to succeed. Because the subgoals are working on the same problem, they share a common set of data and variables. A change to one variable by one subgoal may affect the computation of another subgoal, and may even make it fail. But, in addition, when two subgoals share an unbound variable, and both wish to assign it a value, care must be taken to ensure that they assign a common value to this variable. If they assign different values to the variable, and this error goes undetected, an invalid computation may result. This problem, referred to as the "binding conflict" problem, is examined in detail below.

And-parallelism offers the possibility of producing a single solution more quickly than either sequential or or-parallel Prolog, since to produce the single solution, the different parts of the problem will have been solved in parallel. Further, if a program is largely deterministic (i.e., has few matching clauses for any subgoal), it may continue to exhibit large parallelism where or-parallelism might not.

Or-parallelism and and-parallelism can be effectively combined to take advantage of both, and such models have been proposed (Shapiro 1983). For each type of parallelism, many different forms exist. Some of the major forms of and-parallelism are investigated in the remainder of this chapter.

THE BINDING CONFLICT PROBLEM

Consider a program for calculating the grandfather of a given person, as in Figure 25-4. If a call is made to this program of the form "gfather(tom,Z)," then this subgoal query will be reduced to the two subgoals "father(tom,F)" and "father(F,Z)." When the first subgoal executes, it will unify with "father(tom,sue)," thereby binding F to sue and making the second subgoal change to "father(sue,Z)." No clause can be found to unify with this new second subgoal, and so this subgoal fails. F is unbound, and backtracking to the first subgoal occurs. The first subgoal is retired; this time, it unifies with "father(tom,joe)," binding F to joe, and changing the second subgoal to "father(joe,Z)." Now the second subgoal unifies with "father(joe,jane)." All the subgoals have been solved, and so the original query has been solved. The query is rewritten with the needed variable binding noted; consequently, the reply to the query is "gfather(tom,jane)," and indeed, tom is the father of jane, according to the program.

Notice the flow of variable values in this program. With the given query, Z is unbound upon entry to the grandfather clause. It becomes bound to C, but C also has no value. When the first subgoal executes, a binding for F to "sue" is made, altering the second subgoal. The second subgoal fails with this first binding value and backtracks into the first subgoal. The first subgoal finds another binding for F, and the second subgoal again attempts to execute with this new binding. The second time, it succeeds. In essence, the two subgoals have a typical "producer–consumer" relationship. The first subgoal produces values for F, and the second subgoal consumes them. Remember that they execute sequentially, one after the other.

Suppose instead of executing sequentially, both subgoals are allowed to execute in parallel. After the query "gfather(tom,Z)" is made, the two subgoals become "father(tom,F)" and "father(F,C)." The first subgoal will first unify with "father(tom,sue)," binding F to "sue." The second subgoal will also first unify with "father(tom,sue)," binding F to tom and C to sue. Because both subgoals succeeded, the query will be mistakenly assumed to have succeeded, and the answer to the query will be reported as "gfather(tom,sue)"—clearly a wrong result. The problem is that the two subgoals both bound a value to the variable F, but they bound different values. In order for the computation to be valid, all subgoals sharing a variable must agree on a value to be bound to that variable, if any. (They can all agree simply to leave it unbound, which is valid.) By breaking the "producer–consumer" relationship between the two subgoals, neither checks the values produced by the other, and so different bindings can occur. This problem is called the "binding conflict" problem and is the main problem to be overcome in and-parallel execution models of logic programming.

SETS IN AND-PARALLELISM

One approach to solving the binding conflict problem is to collect all answers from each subgoal and retain only those answers that agree in their assignments to shared variables. The different subgoals may assign many different values to the shared variables, but only those values that agree across all subgoals are returned. Differing values are thrown away, thus avoiding binding conflicts. For example, consider the clause:

$$f(X) \leftarrow g(X) \ \& \ h(X).$$

A parallel Prolog interpreter could invoke $g(X)$ and $h(X)$ in parallel, each of which would return all possible answers instead of simply returning one answer, as in sequential Prolog. After both subgoals are complete, a set-intersection of their two answer sets is computed, and this intersection is returned as the answer for the whole clause—as the answer for $f(X)$. In general, a set-intersection operation is inadequate; instead, a join operation of the relational algebra is required (Codd 1970). The join operation is performed over the set of shared variables.

Consider the grandfather example of Figure 25-4. Suppose a "gfather(tom,Z)" query is made; then the two subgoals become "father(tom,F)" and "father(F,Z)," and the two subgoals share the variable F. Both subgoals are activated in parallel, and both return all possible answers. The first subgoal returns the answer set F = {sue,joe}; the second subgoal unifies with every father clause and so returns the answer set (F,Z) = {(tom,sue), (tom,joe), (john,sam), (joe,jane)}. After the join operation over the variable F, the only answers remaining are (F,Z) = {(joe,jane)}. Because in the original query only the variable Z was unspecified, only the values of Z are returned as answers for the query. (F is a local, or temporary, variable.) Because the only value of Z in the joined answer set is jane, the answer set returned for the query is simply Z = {jane}. So tom is the grandfather of jane, as before.

This join technique can be improved by allowing the join to be computed dynamically instead of statically. As a member of the join result is computed, it can be returned immediately as an answer to the clause. When an answer is returned by one subgoal, a check is made to see if the other subgoal has already returned the same answer. If so, then this answer can be immediately returned as an answer to the whole clause. If not, it is saved for future comparisons. In other words, an answer collection algorithm similar to that in Figure 25-5 is executed for each subgoal activation. This method works by invoking all subgoals in parallel and then collecting the answers from each subgoal as they are produced. Each answer produced must be checked to see if the other subgoals agree that it is an answer. If they do, the answer can be immediately returned; if they do not, the answer must be saved for possible future agreement. This is, of course, an oversimplification, as again a full join operation must in general be used; but instead of being computed statically, the join is computed dynamically.

An important aspect of this "dynamic join" technique is that the answers are pipelined back to the original query from every activated subgoal. As soon as an answer is computed, it is returned. Consequently, the first answer to the query may become available much sooner than in the "static join" technique. The static join technique computes all answers to a subgoal before any are returned, and then they are returned as a single set. As a final, significant optimization, the dynamic join technique can be made demand-driven instead of data-driven, thereby producing even greater efficiencies and response times.

1. start all clause subgoals in parallel;
   set the answer set for each subgoal to nil
2. collect an answer from any ready subgoal;
   if no more answers exist, then quit
3. is this answer in all the other subgoals'
   answer sets? if yes, goto step 5
4. if not, add this answer to the answer
   set for this subgoal; go to step 2
5. all subgoals have returned this answer;
   return this answer as an answer for the
   whole clause; go to step 2

Figure 25-5.   A set-collection algorithm.

The previous example is misleadingly simple. Consider again the clause:

$$f(X) \leftarrow g(X) \ \& \ h(X).$$

Suppose both $g(X)$ and $h(X)$ require a significant amount of computation time. Then if both are activated in parallel, $h(X)$ can be working while $g(X)$ is working. When their answers are produced, a simple join operation is performed (in this case, a set intersection), and the result is passed back as the result of $f(X)$. Clearly the potential for speedups exists. Unfortunately, the potential for great loss of speeds also exists. Suppose $g(x)$ can produce only one answer, but that $h(x)$ can produce many thousands of answers. After g produces its single answer, the system must wait for h to produce its entire answer set. In sequential Prolog, the answer produced by g is "forwarded" to h, and h executes with only this one value. If h is successful with this value, the value is returned as the value of f; if it is unsuccessful, backtracking into g occurs. But because g can produce no further answers, g fails, thereby causing f to fail. Subgoal h does no unnecessary work in sequential Prolog. In the parallel join techniques, however, it may perform lots of extra work, as this example illustrates. Furthermore, suppose g fails outright. It may then become necessary or desirable to track down and destroy h, and all its descendants. This itself may prove costly if h has spread itself out over many processors. This problem does not arise in sequential Prolog.

It must also be stated that the parallel join techniques may suffer from severe storage requirements because all answers from all subgoals must be maintained until the entire join has been computed; only then may these answers be thrown away. And, finally, problems may easily arise in dealing with infinite relations represented by subgoals. For instance, consider the clause:

$$f(X,Y) \leftarrow sum(X,1,Z) \ \& \ times(Z,3,Y).$$

and the query "f(5,A)." The "sum" subgoal will return only the value 6 for Z; but the "times" subgoal may return an infinite set of answers because neither Z nor Y will have values upon its invocation. The possibility for great inefficiencies is clear. Such clauses must clearly execute with at least partial sequentiallity.

CONCURRENT PROLOG

Concurrent Prolog (Shapiro 1983) is a logic programming language with considerably different semantics than sequential Prolog. It embodies and-parallelism, or-parallelism, variable annotations, and guarded-clause nondeterminism (Dijkstra 1975). Because of the and-parallelism, each time a clause is invoked, all subgoals are activated in parallel. However, unlike the previous join technique, Concurrent Prolog retains the producer–consumer relationship between subgoals sharing a variable. Which subgoals can produce values for a variable and which can consume values depends on how the variables are annotated. The specification "X?" denotes that this occurrence of the variable X is read-only; the specification "X" denotes that this occurrence of X may be either read or written. Consider the clause:

$$f \leftarrow g(X) \ \& \ h(X?) \ \& \ k(X?).$$

When activated, all three subgoals will begin executing in parallel. Because the variable X is annotated with a "?" in the calls to subgoals h and k, these subgoals cannot bind values to X during unification—they can only read the value of X. The X in subgoal g(X) is not annotated, and so g can bind a value to X if it so desires. It is possible to unify a read-only annotated variable with another variable, but not with a nonvariable value. If an attempt is made to unify a read-only variable with a nonvariable value, the subgoal attempting the unification is suspended. The subgoal is awakened and allowed to reattempt unification only when and if the read-only variable is finally assigned a value by some other, value-producing subgoal.

Consider the program:

```
f ← g(X) & h(X?) & k(X?)
g(2).
g(3).
h(4).
h(2).
```

k(2).
k(6).
k(3).
← f.

When the clause for f is invoked, all three subgoals begin executing in parallel. At the precise moment they begin execution, X has no value. Consider the subgoal h(X?). As it begins execution, it attempts to unify with every h clause in parallel, as a result of the full or-parallelism. Each of the h clauses in the program contains only nonvariable values, and so the subgoal "h(X?)" suspends because X has no value yet. If execution were not suspended and unification proceeded, X would be bound to a nonvariable value, violating its read-only status. The subgoal "k(X?)" similarly suspends.

The subgoal "g(X)," however, does not suspend, as X is not annotated read-only. Attempts are made in parallel to unify g(X) with each clause for g. Of all those that do unify successfully, exactly one is selected and allowed to succeed; the others are aborted. Suppose the clause "g(2)" is selected to succeed. As a consequence, X becomes bound to the value 2. At this point, the two suspended subgoals, h and k, are awakened and allowed to proceed. Since X is now bound to 2, the subgoals can unify with only the clauses "h(2)" and "k(2)." Because all three subgoals have succeeded, the entire clause succeeds. The subgoal g will have produced the value for X, and h and k will have consumed it.

Suppose that "g(3)" had been selected to succeed instead of "g(2)." When h and k reawaken, they will find X bound to 3 instead of 2. Subgoal k will still succeed, but h will now fail. Consequently the whole clause fails. Backtracking into g to produce another value for X is not allowed because each subgoal is allowed to produce at most one value for any variable. If that value leads to failure, backtracking to produce another value cannot occur. This is clearly different from Prolog.

A subgoal can, however, produce a long, even infinite, "stream" of values. For instance, the clause:

integers (N,N.R) ← sum (N,1,K) & integers (K,R).

generates a single value for the list L when called with "integers (1,L)." This value, however, is the infinite list of positive integers. These integers can be consumed by other clauses. Consider:

squares (N.R) ← times (N,N,Nsq) &
write (Nsq) &
squares (R).

and the query:

$\leftarrow$ integers (1,L) & squares (L).

The "squares" clause should write out the infinite set of squared integers. But does it work? Because no variables are annotated, no subgoal suspensions will occur, and no producer–consumer relationships can be guaranteed. In particular, the squares subgoal does not necessarily consume values produced by integers. And because "squares" calls itself recursively without bound, an unbounded number of "times" and "write" subgoals can be created in parallel. But here too no producer–consumer relationships can be guaranteed, and instead of writing the Nsq value produced by "times," the write subgoal can simply write out an unbound value of Nsq. This is obviously an error. In addition, if "times" is allowed to execute before receiving a value for N, a correct value for Nsq is unlikely to be produced.

To overcome those problems, producer–consumer relations can be reintroduced by rewriting the clause as follows:

squares (N.R) $\leftarrow$ times (N?, N?, Nsq) &
  write (Nsq?) &
  squares (R).

Now "times" cannot execute until a value has been assigned to N by "integers," and "write" will not execute until "times" has completed and assigned a value to Nsq. The program now works correctly, or almost so. The problem is that there is no imposed ordering on all the potentially parallel write statements. Instead of producing "1, 4, 9, 16, 25, . . . " the program might produce instead "49, 4, 16, . . . " or any other ordering due to the timing vagaries of parallel execution. To overcome problems such as these, Concurrent Prolog offers a special "commit" operator. It can be placed between subgoals in the body of a clause. Consider the clause:

squares (N.R) $\leftarrow$ times (N?, N?, Nsq) &
  write (Nsq?) |
  squares (R).

In a clause with a commit, all subgoals preceding the commit are first executed in parallel; if they all succeed, all subgoals following the commit then execute in parallel. In the above "squares" clause, the recursive call to "squares" is held up until the current squared integer is written out. The commit acts as a synchronizing operator, forcing, in this example, the squared values to be written in order. Note that in this last clause, unfortunately, all three subgoals execute sequentially, and no and-parallelism is exploited. There is and-parallelism, however, through the parallel execution of both the "integers" and the "squares" clauses. If the order of the written squared values were unimportant, the commit would not be needed, and "squares" could call itself without bound and without waiting for previous "times" and "write" subgoals to complete. Consequently, an unbounded number of "times" and "write" subgoals could be created in parallel. Each "times" subgoal would wait for a particular integer element in the list of integers created by "integers," and each "write" would wait for an Nsq value from its corresponding "times" subgoal. If it is necessary to restrict this "running ahead" of "squares" in order to conserve memory, for example, the following alternate clause and query could be used:

$$\text{squares (N.R)} \leftarrow \text{times (N,N,Nsq) \&}$$
$$\text{write (Nsq?) \&}$$
$$\text{squares (R?)}$$
$$\leftarrow \text{integers (1,L) \& squares (L?).}$$

Now because a "squares" subgoal cannot proceed until its parameter has at least a next integer, the number of "squares" subgoals can never exceed the number of unconsumed integers.

The subgoals to the left of a commit are called "guards." When a subgoal attempts execution, the clauses with which the subgoal unifies are selected, and their guards are all executed in parallel. The guards of some clauses will succeed, others will fail, and others will suspend. Of those that succeed or later reawaken and later succeed, one is selected to "commit" and is allowed to continue; the others are aborted, even if they have not completed executing their guards. The unified values from the selected, committing clause, if any, are propagated outward to all consuming subgoals; these subgoals are reawakened if they have suspended.

Figure 25-6 illustrates a parallel quicksort algorithm written in Concurrent Prolog. Each invocation of the qsort clause activates three subgoals in parallel. Each of these represents a process in the traditional sense; the processes communicate with each other over the communication channel represented by their shared variables.

Concurrent Prolog has proven itself highly successful as a process-structured programming language. It bears much resemblance to Hoare's Communicating Sequential Processes language (Hoare 1978) but enjoys the many additional benefits of logic programming. It obviously relies upon and-parallelism to activate the different processes in a computation and utilizes shared variables as communication channels between the processes. It should also be noted that Parlog (Clark 1984) is another successful logic programming language that uses variable annotations, guards, and and-parallelism.

RUN-TIME GOAL ORDERING

Conery (1983) developed an and-parallel execution model that attempts to automatically discover the producer–consumer relationships among subgoals in-

```
op('-',rl,50).

quicksort(Unsorted,Sorted) ←
    qsort(Unsorted,Sorted-nil).

qsort(X.Unsorted,Sorted-Rest) ←
    partition(Unsorted?,X,Smaller,Larger) &
    qsort(Smaller?,Sorted-X.Sorted1) &
    qsort(Larger?,Sorted1-Rest).
qsort(nil,Rest-Rest).

partition(X.Xs,A,Smaller,X.Larger) ←
    lt(A,X) | partition(Xs?,A,Smaller,Larger).
partition(X.Xs,A,X.Smaller,Larger) ←
    ge(A,X) | partition(Xs?,A,Smaller,Larger).
partition(nil,,nil,nil).
```

Figure 25-6.   The quicksort program in Concurrent Prolog.

stead of relying upon user annotation of variables, as in Concurrent Prolog (Conery 1983). This model derives at run-time a parallel execution graph for each clause invoked. Because clauses can be invoked in several different ways, several different execution graphs can be derived for any given clause. In addition, as subgoals within a clause execute, the derived execution graph for the clause may have to be modified if certain exit conditions of the subgoals are not met. These points are explained below.

First consider how the producer of a variable is determined. Consider the following three subgoals:

father (tom,bob)
father (tom,X)
father (X,Y)

The first subgoal cannot produce any values because it has no unbound variables as arguments; it is called a "verifier" subgoal because its only job is to verify whether or not tom is indeed the father of bob. The second subgoal, however, is intended to find whom tom is the father of and return that person as the value of X. This subgoal will produce a value of X (if one exists). More than one value may exist, but it is unlikely that there will be many such values. Such is not the case with the third goal because neither of the arguments is bound to value. A potentially large number of X–Y value pairs may be returned by this last subgoal. Because the last two subgoals are supposed to "do" something—return one or more values—they are called "doers." Obviously a "doer" is also a "producer."

Consider again the grandfather clause:

$$\text{gfather } (X,Y) \leftarrow \text{father } (X,Z) \ \& \ \text{father } (Z,Y).$$

and the following query:

$$\leftarrow \text{gfather } (V, \text{ tim}).$$

We would now like to find the grandfather of tim. As a result of the query, the two subgoals become "father(V,Z)" and "father(Z,tim)." The first subgoal, when executed, has neither argument variable bound to a value; hence any father–son pair may be returned as a solution. If the value of Z returned is not the father of Y (tim), the second subgoal will fail, and backtracking into the first subgoal will occur. A second father–son pair will be returned, and so on. Clearly many father–son pairs may be returned by the first subgoal before the second subgoal finds an acceptable value. The potential for great execution inefficiencies is obvious.

Suppose the subgoals could be executed out of order. Then it would appear better to execute the second subgoal first because it will make the actual call:

$$\text{father } (Z,\text{tim}).$$

This subgoal is likely to be able to return only one value for Z since tim most likely has only one father. Suppose Z = john is returned. When the first subgoal is now executed, it makes the actual call:

$$\text{father } (X,\text{john})$$

For the same reasons, this subgoal is also likely to be able to return only a single value for X. If any failure and backtracking occurs, it is unlikely that additional values for either X or Z can be found, and so the query will either quickly fail or succeed. It is clearly more economical to execute the two subgoals in reverse order.

But what about the query:

$$\leftarrow \text{gfather (joe,V)?}$$

Similar reasoning leads us to believe we should now execute the two subgoals of the gfather clause in their original order instead of reversing them. What this goal reordering effectively does is to prevent a "producer" from producing too many values in order to improve efficiency of execution. An obvious goal-ordering heuristic would be to execute first those subgoals with no unbound variables, then those with the fewest unbound variables, and so on. Hopefully, execution costs would thus be minimized.

More precisely, for each variable in the clause, find all subgoals that use that variable and order them by increasing number of unbound variables. Thus the subgoals:

$$f(X,Y) \ \& \ g(2,Y,3) \ \& \ h(Y,V)$$

would be reordered as:

$$g \ (2,Y,3) \ \& \ f \ (X,Y) \ \& \ h \ (Y,V)$$

for the variable Y.

Select the leftmost subgoal and mark it as the producer of the variable; mark the remaining subgoals as consumers of that variable. In the example, g is marked as the producer of Y, and f and h are marked as consumers of Y. This process is carried out for every variable in the clause. After this step, every variable will have been assigned a producer, and certain subgoals will be marked as consumers of values produced by other subgoals. For example, the clause:

$$f \leftarrow g \ (X,Y) \ \& \ h \ (Y,Z) \ \& \ k \ (X,Z)$$

will yield g as the producer of X and Y with h as a consumer of Y, k as a consumer of X, h as a producer of Z, and k as a consumer of Z. These producer–consumer relationships can be represented graphically:

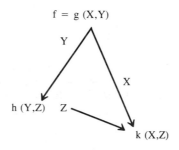

Consider the clause:

$$f (X) \leftarrow g (X) \ \& \ h (X) \ \& \ k (X).$$

We derive the following execution graph:

f (X) = g (X)

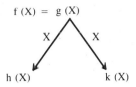

Subgoal g is marked as the producer of X, and subgoals h and k as consumers. But suppose g executes and binds X to a nonground value, such as $X = p(3,q(V))$. Clearly in such cases h and k cannot execute in parallel, as the potential for binding conflicts now exists. The problem lies in the fact that g did not produce a ground value for X; it produced only a partial value.

It can be seen, then, that after a designated producer executes, the value produced must be examined to see if it is ground. If it is not, then this value cannot be forwarded to all consumers in parallel. Instead, a "goal reordering" algorithm must be executed in order to select one of the intended consumers and mark it as the new producer. In the above examples, if g does not produce a ground value for X, h will be designated as the new producer of X, and the execution graph will be reordered as follows:

$$f (X) = g (X)$$
$$\downarrow X$$
$$h (X)$$
$$\downarrow X$$
$$k (X)$$

Suppose now that a query to the above clause is made of the form $\leftarrow f(5)$. Then the three subgoals become g(5), h(5), and k(5). Clearly no subgoals are producers of values. In this case, the clause head is said to be the producer, and the three subgoals are all consumers, yielding the following execution graph:

f (5)

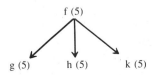

g (5)   h (5)   k (5)

Finally, consider this clause:

$$f(X,Y) \leftarrow g(X) \ \& \ h(Y).$$

The query $\leftarrow f(3,4)$ yields:

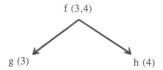

The query $\leftarrow f(V1,V2)$ yields:

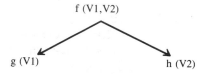

if V1 and V2 are two totally different variables. But the queries:

$$\leftarrow f(X,X)$$
$$\leftarrow f(X,t(2,X))$$

both yield:

$$f(X,Y)$$
$$\downarrow$$
$$g(X)$$
$$\downarrow$$
$$h(Y)$$

since the two input arguments share a common variable.

It should be clear that it is not the source code variables that can be used to derive the execution graphs. Instead, it is the run-time values of those variables. Consequently, the execution graphs cannot be determined until run-time; and because the values of the variables can change as subgoals execute, execution graphs may have to be modified throughout clause execution. The run-time calculations and reorderings of execution graphs are potentially very expensive. In addition, run-time values must be scanned frequently to see if they are ground or contain only unshared variables. This procedure too is potentially very expensive. Finally, it also is not possible to determine which goal a failing subgoal should backtrack to before run-time.

But Conery's model is important for several reasons. First, if parallelism exists, it will be found and correctly utilized. Because of the goal reordering, this model is potentially much more efficient than traditional Prolog. Finally, it retains much of the semantics of traditional Prolog, unlike Concurrent Prolog. Thus it provides the benefit of being able to execute in parallel and without modification the growing body of traditional Prolog programs (or to do so with only minimal modifications).

STATIC DATA-DEPENDENCY GRAPHS

In the "static data-dependency (SDD) graphs" model (Chang et al, 1988), each procedure is assigned an entry mode and an exit mode. The entry mode specifies the state of each head variable upon entry, whether it is ground (G), independent (I), or dependent on some other head variable in the clause (D). Once the procedure completes, the variables may have changed, so that an unground variable may have become ground, two previously independent variables may have become dependent on each other, and so on. The exit mode specifies the state of all variables on exit from a clause. Once the entry and exit modes of each procedure have been calculated, the entry and exit modes of each subgoal in a given clause are known, and the data-dependency graph for the clause may be computed.

Consider the first clause of the append procedure:

$$append(nil,X,X).$$

Depending on how the procedure call to append is made, that is, depending on the state of the variables at call-time (ground, unbound, etc.), different entry modes may result. However, for append, the first two arguments are usually bound to ground terms (lists), and the third argument is usually unbound. In such a case, the entry mode to append will be:

$$mode(append,3,G\text{-}G\text{-}I).$$

Here, "3" refers to the arity of append. The "G-G-I" states that the first two arguments are ground, and that the third is independent. (It is impossible to have G-G-D.) Given this entry mode, the program analyzer determines that the exit mode of the clause:

$$append(nil,X,X).$$

becomes G-G-G through unification. Other entry modes may give rise to different exit modes.

Consider the second clause of the append procedure:

$$append(X.Y,Z,X.L) \leftarrow append(Y,Z,L).$$

The analyzer begins by assigning each head variable to one of three sets: the G (grounded) set, the D (dependent) set, or the I (independent) set. The initial assignment ({G}, {D}, {I}) is ({X, Y, Z}, {−}, {L}). From this variable assignment, the entry mode of the append subgoal (on the left-hand side of the second clause) can be calculated: it is "mode(append,3,G-G-I)," as before. Because this clause is a tail-recursive clause, we know it cannot terminate on itself and must therefore terminate on some other clause in the same procedure. In this case, there is only one such clause—the first one. Because the exit mode of this first append clause is now known to be G-G-G for entry mode G-G-I, the exit mode of the left-hand-side append subgoal can be determined; it too is G-G-G, indicating that the variable L has now become ground. The exit mode of the entire second clause can now be computed to be G-G-G. There are now two exit modes for the append procedure, one for the first clause and one for the second. In this case they are the same, so the exit mode of the entire append procedure is G-G-G if the entry mode is G-G-I. Different entry modes may give rise to different exit modes. In such cases, the analyzer takes the worst of all exit modes as the exit mode of the procedure, irrespective of the call. For calculating worst case, I is worse than G, and D is worse than I.

Consider the following procedure:

f(X,3).
f(4,Z).
f(X,Y) ← f(Y,X).

Suppose the initial entry mode to the procedure f is to be I-I, as in the call f(A,B), where both A and B are unbound, different variables. The exit mode of the first clause is I-G, and the exit mode of the second is G-I. Now consider the third clause, entered with mode I-I. This rule cannot terminate upon itself, but must terminate on one of the other two clauses. Depending upon which clause it terminates on, the exit mode of the subgoal and hence of the third clause itself is either I-G or G-I. The analyzer must make the worst-case assumption, however, and claim that the exit mode is in fact I-I. In other words, given the call f(A,B), upon completion of f, sometimes A will be ground and B unbound, while other times B will be ground and A unbound. But because the analyzer cannot predict which, it assumes the worst case and decides that both will be unground on exit. In addition, it easily determines that the two terms are independent; hence the exit mode I-I.

Now consider adding the following clause to the above clauses for f:

p(X) ← f(X,Y) & g(X,Y) & h(Y).

If p is entered with X unbound, then the entry mode of f will be I-I. (Clearly X and Y cannot be dependent because they can share no variables; this is due to Y's being strictly a local variable. So the entry mode must be I-I and not D-D.) Given the entry mode of I-I for f, the actual exit of mode of f might be either G-I or I-G, as explained above. If it is I-G, then Y will be ground upon

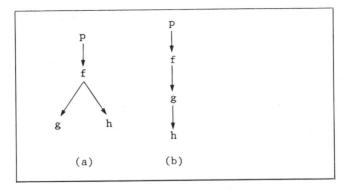

Figure 25-7.   Best- and worst-case execution graphs.

f's termination, and both g and h should be allowed to execute in parallel. However, as stated before, the worst-case mode of f is computed to be I-I, and thus g and h are prevented from executing in parallel, as they might share an unground Y. Thus the best possible actual execution graph is that shown in Figure 25-7(a), whereas the SDD model will achieve only the execution graph of Figure 25-7(b) because of its static, worst-case analysis.

The analyzer processes each clause and every call to determine the worst case of all possible exit modes for all given entry modes. This analysis is nontrivial but can be performed at compile-time (Chang et al. 1985). It is comparable to Mellish's solution for automatically generating I/O modes in Dec-10 Prolog (Mellish 1981). Given the worst-case exit modes for each procedure, a single, worst-case data-dependency graph can be calculated for each clause. Given such a worst-case data-dependency graph, it can be ensured that if any two subgoals execute in parallel, they share no unbound variables whatsoever. Binding conflicts can thus be avoided. Significantly, only one data-dependency graph is computed, and that at compile-time. This assumption of worst-case exit mode is the fundamental and-parallelism restriction mechanism of the SDD model.

RESTRICTED AND-PARALLELISM

In the "Restricted And-Parallelism" (RAP) model (DeGroot 1984), each logic programming clause is transformed into a single execution-graph expression. Each expression may contain several nested subexpressions. Many of these subexpressions can be executed either sequentially or in parallel, depending on the outcome of certain simple run-time tests. As a result, even though each clause is compiled into a single execution-graph expression, more than one run-time execution graph may actually result from a given execution-graph expression. The RAP model attempts to predict a sort of "best case" scenario, derive an execution-graph expression to represent this best case, and embed tests in the code to ensure that the best case actually exists on a given call. If these tests fail, the best case does not exist, and execution reverts to one of several less desirable cases. In the extreme, the worst-case execution might actually result. In fact, the worse-case execution of the RAP model might actually be worse than the assumed worst case of the SDD model since the SDD model examines all calling environments of a clause; the RAP model requires no such examination. Consequently, in a given program the actual calls to a given procedure may all be much better than the RAP model can detect. For instance, suppose a given procedure g(X,Y) is always called with both X and Y independent; the SDD model might detect this, whereas the RAP model might not because the RAP model assumes any two complex, nonground terms are dependent on each other. Clearly this will frequently not be so. In such an event, the RAP model might exhibit less parallelism than the SDD model.

The RAP model consists of three basic components: (1) the typing algorithm, (2) the independence algorithm, and (3) the execution-graph expressions. Each is described below.

THE TYPING ALGORITHM

The typing algorithm is responsible for monitoring the types of all run-time terms (the data items of the program). The goal is to determine when a term has become ground. Each term is assigned one of three type codes: G, for ground; V, for variable (or unbound); and U, for unground and nonvariable. Terms are recursively typed in all their inner components. These type codes are in addition to the normal type codes used by Prolog compilers, such as list, variable, constant, and so on (Warren 1983). On entry into a clause, each U-type (unground) term is flagged as "pending"; then upon successful completion of the clause and just before exiting, these pending terms are examined to see if they have become ground. This is done simply—by inspecting all the top-level components of the term. If all the top-level components have become ground (type G), then the entire term has become ground and of type G. But if at least one top-level component is not ground, the term as a whole is unground—type U. When a variable becomes bound to a term, the variable assumes the type of the unifying term.

The typing algorithm clearly can overlook a ground term and set its type to U when it might have been able to set the type to G if it had been willing to do a more detailed examination of the term and found that the term was ground. There are many ways in which this can occur, although it remains to be seen how frequently these conditions arise in actual programs. Nevertheless, the typing algorithm is approximate. It is extremely efficient and simple instead of being thorough. Any inaccuracies introduced by the approximation technique of the typing algorithm may result in a loss of parallelism, as explained below. However, when it errs, it errs on the side of safety. It is much better to assume a term is nonground and force sequential execution than to assume incorrectly that it is ground and let two subgoals execute in parallel and inconsistently modify the term.

THE INDEPENDENCE ALGORITHM

The independence algorithm is used to determine when two terms are independent, that is, when they share no unbound variables. By definition, two ground terms cannot share any variables, nor can any ground term and any other term. Two different, unbound variables can share no variables because they have no values whatsoever. Two unground, complex terms might share an inner variable, however. To determine whether they do, run-time traversals of these two terms might be required, as in Conery's scheme. The independence algorithm of the RAP model avoids the run-time traversal and simply assumes the two terms are dependent, even though they may not be. A variable and an unground, complex term will similarly be assumed dependent. Clearly the independence algorithm is also an approximate one. This algorithm is shown in Figure 25-8.

The independence algorithm may err because of its simplistic assumption of dependence when unable to prove independence. In addition, when the typing algorithm errs, it may also cause the independence algorithm to err by assuming dependence when independence might otherwise have been detected. The effect of these errors on parallelism is described below.

EXECUTION-GRAPH EXPRESSIONS

The execution-graph expressions are used to express the potential parallelism, if any, inherent in a logic program clause. Certain tests are embedded within the expressions to ensure at run-time that the proper conditions for parallelism exist. Two utility predicate routines are provided for testing terms at run-time: GPAR(X1, . . . ,Xn) and IPAR({V1, . . . ,Vk}, {Vm, . . . ,Vp}). The value of GPAR(X1, . . . Xn) is true if all of the arguments to GPAR are ground terms. If any argument is nonground, the value of GPAR is false. Similarly, the value of IPAR({V1, . . . ,Vk}, {Vm, . . . ,Vp}) is true if variables V1 through Vk are all mutually independent from variables Vm through Vp; but if any two are interdependent, the value of IPAR is false. GPAR simply checks the type code of each of its arguments to ensure that they are all G; IPAR uses the independence algorithm. Both take an arbitrary number of arguments, but IPAR requires at least two. Both are extremely simple and inexpensive.

Six types of execution expressions are allowed:

1. G
2. (SEQ E1 . . . En)
3. (PAR E1 . . . En)
4. (GPAR(X1, . . . ,Xk) E1 . . . En)
5. (IPAR(V1, . . . ,Vk)(Vm, . . . ,Vp)E1 . . . En)
6. (IF E1 E2 E3)

G is any single goal. An SEQ expression states that its subexpressions are to execute sequentially; a PAR expression states that they are all to execute in parallel. A GPAR expression states that if the terms in the GPAR call are all

```
if type(arg1) = G or type(arg2) = G
    then independent else
if type(arg1) = type(arg2) = V and
    address(arg1) ≠ address(arg2)
        then independent else
{assume} dependent;
```

Figure 25-8.   Independence algorithm.

ground, the following subexpressions are to execute in parallel; but if any one of them is not ground, the subexpressions are to execute sequentially. Similarly, an IPAR expression states that if the terms in the IPAR call are independent, then the subexpressions are to execute in parallel; otherwise, they must execute sequentially.

For example, consider the following expression:

$$f(X) \leftarrow g(X) \ \& \ h(X) \ \& \ k(X).$$

and the corresponding execution-graph expression:

$$(GPAR(X) \ g(X) \ (GPAR(X) \ h(X) \ k(X))).$$

Depending on the type of term bound to X, if any, upon entry to f and exit from g, any of the three execution graphs of Figure 25-9 can occur.

If X is ground upon entry to f, all three subgoals will execute in parallel. But if X is unground, g must first execute sequentially. Upon completion of g, another test is made to see if X is now ground; if so, the two remaining subgoals execute in parallel, but they execute sequentially otherwise. Thus with the one graph expression above, three different run-time execution graphs may result. Unlike the SDD model, here the best case and the worst case are both achievable.

LIMITING THE AND-PARALLELISM

In this section, several ways in which the two models restrict the and-parallelism of logic programs are described and compared.

First, consider the clause:

$$f(X,Y) \leftarrow g(X,Y) \ \& \ h(X) \ \& \ k(Y).$$

In the SDD model, if the analyzer calculates that in the worst case the exit mode of g is D-D, then the two subgoals will execute sequentially. An exit mode of D-D means the two terms represented by X and Y "may" be unground upon exit and "may" share a variable. (Remember that just because the exit mode is D-D, it is not necessarily the case that either term is unground, or that they are independent. The exit mode represents the worst case that could arise.) If X and Y do share a variable, then a binding conflict might possibly arise if h and k

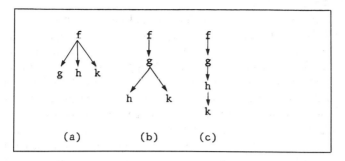

Figure 25-9. Possible execution graphs.

execute in parallel. Consequently, the SDD model forces them to execute sequentially. Many other entry modes for f might be possible and exist within the given program. In fact, potentially 99.9% of all calls to f might be made with X and Y both ground. Then the exit mode of g will usually be G-G (both X and Y still ground), and so h and k should be allowed to execute in parallel in these cases. The SDD model, however, forces h and k to execute sequentially if even one call to f forces the exit mode to be D-D. This is truly a worst-case analysis.

The RAP model will compile the clause into the expression:

$$(GPAR(X,Y)\ g(X,Y)\ (IPAR(X,Y)\ h(X)\ k(Y))).$$

Now if f is called with both X and Y ground, all three subgoals will execute in parallel. In all other cases, g will first execute alone; a subsequent test will then determine whether or not h and k can execute in parallel, in other words, whether the exit mode of g is I-I, G-I, I-G, or even G-G. If not, then h and k must execute sequentially. Again, although only one execution-graph expression is computed, three different run-time execution graphs may result, from the worst case to the best case.

First, notice that the additional flexibility of the RAP model is gained at the expense of additional run-time support. The RAP model requires run-time support in the form of the GPAR and IPAR tests as well as the continual execution of the typing algorithm. Fortunately, these tests are extremely simple, each requiring only a few machine instructions. But they are overhead, nevertheless. Also, sometimes redundant tests may be made in a graph expression, as in the previous graph expression when all head variables are ground upon entry. The cost of these additional tests is, however, quite nominal and may be more than compensated for by the increased parallelism obtained.

The execution-graph expressions of the RAP model themselves limit the obtainable parallelism. Consider the clause:

$$f(X,Y) \leftarrow p(X)\ \&\ q(Yu)\ \&\ s(X,Y)\ \&\ t(Y).$$

The compiler might produce the following execution-graph expression:

$$(GPAR(X,Y)\ (IPAR(X,Y)\ p(X)\ q(Y))$$
$$(GPAR(Y)\ s(X,Y)\ t(Y))$$

Suppose that X and Y are both unbound but different variables upon entry to f. Then the first GPAR test fails, and the two subexpressions will execute sequentially. The IPAR test of the first subexpression succeeds, and so the two subgoals p(X) and q(Y) begin executing in parallel. Suppose that q(Y) is completed first, and binds Y to a ground value. Because of the semantics of the GPAR expression, no part of the second subexpression can begin executing before the entire first subexpression is completed. So both s(X,Y) and t(Y) must wait for both p(X) and q(Y) to finish before either of them may begin. Because q bound Y to a ground variable, however, there is no reason for t(Y) to wait. Both Conery's model and the SDD model may allow t(Y) to begin executing as soon as q(Y) completes; the RAP model will force it to wait. In all cases, the execution graph of Figure 25-10 will be achieved. But this graph is misleading for the RAP model. In the RAP model, subgoals in a given level of the graph might be able to execute only when all subgoals in higher levels are complete, even when those subgoals are not connected to some lower subgoal. The result is frequently a "layered" execution of the graph. This limitation of parallelism is a consequence only of the limited nature of the RAP graph expressions.

As mentioned before, the typing algorithm of the RAP model is an approximation technique. It sacrifices completeness for efficiency. As a result, it may occasionally mistype a term. As discussed above, any mistypings are always on the side of safety and so cause no errors. They may, however, cause loss of potential parallelism. For example, suppose that X is ground but the typing algorithm can detect only that it is nonvariable, nonground, and so sets its type to U; then a GPAR(X) test will fail, and the associated subexpressions will execute sequentially when they would have executed in parallel if the type had been set correctly to G. The SDD model does not have to perform any run-time typing. Because the entire program is analyzed, any calls to a given procedure are always guaranteed to be within the bounds of the worst-case entry mode declared for the procedure by the analyzer.

Finally, it should be noticed that in the RAP model, PAR, GPAR, and IPAR expressions can all be executed sequentially if desired. When all the processors

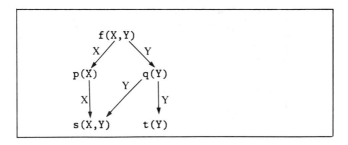

Figure 25-10. A layered execution graph.

in a system become adequately busy, it may prove desirable to discontinue testing for additional parallelism. If requests for work arrive to some processor from other, idle processors, the processor may resume testing for parallelism and distribute any independent pieces it finds. This ability to restrict the parallelism by sequentially executing a parallel expression will prove beneficial in reducing overhead and unnecessary interprocessor communication, as well as increase the benefits of locality.

SUMMARY

Several major models of and-parallel execution of logic programming have been described. These models have been compared and contrasted to each other in order to examine the problems, promises, and limitations of each. All but the join models are actively being researched in major research projects. Which of them may eventually lead to high-performance, robust parallel logic programming systems remains to be seen. Hopefully, all will.

REFERENCES

Cepielewski, Andrzej and Seif Haridi, "Control of Activities in the Or-Parallel Token Machine," *Procs. of the 1984 International Symposium on Logic Programming*, Atlantic City, New Jersey, IEEE, 1984, pp. 49–57.

Chang, Jung-Herng; Alvin Despain, and Doug DeGroot. "And-Parallelism of Logic Programs Based on a Static Data Dependency Analysis," *Procs. of the Spring Compcon 85*, pp. 218–225, IEEE, 1985.

Clark, Keith L., and Steve Gregory, "PARLOG: A Parallel Logic Programming Language," Research Report DOC 83/5, Imperial College, Apr. 1984.

Codd, A. "A Relational Model of Data for Large Shared Data Banks," *Comm. of the ACM*, Vol. 13, No. 6, June 1970, pp. 377–387.

Conery, John S., *The AND/OR Process Model for Parallel Execution of Logic Programs*, dissertation, Univ. of California, Irvine, Tech. Report 204, Information and Computer Science, 1983.

Crammond, Jim "A Comparative Study of Unification Algorithms For Or-Parallel Execution of Logic Languages," to appear in *Procs. of the 1985 Int'l Conference on Parallel Processing*, IEEE.

DeGroot, Doug, "Restricted And-Parallelism," *Procs. of the Int'l Conf. on Fifth Generation Computer Systems 1984*, Tokyo, Japan, Nov. 1984, pp. 471–478.

Dijkstra, "Guarded Commands, Nondeterminancy, and Forma; Derivations of Programs," *Comm. of the ACM*, Vol. 8, No. 8, Aug. 1975, pp. 453–457.

Hoare, C. A. R., "Communicating Sequential Processes," *Comm. of the ACM*, Vol. 21, No. 8., Aug. 1978, pp. 666–667.

Kowalski, Robert, *Logic for Problem Solving*, North Holland, New York, 1979.

Mellish, Chris, "The Automatic Generation of Mode Declarations for Prolog Programs," DAI Research Paper, Dept. of Artificial Intelligence, Univ. of Edinburgh, Scotland, Aug. 1981.

Robinson, Alan, "A Machine-Oriented Logic Based on the Resolution Principle," *Journal of the ACM*, Vol. 12, No. 1, Jan. 1965, pp. 23–41.

Shapiro, Ehud Y., "A Subset of Concurrent Prolog and Its Interpreter," ICOT Tech. Report TR-003, Feb. 1983, Tokyo, Japan.

Warren, David H. D., "An Abstract Prolog Instruction Set," Technical Note 309, Oct. 1983, SRI International.

26. Organization and Operation of Massively Parallel Machines*

David Elliot Shaw

Columbia University

Two observations regarding the evolution of computer systems have, during the past decade or so, become so commonplace as to require little discussion. First, the cost of digital hardware has dropped to the point where, in many applications, processors need no longer be considered a scarce resource. Second, the cost of computer software is increasing, both in absolute terms and, even more dramatically, by comparison with that of the hardware on which it executes.

The design of highly parallel machines is more commonly associated with the first of these observations than the second. Indeed, the availability of large numbers of inexpensive processing elements quite naturally suggests the possibility of constructing highly concurrent systems capable of very rapid execution. The NON-VON machine, which incorporates a large number (between 100,000 and 1,000,000, within the target time frame) of unusually simple processors, is one of the most ambitious proposals to date for the realization of very large scale parallelism using current integrated circuit technology. It should be emphasized, however, that issues related to software are as central to the goals of the NON-VON project as is the achievement of unprecedented processing power.

NON-VON was designed to apply computational parallelism on a rather massive scale to a wide range of data processing tasks. Although the machine would appear to have applications in certain numerical domains as well, the NON-VON architecture is designed to be especially effective in the rapid execution of computationally intensive symbolic information processing tasks. In particular, NON-VON is expected to offer extremely high performance in a number

*This research was supported in part by the Defense Advanced Research Projects Agency, under contract N00039-84-C-0165, by the New York State Center for Advanced Technology in Computers and Information Systems at Columbia University, and by an IBM Faculty Development Award.

of artificial intelligence tasks, including applications involving computer vision, the execution of production systems, and the management of large knowledge bases.

The general NON-VON architecture is heterogeneous in each of the three dimensions that are commonly used to classify parallel architectures: granularity, topology, and synchrony. The machine includes an "active memory" containing as many as a million simple 8-bit *small processing elements* (SPEs), implemented using custom VLSI chips, each containing a number of processing elements. A smaller number (up to a thousand or so) of more powerful *large processing elements* (LPEs) are used to broadcast independent instruction streams for execution in "multiple-SIMD" mode by the small processing elements. Tree- and mesh-structured physical connections, along with linear logical connections, are used to interconnect the SPEs, whereas the LPEs are interconnected through a high-bandwidth, multistage interconnection network. The complete architecture also includes a *secondary processing subsystem* (SPS), based on a bank of intelligent disk drives and connected to the *primary processing subsystem* (PPS) through a high-bandwidth parallel interface.

The first prototype of the machine, however, called NON-VON 1, incorporates only some of these architectural features. In particular, a simplified active memory incorporating only tree connections has been implemented using custom nMOS VLSI chips; it is controlled by a single LPE, which will be referred to here as the *control processor*. In this chapter, we describe the architecture of the NON-VON 1 prototype, and illustrate the manner in which it achieves its unusually high degree of parallelism. Most features of the general architecture that were not incorporated in the NON-VON 1 prototype (the mesh connections at the leaves of the tree, for example, along with the high-bandwidth that would be used to interconnect the LPEs in a multiple instruction stream configuration) will not be discussed in this chapter. In order to convey some idea of the operation of a realistic system configuration, however, the organization and operation of the SPS will also be briefly described, despite the fact that no secondary storage is actually included in the NON-VON 1 prototype.

The chapter is divided into three sections. This introductory section briefly reviews the history and current status of the NON-VON Project, and provides an informal comparison between the essential elements of a conventional computer system and the analogous components of the NON-VON machine. The following section describes NON-VON's physical organization at several levels. A final section describes the instruction set of the NON-VON *processing element* (PE), and introduces some of the most important paradigms for the implementation of NON-VON software.

Project History and Current Status

The theoretical basis for the NON-VON machine was established in the course of a doctoral research project at Stanford University (Shaw 1979, 1980a). Asymptotic improvements in the evaluation of a number of relational database operations were reported. These results employed a highly general technique known as *hash partitioning,* by which many large-scale data processing operations having O(n log n) time complexity on a von Neumann machine may be implemented in linear time on a different type of machine that has the same hardware complexity. The interested reader is referred to these earlier results for a rigorous analysis of the complexity of algorithms, to which the current chapter will make only casual reference.

Detailed design of the NON-VON hardware began in the latter part of 1981. In January 1985, the NON-VON 1 prototype was completed and tested using several simple computer vision application tasks, and was found to be 100% functional. The prototype was implemented using custom-designed nMOS VLSI circuits, which were fabricated remotely using DARPA's "silicon brokerage" system, MOSIS.

The development of software for the NON-VON machine has proceeded in parallel with our hardware implementation efforts. A simulator for the NON-VON instruction set was implemented in the fall of 1981, and has since been enhanced to provide a user-convenient vehicle for the development of NON-VON software. Higher-level linguistic constructs have been implemented as part of an evolving LISP-based programming environment, and a number of applications have been coded and evaluated. About 30 individuals thus far have written NON-VON programs, and have tested their execution using the instruction set simulator. Although no large-scale applications have been implemented yet, our experience with this modest corpus of simple NON-VON programs has already led to major improvements in the architecture and instruction set, which are being incorporated in a later prototype called NON-VON 4.

Comparison with von Neumann Machines

If pressed to identify a single principle underlying the essential "philosophy" of the NON-VON architecture, we probably would choose to highlight the strategy of extensively intermingling processing and storage resources. This strategy is employed at several levels within the NON-VON machine, and perhaps is best appreciated by contrast with the organization of a conventional computer system.

In an ordinary von Neumann machine, a single (often quite powerful) *central processing unit* is connected to a single (often quite large) *random access memory,* which is used for the storage of both programs and data. The CPU and RAM communicate in a serial (or, at best, weakly parallel) fashion through a narrow conduit that Backus (1978) has called the "von Neumann bottleneck." Moreover, the limitations of this organization are becoming more serious as technological progress increases both the potential power of processing hardware and the realizable size of computer memories.

In the NON-VON *primary processing subsystem* (PPS), on the other hand, a large number of very simple, highly area-efficient processing elements (PEs) are, in effect, distributed throughout the memory. In particular, each integrated circuit in the PPS contains a number of PEs (eight, in our planned prototype version, which is based on typical 1982 nMOS device dimensions and die sizes). Each PE is associated with a small amount of locally accessible random access memory (32 bytes, in NON-VON 1, and up to 256 bytes in a production version of the machine). The potential processor/memory bandwidth in NON-VON is thus many orders of magnitude higher than in conventional machines. (Because of the extensive distribution of processing logic throughout NON-VON's memory, the PPS may be regarded as a form of "active memory," capable of manipulating its contents in far more complex ways than is possible in a single conventional RAM.)

In practice, many or all of these tiny PEs often can operate concurrently on data stored in their respective local memories, supporting effective execution speeds far exceeding those of today's fastest supercomputers. Because of their small size, however, the PPS is expected to be only slightly more expensive than an equivalent amount of ordinary static RAM. (In our latest version, NON-VON 3, which supports up to 256 bytes of local RAM, the "active memory," including embedded RAM, should occupy approximately 2.1 times as much silicon area as the embedded RAM alone.) From the viewpoint of performance, the PPS thus may be regarded as an ultra-high-speed parallel processing ensemble; from a cost perspective, though, it is better viewed as a (slightly overpriced) random access memory unit.

A similar comparison between the mass storage facilities of a conventional computer system and the analogous subsystem within the NON-VON machine may also prove instructive. In the typical large-scale data processing system, a large bank of disk drives is charged with the task of responding "mindlessly" to a sequence of requests for data posed by the CPU. In practice, most of this data in fact proves irrelevant to the task at hand. The secondary storage subsystem— a husky and obedient, but rather dim-witted brute—is generally incapable of separating wheat from chaff, and must pass both along to its more intelligent master.

As in the case of the von Neumann bottleneck, the pathway between the "thinking part" and the "remembering part" of such a system is a relatively narrow one, even in the most sophisticated contemporary systems. Although modest degree of parallelism sometimes is achieved in the disk-to-computer interface, the process of transferring data between primary and secondary processing hardware remains, for the most part, an essentially sequential function.

In the NON-VON secondary processing subsystem (SPS), on the other hand, a small amount of processing hardware is associated with each disk head. This hardware allows records to be inspected "on the fly" to determine whether a given record is relevant to the operation at hand. The NON-VON SPS is thus able to be more discriminating in the data it passes along to the primary processing hardware. Furthermore, the topology of the PPS supports a massively parallel interface between primary and secondary storage, allowing data transfer between the subsystems to keep pace with the greatly accelerated execution possible within the PPS. In short, the SPS is able to "filter" data before it is sent to the PPS, and to transfer the "filtrate" in a highly parallel manner.

ORGANIZATION OF THE NON-VON MACHINE

In this section, we describe the physical structure of the NON-VON 1 prototype. The top-level organization of the system is outlined in the first subsection. Our principal concern in this chapter, however, will be with the primary processing subsystem (the "active memory" of the NON-VON architecture), which is described in more detail in the second subsection. In the third subsection, we discuss certain topological considerations that influenced the design of the PPS. The section concludes with a detailed description of the individual processing elements from which the NON-VON PPS is constructed.

System Organization

The top-level organization of the NON-VON machine is illustrated in Figure 26-1.

The PPS is configured as a *binary tree* of processing elements. (It should be recalled that in the general NON-VON architecture, the leaves are also connected to form a rectangular mesh; although these connections have been incorporated

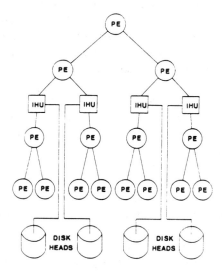

Figure 26-1. Organization of the NON-VON 1 prototype.

in the NON-VON 3 chip design, however, no mesh connections were implemented in the NON-VON 1 prototype.) In addition, the subsystem can be reconfigured to provide for linear, tree-structured or global bus communication by dynamically altering certain switch settings within the PEs. With the exception of minor differences in the "leaf nodes," each PE is laid out identically, and comprises a small random access memory, a modest amount of processing logic, and an I/O switch supporting the various modes of inter-PE communication.

At the root of the tree is a specialized von Neumann machine called the *control processor* (CP), which is responsible for coordinating various activities within the PPS. The CP consists of a conventional uniprocessor (a VAX 11/750 in the NON-VON 1 prototype) that is connected to the PPS through an interface called the *active memory controller*. (In the general NON-VON architecture, the single CP would be replaced by a number of LPEs, each based on an "off-the-shelf" 32-bit microprocessor, and the set of all LPEs would be interconnected by a high-bandwidth interconnection network). Although certain sequences of instructions are executed sequentially within the CP, it is also capable of broadcasting instructions to be simultaneously executed by all enabled PEs in the tree on a single instruction stream, multiple data stream (SIMD) basis (Flynn 1972).

The SPS is based on a number of rotating storage devices, which might in practice be realized using either slightly modified multiple-head disk drives or unmodified single-head drives. Associated with each disk head in the SPS is a separate sense amplifier and a small amount of logic capable of dynamically examining the data passing beneath it. These *intelligent head units* (IHUs) are also capable of performing simple computations (hash coding, for example) and serving a control function similar to the role played by the CP.

Assuming that the number of intelligent disk heads is equal to 2^k, for some integer k, the k-th level of the PPS tree (where the root is considered to be at level zero) is used to interface the PPS and SPS. Specifically, each of the k internal PPS nodes at this level is associated with a different IHU. Physically, this connection is made by interposing the IHU between the interface level PE and its parent PE, as illustrated in Figure 26-1. In its *passive* state, the IHU acts as a simple bus, passing information in both directions without change.

In certain algorithms, though, each IHU serves as an active control processor for the subtree it roots, allowing independent, asynchronous computations within the various *interface-rooted subtrees*. (NON-VON thus is not, strictly speaking, a SIMD machine; in practice, however, it often functions as either a single SIMD machine or a collection of such machines.) The most common application of this capability is in the concurrent loading of each interface-rooted subtree from its respective disk drive. Such parallel transfers between SPS and PPS account for the unusually high effective I/O bandwidth achieved in a wide range of applications. Other algorithms make use of the "top part" of the PPS tree—more precisely, the portion consisting of all PEs lying above the interface level. Among other things, this portion of the tree can be used for the efficient *synchronization* of interface-rooted subtrees following asynchronous operation.

A more thorough discussion of the SPS, its interface to the PPS, and the kinds of algorithms that make explicit use of the upper and lower portions of the tree is beyond the scope of this chapter. The reader may, however, find the discussion of *hash partitioning* presented in Shaw (1980a) to be useful in gaining some appreciation for the way NON-VON–like architectures provide support for at least one important family of highly parallel algorithms involving large amounts of data.

The Primary Processing Subsystem

Although physically structured as a binary tree, the NON-VON 1 PPS can be dynamically reconfigured to support communication patterns characteristic of two other topologies in a highly efficient manner. In this subsection, we describe the physical organization of the NON-VON 1 PPS and discuss the three modes of communication it supports.

The PPS is implemented using a number of identical *PPS chips*. Our use of a single circuit is made possible by the adoption of a tree partitioning scheme first suggested by Leiserson (1981). This approach embeds both a complete subtree (containing $2^c - 1$ constituent PEs, for some c depending on device dimensions) and a single interior node on each chip. Four 9-bit buses (8 bits for data and one bit for a control function, which will not be discussed in this chapter) enter the chip. One, called the T connection, leads to the root of the chip's subtree, while the other three, called the F, L, and R connections, attach the single interior node to its father, left child, and right child, respectively, within the tree.

A simple recursive procedure allows the construction of a complete binary tree of arbitrary size using only chips of this type. This construction is illustrated for the case of two chips in Figure 26-2. Note that the resulting circuit consists

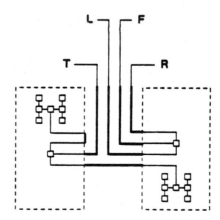

Figure 26-2. Interconnection of two Leiserson chips.

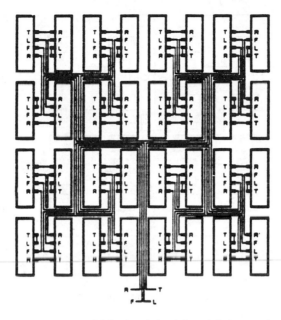

Figure 26-3.   The PPS printed circuit board (Leiserson layout).

of a larger complete binary subtree (in this case rooted by the interior node of the chip on the left side of Figure 26-2), together with a single unconnected interior node (the interior node of the chip on the right). This circuit has the same four external connections—T, F, L, and R—that a single chip had.

The interconnection scheme shown in Figure 26-2 easily may be extended to allow the construction of a simple, planar printed circuit board layout (also due to Leiserson), which is illustrated in Figure 26-3. The regularity of this PC board layout scheme has greatly simplified the task of designing the NON-VON 1 PPS. Furthermore, the area required for routing wires within the PC board is strictly proportional to the number of chips, allowing the efficient implementation of boards of arbitrary size.

The PPS is simply a collection of these PC boards, interconnected in precisely the same manner as are the constituent PPS chips. This scheme is suitable for the construction of a PPS comprising $2^b - 1$ PEs, for arbitrarily large b, and leaves only a single interior PE unused.

The subtree incorporated within each PPS chip is configured geometrically according to a "hyper-H" embedding (Browing 1978), as illustrated in Figure 26-4. This construction is highly regular, is area-optimal (in the sense that the amount of silicon area occupied by the tree is proportional to the number of PEs), and is easily extended to incorporate larger numbers of PEs as device dimensions scale downward.

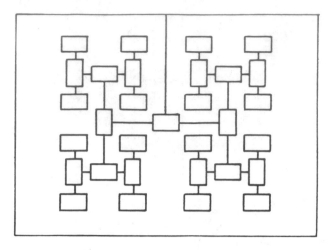

Figure 26-4. Hyper-H embedding of the binary tree.

The tree structured inter-PE bus structure supports three distinct modes of communication:

1. *Global bus communication,* supporting both broadcast by the CP to all PEs in the PPS and data transfers from a single selected PE to the CP.
2. *Physically adjacent (tree) communication* to the *parent* (P), *left child* (LC), and *right child* (RC) PE within the physical PPS tree.
3. *Linearly adjacent neighbor communication* to the *left neighbor* (LN) or *right neighbor* (RN) PE in a particular logical linear sequence.

The global broadcast function supports the rapid parallel communication of instructions and data from the CP to the individual PEs, as required for SIMD execution. As will be seen in the section on "Programming NON-VON," it is also possible for a selected PE to send data to the CP. Using the CP as an intermediary, any PE can thus send data to any other PE. No communication concurrency is achieved, when data is passed from one PE to another using the global mode primitives.

The physically and linearly adjacent communication modes, on the other hand, support fully parallel communication. The former mode is used in many tree-based algorithms. (Parallel sorting and the logarithmic-time addition of n numbers are two examples.) The linear mode is used in algorithms in which many PEs simultaneously exchange data or control information with their immediate predecessor or successor PEs in some predefined total ordering. Several mappings between the linear logical sequence and the tree-structured physical topology of the PPS are possible; these alternatives are discussed in the following subsection.

By way of summary, each PE can communicate with five other PEs, which are referred to within its own local context as P, RC, LC, RN, and LN.

Topological Considerations

The choice of a tree-structured topology for the NON-VON 1 PPS was based on considerations involving such factors as the efficient use of silicon area, favorable pinout properties, and suitability for the rapid broadcasting of data. Another important factor was the ability to efficiently emulate a linear array (a sequence of PEs, each connected only to its immediate precessor and successor), which, among other things, plays a central role in one of our techniques for manipulating records too large to fit within a single PE.

First, we observe that each NON-VON 1 PPS chip has exactly four external connections, each 9 bits wide, regardless of the number of PEs contained within its subtree. Because of its fixed pinout requirements, independent of the size of the embedded subtree, the realizable capacity of the PPS chip will increase quadratically with decreases in minimum feature width. This will permit dramatic increases in the computational power of the NON-VON PPS unit as device dimensions are scaled downward with continuing advances in VLSI technology. (During the target time frame for a production version of a NON-VON–like machine, a c value of 7 or 8, corresponding to several hundred processing elements per PPS chip, would seem feasible.)

It is worth mentioning that, with the notable exceptions of linear arrays and such closely related architectures as simple rings, most topologies proposed for parallel computation in VLSI do not share the area and pinout properties we have just outlined. A homogeneous implementation of the orthogonal and hexagonal mesh-connected topologies proposed for the implementation of *systolic arrays* (Kung and Leiserson 1979), for example, would require a number of pins proportional to the square root of the number of PEs embedded within a chip. This is also true of such "nearly equivalent" architectures as toroidal meshes (Hewitt 1980) and the chordal ring (Arden 1981). In the absence of a breakthrough in packaging technology allowing a dramatic increase in the number of pins per chip, such architectures will thus become progressively more "I/O-bound" as device dimensions continue to scale downward.

Although the NON-VON 3 architecture employs a two-dimensional mesh to connect the leaves, thus sacrificing the strict scalability properties enjoyed by NON-VON 1, the mesh connections are only one bit wide, thus mitigating the import of this disadvantage. The tree connections, on the other hand, which are fully scalable, are 9 bits wide, as in NON-VON 1.

A large family of closely interrelated architectures exemplified by the shuffle-exchange network (Leighton 1981) and cube-connected cycles (Preparata and Vuillemin 1981) are even more limited in this regard. The pinout requirements of this family of architectures grow considerably faster than those of the two- and three-dimensional meshes. Furthermore, area proportional to $n^2/\log^2 n$ is (provably) required to embed in PEs within a single chip using such schemes (Thompson 1980). Thus, such architectures are subject to quickly decreasing returns to scale as improvements are made in logic densities.

Another topological consideration in designing a machine having as many processing elements as in envisioned for NON-VON is the manner in which global communication is handled. If a "processor density" comparable to that of the NON-VON machine is to be achieved, only a very small amount of local memory can be associated with each PE. The extremely fine "granularity" of such a massively parallel machine is thus inconsistent in principle with the replication of substantial programs within each PE. For this reason, the realization of very high processor densities would seem to be inextricably tied to the efficient global broadcasting of instructions.

What are the implications of this requirement for rapid global broadcasting capabilities? First, we note that the "bounded valence assumption" (the restriction that no "node" be connected to more than a fixed maximum number of "wires"), which is central to all contemporary models of computation in VLSI, precludes the possibility of broadcasting in time less than logarithmic in the number of recipients. Whereas this lower bound is realized by members of the tree-structured and shuffle-based families, most other topologies do not share this property. The two-dimensional meshes, for example, are incapable of broadcasting in time less than proportional to the square root of the number of recipients. In the linear array, broadcast requires linear time. The same is true of the ring network, which may be considered "almost equivalent" to the linear array in the context of these concerns.

In the NON-VON PPS, broadcast communication is effected not only in asymptotically optimal time, but with extremely small constants as well. In the NON-VON prototype, information is broadcast through all levels of the tree multiple levels in an unclocked "flow-through" mode, passing through a very small amount of combinational logic embedded within each PE. The alternative option of pipelining instructions through successive levels of the tree has since been investigated in detail for possible application in NON-VON 3, but was rejected in favor of a high-speed flow-through scheme based on a separate, higher-fanout tree implemented using fast bipolar logic. Even NON-VON 1, however, provides relatively efficient support for the global broadcasting of instructions and data to all processing elements.

By way of summary, the meshes are as area-efficient as the binary tree, but pinout limitations and broadcast inefficiencies will make them progressively less attractive as the sole scheme for interconnecting extremely large numbers of PEs as device dimensions continue to scale downward, resulting in an increasing justification for the use of the binary tree. Such architectures as the sole scheme for interconnecting extremely large numbers of PEs as device dimensions continue to scale downward, resulting in an increasing justification for the use of the binary tree. Such architectures as the shuffle-exchange network and cube-connected cycles, while matching the optimal broadcast time of the tree, have area complexity and pinout characteristics that would be incompatible with this degree of parallelism. Of the architectures we have considered, only the linear array and the tree may be considered *indefinitely scalable,* in the sense that their pinout is fixed, and their area proportional to the number of embedded processors.

There are two reasons for our selection of the tree, and not the linear array, as the topology for the NON-VON 1 PPS. First, a strictly linear interconnection network requires time proportional to the number of processors for broadcast. Second, the NON-VON PPS tree is in fact capable of dynamically reconfiguring to *emulate* the behavior of a linear array with only a minor constant-factor degradation in speed, as shown below. (It should be clear that the converse is not true.) Thus, we are giving up very little by choosing the tree over the linear array.

There are several ways in which a binary tree can be used to emulate the behavior of a linear array. The most obvious possibility is to map the nodes of the tree onto a linear sequence according to a standard preorder, in-order, or postorder traversal scheme (Knuth 1969). The nodes of the tree shown in Figure 26-5, for example, are mapped onto these of a linear array by in-order enumeration.

Let us now consider what data would have to pass along each tree edge in

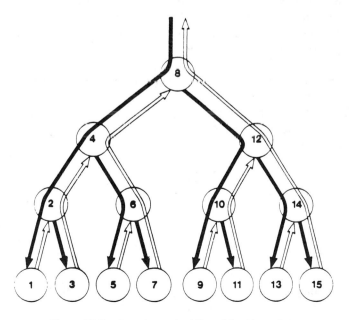

Figure 26-5.   In order embedding of the binary tree.

order to simultaneously transfer a single data element along the path from each tree node to its successor in the linear sequence. These paths are indicated in Figure 26-5 by arrows extending from each node to its linear successor, in general passing through intermediate nodes along the way. It should be noted that because every other element in the in-order sequence is a leaf node, half of these arrows (which we have colored black) originate in internal nodes and terminate in leaf nodes, while the other half (colored white) extend from leaf nodes to internal nodes. Note further that each tree edge is associated with exactly one black and one white arrow. If the communication cycle is divided into separate phases for communication to and from leaf nodes, all nodes in the tree can thus communicate with their respective successors within a single communication cycle.

The in-order embedding scheme, however, has the property that the maximum number of physical tree edges between two nodes that are adjacent in the linear logical sequence grows logarithmically with the size of the tree. This drawback is present in the preorder and postorder enumeration schemes as well because both mappings contain paths extending from root to leaf. As each phase of the communication cycle must be at least as long as the maximum time required for communication between any two linearly adjacent neighbors, it is worth investigating whether a linear array can be embedded in the binary tree in such a way that the maximum path between linearly adjacent nodes is bounded by a constant.

As it happens, we have found a way to configure NON-VON's simple I/O switches so that the longest path between linear neighbors is exactly three. Based on a mathematical result first reported by Sehanina (1960), our scheme requires that the I/O with settings at successive levels of the tree alternate between those that would be employed in a preorder configuration and those that would be used for a postorder mapping. This "bounded neighborhood" embedding is illustrated in Figure 26-6.

In practice, however, the relative advantage of bounded neighborhood embedding over in-order mapping is not so great as it might first appear. The reason for this has to do with the fact that the delay between physically adjacent PEs is not constant throughout the PPS tree. In particular, while most pairs of physically adjacent PEs reside on the same chip, many such pairs are located on different chips, some on different printed circuit boards, and (in a large-scale system) a few in different cabinets. In a realistic large-scale system, the delays encountered between chips, boards, and cabinets typically would be considerably larger than those experienced within a given chip. Because the speed of the communication cycle is limited by the slowest data transfer between linearly adjacent neighbors, each communication phase must be slow enough to allow for the transfer of data between cabinets.

Rough calculations based on estimates on intrachip, interchip, interboard, and intercabinet delays suggest that the relative advantage of the bounded neighborhood mapping over a simple in-order embedding, while not negligible, is not overwhelming for PPS trees of the sizes likely to be encountered in practice. In

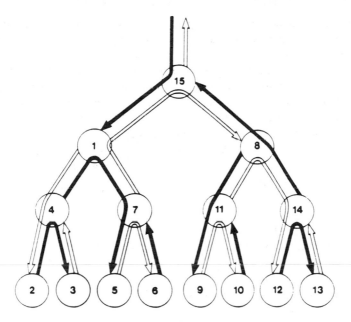

Figure 26-6. Bounded neighborhood embedding of the linear array.

the interest of simplicity, we have thus decided to adopt the in-order embedding
for use in the NON-VON 1 prototype. Future versions of NON-VON may
ultimately be capable of supporting any of the four orderings discussed above,
and of dynamically switching among these orderings.

Having presented arguments for the inclusion of tree-structured physical connections in systems exhibiting parallelism on the massive scale attempted in
NON-VON, we must emphasize that the alternative architectures discussed in
this subsection may in fact prove well suited to applications amenable to coarser
granularities, especially in the short term. In particular, the superficially com-
pelling asymptotic arguments advanced above must be considered in the context
of Larry Snyder's well-phrased reminder that "we don't live in Asymptopia."
On the other hand, if device dimensions continue to decrease, the NON-VON
approach to large-scale parallelism may soon have us "living in the suburbs."

The Processing Element

The NON-VON PE is much simpler, smaller, and less powerful than the processing elements incorporated in previously proposed tree machines (Browning
1980; Seguin et al. 1978). In large part, this difference reflects the SIMD execution of globally broadcast instructions, which characterizes NON-VON's typical operation. By avoiding extensive reliance on MIMD (multiple instruction
stream, multiple data stream) operation, NON-VON obviates the need for large

local program memories and area-expensive processing and communication hardware, and amortizes the cost of most of its control logic over a large number of independent data paths.

The result is a PE that occupies a small fraction of the area required for an ordinary microcomputer, supporting a "processor density" far greater than that of most parallel machines. From an applications viewpoint, the extreme area-efficiency of the NON-VON PE makes it economically feasible to divide primary storage into roughly "record-sized" units, and to associate a separate processing element with each such unit. This aspect of the NON-VON design is central to its processing power in large-scale data processing applications, as we shall see in the remainder of this chapter.

The NON-VON 1 PE comprises:

1. A 32 word X 8 bit *random access memory*.
2. A set of eight 8-bit *byte registers*.
3. A set of eight 1-bit *flag registers*.
4. A byte-wide *arithmetic comparison unit* (ACU).
5. A bit-wide *arithmetic logical unit* (ALU).
6. A byte-wide *I/O switch*.
7. A *programmable logic array* (PLA).

A top-level block diagram of the PE is presented in Figure 26-7.

The data path is organized around two data buses—one 8 bits wide, the other one bit wide. The local RAM, byte-wide registers, and ACU all communicate through the 8-bit bus. One of the 8 byte registers serves as a *memory address register* (MAR), into which addresses are latched in the course of accessing the local RAM. (Although the NON-VON 1 PE contains only 32 bytes of RAM, a production version of the NON-VON machine would probably contain the maximum amount of storage addressable by the instruction set, which is 256 bytes.)

Two of the other registers, labeled A8 and B8, are distinguished as *byte accumulators,* and include special hardware for performing circular shifts. In the course of such shift operations, the bits of A8 and B8 may be rotated through two distinguished flag registers, A1 and B1, which are referred to as the *bit accumulators*. This feature provides a bit-serial link between the byte-wide and bit-wide portions of the data path. In addition, the ACU is capable of comparing the contents of A8 and B8 and latching the results into the bit accumulators. Specifically, A1 is set if and only if the contents of A8 and B8 are identical, whereas B1 is set if and only if the A8 value is greater than that of B8. Another distinguished byte register is IO8, serves a special role, discussed below, involving the latching of data to be transmitted between PEs. The remaining byte registers (labeled C8, X8, Y8, and Z8 in Figure 26-7) are available for general use.

The 1-bit data bus is used to transfer data among the single-bit flag registers,

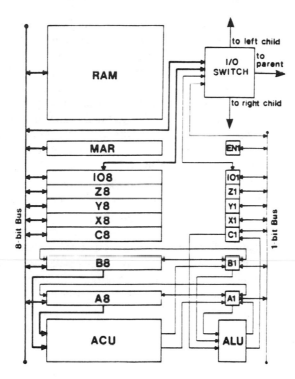

Figure 26-7.   Block diagram of the processing element.

and to supply operands to, and obtain results from, the bit-wide ALU. As noted above, two of the flag registers, called A1 and B1, serve special roles as accumulators. In particular, the bit accumulators serve as inputs to the ALU, along with the contents of a third flag register, C1, which is used to store the carry bit in the course of bit-serial addition and subtraction. Upon execution of one of the *logical function* instructions (described below), the ALU is capable of computing one of the 16 possible Boolean functions of A1 and B1, and storing the result in A1. In response to an ADD1 instruction, the ALU functions as a full adder, computing sum and carry bits for the three inputs A1, B1, and C1. The sum bit is stored in A1 and the new carry bit in C1. Analogous results are produced during a SUB1 instruction.

Another flag register, EN1, is distinguished as an *enable flag*. This flag is used to activate and deactivate individual PEs within the PPS. In general terms, only those PEs whose enable flags are asserted will respond to instructions broadcast by the CP. If EN1 is set to zero in a particular PE, all instructions except one (the ENABLE instruction, discussed below) will be ignored. A number of tricky issues arise in considering the behavior of enabled and disabled PEs, particularly in the case of inter-PE communication operations. These issues will be examined as part of our detailed discussion of the instruction set. Finally, another flag register, IO1, is the Boolean analogue of IO8, serving as an I/O latch in the transmission of single-bit values between PEs. The other flag registers (labeled X1, Y1, and Z1 in Figure 26-7) may be used to store arbitrary Boolean values.

The I/OI switch is connected to both the 8-bit and 1-bit buses, allowing the transfer of byte and flag data to the parent, left child, and right child PEs (and, depending on the switch settings, to other PEs as well). Finite-state control for the I/O switch and data path are provided by a common PLA. Consideration has been given to the possibility of "factoring out" a portion of the PLA associated with each PE on a given chip into a single PLA shared by all such PEs. This approach might ultimately allow the "amortization" of part of the control logic over a large (and increasing, with reductions in device dimensions) number of PEs. Although we have not employed a "PLA factorization" strategy in designing NON-VON 1, this approach is likely to be incorporated in future versions.

In order to keep the area of the PE many times smaller than that of a conventional microprocessor, many decisions have been made in which execution speed is sacrificed for silicon area. It is difficult to rigorously defend such complex and interacting design decisions, but an intuitive justification for this strategy may prove illuminating. First, it is worth mentioning that, in our experience, the savings in area made possible by such decisions in practice often vary as the square of the associated degradation in speed. While such a relationship is observed in many aspects of processor design, the routing of an ordinary n-bit data bus through a 90° turn provides a simple example. Note that the area required to "turn the corner" is proportional not to n, but to the square of n, as illustrated in Figure 26-8. More substantive examples abound.

Because the chip- and board-level layouts employed in the PPS consume area proportional to the number of PEs, the number of PEs realizable in a system containing a fixed number of chips varies inversely ith the area of a single PE. In the critical sections of typical NON-VON programs, all available PEs typically are performing useful computational work in parallel. Thus, NON-VON's maximum achievable execution speed is in some sense inversely proportional to PE area. This being the case, we have found it counterproductive in many cases to achieve a given speedup at the expense of a quadratic penalty in area.

The PE instruction set provides another example of the sacrifice of execution speed within the individual PE in the interest of minimizing area, thus increasing the realizable throughput of the PPS as a whole. As we shall see shortly, the NON-VON PE executes a very small, narrow, and rather low-level set of instructions by comparison with the current generation of powerful 16- and 32-bit microprocessors. In particular, all PE instructions are 8 bits long, including register operands and logical function codes. (In one case, however, the instruction is followed by a byte of data.) In place of a rich set of relatively powerful

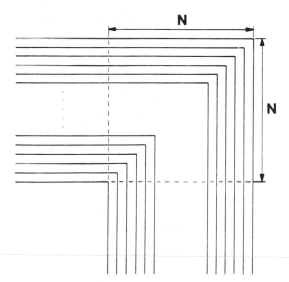

Figure 26-8.   Routing of an N-bit data bus through a 90° trun.

instructions, we have chosen a few low-level operations having extremely simple realizations in hardware.

A single instruction typical of a contemporary 16-bit microprocessor might be implemented in NON-VON using a sequence of between one and four PE instructions. At the cost of a modest degradation in local execution speed, this strategy dramatically simplifies the complexity of (hence, the area required for) the data path and PLA, and reduces the number of pins required to route instructions through the PPS chips.

PROGRAMMING NON-VON

In this section, we introduce the NON-VON instruction set and describe the manner in which it is typically used in the course of programming. While a detailed discussion of each of the applications we have explored is beyond the scope of the current chapter, some feeling for the kinds of techniques employed in constructing NON-VON programs is necessary to understand the basis for our architectural decisions. The remainder of this chapter is thus devoted to an exposition of some of the techniques that characterize the NON-VON approach to parallel programming. As the present chapter is limited to the NON-VON 1 prototype, however, and thus to a strictly SIMD mode of instruction execution, we shall not consider those aspects of NON-VON programming that involve the use of multiple instruction streams.

One conceptual metaphor we have found particularly useful in describing the principles underlying most NON-VON algorithms involves the notion of "intelligent records." This construct is explicated immediately after our description of the instruction set. Next, we discuss the associative operations used to access intelligent records. In a fourth subsection, we describe and compare alternative techniques for the allocation and manipulation of records of various sizes (relative to the local storage capacity of a single PE). Finally, we illustrate typical use of the techniques introduced in this section by informally describing NON-VON algorithms for a few simple symbolic and numerical applications.

The PE Instruction Set

The set of instructions executed by the NON-VON 1 PE may be divided into six categories. The complete instruction set, grouped by category, is described below. Each instruction is followed by a brief specification of its semantics. The following symbols are employed:

⟨byte reg⟩	One of the eight 8-bit registers (A8, B8, C8, X8, Y8, Z8, IO8, or MAR)
⟨flag reg⟩	One of the eight 1-bit registers (A1, B1, C1, X1, Y1, Z1, IO1, or EN1)
⟨PE⟩	One of the physically or linearly adjacent PEs (P, LC, RC, LN, or RN)
⟨address⟩	An 8-bit address in the local RAM
⟨bit⟩	A 1-bit constant
⟨byte⟩	An 8-bit constant

After the presentation of all instruction in a given group, a narrative description of the typical use of each instruction is provided.

1. Register Transfer Group.

Opcode	Operand	Semantics
LOADA8	⟨byte reg⟩	A8 ← ⟨byte reg⟩
LOADB8	⟨byte reg⟩	B8 ← ⟨byte reg⟩
LOADA1	⟨flag reg⟩	A1 ← ⟨flag reg⟩
LOADB1	⟨flag reg⟩	B1 ← ⟨flag reg⟩
STOREA8	⟨byte reg⟩	⟨byte reg⟩ ← A8
STOREB8	⟨byte reg⟩	⟨byte reg⟩ ← B8
STOREA1	⟨flag reg⟩	⟨flag reg⟩ ← A1
STOREB1	⟨flag reg⟩	⟨flag reg⟩ ← B1

The register transfer instructions are used to move data between the four accumulators (A8, B8, A1, and B1) and any of the other registers of compatible length. Note that the MAR may serve as the destination of a 8-bit STORE instruction, allowing different addresses to be stored in the MARs of different PEs, and thus permitting simultaneous access to different locations in the local memories of different PEs, as described below. Similarly, it is worth noting that the value of EN1 may be changed from one to zero using an ordinary STORE instruction, allowing selected PEs to be disabled.

Note that transfers between arbitrary registers must be mediated by one of the accumulator registers, requiring two instructions instead of one. In the context of a massively parallel system, however, the fact that single-operand register transfer instructions are conveniently implemented in an 8-bit instruction word with very little area expended for control logic represents a significant compensating advantage.

2. Memory Access Group.

$$\begin{array}{lll} \text{READRAM} & \langle\text{address}\rangle & \text{A8} \leftarrow \text{RAM(MAR)} \\ \text{WRITERAM} & \langle\text{address}\rangle & \text{RAM(MAR)} \leftarrow \text{A8} \end{array}$$

In order to transfer data between the local RAM and the A8 accumulator, the address of the RAM to be accessed must first be written into the 8-bit MAR register using an ordinary STORE instruction. Note that different PEs may access different RAM locations simultaneously, as the values in their respective MARs need not be the same. This feature is essential to such applications as the parallel processing of variable-length records. The starting addresses of three variable-length fields might be stored in the first, second, and third RAM locations within each PE, for example. In order to access the first byte of the second field of each record in parallel, the contents of RAM location two would be moved (by way of A8) into the MAR, and a READRAM instruction executed. Successive bytes in this field could then be accessed by performing parallel arithmetic on the address stored in the MAR.

3. Arithmetic and Shift Group.

ADD1   A1 ← A1 xor B1 xor C1
       C1 ← (A1 and B1) or (A1 and C1)
       or (B1 and C1)

SUB1   A1 ← A1 xor (not B1) xor C1
       C1 ← (A1 and (not B1)) or (A1 and C1)
       or ((not B1) and C1)

ROTRA   Rotate A8 right by one bit through A1
ROTLA   Rotate A8 left by one bit through A1
ROTRB   Rotate B8 right by one bit through B1
ROTLB   Rotate B8 left by one bit through B1

While we have recently become quite interested in the implementation of parallel numerical algorithms on NON-VON–like machines, the rapid execution of purely numerical problems was not among the primary motivations for the NON-VON machine. Thus, although certain operations critical to NON-VON's typical modes of operation (data transfer and arithmetic comparison operations, for example) are performed 8 bits at a time in NON-VON 1, all arithmetic operations other than comparison are performed in a bit-serial fashion.

Specifically, the ADD1 and SUB1 instruction perform 1-bit addition and subtraction operations, respectively, as described earlier. Arithmetic on operands of arbitrary width is performed by repeated execution of these instructions. (Macros for 8-bit addition and subtraction, along with a number of other common sequences of PE instructions, are provided as part of the NON-VON 1 stimulator.) The result is an ALU that, while fully general and extremely compact, is rather slow by comparison with conventional microprocessors in the performance of standard arithmetic operations.

(Parenthetically, it is worth mentioning that the NON-VON 3 PPS chip contains a full 8-bit ALU, together with a complement of instructions for performing 8-bit arithmetic, including a combined "add/shift" operation useful in performing multiplication. Combined with the faster cycle time and other refinements in the instruction set, these changes should make NON-VON 3 much faster than NON-VON 1 in performing arithmetic and other functions. Using detailed area measurements taken from the NON-VON 1 chip, it was possible to make these architectural changes without a prohibitively expensive increase in silicon area.)

The four NON-VON 1 rotate instructions treat the A8 and A1 registers (and, similarly, the B8 and B1 registers) together as a 9-bit circular shift register. Specifically, ROTRA shifts all but the low-order bit of A8 into the next lowest bit position within A8; the low-order bit of A8 is moved into A1, and the value previously stored in A1 is moved into the high-order bit of A8. ROTLA similarly performs a left circular shift of the combined A8 and A1 registers, while ROTRB and ROTLB perform analogous shifts on the B8 and B1 registers. In combination with the 1-bit logical function operations (discussed below), these instructions permit the execution of arbitrary operations involving 8-bit operands on a bit-serial basis.

4. Logical Function Group.

LOGICAL ⟨operation⟩ A1 ← (A1 ⟨operation⟩ B1) (where ⟨operation⟩ is a 4-bit code specifying one of the 16 possible Boolean functions of two single-bit variables)

CLEAR	A1 ← 0
SET	A1 ← 1
NEGATE	A1 ← not A1
AND	A1 ← A1 and B1
OR	A1 ← A1 or B1
XOR	A1 ← (A1 and (not B1))
	or ((not A1) and B1)
EQU	A1 ← (A1 and B1)
	or ((not A1) and (not B1))
NAND	A1 ← not (A1 and B1)
NOR	A1 ← not (A1 or B1)
NOP	A1 ← A1

The logical function instructions set the flag accumulator A1 to the value of one of the 16 possible Boolean functions of A1 and B1. The desired function can be specified in one of two ways. The first uses the general logical instruction name LOGICAL, together with a 4-bit operation code, each bit of which specifies one product term in a sum-of-products representation of the desired Boolean function. In particular, if the first bit of this code is 1, the function returns true whenever both A1 and B1 are true (and possibly in other cases as well). The second through fourth bit positions similarly correspond to the terms (A1 and (not B1)), ((not A1) and B1), and ((not A1) and (not B1)), respectively. If the condition corresponding to any bit position in which a one appears is satisfied, the function returns true.

For convenience, nine of the most commonly used logical function operations are given special instruction names. (These named instructions have the same machine language representations as the corresponding LOGICAL instructions, however.) Note that two of these named logical operations (NEGATE and NOP) correspond to functions of A1 alone, while the values of two others (SET and CLEAR) are independent of *both* A1 and B1.

5. Communication Group.

BROADCAST8	⟨byte⟩ A8 ← ⟨byte⟩
BROADCAST1	⟨bit⟩ A1 ← ⟨bit⟩
REPORT8	logical register GG8 (in CP) ← A8
REPORT1	logical register GG1 (in CP) ← A1
SEND8	⟨PE⟩ IO8(⟨PE⟩) ← A8
	(⟨PE⟩ may not be P)
SEND1	⟨PE⟩ IO1(⟨PE⟩) ← A1
	(⟨PE⟩ may not be P)

RECV8	$\langle PE \rangle$ A8 $\leftarrow$ IO8($\langle PE \rangle$)
RECV1	$\langle PE \rangle$ A1 $\leftarrow$ IO1($\langle PE \rangle$)

The BROADCAST instructions are used to communicate a specified 8- or 1-bit constant from the CP to all (enabled) PEs in the PPS. (Note that this constant may be the value of a variable in the language being executed on the CP, which has its own instruction set.) The REPORT instructions, on the other hand, provide the means for the contents of the A8 or A1 register of a single enabled PE to be communicated to the CP. In the simulator, the result is to alter the contents of a "logical register" in the CP, called GG8 (or GG1, for REPORT1). The REPORT instructions are intended for use only when it is known that at most one PE is currently enabled—for example, immediately following execution of a RESOLVE instruction (discussed below). The effect of a REPORT instruction is not defined in the case where more than one PE is enabled.

The SEND and RECV instructions are used for communication among phys-ically and linearly adjacent PEs—that is, between PEs that are either physically adjacent within the PPS tree or logically adjacent with respect to the total ordering imposed on the nodes in the linearly adjacent neighbor communication mode. When a PE executes a RECV8 instruction having either P, LC, RC, LN, or RN as its operand, its A8 register takes on the value stored in the IO8 register of the specified (physically or logically) adjacent PE. When a particular PE executes a SEND8 instruction, on the other hand, the contents of its A8 register is transferred to the IO8 register of the adjacent PE specified as its operand.

Unlike the RECV instructions, however, a PE cannot SEND data to its parent because the semantics of this operation would be undefined if both children of that parent were enabled. Thus, only LC, RC, LN, and RN are legal operands for the SEND8 instruction. It should be noted, however, that the parent is capable of receiving data from its children through the use of RECV8 LC and RECV8 RC instructions. The semantics of the SEND8 and RECV8 instructions are not immediately apparent in the case where the operand PE is currently disabled. In such cases, it is the recipient's status, and not that of the originator, that determines whether data is transferred. Specifically, it is always possible to RECV data from a PE, regardless of whether it is enabled, but an attempt to SEND data to a disabled PE will not result in a transfer of data.

The SEND1 and RECV1 instructions function in precisely the same way as SEND8 and RECV8, but operate on flag operands instead of byte-wide values.

6. No Operand Group.

ENABLE	EN1 $\leftarrow$1 in all PEs, including those previously disabled
COMPARE	if A8 = B8 then A1 $\leftarrow$ 1; otherwise A1 $\leftarrow$ 0

$$\text{if } A8 > B8 \text{ then } B1 \leftarrow 1;$$
$$\text{otherwise } B1 \leftarrow 0$$

RESOLVE $A1 \leftarrow 0$ in all PEs except "first" PE
where $A1 = 1$
if no PE has $A1 = 1$,
logical register R1 (in CP) $\leftarrow 0$;
    otherwise $R1 \leftarrow 1$

A PE may be disabled by transferring a 0 into its EN1 register using an ordinary STOREA1 EN1 (or STOREB1 EN1) instruction. In a typical application, the contents of A1 (or B1) will be set to the result of some Boolean test prior to the execution of such a store instruction, resulting in the selective disabling of all PEs for which the test fails. This technique supports the "conditional" execution of a particular code sequence. Following the execution of such a sequence, an ENABLE instruction is used to "awaken" all disabled PEs. In combination with appropriate register transfer and logical operations, this approach may be used to implement more complex conditionals, including nested "IF-THEN-ELSE" constructs.

The COMPARE instruction sets the A1 flag to 1 if the contents of A8 and B8 are the same, and the B1 register to 1 if the contents of A8 exceed that of B8. By combining the 2-bit accumulator values using the appropriate logical instructions, it is thus possible to perform any of the six possible arithmetic relational tests ("equal to," "not equal to," "greater than," "greater than or equal to," "less than," or "less than or equal to") on the values in the byte accumulators. The result may then be used to selectively disable certain processors, allowing the use of general arithmetic tests within a conditional.

The most common use of the COMPARE instruction, however, is in the execution of *content-addressable* operations. As we shall see shortly, such operations are realized by broadcasting character strings or numeric values throughout the PPS, comparing them in parallel with the contents of all enabled PEs, and disabling those for which the match criteria are not satisfied. The decision to implement the COMPARE instruction using byte-wide comparator hardware was based in large part on the central role played by such content-addressable operations in most NON-VON algorithms.

The RESOLVE instruction is used in practice to disable all but a single PE, chosen arbitrarily from among a specified set of PEs. First, the A1 flag is set to one in all PEs to be included in the candidate set. The RESOLVE instruction is then executed, causing all but one of these flags to be changed to zero. (Upon execution of a RESOLVE instruction, one of the inputs to the CP will become high if at least one candidate was found in the tree, and low if the candidate set was found to be empty. In our simulator, this condition code is stored in the "logical register" R1, which may be thought of as existing within the CP.) By issuing a STOREA1 EN1 command, all but the single, chosen PE may be disabled, and a sequence of instructions may be executed on the chosen PE alone. In particular, data from the chosen PE may be communicated to the CP through a sequence of LOAD and REPORT commands.

If the candidate set is first saved (using another flag register in each PE), each of the candidates can be chosen in turn, subjected to individual processing, and removed from the candidate set, allowing the sequential processing of all candidates. Typically, the individual processing performed for each chosen candidate involves the broadcasting of information contained in, or derived from, that candidate to other PEs within the PPS. This paradigm for sequential enumeration is thus employed as a sort of "outer loop" in a number of highly parallel NON-VON algorithms, including the algorithm for set intersection described below (in "Examples of Symbolic and Numerical Algorithms").

In the NON-VON 1 prototype, the A1 flag is preserved in the PE that would be assigned the lowest number in an *in-order enumeration* of all nodes in the PPS tree. The use of in-order enumeration as a criterion for selecting a single PE is an artifact of the NON-VON 1 hardware design, however, and is not guaranteed by the instruction set. The RESOLVE function is implemented using special combinational hardware, embedded within the I/O switch, that propagates a series of "kill" signals in parallel from all candidate PEs to all higher-numbered PEs in the tree. As is the case for all of the global communication functions, the RESOLVE operation is very fast; hundreds of thousands of candidates might be "killed" in less than a microsecond in NON-VON 1, for example.

The "Intelligent Record" Metaphor

A large share of the data processing applications for which computers are now used involves operations on files that consist of a relatively large number of comparatively small records. In many such applications, the relevant files may greatly exceed the capacity of the primary storage device. Whereas the design of NON-VON's SPS and its interface to the PPS were based largely on the essential characteristics of such large-scale data processing tasks, our concern in the following discussion will be with the case in which all records are stored in the PPS. Briefly stated, the NON-VON approach to parallelizing this sort of record-processing application is based on a "nearly one-to-one" physical association of PEs and records. In such applications individual records are often, in effect, capable of manipulating their own contents in parallel. This observation suggests the notion of an "intelligent" record.

As we shall see, NON-VON is designed to support the massively parallel manipulation of records that may be considerably larger or smaller than the local storage available within each PE. Furthermore, the high-level languages we are now developing for use on NON-VON permit the precise mapping between records and PEs to be made *invisible* to the user in most applications. The user-transparency of this mapping is a critical aspect of NON-VON's support for the intelligent record concept because it insulates the programmer from the details of the hardware, allowing each user-defined logical record to be treated as if it had its own private processor.

As an alternative to the intelligent record metaphor, the reader may wish to think in terms of the equivalent notion of "virtual PEs", each consisting of a single processor and an amount of local memory just sufficient to store a single record of arbitrary size.

Associative Operations on the NON-VON Machine

Before examining the manner in which NON-VON's hardware supports records of arbitrary size, let us consider the fundamental mechanisms employed in accessing and manipulating intelligent records. In contrast to a conventional *co-ordinate-addressable* computer, whose primitive instructions access its data by address, NON-VON may be considered a *content-addressable* machine, in which data is accessed on an *associative* basis. In order to illustrate the manner in which records may be accessed by content, let us consider an example in which each PE stores a single "employee record" containing fields for the name, department, years of service, and salary of the employee in question. (Some of these fields will be used in a later example.)

Suppose we wish to associatively identify the records of all employees in the sales department, and to perform some operation on all such records (either concurrently or in succession). Let us assume that the department name is stored in a five-character field beginning in the 17th location within each local RAM, and that all PEs containing an employee record are initially enabled. We now broadcast the first character in the specified department name, which is an "S," to all PE's. Each PE compares this character with the contents of its 17th RAM location, and disables itself if the two are not equal. The precise sequence of PE instructions follows:

BROADCAST8 "S"	;	Send the pattern character
STOREA8 B8	;	and save it in B8
READRAM 17	;	Get the data character
COMPARE	;	Do they match?
STOREA1 EN1	;	If not, disable this PE

Using a similar set of instructions, the second character is broadcast and compared with the 18th location in the local RAM of each enabled PE. After the execution of five such code sequences, only those PEs whose DEPARTMENT fields contain the string "SALES" will remain enabled. It should be noted that this process of *associative marking* is time-dependent only on the length of the pattern string, and is independent of the number of employee records. Furthermore, the values of any combination of fields may be used as criteria for success of the associative marking operation.

In the case where different PEs are used for the storage of different types of records, operations on a given record type must be preceded by the disabling of all PEs but those containing records of that type. To facilitate this process, each record is "tagged" internally to indicate its record type. If there are only a few distinct record types, the records can be tagged by associating a different 1-bit register with each record type, and setting its value to 1 in exactly those PEs containing records of the type in question. In order to enable all records of a given type, the bit contained in the appropriate flag register is simply transferred to EN1 using two register transfer instructions. For a larger number (up to 256) of record types, a distinct "tag byte" is associated with each record type, and stored in the same way as the fields of the record itself. A single BROADCAST and COMPARE sequence, followed by STOREA1 EN1 instruction, may be used to disable all PEs except those containing records of the desired type.

Depending on the application, associative marking is typically followed by one of two operations. The first, and most common, is to perform a sequence of operations in parallel on the records contained in each of the associatively identified PEs. The second involves sending the "marked" records (or selected fields thereof), one at a time, to the CP in an arbitrary sequence, using the RESOLVE and REPORT instructions. The latter operation, when applied to associatively identified records, is called *associative enumeration*. It should be noted that the time required for associative enumeration, while proportional to the number of "matching" records, is independent of the *total* number of records in the file. Both of the above applications of associative marking will be illustrated shortly in the context of particular NON-VON algorithms.

Either a conventional computer or a NON-VON–like machine (and indeed, any device with the power of a Turing machine) is capable of emulating the behavior of either a content- or coordinate-addressed machine. In particular, a conventional system can implement associative operations using only coordinate-addressable primitives by employing one of several well-understood *partial match* algorithms. Because they must provide for retrieval based on any of the 2^k possible combinations of k fields, though, such algorithms are associated with significant costs in time, space, and conceptual complexity.

Conversely, NON-VON is capable of addressing data on a coordinate basis whenever the data under consideration is best understood in terms of an "address-like" numbering scheme. In such applications, coordinate values are explicitly stored as part of each intelligent record and associatively probed to obtain the record corresponding to a given address. This technique is employed in a number of parallel matrix algorithms, for example.

What, then, are the essential differences between NON-VON's addressing capabilities and those supported by a conventional von Neumann computer? From a software perspective, the critical point is that NON-VON uses a numerical addressing scheme only when the problem at hand is most easily *described* in terms of a coordinate system. In the case where records are more naturally identified by content, the programmer is relieved of the responsibility of translating his or her intentions into an artificial coordinate-based descriptive formalism.

In it our contention that a large share of the computer applications encountered to date are most naturally described in terms of content-addressable, as opposed to coordinate-addressable, primitives. Although our argument is perhaps strongest for the kinds of "business-oriented" data processing tasks that presently account for most of our society's expenditures for large-scale computing, we believe that a surprising number of "scientific" applications might also be more easily specified in content-addressable terms. By providing direct, low-level support for associative operations, NON-VON effectively shortens the path between the *description* and *implementation* of many common computational tasks, thus simplifying the task of programming.

The other essential advantage of NON-VON's hardware support for content-addressability relates to the time required for associative operations. In practice, NON-VON might provide as much as several orders of magnitude improvement over the fastest associative retrieval operations on a conventional computer system, without the need for complex, time-consuming, and area-expensive indexing or hashing operations.

Packed and Spanned Records

Up to this point, we have considered the case in which exactly one record is stored in each PE. Let us now consider the manner in which records considerably smaller or larger than the capacity of a single local RAM may be efficiently stored and manipulated within the NON-VON PPS. The former case involves the allocation of more than one record per PE, a scheme we call *packed record allocation*. To illustrate the manner in which small records may be packed, let us consider an application in which it is desirable to pack as many 15-byte records as possible into the PPS at once. (Although records of this size would be uncommon in most symbolic applications, they might well occur in, say, a sparse matrix manipulation or signal processing problem.)

Four such records might be stored in each PE, beginning in local RAM locations 1, 16, 31, and 46. We shall use the term *record slice* to refer to a set of packed records stored in the same position within their respective PEs. (In our example, four record slices are defined.) In general terms, each operation to be performed on a packed record is carried out by issuing a separate set of PE instructions for each record slice. In order to move a single byte from the fifth to the seventh location of each of our 15-byte packed records, for example, we would first execute the sequence:

READRAM 5
WRITERAM 7

followed by the sequence:

READRAM 20
WRITERAM 22

and then by analogous sequences of instructions corresponding to the last two record slices. The high-level languages now under development for use on NON-VON are intended to relieve the programmer of the responsibility for such operations. In our Pascal-based language, for example, the user would simply declare the collection of records to be of type PACKED MULTIPLE RECORD; a subsequent assignment statement involving two fields of that record would be compiled into the four sequences of instructions discussed above.

Not all operations on packed records, though, are so simply handled. In the above example, the A8 register is used only for temporary storage of the value to be transferred, and need not be preserved after the transfer is completed for a given record slice. In general, however, the contents of certain flag and byte registers may have to be saved prior to operations on successive record slices. The question of how best to reduce the overhead involved in such "state-saving" operations is one of the more interesting considerations involved in the design of compilers for NON-VON.

Whereas packed records may be quite useful in some applications, it should be noted that the space saved by packing is at best proportional to the increased time required to broadcast each instruction to all slices. An additional disincentive is provided by the significant compile- and execution-time overhead required for the support of operations on packed records. For these reasons, small records are packed only when this option is explicitly chosen by the programmer, based on the relative importance of time and space in the context of a given application.

In the case of records too large to fit within a single PE, each record is split among several PEs according to one of two schemes. The first, referred to as the *linear allocation* method, splits each record among several linearly adjacent (logical) neighbor PEs. The other, which we call *bush allocation,* stores each record in a distinct "tree-shaped" cluster of physically proximate PEs called a *bush*. In order to illustrate these schemes, let us consider an example involving records 150 bytes in length. Under either allocation scheme, each spanned record is split among three physical PEs. We will refer to the first part of each record a segment A, the second as segment B, and the third as segment C.

Using one of the "tagging" techniques introduced above, all PEs containing the A segment of a record are marked with one tag, those containing B segments with another tag, and those containing C segments with a third. In algorithms requiring no parallel communication between different segments of a spanned record, the A, B, and C segments are treated as if they were distinct record types, only one of which is enabled at any given point in time. As we shall see, algorithms in which activation (the state of being enabled) and data must be transferred in parallel between one segment and another within each record raise a number of more interesting issues. Parallel intersegmental transfers are handled differently (and with different average-case time complexity) in the case of linear and bush allocation. We begin with a discussion of the former technique.

In a linear allocation of our hypothetical 150-byte records, segment A might be assigned to the first PE in the linear sequence used for linearly adjacent neighbor communication (as described above in the subsection on "The Primary Processing Subsystem"). Segment B of the first record would be stored in that PE having linear number two, while segment C would be stored in the "linear three" PE. Segments A, B, and C of the second record would then be assigned to the linear four, five, and six PEs, respectively. The third record would be similarly split among the linear seven, eight, and nine PEs, and so on. It should be recalled that two PEs that are logically adjacent in the linear sequence are not necessarily physically adjacent in the PPS tree. Thus, a single record may be split among PEs that are not physically contiguous, leading to a physical interleaving of records within the PPS. The in-order embedding employed in NON-VON 1, for example, would lead to the allocation shown in Figure 26-9. (The PEs are labeled with the record number and segment of the data; segment B of record 3, for example, is labeled 3B.)

To see how linearly allocated spanned records might be manipulated in the course of an actual application, let us suppose that our sample records each describe one of the employees in our earlier example. Assume also that the first

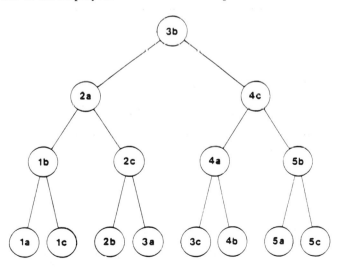

Figure 26-9.   Linear allocation of spanned records.

two characters of the DEPARTMENT field are stored in segment A and the remainder in segment B, and that the salary field is stored entirely within segment C. Now suppose that we wish to raise the salary of all employees in the sales department by 10% in a single parallel operation. Earlier in this section, we presented an informal description of an algorithm for associatively marking each such employee record in the case of one-of-one allocation. After disabling all PEs except those containing A segments, we employ this algorithm to disable all enabled PEs except those having "SA" as the first two characters of their DEPARTMENT field.

At this point, each PE that remains enabled transfers activation to its right linear neighbor. This step is realized through the use of a code sequence that includes a SEND1 RN instruction, which concurrently communicates a Boolean value for each PE to its linear neighbor. At the end of this sequence, which will not be detailed here, the B segments of all records whose DEPARTMENT fields begin with "SA" are enabled, and all A (and C) segments are disabled. The characters "LES" are now matched against the corresponding characters in all enabled records, leaving enabled only the B segments of all records corresponding to employees in the sales department. Activation is now propagated to the C segments of all such PEs, and a sequence of instructions issued to increase the salary fields of all such records by 10%.

In contrast with the linear allocation scheme, the technique of bush allocation groups all segments of a given record together physically within the PPS, as shown in Figure 26-10.

Each of the "tree-shaped" clusters of PEs enclosed within a rectangle in Figure

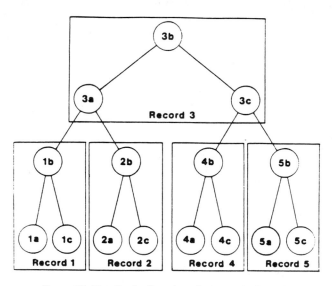

Figure 26-10.   Bush allocation of spanned records.

26-10 is called a *bush*. Within a given bush, successive record segments are assigned to PEs according to the bounded-neighborhood mapping introduced above under "Topological Considerations." The precise manner in which record segments are allocated within a bush, and bushes within the PPS tree, is presented elsewhere (Shaw and Hillyer 1982).

Bush-allocated spanned records are manipulated in much the same way as their linearly allocated counterparts, but using the direct physical tree connections in place of the indirect linear pathways for the parallel propagation of data and activation. In our example application, the first two characters of the string "SALES" are matched concurrently in all of the "A" PEs shown in Figure 26-10. Each matching PE then enables its parent (a "B" PE) using a RECV1 LC instruction. Upon completion of this matching operation, each PE still enabled executes a code sequence including a SEND1 RC instruction to enable its right child (a "C" PE), which then increases its salary field by 10%. As in the case of linear allocation, the transfer of data and activation between segments is fully parallel.

There are certain time/space trade-offs involved in the choice of linear or bush allocation for spanned records, however. Let us first compare the space required for these two allocation methods. The linear allocation method makes progressively more efficient use of the available local RAM as the number of PEs spanned by each record increases. In particular, we would expect to waste only half the space of a single local RAM (16 bytes, in NON-VON 1) per stored record in the average case. This small amount of waste is due to the requirement that the beginning of each record be aligned with the beginning of some PE's local RAM, at least in the method for parallel memory accesses we have outlined. Asymptotically (with increasing record length), the proportion of total available RAM wasted due to alignment thus approaches zero.

By way of comparison, this "waste factor" approaches 25% in the case of bush allocation. To gain an intuitive appreciation for the reason for this comparative inefficiency, consider the case of a spanned record just large enough to require 2^m PEs for storage. The smallest bush capable of storing such a record would contain $2^{m+1} - 1$ PEs, resulting in a waste of $2^m - 1$ PEs' worth of RAM (in addition to an alignment penalty), or approximately half of the total available RAM, for large records. It is easily seen that the average-case waste factor must fall midway between this 50% asymptotic worst-case value and the best-case value of no waste, which occurs for records consuming $2^m - 1$ PEs' worth of RAM. Thus, linear allocation is more space-efficient than bush allocation, particularly in the case of large spanned records.

The space advantage offered by the linear allocation scheme, however, comes at the cost of an increase in the time complexity of data and activation transfers among record segments. Note that in the worst case, the data in question must be transferred from the first to the last PE in the record (with respect to the ordering imposed for purposes of linear neighbor communication). The number of instructions required for such a transfer thus varies linearly with record length in the worse (and in the average) case. In the case of bush allocation, on the other hand, the worst case occurs when data must be passed between two leaves of a bush. On the average, such transferse require time to be logarithmic in the size of the record, a significant advantage in the case of large records. In the case of transfers between *successive* record segments, the bounded-neighborhood ordering reduces this time to a constant.

One other point is worthy of mention in connection with the choice of allocation method. First, we note that binary tree algorithms such as those described by Browning (1980) can be *directly* implemented on NON-VON only when one-to-one allocation is possible (that is, where records are no larger than the capacity of a single local RAM, and each is allocated to a different PE). Many of these algorithms, however, can be easily (and, in some cases, "mechanically") adapted to apply to m-ary trees. (One important class of such algorithms will be described shortly.)

If bush allocation is chosen, such transformed algorithms can be applied to spanned records of arbitrary size, provided that the bushes themselves are allocated within the PPS tree in such a way as to preserve an m-ary tree structure for purposes of interrecord communication. This requirement is satisfied by a particular kind of bush allocation called *landscaped allocation* (discussed in Shaw and Hillyer 1982), in which the bushes are configured as an m-ary tree. Although a thorough discussion of algorithms for landscaped bushes is beyond the scope of the present chapter, the basic approach involves choosing m to be the number of leaves per bush, and treating each bush as a single node in an m-ary tree, where $m = 2^k$ for some positive integer k. (The set of bushes depicted in Figure 26-10 is landscaped, forming a five-node quaternary tree.)

In the case of linear allocation, no such transformation is possible because record segments are interleaved throughout the PPS. The ability to execute many parallel algorithms intrinsically tied to a tree-structured topology thus constitutes another significant advantage of bush allocation.

Examples of Symbolic and Numerical Algorithms

To illustrate some of the more important techniques used in the course of applications programming, we now consider a few simple NON-VON algorithms. First, we describe a highly parallel algorithm for computing the intersection of two sets. This algorithm is based on a commonly used NON-VON programming technique involving a combination of associative enumeration and parallel matching, and is closely related to the algorithms for a number of other set theoretic and relational database operations.

Next, we introduce an important technique for the massively parallel execution of algebraically associative operationns. Using this technique, such quantities as the sum, maximum, or mean of n numbers may be computed in $O(\log n)$

time. We then consider NON-VON's application to a rather "un–NON–VON–like" task: the simulation of large-scale physical systems. We conclude by mentioning a few other examples of symbolic and numerical applications we have considered for parallel implementation on the NON-VON machine.

In general terms, the intersection of two sets is performed by sequentially enumerating the elements of the smaller set, and performing one associative probe for each such element to determine if it is also present in the larger set. Suppose, for example, that we wish to intersect two sets of strings, each stored in its own "virtual PE" (which may be realized using either one-to-one, packed, or spanned records). As in most NON-VON algorithms, these strings may be allocated anywhere within the PPS because all accesses are made on a content-addressable basis. The elements of the two sets are distinguished only by tagging, and may be arbitrarily intermingled.

First, we enable all elements of the smaller set of associatively marking those having the the appropriate tag. An arbitrary one of these elements is then sent to the CP using the RESOLVE and REPORT instructions, and marked so that it will not be chosen again. This value is then matched against all elements of the larger set in parallel, and a RESOLVE instruction executed to see if that string is present. If it is, the element is included in the result set. This procedure is repeated for all elements in the smaller set not already marked as having been processed. The running time of this algorithm is linear in the cardinality of the smaller set, and independent of the size of the larger one. The union or difference of two sets may be constructed in a similar manner.

It is interesting that some of the best algorithms known for set intersection on a von Neumann machine [the hashed intersection algorithms described by Trabb-Pardo (1978), for example] may in fact be viewed as software emulations of the associative approach employed in our algorithm. Whereas we have chosen set intersection to illustrate the "enumeration and probing" paradigm for pedagogical reasons, NON-VON offers more significant advantages in the case of certain "more difficult" operations, whose implementation on a von Neumann machine may in practice be quite expensive. One example having particular importance in relational database management applications is the *equi-join* operation (Codd 1972), of which set intersection may be considered a degenerate case.

The tree-structured topology of the PPS is essential to many aspects of NON-VON's operation, and thus plays an important *implicit* role in all of the algorithms we have discussed so far. None of these algorithms, however, has made *explicit* use of the tree connections. A simple example of an algorithm in which explicit physical tree communication plays an important role is the problem of adding a large number of numeric values, each stored in a distinct "logical record." We might wish, for example, to determine the total yearly payroll of our hypothetical firm by adding the salary fields of all employees.

In the interest of simplicity, let us first consider the case in which each PE in the PPS contains exactly one employee record. First, we disable all nodes except those that are the parent of some leaf node. (This is easily accomplished in constant time using an algorithm that exploits the fact that the leaves are the only nodes that cannot receive a message from any descendant node.) Each of these "penultimate" nodes is then (concurrently) instructed to obtain the salary of its left child (using a sequence of RECV8 LC) instructions, and to add this value to its own salary field.

The process is repeated for all right children, at which point each penultimate node holds the sum of its own salary and those of its two children. At this point, the *parents* of all penultimate nodes are enabled, and all other PEs disabled; the entire procedure is then repeated. After (log n − 1) such steps, the root node will contain the sum of the salaries of all (n − 1) employees. In a full-scale NON-VON prototype containing a million PEs, we would expect the effective execution speed for such a problem to be on the order of tens of billions of arithmetic operations per second.

By substituting other algebraically associative operations for addition, this algorithm can be adapted to compute many other values of practical importance. The mean or maximum salary paid to any employee, for example, can be similarly computed in logarithmic time. Such operations can be combined with the techniques described earlier for the associative identification of records satisfying various criteria, allowing, say, the parallel computation of the average salary paid to employees in Department C who have been employed for between 3 and 5 years.

Finally, it should be noted that such algorithms are easily generalized to support packed records. In our example, we would first add the salary fields of all record slices, leaving a single combined salary in each PE, at a cost proportional to the packing factor. The algorithm for one-to-one addition could then be applied without modification.

Spanned records can also be accommodated, but only when landscaped allocation is employed. In order to adapt our algorithm to the case of landscape-allocated spanned records, we treat each k-level bush as a node in a 2^k-ary tree. The descendants of such a node are precisely the children of all leaves of the bush in question. In Figure 26-10, for example, the bush containing the root node is considered to be the root of a two-level tree with a fan-out of four. The four other bushes in the tree are each treated as leaves of this quaternary tree. In the modified algorithm, each bush adds the salaries of *each* of its "descendants" into a running sum; after approximately $\log_k n$ such steps, the bush containing the root node contains the sum of all salaries.

In order to convey some feeling for the diversity of applications for which NON-VON may provide substantial performance improvements, we now consider a problem that might first appear to be poorly suited for execution on a tree-structured machine. This application, while of only modest economic important by comparison with conventional business data processing tasks, has for some time dominated the attention of most designers and users of "conventional" supercomputers. Although NON-VON was designed to have its primary impact within the mainstream of business computing, we shall succumb to the temptation to discuss its application to this more glamourous scientific application. The task to which we allude is the simulation of large-scale three-dimensional physical systems.

One technique employed in many such simulation problems uses a large number of records (often on the order of a million), each corresponding to a small cubical region in the space being simulated. Each record would typically contain a small number of scalar or vector variables (temperature or fluid velocity, for example) whose values are known to change over time according to certain physical laws involving largely local interactions. The behavior of the system is simulated by repeatedly applying the following two-step process:

1. *The communication step.* The values of certain variables at a given point are communicated to adjacent and "nearly adjacent" neighbors.
2. *The computation step.* A new value is computed for each point in the system, based on the values of variables at neighboring points.

Typically, the same numerical operations are performed at all points during each computation step. The two-step cycle is generally repeated many times to simulate the evolution of a physical system over time.

Although it was certainly not designed with this sort of task in mind, the NON-VON architecture would appear to offer significant asymptotic advantages over existing supercomputer designs in the solution of such problems. Not surprisingly, NON-VON permits the computational component of such problems to be solved in time *independent* of n, the number of points being simulated. To do so, each "cube" of the space being simulated is associated with a distinct virtual PE, and the sequence of operations is broadcast to all such cubes for concurrent execution.

More interesting is the fact that NON-VON permits an $O(n^{1/3})$ speedup in the communication component as well. The algorithm used for communication depends on a particular scheme for allocating the primitive cubes among the leaves of the PPS tree in such a way that the nodes at progressively higher levels correspond to progressively larger cubes. Although the details of this algorithm are beyond the scope of the current chapter, NON-VON's asymptotic speedup is based on the fact that the amount of data passing through each internal node is proportional to the surface areas of these recursively constructed cubes, and not to their volumes. The time complexity of a single communication step is thus $O(n^{2/3})$, and not $O(n)$, as in the case of a von Neumann machine.

Although scientific computing applications have not been central to our design goals, we have investigated the potential application of the NON-VON architecture to a number of numerical problems. Among the applications we have explored are a number of signal processing, matrix manipulation, graphics, and image processing problems. NON-VON's content-addressable primitives permit significant absolute and asymptotic speedups in a number of array processing applications, but provide particularly natural and efficient support for problems involving the manipulation of sparse matrices.

If numerical applications were expected to constitute a large share of the workload of such a machine, the incorporation of a full 8-bit ALU within each PE would almost certainly be warranted, even at the expense of a modest decrease in processor density. Such a change would alter neither the basic NON-VON architecture nor the essential structure of the algorithms we have developed.

Space does not permit a detailed discussion of all of the applications for which we have designed algorithms (at various levels of detail) for NON-VON. It is worth mentioning that the NON-VON PPS supports the execution of several linear-time sorting algorithms, and that at least one promising technique for rapidly sorting very large files is currently under investigation. Highly efficient parallel algorithms for simple transaction processing, and for a number of other operations critical to large-scale commercial data processing, have also been explored. Although we have thus far attacked only a small sampling of the problems to which "real world" computer systems are applied, it has been our experience that *most* such largely symbolic applications prove amenable to massive parallelization on the NON-VON machine.

REFERENCES

Arden, B., "Analysis of Chordal Ring Networks," *IEEE Trans. Computers, C-30,* 291–301 (Apr. 1981).

Backus, J., "Can Programming be Liberated From the von Neumann Style? A Functional Style and Its Algebra of Programs," *Communications of the ACM, 21* (8), 613–641 (Aug. 1978).

Browning, S., "Hierarchically Organized Machines," in C. Mead and L. Conway, Editors, *Introduction to VLSI Systems,* Addison-Wesley, Reading, MA, 1978.

Browning, Sally, "The Tree Machine: A Highly Concurrent Computing Environment," Ph.D. thesis and Technical Report #3760, California Institute of Technology, Jan. 1980.

Codd, E. F., "Relational Completeness of Data Base Sublanguages," in R. Rustin, Editor, *Courant Computer Science Symposium 6: Data Base Systems,* Prentice-Hall, Englewood Cliffs, NJ, 1972.

Flynn, M., "Some Computer Organizations and Their Effectiveness," in *IEEE Trans. Computers, C-21,* 948–960 (Sept. 1972).

Hewitt, C., "Design of the APIARY for Actor Systems," in *Conference Record of the 1980 LISP Conference,* pp. 107–118, Aug. 1980.

Knuth, D. E., *The Art of Computer Programming,* Vol. 1: *Fundamental Algorithms,* Addison-Wesley, Reading MA, 1969.

Kung, H. T., and Leiserson, C. E., "Systolic Arrays (for VLSI)," in *Sparse Matrix Proc. 1978,* Society for Industrial and Applied Mathematics, 1979, pp. 256–282.

Leighton, F., "Layouts for the Shuffle-Exchange Graph and Lower Bound Techniques for VLSI," Ph.D. thesis, Massachussetts Institute of Technology, Aug. 1981.

Leiserson, C. E., "Area-Efficient VLSI Computation," Ph.D. thesis, Dept. of Computer Science, Carnegie-Mellon University, Oct. 1981.

Preparata, F. P., and Vuillemin, J., "The Cube-Connected Cycles: A Versatile Network for Parallel Computation," *Communications of the ACM, 24* (5) 300–309 (May 1981).

Sehanina, M., "On an Ordering of the Set of Vertices of a Connected Graph," in *Publications of the Faculty of Science of the University of Brno,* no. 412, pp. 137–142, 1960.

Sequin, C. H., Despain, A. M., and Patterson, D. A., "Communication in X-Tree, a Modular Multiprocessor System," in *Proc.* Annual Conference of the ACM, Washington, DC, Dec. 1978.

Shaw, D. E., "A Hierarchical Associative Architecture for the Parallel Evaluation of Relational Algebraic Database Primitives," Stanford Computer Science Department Report STAN-CS-79-778, Oct. 1979.

Shaw, D. E., "A Relational Database Machine Architecture," in *Proc.* 1980 Workshop on Computer Architecture for Non-numeric Processing, Asilomar, CA, Mar. 1980.

Shaw, D. E., "Knowledge-Based Retrieval on a Relational Database Machine," Ph.D. thesis, Department of Computer Science, Stanford University, 1980b.

Shaw, D. E., and Hillyer, B. K., "Allocation and Manipulation of Records in the NON-VON Supercomputer," Columbia Computer Science Department Report, Aug. 1982.

Thompson, C., "A Complexity theory for VLSI," Ph.D. thesis, Carnegie-Mellon University, Aug. 1980.

Trabb-Pardo, L., "Set Representation and Set Intersection," Ph.D. thesis and Stanford Computer Science Department Report STAN-CS-78-681, Dec. 1978.

27. Advances in Signal Processor Architecture

Peter N. Marinos

Department of Electrical Engineering Duke University

Nick Kanopoulos

Center for Digital Systems Research Research Triangle Institute

A digital signal processor (DSP) is a discrete parameter system specialized to operate upon and react to externally originated streams of data on a real-time basis. The external sources generating the data streams are usually sensors such as radar, sonar, optical and infrared receivers, medical electronic probes, seismic arrays, and the like.

Some of the most important characteristic features of modern digital signal processors are:

1. Their extensive use of parallelism/concurrency and pipelining in their control, arithmetic, and memory sections.
2. The specialized nature of their instruction set (if programmable).
3. Their interconnection structures, which must be capable of handling not only high data rates but also very complex data patterns.
4. Their use of high-performance arithmetic elements and multiport memories.
5. Their modularity.

Although this chapter is mainly concerned with programmable signal processors utilizing sophisticated, high-level instruction sets, the use of high-functionality hardwired units, such as complex multipliers/accumulators, FFT-butterflies, or even N-point (complex) FFT processors is not excluded.

BACKGROUND

The rapid advances taking place in the field of microelectronics are having a profound effect on digital system design, in general, and on high-performance digital signal processors in particular. In fact, recent improvements in Si and GaAs technologies have been so phenomenol that rethinking of important traditional architectural approaches to signal processor design is both timely and necessary. In the section on "Digital Signal Processor Architectures" we offer a review of important signal processor architectures that have been advanced and implemented over the past 15 years, and we endeavor to identify those architectural attributes that may materially benefit future signal processor design. The particular section of signal processor systems to be reviewed was made on the basis of choosing systems that represented well their respective system design philosophy. Some systems, for instance, exploit parallelism at a very low level (i.e., instruction level), whereas others take advantage of it at a much higher algorithmic level that may permit multiple tasks (i.e., multiply's FFTs, etc.) to proceed concurrently.

As high-performance processors are usually faced with computationally stressful workloads, the need for a high speed arithmetic unit, a memory of large bandwidth, and appropriate interconnection structures that allow efficient address generation and data transfer is of paramount importance. Most programmable digital signal processors are organized according to the functional block diagram given in Figure 27-1. This processor displays several important attributes:

1. Memory specialization.
2. A multiport, parallel, and/or pipeline arithmetic processor.
3. A multiport address generator.
4. Direct access to I/O ports.

Memory specialization implies that (a) data, (b) constants (i.e., coefficients/weights/twiddle factors), and (c) application programs are placed on separate memory modules, each of which is accessed independently of the others, thus achieving the appropriate large memory bandwidth required by the signal processing application. The program memory module is part of the sequencer/microcontroller unit, and it is not explicitly indicated in Figure 27-1.

The arithmetic processor relies on one or several high-speed parallel multipliers and parallel/pipeline hardware modules such as radix-2 and radix-4 FFT butterflies. Multiple copies of the arithmetic processor may be used in parallel or in pipeline fashion to achieve even higher throughput rates, if necessary.

The multiport address generator is an essential function in that it can conveniently generate the complex address sequences that are frequently encountered in signal processing computations (i.e., FFT, convolution, MTI/Doppler processing, etc.).

Finally, provisions for direct I/O access are very important in processing long data streams. Such streams are segmented, and usually only one segment is resident in data memory at any given time; the other data segments reside in fast I/O buffer memory. In such a case, direct access to fast I/O buffer memory

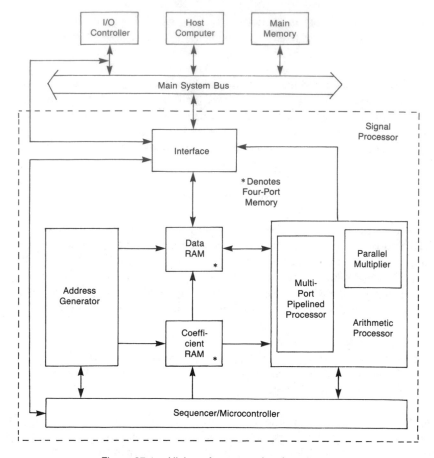

Figure 27-1. High-performance signal processor.

is necessary for maintaining a proper match between I/O and processor bandwidth or speed.

Figure 27-1 illustrates a generic high-performance signal processor organization. The vast majority of the currently available or proposed signal processor designs are simply variations of this basic configuration.

The performance characteristics achieved in a specific signal processor design depend to a great measure, on (a) the technology and algorithms used, (b) how well one matches technology to the selected set of algorithms, and (c) the proper match required between algorithms and hardwired primitives. These conditions are accomplished by careful analysis of the algorithms that the system must execute efficiently in order to satisfy the mission objectives, and by thorough understanding of the potential and the limitations of the technology used.

Given a set of signal processing algorithms, we are usually interested in determining the appropriate available technology that optimizes overall system performance with respect to various system metrics such as speed (throughput), power dissipation, circuit count, cost, size, weight, and reliability. In this chapter we limit our interest to system speed (throughput), power dissipation, and circuit count as we consider various technologies in Si and GaAs. Our specific aim is to investigate several signal processor architectures and existing technologies so that we can determine those choices of architecture and technology that jointly offer superior performance in terms of speed, power, and circuit count.

This approach is supported by many significant recent advances in semiconductor technology that have made possible the fabrication of a wide variety of high-performance signal processing subsystems on a single chip. The availability of such powerful modules offers great opportunities for the design of high performance DSP architectures.

Signal Processing Algorithms

The purpose here is to offer a concise presentation of the signal processing functions and related algorithms believed to be most frequently encountered in high-performance DSP, such as radar signal processing subsystems. For a comprehensive treatment of signal processing algorithms, the interested reader is urged to consult Rabiner and Gold (1975).

The specific signal processing algorithms selected present us with many of the well-known peculiarities of signal processing computations, such as data shuffling and "corner turning," and lead naturally to architectures that are very representative of the ones normally encountered in practice. The signal processing functions considered are:

1. Fast-fourier-transform (FFT).
2. Pulse compression (or fast convolution).
3. Moving target indicator and Doppler processing (MTI/Doppler processing).
4. Constant-false-alarm-rate (CFAR).

These functions display a wide spectrum of computational needs, require unique database arrangements, and rely heavily on specialized hardware modules for their efficient execution.

The type of data acted upon by the signal processor is usually finite sequences of complex points generated by sampling appropriately a real-time signal waveform. We will denote such sequences by $\{x(n)\}$, $n = 0, 1, 2 \ldots N - 1$, where $x_j(n) = x_j(n) + j\,x_j(n)$ and $j = \sqrt{-1}$. The discrete Fourier transform (DFT) of sequence $\{x(n)\}$ is defined as:

$$X(k) = \sum_{n=0}^{N-1} x(n) \cdot \exp\{-(j2\pi/N) \cdot nk\} \qquad [1.1]$$

where $k = 0, 1, 2 \ldots, N - 1$. Expression (1.1) is conveniently rewritten in the form

$$X(k) = \sum_{n=0}^{N-1} x(n)\, w_N^{nk} \qquad [1.2]$$

where $W_N = \exp\{-j2\pi/N\}$. The factors W^{nk} are known as the rotation or phase or twiddle factors, and they display periodicity N.

The inverse DFT is defined in a similar manner, as shown in expression (1.3):

$$x(n) = (1/N) \sum_{k=0}^{N-1} X(k)W_N^{-nk} \qquad [1.3]$$

where $X(k) = X_r(k) + j\, X_i(k)$.

In practical situations, the sequence $\{X(k)\}$ represents the resulting samples in the frequency domain. The importance of DFT in dealing with such sequences stems from the fact that it allows one to perform the necessary signal processing computations in the domain of greater computational convenience.

One of the problems with DFT, however, is that its computational complexity is of order N^2, and thus very slow to compute when N is very large. This impacts adversely real-time signal processing, and the alternate technique of fast-fourier-transform (FFT) for computing DFTs is the algorithm of choice in signal processing applications. The FFT and DFT are equivalent computations in that they generate identical output data sequences from the same input sequence, but the computational complexity of a radix-2 FFT is $(N/2) \cdot \log_2 N$, compared to N^2 for the DFT. This is a substantial improvement in terms of computational speed, and the key reason for the great popularity of FFT over DFT in signal processing applications.

Since its introduction by Cooley and Tukey (1965), the original FFT algorithm has evolved into a large collection of related algorithms specialized to particular applications, various technology options available, and data sequence arrangement that can vary from one-dimensional sequences to multidimensional arrays of data points. In the sequel, we adopt the so-called radix-2 and radix-4, decimation-in-time (DIT) FFT algorithms, and we assume N to be, accordingly, a power of 2 or 4. The choice of N to be a power of 2 (or 4) is not a limiting factor because in practice data sequences may be conveniently adjusted to meet this condition without altering the physical or mathematical characteristics of the problem under consideration (see Rabiner and Gold 1975; Bower and Brown 1982).

The Fast Fourier Transform (FFT). The so-called decimation-in-time (DIT), in-place FFT algorithm reduces the DFT computation given by expression (1.1) or (1.2) into an equivalent computation, as given by expression (1.4):

$$X_k = X_e(k) + W_N^k X_o(k)$$

$$X_{k+N/2} = X_e(k) - W_N^k X_o(k)$$

[1.4]

where $k = 0, 1, 2, \ldots, N/2 - 1$, and $X_e(k)$, $X_o(k)$ are N/2-point sequences obtained from the original N-point data sequence by separating it into its even-indexed and odd-indexed sample points.

This process of dividing by two each of the newly obtained subsequences continues until the new subsequences obtained consist of only two data points. If such a procedure is applied to an $N = 8$-point DFT, expressions (1.4) lead to a sequence of computations as illustrated in Figures 27-2 and 27-3. The terms $X_e(K)$ and $X_o(k)$ represent complex quantities that have resulted from previous computations similar to those given by expressions (1.4). The basic computation implied by (1.4) is referred to as a radix-2 "butterfly," and is a collection of such butterflies as one must compute in order to obtain the FFT of a finite sequence. In a FFT of size N, there are $(N/2) \log_2 N$ radix-2 butterflies. Examination of expressions (1.4) reveals that each radix-2 butterfly involves four real multiplications and six real additions.

An alternate FFT algorithm is the radix-4, decimation-in-time (DIT), in-place FFT. Figures 27-4, 27-5, and 27-6 show the radix-2 butterfly, the radix-4 butterfly, and a 16-point, radix-4 FFT fully developed. Higher radices offer both software and hardware advantages over radix-2, and with the availability of very

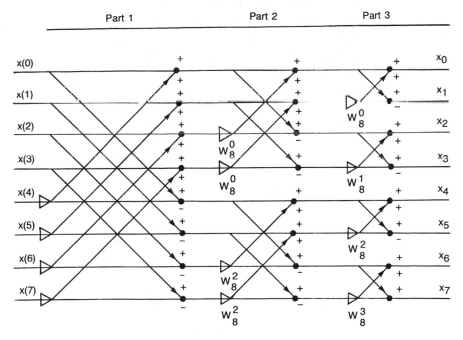

Figure 27-2.   Decimation in time, in place, 8-point FFT.

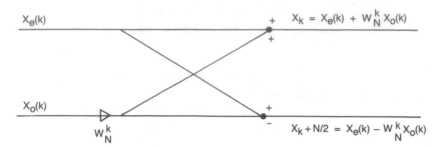

$$X_k = X_e(k) + W_N^k X_o(k)$$

$$X_{k+N/2} = X_e(k) - W_N^k X_o(k)$$

Figure 27-3. Basic butterfly operation.

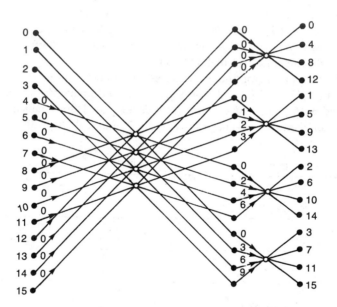

Figure 27-4. Radix-4, DIT, in place, 16-point FFT.

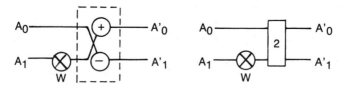

Figure 27-5. Radix-2 butterfly.

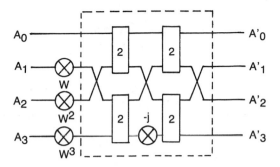

Figure 27-6.   Radix-4 butterfly.

large scale integrated (VLSI) components and very high speed integrated circuits (VHSIC) they are becoming increasingly attractive to designers of signal processing systems.

A complex radix-2 butterfly involves 4 real multiplications and 6 real additions. A complex radix-4 butterfly, on the other hand, requires 12 real multiplications and 16 real additions. Although some of the multiplications are rather trivial (i.e., multiply by ± 1, $\pm j$), programmable signal processors usually deal with all of them in a like manner. For a 1024-complex-point FFT, the number of such operations for radix-2 and radix-4 is as follows:

	Radix-2	Radix-4
Add's	30,720	20,480
Multiply's	20,480	15,360

Another observation is that the radix-4 butterfly requires about three times as many computational resources as radix-2, but it has the potential of achieving twice the throughput rate possible with a radix-2 butterfly. The attribute of the radix-4 butterfly may be used either to double the input data rate for a given processor technology or, given a certain input data rate, to relax the speed requirements of the technology used in the processor relative to a radix-2 FFT implementation.

Another advantage of the radix-4 butterfly over radix-2 is its potential use as an array processor capable of performing multi-operand computations such as four-operand add/subtract, four concurrent multiplications, and so on. As a rule, radix-4, and possibly radix-8, is best utilized in large FFT processors (i.e., $N \geqslant 1024$).

Pulse Compression (via Fast Convolution).   Figure 27-7 illustrates the pulse-compression function when performed via fast convolution. The input data sequence is first processed via a forward FFT of size N, the result of this

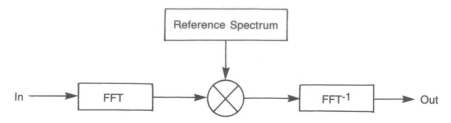

Figure 27-7. Pulse compressor.

transformation is next multiplied by the replica spectrum of the reference signal represented by N complex samples, and the result of this operation is finally acted upon by an inverse FFT of size N. It is assumed throughout this discussion that all computations involve complex data, and that the input data sequences are finite but arbitrarily long. For the importance of this function and related mathematical details, the reader is referred to Bowen and Brown (1982), Blankenship and Hofstetter (1975), Cook and Bernfield (1967), Skolnik (1980), and Nathanson (1969).

The implementation of the pulse-compression system is highly dependent on the length of the input data stream. Very long input sequences cannot be processed in one step as a single pulse-compressor load because of limitations in the size of the FFT one can implement. Therefore, such sequences must be segmented into subsequences of optimal length in a way that minimizes overall processing time.

There are two common approaches for convolving very long sequences: the "overlap-and-save" and "overlap-and-add" techniques (Rabiner and Gold 1975; Bowen and Brown 1982). Both approaches involve essentially the same amount of computation effort, but for purposes of sizing up the time requirements of the pulse-compression function we adopt the "overlap-and-add" method. Figure 27-8 illustrates the manner in which the input sequence $\{x(n)\}$ is segmented and processed. Note that the lengths of the input sequence segments $\{x_j(n)\}$, and reference function sequence, $\{h(n)\}$, are N_2 and N_1, respectively. If the size of the FFT mechanized is N points, this method requires that N_2 be chosen so that:

$$N \geq (N_1 + N_2 - 1)$$

Ideally, we may wish to choose N_2 so that $N = N_1 + N_2 - 1$. However, other overall system requirements may dictate the choice of N_2 (and, for that matter, N_1), thus making adjustments in the choice of FFT size, N, the only option left to exercise.

To illustrate how one may perform sequence segmentation, assume an input sequence $\{x(n)\}$ of length L = 5000 data points., Furthermore, allow available FFT size to be N = 1024 points and the reference function to require N_1 = 120

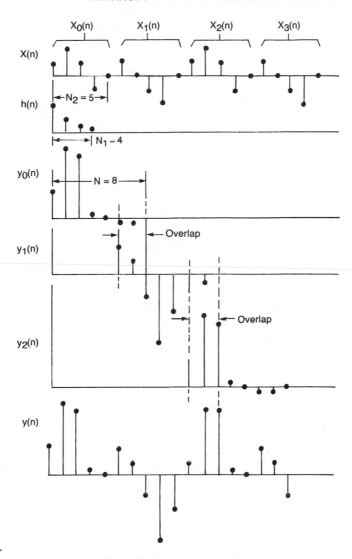

Figure 27-8. Fast convolution.

points for its proper representation. The objective in this case should be to subdivide the input sequence of 5000 data points to be as few subsequences as possible. One possible choice is to form five segments of 905 points each and a sixth segment of 475 points. For FFT purposes the segments and the reference sequence must be zero-extended to $N = 1024$ points. In the last convolution, the points to be discarded are the last $N - (N_1 + 475 - 1 = 1024 - (120 + 474) = 430$ points. Figure 27-9 illustrates graphically the nonrelevant points resulting from an "overlap-and-add" step.

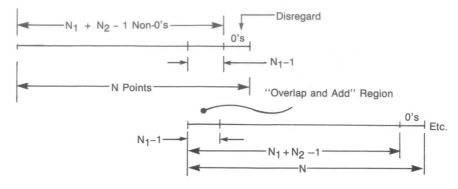

Figure 27-9.   "Overlap and add" correction.

The computational effort generated by a single pulse-compressor load is as follows: (a) $2 \times (N/2 \cdot \log_2 N)$ butterflies (radix-2), and (b) N complex multiplications.

Moving Target Indicator and Doppler Processing—MTI/Doppler Processing.   Figure 27-10 illustrates the manner in which digital MTI/Doppler processing is implemented. It involves first pulse-cancellation with $K(=4)$ pulses followed by a forward FFT of size $N(=64)$. MTI/Doppler is a batch process, and Figure 27-10 shows clearly a batch of data corresponding to several successive observation intervals and ready for processing. Each row represents data collected over a complete observation interval with row-1 being the first interval and row-r the last one. The MTI/Doppler processing function operates on data found along columns in the two-dimensional data arrays, whereas the pulse-compression function requires data arranged along rows. This peculiarity in data accessing is known as "corner turning" and has the potential of introducing serious processing inefficiencies. For a thorough discussion concerning the physical importance and mathematical details of the MTI/Doppler processing functions, the reader is referred to Cook and Bernfield (1967), Skolnik (1980), and Nathanson (1969). Figures 27-11 and 27-12 indicate that pulse cancellation (MTI) may either precede or follow the pulse-compression function. The overall computational effort is the same under either one of the two schemes, and the particular choice is usually dictated by other system performance characteristics.

The weights associated with an $(n + 1)$-pulse canceller are given by expression (1.5) (see Skolnick 1980, p. 110):

$$W = (-1)^{j-1} \frac{n!}{(n - i + 1)!(i - 1)!} \tag{1.5}$$

where $i = 1, 2, \ldots, n + 1$. Other choices for weights are also possible.

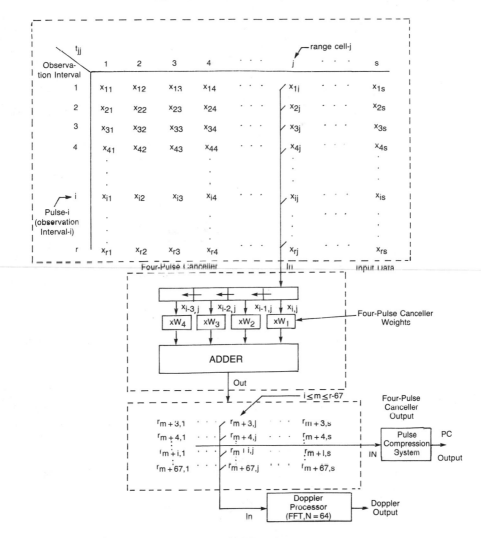

Figure 27-10.   MTI/Doppler processor.

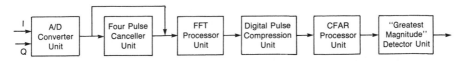

Figure 27-11.   Radar digital signal processor.

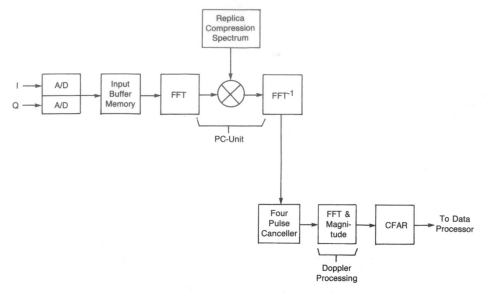

Figure 27-12. Radar digital signal processor.

The computational load generated by the MTI/Doppler processing functions includes the following components:

1. MTI via a 4-pulse canceller. It requires r memory accesses per column; four complex multiplications per point produced by the canceller; and (r − 3) memory accesses per column to place the results into memory.
2. Doppler processing. It requires a 64-complex point FFT per column. The size of this FFT is dictated by the number of Doppler channels desired; the choice of 64 is arbitrary but more than adequate for sizing purposes.

Many of the computational times are reduced to times required to process a column. Therefore, total processing time will depend very heavily on the number of columns of data involved in the two-dimensional data arrays shown in Figure 27-10).

Constant False Alarm Rate (CFAR). Figure 27-13 illustrates the manner in which the digital CFAR function is implemented. Its function consists of averaging a predetermined number of data samples, k, before and after the data point of interest, and comparing the two averages (magnitude-wise), and choosing the larger of the two to set a threshold. The data point of interest is then compared against the threshold value set. For more information concerning the physical meaning and mathematical details of this function, Skolnick (1980) and Nathanson (1969) should be consulted.

The computational effort involved consists of $\{2x(k-1)+1\}$ complex additions, four real multiplications and one subtraction for each data point of interest. Assuming s data points per observation period (i.e., interpulse period), the total computational load is practically s times that estimated for one data point.

Technology Considerations (Marinos 1980)

Advances in signal processing architectures follow very closely the evolution of component technologies. During the past 25 years, the fabrication of solid-state devices has been dominated by the elemental semiconductor material silicon (Si). The associated silicon-based technology has brought us VLSI circuits capable of supporting hundreds of thousands of devices on a chip. This "silicon revolution" is not over, but newer technologies based on the semiconductor compound GaAs have advanced to the point where their availability is beginning to impact digital system design in terms of both performance and system architecture.

With respect to Si-based technologies, bipolar technologies have in the past provided the devices necessary for the design of high-speed digital signal processors at the expense of high power. Emitter-coupled logic (ECL), transistor-transistor logic (TTL), low power Schottkey transistor-transistor logic (STTL), and integrated injection logic (I²L) are the bipolar technologies that have contributed in the past, and will continue to do so in the foreseeable future, the building blocks for the implementation of high-performance digital signal processors. However, parallel advances in silicon-based metal-oxide semiconductor (MOS) technologies have also made possible the design of VHSIC components

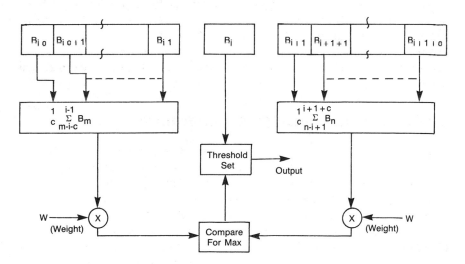

Figure 27-13. CFAR processing.

(Barna 1981) with performance characteristics that compare very favorably to those offered by bipolar technologies. This is a very important development because MOS circuits are less power-consuming than bipolar circuits, and their device densities far exceed those achievable with bipolar technologies. In terms of speed, bipolar circuits are still superior to MOS circuits, but this difference is diminishing very rapidly. When power and size are important design parameters, MOS seems to offer an advantage over bipolar ones, particularly when the problem leads naturally to parallel architectures that require large numbers of components.

GaAs circuits are presently available commercially with device densities at the lower end of the LSI range (i.e., 1000 to 3000 or so gates per chip). Advances in this technology are not occurring at the rate experienced with Si, and projections of GaAs-based logic circuits approaching VLSI densities (i.e., more than 10,000 gates per chip) by the year 1900 may not materialize. Nevertheless, commercial availability of of GaAs logic circuits with even 5000 or so gates will allow us to consider architectures not yet possible. At present, our inability to accommodate on a single chip, or a single board, the logic and memory required to realize a complete processor leads to interconnection delays that are so large that they practically nullify the performance gains that are possible with the superior "power × speed" characteristics of GaAs technologies.

An interesting use of technologies is the one that combines Si multiported memories with GaAs-based processors or fast GaAs memories with several Si-based processors operating in parallel. Neither of these options now leads to cost-effective implementations because GaAs memory is not available in very large sizes or in multiported versions. Similarly, present low device densities in GaAs do not lead to chips with significant computational capability to match the large memory bandwidths that are now possible, with the large Si multiport memory chips that are technologically feasible today.

Both Si and GaAs have much to offer toward improving signal processor performance, and advances in either technology should enhance the potential use of the other.

DIGITAL SIGNAL PROCESSOR ARCHITECTURES

A quick review of developments in signal processing architectures over the past 15 years offers a useful frame of reference in searching either for refinements to existing architectures or for completely new architectural schemes made possible by advances in technology and algorithms.

System architecture is greatly influenced by the problem domain for which it is intended, but the application or problem domain is not the only consideration. Other important factors such as choice of technology, system reliability, maintainability, speed, power, cost, size, functional flexibility, and environmental constraints must be combined with system mission to define the viable architectural options for the system under consideration.

This chapter is concerned with high-performance signal processor architectures intended for missions resulting in data rates of 5 MHz or higher, and thus excludes many interesting implementations identified with less computationally intensive applications. Radar or radarlike signal processing environments are central to this viewpoint, and the architectures to be presented are intended for use in computationally intensive signal processing applications.

In the early 1970s, available technology permitted digital signal processor designs that were suitable for use in ground-based applications or on platforms where weight, size, and power requirements could be easily satisfied. This, however, limited their potential uses considerably, and many important applications related to radar, sonar, and communications remained outside the envelope of practical feasibility.

What follows is a presentation of key digital signal processor architectures utilized over the past 15 years, with special attention paid to architectural features that were the major contributors to these designs' respective high-performance characteristics. The particular choice of design examples was dictated by availability of information, and by our desire to provide a complete coverage of basic architectural approaches used in the past in the actual design of digital signal processors. This includes conventional von Neumann (serial), pipeline, and parallel architectures implemented either as dedicated hardware units or as systems with varying degrees of programmability and thus varying levels of functional flexibility.

Review of High-Performance Digital Signal Processor Architectures

The following examples are the precursors of modern signal processor architectures, and offer important insights into the evolutionary thinking that has brought us to present-day architectures and design practice. Specifically, they reveal the interplay among important factors such as technology, algorithmics, data, programming languages, and availability of specialized hardware units.

The Lincoln Laboratory Fast Digital Processor (FDP) (Rabiner and Gold 1975; Bowen and Brown 1982; Allen 1975). The FDP machine is historically as well as architecturally significant because it was one of the very early serious undertakings in signal processor design intended for use in areas such as radar, sonar, seismology, communications, and speech processing. It was completed in 1971 and included many architectural innovations such as parallelism in its memory, control and arithmetic, data/instruction cycle overlap, and multipurpose instructions.

The technology used was high-speed ECL, which is still the technology of choice for high-speed signal processor implementations although high-speed CMOS and GaAs are rapidly evolving into attractive alternatives.

The FDP was intended as a high-speed arithmetic unit attached to a Univac-1219 computer rather than as a stand-alone processor. Figures 27-14, 27-15, and 27-16 provide an overview of the FDP organization and architecture. Its major subsystems are:

1. Two independently addressable, high-speed (150 ns) memories (M_A, M_B) of 1024 18-bit words each expandable to 4096.
2. Four independently controlled, identical arithmetic elements (AE), each with a buffered multiplier, adder, and three programmable registers.
3. A separate instruction memory that allows overlap of instructions and data cycles.
4. A control unit capable of executing two instructions (i.e., a double-length instruction) simultaneously.
5. Separate I/O buffers that connect the two independently addressable, high-speed memories to the host machine (i.e., Univac-1219) or to external ADC and DAC units.
6. An extra, auxiliary arithmetic unit, XAU, for address generation purposes.

The instruction cycle of FDP is 150 ns, and the number system used throughout is 2's complement, fixed-point arithmetic. Because multiplication and addition are fundamental computations in digital signal processing, their mechanization must be efficient in terms of both speed and their use of resources. In the case of FDP, the multiplier is an 18 × 18-bit array multiplier with an execution time equal to three instruction cycles or 450 ns. In order to make better utilization of resources in the (AEs), and to allow independent arithmetic computations to proceed concurrently, the multiplier is designed as a special hardware attachment to its respective AE. Operationally, the four AEs are interconnected in a ringlike structure in a way that facilitates convenient access to intermediate results (i.e., partial sums and products) usually required in recursive filtering, correlation, and FFT-based computations. It should be pointed out that with the exception of multiplication, which requires 450 ns, all memory instructions require 400 ns for their execution. The effective instruction execution rate is one instruction every 150 ns. Furthermore, a single instruction may activate all four AEs, or any subset of the four, to perform either the same operation or distinctly different tasks.

Normally, executable code resides in memory M_C capable of storing 256 double-length, 36-bit words. If the program is longer, part of it is stored in M_A, M_B. For large programs, part of the code maybe held in the host computer's

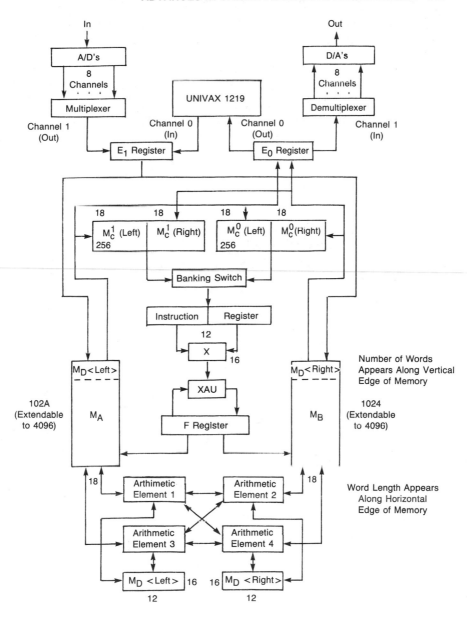

Figure 27-14.   FDP system.

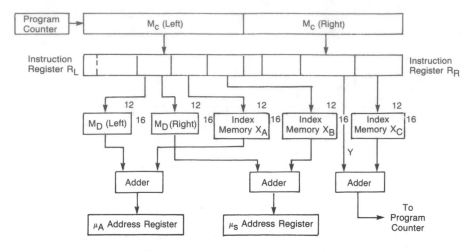

Figure 27-15. FDP address generator.

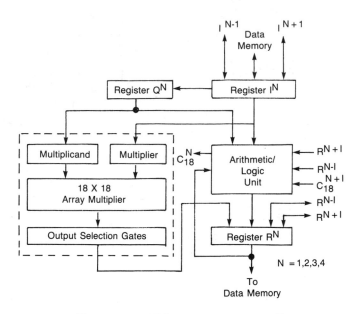

Figure 27-16. FDP arithmetic element (AE).

memory, but this introduces serious programming complications and should be avoided by increasing, if necessary, the size of M_A, M_B.

The FDP allows the following modes of operation:

1. As an independent real-time processor once it has been loaded with executable code from its Univac-1219 host machine.
2. As a real-time processor with the A/Ds and the Univac-1219 computer serving as its peripherals.
3. As a post-processor performing real-time processing under the control of Univac-1219.
4. As a fast peripheral processor to Univac-1219.

Because the FDP was intended for computationally intensive signal processing applications, its hardware, organization, and architecture were finely tuned to carry out, with high efficiency and speed, computations based on formulations such as linear difference equations and fast-Fourier-transform (FFT) butterflies. Expressions (2.1) and (2.2) typify these computations:

$$y(nT) = [2 \cdot \exp(-aT) \cdot \cos(bT)] \cdot y(nT - T) \qquad [2.1]$$
$$- \exp(-2aT) \cdot y(nT - 2T) + A \cdot x(nT) - B \cdot x(nT - T)$$

$$C' = C + D \cdot \exp[j(2\pi M/N)] \qquad [2.2]$$
$$D' = C - D \cdot \exp[j(2\tau M/N)]$$

Expression (2.1) represents the response of a one-zero/two-pole resonator due to input excitation $x(nT)$. Expressions (2.2), on the other hand, represent the radix-2, decimation in time, FFT butterfly, which constitutes the basic computation in evaluating the discrete Fourier transform of a sequence $\{x(n)\}$ given by expression (2.3):

$$X(k) = \sum_{n=0}^{N-1} x(n) \cdot \exp[-j(2\pi nk/N)] \qquad [2.3]$$

For a thorough discussion of FFT and its applications see Rabiner and Gold (1975).

The computations implied by expressions (2.1) and (2.2) constitute the basic elements of the most sophisticated signal processing algorithms, and their actual mechanization uniquely defines the performance characteristics of a digital signal processor. As quantities C and D in expressions (2.2) are usually complex, one can easily verify that both expressions (2.1) and (2.2) can benefit greatly from the presence of four real multipliers when performing computations involving complex multiplication. It is, therefore, not accidental that FDP has four AEs, each capable of performing real multiplication and addition among other operations. In order to perform the computations implied by expressions (2.1) and (2.2), one must: (a) access the operands involved in the computation, (b) perform the four real multiplications implied by a complex multiplication, (c) carry out the associated additions required to arrive at the final result, and (d) store the result. These four tasks are overlapped in FDP, and their concurrent processing materially improves the throughput capabilities of the machine.

The FFT is normally taken as the standard computation when one is evaluating a high-performance signal processor architecture, so we also choose to evaluate FDP's capabilities using an N-complex-point FFT, where N is selected in a way consistent with FDP's memory capabilities. For an N-complex-point FFT, including twiddle factors and filter coefficients, we must provide 4N words of data storage. The choice of $N = 1024$ complex samples is within the maximum FDP memory addressing capability of $M_A = M_B = 4096$ 18-bit words. The data enters directly through the A/D complex and is buffered in M_A and M_B while the previous block of data is being processed. The total time needed to carry out a radix-2, 1024-complex-point FFT is:

$$T \langle FFT - 1024 \rangle = [(N/2)\log_2 N] \times q \times T \langle FDP \rangle \qquad [2.4]$$

or:

$$T \langle FFT - 1024 \rangle = 512 \times 10 \times 8 \times 150ns = 6.144ms$$

where: q = effective number of FDP instruction cycles per butterfly (≈ 8), and $T \langle FDP \rangle$ = time per FDP instruction cycle = 150 ns.

Additionally, one must account for bit-reversal time, as well as time consumed whenever an overflow condition occurs that must be checked and corrected. For a 1024-point FFT, FDP requires about 0.8 ms to carry out bit-reversal; overflow conditions are not predictable, but one can estimate a worst case in which overflow occurs at each step in the $\log_2 N$ iterations, and thus compute an upper bound. For $N = 1024$, the estimated worst-case overflow condition may contribute as much as 0.7 ms. Therefore, the total time required by the FDP to do a 1024-complex-point FFT is:

$$T \langle FFT \rangle = 6.144 + 0.8 + .7 \text{ ms}$$
$$= 7.64 \text{ ms}$$

Next, we turn to the case in which the transform exceeds memory capacity, and it becomes necessary to make use of external memory. Slow external memory has an adverse effect on the performance of FDP, and an 18-to-1 degradation in speed is possible. For illustrative purposes, assume N-4096 complex points and that FDP is equipped to handle only 1024-complex-point FFTs at a time. We have two options:

(a) Treat the 4096 points as a linear array residing in the memory of the host computer (i.e., Univac-1219), and proceed to perform a 4096-point FFT using FDP as a high-speed peripheral processor.
(b) Recast the linear 4096-point array as a two-dimensional 4 × 1024 array (Rabiner and Gold 1975), and use the 1024-point FFT capability of FDP to obtain the 4096-point FFT desired.

Under Option-(a), the time required to carry out the necessary "read-and-write's" into Univac-1219 memory is given by the expression:

$$T\langle I/O \rangle = 3 \times N \times [\log_2 N] \times T \langle R/W \rangle \qquad [2.5]$$

where $T\langle R/W \rangle$ is the time required to read or write a complex word into Univac-1219 memory, and N is the size of the FFT being processed. Note that there are $(N/2) \cdot \log_2 N$ butterflies, each using two complex points and each complex point requiring a "read" and a "write." Additionally, for each complex point read, we must also supply (i.e., "read") a corresponding twiddle factor. Based on this information and the fact that $T\langle R/W \rangle = 3.8$ μs, expression (2.5) yields $T\langle I/O \rangle = 560.33$ ms for 4096 points or $560.33/4 = 140.08$ ms per 1024 points.

Noting that it took 7.64 ms (including overhead) to process a 1024-complex-point FFT using the FDP as the main processor, we conclude that I/O inefficiencies, under Option-(a), result in 140.08/7.64 or, approximately, an 18-to-1 degradation (loss) in speed. This degradation points out the potential impact of slow external memory on the performance of the FDP unit.

Next, we turn to Option-(b), which allows the 4096 points to be arranged in four rows of 1024 points per row and subsequently processed according to Figure 27-17. The 4096 points are stored in a large memory, M_L, accessible by the FDP through the I/O channel. The "L" and "M" computational loops shown in Figure 27-17 require processing times $T\langle L \rangle$ and $T\langle M \rangle$, respectively, as given by expressions (2.6) and (2.7):

$$T\langle L \rangle = Lx[M/2)\log_2 M] \times q \times T\langle FDP \rangle \qquad [2.6]$$
$$+ L \times M [q \times T\langle FPD \rangle] + 3NT_m$$

$$T\langle M \rangle = M \times [L/2)\log_2 L] \times q \times T\langle FDP \rangle + 2NT_m \qquad [2.7]$$

where $N = L \times M$, T_m is the FDP I/O time needed to read or write information into memory M_L and q, $T\langle FDP \rangle$ are as defined previously. Note that the term $3NT_m$ in (2.6) implies that the N points are first read, along with N twiddle factors, and when the L-loop processing is completed, the resulting N complex points are written back into M_L for a total of 3N memory accesses (read/write's); and similarly for expression (2.7) except that the absence of twiddle factors reduces the read/write's to 2N. An implicit assumption used in expression (2.6) is that twiddling takes as much time as a butterfly. The sum of $T\langle L \rangle$ and $T\langle M \rangle$ gives the total FFT time as follows:

$$T\langle FFT - N \rangle = T\langle L \rangle + T\langle M \rangle$$
$$= (N/2)\{\log_2 N + 2\} \times q \times T\langle FDP \rangle + 5NT_m$$

For $N = 4096$, $q = 8$, $T\langle FPD \rangle = 150$ ns and $T_m = 1.6$ μs, one has:

$$T\langle FFT - 4096 \rangle = 2048 \times 14 \times 8 \times 150 + 5 \times 4096 \times 1600 \text{ ns}$$
$$= 34.4 \text{ ms} + 32.768 \text{ ms} = 67.17 \text{ ms.}$$

In this case, the memory access time (32.768 ms) is less than the actual processing time (34.4 ms), and is not the dominant factor that it was in Option-(a), where memory access time was $T\langle I/O \rangle = 560.33$ ms. (In fact, 560.33 ms was all the time necessary to FFT − 4096 complex points.) Therefore, under Option-(b), we achieve a 560.33/67.17 or, approximately, an 8-to-1 improvement in speed over Option-(a). Comparing the three modes of operation represented by Option-(a), Option-(b), and use of FDP without external memory on the basis of a 1024-complex-point FFT, we arrive at the following FFT processing times:

Option-(a): 140.08 ms
Option-(b): 16.79 ms
FDP without external memory: 7.64 ms

It becomes clear that use of slow external memory causes degradation in performance, but when it becomes necessary, Option-(b) is obviously the better option of the two to select.

A few remarks concerning FDP hardware requirements are in order. The technology used for implementation of the FDP system was basically Motorola MECL II, ECL dual-in-line SSI ICs mounted on large wire-wrapped boards (8 × 17 in.) with each board holding up to 180 dual-in-line packages. Each arithmetic element uses ten boards with a total of 1200 packages; the control section of FDP uses 28 boards with a total of 2800 packages. The memory used is also MECL II, 16-bit ICs mounted on multilayer PCBs.

In terms of size and implied power requirements, FDP compares very poorly to present-day designs, but with respect to speed and functional flexibility, the FDP architecture offers performance comparable to that of current signal processor architectures. The most distinguishing characteristics of the FDP are:

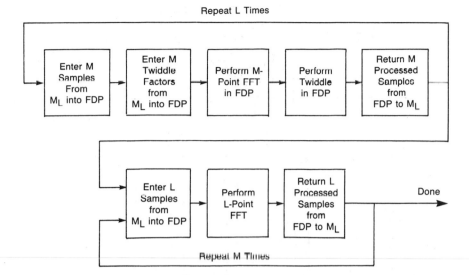

Figure 27-17. Large FFT done by using the FDP's large core memory.

parallelism in the arithmetic section through the use of four arithmetic elements; concurrency in the control section via tightly timed instruction overlap; internal main-memory organization resulting, effectively, in a multiported memory; and sophisticated address generation capabilities. Modern signal processor design draws very heavily on the FDP experience, and the FDP architecture is a classic example of how to exploit parallelism in signal processing efficiently at a low functional level without the use of specialized high-speed functional and storage units.

The Lincoln Signal Processor-2 (LSP/2) (Rabiner and Gold 1975; Bowen and Brown 1982). The LSP/2 signal processor was an outgrowth of the FDP design experience, which demonstrated that the proper mix of functional flexibility (i.e., programmability) and speed in signal processor design is more beneficial than placing emphasis on one at the expense of the other.

Figure 27-18 offers an overview of the LSP/2 structure. This processor relies heavily on the use of several specialized high-speed functional and storage units interconnected through a group of buses. This is in contrast to the FDP system, which utilized four arithmetic units and storage units with direct interconnections to carry out all required computations. This change in interconnection strategy made selective modular enhancements in the LSP/2 system possible, and resulted in a high-performance architecture capable of tracking future technological advances. The distinguishing features of LSP/2 are:

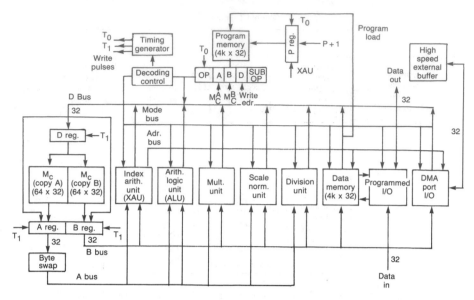

Figure 27-18. LSP/2 signal processor.

1. Separate data and program memories.
2. A multibus architecture that allows mapping of independent sequential instructions into concurrently executable tasks (i.e., computations as well as transfers).
3. Availability of resources that make floating-point arithmetic and high-precision computations easily implementable.
4. A high-speed external buffer and DMA I/O port for speeding up I/O transfers.

All four features listed above were the result of lessons learned from the FDP design. The technology intended for LSP/2 was still ECL, but at least twice as fast, and certainly more densely packed than that used in FDP. The result was an LSP/2 design that achieved about four times the performance of FPD with only one-third as many IC packages.

The LSP/2 architecture is even today a very viable option, and is implementation using high-performance CMOS or GaAs circuits should yield at least an order of magnitude improvement over the original LSP/2 design.

Signal Processing Systems-41/81 (SPS-41/81) (Rabiner and Gold 1975; Bowen and Brown 1982; Fisher 1974). Whereas FDP and LSP/2 achieved their high performance by exploiting parallelism at relatively low functional levels, SPS-41/81 was an architecture designed to achieve its high performance through the use of specialized processors rather than low-level functional units, thus exploiting parallelism at higher functional levels.

The original SPS-41 design utilized three different processors, each optimized to perform a specialized function efficiently. Figure 27-19 offers a system overview and shows clearly the three major processing elements. The IOP processor is a complex of 16 virtual I/O processors, and is capable of retaining complete state information for each of the virtual I/O processors in high-speed memory-A. Memory-A is so designed that it permits access to all elements (i.e., words) of the state vector directly and in parallel as if they were always in the processor. This allows context-switching among the 16 virtual I/O processes without the penalty of overhead. The initial TTL version of this system allowed a different I/O processor instantiation every 200 ns. Which one of the 16 virtual I/O processors is operative at a given time is determined by associative matching against existing I/O conditions. Once a match is established of a virtual processor and a requesting process (such as "read a converted word from the A/D"), the selected IOP provides the appropriate set of I/O routines that may be invoked without incurring any overhead. A bus-oriented interconnection structure provides the paths for passing data between the IOP and the other processors. Interprocessor synchronization is achieved via handshaking or protocol-enforcing logic circuitry.

The second process is the so-called index-section processor (IS), which pro-

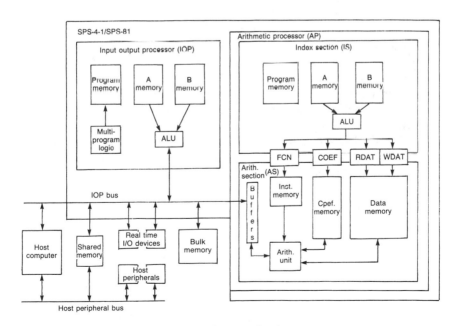

Figure 27-19. SPS-41/81 signal processor.

vides the main control for arithmetic computations. It is essentially a sophisticated address generator responsible for program control and operand selection. It should be pointed out that control tasks and arithmetic computations may proceed concurrently, thus improving overall system performance.

The third processor, referred to as the arithmetic section (AS), operates under the control of the IS processor, which provides the functions, constants, and read/write addresses. The function code identifies one of the sixteen 64-bit registers in program memory, which, in turn, determines the corresponding data-flow configuration capable of carrying out that function. Therefore, the processor is dynamically configurable on a cycle-by-cycle basis. The resources available to the AS processor consist of:

1. Four multipliers.
2. Six adders.
3. Three data memories.
4. A read-only sine/cosine table.

These resources, through the use of multiplexing, may be configured to perform recursive filtering, complex multiplication, and fast-Fourier-transforms, as well as other computations commonly found in signal processing applications.

One of the difficulties associated with this multiple-instruction/multiple-data (MIMD) architecture is the fact that it cannot be easily programmed. This forces one to write interprocess control and local control microcode either at the machine level or, at best, in assembly language. The resulting level of performance is about half that of FDP.

Signal Processing Systems-1000 (SPS-1000) (Booth 1983). The SPS-41/81 architecture is distinguished by the fact that it utilized specialized processors capable of being reconfigured dynamically to accommodate the function desired. In other words, the function code set up appropriate connections that facilitated the data-flow through the available arithmetic resources. However, lack of software support made the use of SPS-41/81 rather difficult and therefore its acceptance rather limited. The SPS-1000 was subsequently introduced as a replacement for the SPS-41/81, and it relies on a FORTRAN-compatible block-diagram high-level language for programming purposes. This choice of high-level language makes SPS-1000 user-friendly and constitutes a definite improvement over the SPS-41/81.

The SPS-1000 system is a data-flow architecture, similar to SPS-41/81, but its large collection of specialized processors is organized somewhat differently, as shown in Figure 27-20. The SPS-1000 architecture differs from other digital signal processor architectures in that an operation is enabled upon acquisition of its operand resources, which include both domain and range resources (i.e., input and output space in memory). Therefore, the user does not need to worry about synchronization of the various processors.

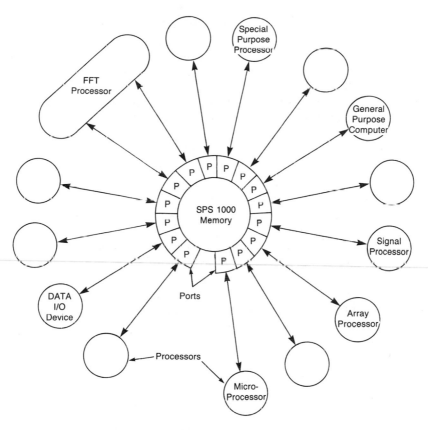

Figure 27-20.   SPS-1000 architecture showing the memory, prots, and various types of processors.

The SPS-1000 uses memory interleaving and parallel bus structures to achieve memory bandwidths of up to 48-MHz word rates. Figure 27-20 shows also the multiported SPS-1000 central memory that each of the attached processors is capable of accessing. In addition, each processor has its own private memory that facilitates local storage and thus immediate access to important data and program instructions. The SPS-1000 central memory is endowed with built-in control and self-addressing capability in addition to being addressable randomly by the individual processors. The memory maintains dynamic protection boundaries with releases of buffer space from one processor to another in increments as small as one word or as large as needed. The memory address generator implements complex nested DO-loop structures that facilitate various ways of accessing multidimensional arrays.

The functions in an SPS-1000 system operate in a data-flow manner that parallels very closely modular implementations of custom-designed, high-performance signal processors in which data is transferred from one module to another via intervening buffers. In the SPS-1000, however, the buffers and process modules are located in the memory and are all defined via software.

As noted above, the language used for programming the SPS-1000 is a FORTRAN-compatible block diagram language. In order to define a system, one begins with a block diagram as shown in Figure 27-21, and then proceeds with the description of the blocks and their interconnections using the FORTRAN-compatible block diagram language. For instance, to describe a process one needs only one statement, and the same holds for the description of a buffer.

The arrows between a process and a buffer are also described by a single statement. The process statements are in effect subroutine calls except that their order of execution is determined by the data-flow and not by the order in which the statements are encountered in the program. Furthermore, the host computer is not involved in the routine operation of the system unless, of course, it is one of the processors performing a function. The FORTRAN-compatible compiler is responsible for generating a downloader table for transfer to the SPS-1000. At run-time, the SPS-1000 executes its workload in the manner defined by the block diagram and does not need subroutine calls from the host computer. However, FORTRAN subroutine calls are used for (a) defining and allocating buffer in the SPS-1000 main memory, (b) defining and allocating processes to one or more processors, and (c) defining the data paths between processes and buffers.

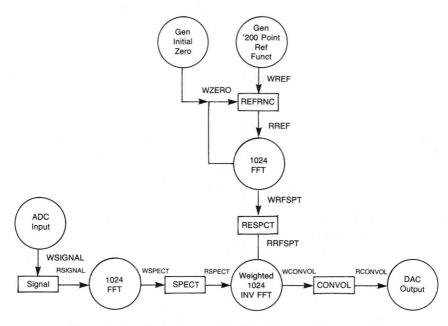

Figure 27-21. Typical system data flow diagrmam (convolution).

It should be noted that proper data-flow for the various processes is established by the interconnection of the buffers and processes in the block diagram. Furthermore, the data-flow features of the system are designed into the ports and the memory, an arrangement that facilitates the use of non-data-flow machines in a data-flow sense.

The Advanced Micro-Devices Am29500 Family of Signal Processing Chips (Advanced Micro Devices 1982; New AND Pittroff 1982). The Am29500 family of signal processing chips combines internal ECL circuit-design for speed with TTL outputs for compatability with the outside world. The general-purpose building blocks of the Am29500 series include the following chips:

1. A byte-slice, multiport microprogrammable signal processor (Am29501).
2. A 16 × 16-bit multiplier with programmable input/output (Am29516/17).
3. A multilevel pipeline register for data and address pipelining (Am29520/21).
4. A fast-Fourier transform address sequencer (Am29540).

The Am29500 series is an outgrowth of the older Am2900 family that lacked parallel channels, which are required for high-speed array or signal processing environments. This deficiency was removed by introduction of a new bus structure, a very efficient resource management scheme, and a new arithmetic unit capable of handling complex numbers efficiently via parallel processing techniques. In other words, the Am29500 family emphasizes parallel and pipelined processing, and uses high-speed buses to support high-performance signal processing architectures in a cost-effective way.

The resource management scheme used in Am29500-based architectures allows the building blocks to be interconnected in a variety of ways, thus realizing conveniently any algorithmic mechanization of interest in a signal processing environment. Figure 27-22 illustrates the use of the Am29501 chip in a dedicated-function involving multiple-algorithm processing. The Am29501 operates under the control of a host computer system that downloads large blocks of data in the memory of the Am29501-based processor through DMA transfers. Once the downloading is completed, the Am29501 operates under local program control. Each algorithm is executed by its own software routine stored in its own local memory, and without any interference from the host computer.

It should be pointed out that in signal processing there are many more memory accesses than there are data points, but the memory-access sequence, although long, is very well-structured and makes possible the design and use of dedicated address sequencers (generators). The FFT is a good example of an algorithm that can use an address generator very efficiently. The use of address generators relieves the CPU from the task of address generation, and permits the CPU to use all of its machine cycles for arithmetic computations, thus boosting overall system performance.

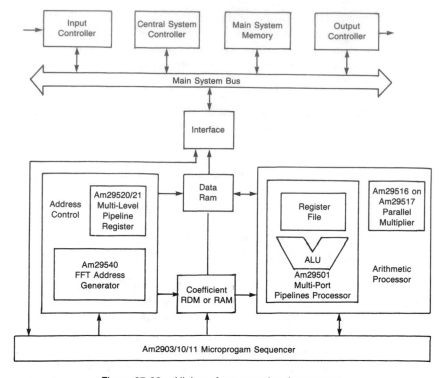

Figure 27-22. High-performance signal processor.

The multiport organization of Am29501, which includes a data-bus port, an output port to a multipliers, and an input port from a multiplier, also contributes to the overall performance of Am29501-based systems. A typical cycle on the Am29501, for instance, consists of:

1. Data input from memory.
2. Data output to the multiplier.
3. Retrieving a previous product from the multiplier.
4. Register-to-register ALU operations and data moves.

(All these operations occur during the same cycle!)

The heart of any high-performance signal processor is its multiplier. The Am29500 family uses high-speed, parallel 16 × 16-bit multipliers (Am29516 and Am29517) capable of delivering a product every 40 ns. The Am29516 chip is compatible with TRW's MPY16H J multiplier chip except that the Am29516 is twice as fast and has an output multiplexor. One may select either the least or the most significant part of the product at the multiplexor output for subsequent use in many pipelined architecture calculations.

Address sequencing complexity for array and signal processors can range from integer counting to the more complex sequences of data-point addresses occurring in FFTs. The Am29540 chip handles the most common FFT formats. The designer can choose, for instance, bit-reversed output order or bit-reversed input order; radix-2 or radix-4 address sequences; and decimation-in-frequency (DIF) or decimation-in-time (DIT). Any transform from 2 to 65,536 points long (requiring a 16-bit address) can be selected. The high-order bits not required for the specified transform can be preloaded through a bidirectional address port to access the next data block. For example, a 1024-point FFT requires only 10-bit addresses, which leaves the remaining 6 bits for accessing data blocks.

While parallel data processing by the Am29500 family is the major contributor to its high-throughput capability, another fairly significant factor is the architecture's ability to facilitate address pipelining, as shown in Figure 27-23, which in effect accelerates address availability. The chips responsible for this (i.e., chips Am29520/1) may be configured as either dual or single pipelines, thus facilitating concurrent access to real and imaginary data, as in the case of complex-point FFTs.

The Am29500 family of chips operates at 10-MHz instruction rates and pro-

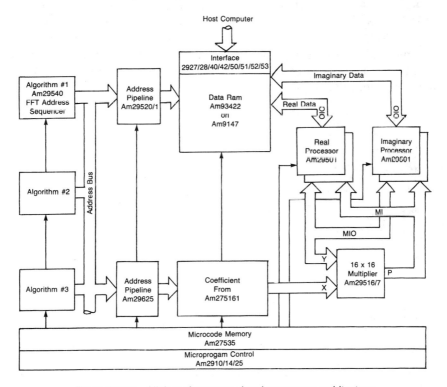

Figure 27-23.   High-performance signal processor architecture.

cesses a radix-2 butterfly in 400 ns. This speed allows a 1024-complex-point FFT to be completed in about 2.0 ms.

The Advanced Micro-Devices (AMD) Am29300 Family of Chips (Advanced Micro Devices n.d.).

This is a 32-bit family of basic building blocks, similar to the Am29500 family of chips, suitable for signal processing as well as other applications that are computationally intensive and could benefit from its floating-point processor (Am29325) capabilities. It requires only one cycle (80 to 100 ns) to execute all instructions including 32 × 32-bit multiplication. Figure 27-24 provides an arrangement of Am29300 parts presently available in very limited quantities. A common feature of all chips in this family is their three-bus architecture, which allows two of them for use as input buses and the third as output bus, thus eliminating the need for bidirectional data buses. This yields increased I/O bandwidth capability and design flexibility, two extremely important characteristics of signal processing systems.

The devices use a scaled, ion-implanted, oxide-isolated bipolar process that features three-layer metal interconnections. Internal ECL circuitry is used for high speed with I/O implemented in TTL.

Signal processors based on the Am29300 family of chips should be capable of signal processing throughput capability comparable to that achievable with the Am29500 family of chips. The former, however, offers floating-point capability and improved computational accuracy not available in the Am29500-based signal processor implementations, and these two added qualities should prove very attractive to system designers.

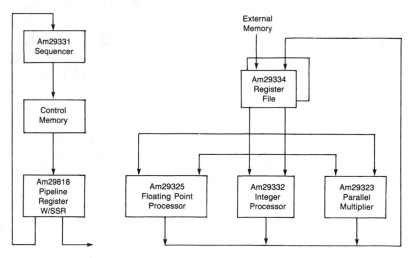

Figure 27-24.   AM29300 32-bit family.

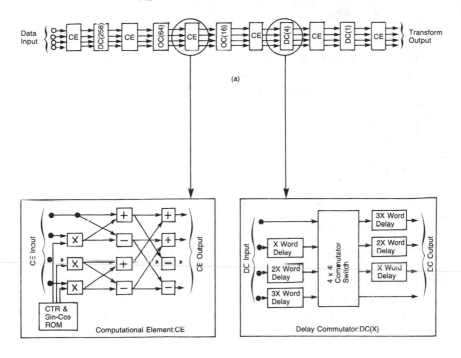

(a)

(b)

Figure 27-25.   Frequency domain adaptive digital filter.

The TRW Fast Transform Processor (Swartzlander and Hallnor 1984; Swartzlander et al. 1984).

It was reported at the IEEE International Conference on Acoustics, Speech and Signal Processing, March 19–21, 1984, that TRW had implemented a 40-MHz (complex) data rate, frequency-domain adaptive digital filter (Swartzlander and Hallnor 1984). The filter uses multiple time overlapped channels each consisting of an FFT, a frequency domain multiplier, and an inverse FFT. The 4096-complex-point FFT and inverse FFT processors realize the McClellan and Purdy radix-4 pipeline FFT algorithm with 22-bit floating-point arithmetic (McClellan and Purdy 1978). The arithmetic is performed with single-chip floating-point adders and multipliers. The interstage reordering is performed with a delay commutator implemented with semicustom VLSI. By using state-of-the-art arithmetic components and semicustom circuit design, an FFT 6-stage pipeline processor was implemented that computes a 4096-complex-point FFT in 102 μs. Figure 27-25 provide an overview of the complexity of the system, which operates at a clock rate of 10-MHz but realizes a 40-MHz data rate because of the radix-4 butterfly used in the implementation of the processor.

The two basic modules of this pipeline FFT architecture are: the computational element (CE) and the delay commutator (DC(X)). The computational element was built around state-of-the-art arithmetic components, whereas the delay commutator was implemented using the 2.5 μm Bell Laboratory polycell (standard cell) CMOS technology. The delay commutator is a single-chip unit involving 12,288 shift register stages and about 2000 gates of random logic, for a total of 108,000 transistors (Swartzlander et al. 1984). Its power dissipation is less than 500 mW. The VLSI implementation of the delay commutator has resulted in a dramatic reduction in the number of ICs needed to implement a 40-MHz, 4096-complex-point FFT. Table 27-1 gives a quick comparison of two FFT implementations with and without the delay commutator circuit (Swartzlander and Hallnor 1984).

It should be pointed out that the resulting commutator uses programmable length shift registers, thus making possible the realization of the various interstage delays required by the FFT.

The implications of this design go beyond the mere implementation of the 4096-complex-point FFT. It demonstrates the power of semicustom IC design in implementing high-performance signal processing systems. The key requirement is that we make our choice of algorithms, architecture, and technology in a way that matches the requirement of the signal processing function.

Recent Developments in VLSI Building Blocks for Digital Signal Processing (Silicon Technologies) (Marinos 1980).

Table 27-2 offers a listing of VLSI digital signal processing building blocks. Some of them are available only in limited quantities at this time. Table 27-3 indicates technological trends for single-chip programmable digital signal processors.

Recent Advances in GaAs-Based Building Blocks for Digital Signal Processing (Marinos 1980).

Recent advances in GaAs technologies have resulted in LSI circuits capable of serving as basic building blocks in the design of high-performance signal processing systems. Table 27-4 offers a fairly complete list of commercially available GaAs components. However, there is no capability for realizing, at this time, single-chip, programmable functional units such as computational elements, delay commutators, address generators, and the like. We are at least three to five years away from realizing GaAs logic circuits of 10,000 or so gate complexity.

GaAs-based gate arrays are also attracting a great deal of interest because of their cost-effectiveness in system design. Table 27-5 gives a listing of current offerings in GaAs gate arrays. This technology is evolving very rapidly with projection of 3000- to 5000-gate chips becoming commercially available in the very near future.

Table 27-1. System complexity (integrated circuit count).

	WITHOUT DELAY COMMUTATOR CIRCUIT		WITH DELAY COMMUTATOR CIRCUIT	
4096 Point FFT				
Computational Element	6 cards at 80 ckts/card =	480 ckts	6 cards at 91 ckts/card = 546 ckts	
Delay Commutator	5 cards at 179 ckts/card =	895 ckts	(included on CE)	
TOTAL	11 cards	1375 ckts	6 cards	546 ckts
16384 Point FFT				
Computational Element	7 cards at 80 ckts/cards =	560 ckts	7 cards at 91 ckts/card = 637 ckts	
Delay Commutator	6 cards at 179 ckts/cards =	1074 ckts	1 extended DC (1024) = 33 ckts	
TOTAL	13 cards	1634 ckts	8 cards	670 ckts

PROPOSED HIGH-PERFORMANCE DSP ARCHITECTURES

The major contributing factors in realizing high-performance DSP architectures within ever decreasing power and space budgets are: (a) the advancement of semiconductor technologies and (b) the development of appropriate tools and methodologies that make possible the use of these technologies in the implementation of high-performance circuits and subsystems.

This section presents DSP architectures that take full advantage not only of existing high-performance integrated circuits, but also of technologies sufficiently developed to make five-year projections of circuit availability and performance relatively accurate and safe.

An important consideration throughout this section is the assumption that each of the signal processor architectures proposed must be capable of performing a postulated set of signal processing functions under specified input signal conditions. Such an assumption permits us to compare the various architectures under identical conditions, and makes processor characteristics such as speed, circuit count, and power consumption important parameters in arriving at a meaningful figure of merit for each of the proposed architectures.

Although the degree of design detail offered throughout this section makes comparisons among the various architectures very meaningful, it lacks the level of detail necessary for accurate system sizing (i.e., determination of circuit count and power consumption) and throughput projections. This presentation is consistent with the main objective of this chapter, which is to establish the relative merits of various architectures rather than their absolute levels of complexity and performance.

It should be reiterated that the choice of signal processing functions and workload environment were motivated by the potential uses of the proposed signal processor architectures and our desire to be as realistic as possible in assessing various system design options. System scaling for meeting different application requirements, however, is achieved by making appropriate choices in either technology or circuit count, or both.

Table 27-2. High-performance silicon-based circuits for signal processing applications.

MANUFACTURER	FUNCTION	TECHNOLOGY	PERFORMANCE	POWER DISSIPATION
AT&T BELL LABS.	Dual Port 2k × 9-bit SRAM	DOUBLE POLY CMOS	150 ns ACCESS TIME PER PORT ASYNCHRONOUS PORT OPERATION AVAILABLE	330 mW ACTIVE 550 μW STANDBY
FAIRCHILD RES.	64k SRAM (8k × 8-bit)	2μ, BIPOLAR	15 ns ACCESS TIME	900 mW
MOTOROLA	64k SRAM (8k × 8-bit)	1.5μ CMOS	37 ns ACCESS TIME	270 mW ACTIVE 300 mW STANDBY
HUGHES RES.	8 × 8-bit MULTIPLIER	DOUBLE METAL 0.85μ, NMOS 1.5 μ, NMOS	9.5 ns (16-bit OUTPUT)	600 mW
AT&T BELL LABS.	32-BIT FLOATING-POINT PROGRAMMABLE SIGNAL PROCESSOR		4 MIPS 8 MFLOPS 32-BIT FLOATING-POINT (24-bit MANTISSA 8-bit EXPONENT)	N/A
ADVANCED MICRO DEVICES- Am29323	32 × 32-bit MULTIPLIER	ECL	80 ns	N/A
Am29334	64 × 18-bit REGISTER FILE WITH FOUR	ECL	20 ns PER PORT/ALL PORTS	N/A

Am9150	NMOS	1k × 4-bit SRAM	PORTS (TWO READ, TWO WRITE) CAN BE ACCESSED CONCURRENTLY 25 ns ACCESS TIME	900 mW
Am29500 NTT	CHIPS 1.2 μ CMOS	FAMILY OF 80-bit FLOATING-POINT PROCESSOR	5.6 MFLOPS CLOCK CYCLE = 60 ns MULTIPLY TIME = 180	1 W
INTEGRATED DEVICE TECHNOLOGIES, INC. IDT-7132	CMOS	16k DUAL PORT SRAM (2k × 8-bit)	90 ns ACCESS PER PORT BOTH PORTS ACCESSABLE CONCURRENTLY	320 mW
FAIRCHILD RES. 54F/74 F -784	BIPOLAR	8-bit SERIAL/ PARALLEL MULTIPLIER (THE 8-bit OPERAND ENTERED IN PARALLEL, THE SECOND OPERAND IS ARBITRARILY LONG AND IS ENTERED SERIALLY)	50 MHz RATE FOR THE SERIAL INPUT	500 mW

Table 27-3. Technology trends for single-chip programmable digital signal processors.

	AMI	AMI	BELL	FUJITSU	HITACHI	INTOL	NBC	NBC	TI	TOSHIBA
Device	S2811	28211/2	DSP-1	MB 8764	61810	2920/21	7720/F20	77220	320	T6386/7
Process	aMOS	aMOS	aMOS	CMOS	CMOS	EPROM/ aMOS	aMOS EPROM	CMOS	aMOS	CMOS
Minimum feature size	4.5 microns	3 microns	4.5 microns	2.3 microns	3 microns	6/4 microns	3 microns	2 microns	2.7 microns	2 microns
Year described	1978	1983	1980	1983	1982	1978/81	1980	1984	1982	1983
Area (sq. mils)	41,000	—	106,000	145,000	79,000	47,000	44,000	—	70,000	76,000 74,000
Number of Transistors	30,000	—	45,000	91,000	55,000	20,000	40,000	—	—	66,000‡ 48,000
Pins	28	28/64	40	88	40	28	28	—	40	28/64
Power	1 W	0.7 W	1.25 W	0.290 W	0.200 W	1 W	1 W	—	0.9 W	0.360 W
Precision	16	16	20	16	16 FP	25	16	24	16	16
Multiplier	12×12=16	12×12=16	4(4×20)=36	16×16=26	12×12=16	25×1=25	16×16=31	24×24=48	16×16=31	16×16=31
Speed	300 ns	300 ns	4(200) ns	100 ns	250 ns	600/400 ns	250 ns	100 ns	200 ns	200 ns
Program Memory	256×17 ROM	512×18 ROM/EXT	1K×16 ROM,EXT	1K×24 ROM,EXT	512×22 ROM	192×24 EPROM/ROM	512×23 ROM/EPROM	4096 ROM	4K×16 ROM,EXT	512×16 ROM,EXT
Data RAM	128×16	256×16	128×20	200×16	40×25	128×16	1024×24	144×16	128×16	512×16
Data ROM	128×16	128×16	In program	In program	128×16	16×4	512×13	1024	In program	512×16 ROM/EXT

Table 27-4. Digital GaAs IC's.

FUNCTION	TECHNOLOGY	SPEED	POWER DIS	POWER SUP	IC MAKER
32-bit adder	BFL	2.9 ns	1.2 W	2, −1 (v)	NEC
		7.6 ns	340 mW	2.16, −1.02	
4-bit ripple	lp-BFL	1.9 ns	45 mW	5, −3	Thomson
Carry adder		1.25 ns	180 mW		CSF
16 × 16 multiplier	DCFL	10.5 ns	952 mW	1.6, 0	Fujitsu
10-stage linear	BFL	2GHz clock	n/av	5, −3.5	H.P.
Feedback S/R					
8-bit Mux	BFL	250 MHz	250 mW	−5.2	British/ Telecom
1-KBit SRAM	DCFL	4 ns access	68 mW	1.5	Fujitsu
1-KBit SRAM	DCFL	6 ns access	30 mW	1	NEC
4 × 4 multiplier	DCFL	5.5 ns	39 mW	1	Toshiba
8-bit MUX,DMUX	DFL	2GHz clock	n/av	3, −2.5	Tektronix
1-KBit SRAM	DCFL	6 ns access	38 mW	1	NTT
4-KBit SRAM	DCFL	3 ns access	700 nW	1	Fujitsu
4-KBit SRAM	DCFL	2.8 ns access	1.2 W	1.3 cell, 1.8 peripheral	NTT
1-KBit SRAM	DCFL(HEMT)	3.4 ns (300 K)	290 mW	.13	Fujitsu
		0.9 ns (77 K)	360 mW	1.6	
8 × 8 multiplier	BFL	5.2 ns	2.2	n/av	Rockwell
6 × 6 multiplier	DCFL	6.4 ns	173 mW	1.5	Fujitsu

BFL: buffered FET logic.
DCFL: direct coupled FET logic
lp BFL: low power BFL.
HEMT: high electron mobility transistor.
n/av: not available.

Table 27-5. Gate arrays.

# GATES	I/O	GATE DELAY	IC MAKER	TECHNOLOGY
320	40	184 ps	Lockheed	BFL
1-K	56	100 ps (no load)	Toshiba	DCFL
1 K	64	1.5 ns, 0.4 ns	T.I.	Heterojunction Bipolar
432	n/av.	n/av.	Honeywell	BFL

As previously stated, the three signal processing functions considered here are pulse compression, MTI/Doppler processing, and CFAR. Furthermore, we assume real-time signal processing with information bandwidths of 5 MHz or higher. Because the FFT is a fundamental operation in performing signal processing functions such as pulse compression and Doppler processing, its efficient computation is of paramount importance. This necessitates the use of fast complex multipliers/adders, address generators, and fast multiport memory. In order to determine the hardware requirements of the processor, we assume that the maximum size FFT to be handled by the processor at any given time is 1024 complex points. Data-point sequences consisting of more than 1024 complex points are processed (i.e., filtered) by subdivision into sequences of 1024 (or fewer) complex points and appropriate processing of these subsequences.

With regard to programming the proposed DSP architectures, it is assumed that programming is carried out at a very high level through the use of a minimal instruction set. The instruction set includes I/O instructions for moving data in and out of the processor as well as data manipulative instructions for performing pulse compression, MTI/Doppler processing, CFAR, and so on. These very high level instructions differ from conventional instruction sets in that each instruction configures the processor to perform the respective function and leaves it in that configuration until another instruction is invoked.

In the sequel, several signal processor architectures are presented utilizing either silicon-based technologies or GaAs technologies of MSI and LSI complexity. Processor designs using combinations of these technologies are also included.

Si-Based DSP Architectures

The DSP architectures that follow utilize Si-based technologies and related VLSI and VHSIC (Barna 1981) components. When commercially available components meet the needs of a proposed architecture, their use is recommended. In those cases in which the components required are not available, but there is available technology to produce them, their existence is assumed, and their performance and complexity are conservatively extrapolated.

DSP Architecture-A.   This architecture, which relies on VLSI and VHSIC components for its implementation, is illustrated in Figure 27-26.

The process is configured using the Advanced Mirco Devices (or equivalent) bipolar technology and includes the Am29516 multiplier chip.

The memory utilizes CMOS very high speed integrated circuits configured in a four-port RAM type memory system with two input and two output ports per chip. The availability of multiport memory improves overall system performance by allowing concurrent READS and WRITES and by permitting retrieval of

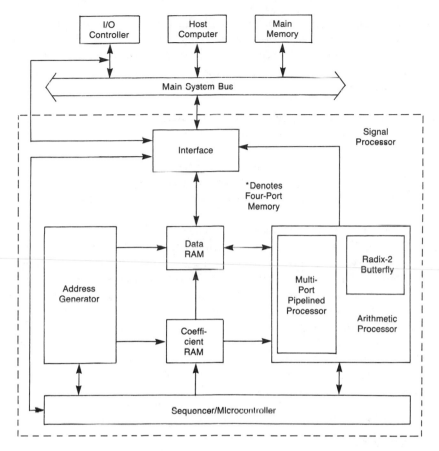

Figure 27-26.   High performance signal processor (Architecture-A).

multiple complex data points concurrently when performing fast-Fourier-transforms (FFTs). RAM memory is used for both data and coefficient storage.

The address generator is also a very high speed integrated circuit, and serves as a slave processor. It generates sequences of 16-bit addresses, one new address with each successive clock cycle, for quick access of data structures with minimal program control.

The host computer shown in Figure 27-26 is either the central system processor or a special-purpose processor chip capable of interpreting the high-level instruction set associated with the signal processor. Its presence endows the overall system with the necessary flexibility to accommodate variable-size FFTs and the use of arbitrary filter transfer functions when performing filtering operations. For purposes of achieving high throughput and better data management, the memory is divided into Data-memory and Coefficient-memory. The size of these memories depends primarily on the size of FFT desired and the number of coefficients used in filtering operations. Memory sizing as well as overall system hardware requirements are based on system ability to perform 1024-complex-point FFTs.

DSP Architecture-A assumes:

1. 16-bit operands (i.e., 16-bit real, 16-bit imaginary).
2. 2's complement, fixed-point arithmetic.
3. Up to 1024-complex-point FFTs.
4. The use of radix-2, decimation-in-time (DIT), in-place FFT.
5. That convolution of long input sequences is performed via the "fast-convolution" algorithm using the overlap-and add technique.

In order to meet the computational requirements of signal processing functions performed via "fast convolution," the memory identified with this architecture should utilize twenty-eight 1024 × 4-bit (i.e., 1k × 4-bit) static RAM chips, and be organized for concurrent read/write of two complex operands via separate 64-bit output and input buses. That is, the memory should be large enough to accommodate: (a) up to 1024 complex data points; (b) up to 1024 twiddle factors required by a 1024-complex-point FFT (it is possible to use only 512 twiddle factors by eliminating inherent redundancies); (c) up to 1024 complex filter coefficients; and (d) up to 512 complex data points generated by the "overlap-and-add" technique when performing fast convolution.

When computing a complex, radix-2 FFT butterfly, one performs four real multiplications and six real additions. This necessitates the use of four 16 × 16-bit multipliers and six 16-bit adders. An additional adder is provided to perform the appropriate correction arising in the computation of the "fast convolution" using the "overlap-and-add" technique.

Taking into account all the computational needs outlined previously, one arrives at the set of hardware requirements for DSP Architecture-A shown in Table 27-6. This architecture can be implemented on a single circuit board, and requires one input bus to transfer input data to the RAMS, and one output bus to transfer off-board the results of completed signal processing tasks. The use

Table 27-6. Hardware requirements of Architecture-A.

1. Four (4) Multiplier Chips (16 × 16-bit)
2. Seven (7) Adder Chips (16-bit)
3. Twenty-eight (28) 1k × 4-bit RAM Chips
4. One (1) Address Generator Chip (16-bit)
5. One (1) Sequencer/Microcontroller Chip
   Total Number of Chips = 41

Table 27-7. Power requirements of Architecture-A.

1. Multipliers (Bipolar, ECL)	4 chips × 3.0 watts/chip = 12.0 watts
2. Adders (Bipolar, TTL)	7 chips × 0.5 watt/chip = 3.5 watts
3. Memory (SRAM, CMOS)	28 chips × 0.3 watt/chip = 8.4 watts
4. Address Generator (Bipolar)	1 chip × 3.0 watts/chip = 3.0 watts
5. Sequence/Microcontroller (Bipolar)	1 chip × 1.3 watts/chip = 1.3 watts
	Totals: 41 chips 28.2 watts

of two separate buses permits concurrent I/O operations, and makes possible the use of this architecture in a pipelined fashion on arbitrarily long streams of data.

Determining the power requirements of this architecture involves estimation of power quantities for the sequencer/microcontroller, address generator, and two-port memory chips that are not presently readily available. Nevertheless, one can arrive at sound, conservative estimates of power consumption for each chip, as shown in Table 27-7, based on knowledge of the technology used, a good estimate of device count, and preliminary available data.

The performance of this architecture is limited by the speed of its various component chips because its memory organization (i.e., two input and two output ports) permits inputting and outputting operands concurrently at a rate compatible with the processing capability of the system. Table 27-8 shows the speed characteristics of the components used.

Given the manner in which DSP Architecture-A is organized and the speed characteristics of its component parts, one can conclude that the time required to compute a radix-2 butterfly is 40 ns with a 25-MHz (100 ns with a 10-MHz) clock. Note that a radix-2 butterfly processes two complex points every 40 ns by overlapping the "multiply" and "add" times required by successive pairs of such points, as shown in Figure 27-27. This throughput capability is consistent

Table 27-8. Speed characteristics of chips used in Architecture-A.

1. Array Multiplier, 16 × 16-bit (Bipolar, ECL)	16 or 32-bit output in 40 ns
2. Adder (Carry Look-Ahead), 16-bit (Bipolar, TTL)	Add time = 10 ns
3. Memory (SRAM, CMOS), Four-Port	Access time = 40 ns
4. System Clock	10 MHz/25MHz

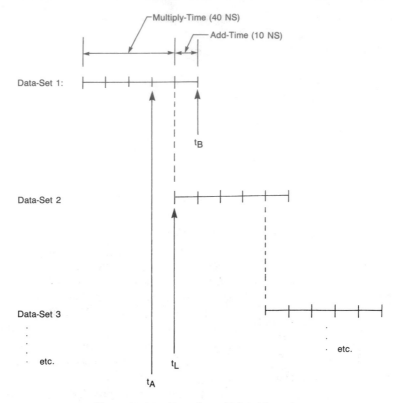

Figure 27-27.   Butterfly multiply/add overlap.

with the 40 ns memory access time, the speed characteristics of the butterfly, and the fact that the address generator produces a new address every clock period (i.e., every 40 ns for a 25-MHz clock). Referring to Figure 27-27, points designated at t_A denote times at which a new address is generated; they are immediately followed (i.e., 10 ns later) by t_L, which represents times at which a new pair of complex points is received by the butterfly for processing; finally, at points in time denoted by t_B, a butterfly output is produced and stored in memory using the address associated with the corresponding complex points just processed by the butterfly. Therefore, the processor keeps track of two addresses at any given time: the address used to load a new pair of complex points to the butterfly and the address used for storing the current butterfly output in memory. These "read" and "write" addresses related in the sense that the current "write" address is the same as the "read" address used one computational cycle (i.e., butterfly) earlier.

When the processor is operated in a pipelined configuration and the I/O buses are assumed to be 64-bits wide, it is only necessary that I/O transfers be carried out at 25 MHz. This is the same as the internal processors clock rate. If, on the other hand, the I/O buses are limited to a 16-bit width, then I/O transfers must take place at 100 MHz. This high I/O transfer rate is not possible with the 40 ns access-time memory used without appropriate buffering.

We turn, next, to the throughput capabilities of DSP Architecture-A. One of the basic computations the system should be capable of performing is the FFT with as many as 1024 complex points. Such a computation involves $(N/2)\log_2 N$ complex radix-2 butterflies, where N is the number of complex points to be processed. Assuming a 25-MHz system clocks, the time required to compute a radix-2 butterfly is 40 ns, and the total time needed to FFT 1024 complex points is:

$$T\langle FFT - 1024\rangle = [(N/2)\log_2 N] \times T\langle Butterfly\rangle$$
$$= 512 \times 10 \times 40 \text{ ns} \qquad\qquad [3.1]$$
$$= 204.8 \text{ μs}$$

The FFT along with complex multiplication and addition form the basic operations in carrying out the three signal processing functions: pulse compression, MTI/Doppler processing and CFAR.

Pulse Compression. It was stated earlier that pulse compression is best carried out via fast convolution. This implies processing of input data streams of arbitrary length using techniques such as the overlap-and-add method, which requires segmentation of the input data stream into groups of N points, and convolution of each segment separately. (For more details the reader is referred to the first section of the chapter.) DSP Architecture-A is so organized that it can support FFTs of up to 1024 complex points, and general filtering operations utilizing unsampled filter transfer functions with as many as 1024 complex points.

Pulse compression requires an FFT, a complex multiplication, and an inverse FFT to be carried out sequentially. One may use either a single processor to do all three computations sequentially or two pipelined processors operating concurrently. The latter scheme takes advantage of the inherent concurrency possible with successive pulse compression computations when long streams of input data are processed.

Option-1 (Pulse Compression Using a Single Processor). Assuming N_1 and N_2 to be the number of replica spectrum and input samples, respectively, the size of FFT, N, required for pulse compression is $N \geqslant N_1 + N_2 - 1$ (see first section of this chapter). The FFT and inverse FFT result in a total of:

$$2 \times (N/2)\log_2 N = 2 \times 512 \times 10 = 10{,}240 \text{ butterflies}$$

and require, collectively, for this processing:

$$10{,}240 \text{ butterflies} \times 40 \text{ ns/butterfly} = 409.6 \text{ μs}$$

assuming a 25-MHz system clock. Complex multiplication with the replica spectrum results in N = 1024 multiplies, and requires:

$$1024 \text{ multiplications} \times (40 \text{ ns/multiplication}) = 40.96 \text{ } \mu s$$

Thus, the total time required for fast convolution of N = 1024 points is 450.56 μs.

For long input data streams, appropriate stream segmentation and use of the overlap-and-add technique must be invoked to perform pulse compression, as discussed in the first section of the chapter. The overlap-and-add operation is an overlapped computational activity easily accommodated by the presence of the seventh adder in the processor.

Option-2 (Pulse Compression Using Two Pipelined Processors). The two stages of the pipeline are assigned to computations, as shown in Figure 27-28.

Assuming the same parameters as in Option-1 and noting that loading of Processor-1 with anew set of N_2 complex samples is carried out concurrently with the replica-spectrum multiplication, one obtains a throughput rate equivalent to one fast convolution every

$$T\langle N - CONVOL.\rangle = [(N/2)\log_2 N] \times T\langle Butterfly\rangle \qquad [3.2]$$
$$+ N \times T\langle Multiply\rangle \text{ sec}$$

For N = 1024, $T\langle Butterfly\rangle = T\langle Multiply\rangle = 40$ ns, one has:

$$T\langle 1024 - CONVOL.\rangle = 245.76 \text{ } \mu s$$

In general, an input stream of N complex samples will require the execution of M fast convolutions where:

$$M \geq [N_S/N_2] \text{ and } N_2 \leq N - (N_1 - 1) \qquad [3.3]$$

Assuming $N_S = 5000$, N = 1024, and $N_1 = 120$, one obtains M = 6. This implies that the input stream of 5000 points is segmented into six subsequences; the first five are selected to be 834 samples each, with the sixth carrying the remaining 830 samples. All these sequences, including $N_1 = 120$, are zero-extended to N = 1024, and processed as described in the first section of the chapter.

MTI/Doppler Processing. The computational load associated with the MTI/Doppler processing function was analyzed in the first section. It consists of data "corner-turning," four-pulse cancellation, and FFT. Note that Doppler processing is a batch process, as we must acquire target information in a given range cell from every one of the pulses we have illuminated the target with before we can begin Doppler processing.

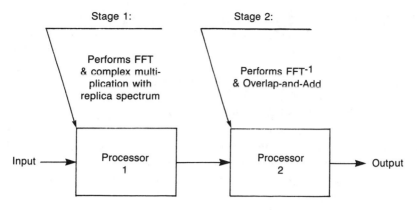

Figure 27-28.   Pulse-compression pipeline.

Because Architecture-A has four real multipliers and seven adders, the four-pulse canceller can generate a new column of samples, r_{ij}, every

$$T\langle\text{Column} - j\rangle - 4 \times (P - 3) \times T\langle\text{Multiply}\rangle \text{ sec} \qquad [3.4]$$

where P is the number of pulses transmitted in a dwell, $T\langle\text{Multiply}\rangle$ is the time required to perform a complex multiply, and the factor 4 denotes the number of multiplications per complex point generated. This time is sufficiently long to absorb the add-time incurred in the adder and also the time needed to (a): load the four-pulse canceller with new samples and (b) store the result in processor memory. Assuming P − 67 and $T\langle\text{Multiply}\rangle$ − 40 ns, one obtains:

$$T\langle\text{Column} - j\rangle = 10.24 \text{ }\mu\text{s}$$

If the number of range cells of interest is k, then the total time involved in four-pulse cancellation is:

$$T\langle 4 - \text{pulse CANC.}\rangle = k \times T\langle\text{Column} - j\rangle \text{ sec} \qquad [3.5]$$

If k is very large, the time $T\langle 4 - \text{pulse CANC.}\rangle$ may be unacceptably high, and one may be forced to subdivide the range cells of interest in smaller subgroups and utilize a separate four-pulse canceller (i.e., processor) with each of the subgroups. Architecture-A is very well-suited for such a computation.

Inspection of the output resulting from four-pulse cancellation, and shown in Figure 27-10, clearly suggests the manner in which one must perform Doppler processing. Each column represents an FFT load and corresponds to a specific range cell. The time required to FFT each column, j, is:

$$T\langle\text{FFT} - j\rangle = \{(L/2)\log_2 L\} \times T\langle\text{Butterfly}\rangle \text{ sec} \qquad [3.6]$$

where L is the number of pulses per dwell, and T⟨Butterfly⟩ is the time required to perform a complex butterfly. For typical values, L = 64 and T⟨Butterfly⟩ = 40 ns, one obtains:

$$T\langle FFT - j\rangle = 7.68\mu s$$

Usually the FFT computation needed for Doppler processing and the four-pulse cancellation task are overlapped, thus resulting in increased processor throughput. This, however, necessitates the use of two Architecture-A processors. Under overlapped operation, and using the typical parameters given above, the total time needed to generate a column of the output matrix shown in Figure 27-10, and its subsequent processing via FFT to obtain Doppler information, is

$$T\langle MTI/DOPPL - j\rangle = 10.24 \ \mu s \qquad [3.7]$$

which is simply the time needed to do MTI via four-pulse cancellation because a 64-point FFT used in Doppler processing requires only 7.68 μs, and the two are overlapped, as shown in Figure 27-29. In the case of a single-processor system, the total time required for MTI and Doppler processing per range cell is 10;24 μs + 7.68 μs = 17.92 μs, using, or course, the same typical values for various parameters as given previously.

CFAR—Constant False Alarm Rate. The computational load associated with the CFAR function was analyzed in the first section of the chapter. Figure 27-13

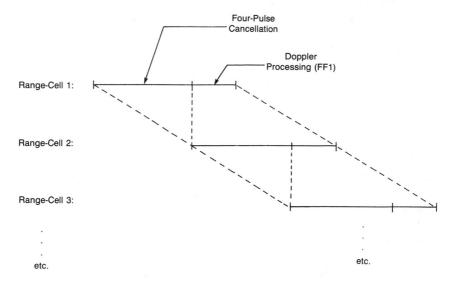

Figure 27-29. MTI/DOppler processing overlap.

suggests the operations to be performed. The processor of Architecture-A has sufficient arithmetic resources from the entire CFAR computation in time

$$T\langle CFAR \rangle = 2 \times \{ [k/a] + [k/a^2] + \ldots + [k/a^m] \} \times \{T\langle ADD \rangle\}$$
$$+ 2 \times \{T\langle MULT \rangle + T\langle ADD \rangle\} \qquad [3.8]$$
$$+ 1 \times T\langle SUBT \rangle \text{ sec}$$

where the first term denotes the addition time required by the two segments of k range cells per segment; the second is the time needed to compute the magnitude of each accumulated sum; and the last denotes the time associated with the comparison of the two segment magnitudes. Note that the first term accounts for the fact that there will be m passes over partially formed sums before the complete sum of each segment is obtained. Therefore, m is the smallest integer such that $[k/a^m] = 1$, where "a" denotes the number of adders available to do CFAR. It is also true that during each pass, i, we use an amount of time equal to

$$[k/a^i] \times \{T\langle ADD \rangle\} \text{ sec}$$

Assuming k = 20, and given that a = 6, $T\langle MULT \rangle$ = 40 ns, and $T\langle ADD \rangle$ = $T\langle SUBT \rangle$ = 40 ns (for a 25-MHz clock), one concludes that m = 2. This leads to:

$$T\langle CFAR \rangle = 2 \times \{4 + 1\} \times \{40\}$$
$$+ 2 \times \{40 + 40\}$$
$$+ 1 \times 40 \text{ ns}$$

or $T\langle CFAR \rangle = 600$ ns per range cell. When this time for CFAR is compared to the time necessary for MTI/Doppler processing of one range cell, which was 17.92 μs, we conclude that CFAR is not as time-consuming a function as MTI/Doppler processing.

Table 27-9 offers a summary of processor performance characteristics in terms of the signal processing functions considered. The parenthetical entries in the table represent performance due to two processsors used in a pipelined fashion. All other entries are based on the use of a single processor.

DSP Architecture-B.   This architecture is a variation of Architecture-A. The processor uses a radix-4 butterfly to perform the FFT, instead of the radix-2 used in Architecture-A, but relies on the same technology and chips for its implementation that were used in Architecture-A.

The time required to execute a radix-4 butterfly is 80 ns because the memory is still organized as in Architecture-A, and can supply only two complex data

Table 27-9. Throughput characteristics of Architecture-A.

	CLOCK	
FUNCTION	25 MHz	10 MHz
FFT-1024 Complex Points	204.8 μs	512 μs
Pulse Compression	450.56 μs (245.76 μs)	1126.4 μs (614.4 μs)
MTI/Doppler (Per Range-Cell)	17.92 μs (10.24 μs)	44.8 μs (25.6 μs)
CFAR (Per Range-Cell)	0.60 μs	1.50 μs

samples at a time at the rate of 25 MHz. Nevertheless, the achievable throughput rate is improved over Architecture-A, as the time required to perform an N-complex-point FFT, where N is a power of 4, is:

$$T\langle FFT - N\rangle = \{(N/4)\log_4 N\} \times T\langle Butterfly\rangle \qquad [3.9]$$

or for $N = 1024$ and $T\langle Butterfly\rangle = 80$ ns, $T\langle FFT - 1024\rangle = 102.4$ μs. Thus, the time required to perform a radix-4 FFT using Architecture-B is one half that taken by Architecture-A.

A radix-4 butterfly requires 12 real multipliers and 16 real adders, analyzed in the first section. The memory requirements remain the same as in Architecture-A, at twenty-eight 1k × 4-bit static RAM chips. The same is true with respect to the address generator and sequencer/microcontroller chips. An extra adder is provided in Architecture-B, as it was in Architecture-A, to facilitate concurrent execution of the "overlap-and-add" technique used in "fast convolution" with other computations of the convolution algorithm. All these hardware requirements are summarized in Table 27-10. This architecture can be implemented on

Table 27-10. Hardware requirements of Architecture-B.

1. Twelve (12) Multiplier Chips (16 × 16-bit)
2. Seventeen (17) Adder Chips (16-bit)
3. Twenty-eight (28) 1k × 4-bit RAM Chips
4. One (1) Address Generator Chip (16-bit)
5. One (1) Sequencer/Microcontroller Chip
   Total Number of Chips = 59

a single board, and requires one input bus to transfer input data to the RAMS and one output bus to transfer off-board the results of completed signal processing tasks. The power requirements for this architecture are summarized in Table 27-11.

The speed characteristics of the chips used in Architecture-B are summarized in Table 27-8, given earlier.

We turn next to the throughput capabilities of DSP Architecture-B. The processor in this case can computer an N-point FFT using as few as $(N/4)\log_4 N$, radix-4 butterflies when N is a power of 4. If N is not a power of 4, one must consider either zero-extending the N-point sequence to such a value or combining radix-4 and radix-2 computations. For instance, N = 1024 is a power of 4 (i.e., $1024 = 4^5$), whereas N = 2048 is not (i.e., $2048 = 4^5 \cdot 2^1$); the former may use radix-4 butterflies throughout the FFT computation, while the latter is carried out using five stages of radix-4 butterfly computations followed by a sixth stage using the radix-2 butterfly. The times required to perform pulse compression, MTI/Doppler processing, and CFAR are as follows

Pulse Compression.

Option-1 (Pulse Compression Using a Single Processor). The forward and inverse FFT result in $2 \times \{(N/4)\log_4 N\}$, radix-4 butterflies. Assuming N = 1024 and given that T⟨Butterfly⟩ = 80 ns, one obtains 204.8 µs as the time required to carry out the two FFTs.

The complex multiplication with the replica spectrum utilizes 12 multipliers instead of the 4 available in Architecture-A. This still results in a multiplication time of 40 ns due to memory bandwidth limitations, and a total replica spectrum multiplication time of 1024×40 ns = 40.96 µs.

Therefore, the total time required to perform fast convolution with N = 1024 is 245.76 µs.

Table 27-11. Power requirements of Architecture-B.

1. Multipliers (Bipolar, ECL)	12 chips × 3.0 watts/chip = 36.0 watts
2. Adders (Bipolar, TTL)	17 chips × 0.5 watt/chip = 8.5 watts
3. Memory (SRAM, CMOS)	28 chips × 0.3 watt/chip = 8.4 watts
4. Address Generator (Bipolar)	1 chip × 3.0 watts/chip = 3.0 watts
5. Sequencer/Microcontroller (Bipolar)	1 chip × 1.3 watts/chip = 1.3 watts
	Totalsd: 59 chips 51.2 watts

Option-2 (Pulse Compression Using Two Pipelined Processors). Assuming the same parameters as in Option-1, and following a procedure similar to the one given for DSP Architecture-A, one obtains:

$$T\langle N - CONVOL.\rangle = \{(N/4)\log_4 N\} \quad\quad [3.10]$$
$$\times \ T\langle Butterfly\rangle + N \times T\langle Multiply\rangle \ sec$$

or:

$$T\langle 1024 - CONVL.\rangle = 256 \times 5 \times 80 \ ns + 1024 \times 40.0 \ ns$$
$$= 143.36 \ \mu s$$

MTI/Doppler Processing. Making the same assumptions and following the same procedure given for DSP Architecture-A, one obtains:

$$T\langle Column - j\rangle = 4 \times (P - 3) \times T\langle Multiply\rangle$$
$$= 4 \times 64 \times (40/3) \ ns = 3.41 \ \mu s$$

Similarly:

$$T\langle FFT - j\rangle = \{L/4\} \ \log_4 L\} \times T\langle Butterfly\rangle \quad\quad [3.11]$$
$$= \{(64/4) \ \log_4 64\} \times 80 \ ns = 3.84 \ \mu gs$$

In the case of a single-processor system, the total time required for MTI and Doppler processing per range cell is $3.41 \ \mu s + 3.84 \ \mu s = 7.25 \ \mu s$.

For a two-processor system, the MTI and Doppler processing time per range cell is simply $3.84 \ \mu s$.

CFAR—Constant False Alarm Rate. The time needed to perform CFAR is the same as in Architecture-A even though many more resources (adders, multipliers) are available than was the case in Architecture-A. This is due to the fact that its bud bandwidth allows accessing only two complex samples at a time.

Table 27-12 offers a summary of processor performance characteristics in terms of the signal processing functions considered. The parenthetical entries in the table represent performance due to two processors used in a pipelined fashion. All other entries assume the use of a single processor.

The performance of Architecture-B could be improved by a factor of two if the I/O buses were enlarged from the present 64-bit size to a 128-bit width. This would allow read/write's of four complex samples at a time, thus matching perfectly the processing capabilities of the radix-4 butterfly processor. It also allows better utilization of resources during pulse compression, MTI/Doppler, and CFAR processing. Power requirements would be somewhat higher.

Table 27-12. Throughput characteristics of Architecture-B.

FUNCTION	CLOCK	
	25 MHz	10 MHz
FFT-1024	102.4 μs	256 μs
Complex		
Points		
Pulse	245.76 μs	614.4 μs
Compression	(143.36 μs)	(358.4 μs)
MTI/Doppler	7.42 μs	18.55 μs
(Per Range-Cell)	(3.84 μs)	(9.6 μs)
CFAR	0.6 μs	1.50 μs
(Per Range-Cell)		

DSP Architecture-C (Kanopoulos and Marinos 1984).   This architecture, which utilizes FIFO memory in place of RAMS and a bit-serially configured arithmetic processor, is proposed in an attempt to significantly reduce circuit count and power dissipation. It is illustrated in Figure 27-30. A similar arithmetic processor has been previously proposed (Kanapoulos and Marinos 1984; Kanapoulos 1984), and it is possible to implement using CMOS-based very high speed integrated circuits. Such a processor can perform bit-serial, pipeline multiplication based on Booth's algorithm in 16 clock pulses or 640 ns with a 25-MHz clock. The product is rounded to the 16 most significant bits.

Use of the FIFO memory eliminates the need for an address generation chip, and thus reduces circuit count and overall power dissipation. The resulting output data, however, may be provided in an order not suitable for immediate subsequent use. When shuffling of the data is required, appropriate shuffling networks or an address generator must be utilized. Therefore, this architecture makes use of address generators (or shuffling networks) only as needed.

Compared to the processor of Architecture-A, which can effectively perform one real multiplication in 40 ns, the multiplication time of Architecture-C is 16 times longer. Another important observation is that a bit-serial multiplier in this case is 1/16th the size of an array multiplier, and this, using very-high-speed-integrated-circuit technology, can lead to greatly increased functional capability on a single chip.

The proposed processor is to provide a hardwired radix-4 butterfly, and a multiplier bank capable of performing four complex multiplications concurrently in a bit-serial manner. Such a multiplier bank facilitates filter implementations in which complex multiplication of FFT output with filter coefficients is necessary. This leads to a processor with a total of 12 multipliers in the radix-4 section and 16 multipliers in the "multiplier bank," for a grand total of 28 16-

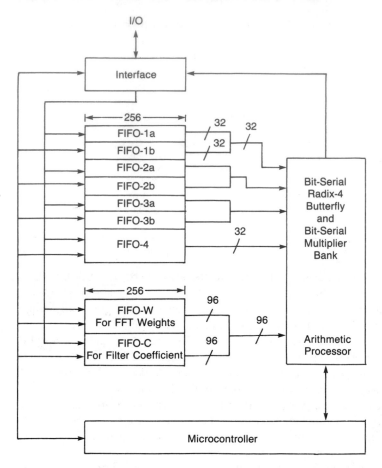

Figure 27-30. High-performance signal processor with FIFI memory and bit-serial arithmetic processor (Architecture-C).

bit serial multipliers. This results in $28 \times 16 = 448$ multiplier stages, and by assuming 60 devices per multiplier stage one obtains 26,880 devices needed to implement the multipliers alone. If we conservatively assume that the rest of the circuitry involving adders, latching, and control requires an equal number of devices, we reach a total of about 53,000 devices for the entire arithmetic processor. For a 20 μ^2m average device size and a gate capacitance of 3.68×10^{-4} pf/μ^2m, a single chip implementation of 53,000 devices will display 390 pf capacitance. Assuming all the devices to switch every clock cycle, and 50% additional capacitance due to parasitics, then for a power supply of 5 V and a clock of 25 MHz dynamic power dissipation is approximately equal to 400 mW.

The processor performs arithmetic operations bit-serially, which implies that

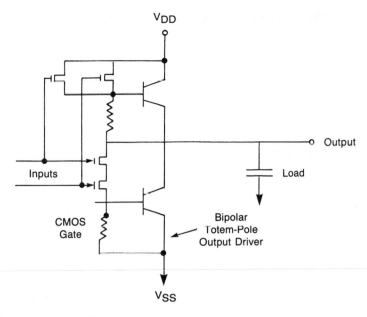

Figure 27-31.   CMOS output driver with bipolar devices in totem-pole configuration.

a radix-4 butterfly (or a complex multiplication) will be performed in 640 ns with a 25-MHz clock. However, the multiplier bank permits four complex multiplications at a time, resulting in an effective multiplication rate of one complex multiplication per 160 ns. Assuming the processor's ports to operate at the clock rate, one ensures adequate availability, without the need for wait states, to keep the processor busy continuously. There are only four clock cycles needed for transferring input data into the chip, and similarly for taking data off the chip.

A potential problem for the CMOS processor may arise if this chip is required to drive large capacitive loads off-chip when technologies other than CMOS are used (e.g., for interfacing). In this case, the output drivers of the processor can be made with bipolar devices, as shown in Figure 27-31. This permits the processor to interface with a wide range of technologies at the 25-MHz clock rate. Hitachi has already demonstrated the feasibility of incorporating bipolar drivers in CMOS gate arrays. This feature will increase the processor power requirements to about 1 watt.

Tables 27-13 through 27-16 summarize the relevant features of Architecture-C.

DSP Architecture-D.   This architecture employs the same arithmetic processor used in Architecture-C, but instead of FIFO memory it makes use of CMOS four-port memory chips. Because the memory can supply two words every 40

Table 27-13. Hardware requirements of Architecture-C.

1. One (1) Arithmetic Processor Chip (CMOS)
2. Twenty-Eight (28), 256 × 16-bit, FIFO Chips
3. One (1) Controller Chip
   Total Number of Chips = 30

Table 27-14. Power requirements of Architecture-C.

1. Arithmetic Processor Chip (CMOS)	1 chip × 0.4 watt/chip = 0.4 watt
2. FIFO (CMOS)	28 chips × 0.3 watt/chip = 8.4 watts
3. Controller (CMOS)	1 chip × 0.3 watt/chip = 0.3 watt
	Totals: 30 chips 9.1 watts

Table 27-15. Speed characteristics of chips used in Architecture-C.

1. Arithmetic Processor—	Effective Multiplication Time 160 ns per multiply (complex) Radix-4 Butterfly Time = 640 ns (Assume 25 MHz System Clock)	
2. FIFO Memory		Clocked at 25-MHz
3. Controller Chip		Clocked at 25-MHz.

Table 27-16. Throughput characteristics of Architecture-C.

	CLOCK	
FUNCTION	25 MHz	10 MHz
FFT-1024 Complex Points	819.2 μs	2048 μs
Pulse Compression	638.4 μs (819.2 μs)	4096 μs (2048 μs)
MTI/Doppler (Per Range-Cell)	71.68 μs (40.96 μs)	179.2 μs (102.4 μs)
CFAR (Per Range-Cell)	3.36 μs	8.4 μs

ns, and the arithmetic processor's computational cycle is 640 ns, one can generate within 640 ns eight "radix-4 arithmetic processor loads." This offers us, equivalently, the option of using eight copies of the arithmetic processor concurrently, resulting in a significant improvement in throughput performance. [Note that complex data samples and relevant twiddle factors (or filter coefficients) are loaded concurrently.] Figure 27-32 illustrates Architecture-D in block diagram form. Tables 27-17 through 27-19 present the requirements and performance of Architecture-D.

Note that from the standpoint of throughput, Architecture D has eight times the computational power of Architecture-C. This increased computational power

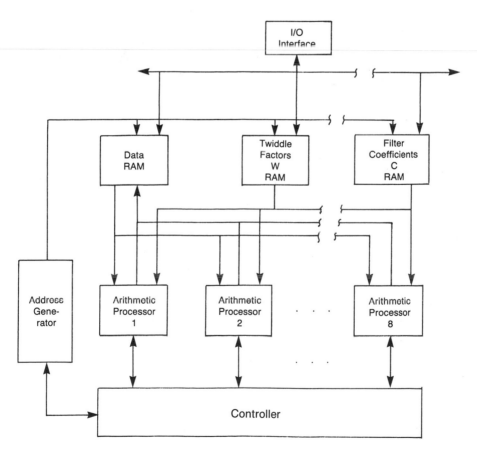

Figure 27-32.   High-performance signal processor with multiple arithmetic processors (Architecture-D).

Table 27-17. Hardware requirements of Architecture-D.

1. Eight (8) Arithmetic Processor Chips
2. Twenty-Eight (28) (1k × 4-bit) Four-Port Memory Chips
3. One (1) Address Generator Chip
4. One (1) Controller Chip
   Total Number of Chips = 38

can be fully utilized in performing FFT, pulse-compression, MTI/Doppler, and CFAR processing.

DSP Architecture-E (Marinos 1981).   This architecture, illustrated in Figure 27-33, utilizes the same arithmetic processor chip used in Architecture-D. The significant difference here is that Architecture-E relies on two ring-buses for resource interconnection and data transfers. This is a high-performance architecture in terms of both throughput and system reliability. A single failure on a ring bus does not normally led to overall system failure. Furthermore, technological improvements are very conveniently accommodated by this architecture as long as bus speed does not become a limiting factor. Similarly, one can vary the number and type of processors attached to the buses. Note that failure of a processor causes only system degradation, not system failure. This is not necessarily true in Architecture-D.

Tables 27-20 and 27-21 give relevant features of Architecture-E. The throughput characteristics of Architecture-E are identical to those given for Architecture-D.

GaAs-Based DSP Architectures

The following DSP architectures utilize GaAs technologies. When commercially available components meet the needs of a proposed architecture, their use is recommended. In those cases in which the components required are not available, but there is available technology to produce them, their existence is assumed, and their performance and complexity are conservatively extrapolated.

Table 27-18. Power requirements of Architecture-D.

1. Arithmetic Processor Chip (CMOS)	8 chips × 0.4 watt/chip = 3.2 watts
2. Memory, 1k × 4-bit SRAM (CMOS)	28 chips × 0.3 watt/chip = 8.4 watts
3. Address Generator (Bipolar)	1 chip × 3.0 watts/chip = 3.0 watts
4. Controller (CMOS)	1 chip × 0.3 watt/chip = 0.3 watt
	Totals: 38 chips 14.9 watts

Table 27-19. Throughput characteristics of Architecture-D.

FUNCTION	CLOCK	
	25 MHz	10 MHz
FFT-1024 Complex Points	102.4 μs	256 μs
Pulse Compression	204.8 μs (102.4 μs)	512 μs (256 μs)
MTI/Doppler (Per Range-Cell)	8.96 μs (5.12 μs)	22.4 μs (12.8 μs)
CFAR (Per Range-Cell)	0.42 μs	1.05 μs

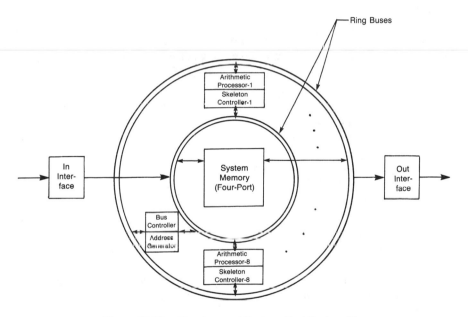

Figure 27-33.   Ring-bus architecture (Architecture-D).

Table 27-20. Hardware requirements of Architecture-E.

1. Eight (8) Arithmetic Processor/Skeleton Controller Chips
2. Twenty-eight (28), Four-Port (1k × 4-bit) Memory Chips
3. One (1) Address Generator Chip
4. One (1) Bus Controller Chip
        Total Number of Chips = 38 chips

Table 27-21. Power requirements of Architecture-E.

1. Arithmetic Processor and Skeleton Controller (CMOS)	8 chips × 0.5 watt/chip = 4.0 watts
2. Memory, 1k × 4-bit RAM (CMOS)	28 chips × 0.3 watt/chip = 8.4 watts
3. Address Generator (Bipolar)	1 chip × 3.0 watts/chiop = 3.0 watts
4. Bus Controller (CMOS)	1 chip × 0.3 watt/chip = 0.3 watt
	Totals: 38 chips 15.7 watts

DSP Architecture-1. This architecture is configured to perform the radix-2, DIT, in-place FFT algorithm relying entirely on existing GaAs components. Figure 27-34 illustrates Architecture-1 in block diagram form. The arithmetic processor consists essentially of a hardwired circuit designed to perform efficiently the radix-2 complex butterfly. This processor is further augmented with a "multiplier bank" that in conjunction with the radix-2 butterfly unit facilities the implementation of various signal processing functions such as convolution, MTI/Doppler, and CFAR. The multiplier bank can receive FFT output, act upon it, and write the results into data memory. Using this scheme, FFT computations and spectrum multiplications can be pipelined in such a way that spectrum multiplication is completely overlapped with the last FFT pass when convolution or other filtering operations are carried out.

The multipliers used in this architecture are 16-bit × 16-bit array multipliers by Fujitsu (or equivalent) with 10.5 ns multiplication time. The adder is the 32-bit adder by NEC (or equivalent) with 4 ns addition time. The memory is assumed to be configured with NTT 4k SRAM chips (or equivalent) whose access time is less than 4 ns.

The hardware requirements for implementing Architecture-1 are given in Table 27-22. Four of the multipliers are used to implement the radix-2 butterfly unit, and the remaining eight form the "multiplier bank," The latter facilitates multiplication of two complex samples with appropriate filter coefficients, concurrently. The eleven adders are distributed as follows: six in the radix-2 butterfly, four in the "multiplier bank," and one provided to facilitate the "overlap-and-add" computation.

The power requirements of Architecture-1 are given in Table 27-23 and the speed characteristics of the various chips used in Architecture-1 are given in Table 27-24.

As there are two memory operations (i.e., one read and one write) associated with each of the two operands in every radix-2 FFT butterfly computation, there will be 16 ns consumed in memory accesses. This is in addition to multiply and add times of 10.4 ns and 4 ns, respectively. However, a timing scheme overlapping fetch and execution periods as shown in Figure 27-35 is possible, and the result is a radix-2 butterfly every 20 ns. It should be pointed out that twiddle

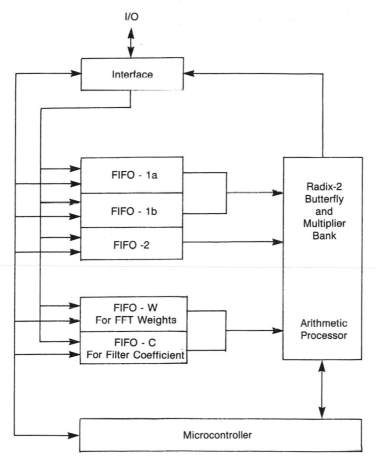

Figure 27-34. high-performance signal processor with FIFO memory (ARchitecture-1).

Table 27-22. Hardware requirements of Architecture-1.

---

1. Twelve (12) Multiplier Chips
2. Eleven (11) Adder Chips
3. Twenty-eight (28), Single-Port
   1k × 4-bit, Memory Chips
4. One (1) Address Generator Chip
5. One (1) Controller Chip
   Total Number of Chips = 53

Table 27-23. Power requirements of Architecture-1.

1. Multiplier (GaAs)	12 chips $\times$ 0.9532 watt/chip = 11.42 watts
2. Adders (GaAs)	11 chips $\times$ 1.2 watts/chip = 13.20 watts
3. Memory (SRAM, GaAs)	28 chips $\times$ 0.9 watt/chip = 25.20 watts
4. Controller (GaAs)	1 chip $\times$ 1 watt/chip = 1.00 watt
5. Address Generator (GaAs)	1 chip $\times$ 1 watt/chip = 1.00 watt
	Totals: 53 chips 51.82 watts

factors (or filter coefficients) are accessed independently through a bus other than the one used to access data samples. Furthermore, they are only read, and thus they are never a factor in determining system throughput.

Based on a 250-MHz clock and a radix-2 butterfly time of 20 ns, one obtains the throughput performance for Architecture-1 given in Table 27-25. Parenthetical entries in Table 27-25 are based on the availability of two-port memories. In computing the CFAR function, only two multipliers can be used; therefore, the parameter "a" in expression (3.8) is a = 2. Pipelining processor boards based on Architecture-1 should prove costly in terms of performance degradation, so it is not considered.

DSP Architecture-2.   This architecture is proposed as a means for reducing circuit count and power dissipation relative to Architecture-1 without significant degradation in throughput performance. Again, we propose a radix-2 butterfly as in Architecture-1 except that all arithmetic computations are to be carried out bit-serially. This reduces by a factor of 16 (i.e., the word length) the amount of hardware required for arithmetic, but it will, of course, increase computation time proportionately.

Assuming bit-serial arithmetic, one should be able to integrate comfortably a hardwired radix-2 complex butterfly and a "multiplier bank" of eight multipliers on the same chip. This is not an unreasonable expectation if one considers that the circuitry involved (i.e., 12 bit-serial multipliers, and 11 single-bit adders) is less than the circuitry used to realize Fujitsu's 16-bit array multiplier. Figure

Table 27-24. Speed characteristics of chips used in Architecture-1.

1. Array Multiplier, 16 $\times$ 16-bit (Direct Coupled FET Logic)	Multiplication time = 10.5 ns
2. Adder, 32-bit (Buffered FET Logic)	Add time = 4 ns
3. Memory (SRAM), 1k $\times$ 4-bit (DCFL)	Access time = 2.8 ns
4. System Clock	250 MHz
(Address Generator produces a new address every 4 ns)	

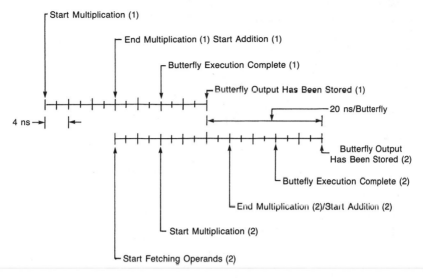

Figure 27-35.   Butterfly fetch/execution overlap.

27-36 illustrates in block diagram form the arithmetic processor of Architecture-2. The computational cycle of this processor is (16-bit operand) × (4 ns/bit processing time) = 64 ns. As there are four memory references (i.e., two reads and two writers) per computational cycle, and memory access time is 4 ns, one should able able to support {(64 ns per comput. cycle)/(16 ns memory referencing time per processor)} = 4 processors. Figure 27-37 shows Architecture-2 utilizing four copies of the arithmetic processor given in Figure 27-36. The radix-2 butterfly of each processor requires 16-bits × 4 ns/bit = 64 to perform multipli-

Table 27-25. Throughput characteristics of Architecture-1.

| | CLOCK | |
FUNCTION	250 MHz	100 MHz
FFT-1024	102.40 μs	256.0 μs
Complex	(81.92 μs)	(204.8 μs)
Points		
Pulse	204.8 μs	512.0 μs
Compression	(163.84 μs)	(409.60 μs)
MTI/Doppler	6.53 μs	16.32 μs
(Per Range-Cell)	(5.76 μs)	(14.40 μs)
CFAR	0.201 μs	0.502 μs
(Per Range-Cell)		

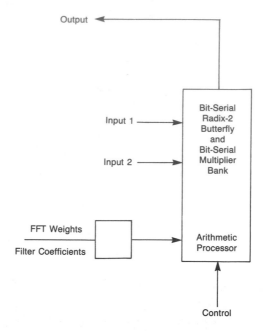

Figure 27-36.   Bit serial arithmetic processor.

cation with the two additions which follow the multiplication executed in a pipelined fashion at no additional cost in time. Thus, we only need to provide a new set of operands to the butterfly every 64 ns. Because we use four processors, the result is four butterflies every 62 ns for an *effective rate* of one butterfly per 16 ns.

The hardware requirements of this architecture are given in Table 27-26.

In estimating the power requirements of this architecture, we assume that the bit-serial arithmetic processor chip is comparable in complexity and power dissipation to the 16b × 16b array multiplier offered by Fujitsu. Table 27-27 gives the power requirements of Architecture-2.

Table 27-28 gives the speed characteristics of chips used in Architecture-2.

Based on a 250-MHz clock and an effective radix-2 butterfly time of 16 ns (i.e., four arithmetic processors complete four butterflies every 64 ns), one obtains for Architecture-2 the throughput characteristics given in Table 27-29.

In computing the MTI/Doppler function, we use an effective addition time of 16 ns because four arithmetic processors operate in parallel. Similarly, in computing the CFAR function, use is made of expression (3.8) with a = 4.

Memory bandwidth limitations are the determining factor of system performance in Architecture-2, and, in fact, in Architecture-1 as well. Until further progress is made toward realizing multiport memory chips and achieving greater device densities, system performance of GaAs-based signal processors will re-

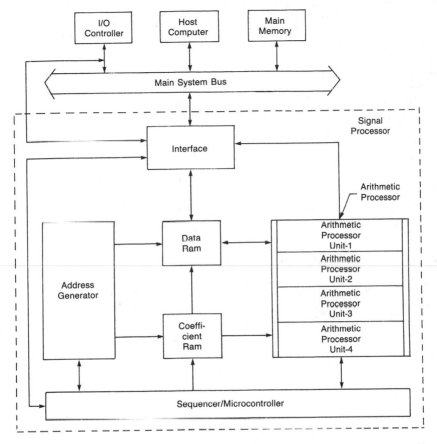

Figure 27-37. High-performance signal processor using multiple arithmetic processors (Architecture-2).

main comparable to that possible with mature, state-of-the-art Si technologies. If Architecture-B (or D) could be implemented in GaAs, a "1024-complex-point FFT" would require only 10.24 μs to complete, assuming a 250-MHz clock. This would clearly make GaAs the technology of choice in computationally intensive situations, such as electronic intelligence, and in applications with

Table 27-26. Hardware requirements of Architecture-2.

1. Four (4), Bit-Serial Arithmetic Processor Chips
2. Twenty-Eight (28), 1 k × 4-bit, Memory Chips
3. One (1) Address Generator Chip
4. One (1) Controller Chip
   Total Number of Chips = 34

Table 27-27. Power requirements of Architecture-2.

1. Bit-Serial GaAs Arithmetic Processor:

   4 chips × 0.952 watt/chip = 3.8 watts

2. Memory (SRAM, GaAs):

   28 chips × 0.9 watt/chip = 25.2 watts

3. Address Generator GaAS   1 chip × 1 watt/chip = 1.0 watts
4. Controller GaAs   1 chip × 1 watt/chip = 1.0 watts

   Totals: 34 chips   31 watts

system power and size constraints. The assumption of a 250-MHz clock is rather conservative, and any improvements in switching speeds will contribute further to overall system performance.

GaAs/Sl-Based DSP Architectures

GaAs technology may be used advantageously in Si-based systems through selective replacement of silicon-based components. The memory and address generator chips are prime candidates for replacement with GaAs parts. In order to match the speed of the memory and address generator complex to the rest of the Si-based system, extensive data buffering becomes necessary. The buffers will be GaAs circuits interacting with main (GaAs) memory at a clock rate of 250 MHz and with the Si-based arithmetic unit at 25 MHz.

DSP Architecture-X.   This architecture is in effect Architecture-D, presented earlier, except its Si four-port memory has been replaced with GaAs, single-port memory chips, and the address generator is implemented in GaAs. Figure 27-38 illustrates how these changes are incorporated in Architecture-D and shows the rather extensive buffering required to realize Architecture-X.

The computational cycle in Architecture-D was 640 ns and Read/Write's were

Table 27-28. Speed characteristics of chips used in Architecture-2.

1. Bit-Serial Arithmetic Processor (GaAs-DCFL)
   —Multiplication Time = Add/Subtract Time =   64 ns.
   —Radix-2 Butterfly in Pipelined Fashion
     Requires
2. Memory (SRAM), 1k × 4-bit (DCFL)
   —Access Time =   2.8 ns.
3. System Check —   250 MHz
   (Address Generator produces a new address every 4 ns)

Table 27-29. Throughput characteristics of Architecture-2.

| | CLOCK | |
FUNCTION	25 MHz	10 MHz
FFT-1024 Complex Points	81.92 μs	204.8 μs
Pulse Compression	163.84 μs	409.6 μs
MTI/Doppler (Per Range-Cell)	7.17 μs	17.0 μs
CFAR (Per Range-Cell)	1.34 μs	3.36 μs

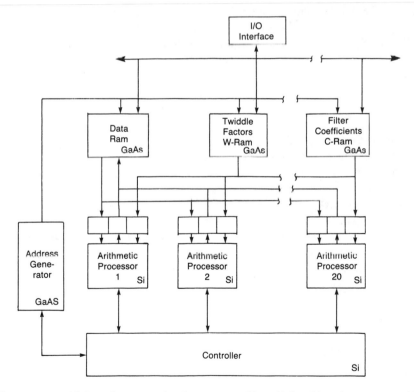

Figure 27-38.   High-performance signal processor with multiple arithmetic processors (Architecture-X).

carried out concurrently due to availability of multiported memory. In Architecture-X, it is not possible to Read and Write in main memory concurrently because of the use of single-port GaAs memory chips. Therefore, of the 640 ns period, one can use only 320 ns to Read from main memory. Assuming 4 ns main-memory access time, one concludes that we can fetch a full processor load (i.e., four complex samples for the radix-4 butterfly) in 16 ns or 320/16 = 20 such loads during a complete (i.e., 640 ns) computational cycle. This leads to the choice of 20 arithmetic processors operating concurrently, as shown in Figure 27-38.

The hardware requirements of Architecture-X are given in Table 27-30. A single-board implementation of Architecture-X is possible.

The power requirements of Architecture-X are given in Table 27-31.

The speed characteristics of the chips used are similar to those given in conjunction with Architecture-D and Architecture-1, earlier. Table 27-32 provides the throughput characteristics of Architecture-X.

In this architecture we use 20 arithmetic processors in parallel, and each processor executes a radix-4 butterfly every 640 ns. Therefore, the effective rate at which a radix-4 butterfly is completed is 32 ns; and the time required to process a 1024-complex-point FFT is 40.96 μs. Overall, the resulting throughput performance is 2.5 times that of Architecture-D, which uses eight processors identical to those used by Architecture-X.

CONCLUSION

The objectives of this chapter were: (a) to review relevant, high-performance signal processor architectures used in actual system implementations over the past 15 years; (b) to propose new high-performance architectures suitable for use in computationally intensive signal processing environments (i.e., radar, sonar, seismic arrays, and the like); and (c) to recommend specific uses of advanced Si and GaAs technologies in the implementation of high-performance DSP architectures. The emphasis here is placed on optimal matching of technology to architecture at the memory–processor and I/O–processor interfaces.

Table 27-30. Hardware requirements of Architecture-X.

1. Twenty (20) Arithmetic
   Processor Chips (Si)
2. Twenty-Eight (28), Single-Port,
   1k × 4-bit Memory Chips (GaAs)
3. Sixty (60), 4 × 32-bit, Buffer Chips (GaAs)
4. One (1) Address Generator Chip (GaAs)
5. One (1) Controller Chip (Si)
   Total Number of Chips = 110

Table 27-31. Power requirements of Architecture-X.

1. Arithmetic Processor
   (CMOS):   20 chips $\times$ 0.4 watts/chip = 8.0 watts

2. Memory, 1k $\times$ 4-bit (GaAs):
                            28 chips $\times$ 0.9 watt/chip = 25.2 watts

3. Buffer, 4 $\times$ 32-bit (GaAs):
                            60 chips $\times$ 0.2 watt/chip = 6.0 watts

3. Address Generator GaAS:
                            1 chip $\times$ 1 watt/chip = 1.0 watts

4. Controller (CMOS):
                            1 chip $\times$ 0.3 watt/chip = 0.3 watts

                            Totals: 110 chips   40.5 watts

With respect to architecture and technology selections, we observe the following:

Selection of Architecture

Signal processor designers have at their disposal a wide variety of high-performance signal processing architectures, but a particular choice of architecture is heavily dependent on: (a) mission-imposed constraints such as throughput requirements, power needs, size, reliability, and so on; (b) the technology available at the time of actual system implementation; and (c) system cost.

Of the several signal processing architectures investigated, some emphasized throughput performance, and others minimal use of components (circuits) and power. The common characteristic of all these architectures is their use of massive

Table 27-32. Throughput characteristics of Architecture-X.

FUNCTION	CLOCK	
	25 MHz	10 MHz
FFT-1024 Complex Points	40.96 μs	102.4 μs
Pulse Compression	81.92 μs	204.8 μs
MTI/Doppler (Per Range-Cell)	3.58 μs	8.95 μs
CFAR (Per Range-Cell)	0.168 μs	0.42 μs

parallelism at all functional levels and across all subsystems. Functional parallelism is exploited at the instruction, algorithm, and program level, whereas parallelism at the subsystem level is possible via appropriate organization and design of the memory, arithmetic, and control unit.

It was determined that certain architectures can benefit from the use of serially (bit-wise) operating arithmetic units, particularly when circuit count and power dissipation are important design parameters.

All architectures studied are computationally powerful enough in a single-board configuration to satisfy most of the problem needs arising in radar, sonar, and similar applications without the need for sophisticated software support.

One of the architectures analyzed was the so-called ring-bus architecture (Architecture-E). This architecture is of particular importance because of its fault-tolerant attributes and its ability to readily accommodate system growth, or other system improvements, in a graceful manner. It is an architecture with generic characteristics and thus worthy of further serious consideration.

Selection of Technology

Although technologies are evolving very rapidly, the object of this chapter was to use state-of-the-art, "commercially available" parts to perform parametric trade-off analysis of various signal processor architectures for determination of their performance characteristics. Emphasis was placed on high-performance Si and GaAs technologies.

Silicon (Si) dominates presently digital signal processor design, and this trend should continue unabated in the foreseeable future. Availability of high-performance VLSI and very high speed integrated circuits should considerably expand the potential uses of Si technologies in signal processing hardware.

GaAs is clearly a signal processing technology of great promise, but its contributions to significant signal processing system designs are not likely to be substantial before 1990, at the earliest. At the present time, its low device density circuits prevent its use in sophisticated processor and memory designs that are essential in high-performance signal processing architectures. This situation is not likely to change very significantly over the next five years or so. There is, however, a place where GaAs now can play a significant role, namely, its selective insertion in the various subsystems of existing or currently designed signal processing systems. We have demonstrated that we can utilize architectures that exploit the strengths of Si and GaAs technologies to realize processors with substantially improved performance over Si-based designs. This attribute of the GaAs technology needs to be exploited fully in signal processing during this transition period.

In the immediate future, and while GaAs technologies are advancing toward LSI and VLSI levels of complexity, more aggressive use of high-performance Si and available GaAs components should be made in signal processor design. Architectures that can benefit from such a technology mix should be identified, investigated, and implemented. There is, in fact, a need for such a signal processor design to serve as a testbed for future developments in GaAs and Si technologies.

REFERENCES

Advanced Micro Devices, Inc., "The AM 29500 Family of Chips," *Array and Digital Signal Processing Products Manual,* Aug. 1982.

Advanced Micro Devices, Inc., "AM29300 Family of 32 Bit Building Blocks (data Sheets and personal communication).

Allen, Jonathan, "Computer Architecture for Signal Processing," *Proc. IEEE, 63* (4), 624–633 (Apr. 1975).

Barna, Arpad, *VHSIC—Very High Speed Integrated Circuits/Technologies and Tradeoffs,* John Wiley and Sons, New York, 1981.

Blankenship, P. E., and Hofstetter," E. M., "Digital Pulse Compression via Fast Convolution," *IEEE Trans. Acoustics, Speech, and Signal Processing, ASSP-23* (2), 189–201 (Apr. 1975).

Booth, William, "Approaches to Radar Signal Processing," *IEEE Computer, 16* (6), 32–42 (June 1983).

Bowen, B. A., and Brown, W. R., *VLSI Systems Design for Digital Signal Processing,* Vol. 1, Prentice-Hall, Englewood Cliffs, NJ, 1982.

Cook, C. E., and Bernfeld, M., *Radar Signals—An Introduction to Theory and Applications,* Academic Press, New York, 1967.

Cooley, J. W., and Tukey, J. W., "An Algorithm for the machine Calculation of Complex Fourier Series," *Mathematics of Computation, 19,* 297–302 (Apr. 1965).

Fisher, J. R., "Architecture and Applications of the SPS-41 and SPS 81 Programmable Digital Signal Processors," *EASCON 74 Record,* Oct. 1974, pp. 674–678.

Kanopoulos, Nick, "A Single Chip VLSI Architecture for Radar Signal Processing," Ph.D. dissertation, Department of Electrical Engineering, Duke University, Durham, NC, 1984.

Kanopoulos, Nick, and Marinos, Peter N., "A High-Performance VLSI Signal Processor Architecture," *Proc.* IEEE International Conference on Computer Design: VLSI in Computers, Oct. 1984, pp. 526–531.

Marinos, Peter N., "Present and Future Trends in Signal Processing Technologies," NRL Memorandum Report no. 4299, Aug. 1980.

Marinos, Peter N., "An Integrated Signal/Data Processor Architecture for Use in High Availability Radar Systems," NRL Memorandum Report 4693, Dec. 1981.

McClellan, J. H., and Prudy, R. J., "Applications of Digital Signal Processing to Radar," in *Applications of Digital Signal Processing,* A. V. Oppenheim, Editor, Prentice-Hall, Englewood Cliffs, NJ, 1982.

Nathanson, F. E., *Radar Design Principles,* McGraw-Hill Book Co., New York, 1969.

New, B., and Pittroff, Lyle, "Record Signal-Processing Rates Spring from Chip Refinements," *Electronics,* 114–116 (July 28, 1982).

Rabiner, L. R., and Gold, Bernard, *Theory and Application of Digital Signal Processing,* Prentice-Hall, Englewood Cliffs, NJ, 1975.

Skolnik, M. I., *Introduction to Radar Systems,* 2nd ed., McGraw-Hill Book Co., New York, 2nd ed., 1980.

Swartzlander, Earl E., Jr., and Hallnor, George, "Fast Transform Implementation," *Proc.* IEEE International Conference on Acoustics, Speech, and Signal Processing, March 19–21, 1984, pp. 25 A.5.1–25A.5.4.

Swartzlander, Earl E., Jr., Young, Wendell K. W., and Joseph, Saul J., "A VLSI Delay Commutator for FFT Implementation," *Proc. IEEE,* International Solid-State Circuits Conference, Feb. 24, 1984, pp. 266–267.

Index